Advanced Mathematics for Engineers and Physicists

Sever Angel Popescu • Marilena Jianu

Advanced Mathematics for Engineers and Physicists

 Springer

Sever Angel Popescu
Department of Mathematics and Computer Science
Technical University of Civil Engineering of Bucharest
Bucharest, Romania

Marilena Jianu
Department of Mathematics and Computer Science
Technical University of Civil Engineering of Bucharest
Bucharest, Romania

ISBN 978-3-031-21501-8   ISBN 978-3-031-21502-5 (eBook)
https://doi.org/10.1007/978-3-031-21502-5

© The Editor(s) (if applicable) and The Author(s), under exclusive license to Springer Nature Switzerland AG 2022

This work is subject to copyright. All rights are solely and exclusively licensed by the Publisher, whether the whole or part of the material is concerned, specifically the rights of translation, reprinting, reuse of illustrations, recitation, broadcasting, reproduction on microfilms or in any other physical way, and transmission or information storage and retrieval, electronic adaptation, computer software, or by similar or dissimilar methodology now known or hereafter developed.

The use of general descriptive names, registered names, trademarks, service marks, etc. in this publication does not imply, even in the absence of a specific statement, that such names are exempt from the relevant protective laws and regulations and therefore free for general use.

The publisher, the authors, and the editors are safe to assume that the advice and information in this book are believed to be true and accurate at the date of publication. Neither the publisher nor the authors or the editors give a warranty, expressed or implied, with respect to the material contained herein or for any errors or omissions that may have been made. The publisher remains neutral with regard to jurisdictional claims in published maps and institutional affiliations.

This Springer imprint is published by the registered company Springer Nature Switzerland AG
The registered company address is: Gewerbestrasse 11, 6330 Cham, Switzerland

Preface

This book is designed to be an introductory course to some important chapters of Advanced Mathematics for Engineering and Physics students: Differential Equations, Fourier Series, Laplace and Fourier Transforms, Calculus of Variations, and Probability Theory. It requires a complete course in Mathematical Analysis and in Linear Algebra. We mostly tried to use the engineering intuition instead of insisting on mathematical tricks. Even if the book was first of all intended for engineering students, it may be successfully used by other students and scientists as an introduction to Applied Mathematics. The main feature of the material presented here is its clarity, motivation, and the genuine desire of the authors to make extremely transparent the "mysterious" mathematical tools which are used to describe and organize the great variety of impressions of our searching Mind in the infinite hidden corners of Nature. We have always had in mind the common student who really tries to understand the process of mathematical modeling of various physical phenomena, because we are fully convinced that mathematics education does not mean learning (lots of) formulas by heart, but the conscious understanding of the meaning of these formulas. We believe that this course is also useful for mathematics students because the presentation is elementary, but rigorous enough to satisfy the taste of a mathematician.

In the first three chapters, we offer a detailed study of ordinary differential equations. In Chap. 3, we begin by presenting the general framework, focusing on first-order ordinary differential equations. We show by several examples how differential equations can describe (model) physical phenomena in mathematical terms. The most common types of first-order differential equations are presented, together with the techniques for obtaining their solutions (either in implicit or explicit form). The last section is dedicated to the theorem of existence and uniqueness of the solution of the Cauchy problem for first-order differential equations. Chapter 2 is devoted to higher-order differential equations, mainly linear differential equations. We introduce the vector space of the solutions of a homogeneous linear differential equation and the affine space of the solutions for the non-homogeneous case. Linear equations with constant coefficients are placed in the foreground, because in this case we have a closed-form solution for homogeneous equations as well as for a wide class of non-homogeneous equations. We also present the Euler-type equations which have variable coefficients but can be transformed into linear equations with constant coefficients. The systems of differential equations are presented in Chap. 3. We state and prove the existence and uniqueness theorem for the Cauchy problem in this case. Particular attention is paid to linear systems of differential equations and to the relation between a linear system of n equations and an n-th order linear equation. Autonomous systems are also studied, and since first-order partial differential equations are closely related to autonomous systems, we have chosen to present them in this chapter.

In Chap. 4 (Fourier series), we discuss the expansion of periodic functions in infinite series of sines and cosines and introduce the general orthogonal systems of functions. Fourier series will be used in Chap. 7 to solve partial differential equations such as the wave equation, the heat equation, or the Laplace equation. Fourier series are used to represent periodic functions (or functions defined on a finite interval, which can be extended by periodicity to the entire axis). The Fourier transform generalizes this idea to the integral representation of nonperiodic functions defined on the whole set of real numbers. We present the Fourier transform and some of its applications in Chap. 5. The discrete Fourier transform, which can be used when a function is given only in terms of values at a finite number of points, is also introduced. Another important mathematical tool is the Laplace transform, presented in Chap. 6. As in the case of the Fourier transform, the Laplace transform simplifies the solution of linear differential equations by transforming them into algebraic equations. It is also applied for the solution of partial differential equations—in Chap. 7, we use the Laplace transform for the solution of the finite vibrating string equation. By using the Heaviside step function or the Dirac delta function, the Laplace transform can be applied in problems where the free term has some discontinuities or represents short impulses. Chapter 7 is an introduction to the field of the "Equations of Mathematical Physics," the most important partial differential equations used in Physics and Engineering. We begin by classifying these equations and introduce the characteristics method to find the canonical form of a quasilinear second-order partial differential equation. Then we present the solution of the one-dimensional and two-dimensional wave equation, the heat flow equation, and the Dirichlet problem for Laplace's equation. Although this chapter is mainly devoted to second-order PDE, we have decided to include here also a fourth-order partial differential equation—the simply supported beam equation.

Chapter 8 contains the basic theory of Calculus of Variations applied to fundamental types of variational problems with applications in physics and engineering. We begin by stating several classical problems (such as the brachistochrone problem, the minimal surface of revolution, Hamilton's principle of the least action, and Dido's problem). Then we introduce the general frame of calculus of variation, focusing on necessary conditions of extremum of a functional. We deduce the basic differential equations of calculus of variations and apply them to solve some classical variational problems.

In Chap. 9, we make an elementary introduction to probability theory. We present Laplace's and Kolmogorov's definitions of probability and show how they can be applied to solve practical problems. We introduce the conditional probability and the Bayes formula, providing also numerous applications. We define the discrete random variables and the continuous random variables, emphasizing the most used distributions of discrete type (Bernoulli, binomial, Poisson), and of continuous type (normal, gamma, chi-squared, student). We also present the most important limit theorems: the weak law and the strong law of large numbers and the central limit theorem, highlighting the role played by the normal distribution. Each chapter has a final section with exercises. The complete solutions of these problems are presented step by step in Chap. 10.

Considering the great diversity of topics covered in this book, we have tried to make it as self-contained and unitary as possible. This is why we have added a supplementary chapter—Chap. 11—which contains the basic theory in some areas of Linear Algebra, Calculus, and Complex Analysis, necessary for a deep understanding of the material presented in the book. Thus, the first section of this chapter contains elementary results on metric, normed, Banach, and Hilbert spaces, while the second section provides a brief introduction to complex analysis, with a special focus on the calculus of residues (used in the calculation of the Fourier and Laplace transforms).

It is not possible to express our gratitude to all those who contributed indirectly to the writing of this book. A special role in the formation of the first author as a mathematician was played by his former Professor of Algebra and Number Theory, his PhD supervisor, and his spiritual master, Dr. doc. Nicolae Popescu.

We would like to express our sincere gratefulness to Professor Dr. Octav Olteanu, Professor Dr. Gavriil Păltineanu, Professor Dr. Ghiocel Groza, and the reviewers for their considerable attention and patience in reading the manuscript and for the valuable suggestions and comments that greatly improved the initial version of this book.

Our families were so patient during the complicated and lengthy conception of this writing. We thank them all for their loving support.

We shall be very grateful to all the readers who will let us know about possible mistakes or some of their particular opinions related to the topics of this book, because improvement is the most important human activity.

Bucharest, Romania
September 2022

Sever Angel Popescu
Marilena Jianu

Contents

1 First-Order Differential Equations 1
    1.1   Introduction to Ordinary Differential Equations 1
    1.2   Separable Equations... 7
    1.3   Homogeneous Equations... 13
    1.4   First-Order Linear Differential Equations.......................... 16
    1.5   Bernoulli Equations 20
    1.6   Riccati Equations 22
    1.7   Exact Differential Equations.. 24
    1.8   Lagrange Equations and Clairaut Equations 29
    1.9   Existence and Uniqueness of Solution of the Cauchy Problem 34
    1.10   Exercises... 42

2 Higher-Order Differential Equations 45
    2.1   Introduction.. 45
    2.2   Homogeneous Linear Differential Equations of Order n 57
    2.3   Non-Homogeneous Linear Differential Equations of
           Order n... 64
    2.4   Homogeneous Linear Equations with Constant Coefficients 68
    2.5   Nonhomogeneous Linear Equations with Constant
           Coefficients 82
    2.6   Euler Equations... 93
    2.7   Exercises... 97

3 Systems of Differential Equations 101
    3.1   Introduction.. 101
    3.2   First-Order Systems and Differential Equations of Order n 109
    3.3   Linear Systems of Differential Equations 114
    3.4   Linear Systems with Constant Coefficients 125
           3.4.1   The Homogeneous Case (the Algebraic Method)......... 125
           3.4.2   The Non-Homogeneous Case (the Method of
                    Undetermined Coefficients) 138

		3.4.3	Matrix Exponential and Linear Systems with Constant Coefficients	148

- 3.4.3 Matrix Exponential and Linear Systems with Constant Coefficients 148
- 3.4.4 Elimination Method for Linear Systems with Constant Coefficients 164
- 3.5 Autonomous Systems of Differential Equations 170
- 3.6 First-Order Partial Differential Equations 178
  - 3.6.1 Linear Homogeneous First-Order PDE 179
  - 3.6.2 Quasilinear First-Order Partial Differential Equations 183
- 3.7 Exercises 186

4 Fourier Series 191
- 4.1 Introduction: Periodic, Piecewise Smooth Functions 191
  - 4.1.1 Periodic Functions 192
  - 4.1.2 Piecewise Continuous and Piecewise Smooth Functions 195
- 4.2 Fourier Series Expansions 198
  - 4.2.1 Series of Functions 199
  - 4.2.2 A Basic Trigonometric System 201
  - 4.2.3 Fourier Coefficients 203
- 4.3 Orthogonal Systems of Functions 205
  - 4.3.1 Inner Product 205
  - 4.3.2 Best Approximation in the Mean: Bessel's Inequality 208
- 4.4 The Convergence of Fourier Series 212
- 4.5 Differentiation and Integration of the Fourier Series 225
- 4.6 The Convergence in the Mean: Complete Systems 227
- 4.7 Examples of Fourier Expansions 235
- 4.8 The Complex form of the Fourier Series 241
- 4.9 Exercises 246

5 Fourier Transform 249
- 5.1 Improper Integrals 249
- 5.2 The Fourier Integral Formula 258
- 5.3 The Fourier Transform 265
- 5.4 Solving Linear Differential Equations 284
- 5.5 Moments Theorems 288
- 5.6 Sampling Theorem 297
- 5.7 Discrete Fourier Transform 298
- 5.8 Exercises 302

6 Laplace Transform 305
- 6.1 Introduction 305
- 6.2 Properties of the Laplace Transform 312
- 6.3 Inverse Laplace Transform 332
- 6.4 Solving Linear Differential Equations 344
- 6.5 The Dirac Delta Function 352
- 6.6 Exercises 358

7	**Second-Order Partial Differential Equations**		359
	7.1	Classification: Canonical Form	359
	7.2	The Wave Equation	378
		7.2.1 Infinite Vibrating String: D'Alembert Formula	381
		7.2.2 Finite Vibrating String: Fourier Method	385
		7.2.3 Laplace Transform Method for the Vibrating String	398
		7.2.4 Vibrations of a Rectangular Membrane: Two-Dimensional Wave Equation	400
	7.3	Vibrations of a Simply Supported Beam: Fourier Method	404
	7.4	The Heat Equation	409
		7.4.1 Modeling the Heat Flow from a Body in Space	409
		7.4.2 Heat Flow in a Finite Rod: Fourier Method	411
		7.4.3 Heat Flow in an Infinite Rod	415
		7.4.4 Heat Flow in a Rectangular Plate	417
	7.5	The Laplace's Equation	422
		7.5.1 Dirichlet Problem for a Rectangle	423
		7.5.2 Dirichlet Problem for a Disk	425
	7.6	Exercises	432
8	**Introduction to the Calculus of Variations**		435
	8.1	Classical Variational Problems	435
	8.2	General Frame of Calculus of Variations	443
	8.3	The Case $\mathcal{F}[y] = \int_a^b F(x, y, y')dx$	448
	8.4	The Case $\mathcal{F}[y] = \int_a^b F(x, y, y', \ldots, y^{(n)})dx$	456
	8.5	The Case $\mathcal{F}[y_1, \ldots, y_n] = \int_a^b F(x, y_1, \ldots, y_n, y_1', \ldots, y_n')dx$	460
	8.6	The Case $\mathcal{F}[z] = \iint_D F\left(x, y, z, \frac{\partial z}{\partial x}, \frac{\partial z}{\partial y}\right)dxdy$	467
	8.7	Isoperimetric Problems and Geodesic Problems	472
		8.7.1 Isoperimetric Problems	472
		8.7.2 Geodesic Problems	477
	8.8	Exercises	482
9	**Elements of Probability Theory**		485
	9.1	Sample Space: Event Space	485
	9.2	Probability Space	492
	9.3	Conditional Probability: Bayes Formula	506
	9.4	Discrete Random Variables	511
		9.4.1 Random Variables	511
		9.4.2 Expected Value; Moments	518
		9.4.3 Variance	529
		9.4.4 Discrete Uniform Distribution	535
		9.4.5 Bernoulli Distribution	536
		9.4.6 Binomial Distribution	536
		9.4.7 Poisson Distribution	540
		9.4.8 Geometric Distribution	543

	9.5	Continuous Random Variables	545
		9.5.1 The Probability Density Function; The Distribution Function	545
		9.5.2 Expected Value, Moments and Variance for Continuous Random Variables	548
		9.5.3 Characteristic Function	559
		9.5.4 The Uniform Distribution	568
		9.5.5 The Exponential Distribution	572
		9.5.6 The Normal Distribution	575
		9.5.7 Gamma Distribution	584
		9.5.8 Chi-Squared Distribution	588
		9.5.9 Student **t**-Distribution	589
	9.6	Limit Theorems	595
	9.7	Exercises	606
10	**Answers and Solutions to Exercises**		**611**
	10.1	Chapter 1	611
	10.2	Chapter 2	622
	10.3	Chapter 3	636
	10.4	Chapter 4	663
	10.5	Chapter 5	670
	10.6	Chapter 6	675
	10.7	Chapter 7	679
	10.8	Chapter 8	696
	10.9	Chapter 9	706
11	**Supplementary Materials**		**723**
	11.1	Normed, Metric and Hilbert Spaces	723
		11.1.1 Normed Vector Spaces	723
		11.1.2 Sequences and Series of Functions	727
		11.1.3 Metric Spaces. Some Density Theorems	734
		11.1.4 The Fields $\mathbb{Q}$, $\mathbb{R}$ and $\mathbb{C}$	747
		11.1.5 Hilbert Spaces	762
		11.1.6 Continuous Functions and Step Functions	772
		11.1.7 Orthonormal Systems in a Hilbert Space	774
	11.2	Complex Function Theory	785
		11.2.1 Differentiability of Complex Functions	785
		11.2.2 Integration of Complex Functions	791
		11.2.3 Power Series Representation	803
		11.2.4 Residue Theorem and Applications	810
Bibliography			**823**
Index			**825**

Basic Notations

Sets

$\mathbb{N}$ Natural numbers
$\mathbb{Z}$ Integer numbers
$\mathbb{Q}$ Rational numbers
$\mathbb{R}$ Real numbers
$\mathbb{C}$ Complex numbers

If $z = x + iy$ is a complex number, then:

$\text{Re}(z) = x$ (the real part of z)
$\text{Im}(z) = y$ (the imaginary part of z)
$\bar{z} = x - iy$ (the conjugate of z)
$|z| = \sqrt{x^2 + y^2}$ (the modulus of z)
$\exp(z) = e^x (\cos y + i \sin y)$

$\mathbb{R}^n = \{(x_1, \ldots, x_n) : x_1, \ldots, x_n \in \mathbb{R}\}$ the Euclidean n-dimensional space.

$\{\mathbf{i} = (1, 0, 0), \mathbf{j} = (0, 1, 0), \mathbf{k} = (0, 0, 1)\}$ the canonical base of $\mathbb{R}^3$

$\text{grad } U = \nabla U = \dfrac{\partial U}{\partial x}\mathbf{i} + \dfrac{\partial U}{\partial y}\mathbf{j}$ the gradient of the function $U(x, y)$

$\text{grad } U = \nabla U = \dfrac{\partial U}{\partial x}\mathbf{i} + \dfrac{\partial U}{\partial y}\mathbf{j} + \dfrac{\partial U}{\partial z}\mathbf{k}$ the gradient of the function $U(x, y, z)$

$\Delta U = \nabla^2 U = \dfrac{\partial^2 U}{\partial x^2} + \dfrac{\partial^2 U}{\partial y^2}$ the Laplacian (Laplace operator) of the function $U(x, y)$

$\Delta U = \nabla^2 U = \dfrac{\partial^2 U}{\partial x^2} + \dfrac{\partial^2 U}{\partial y^2} + \dfrac{\partial^2 U}{\partial z^2}$ the Laplacian (Laplace operator) of the function $U(x, y, z)$

$\text{div } V = \dfrac{\partial V_1}{\partial x} + \dfrac{\partial V_2}{\partial y} + \dfrac{\partial V_3}{\partial z}$ the divergence of the vector function $V(x, y, z) = V_1(x, y, z)\mathbf{i} + V_2(x, y, z)\mathbf{j} + V_3(x, y, z)\mathbf{k}$

C^n A function is said to be of class C^n if it is n times continuously differentiable
C^∞ A function is said to be of class C^∞ if it is infinitely differentiable

$$\delta_{i,j} = \begin{cases} 1 \text{ if } i = j \\ 0 \text{ if } i \neq j \end{cases} \text{ (Kronecker's symbol)}$$

$$C_n^k = \binom{n}{k} = \frac{n!}{k!(n-k)!} \quad k = 0, 1, \ldots, n \text{ binomial coefficients}$$

Hyperbolic Functions

$$\cosh(x) = \frac{e^x + e^{-x}}{2} \text{ (hyperbolic cosine)}$$

$$\sinh(x) = \frac{e^x - e^{-x}}{2} \text{ (hyperbolic sine)}$$

Euler Integrals

$$\Gamma(x) = \int_0^\infty e^{-t} t^{x-1} dt \text{ the gamma function (Euler integral of the second kind)}$$

$$B(x, y) = \int_0^1 t^{x-1}(1-t)^{y-1} dt \text{ the beta function (Euler integral of the first kind)}$$

Chapter 1
First-Order Differential Equations

In the first chapter we introduce the notion of differential equation, focusing on first-order ordinary differential equations. We show by several examples how differential equations can describe (model) physical phenomena in mathematical terms. The most common types of first-order differential equations are presented, together with the techniques for obtaining their solutions (either in implicit or explicit form). The last section is dedicated to the Theorem of existence and uniqueness of the solution of the Cauchy problem for first-order differential equations.

1.1 Introduction to Ordinary Differential Equations

Generally speaking, a *differential equation* is an equation containing one or more derivatives of a function. An *ordinary differential equation (ODE)* contains one or more *ordinary* derivatives of an unknown function, usually denoted by $y(x)$, where the *independent variable* x is running over a given interval $I \subset \mathbb{R}$. The unknown function, y, is also called the *dependent variable*. Its derivatives, $y'(x)$, $y''(x), \ldots, y^{(n)}(x)$, can be also denoted using the Leibniz notation: $y' = \frac{dy}{dx}$, $y'' = \frac{d^2y}{dx^2}$, etc. For instance, the following ordinary differential equations

$$xy' + y - 1 = 0, \tag{1.1}$$

$$y'' + 4y = 0 \tag{1.2}$$

can be written in the form:

$$x\frac{dy}{dx} + y - 1 = 0$$

© The Author(s), under exclusive license to Springer Nature Switzerland AG 2022
S. A. Popescu, M. Jianu, *Advanced Mathematics for Engineers and Physicists*,
https://doi.org/10.1007/978-3-031-21502-5_1

and, respectively,

$$\frac{d^2y}{dx^2} + 4y = 0.$$

An equation involving *partial derivatives* of a function of two or more independent variables, $u(x, y, z, \ldots)$, $\frac{\partial u}{\partial x}$, $\frac{\partial u}{\partial y}$, ..., $\frac{\partial^2 u}{\partial x^2}$, $\frac{\partial^2 u}{\partial x \partial y}$, ..., is called a *partial differential equation (PDE)* (see the example below).

$$\frac{\partial^2 u}{\partial x^2} + \frac{\partial^2 u}{\partial y^2} = 0. \tag{1.3}$$

Partial differential equations will be considered in Chap. 7 and in Sect. 3.6. In this chapter we study only ordinary differential equations.

The *order* of a differential equation is the order of the highest derivative in the equation. For example, (1.1) is a first-order ODE, (1.2) is a second-order ODE, while (1.3) is a second-order PDE.

The general form of an *n*th-order ODE is:

$$F(x, y, y', \ldots, y^{(n)}) = 0, \tag{1.4}$$

where F is a real-valued continuous function of $n + 2$ variables. If Eq. (1.4), can be solved for the highest derivative, as

$$y^{(n)} = f(x, y, y', \ldots, y^{(n-1)}), \tag{1.5}$$

where f is a continuous function of $n+1$ variables, then (1.5) is said to be the *normal form* of Eq. (1.4). For example, the normal form of the first-order equation (1.1) is

$$y' = -\frac{y-1}{x}, \quad x \neq 0, \tag{1.6}$$

while the normal form of the second-order equation (1.2) is

$$y'' = -4y. \tag{1.7}$$

Let I be a real interval and k a nonnegative integer. We say that a function $f : I \to \mathbb{R}$ is *of (differentiability) class C^k* if the derivatives $f', f'', \ldots, f^{(k)}$ exist and are continuous. A function of class C^1 is said to be *continuously differentiable*. If the function f has derivatives of any order, then it is said to be *of class C^∞*.

Definition 1.1 A solution of Eq. (1.4) on the interval I is a function $\varphi : I \to \mathbb{R}$ of class C^n that satisfies the equation for all x in I:

$$F(x, \varphi(x), \varphi'(x), \ldots, \varphi^{(n)}(x)) = 0, \text{ for all } x \in I. \tag{1.8}$$

1.1 Introduction to Ordinary Differential Equations

The graph of φ is called an *integral curve* (or a *solution curve*) of the ODE (1.4).

In this chapter we study *first-order* differential equations. The general form of a first-order ODE is

$$F(x, y, y') = 0, \tag{1.9}$$

while the *normal form* of a first-order differential equation is

$$y' = f(x, y), \tag{1.10}$$

A solution of Eq. (1.9) on the interval I is a function $\varphi : I \to \mathbb{R}$ of class C^1 that satisfies the equation for all x in I:

$$F\left(x, \varphi(x), \varphi'(x)\right) = 0, \text{ for all } x \in I. \tag{1.11}$$

For instance, as one can directly verify, $y = \frac{1}{x} + 1$ is a solution of the first-order equation (1.1) on the interval $I = (0, \infty)$. But we can see that the functions

$$y_1 = \frac{2}{x} + 1, \quad y_2 = -\frac{1}{x} + 1, \quad y_3 = 1, \text{ etc.} \tag{1.12}$$

also satisfy Eq. (1.1) and, as a matter of fact, any function of the form

$$y = \frac{C}{x} + 1, \tag{1.13}$$

where $C \in \mathbb{R}$ is an arbitrary constant, is a solution of Eq. (1.1) on the interval I. We say that (1.13) is the *general solution* of the differential equation (1.1).

The *general solution* of a first-order ODE is a *family of solutions* which depend on an arbitrary constant C. Each choice of the constant in the general solution yields a *particular solution* of the differential equation. For example, in the general solution (1.13) of Eq. (1.1) if we take $C = 2$, $C = -1$ and $C = 0$, the particular solutions (1.12) are obtained. The graphs of all the particular solutions form an infinite family of integral curves, one for each choice of the constant C (see Fig. 1.1).

Some equations may have solutions which cannot be obtained by choosing a value of the constant C in the general solution. These solutions are called *singular solutions*.

Example 1.1 The first-order differential equation

$$y'^2 - xy' + y = 0 \tag{1.14}$$

has the general solution

$$y = Cx - C^2 \tag{1.15}$$

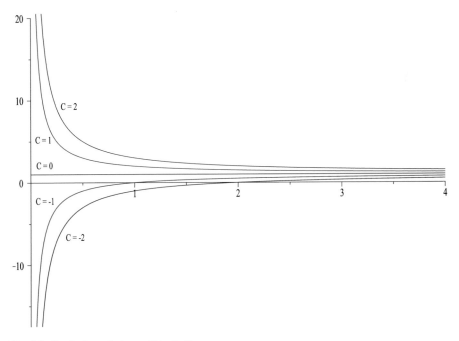

Fig. 1.1 Particular solutions of Eq. (1.1)

(a family of straight lines), but it also has the *singular solution*

$$y_s = \frac{x^2}{4} \tag{1.16}$$

(a parabola). You can verify by direct differentiation that (1.15) and (1.16) are solutions of Eq. (1.14). In Sect. 1.8 we will see that any solution of Eq. (1.14) is either (1.16), or of the form (1.15).

In many cases we have to find the particular solution $y(x)$ that passes through a given point, (x_0, y_0), so the function $y(x)$ must satisfy the *initial condition*:

$$y(x_0) = y_0. \tag{1.17}$$

This is called a *Cauchy problem* or an *initial value problem*. Thus, a first-order Cauchy problem is of the form

$$F(x, y, y') = 0, \quad y(x_0) = y_0, \tag{1.18}$$

where x_0 and y_0 are given numbers.

1.1 Introduction to Ordinary Differential Equations

The terms *initial condition*, *initial value problem* originate in physical systems where the independent variable is time t and where $y(t_0) = y_0$ represents the value of the function $y(t)$ at the initial time, t_0 (see Example 1.3 where the initial time is $t_0 = 0$).

Consider, for instance, the following Cauchy problem for Eq. (1.1):

$$xy' + y - 1 = 0, \quad y(1) = 3. \tag{1.19}$$

We saw that the general solution of the equation is given by (1.13). To solve the Cauchy problem, we impose the initial condition $y(1) = 3$ and find the constant C:

$$y(1) = C + 1 = 3 \Rightarrow C = 2.$$

Hence the function $y(x) = \frac{2}{x} + 1$ is the solution of the Cauchy problem (1.19).

The effect of the initial condition in this example is to pick out from the family of integral curves (1.13) the one that passes through the point $(1, 3)$. This suggests that a Cauchy problem may be expected to have a unique solution. The existence and uniqueness of the solution of an initial value problem will be analyzed in Sect. 1.9. The example below shows that a Cauchy problem may have more than one solution or even no solution.

Example 1.2 Consider the following Cauchy problems (for the ODE (1.14)):

$$y'^2 - xy' + y = 0, \quad y(2) = 1; \tag{1.20}$$

$$y'^2 - xy' + y = 0, \quad y(1) = 2. \tag{1.21}$$

To solve (1.20) we need to find the constant C such that

$$y(2) = 2C - C^2 = 1.$$

It follows that $C = 1$, hence $y(x) = x - 1$ is a solution of the Cauchy problem (1.20). But we also notice that the singular solution of the equation, (1.16), satisfies the condition $y_s(2) = 1$, so we have found two solutions for the initial value problem (1.20).

On the other hand, we can see that the parabola (1.16) does not pass through the point $(1, 2)$ and there is no real constant C such that

$$y(1) = C - C^2 = 2,$$

so the Cauchy problem (1.21) has no solution.

Many laws that describe the behavior of the natural world can be expressed as mathematical relations between some physical quantities and their rates of change. These mathematical relations are *differential equations* (since the rates of change are *derivatives*). A differential equation that describes some physical process (such as the motion of a body, the flow of current in electrical circuits, the dissipation of heat in solid objects, or the dynamic of population) is called a *mathematical model* of the process (see [3, 20, 21, 26, 44] for detailed presentations of such models).

Example 1.3 **Newton's law of cooling** states that the temperature of an object changes at a rate proportional to the difference between the temperature of the object itself and the temperature of the surrounding medium. If we denote by $T(t)$ the temperature of the object at time t and by T_m the temperature of the surrounding medium (considered to be constant), then the mathematical expression of the Newton's law of cooling (or warming) is the following differential equation:

$$\frac{dT}{dt} = -k(T - T_m), \tag{1.22}$$

where k is a positive constant of proportionality. Notice that if $T > T_m$, then $\frac{dT}{dt} < 0$, so the temperature is a decreasing function (the object is cooling). Otherwise, if the temperature of the object is lower then the temperature of the surrounding medium ($T < T_m$), then $\frac{dT}{dt} > 0$, so the temperature is an increasing function (the object is warming).

If we know the initial temperature of the object,

$$T(0) = T_0, \tag{1.23}$$

then we can predict the temperature of the object at any time t by solving the Cauchy problem (1.22)–(1.23).

Equation (1.22) can be written in the following equivalent form:

$$\frac{T'(t)}{T(t) - T_m} = -k.$$

We integrate with respect to t

$$\int \frac{T'(t)}{T(t) - T_m} dt = -\int k \, dt$$

and obtain:

$$\ln |T(t) - T_m| = -kt + c,$$

where c is a real constant. Hence

$$T(t) - T_m = \pm e^{-kt+c}$$

and, by denoting $\pm e^c = C \in \mathbb{R}$, the *general solution* of Eq. (1.22) is:

$$T(t) = T_m + Ce^{-kt}. \tag{1.24}$$

1.2 Separable Equations

We have to find the *particular solution* that satisfies the initial condition (1.23), i.e. to find the real constant C such that $T(0) = T_0$. By replacing $t = 0$ in Eq. (1.24), we obtain the value of constant: $C = T_0 - T_m$. Hence the solution of the Cauchy problem (1.22)–(1.23) is:

$$T(t) = T_m + (T_0 - T_m)e^{-kt}. \qquad (1.25)$$

In the next sections we will consider some classes of first-order ODEs that can be integrated in a finite number of steps involving elementary operations, quadratures (integral operations) and appropriate substitutions. For more details regarding first-order differential equations, we refer the reader to [2, 3, 6, 13, 21, 44]. A brief review of the major trends in the historical development of the subject is given in [3].

1.2 Separable Equations

A first-order differential equation $y' = f(x, y)$ is a *separable equation* if the function $f(x, y)$ is a product of a function of x and a function of y:

$$\frac{dy}{dx} = g(x)h(y). \qquad (1.26)$$

When $h(y) \neq 0$, the separable equation can be written in the *differential form*:

$$\frac{1}{h(y)} dy = g(x) dx. \qquad (1.27)$$

We integrate:

$$\int \frac{1}{h(y)} dy = \int g(x) dx \qquad (1.28)$$

and obtain:

$$H(y) = G(x) + C, \qquad (1.29)$$

where $H(y)$ is a primitive of the function $\frac{1}{h(y)}$, $G(x)$ is a primitive of the function $g(x)$, and $C \in \mathbb{R}$ is an arbitrary constant. Notice that we do not need to use two constants in the integration of Eq. (1.28), because if we write $H(y) + C_1 = G(x) + C_2$, then the difference $C_2 - C_1$ can be replaced by a single constant C, as in (1.29). Sometimes it is convenient to use instead of the constant C, the natural logarithm of a positive constant, $C = \ln C_1$ (see Example 1.4), or another expression (for instance, $C = \frac{C_1^2}{2}$, as in Example 1.5).

Equation (1.29) implicitly defines the general solution $y(x)$. Sometimes it is possible to solve it explicitly for $y(x)$, other times we may leave it in the implicit form (see Example 1.5).

Example 1.4 Find the general solution of the equation

$$y' = -xy. \tag{1.30}$$

We notice that (1.30) is a separable equation and can be written:

$$\frac{1}{y}dy = -xdx. \tag{1.31}$$

We integrate and obtain:

$$\int \frac{dy}{y} = -\int xdx$$

if $y > 0$ or $y < 0$. In both cases we have:

$$\ln|y| = -\frac{x^2}{2} + k,$$

where k is a real constant. If we replace the constant k with $\ln C$, $C > 0$, we get:

$$|y| = C \exp\left(-\frac{x^2}{2}\right),$$

hence

$$y = \pm C \exp\left(-\frac{x^2}{2}\right).$$

We denote by $C_1 = \pm C$, so the general solution of Eq. (1.30) is

$$y = C_1 \exp\left(-\frac{x^2}{2}\right), \tag{1.32}$$

where $C_1 \in \mathbb{R}$ is an arbitrary constant (notice that we may have $C_1 = 0$; the solution in this case is the constant function $y(x) = 0$, which satisfies the initial Eq. (1.30), although it is not a solution of (1.31)). Except for the case $C_1 = 0$, the solutions are *bell-shaped curves* (as can be seen in Fig. 1.2).

1.2 Separable Equations

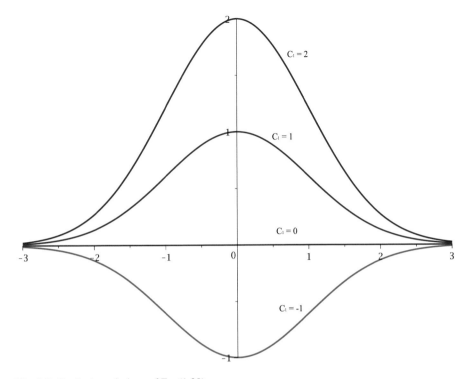

Fig. 1.2 Particular solutions of Eq. (1.30)

Example 1.5 Find the general solution of the equation

$$y' = -\frac{x}{y}. \tag{1.33}$$

We separate the variables:

$$y\,dy = -x\,dx, \tag{1.34}$$

integrate:

$$\int y\,dy = -\int x\,dx$$

and obtain:

$$\frac{y^2}{2} = -\frac{x^2}{2} + C.$$

If we replace the constant C with $\frac{C_1^2}{2}$, $C_1 > 0$, the general solution is written:

$$x^2 + y^2 = C_1^2$$

(a family of circles with the center at the origin).

Now, let us consider the first-order differential equation of the form

$$y' = g(ax + by + c), \qquad (1.35)$$

where g is a continuous function. It is not a separable equation, but, using an adequate substitution, it may be transformed into a separable equation We denote:

$$z(x) = ax + by(x) + c.$$

Differentiating this last equality w.r.t. x, we get:

$$z' = a + by' = a + bg(z),$$

so we obtain the following separable equation:

$$\frac{dz}{a + bg(z)} = dx,$$

(if $a + bg(z) \neq 0$). We integrate,

$$\int \frac{dz}{a + bg(z)} = x + C_1,$$

and obtain that the general solution (in an implicit form) can be written as $G(x, z, C_1) = 0$. If we replace here z by $ax + by + c$, we find the general solution of the initial equation.

Example 1.6 Let us integrate the equation $y' = (2x + 3y + 1)^2$ and solve the Cauchy problem

$$\begin{cases} y' = (2x + 3y + 1)^2 \\ y(-\frac{1}{2}) = 0. \end{cases}$$

It is easy to see that we cannot separate the variables. Make the substitution $z = 2x + 3y + 1$, so

$$z' = 2 + 3y' = 2 + 3(2x + 3y + 1)^2 = 2 + 3z^2.$$

1.2 Separable Equations

Hence $\frac{dz}{2+3z^2} = dx$ and $\int \frac{dz}{2+3z^2} = x + C_1$, i.e.

$$x + C_1 = \frac{1}{\sqrt{3}} \int \frac{d\left(\sqrt{3}z\right)}{2+\left(\sqrt{3}z\right)^2} \overset{\sqrt{3}z=u}{=} \frac{1}{\sqrt{3}} \int \frac{du}{2+u^2} =$$

$$= \frac{1}{\sqrt{6}} \tan^{-1} \frac{u}{\sqrt{2}} = \frac{1}{\sqrt{6}} \tan^{-1} \frac{\sqrt{3}z}{\sqrt{2}}.$$

Now, by replacing z with $2x + 3y + 1$, we find the general solution, in an implicit form:

$$\frac{1}{\sqrt{6}} \tan^{-1} \frac{\sqrt{3}(2x+3y+1)}{\sqrt{2}} - x = C_1. \qquad (1.36)$$

It is easy to find the explicit form. For this, let us multiply the last equality by $\sqrt{6}$ and then apply the tangent function:

$$\frac{\sqrt{3}(2x+3y+1)}{\sqrt{2}} = \tan\left[\sqrt{6}(x+C_1)\right],$$

hence

$$y = \frac{\sqrt{2}}{3\sqrt{3}} \tan\left[\sqrt{6}(x+C_1)\right] - \frac{2x+1}{3}.$$

From $y(-\frac{1}{2}) = 0$ we get $C_1 = \frac{1}{2} + \frac{k\pi}{\sqrt{6}}$, where $k \in \mathbb{Z}$ and, since $\tan x = \tan(x+k\pi)$, we obtain that the solution of the Cauchy problem is:

$$y = \frac{\sqrt{2}}{3\sqrt{3}} \tan\left[\sqrt{6}\left(x+\frac{1}{2}\right)\right] - \frac{2x+1}{3}$$

Example 1.7 (Dynamics of Population Models) By "population", in general, we mean a multitude of individuals from a species of plants, animals, humans, or a set of radioactive particles etc. For instance, if we denote by $y(t)$ the amount of a radioactive substance at the moment t, since the rate of radioactive decay at the same moment t is proportional to the amount $y(t)$, we obtain the following differential equation with separable variables:

$$y' = -ky, \qquad (1.37)$$

where $k > 0$ is the decay constant. The sign "−" means that the amount y decreases with time. Suppose that at the moment $t = t_0$ we know that

$$y(t_0) = y_0. \tag{1.38}$$

Then, by solving the Cauchy problem (1.37)–(1.38) we get:

$$y(t) = y_0 \exp(-k(t - t_0)). \tag{1.39}$$

If we need to find the mathematical model for a multiplication process, e.g. the multiplication of neutrons in chain reactions or the multiplication of bacteria on the assumption that the rate of multiplication is proportional to the number of bacteria at the considered moment, we obtain the equation

$$y' = ky,$$

where $k > 0$. The solution in this case is $y(t) = y_0 \exp(k(t - t_0))$. Hence, if k is an arbitrary real number, all the above phenomena have the same mathematical model:

$$y' = ky. \tag{1.40}$$

This is the simplest model for the dynamics of populations. Assume now that $k = m - n$, where m is the coefficient of the relative birth rate and n is the coefficient of the relative death rate. It is obvious that if $m > n$ the population increases to infinity and if $m < n$ the population decreases to zero. But, for a large population, the resources available become fewer and fewer, leading to a decrease in the birth rate and an increase in the death rate. This means that there exist four positive constants b_1, b_2, b_3 and b_4 such that $m = b_1 - b_2 y$ and $n = b_3 + b_4 y$. Then

$$k = b_1 - b_3 - (b_2 + b_4)y = \alpha(A - y),$$

where $\alpha = b_2 + b_4$ and $A = \frac{b_1 - b_3}{b_2 + b_4}$. Thus, Eq. (1.40) becomes

$$y' = \alpha(A - y)y, \tag{1.41}$$

which is a separable ODE, called the *logistic equation*. It is used in demography, ecology, medicine, psychology, sociology etc. By solving Eq. (1.41) with the initial condition (1.38), we obtain the solution

$$y(t) = \frac{A}{1 + \left(\frac{A}{y_0} - 1\right)\exp(-A\alpha(t - t_0))},$$

where A and α are two parameters which can be computed by measuring the values of $y(t)$ at two distinct moments t_1 and t_2.

1.3 Homogeneous Equations

A function $f(x, y)$ is said to be *homogeneous of degree k* if it satisfies the equality $f(tx, ty) = t^k f(x, y)$, for any positive number t. For instance, $f(x, y) = \frac{\sqrt[3]{x^3+y^3}}{x+y}$ is homogeneous of degree 0. An ODE of the first order $y' = f(x, y)$ is *homogeneous* if $f(x, y)$ is homogeneous of degree 0.

We assume in the following that $x \in I$, where I is an interval which does not contain 0. It is very easy to see that the homogeneous function $f(x, y)$ can be written as a function of $\frac{y}{x}$:

$$f(x, y) = f\left(1, \frac{y}{x}\right) =: g\left(\frac{y}{x}\right),$$

so it is natural to substitute the initial function $y(x)$ with a new function $u(x) = \frac{y(x)}{x}$. From $y(x) = xu(x)$, by differentiation with respect to x, we get $y' = u + xu'$. Substituting this form of y' in

$$y' = f(x, y) = g\left(\frac{y}{x}\right) = g(u),$$

we obtain

$$u' = \frac{du}{dx} = \frac{1}{x}[g(u) - u],$$

or

$$\frac{du}{g(u) - u} = \frac{dx}{x},$$

where $g(u) - u \neq 0$. This is an equation with separable variables, which can be solved by a simple quadrature

$$\int \frac{du}{g(u) - u} = \ln|x| + \ln|C_1|,$$

because any constant C can be written as $\ln|C_1|$. Finally we come back and replace u with y/x etc.

Example 1.8 Let us consider the following Cauchy problem:

$$\begin{cases} y' = \dfrac{x + 2y}{2x + y} \\ y(1) = 2 \end{cases}.$$

Since the function $f(x, y) = \frac{x+2y}{2x+y}$ is homogeneous:

$$f(tx, ty) = \frac{tx + 2ty}{2tx + ty} = \frac{x + 2y}{2x + y} = f(x, y),$$

we can use the substitution $u(x) = \frac{y(x)}{x}$ and the ODE becomes:

$$u + xu' = \frac{1 + 2u}{2 + u},$$

or

$$\frac{2+u}{u^2 - 1} du = -\frac{1}{x} dx.$$

By integration we get

$$\int \frac{2+u}{u^2 - 1} du = -\ln|x| + \ln|C_1|.$$

Since

$$\frac{2+u}{u^2 - 1} = \frac{3}{2(u-1)} - \frac{1}{2(u+1)},$$

we have

$$\frac{3}{2} \ln|u - 1| - \frac{1}{2} \ln|u + 1| = \ln\left|\frac{C_1}{x}\right|,$$

or, from $u = y/x$,

$$(y - x)^3 - C(y + x) = 0,$$

where $y > x$ and $C > 0$. Since $y(1) = 2$ we get $C = \frac{1}{3}$, thus the unique solution of the above Cauchy problem is the plane curve

$$3(y - x)^3 - y - x = 0, \ y > x, \qquad (1.42)$$

considered in a sufficiently small neighborhood of the point $(1, 2)$.

The Case $y' = f\left(\frac{a_1 x + b_1 y + c_1}{a_2 x + b_2 y + c_2}\right)$

In this case, if at least one of c_1, c_2 is nonzero, the equation is not homogeneous. But, using an adequate change of variables (by changing the independent variable, x, as well as the dependent variable, y) the equation can be transformed into a homogeneous one.

1.3 Homogeneous Equations

There are two distinct situations:

(1) The lines $a_1 x + b_1 y + c_1 = 0$, $a_2 x + b_2 y + c_2 = 0$ have a common point, $M_0(x_0, y_0, z_0)$. Since $a_1 x_0 + b_1 y_0 + c_1 = 0$ and $a_2 x_0 + b_2 y_0 + c_2 = 0$, we get:

$$y' = f\left(\frac{a_1(x-x_0) + b_1(y-y_0)}{a_2(x-x_0) + b_2(y-y_0)}\right). \tag{1.43}$$

Let us make the following substitutions: $x - x_0 = u$, $y - y_0 = v(u)$ (each function of x is also a new function of $x - x_0$). Then,

$$y' = v'(u) \cdot u'(x) = v'(u)$$

and the ODE (1.43) becomes:

$$v'(u) = f\left(\frac{a_1 u + b_1 v}{a_2 u + b_2 v}\right),$$

which is a homogeneous equation in u and v. We solve it as above and finally put $x - x_0$ instead of u and $y - y_0$ instead of v, respectively.

(2) The lines $a_1 x + b_1 y + c_1 = 0$ and $a_2 x + b_2 y + c_2 = 0$ are parallel. This means that $a_2 x + b_2 y = \gamma(a_1 x + b_1 y)$ and the equation becomes:

$$y' = f\left(\frac{a_1 x + b_1 y + c_1}{\gamma(a_1 x + b_1 y) + c_2}\right).$$

If $b_1 = 0$, one can find the solution by a simple quadrature (integration). Assume that $b_1 \neq 0$. Let us make the substitution $a_1 x + b_1 y(x) = w(x)$, a new function of x, and find:

$$w' = b_1 f\left(\frac{w + c_1}{\gamma w + c_2}\right) + a_1 = g(w),$$

a separable equation which can be easily integrated:

$$\int \frac{dw}{g(w)} = \int dx = x + C_1 \Rightarrow G(x, w, C_1) = 0.$$

We replace w by $a_1 x + b_1 y$ and find the general solution of the initial equation.

1.4 First-Order Linear Differential Equations

A *linear differential equation of the first order* is an equation of the form:

$$y' + a(x)y = b(x), \tag{1.44}$$

where $a(x)$ and $b(x)$ are continuous functions defined on a given interval I. If the function $b(x)$ is identically zero, then the equation

$$y' + a(x)y = 0 \tag{1.45}$$

is said to be a *homogeneous linear equation*. Otherwise, Eq. (1.44) is called *non-homogeneous* or *inhomogeneous* (there is no relation between *homogeneous* linear equations and *homogeneous* equations studied in Sect. 1.3, the word *homogeneous* has two different meanings). Equation (1.44) is a linear equation because the "variables" y and y' appear to power one, i.e. the left side of the equation is linear in the variables y and y'.

Let $y_P(x)$ be a fixed *particular* solution of Eq. (1.44) and let $y(x)$ be an arbitrary solution of this equation. We have to find the *general* solution of Eq. (1.44). So, we have:

$$y'(x) + a(x)y(x) = b(x),$$
$$y'_P(x) + a(x)y_P(x) = b(x)$$

and by subtraction we get:

$$[y(x) - y_P(x)]' + a(x)[y(x) - y_P(x)] = 0,$$

i.e. the function $y(x) - y_P(x)$ is a solution of the homogeneous equation (1.45). Hence, for solving Eq. (1.44) it is sufficient to find the general solution of its associated homogeneous equation (1.45) and a particular solution for the initial, non-homogeneous equation (1.44).

The homogeneous equation (1.45) can be also written in the form:

$$\frac{dy}{y} = -a(x)dx,$$

so it is a *separable* differential equation. By integrating both sides, we obtain:

$$\ln|y| = -\int a(x)\,dx + \ln|C|. \tag{1.46}$$

1.4 First-Order Linear Differential Equations

Recall that $\int a(x)\,dx$ is a fixed primitive (antiderivative) of $a(x)$. The function

$$y_{GH}(x) = C \exp\left(-\int a(x)\,dx\right), \tag{1.47}$$

where C is an arbitrary constant, is the general solution of the associated homogeneous equation (1.45) (we also considered here the case $C = 0$ which corresponds to $y(x) \equiv 0$).

The method used to find a particular solution of the non-homogeneous equation (1.44) was developed by the Italian-French mathematician Joseph-Louis Lagrange and it is known as the *method of variation of constants* or the *method of variation of parameters*. Lagrange substituted the arbitrary constant C with a function of x, $C(x)$, and forced the new function

$$y_P(x) = C(x) \exp\left(-\int a(x)\,dx\right) \tag{1.48}$$

to be a particular solution for the non-homogeneous equation (1.44).

Since

$$y'_P(x) = C'(x) \exp\left(-\int a(x)\,dx\right) + C(x) \cdot [-a(x)] \exp\left(-\int a(x)\,dx\right),$$

by substituting (1.48) into (1.44) we obtain that

$$C'(x) \exp\left(-\int a(x)\,dx\right) = b(x),$$

so it follows that

$$C(x) = \int b(x) \exp\left(\int a(x)\,dx\right) dx,$$

i.e.

$$y_P(x) = \left[\int b(x) \exp\left(\int a(x)\,dx\right) dx\right] \exp\left(-\int a(x)\,dx\right). \tag{1.49}$$

Since $y(x) - y_P(x)$ is the general solution of the homogeneous equation (1.45), the general solution of the initial non-homogeneous equation (1.44) is

$$y(x) = \left[\int b(x) \exp\left(\int a(x)\,dx\right) dx\right] \exp\left(-\int a(x)\,dx\right) + \tag{1.50}$$
$$+ C_1 \exp\left(-\int a(x)\,dx\right),$$

where C_1 is an arbitrary real constant.

Now, suppose that we have to solve the Cauchy problem

$$y' + a(x)y = b(x), \quad y(x_0) = y_0. \tag{1.51}$$

In the formula (1.50) we can take instead of $\int a(x)\,dx$ the particular primitive $\int_{x_0}^{x} a(t)\,dt$, etc., and find:

$$y(x) = \left[\int_{x_0}^{x} b(t) \exp\left(\int_{x_0}^{t} a(u)\,du\right) dt\right] \exp\left(-\int_{x_0}^{x} a(t)\,dt\right) +$$
$$+ C_1 \exp\left(-\int_{x_0}^{x} a(t)\,dt\right).$$

The initial condition $y(x_0) = y_0$ implies that $C_1 = y_0$ and the next theorem is proved (see also [3, 6]):

Theorem 1.1 *The Cauchy problem for the linear differential equation (1.51) has a unique solution, given by the formula:*

$$y(x) = \left[\int_{x_0}^{x} b(t) \exp\left(\int_{x_0}^{t} a(u)\,du\right) dt\right] \exp\left(-\int_{x_0}^{x} a(t)\,dt\right) + \tag{1.52}$$
$$+ y_0 \exp\left(-\int_{x_0}^{x} a(t)\,dt\right).$$

In practice, the formula (1.52) is used by computers or by those who like very much to use "magic" formulas. Usually, at least for teaching purposes, it is desired to apply the above Lagrange's reasoning.

Example 1.9 Let us solve the following Cauchy problem:

$$y' + 3y = \exp(2x), \quad y(0) = 2.$$

1.4 First-Order Linear Differential Equations

First of all let us use the sophisticated formula (1.52).

$$y(x) = \left[\int_0^x \exp(2t) \exp\left(\int_0^t 3\,du\right) dt\right] \exp\left(-\int_0^x 3\,dt\right) + \quad (1.53)$$
$$+ 2\exp\left(-\int_0^x 3\,dt\right).$$

Let us compute the brackets.

$$\int_0^x \exp(2t) \exp\left(\int_0^t 3\,du\right) dt = \int_0^x \exp(5t)\,dt = \frac{1}{5}\left[\exp(5x) - 1\right].$$

So, (1.53) becomes

$$y(x) = \frac{1}{5}\left[\exp(5x) - 1\right] \cdot \exp(-3x) + 2\exp(-3x) =$$
$$= \frac{1}{5}\exp(2x) + \frac{9}{5}\exp(-3x).$$

Now, let us use directly the Lagrange method only, without "magic" formulas.

Firstly, we find the general solution of the homogeneous associated equation $y' + 3y = 0$ or $\frac{dy}{y} = -3dx$. So the general solution of the associated homogeneous equation is

$$y_{GH} = C\exp(-3x).$$

The second step is to search for a particular solution of the non-homogeneous equation,

$$y_P = C(x)\exp(-3x).$$

By replacing this function in the non-homogeneous equation $y' + 3y = \exp(2x)$, we obtain

$$C'(x)\exp(-3x) - 3C(x)\exp(-3x) + 3C(x)\exp(-3x) = \exp(2x),$$

or

$$C'(x) = \exp(5x),$$

i.e. $C(x) = \frac{1}{5}\exp(5x)$. Thus, $y_P = \frac{1}{5}\exp(2x)$. Since the general solution of the non-homogeneous equation is $y = y_{GH} + y_P$, we obtain

$$y = C\exp(-3x) + \frac{1}{5}\exp(2x).$$

From the initial condition $y(0) = 2$, we get $C = \frac{9}{5}$ and the solution of our Cauchy problem is

$$y = \frac{9}{5}\exp(-3x) + \frac{1}{5}\exp(2x).$$

The readers may find that the latter direct method is much easier.

1.5 Bernoulli Equations

A *Bernoulli equation* is an equation of the form:

$$y' + a(x)y = b(x)y^\alpha, \tag{1.54}$$

where $a(x)$ and $b(x)$ are continuous functions defined on an interval I and α is a real constant. Since for $\alpha = 0$ Eq. (1.54) is a linear equation and since for $\alpha = 1$ the equation becomes

$$y' + [a(x) - b(x)]y = 0,$$

i.e. a homogeneous linear equation, we can assume in the following that $\alpha \neq 0$, and $\alpha \neq 1$.

Dividing the equality (1.54) by y^α we get

$$y'y^{-\alpha} + a(x)y^{1-\alpha} = b(x). \tag{1.55}$$

Let us substitute $z(x)$ for $y^{1-\alpha}(x)$. By differentiating the equality $z = y^{1-\alpha}$ we get $z' = (1-\alpha)y^{-\alpha}y'$, so $y'y^{-\alpha} = \frac{z'}{1-\alpha}$. Thus, Eq. (1.55) becomes

$$z' + (1-\alpha)a(x)z(x) = (1-\alpha)b(x),$$

which is a linear equation in the unknown function $z(x)$. We integrate it by following the method described above and then compute $y(x) = z^{\frac{1}{1-\alpha}}$.

Another method is to directly use the Lagrange method of variation of constant after finding the general solution of the homogeneous associated equation $y' + a(x)y = 0$. (See Example 1.10).

Example 1.10 Let us consider the following Cauchy problem

$$\begin{cases} y' + y = 2\exp(-x)y^2 \\ y(0) = 1. \end{cases} \tag{1.56}$$

1.5 Bernoulli Equations

Method I (Using Substitution) Since this ODE is a Bernoulli equation with $\alpha = 2$, we make the substitution $z = y^{-1}$ and obtain $z' = -y^{-2}y'$. Since, by division by y^2 our equation becomes $y'y^{-2} + y^{-1} = 2\exp(-x)$, introducing z, one obtains the ODE in the unknown function $z(x)$:

$$-z' + z = 2\exp(-x), \tag{1.57}$$

which is a linear equation. The general solution of the homogeneous associated equation $-z' + z = 0$ is $z_{GH}(x) = C\exp(x)$, where C is an arbitrary constant. Let us search for a particular solution of the form: $z_P(x) = C(x)\exp(x)$, where this time $C(x)$ is a function of x (we just varied the constant C). Now we force this z_P to be a solution of (1.57).

$$-C'(x)\exp(x) - C(x)\exp(x) + C(x)\exp(x) = 2\exp(-x),$$

or

$$C'(x) = -2\exp(-2x),$$

i.e. $C(x) = \exp(-2x)$. Thus,

$$z_P(x) = \exp(-2x) \cdot \exp(x) = \exp(-x).$$

Since the general solution of the non-homogeneous equation (1.57) is the sum of a particular solution of this last equation and the general solution of its associated homogeneous equation, we finally get:

$$z(x) = \exp(-x) + C\exp(x).$$

Coming back to y, from $y = z^{-1}$ we find:

$$y(x) = \frac{1}{\exp(-x) + C\exp(x)}, \tag{1.58}$$

which is the general solution of our initial Eq. (1.56). From the initial condition $y(0) = 1$ we get $C = 0$ thus, the solution of the initial Cauchy problem (1.56) is $y(x) = \exp(x)$.

Method II (Using Directly the Variation of Constant) The associated homogeneous linear equation is $y' + y = 0$, so $y_{GH}(x) = C\exp(-x)$. To find the general solution of the Bernoulli equation from (1.56), we vary the constant C: we search for a solution of the form $y(x) = C(x)\exp(-x)$ and obtain:

$$C'(x)\exp(-x) - C(x)\exp(-x) + C(x)\exp(-x)$$
$$= 2\exp(-x) \cdot C^2(x)\exp(-2x) = 2\exp(-3x)C^2(x),$$

or

$$\frac{dC}{C^2} = 2\exp(-2x)dx,$$

which is an ODE with separable variables. By integrating it directly we get

$$-\frac{1}{C(x)} = -\exp(-2x) - K,$$

where K is an arbitrary constant. So $C(x) = \frac{1}{\exp(-2x)+K}$. Thus, the general solution of (1.56) is

$$y(x) = \frac{\exp(-x)}{\exp(-2x)+K} = \frac{1}{\exp(-x)+K\exp(x)},$$

and it is easy to see that this one coincide with (1.58).

1.6 Riccati Equations

A *Riccati equation* is an equation of the form:

$$y' = a(x)y^2 + b(x)y + c(x), \tag{1.59}$$

where $a(x)$, $b(x)$, $c(x)$ are continuous functions on a given interval I. For some particular functions $a(x)$, $b(x)$, $c(x)$ this equation becomes an equation of a previous type. For instance, if $a(x)$, $b(x)$, $c(x)$ are constant numbers a, b, c, the equation is a separable equation. If $a(x) = 0$ for all x in I, it becomes a linear equation, if $c(x) = 0$, it is a Bernoulli equation.

In general, if we know a particular solution of it, $y_1(x)$, the Riccati equation can be transformed into a Bernoulli equation by the substitution

$$z(x) = y(x) - y_1(x). \tag{1.60}$$

Thus, since $y_1(x)$ is a particular solution of (1.59), one can write:

$$y_1' = a(x)y_1^2 + b(x)y_1 + c(x).$$

By subtracting this last equality from the equality (1.59) we get:

$$y' - y_1' = a(x)(y - y_1)(y + y_1) + b(x)(y - y_1).$$

1.6 Riccati Equations

Since $y = y_1 + z$, we finally obtain:

$$z' - [2a(x)y_1(x) + b(x)]z = a(x)z^2,$$

which is indeed a Bernoulli equation in z.

Since always this last Bernoulli equation has $\alpha = 2$, we can use even from the beginning the substitution $y(x) = y_1(x) + \frac{1}{u(x)}$ to obtain directly a linear equation in u.

Example 1.11 Verify that $y_1(x) = x$ is a solution of the ODE

$$xy' - y^2 + (2x+1)y = x^2 + 2x \tag{1.61}$$

and then solve it (find its general solution).

It is clear that the equation is a Riccati equation and that $y_1(x) = x$ is a solution of it. Let us use the substitution (1.60): we put $y(x) = x + z(x)$, where z is a new function of x. Thus, (1.61) becomes:

$$xz' + z = z^2, \tag{1.62}$$

which is a Bernoulli equation. To solve it, we divide (1.62) by z^2 and make the substitution $z^{-1} = u$. Hence Eq. (1.62) becomes:

$$-xu' + u = 1, \tag{1.63}$$

a linear differential equation of the first order. The general solution of its associated homogeneous equation $-xu' + u = 0$ is

$$u_{GH} = Cx$$

and we continue by using the Lagrange method in order to find a particular solution for the non-homogeneous equation (1.63). For this, let us force

$$u_P = C(x)x$$

to be a solution of (1.63): $-xu'_P + u_P = 1$. We obtain $C' = -x^{-2}$, hence $C(x) = \frac{1}{x}$ and $u_P = 1$. Finally, the general solution of the non-homogeneous equation (1.63) is $u = Cx + 1$. Hence, $z = \frac{1}{Cx+1}$ is the general solution of Eq. (1.62), and $y = x + \frac{1}{Cx+1}$ is the general solution for the initial Riccati equation (1.61).

1.7 Exact Differential Equations

We have seen that any first-order ODE, $y' = f(x, y)$, can be written as $dy = f(x, y) dx$, or, equivalently, $f(x, y) dx - dy = 0$. The left side of this equality is nothing else but a particular first-order *differential form*. The general expression of such a differential form is $P(x, y) dx + Q(x, y) dy$, where P and Q are continuous functions in x and y. A differential form is said to be *exact* if there exists a function $U(x, y)$ of class C^1 such that

$$dU = P(x, y) dx + Q(x, y) dy. \tag{1.64}$$

In this case, the corresponding differential equation,

$$P(x, y) dx + Q(x, y) dy = 0. \tag{1.65}$$

is said to be an *exact equation*, and a function $U(x, y)$ that satisfies Eq. (1.64) is called a *potential function* for the differential equation (1.65) ([26], Definition 1.3).

Since the expression of the differential dU of the potential function is

$$dU = \frac{\partial U}{\partial x} dx + \frac{\partial U}{\partial y} dy,$$

Equation (1.64) is equivalent to

$$\frac{\partial U}{\partial x} = P(x, y), \quad \frac{\partial U}{\partial y} = Q(x, y).$$

If we know a potential function $U(x, y)$ for the exact differential equation (1.65), the solution (in an *implicit form*) follows from $dU = 0$:

$$U(x, y) = C, \tag{1.66}$$

where C is a real constant. Sometimes $y(x)$ can be written in an explicit form from (1.66), sometimes, not.

The question that arises is how can we recognize an exact equation (because not *any* differential equation written in the form (1.65) is an exact equation!). A convenient test for *exactness* is provided by the following theorem (see also [3], Theorem 2.6.1 or [29], Theorem 68).

Theorem 1.2 *Let $P(x, y)$ and $Q(x, y)$ be two functions of class C^1 defined on a simply connected open set $D \subset \mathbb{R}^2$. Then, the differential form $P(x, y) dx + Q(x, y) dy$ is exact if and only if*

$$\frac{\partial P}{\partial y} = \frac{\partial Q}{\partial x}. \tag{1.67}$$

1.7 Exact Differential Equations

Proof (Necessity) If the form $P(x, y)\, dx + Q(x, y)\, dy$ is exact, then there exists a *potential function* $U(x, y)$ such that $\dfrac{\partial U}{\partial x} = P(x, y)$ and $\dfrac{\partial U}{\partial y} = Q(x, y)$. Since P and Q are functions of class C^1, the function U is of class C^2 and so, by Schwarz Theorem (also known as Clairaut Theorem—see [28], Theorem 71, or [40], Theorem 17.5.4), we have:

$$\frac{\partial^2 U}{\partial x \partial y} = \frac{\partial^2 U}{\partial y \partial x} \tag{1.68}$$

But

$$\frac{\partial P}{\partial y} = \frac{\partial}{\partial y}\left(\frac{\partial U}{\partial x}\right) = \frac{\partial^2 U}{\partial y \partial x}$$

and

$$\frac{\partial Q}{\partial x} = \frac{\partial}{\partial x}\left(\frac{\partial U}{\partial y}\right) = \frac{\partial^2 U}{\partial x \partial y},$$

so, by (1.68), the equality (1.67) follows.

(Sufficiency) Assume now that the equality (1.67) is true and let us search for a function $U(x, y)$ such that

$$\frac{\partial U}{\partial x} = P(x, y) \tag{1.69}$$

and

$$\frac{\partial U}{\partial y} = Q(x, y). \tag{1.70}$$

Thus, from (1.69), the function $U(x, y)$ must be of the form:

$$U(x, y) = \int_{x_0}^{x} P(t, y)\, dt + C(y), \tag{1.71}$$

where $\int_{x_0}^{x} P(t, y)\, dt$ is a primitive (antiderivative) of $P(x, y)$ with respect to the variable x and $C(y)$ is a constant with respect to the variable of integration x, but it can depend on y, i.e. it is a function of y.

By *Leibniz' formula* for differentiation under the integral sign (see [22], Theorem 25.2), we have:

$$\frac{d}{dx}\int_{a}^{b} f(x, t)\, dt = \int_{a}^{b} \frac{\partial f}{\partial x}(x, t)\, dt \tag{1.72}$$

Thus, since the function $U(x, y)$ from (1.71) must satisfy Eq. (1.70), we have:

$$\int_{x_0}^{x} \frac{\partial P}{\partial y}(t, y)\, dt + C'(y) = Q(x, y).$$

In order to find the function $C(y)$ from here, we must prove that the expression

$$Q(x, y) - \int_{x_0}^{x} \frac{\partial P}{\partial y}(t, y)\, dt \tag{1.73}$$

does not depend on x, i.e. its derivative with respect to x is zero.

By the *Fundamental Theorem of Calculus* ([38], 5.3) we have that, for any continuous function $f(x)$,

$$\frac{d}{dx} \int_{x_0}^{x} f(t)\, dt = f(x). \tag{1.74}$$

Hence we obtain that the derivative of (1.73) is

$$\frac{\partial Q}{\partial x}(x, y) - \frac{\partial P}{\partial y}(x, y),$$

which is equal to 0 by the hypothesis (1.67), so the proof is completed. □

Example 1.12 Let us solve the following Cauchy problem:

$$\begin{cases} 2xy\, dx + x^2\, dy = 0,\, x > 0 \\ y(1) = 1. \end{cases}$$

Here $P(x, y) = 2xy$ and $Q(x, y) = x^2$. They are clearly functions of class C^1 on the entire plane $\mathbb{R}^2$, thus one can apply Theorem 1.2. Since

$$\frac{\partial P}{\partial y} = 2x = \frac{\partial Q}{\partial x},$$

there exists a function $U(x, y)$ such that $\dfrac{\partial U}{\partial x} = 2xy$ and $\dfrac{\partial U}{\partial y} = x^2$. Let us find this potential function $U(x, y)$. We have:

$$\frac{\partial U}{\partial x} = 2xy \Rightarrow U(x, y) = x^2 y + C(y), \tag{1.75}$$

so $\dfrac{\partial U}{\partial y} = x^2 + C'(y)$. But we know that $\dfrac{\partial U}{\partial y} = x^2$, hence $C'(y) = 0$, so $C(y)$ is a constant. By considering this constant 0, we obtain that $U(x, y) = x^2 y$ is a potential function of our exact differential equation. Consequently, since the equation can be

1.7 Exact Differential Equations

written $dU = 0$, the general solution of this equation can be written in the implicit form

$$x^2 y = C$$

or in the explicit form

$$y = \frac{C}{x^2}$$

(where C is an arbitrary constant). Since $y(1) = 1$, the constant C must be 1. Hence, the unique solution of the above Cauchy problem is $y(x) = \frac{1}{x^2}$.

Sometimes, the equation $P(x, y)\,dx + Q(x, y)\,dy = 0$ is not exact but, after the multiplication with an appropriate function $\mu(x, y)$ (nonzero at any point of the considered domain D), called an *integrating factor*, this equation becomes exact and thus it can be integrated. Let $\mu(x, y)$ be an integrating factor. It must satisfy:

$$\frac{\partial(\mu P)}{\partial y} = \frac{\partial(\mu Q)}{\partial x}, \tag{1.76}$$

or, equivalently,

$$\frac{\partial \mu}{\partial y} P - \frac{\partial \mu}{\partial x} Q = \mu \left(\frac{\partial Q}{\partial x} - \frac{\partial P}{\partial y} \right). \tag{1.77}$$

This is a quasilinear partial differential equation in the unknown function $\mu(x, y)$ (see Chap. 3.6).

Usually we try to find integrating factors which depend only on x or only on y, by checking the equality (1.77). For instance, if we want to find an integrating factor $\mu = \mu(x)$, then the expression $\left(\frac{\partial Q}{\partial x} - \frac{\partial P}{\partial y} \right) \frac{1}{Q}$ must be a function of x only and, conversely, if this expression depends only on x, then one can find an integrating factor $\mu = \mu(x)$ by solving the ODE

$$-\mu' = \mu \left(\frac{\partial Q}{\partial x} - \frac{\partial P}{\partial y} \right) \frac{1}{Q}. \tag{1.78}$$

Example 1.13 Find an integrating factor $\mu = \mu(x)$ for the following ODE and then solve it:

$$(y \exp(x) - 1)\,dx + (\exp(x + y) + x \exp(x))\,dy = 0.$$

Here $P(x, y) = y\exp(x) - 1$ and $Q(x, y) = \exp(x + y) + x\exp(x)$. Since $\frac{\partial Q}{\partial x} = \exp(x + y) + \exp(x) + x\exp(x)$ and $\frac{\partial P}{\partial y} = \exp(x)$, we have

$$\left(\frac{\partial Q}{\partial x} - \frac{\partial P}{\partial y}\right)\frac{1}{Q} = 1$$

and, from (1.78) we get $\mu' = -\mu$ and so we obtain the integrating factor $\mu = \exp(-x)$. Thus, by multiplying both sides of the initial equation with $\exp(-x)$, the following *exact equation* is obtained:

$$(y - \exp(-x))dx + (\exp(y) + x)dy = 0.$$

To solve this exact equation we need to find a function $U(x, y)$ such that

$$dU = (y - \exp(-x))\,dx + (\exp(y) + x)\,dy.$$

$$\frac{\partial U}{\partial x} = y - \exp(-x) \implies U(x, y) = \int (y - \exp(-x))\,dx = xy + \exp(-x) + C(y),$$

so $\frac{\partial U}{\partial y} = x + C'(y)$. But, since $U(x, y)$ is a potential function, it must satisfy

$$\frac{\partial U}{\partial y} = \exp(y) + x.$$

It follows that $C(y) = \exp(y)$, hence $U(x, y) = xy + \exp(-x) + \exp(y)$ and the general solution of our differential equation is

$$xy + \exp(-x) + \exp(y) = C,$$

where C is an arbitrary constant.

Example 1.14 Find an integrating factor $\mu = \mu(y)$ for the ODE

$$2xy \ln y\, dx + \left(x^2 + y^2\sqrt{y^2 + 1}\right) dy = 0,$$

then solve it and find the particular solution that satisfies $y(0) = \sqrt{2}$.

Here $P(x, y) = 2xy \ln y$ and $Q(x, y) = x^2 + y^2\sqrt{y^2 + 1}$. Since $\frac{\partial Q}{\partial x} = 2x$ and $\frac{\partial P}{\partial y} = 2x(\ln y + 1)$, the equation is not exact. For $\mu = \mu(y)$ Eq. (1.77) becomes

$$\mu' = \mu \left(\frac{\partial Q}{\partial x} - \frac{\partial P}{\partial y}\right)\frac{1}{P}$$

and we obtain $\frac{\mu'}{\mu} = -\frac{1}{y}$. Thus, $\ln|\mu| = -\ln|y| + \ln C_1$, so $\mu = \frac{1}{y}$ is an integrating factor and the following exact equation is obtained:

$$2x \ln y \, dx + \frac{x^2}{y} dy + y\sqrt{y^2+1} \, dy = 0,$$

or, equivalently,

$$d\left(x^2 \ln y\right) + d\left(\frac{1}{3}(y^2+1)^{\frac{3}{2}}\right) = 0.$$

Hence the general implicit solution is:

$$x^2 \ln y + \frac{1}{3}(y^2+1)^{\frac{3}{2}} = C$$

and the particular solution that satisfies the initial condition $y(0) = \sqrt{2}$ is obtained for $C = \sqrt{3}$:

$$x^2 \ln y + \frac{1}{3}(y^2+1)^{\frac{3}{2}} = \sqrt{3}.$$

Remark 1.1 It is easy to see that $\mu = \exp\left[\int p(x)\,dx\right]$ is an integrating factor for the linear differential equation $\frac{dy}{dx} + p(x)y = q(x)$. Indeed, let us write the equation as: $[p(x)y - q(x)]\,dx + dy = 0$. Here $P = p(x)y - q(x)$ and $Q = 1$. Since $\frac{\partial Q}{\partial x} = 0 \neq \frac{\partial P}{\partial y} = p(x)$ in general, the relation (1.78) becomes: $-\frac{\mu'}{\mu} = -p(x)$, so $\mu = \exp\left[\int p(x)\,dx\right]$. Hence, the general solution of the linear differential equation $\frac{dy}{dx} + p(x)y = q(x)$ can be given as:

$$y \exp\left[\int p(x)\,dx\right] - \int q(x) \exp\left[\int p(x)\,dx\right] dx = C.$$

This is another way to solve first-order linear differential equations.

1.8 Lagrange Equations and Clairaut Equations

A *Lagrange equation* is a differential equation of the form:

$$y = xa(y') + b(y'), \qquad (1.79)$$

where $a(z)$ and $b(z)$ are continuously differentiable functions and $a(z)$ is not the identity function. Let us introduce a parameter $p = y' = \frac{dy}{dx}$ and try to find solutions in a parametric form $\begin{cases} x = x(p) \\ y = y(p) \end{cases}$. So (1.79) becomes

$$y = xa(p) + b(p).$$

By differentiating this last equality with respect to x, we get:

$$p = a(p) + xa'(p)\frac{dp}{dx} + b'(p)\frac{dp}{dx},$$

or, equivalently,

$$p - a(p) = [xa'(p) + b'(p)]\frac{dp}{dx}. \tag{1.80}$$

If $\frac{dp}{dx} = 0$, then p is a constant and, from (1.80), this constant must satisfy the equality $p - a(p) = 0$, i.e p is a root of the algebraic equation $z = a(z)$. Let $c_1, c_2, \ldots, c_n$ be the roots of this equation. Then the straight lines

$$y = c_i x + b(c_i), i = 1, \ldots, n$$

are singular solutions for the Lagrange equation (1.79).

If $\frac{dp}{dx} \neq 0$, then we can rewrite (1.80) as:

$$[p - a(p)]\frac{dx}{dp} - xa'(p) = b'(p), \tag{1.81}$$

which is a linear differential equation in the unknown function $x(p)$ and the independent variable p. We solve it and find its general solution $x = x(p, C)$ (where C is an arbitrary constant). Since $y = xa(p) + b(p)$, we obtain the parametric equations of the general solution for the Lagrange equation:

$$\begin{cases} x = x(p, C) \\ y = x(p, C)a(p) + b(p). \end{cases} \tag{1.82}$$

Example 1.15 Let us study the set of solutions of the following Lagrange equation:

$$y = xy'^2 - \frac{1}{y'}.$$

Here $a(z) = z^2$, $b(z) = -\frac{1}{z}$. The algebraic equation $c = a(c)$ has the solutions: $c_1 = 0$ and $c_2 = 1$. Since $y' \neq 0$, we get only one singular solution: the straight line $y = x - 1$.

1.8 Lagrange Equations and Clairaut Equations

Now, let us find the general solution, by solving Eq. (1.81):

$$(p - p^2)\frac{dx}{dp} - 2px = \frac{1}{p^2}. \tag{1.83}$$

The homogeneous associated equation ($p \neq 0, 1$) has the general solution: $x_{GH}(p) = \frac{C}{(1-p)^2}$. By using Lagrange method of variation of constants, we find $C(p) = \frac{1}{p} - \frac{1}{2p^2}$, so a particular solution for (1.83) is $x_P(p) = \frac{p-0.5}{p^2(1-p)^2}$. Thus, the general solution of (1.83) is

$$x(p) = \frac{Cp^2 + p - 0.5}{p^2(p-1)^2},$$

and, using the initial equation, we obtain:

$$y(p) = \frac{Cp^2 + p - 0.5}{(p-1)^2} - \frac{1}{p}.$$

It is easy to see that there is no value of the arbitrary constant C such that $y(p) = x(p) - 1$ for any $p \neq 0, 1$. Hence, $y = x - 1$ is a singular solution indeed for the above Lagrange equation.

A *Clairaut equation* is a differential equation of the form:

$$y = xy' + b(y'), \tag{1.84}$$

where $b(z)$ is a continuously differentiable function. Notice that it has the same form as the *Lagrange equation*, but here $a(z)$ is the identity function, so the above theory does not work in this case. However, we can use the same idea: we denote $p = y'$ and the equation becomes

$$y = xp + b(p).$$

After differentiating this last equation with respect to x, we obtain: $p = p + xp' + b'(p)p'$, or, equivalently,

$$p'\left[x + b'(p)\right] = 0.$$

If $p' = 0$, then $p = C$, a constant, and the general solution of the Clairaut equation is a family of straight lines:

$$y = Cx + b(C). \tag{1.85}$$

If $x = -b'(p)$, then $y = -pb'(p) + b(p)$, so we have the (*singular*) parametric solution:

$$\begin{cases} x = -b'(p) \\ y = -pb'(p) + b(p) \end{cases}. \tag{1.86}$$

It is easy to see that (1.86) is an envelope curve for the family of lines (1.85), i.e. for the general solution. Indeed, the envelope of a family of curves with one parameter C, $F(x, y, C) = 0$, is given by the parametric equations:

$$\begin{cases} F(x, y, C) = 0 \\ \frac{dF}{dC}(x, y, C) = 0 \end{cases}.$$

In our case, these equations are:

$$\begin{cases} y = Cx + b(C) \\ 0 = x + b'(C) \end{cases},$$

equivalent to Eqs. (1.86) (using the parameter p instead of C).

Example 1.16 Let us solve the Clairaut equation $y = xy' + y'^2$.

We denote $y' = p$, so $y = xp + p^2$. After differentiating with respect to x this last equality we obtain: $p'(x + 2p) = 0$. Thus, either $p = C$, a constant, and we obtain the general solution, a family of straight lines $y = Cx + C^2$, or $x = -2p$, $y = -p^2$, and we obtain a parabola, $y = -\frac{x^2}{4}$, which is the singular solution. Any straight line of the family $y = Cx + C^2$ is tangent to this parabola (see Fig. 1.3).

Orthogonal Trajectories

Assume that $F(x, y, C) = 0$ is a family of plane curves with one parameter C, where F is a continuously differentiable function such that $\frac{\partial F}{\partial y} \neq 0$. We regard it as the general solution of a first-order differential equation. Let us find this differential equation. For this we differentiate the equality $F(x, y, C) = 0$ with respect to x (considering $y = y(x)$) and eliminate C from the following equalities:

$$\begin{cases} F(x, y, C) = 0 \\ \frac{\partial F}{\partial x}(x, y, C) + \frac{\partial F}{\partial y}(x, y, C)y' = 0 \end{cases}. \tag{1.87}$$

We obtain a first-order ODE which has the general solution $F(x, y, C) = 0$. If C does not appear in the second equality then this equality itself is the sought equation. For instance: $y^2 + 4x - C = 0$ is the general solution of $2yy' + 4 = 0$. Indeed, this differential equation can be written as $(y^2)' = -4$, so its general solution is $y^2 + 4x - C = 0$.

Let us find now the differential equation of the family of circles tangent to Oy-axis at the origin: $x^2 + y^2 - 2Cx = 0$. For this we must eliminate C between the

1.8 Lagrange Equations and Clairaut Equations

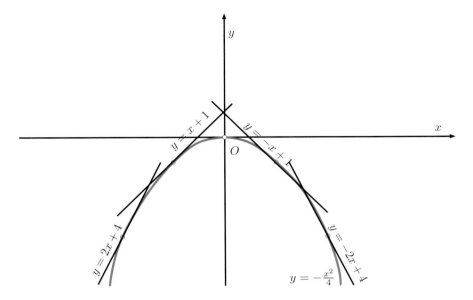

Fig. 1.3 Solutions of the equation $y = xy' + y'^2$

equalities:

$$\begin{cases} x^2 + y^2 - 2Cx = 0 \\ 2x + 2yy' - 2C = 0. \end{cases}$$

Thus we get: $2xyy' + x^2 - y^2 = 0$, the first order ODE that has the general solution $x^2 + y^2 - 2Cx = 0$ (prove it!).

Two curves $y = f(x)$ and $y = g(x)$ are said to be *orthogonal* at the (common) point $M_0(x_0, y_0)$ if their tangent lines at M_0 are perpendicular (orthogonal), i.e. if $f'(x_0)g'(x_0) = -1$.

Let $F(x, y, C) = 0$ be a family of plane curves of class C^1 (each curve in the family has a unique tangent line at each one of its points). Another family of plane curves of class C^1, $G(x, y, K) = 0$ is called a *family of orthogonal trajectories* for $F(x, y, C) = 0$ if for any point $M_0(x_0, y_0)$ on a curve γ from the first family $F(x, y, C) = 0$, there exists a curve δ from the second family $G(x, y, K) = 0$ which passes through M_0 and is orthogonal to γ at this common point M_0. Let $\psi(x, y, y') = 0$ be the differential equation of the first family $F(x, y, C) = 0$. Then $\psi(x, y, -\frac{1}{y'}) = 0$ is the differential equation of the second family $G(x, y, K) = 0$, i.e. the differential equation of its family of orthogonal trajectories.

Example 1.17 Let $x^2 + y^2 - C = 0$, $C > 0$, be the family of all concentric circles with centers at the origin $(0, 0)$. Its differential equation is: $x + yy' = 0$. In order to find the orthogonal trajectories of this family of concentric circles we simply substitute y' with $-\frac{1}{y'}$ in $x + yy' = 0$. So we get: $x - \frac{y}{y'} = 0$, or $\frac{y'}{y} = \frac{1}{x}$ and, by integration, we finally obtain $y = Kx$, which is the family of all the straight lines which pass through the origin.

1.9 Existence and Uniqueness of Solution of the Cauchy Problem

Theorem 1.1 states that any Cauchy problem for a *linear* ODE has a unique solution. On the other hand, this is not true for *any* differential equation. For instance, consider the following two Cauchy problems for the differential equation from Example 1.16:

$$y = xy' + y'^2, \quad y(2) = -1,$$

$$y = xy' + y'^2, \quad y(1) = -2.$$

It is easy to see that the first one has two solutions: $y = -x + 1$ and $y = -\frac{x^2}{4}$, while the second one has no solution (see Fig. 1.3).

The questions that arise naturally are: what are the conditions that make a Cauchy problem to have *at least one solution* and what are the conditions that make it to have *at most one solution*. The next theorem (see [7], Theorem 3.1, or [22], Theorem 17.1) gives *sufficient conditions* such that the Cauchy problem for a first-order differential equation in the normal form,

$$y' = f(x, y) \qquad (1.88)$$

with the initial condition

$$y(x_0) = y_0 \qquad (1.89)$$

has a solution and this solution is unique.

Theorem 1.3 (Existence and Uniqueness Theorem) *If $f(x, y)$ is a continuous function in the rectangle $D = [x_0 - a, x_0 + a] \times [y_0 - b, y_0 + b]$ ($a, b > 0$) and it has a bounded partial derivative with respect to y, then there is some interval $I = [x_0 - h, x_0 + h]$, $h > 0$, in which there exists a unique solution $y = y(x)$ of the differential equation (1.88) such that $y(x_0) = y_0$.*

Proof **Existence.** First of all, we prove that our Cauchy problem (1.88)–(1.89) is equivalent to the following integral equation:

$$y(x) = y_0 + \int_{x_0}^{x} f(t, y(t)) \, dt. \qquad (1.90)$$

1.9 Existence and Uniqueness of Solution of the Cauchy Problem

Let $y(x)$ be a solution of the Cauchy problem (1.88)–(1.89). We can write Eq. (1.88) in the independent variable t,

$$y'(t) = f(t, y(t))$$

and, since the function $f(t, y(t))$ is continuous, we can integrate it on the interval $[x_0, x]$. Using the initial condition (1.89), we obtain (1.90).

Conversely, if $y(x)$ satisfies the integral equation (1.90), then $y(x_0) = y_0$ and, by differentiation (using the Fundamental Theorem of Calculus (1.74)) we obtain that $y'(x) = f(x, y(x))$.

The method we use to prove that the integral equation (1.90) has a solution is called the *method of successive approximations* and was discovered by the French mathematician *C.E. Picard*. It is based on the construction of a sequence of functions, $\{y_n(x)\}$ which is uniformly convergent to the solution of (1.90).

Both functions f and $\frac{\partial f}{\partial y}$ are bounded (since f is continuous on the closed rectangle D). We denote:

$$M = \|f\| = \sup_{(x,y) \in D} |f(x, y)| \quad \text{and} \quad L = \left\|\frac{\partial f}{\partial y}\right\| = \sup_{(x,y) \in D} \left|\frac{\partial f}{\partial y}(x, y)\right|.$$

From the Mean Value Theorem ([38], 4.2), for any $x \in [x_0 - a, x_0 + a]$ and $y_1, y_2 \in [y_0 - b, y_0 + b]$, there exists $z \in [y_1, y_2]$ such that

$$f(x, y_1) - f(x, y_2) = \frac{\partial f}{\partial y}(x, z)(y_1 - y_2),$$

hence the following inequality holds:

$$|f(x, y_1) - f(x, y_2)| \leq L |y_1 - y_2|. \tag{1.91}$$

Let $h = \min\left\{a, \frac{b}{M}\right\}$. We consider the following sequence of functions defined on the interval $I = [x_0 - h, x_0 + h]$ (see Fig. 1.4):

$$y_0(x) = y_0$$

and, for $n = 1, 2, \ldots,$

$$y_n(x) = y_0 + \int_{x_0}^{x} f(t, y_{n-1}(t)) \, dt. \tag{1.92}$$

Since

$$|y_n(x) - y_0| \leq \left|\int_{x_0}^{x} |f(t, y_{n-1}(t))| \, dt\right| \leq M|x - x_0|, \tag{1.93}$$

it follows that $y_n(x) \in [y_0 - b, y_0 + b]$, for any $x \in I$.

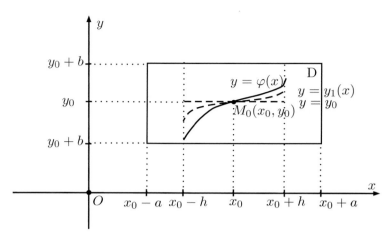

Fig. 1.4 The first two approximations: $y_0(x)$ and $y_1(x)$

By the definition of the sequence (1.92) we have:

$$|y_{n+1}(x) - y_n(x)| \leq \left| \int_{x_0}^{x} |f(t, y_n(t)) - f(t, y_{n-1}(t))| \, dt \right|$$

and, from the inequality (1.91), we obtain:

$$|y_{n+1}(x) - y_n(x)| \leq L \left| \int_{x_0}^{x} |y_n(t) - y_{n-1}(t)| \, dt \right|, \tag{1.94}$$

for any $x \in I$ and $n = 1, 2, \ldots$.

From (1.93) we can write:

$$|y_1(x) - y_0| \leq M|x - x_0|$$

and, using (1.94), the following inequality can be easily proved by mathematical induction:

$$|y_{n+1}(x) - y_n(x)| \leq \frac{ML^n |x - x_0|^{n+1}}{(n+1)!}, \tag{1.95}$$

for any $x \in I$ and $n = 0, 1, \ldots$. It follows that

$$|y_{n+1}(x) - y_n(x)| \leq \frac{ML^n h^{n+1}}{(n+1)!}, \tag{1.96}$$

1.9 Existence and Uniqueness of Solution of the Cauchy Problem

for any $x \in I$ and $n = 0, 1, \ldots$. Hence, by *Weierstrass M-test* (see Theorem 4.1), since the series

$$\sum_{n=0}^{\infty} \frac{ML^n h^{n+1}}{(n+1)!} \tag{1.97}$$

is convergent (it converges to $\frac{M}{L}[\exp(Lh) - 1]$, we obtain that the series

$$y_0(x) + \sum_{n=0}^{\infty} (y_{n+1}(x) - y_n(x)) \tag{1.98}$$

uniformly converges on the interval I to a continuous function, $y(x)$. But the partial sums of the series (1.98) are the terms of the sequence $\{y_n\}$:

$$y_0(x) + \sum_{k=0}^{n-1} (y_{k+1}(x) - y_k(x)) = y_n(x),$$

so the sequence $\{y_n\}$ constructed by (1.92) uniformly converges to $y(x)$, that is:

$$\lim_{n \to \infty} \sup_{x \in I} |y_n(x) - y(x)| = \lim_{n \to \infty} \|y_n - y\| = 0. \tag{1.99}$$

Since $y_n(x) \in [y_0 - b, y_0 + b]$ for any $x \in I$ and $n = 1, 2, \ldots$, it follows that $y(x) \in [y_0 - b, y_0 + b]$ for any $x \in I$. We prove that the function $y(x)$ satisfies the integral equation (1.90).

For any $x \in I$ and $n = 1, 2, \ldots$ we can write (using the inequality (1.91)):

$$\left| \int_{x_0}^{x} f(t, y_n(t)) \, dt - \int_{x_0}^{x} f(t, y(t)) \, dt \right| \leq \left| \int_{x_0}^{x} |f(t, y_n(t)) - f(t, y(t))| \, dt \right|$$

$$\leq L \left| \int_{x_0}^{x} |y_n(t) - y(t)| \, dt \right| \leq Lh \, \|y_n - y\| \to 0,$$

and it follows that

$$\lim_{n \to \infty} \int_{x_0}^{x} f(t, y_n(t)) \, dt = \int_{x_0}^{x} f(t, y(t)) \, dt.$$

Thus, when $n \to \infty$, the relation (1.92) yields (1.90), so $y(x)$ satisfies the integral equation (1.90) and, as we have proved above, it is a solution of the Cauchy problem (1.88)–(1.89).

Uniqueness. Let $z : I \to \mathbb{R}$ be another solution of the Cauchy problem (1.88)–(1.89). Then z must also satisfy the integral equation (1.90):

$$z(x) = y_0 + \int_{x_0}^{x} f(t, z(t))\, dt, \tag{1.100}$$

for all $x \in I$. It follows that:

$$|z(x) - y_0| \leq \left| \int_{x_0}^{x} |f(t, z(t))|\, dt \right| \leq M|x - x_0|. \tag{1.101}$$

On the other hand, by (1.92), (1.100) and (1.91), we have that:

$$|z(x) - y_n(x)| \leq \left| \int_{x_0}^{x} |f(t, z(t)) - f(t, y_{n-1}(t))|\, dt \right| \leq L \left| \int_{x_0}^{x} |z(t) - y_{n-1}(t)|\, dt \right|, \tag{1.102}$$

for all $x \in I$ and $n = 1, 2, \ldots$. For $n = 1$ the above equation is written:

$$|z(x) - y_1(x)| \leq L \left| \int_{x_0}^{x} |z(t) - y_0|\, dt \right|$$

and, by Eq. (1.101), we obtain that

$$|z(x) - y_1(x)| \leq LM \frac{(x - x_0)^2}{2}, \tag{1.103}$$

for all $x \in I$. By writing Eq. (1.102) for $n = 2$ and using (1.103), we obtain:

$$|z(x) - y_2(x)| \leq L \left| \int_{x_0}^{x} |z(t) - y_1(t)|\, dt \right| \leq ML^2 \frac{|x - x_0|^3}{3!}$$

and, by mathematical induction it can be shown that

$$|z(x) - y_n(x)| \leq ML^n \frac{|x - x_0|^{n+1}}{(n+1)!}.$$

It follows that, for all $x \in I$ and $n = 1, 2, \ldots$, the following inequality holds:

$$|z(x) - y_n(x)| \leq ML^n \frac{h^{n+1}}{(n+1)!}.$$

1.9 Existence and Uniqueness of Solution of the Cauchy Problem

Since the series (1.97) is convergent, the sequence $ML^n \dfrac{h^{n+1}}{(n+1)!}$ tends to 0. Therefore,

$$\lim_{n \to \infty} y_n(x) = z(x),$$

which means that $y = z$ and the uniqueness of the solution is proved. □

Remark 1.2 Using the inequality (1.96), we can compute an upper bound of the error in approximating the solution $y(x)$ by $y_n(x)$:

$$|y_n(x) - y(x)| \leq \sum_{k=n}^{\infty} |y_{k+1}(x) - y_k(x)|$$

$$\leq \frac{M}{L} \sum_{k=n}^{\infty} \frac{(Lh)^{k+1}}{(k+1)!} < \frac{M}{L} \frac{(Lh)^{n+1}}{(n+1)!} \sum_{k=0}^{\infty} \frac{(Lh)^k}{k!},$$

so

$$|y_n(x) - y(x)| < \frac{M}{L} \frac{(Lh)^{n+1}}{(n+1)!} \exp(Lh), \qquad (1.104)$$

for all $x \in [x_0 - h, x_0 + h]$ and $n = 1, 2, \ldots$.

Example 1.18 The general solution of the (linear) ODE $y' = y$ is $y(x) = C_1 \exp(x)$ with C_1 an arbitrary constant. The Cauchy problem

$$y' = y, \quad y(0) = 1$$

has the unique solution $y(x) = \exp(x)$.

Let us apply in this case the method of successive approximations described in the proof of Theorem 1.3. We construct the sequence $\{y_n(x)\}$:

$$y_0(x) = 1,$$

$$y_1(x) = 1 + \int_0^x 1 \, dx = 1 + \frac{x}{1!},$$

$$y_2(x) = 1 + \int_0^x \left(1 + \frac{t}{1!}\right) dt = 1 + \frac{x}{1!} + \frac{x^2}{2!}.$$

Using mathematical induction it is easy to prove that, for any $n = 0, 1, \ldots$,

$$y_n(x) = 1 + \frac{x}{1!} + \frac{x^2}{2!} + \ldots + \frac{x^n}{n!},$$

i.e. the nth partial sum of the well known series

$$\exp(x) = \sum_{n=0}^{\infty} \frac{x^n}{n!}.$$

So, we obtained by the method of successive approximations the same solution, $y(x) = \exp(x)$. We let the reader to evaluate the constants L, M, and h for an arbitrary rectangle $D = [-a, +a] \times [1-b, 1+b]$, $a, b > 0$ ($x_0 = 0$ and $y_0 = 1$ in our case).

Example 1.19 Using the method of successive approximations, let us find an approximate solution with an error less than 0.01 for the following Cauchy problem:

$$\begin{cases} y' = x + y \\ y(0) = 0, \end{cases} \quad (x, y) \in \left[-\frac{1}{2}, \frac{1}{2}\right] \times [-1, 1].$$

In our case $f(x, y) = x + y$, so $L = 1$, $M = \frac{3}{2}$ and $h = \min\left\{\frac{1}{2}, \frac{2}{3}\right\} = \frac{1}{2}$. By the formula (1.104) we have:

$$|y_n(x) - y(x)| < \frac{3}{2} \cdot \frac{\exp(\frac{1}{2})}{2^{n+1}(n+1)!}.$$

So, to obtain an error less than 0.01, it is sufficient to find the smallest integer n such that

$$\frac{3 \exp(\frac{1}{2})}{2^{n+2}(n+1)!} \leq 0.01,$$

or, equivalently, such that

$$150\sqrt{e} \leq 2^{n+1}(n+1)!$$

It is easy to see that the smallest n which satisfies this last inequality is $n = 3$. We calculate the approximate solution $y_3(x)$:

$$y_0(x) = 0, \quad y_1(x) = \int_0^x (t + y_0(t))\, dt = \frac{x^2}{2},$$

$$y_2(x) = \int_0^x (t + y_1(t))\, dt = \int_0^x \left(t + \frac{t^2}{2}\right) dt = \frac{x^2}{2!} + \frac{x^3}{3!},$$

$$y_3(x) = \int_0^x (t + y_2(t))\, dt = \int_0^x \left(t + \frac{t^2}{2} + \frac{t^3}{3}\right) dt = \frac{x^2}{2!} + \frac{x^3}{3!} + \frac{x^4}{4!}.$$

1.9 Existence and Uniqueness of Solution of the Cauchy Problem

We notice that it can be easily proved (by mathematical induction) that the nth term of the sequence is:

$$y_n(x) = \frac{x^2}{2!} + \frac{x^3}{3!} + \ldots + \frac{x^{n+1}}{(n+1)!},$$

hence the exact solution (the limit of this sequence) is $y(x) = \exp(x) - x - 1$.

Remark 1.3 In Theorem 1.3, if the function f is just continuous on the rectangle D, it is possible to show the existence of a solution of the Cauchy problem (1.88)–(1.89) (see [7], *Cauchy-Peano Existence Theorem*).

On the other hand, the condition that the partial derivative $\frac{\partial f}{\partial y}$ exists and is bounded on D can be replaced with a weaker condition, namely, the *Lipschitz condition*: the function $f(x, y)$ is said to satisfy a *Lipschitz condition* (with respect to y) in D if there exists a constant $L > 0$ such that

$$|f(x, y_1) - f(x, y_2)| \leq L |y_1 - y_2| \tag{1.105}$$

for every (x, y_1) and (x, y_2) in D (see [7]).

As shown in the proof of Theorem 1.3 (see inequality (1.91)), if $\frac{\partial f}{\partial y}$ exists and is bounded on D, then the function $f(x, y)$ satisfies a Lipschitz condition (with respect to y) with the constant $L = \sup_{(x,y) \in D} \left| \frac{\partial f}{\partial y}(x, y) \right|$. The converse is not true: $f(x, y) = |y|$ satisfies a Lipschitz condition (with the constant $L = 1$) but it is not differentiable with respect to y at 0.

For the existence of a solution, it is sufficient that $f(x, y)$ is continuous. But if the Lipschitz condition is not fulfilled, then the uniqueness of the solution of the initial value problem may fail. For example, the following Cauchy problem:

$$y' = 3y^{2/3}, \quad y(0) = 0$$

has two distinct solutions: $y(x) = 0$ and $y = x^3$, $x \in \mathbb{R}$. As one can readily verify, the function $f(x, y) = y^{2/3}$ does not satisfy the Lipschitz condition since the ratio

$$\frac{f(x, y) - f(x, 0)}{y - 0} = \frac{y^{2/3} - 0}{y} = \frac{1}{y^{1/3}}$$

is unbounded in any rectangle $[-a, a] \times [-b, b]$.

Remark 1.4 Non-local existence of solutions (see [6], p. 217–219). Theorem 1.3 is local in nature: it guarantees a solution to the Cauchy problem only for x in a neighborhood of x_0. However, in many cases (for instance, in the case of linear differential equations), there exists a solution to the initial-value problem on the whole interval $[x_0 - a, x_0 + a]$ and in such cases we say that a solution exists *non-locally*. Thus, it can be proved that, if the function $f(x, y)$ is continuous on $I \times \mathbb{R}$ (where I is a real interval, bounded or not, containing the point x_0) and satisfies the

Lipschitz condition w.r.t y on every strip $[\alpha, \beta] \times \mathbb{R}$ (where $x_0 \in [\alpha, \beta] \subset I$), that is: there exists a positive constant $L = L_{[\alpha,\beta]} > 0$ such that the inequality (1.105) holds for all $x \in [\alpha, \beta]$ and $y_1, y_2 \in \mathbb{R}$, then the Cauchy problem (1.88)–(1.89) has a (unique) solution $y(x)$ defined on the whole interval I.

1.10 Exercises

1. Prove that the function $y(x) = Cx - x\cos x$ is a solution of the first-order differential equation $xy' - y = x^2 \sin x$, for any arbitrary real constant C. Find the particular solution of the equation that satisfies the initial condition $y(\pi) = 4$.
2. Prove that $y(x) = \frac{1}{C\exp(-x)+1}$ is a solution of the first-order differential equation $y' = y - y^2$, for any real constant C. Find the particular solution of the equation that satisfies the initial condition $y(0) = 0.2$.
3. Find the general solution for the following ODEs:
   (a) $y' = \frac{3x-y}{x+y}$; (b) $2xyy' = x^2 + y^2$; (c) $(x^2 + xy + y^2)dx - x^2 dy = 0$;
   (d) $y' = \frac{4x-y}{2x+y-3}$; (e) $(x-2y+3)y' = 2x - y$; (f) $(x+y)(dx - dy) = -dy$;
   (g) $\frac{dy}{dx} = \frac{2x+y-4}{4x+2y+2}$; (h) $xdx + ydy = 0$; (i) $xdy - ydx = y^2 dx$; (j)
   $xy' = y^3 + y$; (k) $y'' = t^{100}$; (l) $y^{(5)} = 1$; (m) $\frac{d^4x}{dt^4} = \cos 3t$; (n) $\frac{dy}{dx} + \frac{2y}{x} = x^3$;
   (o) $y' + xy = 1 + x^2$; (p) $\frac{dy}{dx} - \frac{y}{x} = -x(\sin x + \cos x)$; (q) $ydx - (2x+y^3)dy = 0$
   (Hint: take $x = x(y)$); (r) $y' + \frac{y}{x} = -xy^2$; (s) $y' = \frac{4}{x}y + x\sqrt{y}$; (t) $3xy^2 y' + y^3 = 2x$; (u) $(2x + 3x^2 y)dx + (x^3 - 3y^2)dy = 0$; (v) $(2x + \exp(-y))dx - (2y + x\exp(-y))dy = 0$; (w) $(xy^2 - y^3)dx + (1 - xy^2)dy = 0$ (Hint: find an integrating factor $\mu = \mu(y)$); (x) $yy' = 2$; (y) $x^2 y'' = y'^2$; (z) $y'' = 2yy'$.
4. Find α such that the function $y_1(x) = \frac{1}{x}, x > 0$, is a particular solution of the differential equation $2y' + y^2 + \frac{1}{x^\alpha} = 0$. Then, find the general solution of the equation.
5. Solve the following Cauchy problems:
   (a) $xy' + y = 0$, $y(1) = 2$; (b) $(1 + \exp x)y' = y\exp x$, $y(0) = 2$; (c) $y' + (\cos x)y = \cos x$, $y(0) = 1$; (d) $xy' - y + \ln x = 0$, $y(1) = 1$; (e) $y' + xy = 1 + x^2$, $y(0) = 1$; (f) $xy' + y = 2x^2 y^2$, $y(1) = 1$; (g) $xyy' + y^2 = 3x^4$, $y(1) = 1$; (h) $y' + 2y = y^2 \exp x$, $y(0) = \frac{1}{2}$; (i) $y' = \frac{y-x}{x}$, $y(1) = 0$; (j) $y' = \frac{xy-y^2}{x^2}$, $y(1) = 1$.

1.10 Exercises

6. Find the general solutions for the following Clairaut and Lagrange equations:
   (a) $y = xy' - (2 + y'^2)$; (b) $y = xy' + 4\sqrt{y'}$; (c) $y = xy' + \sqrt{1 + y'^2}$; (d) $y = xy' - \ln y'$; (e) $y = 2xy' - 4y'^3$; (f) $y = -xy' + y'^2 - 2y' + 2x$.
7. Use the Picard's method of successive approximations to compute an approximate of order 3 (y_3) for the following Cauchy problems:
   (a) $y' = x^2 + y$, $y(0) = 0$; (b) $xy' + y = x$, $y(1) = 1$.

Chapter 2
Higher-Order Differential Equations

Chapter 2 is devoted to higher order differential equations, mainly to linear differential equations. We introduce the vector space of the solutions of a homogeneous linear differential equation and the affine space of the solutions for the non-homogeneous case. Linear equations with constant coefficients are placed in the foreground, because in this case we have a closed-form solution for homogeneous equations, as well as for a wide class of non-homogeneous equations. We also present the Euler-type equations which have variable coefficients but can be transformed into linear equations with constant coefficients.

2.1 Introduction

As we have seen in the beginning of Chap. 1, the general form of an ordinary differential equation of order n is

$$F(x, y, y', \ldots, y^{(n)}) = 0, \tag{2.1}$$

where $F(x, \xi_1, \xi_2, \ldots, \xi_{n+1})$ is a real-valued continuous function of $n+2$ variables. If the Eq. (2.1), can be solved for the highest derivative, $y^{(n)}$, as

$$y^{(n)} = f(x, y, y', \ldots, y^{(n-1)}), \tag{2.2}$$

where $f(x, \xi_1, \xi_2, \ldots, \xi_n)$ is a continuous function of $n+1$ variables, then (2.2) is said to be the *normal form* of the Eq. (2.1).

A *solution* of the Eq. (2.1) on the interval I is a function $\varphi : I \to \mathbb{R}$ of class C^n that satisfies the equation for all x in I:

$$F(x, \varphi(x), \varphi'(x), \ldots, \varphi^{(n)}(x)) = 0, \text{ for all } x \in I. \tag{2.3}$$

© The Author(s), under exclusive license to Springer Nature Switzerland AG 2022
S. A. Popescu, M. Jianu, *Advanced Mathematics for Engineers and Physicists*,
https://doi.org/10.1007/978-3-031-21502-5_2

For instance, let us find a solution of the third-order ODE $y''' = 24x + 12$. We apply three successive quadratures (integration operations):

$$y'' = 12x^2 + 12x + C,$$

$$y' = 4x^3 + 6x^2 + Cx + C_2,$$

$$y = x^4 + 2x^3 + C_1 x^2 + C_2 x + C_3 \tag{2.4}$$

(we denoted $C_1 = \frac{C}{2}$). As can be seen, the *general solution* (2.4) depends on 3 arbitrary constants (or parameters), C_1, C_2 and C_3. When these constants take some particular values, a *particular solution* of the ODE is obtained.

Example 2.1 The motion of an object is governed by Newton's second law, which states that the mass of the object (m) multiplied by its acceleration (a) is equal to the net force on the object (F):

$$F = ma. \tag{2.5}$$

Consider an object that falls under the action of gravity, neglecting the force due to air resistance, so $F = mg$ (where g is the gravitational acceleration, $g = 9.8$ m/s^2.) If the function $y(t)$ denotes the position of the object at time t, then its acceleration at time t is the second derivative, $y''(t) = \frac{d^2 y}{dt^2}$. So the Eq. (2.5) can be written: $-mg = my''$ (the minus sign indicates that the gravitational force acts in the downward (negative) direction of y-axis). Hence we have to integrate a simple second-order ODE:

$$y'' = -g \tag{2.6}$$

As above, we integrate this equation twice and obtain the general solution:

$$y(t) = -\frac{gt^2}{2} + C_1 t + C_2. \tag{2.7}$$

For *any* real constants C_1, C_2, the function (2.7) is a solution of the Eq. (2.6). To find the values of these constants in our problem, we need to know the *initial conditions*:

$$\begin{cases} y(0) = y_0 \text{ the initial position} \\ y'(0) = v_0 \text{ the initial velocity} \end{cases} \tag{2.8}$$

The differential Eq. (2.6) together with the initial conditions (2.8) form a *Cauchy problem* (or *initial value problem*). By (2.8) we obtain: $y(0) = C_1 = y_0$ and $y'(0) =$

$C_2 = v_0$, so the solution of our problem is:

$$y(t) = -\frac{gt^2}{2} + v_0 t + y_0.$$

Remark 2.1 The *general solution* of an nth-order ODE depends on n constants, $C_1, C_2, \ldots, C_n$. These constants can be found using either *initial conditions*,

$$y(x_0) = y_0, \ y'(x_0) = y_1, \ \ldots, y^{(n-1)}(x_0) = y_{n-1}, \qquad (2.9)$$

(where $y_0, y_1, \ldots, y_{n-1}$ are fixed real numbers) and in this case we have a *Cauchy problem* (*initial value problem*), or *boundary conditions*—that is, conditions specified on the unknown function, and/or on one of its derivatives, at two different points (usually, the endpoints of the interval); in this case we have to solve a *boundary value problem*. We remark that, while a Cauchy problem has a unique solution (assuming some properties of the function f—see Theorem 2.1), a boundary value problem may have either no solution, one solution, or even an infinity of solutions (see Example 2.3).

The next two examples deal with the same equation but with different conditions: initial conditions in Example 2.2, and boundary conditions in Example 2.3.

Example 2.2 Modeling the free oscillations of a mass-spring system (see [3] 3.8). Suppose we have a spring suspended vertically from a fixed support and we attach a body of mass m at its lower end (see Fig. 2.1). The spring will suffer an elongation s_0 and the system will be at rest in the equilibrium position. We let $y = 0$ denote this position and choose the downward direction as positive. There are two forces acting on the body: the gravitational force (the weight of the body), $W = mg$, and the spring (restoring) force F_0 that acts upward (in the negative direction) and, by Hooke's law, it is proportional to the elongation s_0:

$$F_0 = -ks_0,$$

where $k > 0$ is the spring constant. Since the system is in equilibrium, we have $W + F_0 = 0$, hence

$$mg = ks_0. \qquad (2.10)$$

From the equilibrium position, the body is displaced vertically a distance y_0 (up: $y_0 < 0$ or down: $y_0 > 0$) and released, possibly with an initial velocity v_0. Let $y(t)$ be the displacement of the object from the equilibrium position at time t (as we saw above, $y = 0$ in the equilibrium position). We consider that the damping (due to air resistance for example) is so small that can be neglected and that no external forces (apart from the gravitational force W and the spring restoring force F_s) act on the

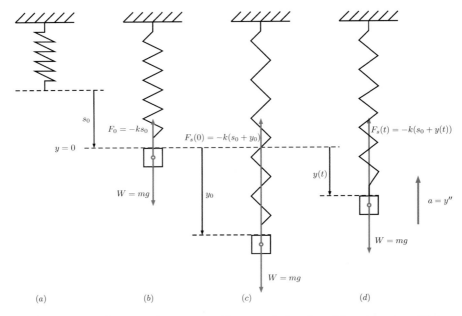

Fig. 2.1 Mechanical mass–spring system: (**a**) unstretched spring; (**b**) system at equilibrium position; (**c**) system at time $t = 0$; (**d**) system in motion at time t

body. Thus, the body has a *free, undamped* motion. By Newton's second law we have that

$$ma = W + F_s,$$

where $a = y'' = \frac{d^2y}{dt^2}$ is the acceleration of the mass m. By Hooke's law, the force exerted by the spring is $F_s = -k(y + s_0)$, so we can write:

$$my'' = mg - k(y + s_0)$$

and using (2.10) we obtain:

$$my'' + ky = 0. \qquad (2.11)$$

We denote the positive constant $\frac{k}{m} = \omega^2$ and obtain the following ODE (called the *harmonic oscillator equation*):

$$y'' + \omega^2 y = 0. \qquad (2.12)$$

As one can easily verify, the function

$$y = C_1 \cos \omega t + C_2 \sin \omega t \qquad (2.13)$$

2.1 Introduction

is a solution of Eq. (2.12), for any real constants C_1 and C_2. Later in this chapter we shall see that *any* solution of the Eq. (2.12) is of this form. Thus, (2.13) is the *general solution* of the ODE (2.12).

Given the initial conditions

$$\begin{cases} y(0) = y_0 \\ y'(0) = v_0 \end{cases}, \qquad (2.14)$$

we can find the constants C_1 and C_2,

$$C_1 = y_0, \quad C_2 = \frac{v_0}{\omega},$$

and write the *unique* solution of the Cauchy problem ((2.12) and (2.14)).

Remark 2.2 The motion described by Eq. (2.13) is called a *harmonic oscillation*. The function $y(t)$ is a periodic function of period $T = \frac{2\pi}{\omega}$. The circular frequency $\omega = \sqrt{\frac{k}{m}}$ (measured in radians per unit time), is called the *natural frequency of the vibration*. We notice that (2.13) can also be written in the form:

$$y = \sqrt{C_1^2 + C_2^2} \left(\frac{C_1}{\sqrt{C_1^2 + C_2^2}} \cos \omega t + \frac{C_2}{\sqrt{C_1^2 + C_2^2}} \sin \omega t \right)$$

$$= \sqrt{C_1^2 + C_2^2} \left(\cos \varphi \cos \omega t + \sin \varphi \sin \omega t \right),$$

or:

$$y = C \cos(\omega t - \varphi), \qquad (2.15)$$

where $C = \sqrt{C_1^2 + C_2^2}$ is the *amplitude* of the motion (the maximum displacement from equilibrium) and the angle φ such that $\cos \varphi = \frac{C_1}{C}$ and $\sin \varphi = \frac{C_2}{C}$ is the *phase* (or *phase angle*) (Fig. 2.2).

Example 2.3 The Euler load (see [44], 3.9). In the eighteenth century Leonhard Euler analyzed how a thin elastic column of uniform cross section buckles under a compressive axial force, P. Let L be the length of the column and $y(x)$ be the deflection of the column at the point $x \in [0, L]$ when the constant vertical load P is applied to its top (as shown in Fig. 2.3). By comparing the bending moments at any point x along the column we obtain the second-order ODE:

$$EI y''(x) = -P y(x),$$

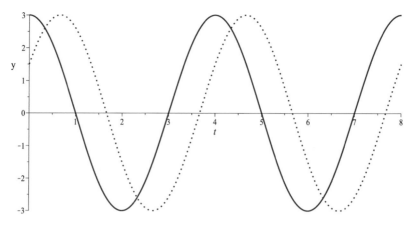

Fig. 2.2 Harmonic oscillation $y = C\cos(\omega t - \varphi)$ (solid line: $\varphi = 0$; dot line: $\varphi \neq 0$)

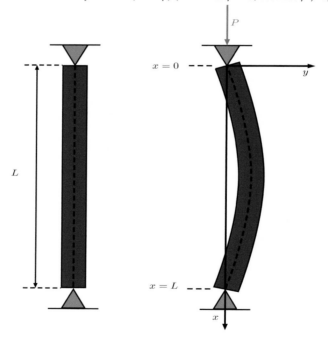

Fig. 2.3 Elastic column buckling under the compressive force P

where E is Young's modulus of elasticity and I is the moment of inertia of a cross section about a vertical line through its centroid.

We want to find the deflection $y(x)$ if the column is hinged at both ends (so $y(0) = y(L) = 0$). By denoting $\lambda = \frac{P}{EI} > 0$, we have to find the solutions of the

2.1 Introduction

equation

$$y'' + \lambda y = 0 \tag{2.16}$$

that satisfies the *boundary conditions*:

$$y(0) = 0, \quad y(L) = 0. \tag{2.17}$$

The Eq. (2.16) is similar to Eq. (2.12), so its general solution is similar to (2.13):

$$y = C_1 \cos \sqrt{\lambda} x + C_2 \sin \sqrt{\lambda} x. \tag{2.18}$$

We can see that, for $C_1 = C_2 = 0$ we obtain the trivial solution $y = 0$. This solution has a simple interpretation: if the load P is not great enough, there is no deflection. The question is: for what values of P will the column bend? That is, what are the values of P such that the boundary-value problem (2.16)–(2.17) has nontrivial solutions?

From the boundary conditions (2.17) we obtain $y(0) = C_1 = 0$ and $y(L) = C_2 \sin \sqrt{\lambda} L = 0$. To have a nontrivial solution we need that $C_2 \neq 0$, hence $\sqrt{\lambda} L = n\pi$ The numbers

$$\lambda_n = \frac{n^2 \pi^2}{L^2}, \quad n = 1, 2, \ldots$$

for which the boundary value problem (2.16)–(2.17) has nontrivial solutions are called *eigenvalues*. The corresponding functions,

$$y_n = \sin \frac{n\pi x}{L}, \quad n = 1, 2, \ldots$$

are called *eigenfunctions*. We notice that any function of the form

$$\tilde{y}_n = C \sin \frac{n\pi x}{L}, \tag{2.19}$$

(for any real constant C) is a solution of the boundary value problem. In our physical problem, since $\lambda = \dfrac{P}{EI}$, this means that the column will buckle or deflect only when the compressive force is one of the values

$$P_n = \frac{n^2 \pi^2 EI}{L^2}, \quad n = 1, 2, \ldots \tag{2.20}$$

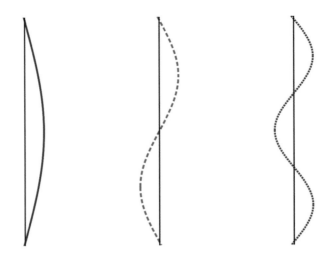

Fig. 2.4 Deflection curves for compressive forces $P_1 = \frac{\pi^2 EI}{L^2}$, $P_2 = \frac{4\pi^2 EI}{L^2}$ and $P_3 = \frac{9\pi^2 EI}{L^2}$

The forces defined by (2.20) are called *critical loads*. The smallest critical load, $P_1 = \frac{\pi^2 EI}{L^2}$, is called the *Euler load* and the corresponding eigenfunction, $y_1 = \sin\frac{\pi x}{L}$ is known as *the first buckling mode*. The graphs of the functions y_i, $i = 1, 2, 3$ are presented in Fig. 2.4.

In conclusion, if the compressive force P is not of the form (2.20), the *boundary value problem* (2.16)–(2.17) has only the trivial solution $y = 0$. If P is of the form (2.20), then the boundary value problem has an infinity of solutions of the form (2.19).

Unlike the *boundary value problem, initial value problem* ((2.2) and (2.9)) has a unique solution, given some hypotheses about the function f. In the same way as Theorem 1.3 for first-order ODEs, the next theorem states sufficient conditions such that the *Cauchy problem* for an ODE of order n has a solution and the solution is unique (see [22], Th. 18.1).

Theorem 2.1 (Existence and Uniqueness Theorem for nth-Order ODEs)
Assume that $f(x, \xi_1, \xi_2, \ldots, \xi_n)$ is a continuous function of $n + 1$ variables defined on the $(n + 1)$-dimensional rectangle

$$\Omega = [x_0 - a, x_0 + a] \times [y_0 - b_0, y_0 + b_0] \times \ldots \times [y_{n-1} - b_{n-1}, y_{n-1} + b_{n-1}].$$

Then there is some interval $I = [x_0 - h, x_0 + h]$ ($0 < h \leq a$) and a function $y = y(x)$ defined on this interval such that $y^{(k)}(x) \in [y_k - b_k, y_k + b_k]$ for all $x \in I$

2.1 Introduction

and $k = 0, 1, \ldots, n-1$, and $y(x)$ is a solution of the Cauchy problem:

$$\begin{cases} y^{(n)}(x) = f(x, y(x), y'(x), \ldots, y^{(n-1)}(x)), \ \forall x \in I \\ y(x_0) = y_0, y'(x_0) = y_1, \ldots, y^{(n-1)}(x_0) = y_{n-1}. \end{cases} \quad (2.21)$$

Furthermore, if the function $f(x, \xi_1, \xi_2, \ldots, \xi_n)$ has bounded partial derivatives with respect $\xi_i, i = 1, 2, \ldots, n$, then the solution $y(x)$ is unique.

Remark 2.3 As in Theorem 1.3 for first-order differential equations, instead of the condition regarding the existence and the boundedness of the partial derivatives, a weaker condition can be used. Thus, for the uniqueness of the solution it is sufficient that the function $f(x, \xi_1, \xi_2, \ldots, \xi_n)$ satisfies the Lipschitz condition w.r.t. $\xi_1, \xi_2, \ldots, \xi_n$, i.e., there exists a positive constant $L > 0$ such that

$$|f(x, \xi_1, \ldots, \xi_i, \ldots, \xi_n) - f(x, \xi_1, \ldots, \xi_i', \ldots, \xi_n)| \leq L|\xi_i - \xi_i'|, \quad (2.22)$$

for all $(x, \xi_1, \ldots, \xi_i, \ldots, \xi_n), (x, \xi_1, \ldots, \xi_i', \ldots, \xi_n) \in \Omega, i = 1, 2, \ldots, n$.

We remark that Theorem 2.1 is a direct consequence of Theorem 3.1 which states the existence and uniqueness of the solution of the Cauchy problem for *systems of differential equations* (see Chap. 3). It is a direct consequence because of the equivalence between the nth-order ODE (2.2) and the system of n first-order equations which is obtained by denoting $y = y_1, y' = y_2, \ldots, y^{(n-1)} = y_n$:

$$\begin{cases} y_1' = y_2 \\ y_2' = y_3 \\ \ldots\ldots\ldots \\ y_{n-1}' = y_n \\ y_n' = f(x, y_1, y_2, \ldots, y_n). \end{cases}$$

Remark 2.4 (Non-Local Existence of Solutions) As we saw in Remark 1.4 for first-order equations, if the function $f(x, \xi_1, \xi_2, \ldots, \xi_n)$ is continuous on $I \times \mathbb{R}^n$ and satisfies the Lipschitz condition on every strip $[\alpha, \beta] \times \mathbb{R}^n$ (where $x_0 \in [\alpha, \beta] \subset I$), that is: there exists a positive constant $L = L_{[\alpha,\beta]} > 0$ such that the inequality (2.22) holds for all $x \in [\alpha, \beta]$ and $\xi_1, \xi_2, \ldots, \xi_i, \xi_i', \ldots, \xi_n \in \mathbb{R}, i = 1, 2, \ldots, n$, then the Cauchy problem (2.21) has a (unique) solution $y(x)$ defined on the whole interval I.

At the end of this section, before moving on to the most used nth-order ODEs, namely to *linear differential equations of order n*, we discuss some particular cases of higher-order equations where the general solution can be found by simple methods (such as successive integration, reducing the order of the equation, etc.).

The Case $y^{(n)} = f(x)$

By successive quadratures, we finally find:

$$y(x) = \underbrace{\int \left[\int \left[\int \left[\cdots \left[\int f(x)\,dx\right]\cdots\right] dx\right] dx\right] dx}_{n-times} +$$

$$+ C_1 \frac{x^{n-1}}{(n-1)!} + C_2 \frac{x^{n-2}}{(n-2)!} + \ldots + C_n.$$

Example 2.4 $y'' = x + 1$. We integrate twice and obtain:

$$y' = \int (x+1)\,dx = \frac{x^2}{2} + x + C_1,$$

$$y = \int \left(\frac{x^2}{2} + x + C_1\right) dx,$$

so the general solution is

$$y = \frac{x^3}{6} + \frac{x^2}{2} + C_1 x + C_2.$$

The Case $F(x, y^{(k)}, y^{(k+1)}, \ldots, y^{(n)}) = 0,\ k \le n$

In this case we can reduce the order of the ODE from n to $n-k$ if we use the substitution $z(x) = y^{(k)}(x)$, solve the ODE of order $(n-k)$, $F(x, z, z', \ldots, z^{(n-k)}) = 0$, and then apply k successive quadratures.

Example 2.5 $y'' - \frac{1}{x} y' = 0$.

By denoting $z = y'$, the equation becomes: $z' = \frac{1}{x} z$, or $\frac{z'}{z} = \frac{1}{x}$. We integrate and obtain: $\ln|z| = \ln|x| + \ln|C_1|$, or $z = C_1 x$. Hence we obtained that $y' = C_1 x$. It follows that $y = C_1 \frac{x^2}{2} + C_2$.

Example 2.6 Hanging cable ([44], 3.11). Consider an inextensible flexible homogeneous cable (or chain) hanging between two fixed points, $M_1(-b, h)$ and $M_2(b, h)$. We take the origin of the y-axis at the lowest point of the cable, $O(0, 0)$. The shape taken by the cable is called a *catenary*, after the Latin word *catena* = chain. We want to find the mathematical expression of this curve, $y(x)$. Let $P(x, y(x))$ be an arbitrary point of the chain. The tension T in the cable at the point P is tangent to the cable and can be decomposed into horizontal and vertical components,

$$T_h = T \cos\theta \quad \text{and} \quad T_v = T \sin\theta,$$

2.1 Introduction

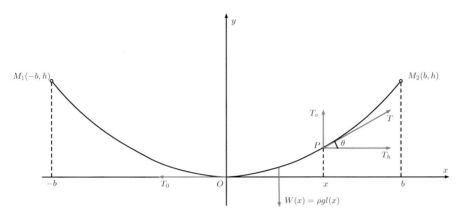

Fig. 2.5 Cable suspended between two fixed points

where θ is the angle between the x- axis and the tangent to the curve at the point P, hence $\tan\theta = y'(x)$ (Fig. 2.5).

Let T_0 be the tension in the cable at the lowest point, O (oriented horizontally) and let $W(x)$ be the weight of the segment of cable OP. Because the system is in equilibrium, we have:

$$T \sin\theta = W(x) \quad \text{and} \quad T \cos\theta = T_0,$$

so we obtain

$$y'(x) = \frac{W(x)}{T_0}. \tag{2.23}$$

Let ρ denotes the linear density of the cable ($\rho = M/L$ where M is the mass and L is the length of the cable). Then $W(x) = \rho g l(x)$, where g is the gravitational acceleration and $l(x)$ is the length of the arc OP, which is given by the formula

$$l(x) = \int_0^x \sqrt{1 + y'(t)^2}\, dt.$$

By the Fundamental Theorem of Calculus it follows that $l'(x) = \sqrt{1 + y'(x)^2}$ and, using Eq. (2.23), we obtain that $y(x)$ satisfies the following *second-order* ODE:

$$y'' = \frac{\rho g}{T_0} \sqrt{1 + y'(x)^2}. \tag{2.24}$$

By denoting $a = \dfrac{T_0}{\rho g}$, and $z(x) = y'(x)$, we obtain the following *first-order* (separable) equation in $z = z(x)$:

$$\frac{dz}{dx} = \frac{\sqrt{1+z^2}}{a},$$

which can be integrated as follows:

$$\int \frac{dz}{\sqrt{1+z^2}} = \frac{1}{a} \int dx,$$

or

$$\sinh^{-1} z = \frac{x}{a} + C_1.$$

Hence the general solution of the first-order equation in z is $z(x) = \sinh\left(\frac{x}{a} + C_1\right)$. Since $z = y'$, we obtain that the general solution of the Eq. (2.24) is:

$$y(x) = a \cosh\left(\frac{x}{a} + C_1\right) + C_2.$$

The last step is to find the constants C_1 and C_2 such that the following initial conditions are fulfilled: $y(0) = 0$ and $y'(0) = 0$ (because $O(0, 0)$ is the minimum point of the curve). Since $y'(0) = \sinh(C_1) = 0$, it follows that $C_1 = 0$ and, because $y(0) = a + C_2 = 0$, we obtain that $C_2 = -a$. Hence, the solution of our *Cauchy problem* is

$$y = a \cosh \frac{x}{a} - a.$$

The Case $F(y, y', \ldots, y^{(n)}) = 0$

If the differential equation does not explicitly contain the independent variable, x, then we denote $y' = p$, where $p = p(y)$ is treated as a new unknown function of the new independent variable y. The higher-order derivatives of y with respect to x can be expressed as follows:

$$y'' = \frac{d}{dx}(y') = \frac{d}{dx}(p(y)) = \frac{d}{dy}(p(y)) \cdot \frac{dy}{dx} = p' \cdot p,$$

$$y''' = \frac{d}{dx}(y'') = \frac{d}{dy}(p'(y)p(y)) \cdot \frac{dy}{dx} = \left(p'' p + p'^2\right) p = p'' p^2 + p'^2 p,$$

and so on. As can be easily noticed, by substituting these derivatives in the initial nth-order equation, we obtain a differential equation of order $n - 1$.

Example 2.7 $y'' + y'^2 = 2\exp(-y)$.

We denote $p = y'$ and obtain the following first-order differential equation in the unknown function $p = p(y)$:

$$p' \cdot p + p^2 = 2\exp(-y)$$

or, equivalently,

$$p'(y) + p(y) = 2\exp(-y)p(y)^{-1},$$

which is a Bernoulli equation with $\alpha = -1$. We make the substitution $z = p^2$ and find the linear equation:

$$z' + 2z = 4\exp(-y)$$

with the general solution $z = 4\exp(-y) + C\exp(-2y)$. Furthermore, since $z = y'^2$, we obtain:

$$y' = \frac{dy}{dx} = \pm\sqrt{4\exp(-y) + C\exp(-2y)}.$$

This is an equation with separable variables, so we can integrate it:

$$\pm\int \frac{4\exp(y)dy}{2\sqrt{4\exp(y) + C}} = \int 2\,dx \iff \pm\sqrt{4\exp(y) + C} = 2x + C_1$$

$$\iff 4\exp(y) = 4x^2 + 4C_1 x + C_1^2 - C$$

and, by denoting $C_2 = \frac{1}{4}(C_1^2 - C)$, we obtain the general solution:

$$y = \ln(x^2 + C_1 x + C_2).$$

2.2 Homogeneous Linear Differential Equations of Order n

If the function $F(x, \xi_1, \xi_2, \ldots, \xi_{n+1})$ from Eq. (2.1) is *linear* in $\xi_1, \xi_2, \ldots, \xi_{n+1}$, the ODE is said to be *linear*. Thus, a linear differential equation of nth order is an equation of the form:

$$a_0(x)y^{(n)} + a_1(x)y^{(n-1)} + \ldots + a_{n-1}(x)y' + a_n(x)y = b(x), \quad (2.25)$$

where $a_0, a_1, \ldots, a_n$ and b are continuous functions on the real interval I, and $a_0(x) \neq 0$ for any $x \in I$.

If $b(x)$ is identically zero, $b(x) = 0$ for all $x \in I$, then (2.25) is written:

$$a_0(x)y^{(n)} + a_1(x)y^{(n-1)} + \ldots + a_{n-1}(x)y' + a_n(x)y = 0 \qquad (2.26)$$

and it is said to be a *homogeneous* linear ODE of order n. Otherwise, the Eq. (2.25) is called *non-homogeneous*.

Since $a_0(x) \neq 0$ for any $x \in I$, the Eq. (2.25) can be written in the normal form:

$$y^{(n)} = \frac{b(x)}{a_0(x)} - \frac{a_n(x)}{a_0(x)}y - \frac{a_{n-1}(x)}{a_0(x)}y' - \ldots - \frac{a_1(x)}{a_0(x)}y^{(n-1)} = f(x, y, y', \ldots, y^{(n-1)}), \qquad (2.27)$$

where f is a function of $n+1$ variables, defined for all $x \in I$ and $\xi_i \in \mathbb{R}$ as:

$$f(x, \xi_1, \xi_2, \ldots, \xi_n) = \frac{b(x)}{a_0(x)} - \frac{a_n(x)}{a_0(x)}\xi_1 - \frac{a_{n-1}(x)}{a_0(x)}\xi_2 - \ldots - \frac{a_1(x)}{a_0(x)}\xi_n.$$

We can see that

$$\frac{\partial f}{\partial \xi} = -\frac{a_{n+1-i}(x)}{a_0(x)},$$

for any $i = 1, 2, \ldots, n$. For every interval $[\alpha, \beta] \subset I$ (containing x_0) we denote by $L_{[\alpha,\beta]} = \max_{i=1,n} \left| \frac{a_{n+1-i}(x)}{a_0(x)} \right|$. By the Mean Value Theorem, it follows that

$$\left| f(x, \xi_1, \ldots, \xi_i, \ldots, \xi_n) - f(x, \xi_1, \ldots, \xi_i', \ldots, \xi_n) \right| \leq L_{[\alpha,\beta]} \left| \xi_i - \xi_i' \right|,$$

for any $x \in [\alpha, \beta]$, $\xi_1, \xi_2, \ldots, \xi_i, \xi_i', \ldots, \xi_n \in \mathbb{R}$. Hence the function f satisfies the Lipschitz condition w.r.t $\xi_1, \xi_2, \ldots, \xi_n$, on every strip $[\alpha, \beta] \times \mathbb{R}^n$ and, by Theorem 2.1 and Remark 2.4, the next theorem follows.

Theorem 2.2 *Let $a_0, a_1, \ldots, a_n$ and b be continuous functions on the real interval I, such that $a_0(x) \neq 0$ for any $x \in I$ and consider $x_0 \in I$ and $y_0, y_1, \ldots, y_{n-1} \in \mathbb{R}$. Then the Cauchy problem*

$$\begin{cases} a_0(x)y^{(n)} + a_1(x)y^{(n-1)} + \ldots + a_{n-1}(x)y' + a_n(x)y = b(x) \\ y(x_0) = y_0, \ y'(x_0) = y_1, \ldots, y^{(n-1)}(x_0) = y_{n-1} \end{cases} \qquad (2.28)$$

has a unique solution $y(x)$, defined on the whole interval I.

Let us consider the homogeneous Eq. (2.26) and denote by L the differential operator defined by:

$$L[u] = a_0(x)u^{(n)} + a_1(x)u^{(n-1)} + \ldots + a_{n-1}(x)u' + a_n(x)u. \qquad (2.29)$$

2.2 Homogeneous Linear Differential Equations of Order n

It is very easy to verify that L is a linear mapping from $C^n(I)$, the real vector space of all functions defined on I of class C^n, to $C(I)$, the real vector space of all continuous functions defined on the interval I. Indeed, for any $u, v \in C^n(I)$ and $\alpha, \beta \in \mathbb{R}$, since $[\alpha u(x) + \beta v(x)]^{(j)} = \alpha u^{(j)}(x) + \beta v^{(j)}(x)$, $j = 0, 1, \ldots, n$, one easily obtain that

$$L[\alpha u + \beta v] = \alpha L[u] + \beta L[v].$$

Let S denote the set of all solutions of the homogeneous Eq. (2.26), which can be also written as $L[y] = 0$. As can be readily verified, for any solutions $u, v \in S$ and $\alpha, \beta \in \mathbb{R}$, we have $\alpha u + \beta v \in S$, so S is a vector subspace of $C^n(I)$ (actually S is the kernel of the linear operator L).

Theorem 2.3 *If $a_0(x) \neq 0$ for any $x \in I$, then the dimension of the vector subspace S of all solutions of the homogeneous equation $L[y] = 0$ is exactly n, the order of the equation. This means that there exist n linearly independent solutions $y_1, y_2, \ldots, y_n \in S$ such that any solution $y(x)$ of $L[y] = 0$ can be uniquely written as:*

$$y(x) = C_1 y_1(x) + \ldots + C_n y_n(x), \qquad (2.30)$$

for any $x \in I$, where $C_1, \ldots, C_n$ are some real constants.

We say that the functions $y_1, y_2, \ldots, y_n$ form a *basis of solutions* (or a *fundamental set of solutions*). Formula (2.30) is called the *general solution* of the homogeneous linear Eq. (2.26). It is also written as:

$$y_{GH}(x) = C_1 y_1(x) + \ldots + C_n y_n(x). \qquad (2.31)$$

Proof To prove this theorem we need some basic linear algebra and Theorem 2.2. The idea is to fix a point $x_0 \in I$ and to construct the isomorphism $T : S \to \mathbb{R}^n$ which associates to every solution $u \in S$ the vector $\bigl(u(x_0), u'(x_0), \ldots, u^{(n-1)}(x_0)\bigr) \in \mathbb{R}^n$. Obviously, T is a linear mapping. Furthermore, the existence part of Theorem 2.2 proves the surjectivity of T, while the uniqueness part of the theorem gives the injectivity of T. Therefore, T is an isomorphism, hence the vector spaces S and $\mathbb{R}^n$ are isomorphic. It follows that $\dim S = \dim \mathbb{R}^n = n$, so there exists a basis of S formed by n solutions of (2.26), $y_1, y_2, \ldots, y_n$. For example, if $e_1 = (1, 0, \ldots, 0)$, $e_2 = (0, 1, \ldots, 0), \ldots, e_n = (0, 0, \ldots, 1)$ is the canonical basis of $\mathbb{R}^n$, we can take $y_1 = T^{-1}(e_1), \ldots, y_n = T^{-1}(e_n)$. □

The next problem is: "How to construct such a basis of solutions?"

First of all we need a test for the linear independence of a set of n functions of $C^n(I)$. The polish mathematician Wronski (J. M. Hoene-Wronski, 1776–1853) discovered such a test, based on the following determinant, called the *Wronskian* of the system of functions $y_1, y_2, \ldots, y_n$:

$$W(x) = W[y_1, y_2, \ldots, y_n](x) = \begin{vmatrix} y_1(x) & y_2(x) & \ldots & y_n(x) \\ y_1'(x) & y_2'(x) & \ldots & y_n'(x) \\ \vdots & \vdots & \ldots & \vdots \\ y_1^{(n-1)}(x) & y_2^{(n-1)}(x) & \ldots & y_n^{(n-1)}(x) \end{vmatrix} \quad (2.32)$$

for all $x \in I$.

Theorem 2.4 *Let $y_1, y_2, \ldots, y_n$ be n functions of class C^n defined on the interval I and $W(x)$ be the Wronskian of this set of functions, defined by (2.32). If there is a point $x_0 \in I$ such that $W(x_0) \neq 0$, then the functions $y_1, y_2, \ldots, y_n$ are linearly independent.*

Proof Assume that $C_1, C_2, \ldots, C_n$ are real numbers such that

$$C_1 y_1(x) + C_2 y_2(x) + \ldots + C_n y_n(x) = 0 \quad (2.33)$$

for any $x \in I$. If we differentiate the equality (2.33) $n - 1$ times and write it for the point $x = x_0$, we get:

$$\begin{cases} C_1 y_1(x_0) + C_2 y_2(x_0) + \ldots + C_n y_n(x_0) = 0 \\ C_1 y_1'(x_0) + C_2 y_2'(x_0) + \ldots + C_n y_n'(x_0) = 0 \\ \vdots \\ C_1 y_1^{(n-1)}(x_0) + C_2 y_2^{(n-1)}(x_0) + \ldots + C_n y_n^{(n-1)}(x_0) = 0. \end{cases}$$

This is a homogeneous linear system of equations (with the unknowns $C_1, C_2, \ldots, C_n$) and its determinant is $W(x_0) \neq 0$. Hence the unique solution of the system is the trivial one: $C_1 = 0, C_2 = 0, \ldots, C_n = 0$. □

We note that the reciprocal of Theorem 2.4 is not always true. As shown in the example below, if $y_1, y_2, \ldots, y_n \in C^n(I)$ such that $W(x) = 0$ for all $x \in I$, it doesn't mean that the functions are linearly dependent. However, Theorem 2.5 states that if the n functions are *solutions of the homogeneous equation* (2.26) (i.e. $y_1, y_2, \ldots, y_n \in S$), then the reciprocal of Theorem 2.4 is true. As a matter of fact, in Theorem 2.5, a stronger statement is expressed: if $y_1, y_2, \ldots, y_n \in S$ are linearly independent, then not only is there a point where their Wronskian is nonzero, but $W(x) \neq 0$ at any $x \in I$.

Example 2.8 Consider the following functions (of class C^2):

$$y_1(x) = \begin{cases} 0, & \text{if } x \in [-1, 0) \\ x^3, & \text{if } x \in [0, 1] \end{cases}, \quad y_2(x) = \begin{cases} x^3, & \text{if } x \in [-1, 0) \\ 0, & \text{if } x \in [0, 1] \end{cases}.$$

2.2 Homogeneous Linear Differential Equations of Order n

It is easy to see that these y_1, y_2 are linearly independent on $[-1, 1]$ and yet the Wronskian $W[y_1, y_2](x) = 0$ for any x in $[-1, 1]$, so the reciprocal of Theorem 2.4 is not true. Anyway, by Theorem 2.5, we can say that the functions $y_1(x)$, $y_2(x)$ cannot be solutions of a 2nd-order homogeneous linear equation with $a_0(x) \neq 0$ for all x in $[-1, 1]$.

Notice that, if we drop the condition $a_0(x) \neq 0$ for all x in $[-1, 1]$, we can find a 2nd-order homogeneous equation which has the solutions y_1, y_2. For example, we can take the equation

$$xy'' - 2y' = 0, \ x \in [-1, 1].$$

However, the condition $a_0(x) \neq 0$ for all $x \in I$ is of the utmost importance for the theory of homogeneous linear equations. If it is not fulfilled, the dimension of the vector space of solutions, dim S, may be greater than n. For example, the equation above has at least three linearly independent solutions: $y_0(x) = 1$, $y_1(x)$ and $y_2(x)$.

Theorem 2.5 *If $y_1, y_2, \ldots, y_n$ are n linearly independent solutions of the homogeneous Eq. (2.26) with $a_0(x) \neq 0$ for all $x \in I$, then*

$$W[y_1, \ldots, y_n](x) \neq 0$$

at any point $x \in I$.

Proof We suppose that there exists a point x_0 such that

$$W[y_1, \ldots, y_n](x_0) = 0.$$

This means that at least one column of the Wronskian

$$W[y_1, \ldots, y_n](x_0) = \begin{vmatrix} y_1(x_0) & y_2(x_0) & \cdots & y_n(x_0) \\ y_1'(x_0) & y_2'(x_0) & \cdots & y_n'(x_0) \\ \vdots & \vdots & \cdots & \vdots \\ y_1^{(n-1)}(x_0) & y_2^{(n-1)}(x_0) & \cdots & y_n^{(n-1)}(x_0) \end{vmatrix}$$

is a linear combination of the others. To be easier, let us assume that the first one is a linear combination of the others, so there exist the real numbers $C_2, \ldots, C_n$ such that:

$$\begin{cases} y_1(x_0) = C_2 y_2(x_0) + \ldots + C_n y_n(x_0) \\ y_1'(x_0) = C_2 y_2'(x_0) + \ldots + C_n y_n'(x_0) \\ \vdots \\ y_1^{(n-1)}(x_0) = C_2 y_2^{(n-1)}(x_0) + \ldots + C_n y_n^{(n-1)}(x_0) \end{cases} \quad (2.34)$$

Let us consider the following solution of the homogeneous equation $L[y] = 0$:

$$y(x) = C_2 y_2(x) + \ldots + C_n y_n(x).$$

Since

$$y_1(x_0) = y(x_0),\ y_1'(x_0) = y'(x_0),\ \ldots,\ y_1^{(n-1)}(x_0) = y^{(n-1)}(x_0),$$

from Theorem 2.2 we obtain that $y(x) = y_1(x)$ for any x in I. Thus, $y_1(x) = C_2 y_2(x) + \ldots + C_n y_n(x)$ for any $x \in I$, which means that $y_1, y_2, \ldots, y_n$ are not linearly independent, a contradiction! Hence, the Wronskian $W[y_1, y_2, \ldots, y_n]$ is nonzero at any point $x \in I$. □

Example 2.9 It is easy to see that $y_1(x) = \sin x$ and $y_2(x) = \cos x$ are solutions of the linear ODE $y'' + y = 0$. Since their Wronskian is obvious $-1 \neq 0$, $\{\sin x, \cos x\}$ is linearly independent, i.e. it is a basis of solutions for our equation or, equivalently, any other solution $y(x)$ is a linear combination of $\sin x$ and $\cos x$: $y(x) = C_1 \sin x + C_2 \cos x$.

Example 2.10 Using Theorem 2.5 we can prove that, for any $n \geq 2$, the functions $\{\sin x, \sin 2x, \ldots, \sin nx\}$, which are linearly independent on $[-\pi, \pi]$ (see Chap. 4), cannot be simultaneously solutions of a homogeneous linear Eq. (2.26) with $a_0(x) \neq 0$ on the interval $I = [-\pi, \pi]$.

If they were so, then their Wronskian, $W(x)$, should be nonzero at any point $x \in [-\pi, \pi]$. But for $x = 0$ we have $W(0) = 0$.

We note that, given a set of n linearly independent functions of class C^n, $\{y_1, y_2, \ldots, y_n\}$, we can always find a homogeneous linear equation of nth order which is satisfied by all these functions. The equation

$$W[y, y_1, y_2, \ldots, y_n] = 0$$

is such an equation. But as can be easily noticed, the coefficient of $y^{(n)}$ in this equation is $a_0(x) = W[y_1, y_2, \ldots, y_n](x)$.

For instance, a linear 2nd-order ODE which has as solutions $\{\sin x, \sin 2x\}$ is:

$$\begin{vmatrix} y & \sin x & \sin 2x \\ y' & \cos x & 2\cos 2x \\ y'' & -\sin x & -4\sin 2x \end{vmatrix} = 0,$$

which is written

$$[2 \sin x \cos 2x - \sin 2x \cos x] y'' + [3 \sin x \sin 2x] y' +$$

$$+[-4 \cos x \sin 2x + 2 \sin x \cos 2x] y = 0,$$

2.2 Homogeneous Linear Differential Equations of Order n

or, equivalently,

$$(\sin^2 x)y'' - (3\sin x \cos x)y' + (2\cos^2 x + 1)y = 0.$$

Let us recall the general rule of differentiating the determinant of a matrix of functions:

$$\frac{d}{dx}\left(\begin{vmatrix} a_{11}(x) & a_{12}(x) & \cdots & a_{1,n-1}(x) & a_{1n}(x) \\ a_{21}(x) & a_{22}(x) & \cdots & a_{2,n-1}(x) & a_{2n}(x) \\ \vdots & \vdots & \cdots & \vdots & \vdots \\ a_{n-1,1}(x) & a_{n-1,2}(x) & \cdots & a_{n-1,n-1}(x) & a_{n-1,n}(x) \\ a_{n1}(x) & a_{n1}(x) & \cdots & a_{n,n-1}(x) & a_{nn}(x) \end{vmatrix}\right) = \qquad (2.35)$$

$$= \begin{vmatrix} a'_{11}(x) & a'_{12}(x) & \cdots & a'_{1,n-1}(x) & a'_{1n}(x) \\ a_{21}(x) & a_{22}(x) & \cdots & a_{2,n-1}(x) & a_{2n}(x) \\ \vdots & \vdots & \cdots & \vdots & \vdots \\ a_{n-1,1}(x) & a_{n-1,2}(x) & \cdots & a_{n-1,n-1}(x) & a_{n-1,n}(x) \\ a_{n1}(x) & a_{n1}(x) & \cdots & a_{n,n-1}(x) & a_{nn}(x) \end{vmatrix} +$$

$$+ \begin{vmatrix} a_{11}(x) & a_{12}(x) & \cdots & a_{1,n-1}(x) & a_{1n}(x) \\ a'_{21}(x) & a'_{22}(x) & \cdots & a'_{2,n-1}(x) & a'_{2n}(x) \\ \vdots & \vdots & \cdots & \vdots & \vdots \\ a_{n-1,1}(x) & a_{n-1,2}(x) & \cdots & a_{n-1,n-1}(x) & a_{n-1,n}(x) \\ a_{n1}(x) & a_{n1}(x) & \cdots & a_{n,n-1}(x) & a_{nn}(x) \end{vmatrix} + \cdots +$$

$$+ \begin{vmatrix} a_{11}(x) & a_{12}(x) & \cdots & a_{1,n-1}(x) & a_{1n}(x) \\ a_{21}(x) & a_{22}(x) & \cdots & a_{2,n-1}(x) & a_{2n}(x) \\ \vdots & \vdots & \cdots & \vdots & \vdots \\ a_{n-1,1}(x) & a_{n-1,2}(x) & \cdots & a_{n-1,n-1}(x) & a_{n-1,n}(x) \\ a'_{n1}(x) & a'_{n1}(x) & \cdots & a'_{n,n-1}(x) & a'_{nn}(x) \end{vmatrix}.$$

We apply this rule to differentiate the Wronskian $W(x) = W[y_1, y_2](x)$ of the functions y_1, y_2, two solutions of the homogeneous linear ODE

$$a_0(x)y'' + a_1(x)y' + a_2(x)y = 0,$$

with $a_0(x) \neq 0$ for any $x \in I$.

$$W'(x) = \frac{d}{dx}\left(\begin{vmatrix} y_1(x) & y_2(x) \\ y'_1(x) & y'_2(x) \end{vmatrix}\right) = \begin{vmatrix} y'_1(x) & y'_2(x) \\ y'_1(x) & y'_2(x) \end{vmatrix} + \begin{vmatrix} y_1(x) & y_2(x) \\ y''_1(x) & y''_2(x) \end{vmatrix} =$$

$$= \frac{1}{a_0(x)} \begin{vmatrix} y_1(x) & y_2(x) \\ a_0(x)y''_1(x) & a_0(x)y''_2(x) \end{vmatrix} =$$

$$= \frac{1}{a_0(x)} \begin{vmatrix} y_1(x) & y_2(x) \\ -a_1(x)y'_1(x) - a_2(x)y_1(x) & -a_1(x)y'_2(x) - a_2(x)y_2(x) \end{vmatrix} =$$

$$= -\frac{a_1(x)}{a_0(x)} \begin{vmatrix} y_1(x) & y_2(x) \\ y'_1(x) & y'_2(x) \end{vmatrix} = -\frac{a_1(x)}{a_0(x)} W(x).$$

Hence

$$\frac{d}{dx} W(x) = -\frac{a_1(x)}{a_0(x)} W(x).$$

Integrating this last equality on a subinterval $[x_0, x] \subset I$ we obtain the Abel's formula:

$$W(x) = W(x_0) \exp\left(-\int_{x_0}^{x} \frac{a_1(t)}{a_0(t)} dt\right). \tag{2.36}$$

In particular, we obtain again that the Wronskian is either identically zero on the interval I, or it is nonzero at every point of I. Note that the formula (2.36) is true for any integer $n \geq 2$. For the general case $n > 2$ the proof can be done in the same manner.

2.3 Non-Homogeneous Linear Differential Equations of Order n

Let us consider now the non-homogeneous Eq. (2.25):

$$a_0(x)y^{(n)} + a_1(x)y^{(n-1)} + \ldots + a_{n-1}(x)y' + a_n(x)y = b(x), \tag{2.37}$$

where $b, a_0, a_1, \ldots, a_n : I \to \mathbb{R}$ are continuous functions and $a_0(x) \neq 0$ for any $x \in I$. Using the differential operator defined by (2.29), the non-homogeneous equation is written as:

$$L[y] = b.$$

2.3 Non-Homogeneous Linear Differential Equations of Order n

Suppose that $y_1, y_2, \ldots, y_n$ is a fundamental set of solutions of the *homogeneous* Eq. (2.26) and y_P is a particular solution of the *non-homogeneous* Eq. (2.37). Then, for any constants $C_1, \ldots, C_n$,

$$L[C_1 y_1 + \ldots + C_n y_n + y_P] = L[y_P] = b,$$

so $y = C_1 y_1 + \ldots + C_n y_n + y_P$ is a solution of the non-homogeneous Eq. (2.37).

Moreover, if $y(x)$ is an arbitrary solution of the non-homogeneous equation $L[y] = b$, then, since

$$L[y - y_P] = L[y] - L[y_P] = 0,$$

it follows that $y - y_P$ is a solution of the homogeneous equation $L[y] = 0$. Hence $y - y_P$ can be written as a linear combination of $y_1, y_2, \ldots, y_n$:

$$y - y_P = C_1 y_1 + C_2 y_2 + \ldots + C_n y_n$$

and the next theorem is proved.

Theorem 2.6 *Let $\{y_1, y_2, \ldots, y_n\}$ be a fundamental set of solutions for the associated homogeneous equation $L[y] = 0$ and let y_P be a particular solution of the non-homogeneous Eq. (2.37) $L[y] = b$. Then the **general solution** of (2.37) is*

$$y(x) = y_P(x) + C_1 y_1(x) + C_2 y_2(x) + \ldots + C_n y_n(x), \tag{2.38}$$

for all $x \in I$, where $C_1, C_2, \ldots, C_n$ are real constants.

The formula of the general solution (2.38) can also be written as

$$y_G = y_P + y_{GH}, \tag{2.39}$$

i.e. the general solution of the non-homogeneous equation $L[y] = b$ is a sum between a particular solution y_P of the same non-homogeneous equation and the general solution of the associated homogeneous equation $L[y] = 0$. Hence, the space of all solutions of the non-homogeneous equation $L[y] = f$ is an *affine space*, i.e. a sum between a "vector" y_P and the vector subspace S of all solutions of the associated homogeneous equation $L[y] = 0$.

As can be seen, in order to solve the non-homogeneous equation $L[y] = b$, we need a fundamental set of solutions $\{y_1, y_2, \ldots, y_n\}$ of S, the vector space of all solutions of the homogeneous equation $L[y] = 0$ and a particular solution $y_P(x)$ for the equation $L[y] = b$. There is no general method to find a fundamental set of solutions (though it can be constructed in some particular cases, for instance, when the equation has constant coefficients). But, to find a particular solution for the equation $L[y] = b$, when we know such a basis of solutions $\{y_1, y_2, \ldots, y_n\}$, is theoretically possible by using the *method of variation of constants*, discovered by Lagrange (in Sect. 1.4 we applied this method for first-order linear differential equations).

Theorem 2.7 (Lagrange's Method of Variation of Constants) *Let $\{y_1, y_2, \ldots, y_n\}$ be a fundamental set of solutions of the homogeneous equation $L[y] = 0$ and let $C_1(x), \ldots, C_n(x)$ be n functions of class $C^1(I)$ whose derivatives $C_1'(x), \ldots, C_n'(x)$ satisfy the following linear algebraic system of equations, for all $x \in I$:*

$$\begin{pmatrix} y_1(x) & y_2(x) & \ldots & y_n(x) \\ y_1'(x) & y_2'(x) & \ldots & y_n'(x) \\ \vdots & \vdots & \ldots & \vdots \\ y_1^{(n-1)}(x) & y_2^{(n-1)}(x) & \ldots & y_n^{(n-1)}(x) \end{pmatrix} \begin{pmatrix} C_1'(x) \\ C_2'(x) \\ \vdots \\ C_n'(x) \end{pmatrix} = \begin{pmatrix} 0 \\ \vdots \\ 0 \\ \frac{b(x)}{a_0(x)} \end{pmatrix}. \tag{2.40}$$

Then the function $y_P : I \to \mathbb{R}$, defined by

$$y_P(x) = C_1(x)y_1(x) + C_2(x)y_2(x) + \ldots + C_n(x)y_n(x) \tag{2.41}$$

is a particular solution of the non-homogeneous equation $L[y] = b$.

Proof We differentiate (2.41) and obtain:

$$y_P'(x) = C_1(x)y_1'(x) + C_2(x)y_2'(x) + \ldots + C_n(x)y_n'(x) +$$

$$+ C_1'(x)y_1(x) + C_2'(x)y_2(x) + \ldots + C_n'(x)y_n(x).$$

But the first equation of the system (2.40) is

$$C_1'(x)y_1(x) + C_2'(x)y_2(x) + \ldots + C_n'(x)y_n(x) = 0,$$

hence

$$y_P'(x) = C_1(x)y_1'(x) + C_2(x)y_2'(x) + \ldots + C_n(x)y_n'(x) \tag{2.42}$$

By differentiating the equality (2.42) and using the second equation of the system (2.40), we get:

$$y_P''(x) = C_1(x)y_1''(x) + C_2(x)y_2''(x) + \ldots + C_n(x)y_n''(x). \tag{2.43}$$

We continue in this way up to the $(n-1)$th derivative,

$$y_P^{(n-1)}(x) = C_1(x)y_1^{(n-1)}(x) + C_2(x)y_2^{(n-1)}(x) + \ldots + C_n(x)y_n^{(n-1)}(x). \tag{2.44}$$

2.3 Non-Homogeneous Linear Differential Equations of Order n

Since the last equation of the system (2.40) is

$$C_1'(x)y_1^{(n-1)}(x) + C_2'(x)y_2^{(n-1)}(x) + \ldots + C_n'(x)y_n^{(n-1)}(x) = \frac{b(x)}{a_0(x)},$$

by differentiating (2.44) we obtain:

$$y_P^{(n)}(x) = C_1(x)y_1^{(n)}(x) + C_2(x)y_2^{(n)}(x) + \ldots + C_n(x)y_n^{(n)}(x) + \frac{b(x)}{a_0(x)}. \quad (2.45)$$

Now, we multiply the equality (2.45) by $a_0(x)$, (2.44) by $a_1(x)$ and so on, up to the equality (2.41) which is multiplied by $a_n(x)$; then we add all these equalities and finally find:

$$L[y_P(x)] = C_1(x)L[y_1(x)] + \ldots + C_n(x)L[y_n(x)] + b(x). \quad (2.46)$$

Since $y_1, y_2, \ldots, y_n$ are solutions of the homogeneous equation $L[y] = 0$, from (2.46), we get: $L[y_P] = b(x)$, i.e. $y_P(x)$ is a solution of the non-homogeneous equation $L[y] = b$. □

Example 2.11 Let us apply the Lagrange's method of variation of constants to solve the following Cauchy problem: $y'' + y = \frac{1}{\cos x}$, $y(0) = 0$, $y'(0) = 1$. It is easy to see that $\{\cos x, \sin x\}$ is a basis of solutions for the homogeneous associated equation $y'' + y = 0$. Let us search for a particular solution of the initial non-homogeneous equation:

$$y_P(x) = C_1(x)\cos x + C_2(x)\sin x, \quad (2.47)$$

where the derivatives $\{C_1'(x), C_2'(x)\}$ constitute the solution of the linear algebraic system (2.40):

$$\begin{pmatrix} \cos x & \sin x \\ -\sin x & \cos x \end{pmatrix} \begin{pmatrix} C_1'(x) \\ C_2'(x) \end{pmatrix} = \begin{pmatrix} 0 \\ \frac{1}{\cos x} \end{pmatrix}.$$

The solution of this system is: $C_1'(x) = -\frac{\sin x}{\cos x}$, $C_2'(x) = 1$. Thus, $C_1(x) = \ln|\cos x|$ and $C_2(x) = x$. Now, we remark that x must be in an interval I such that either $\cos x > 0$, or $\cos x < 0$, for all $x \in I$ (because $\cos x$ appears at the denominator of the free term). Since $0 \in I$, it follows that we have $\cos x > 0$, for all $x \in I$, hence $C_1(x) = \ln \cos x$.

Coming back in (2.47), we find:

$$y_P(x) = \cos x \cdot \ln \cos x + x \sin x.$$

Hence, the general solution of the equation $y'' + y = \frac{1}{\cos x}$ is:

$$y(x) = \cos x \cdot \ln \cos x + x \sin x + C_1 \cos x + C_2 \sin x.$$

Since $y(0) = 0$, $y'(0) = 1$, we find that $C_1 = 0$ and $C_2 = 1$. So, the solution of the above Cauchy problem is: $y(x) = \cos x \cdot \ln \cos x + (x + 1) \sin x$.

Remark 2.5 *Superposition principle.* If the "free term" $b(x)$ (which does not contain y and its derivatives) is too complicated, then Lagrange's method can become extremely difficult. The following superposition law can be useful in order to make easier the searching for a particular solution of our non-homogeneous equation $L[y] = b$. Namely, if $b = b_1 + b_2 + \ldots + b_k$ and if $y_{P_1}, \ldots, y_{P_k}$ are particular solutions for $L[y] = b_1, \ldots, L[y] = b_k$ respectively, then $y_P = y_{P_1} + \ldots + y_{P_k}$ (their superposition) is a particular solution for $L[y] = b$. Indeed,

$$L[y_P] = L[y_{P_1} + \ldots + y_{P_k}] = L[y_{P_1}] + \ldots + L[y_{P_k}] = b_1 + \ldots + b_k = b.$$

2.4 Homogeneous Linear Equations with Constant Coefficients

A linear differential equation of nth order with constant coefficients is an ODE of the following form:

$$a_0 y^{(n)} + a_1 y^{(n-1)} + \ldots + a_{n-1} y' + a_n y = b(x),$$

where the coefficients $a_0 \neq 0, a_1 \ldots, a_n$ are real numbers and $b(x)$ is a function continuous on the real interval I. We may always assume that $a_0 = 1$ (if not, we divide the equation by a_0), so it is sufficient to consider only linear differential equations of the form:

$$y^{(n)} + a_1 y^{(n-1)} + \ldots + a_{n-1} y' + a_n y = b(x). \tag{2.48}$$

First of all, we study how to construct a fundamental set of solutions for the homogeneous equation associated,

$$y^{(n)} + a_1 y^{(n-1)} + \ldots + a_{n-1} y' + a_n y = 0. \tag{2.49}$$

Let us search for a solution of the form

$$y = \exp(\lambda x), \, x \in \mathbb{R}, \tag{2.50}$$

2.4 Homogeneous Linear Equations with Constant Coefficients

where λ is a fixed real number. Since

$$y^{(k)} = \frac{d^k}{dx^k}(\exp(\lambda x)) = \lambda^k \exp(\lambda x), \tag{2.51}$$

for all $k = 1, 2, \ldots$, by replacing in (2.49) we obtain:

$$(\lambda^n + a_1\lambda^{n-1} + \ldots + a_{n-1}\lambda + a_n)\exp(\lambda x) = 0.$$

But $\exp(\lambda x) \neq 0$ for any $x \in \mathbb{R}$, hence $y = \exp(\lambda x)$ is a solution of the homogeneous Eq. (2.49) if and only if λ is a solution of the nth degree polynomial equation

$$\lambda^n + a_1\lambda^{n-1} + \ldots + a_{n-1}\lambda + a_n = 0 \tag{2.52}$$

The polynomial

$$P(\lambda) = \lambda^n + a_1\lambda^{n-1} + \ldots + a_{n-1}\lambda + a_n \tag{2.53}$$

is called the *characteristic polynomial*, while the Eq. (2.52) is called the *characteristic equation* (or the *auxiliary equation*) of the differential Eq. (2.49).

By the *Fundamental Theorem of Algebra*, every nth-degree polynomial has n roots, though not necessarily distinct and not necessarily real.

We first consider the case when all the roots of the polynomial (2.53) are *real* and *distinct*:

$$\lambda_1, \lambda_2, \ldots, \lambda_n \in \mathbb{R}, \ \lambda_i \neq \lambda_j, \ \forall i \neq j.$$

As shown above, the functions $y_i : \mathbb{R} \to \mathbb{R}$, $i = 1, \ldots, n$,

$$y_1 = \exp(\lambda_1 x), y_2 = \exp(\lambda_2 x), \ldots, y_n = \exp(\lambda_n x) \tag{2.54}$$

are solutions of the homogeneous Eq. (2.49). We shall prove that $y_1, \ldots, y_n$ form a fundamental set of solutions.

Lemma 2.1 *If $\lambda_1, \lambda_2, \ldots, \lambda_n$ are real, distinct numbers, then the functions*

$$y_1 = \exp(\lambda_1 x), y_2 = \exp(\lambda_2 x), \ldots, y_n = \exp(\lambda_n x), x \in \mathbb{R},$$

are linearly independent.

Proof In order to prove the linear independence, we compute their Wronskian:

$$W[y_1,\ldots,y_n](x) = \begin{vmatrix} \exp(\lambda_1 x) & \exp(\lambda_2 x) & \ldots & \exp(\lambda_n x) \\ \lambda_1 \exp(\lambda_1 x) & \lambda_2 \exp(\lambda_2 x) & \ldots & \lambda_n \exp(\lambda_n x) \\ \lambda_1^2 \exp(\lambda_1 x) & \lambda_2^2 \exp(\lambda_2 x) & \ldots & \lambda_n^2 \exp(\lambda_n x) \\ \vdots & \vdots & & \vdots \\ \lambda_1^{n-1} \exp(\lambda_1 x) & \lambda_2^{n-1} \exp(\lambda_2 x) & \ldots & \lambda_n^{n-1} \exp(\lambda_n x) \end{vmatrix}$$

$$= \exp[(\lambda_1 + \lambda_2 + \ldots + \lambda_n)x] \begin{vmatrix} 1 & 1 & \ldots & 1 \\ \lambda_1 & \lambda_2 & \ldots & \lambda_n \\ \lambda_1^2 & \lambda_2^2 & \ldots & \lambda_n^2 \\ \vdots & \vdots & & \vdots \\ \lambda_1^{n-1} & \lambda_2^{n-1} & \ldots & \lambda_n^{n-1} \end{vmatrix}$$

The determinant above is a *Vandermonde* determinant ([35], 1.55), so

$$W[y_1,\ldots,y_n](x) = \exp[(\lambda_1 + \lambda_2 + \ldots + \lambda_n)x] \prod_{1 \le i < j \le n} (\lambda_j - \lambda_i) \ne 0,$$

because $\lambda_j - \lambda_i$ for any $i \ne j$. Hence the functions (2.54) are linearly independent and the lemma is proved. □

Corollary 2.1 *If all the roots of the characteristic polynomial* (2.53) *are real and distinct numbers,* $\lambda_1, \lambda_2, \ldots, \lambda_n$, *then* (2.54) *is a fundamental system of solutions for the homogeneous Eq.* (2.48), *so its general solution is:*

$$y_{GH} = C_1 \exp(\lambda_1 x) + C_2 \exp(\lambda_2 x) + \ldots + C_n \exp(\lambda_n x). \tag{2.55}$$

But what happens when (some of) the roots of the characteristic polynomial are complex numbers ?

Recall the *Euler's formula*, which establishes the fundamental relationship between the trigonometric functions and the complex exponential function. It states that, for any real number x, we have:

$$\exp(ix) = \cos x + i \sin x. \tag{2.56}$$

Now, let $\lambda = \alpha + i\beta$ be a *complex root* of the polynomial (2.53) and consider the *complex function* (of a real variable x),

$$y = \exp(\lambda x) = \exp(\alpha x)(\cos \beta x + i \sin \beta x), \ x \in \mathbb{R}. \tag{2.57}$$

A detailed presentation of complex functions can be found in Chap. 11, Sect. 11.2. For the moment, we only mention that a complex function $f(x) = u(x) + iv(x)$

2.4 Homogeneous Linear Equations with Constant Coefficients

with the real part $u(x)$ and the imaginary part $v(x)$ differentiable functions, is also differentiable and $f'(x) = u'(x) + iv'(x)$. It can be readily seen that the formula

$$\frac{d}{dx}(\exp(\lambda x)) = \lambda \exp(\lambda x),$$

as well as the general formula (2.51) hold for complex values of λ as well. Thus, we can say that $y = \exp(\lambda x)$ is a solution of the homogeneous Eq. (2.49) if and only if λ is a (*real* or *complex*) root of the characteristic polynomial (2.53). Moreover, it can be easily seen that Lemma 2.1 holds true even if the roots $\lambda_1, \ldots, \lambda_n$ are complex numbers (they only need to be *distinct*). Nevertheless, in the case with complex roots, the formula (2.55) is not very helpful, because it yields a *complex function*, while we are searching for the *real* solutions of the differential Eq. (2.49). But, if the *complex* function (2.57) is a solution of homogeneous Eq. (2.49), it is compulsory that the equation is also satisfied by its real part $\exp(\alpha x) \cos \beta x$ and its imaginary part $\exp(\alpha x) \sin \beta x$ respectively. It seems like the complex solution (2.57) produces two *linearly independent* real solutions of the equation (you can calculate the Wronskian to check the linear independence of the two functions).

Remember that the complex roots of a polynomial with *real* coefficients are "in pairs", that is, if $\lambda = \alpha + i\beta$ with $\beta \neq 0$ is a root, then its complex conjugate $\bar\lambda = \alpha - i\beta$ is also a root (with the same multiplicity as λ).

Corollary 2.2 *Suppose that the characteristic polynomial* (2.53) *has* **distinct** *roots and* $2p$ *of them are complex,*

$$\lambda_{2k-1} = \alpha_k + i\beta_k, \ \lambda_{2k} = \bar\lambda_{2k-1} = \alpha_k - i\beta_k, \ \beta_k \neq 0, \ k = 1, \ldots, p,$$

and the other $n - 2p$ *are real:* $\lambda_{2p+1}, \ldots, \lambda_n \in \mathbb{R}$.
Then, the (real) functions $\tilde y_1, \ldots, \tilde y_{2p}, y_{2p+1}, \ldots y_n$ *defined by:*

$$\begin{aligned}
\tilde y_{2k-1} &= \exp(\alpha_k x) \cos \beta_k x, \ \tilde y_{2k} = \exp(\alpha_k x) \sin \beta_k x, \ k = 1, \ldots, p, \\
y_{2p+1} &= \exp(\lambda_{2p+1} x), \ldots, y_n = \exp(\lambda_n x)
\end{aligned} \quad (2.58)$$

form a fundamental system of solutions of the Eq. (2.49).

Proof Let us prove that the functions (2.58) are linearly independent. We denote $y_j = \exp(\lambda_j x)$ for any $j = 1, \ldots, n$. It is easy to see that, for any $k = 1, \ldots, p$,

$$\begin{cases} y_{2k-1} = \tilde y_{2k-1} + i\tilde y_{2k} \\ y_{2k} = \tilde y_{2k-1} - i\tilde y_{2k}, \end{cases}$$

so

$$\begin{cases} \tilde y_{2k-1} = \frac{1}{2}(y_{2k-1} + y_{2k}) \\ \tilde y_{2k} = \frac{1}{2i}(y_{2k-1} - y_{2k}). \end{cases} \quad (2.59)$$

Suppose that $\gamma_1, \ldots, \gamma_n$ are real numbers such that

$$\sum_{k=1}^{p}(\gamma_{2k-1}\tilde{y}_{2k-1} + \gamma_{2k}\tilde{y}_{2k}) + \sum_{k=2p+1}^{n}\gamma_k y_k = 0.$$

By using (2.59) we can write this equality as

$$\sum_{k=1}^{p}\left(\frac{\gamma_k - i\gamma_{k+1}}{2}y_k + \frac{\gamma_k + i\gamma_{k+1}}{2}y_{k+1}\right) + \sum_{k=2p+1}^{n}\gamma_k y_k = 0.$$

As shown above, Lemma 2.1 is applicable also for complex numbers, so the functions $y_1, \ldots, y_n$ are linearly independent, hence all the coefficients of the linear combination above are equal to 0. Since $\gamma_j = 0$, for all $j = 1, \ldots, n$, the functions (2.58) are linearly independent.

□

If the roots of the characteristic polynomial (2.53) are not distinct (if it has *multiple* roots), then (2.54) cannot yield n linearly independent solutions of the Eq. (2.49). For example, if the roots are 1, 1, 1 and -1, we obtain only the two functions $\exp(x)$ and $\exp(-x)$. To produce the missing linearly independent solutions, it is convenient to write Eq. (2.49) in the form $L[y] = 0$, where L is the differential operator defined in the same way as in (2.29), but using for derivative the Leibniz notation, $\frac{dy}{dx}$, instead of the "prime" notation, y':

$$L = \frac{d^n}{dx^n} + a_1\frac{d^{n-1}}{dx^{n-1}} + \ldots + a_{n-1}\frac{d}{dx} + a_n I. \tag{2.60}$$

Here I is the identity operator and $\frac{d^k}{dx^k}$, the composition of $\frac{d}{dx}$ by itself, k-times (to produce the kth derivative, $y^{(k)}$) can also be seen as the *multiplication* of $\frac{d}{dx}$ by itself, k-times. We consider the *differential polynomial ring* $\mathbb{R}\left[\frac{d}{dx}\right]$ which consists of all the formal expressions

$$b_0\frac{d^k}{dx^k} + b_1\frac{d^{k-1}}{dx^{k-1}} + \ldots + b_{k-1}\frac{d}{dx} + b_k I, \tag{2.61}$$

where $b_0, \ldots, b_n$ are real numbers. The differential polynomial ring $\mathbb{R}\left[\frac{d}{dx}\right]$ is isomorphic to the usual polynomial ring $\mathbb{R}[r]$, where r is a variable, since $T : \mathbb{R}[r] \to \mathbb{R}\left[\frac{d}{dx}\right]$,

$$T(b_0r^k + b_1r^{k-1} + \ldots + b_{k-1}r + b_k) = b_0\frac{d^k}{dx^k} + b_1\frac{d^{k-1}}{dx^{k-1}} + \ldots + b_{k-1}\frac{d}{dx} + b_k I$$

is an isomorphism.

2.4 Homogeneous Linear Equations with Constant Coefficients

If we consider the polynomials with *complex* coefficients, the rings $\mathbb{C}\left[\frac{d}{dx}\right]$ and $\mathbb{C}[r]$ are also isomorphic (by the same isomorphism T defined above). Therefore, a factorization of a polynomial $P(r) \in \mathbb{R}[r]$ in linear factors over the complex number field $\mathbb{C}$ corresponds to a factorization of the corresponding differential polynomial as an element of $\mathbb{R}\left[\frac{d}{dx}\right]$, in linear differential factors from $\mathbb{C}\left[\frac{d}{dx}\right]$.

For instance,

$$r^7 + 4r^5 + 4r^3 = r^3 \left(r - i\sqrt{2}\right)^2 \left(r + i\sqrt{2}\right)^2$$

in $\mathbb{C}[r]$, corresponds to

$$\frac{d^7}{dx^7} + 4\frac{d^5}{dx^5} + 4\frac{d^3}{dx^3} = \frac{d^3}{dx^3}\left(\frac{d}{dx} - i\sqrt{2}I\right)^2 \left(\frac{d}{dx} + i\sqrt{2}I\right)^2,$$

in $\mathbb{C}\left[\frac{d}{dx}\right]$.

We remark that the multiplication in $\mathbb{C}\left[\frac{d}{dx}\right]$ is commutative (the rings $\mathbb{C}\left[\frac{d}{dx}\right]$ and $\mathbb{C}[r]$ are commutative). By applying this useful property, some computations can be made easier. For instance, let us compute

$$\left(\frac{d^4}{dx^4} - 8\frac{d^2}{dx^2} + 16I\right)[x \exp(2x)].$$

To calculate $\frac{d^4}{dx^4}[x \exp(2x)]$ is not an easy task. Let us factor the operator on the left side into linear factors by using the corresponding polynomial factoring:

$$r^4 - 8r^2 + 16 = (r-2)^2 (r+2)^2.$$

Thus,

$$\frac{d^4}{dx^4} - 8\frac{d^2}{dx^2} + 16I = \left(\frac{d}{dx} - 2I\right)^2 \left(\frac{d}{dx} + 2I\right)^2.$$

By using the commutativity property in $\mathbb{R}\left[\frac{d}{dx}\right]$, we can apply first of all the operator $\left(\frac{d}{dx} - 2I\right)^2$ to our function $x \exp(2x)$, then we will apply the operator $\left(\frac{d}{dx} + 2I\right)^2$ to the resulting function. So,

$$\left(\frac{d}{dx} - 2I\right)^2 [x \exp(2x)] = \left(\frac{d}{dx} - 2I\right)\left(\frac{d}{dx} - 2I\right)[x \exp(2x)] =$$

$$= \left(\frac{d}{dx} - 2I\right)[\exp(2x) + 2x \exp(2x) - 2x \exp(2x)] =$$

$$\left(\frac{d}{dx} - 2I\right)[\exp(2x)] = 0,$$

hence
$$\left(\frac{d}{dx} - 2I\right)^2 \left(\frac{d}{dx} + 2I\right)^2 [x \exp(2x)] = 0.$$

We can prove the following general result.

Lemma 2.2 *Let λ be a real or a complex number, m a positive integer and $P(x) \in \mathbb{C}[x]$ a polynomial of degree less than m. Then*

$$\left(\frac{d}{dx} - \lambda I\right)^m [P(x) \exp(\lambda x)] = 0. \tag{2.62}$$

Proof We prove this statement by mathematical induction on $m \geq 1$. For $m = 1$, since the polynomial $P(x)$ must be a constant, $P(x) = c$, we have:

$$\left(\frac{d}{dx} - \lambda I\right)[c \exp(\lambda x)] = c\left(\frac{d}{dx} - \lambda I\right)[\exp(\lambda x)] =$$

$$c[\lambda \exp(\lambda x) - \lambda \exp(\lambda x)] = 0.$$

Assume now that $m > 1$ and that the statement is true for $m - 1$. We prove it for m:

$$\left(\frac{d}{dx} - \lambda I\right)^m [P(x) \exp(\lambda x)] =$$

$$= \left(\frac{d}{dx} - \lambda I\right)^{m-1} \left(\frac{d}{dx} - \lambda I\right)[P(x) \exp(\lambda x)] =$$

$$= \left(\frac{d}{dx} - \lambda I\right)^{m-1} [P'(x) \exp(\lambda x) + \lambda P(x) \exp(\lambda x) - \lambda P(x) \exp(\lambda x)] =$$

$$= \left(\frac{d}{dx} - \lambda I\right)^{m-1} [P'(x) \exp(\lambda x)].$$

Since the degree of $P(x)$ is less than m, the degree of the polynomial $P'(x)$ is less than $m - 1$. Hence, by the induction hypothesis, we obtain that the expression above equals zero and the lemma is proved. □

Corollary 2.3 *Consider the differential operator with real coefficients, $L \in \mathbb{R}\left[\frac{d}{dx}\right]$,*

$$L = \frac{d^n}{dx^n} + a_1 \frac{d^{n-1}}{dx^{n-1}} + \ldots + a_{n-1}\frac{d}{dx} + a_n I \tag{2.63}$$

2.4 Homogeneous Linear Equations with Constant Coefficients

and its corresponding polynomial $P(r) \in \mathbb{R}[r]$,

$$P(r) = r^n + a_1 r^{n-1} + \ldots + a_{n-1} r + a_n = 0,$$

which can be (uniquely) factored in $\mathbb{C}[r]$ as

$$P(r) = (r - \lambda_1)^{m_1} \ldots (r - \lambda_k)^{m_k}.$$

Then, for any polynomials $P_1(x), \ldots, P_k(x) \in \mathbb{C}[x]$ of degrees less than $m_1, \ldots, m_k$ respectively, the functions

$$y_1(x) = P_1(x) \exp(\lambda_1 x), \ldots, y_k(x) = P_k(x) \exp(\lambda_k x)$$

are solutions of the homogeneous Eq. (2.49): $L[y] = 0$.

Proof The differential operator L can be factored as

$$L = \left(\frac{d}{dx} - \lambda_1 I\right)^{m_1} \ldots \left(\frac{d}{dx} - \lambda_k I\right)^{m_k}. \tag{2.64}$$

Since the multiplication in $\mathbb{C}\left[\frac{d}{dx}\right]$ is commutative, for a fixed $i \in \{1, 2, \ldots, k\}$, we can apply first of all $\left(\frac{d}{dx} - \lambda_i I\right)^{m_i}$ to $y_i(x) = P_i(x) \exp(\lambda_i x)$. Thus, according to Lemma 2.2,

$$\left(\frac{d}{dx} - \lambda_i I\right)^{m_i} [y_i(x)] = 0,$$

so $L[y_i(x)] = 0$. □

Lemma 2.3 Let $Q_1(x), \ldots, Q_k(x)$ be arbitrary non zero polynomials in $\mathbb{C}[x]$ and $\lambda_1, \ldots, \lambda_k$ be real or complex distinct numbers. Then the functions

$$y_1(x) = Q_1(x) \exp(\lambda_1 x), \ldots, y_k(x) = Q_k(x) \exp(\lambda_k x)$$

are linearly independent over $\mathbb{C}$, the field of complex numbers.

Proof We shall prove the lemma by mathematical induction on $m \geq 0$, where m is the sum of the degrees of above polynomials $Q_1(x), \ldots, Q_k(x)$,

$$m = \deg Q_1(x) + \ldots + \deg Q_k(x).$$

For $m = 0$, all the polynomials $Q_1(x), \ldots, Q_k(x)$ are constants and the linear independence over $\mathbb{C}$ of the functions

$$y_1 = \exp(\lambda_1 x), \ldots, y_k = \exp(\lambda_k x),$$

(when $\lambda_1, \ldots, \lambda_k$ are *distinct* numbers) follows by simply extending Lemma 2.1 from *real* numbers to *complex* numbers.

Let $m > 0$. We assume that the statement is true for any polynomials $Q_1(x), \ldots, Q_k(x)$ with the sum of degrees equal to $m - 1$ and prove it for m. Consider the polynomials $Q_1(x), \ldots, Q_k(x)$ such that $\deg Q_1(x) + \ldots + \deg Q_k(x) = m$ and suppose that $\beta_1, \ldots, \beta_k$ are complex numbers such that

$$\beta_1 Q_1(x) \exp(\lambda_1 x) + \ldots \beta_k Q_k(x) \exp(\lambda_k x) = 0, \tag{2.65}$$

for any $x \in \mathbb{R}$. We differentiate this equality:

$$\beta_1 [Q_1'(x) + \lambda_1 Q_1(x)] \exp(\lambda_1 x) + \ldots + \beta_k [Q_k'(x) + \lambda_k Q_k(x)] \exp(\lambda_k x) = 0. \tag{2.66}$$

Since $m > 0$, at least one of the polynomials $Q_1(x), \ldots, Q_k(x)$ is not a constant. Assume that $\deg Q_1(x) > 0$ and $\beta_1 \neq 0$. We multiply the equality (2.65) by λ_1 and subtract the resulting equality from (2.66). We obtain a null linear combination of the form:

$$\beta_1 Q_1'(x) \exp(\lambda_1 x) + \beta_2 [Q_2'(x) + (\lambda_2 - \lambda_1) Q_2(x)] \exp(\lambda_2 x) + \ldots +$$

$$+ \beta_k [Q_k'(x) + (\lambda_k - \lambda_1) Q_k(x)] \exp(\lambda_k x) = 0.$$

Since $\deg Q_1(x) > 0$ and $\lambda_1, \ldots, \lambda_k$ are distinct numbers, we have:

$$\deg Q_1'(x) + \deg \left(Q_2'(x) + (\lambda_2 - \lambda_1) Q_2(x) \right) + \ldots + \deg \left(Q_k'(x) + (\lambda_k - \lambda_1) Q_k(x) \right) =$$

$$= \deg Q_1(x) - 1 + \deg Q_2(x) + \ldots + \deg Q_k(x) = m - 1.$$

By applying the induction hypothesis, we obtain that $\beta_i = 0$, for all $i = 1, 2, \ldots, k$. Hence the functions $y_1(x), \ldots, y_k(x)$ are linearly independent over $\mathbb{C}$. □

Corollary 2.4 *If $\lambda_1, \ldots, \lambda_k$ are real or complex distinct numbers and $m_1, \ldots, m_k$ are positive integers, then the functions*

$$\begin{aligned} \exp(\lambda_1 x), x \exp(\lambda_1 x), \ldots, x^{m_1-1} \exp(\lambda_1 x), \\ \exp(\lambda_2 x), x \exp(\lambda_2 x), \ldots, x^{m_2-1} \exp(\lambda_2 x), \\ \ldots\ldots\ldots\ldots\ldots\ldots\ldots\ldots\ldots\ldots\ldots\ldots\ldots\ldots\ldots\ldots \\ \exp(\lambda_k x), x \exp(\lambda_k x), \ldots, x^{m_k-1} \exp(\lambda_k x) \end{aligned} \tag{2.67}$$

defined for all $x \in \mathbb{R}$, are linearly independent over $\mathbb{C}$.

Proof Any linear combination of the functions (2.67) can be written in the form

$$Q_1(x) \exp(\lambda_1 x) + Q_2(x) \exp(\lambda_2 x) + \ldots + Q_k(x) \exp(\lambda_k x),$$

2.4 Homogeneous Linear Equations with Constant Coefficients

where $Q_1(x), \ldots, Q_k(x) \in \mathbb{C}[x]$ are polynomials of degrees less than or equal to $m_1 - 1, \ldots, m_k - 1$, respectively. If the linear combination above equals 0 for any $x \in \mathbb{R}$, by Lemma 2.3, it follows that all the polynomials must be identically zero, hence the functions (2.67) are linearly independent. □

Remark 2.6 Suppose that the characteristic polynomial attached to the homogeneous Eq. (2.49) has the *real* (multiple) roots $\lambda_1, \ldots, \lambda_k$, $\lambda_i \neq \lambda_j$ for $i \neq j$:

$$P(r) = r^n + a_1 r^{n-1} + \ldots + a_{n-1} r + a_n =$$
$$= (r - \lambda_1)^{m_1} (r - \lambda_2)^{m_2} \ldots (r - \lambda_k)^{m_k}, \quad (2.68)$$

where $m_1 + m_2 + \ldots + m_k = n$. By Lemma 2.2, all the n the functions from (2.67) are solutions of the homogeneous equation $L[y] = 0$ and, by Corollary 2.4, they are linearly independent. Hence (2.67) is a fundamental system of solutions of the homogeneous Eq. (2.49).

Remark 2.7 If $\lambda = \alpha + i\beta$ with $\beta \neq 0$, is a complex root of the characteristic polynomial (with *real* coefficients) $P(r)$, then its conjugate, $\bar{\lambda} = \alpha - i\beta$ is also a root and both roots λ and $\bar{\lambda}$ have the same multiplicity, $\nu \geq 0$. We have seen that the functions

$$\begin{array}{c} \exp(\lambda x), x \exp(\lambda x), \ldots, x^{\nu-1} \exp(\lambda x) \\ \exp(\bar{\lambda} x), x \exp(\bar{\lambda} x), \ldots, x^{\nu-1} \exp(\bar{\lambda} x) \end{array} \quad (2.69)$$

are 2ν linearly independent solutions of the homogeneous equation $L[y] = 0$. Since

$$\exp(\lambda x) = \exp(\alpha x)(\cos \beta x + i \sin \beta x),$$

any linear combination (with complex coefficients) of the *complex* functions (2.69) is also a linear combination of the *real* functions

$$\begin{array}{c} \exp(\alpha x) \cos \beta x, x \exp(\alpha x) \cos \beta x, \ldots, x^{\nu-1} \exp(\alpha x) \cos \beta x \\ \exp(\alpha x) \sin \beta x, x \exp(\alpha x) \sin \beta x, \ldots, x^{\nu-1} \exp(\alpha x) \sin \beta x \end{array} \quad (2.70)$$

and we can see that the functions (2.70) are linearly independent, as well. Thus, if

$$\sum_{j=0}^{\nu-1} \left[\gamma_j x^j \exp(\alpha x) \cos \beta x + \delta_j x^j \exp(\alpha x) \sin \beta x \right] = 0,$$

by the formulas

$$\exp(\alpha x) \cos \beta x = \frac{\exp(\lambda x) + \exp(\bar{\lambda} x)}{2}, \quad \exp(\alpha x) \sin \beta x = \frac{\exp(\lambda x) - \exp(\bar{\lambda} x)}{2i},$$

we obtain:

$$\sum_{j=0}^{\nu-1}\left[\frac{\gamma_j - i\delta_j}{2}x^j \exp(\lambda x) + \frac{\gamma_j + i\delta_j}{2}x^j \exp(\bar\lambda x)\right] = 0$$

and, since the functions (2.69) are linearly independent, it follows that all the coefficients must be zero, so $\gamma_j = \delta_j = 0$, for all $j = 0, 1, \ldots, \nu - 1$.

Thus, a fundamental system of solutions of the homogeneous equation in the general case (when the roots of the characteristic equation are real or complex, simple or multiple roots) is given by the next theorem.

Theorem 2.8 *Suppose that the characteristic polynomial attached to the homogeneous Eq. (2.49) has the real roots $\lambda_1, \ldots, \lambda_k \in \mathbb{R}$ of multiplicity $m_1, \ldots, m_k$, and the complex (conjugate) roots $\alpha_1 \pm i\beta_1, \ldots, \alpha_s \pm i\beta_s$ of multiplicity $n_1, \ldots, n_s$, respectively:*

$$P(r) = (r - \lambda_1)^{m_1} \ldots (r - \lambda_k)^{m_k} (r - \alpha_1 - i\beta_1)^{n_1} (r - \alpha_1 + i\beta_1)^{n_1} \ldots$$

$$\ldots \cdot (r - \alpha_s - i\beta_s)^{n_s} (r - \alpha_s + i\beta_s)^{n_s}. \tag{2.71}$$

where $m_1 + m_2 + \ldots + m_k + 2n_1 + \ldots + 2n_s = n$. Then, a fundamental system of solutions of the homogeneous differential Eq. (2.49) is formed by the following functions:

$$\begin{array}{c}
\exp(\lambda_1 x), x\exp(\lambda_1 x), \ldots, x^{m_1-1}\exp(\lambda_1 x), \\
\exp(\lambda_2 x), x\exp(\lambda_2 x), \ldots, x^{m_2-1}\exp(\lambda_2 x), \\
\ldots\ldots\ldots\ldots\ldots\ldots\ldots\ldots\ldots\ldots\ldots\ldots\ldots\ldots \\
\exp(\lambda_k x), x\exp(\lambda_k x), \ldots, x^{m_k-1}\exp(\lambda_k x) \\
\exp(\alpha_1 x)\cos\beta_1 x, \exp(\alpha_1 x)\sin\beta_1 x, x\exp(\alpha_1 x)\cos\beta_1 x, \\
x\exp(\alpha_1 x)\sin\beta_1 x, \ldots, x^{n_1-1}\exp(\alpha_1 x)\cos\beta_1 x, x^{n_1-1}\exp(\alpha_1 x)\sin\beta_1 x, \\
\ldots\ldots\ldots\ldots\ldots\ldots\ldots\ldots\ldots\ldots\ldots\ldots\ldots\ldots \\
\exp(\alpha_s x)\cos\beta_s x, \exp(\alpha_s x)\sin\beta_s x, x\exp(\alpha_s x)\cos\beta_s x, \\
x\exp(\alpha_s x)\sin\beta_s x, \ldots, x^{n_s-1}\exp(\alpha_s x)\cos\beta_s x, x^{n_s-1}\exp(\alpha_s x)\sin\beta_s x.
\end{array} \tag{2.72}$$

Example 2.12 Let us find $y(x)$ if $y'' + 3y' + 2y = 0$ and $y(0) = 0$, $y'(0) = 2$.

Since the equation is a homogeneous equation with constant coefficients, we apply the above theory and first of all solve the characteristic equation: $r^2 + 3r + 2 = 0$. The roots are $\lambda_1 = -1$ and $\lambda_2 = -2$. Thus, $\{y_1(x) = \exp(-x), y_2(x) = \exp(-2x)\}$ is a fundamental system of solutions for our equation. Hence, the general solution is:

$$y(x) = C_1 \exp(-x) + C_2 \exp(-2x), \tag{2.73}$$

2.4 Homogeneous Linear Equations with Constant Coefficients

where C_1, C_2 are arbitrary constants. To find the particular solution $y_0(x)$ with $y_0(0) = 0$, $y'_0(0) = 2$, we compute the derivative of y in (2.73):

$$y'(x) = -C_1 \exp(-x) - 2C_2 \exp(-2x).$$

But $y(0) = 0$ implies $C_1 + C_2 = 0$ and $y'(0) = 2$ implies $-C_1 - 2C_2 = 2$. So $C_1 = 2$ and $C_2 = -2$, i.e. $y_0(x) = 2\exp(-x) - 2\exp(-2x)$.

Example 2.13 Find the general solution of the equation:

$$\frac{d^5 y}{dx^5} - 4\frac{d^4 y}{dx^4} + 4\frac{d^3 y}{dx^3} = 0.$$

The characteristic equation is: $r^5 - 4r^4 + 4r^3 = 0$. It has two distinct solutions, $\lambda_1 = 0$, of multiplicity 3 and $\lambda_2 = 2$, of multiplicity 2. We apply Theorem 2.8 and find the linearly independent functions $y_1(x) = \exp(0 \cdot x) = 1$, $y_2(x) = x$, $y_3(x) = x^2$, which correspond to $\lambda_1 = 0$ and $y_4(x) = \exp(2x)$, $y_5(x) = x\exp(2x)$, which correspond to $\lambda_2 = 2$. A basis of solutions is:

$$y_1(x) = 1,\ y_2(x) = x,\ y_3(x) = x^2,$$
$$y_4(x) = \exp(2x),\ y_5(x) = x\exp(2x).$$

Hence, the general solution for our equation is:

$$y(x) = C_1 + C_2 x + C_3 x^2 + C_4 \exp(2x) + C_5 x \exp(2x).$$

Example 2.14 Write the differential equation which corresponds to the algebraic equation: $(r^2 + r + 1)^2(r^2 + 9) = 0$ and solve it.

The algebraic equation above can also be written:

$$r^6 + 2r^5 + 12r^4 + 20r^3 + 28r^2 + 18r + 9 = 0.$$

So, its corresponding differential equation is:

$$\frac{d^6 y}{dx^6} + 2\frac{d^5 y}{dx^5} + 12\frac{d^4 y}{dx^4} + 20\frac{d^3 y}{dx^3} + 28\frac{d^2 y}{dx^2} + 18\frac{dy}{dx} + 9y = 0.$$

We look at the initial form of the algebraic equation and easily find its roots: $\lambda_{1,2} = -\frac{1}{2} \pm i\frac{\sqrt{3}}{2}$, of multiplicity 2 and $\lambda_{3,4} = \pm 3i$, of multiplicity 1. By Theorem 2.8, we obtain the fundamental system of solutions:

$$y_1(x) = \exp\left(-\tfrac{1}{2}x\right)\cos\left(\tfrac{\sqrt{3}}{2}x\right),\ y_2(x) = \exp\left(-\tfrac{1}{2}x\right)\sin\left(\tfrac{\sqrt{3}}{2}x\right),$$
$$y_3(x) = x\exp\left(-\tfrac{1}{2}x\right)\cos\left(\tfrac{\sqrt{3}}{2}x\right),\ y_4(x) = x\exp\left(-\tfrac{1}{2}x\right)\sin\left(\tfrac{\sqrt{3}}{2}x\right),$$
$$y_5(x) = \cos 3x,\ y_6(x) = \sin 3x.$$

The general solution is an arbitrary linear combination of these six functions.

Example 2.15 Damped mass-spring system (see [21], p. 64). Let us analyze the behavior of the mass-spring system from Example 2.2 if the damping force F_d (due to air resistance for instance) cannot be neglected. We assume that F_d is proportional to the velocity y'. Since the damping force acts *against* the motion (just like the restoring force of the spring $F_s = -ky$), we can write:

$$F_d = -cy',$$

where c is a positive constant. Thus, instead of Eq. (2.11), we obtain:

$$my'' + cy' + ky = 0. \qquad (2.74)$$

Find the particular solution of Eq. (2.74) that satisfies the initial conditions

$$y(0) = 4, \quad y'(0) = 0,$$

knowing the mass $m = 10$, the spring constant $k = 90$, and taking three cases for the damping constant $c > 0$:

(a) $c = 100$; (b) $c = 60$; (c) $c = 10$.

(a) Dividing everything by 10, the equation can be written:

$$y'' + 10y' + 9y = 0.$$

The characteristic equation $r^2 + 10r + 9 = 0$ has the real distinct roots $r_1 = -1$ and $r_2 = -9$, hence the general solution is

$$y = C_1 \exp(-t) + C_2 \exp(-9t).$$

From the initial conditions we obtain $C_1 + C_2 = 4$ and $-C_1 - 9C_2 = 0$ so $C_1 = \frac{9}{2}$ and $C_2 = -\frac{1}{2}$ and the solution of the Cauchy problem is:

$$y_1 = \frac{9}{2} \exp(-t) - \frac{1}{2} \exp(-9t).$$

(b) The equation is written:

$$y'' + 6y' + 9y = 0.$$

The characteristic equation $r^2 + 6r + 9 = 0$ has one (double) real root $r_1 = r_2 = -3$, hence the general solution is

$$y = (C_1 + C_2 t) \exp(-3t).$$

2.4 Homogeneous Linear Equations with Constant Coefficients

From the initial conditions we obtain $C_1 = 4$ and $C_2 = 12$, so the solution of the Cauchy problem is:

$$y_2 = (4 + 12t)\exp(-3t).$$

(c) The equation is written:

$$y'' + 2y' + 9y = 0.$$

The characteristic equation $r^2 + 2r + 9 = 0$ has the complex conjugate roots $r_{1,2} = -1 \pm 2\sqrt{2}i$, hence the general solution is

$$y = C_1 \exp(-t)\cos(2\sqrt{2}t) + C_2 \exp(-t)\sin(2\sqrt{2}t).$$

From the initial conditions we obtain $C_1 = 4$ and $C_2 = \sqrt{2}$ so the solution of the Cauchy problem is:

$$y_3 = \exp(-t)[4\cos(2\sqrt{2}t) + \sqrt{2}\sin(2\sqrt{2}t)].$$

Figure 2.6 shows the graphs of the solutions $y_1(t)$, $y_2(2)$ and $y_3(t)$. As can be noticed, all of them approach 0 as $t \to \infty$ (the system tends to the equilibrium

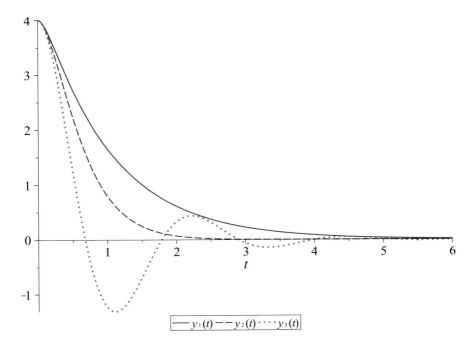

Fig. 2.6 The solutions $y_1(t)$, $y_2(t)$ and $y_3(t)$

position). But in the first two cases the solutions decrease monotonically, while in the last case we have *damped oscillations* (their amplitude goes to 0).

2.5 Nonhomogeneous Linear Equations with Constant Coefficients

If the linear differential equation is not homogeneous, then we can apply Lagrange method of variation of constants described in Sect. 2.3. But this method can involve difficult computations, for instance, it can be difficult to integrate the functions $C'_1(x), \ldots, C'_n(x)$ obtained after solving the system of equations from Theorem 2.7.

Fortunately, in the case of non-homogeneous linear differential equation with *constant coefficients*,

$$L[y] = y^{(n)} + a_1 y^{(n-1)} + \ldots + a_{n-1} y + a_n = f(x), \tag{2.75}$$

if the free term $f(x)$ is a finite sum of the following types of functions:

$$P(x)\exp(\lambda x), \ Q(x)\exp(\alpha x)\cos(\beta x) \text{ or } R(x)\exp(\alpha x)\sin(\beta x), \tag{2.76}$$

where $P(x)$, $Q(x)$ and $R(x)$ are polynomials with real coefficients and λ, α, β are real numbers, then we can find a particular solution for (2.75) of a similar type. This method is known as the *method of undetermined coefficients*. Using Remark 2.5 and the formulas

$$\begin{aligned}\cos(\beta x) &= \tfrac{1}{2}\left[\exp(i\beta x) + \exp(-i\beta x)\right], \\ \sin(\beta x) &= \tfrac{1}{2i}\left[\exp(i\beta x) - \exp(-i\beta x)\right],\end{aligned} \tag{2.77}$$

we can reduce everything to the case

$$L[y] = P(x)\exp(\lambda x), \tag{2.78}$$

where $P \in \mathbb{C}[x]$ and $\lambda \in \mathbb{C}$. Moreover, if $y_{P_1} + iy_{P_2}$ is a particular solution of the equation $L[y] = f$, with $f = f_1 + if_2$ (where $y_{P_1}(x), y_{P_2}(x), f_1, f_2$ are functions with real values), then y_{P_1} is a particular solution of the equation $L[y] = f_1$ and y_{P_2} is a particular solution for $L[y] = f_2$ ("the real part of the left side is equal to the real part of the right side, and the imaginary part of the left side is equal to the imaginary part of the right side").

Consider the characteristic equation

$$r^n + a_1 r^{n-1} + \ldots + a_{n-1} r + a_n = 0, \tag{2.79}$$

2.5 Nonhomogeneous Linear Equations with Constant Coefficients

with the roots $\lambda_1, \ldots, \lambda_k \in \mathbb{C}$ of multiplicities $m_1, \ldots, m_k$, respectively. As we have seen in the previous section, the linear differential operator from Eq. (2.75),

$$L[y] = \frac{d^n}{dx^n} + a_1 \frac{d^{n-1}}{dx^{n-1}} + \ldots + a_{n-1} \frac{d}{dx} + a_n I, \tag{2.80}$$

can be written as a product (composition) of linear first-order operators,

$$L = \left(\frac{d}{dx} - \lambda_1 I\right)^{m_1} \ldots \left(\frac{d}{dx} - \lambda_k I\right)^{m_k}, \tag{2.81}$$

where I is the identity operator.

Lemma 2.4 *Let*

$$\left(\frac{d}{dx} - \lambda_0 I\right)[y] = P(x) \exp(\lambda x), \quad \lambda \neq \lambda_0, \tag{2.82}$$

be a linear first-order differential equation over $\mathbb{C}$ ($P(x) \in \mathbb{C}[x]$, $\lambda, \lambda_0 \in \mathbb{C}$). Then there exists a unique polynomial $Q(x) \in \mathbb{C}[x]$ such that

$$y_P(x) = Q(x) \exp(\lambda x)$$

is a particular solution for the Eq. (2.82). Moreover, $\deg P = \deg Q$ and, if $\lambda, \lambda_0 \in \mathbb{R}$ and $P(x) \in \mathbb{R}[x]$, then $Q(x)$ is also a polynomial with real coefficients.

Proof Let $P(x) = b_0 + b_1 x + \ldots + b_k x^k$, $b_k \neq 0$ be a polynomial of degree k. Let $Q(x) = c_0 + c_1 x + \ldots + c_m x^m$ be the searched polynomial. We introduce $y_P(x) = Q(x) \exp(\lambda x)$ in (2.82) and find the equality:

$$Q'(x) + (\lambda - \lambda_0) Q(x) = P(x).$$

Since $\lambda \neq \lambda_0$, we have $\deg P = \deg Q$, so $k = m$ and

$$\sum_{j=1}^{k} j b_j x^{j-1} + (\lambda - \lambda_0) \sum_{j=0}^{k} b_j x^j = \sum_{j=0}^{k} a_j x^j.$$

By equating the coefficients corresponding to $x^0, x^1, \ldots, x^k$, we obtain:

$$c_1 + c_0 (\lambda - \lambda_0) = b_0$$
$$2c_2 + c_1 (\lambda - \lambda_0) = b_1$$
$$\vdots$$
$$k c_k + c_{k-1} (\lambda - \lambda_0) = b_{k-1}$$
$$c_k (\lambda - \lambda_0) = b_k.$$

Since $\lambda \neq \lambda_0$, from the last equation we obtain $c_k = \frac{b_k}{\lambda - \lambda_0} \neq 0$ and the system can be solved to find $c_{k-1}, c_{k-2}, \ldots, c_0$ by backward substitution. Hence the polynomial $Q(x)$ is *uniquely* determined such that $y_P(x) = Q(x)\exp(\lambda x)$ is a solution of the Eq. (2.82).

Moreover, if λ, λ_0 and $b_0, \ldots, b_k \in \mathbb{R}$, then, obviously, it follows that $c_0, \ldots, c_k \in \mathbb{R}$. □

Corollary 2.5 *Let L be the differential operator defined by* (2.80), $P(x) \in \mathbb{C}[x]$ *be a polynomial with complex coefficients and* $\lambda \in \mathbb{C}$ *be a real or complex number which is not a root of the characteristic Eq.* (2.79). *Then the differential Eq.* (2.78) *has a unique solution of the form*

$$y_P(x) = Q(x)\exp(\lambda x)$$

where $Q(x) \in \mathbb{C}[x]$. *Moreover,* $\deg Q(x) = \deg P(x)$ *and, if* $\lambda \in \mathbb{R}$ *and* $P(x) \in \mathbb{R}[x]$, *then* $Q(x) \in \mathbb{R}[x]$.

Proof We use mathematical induction on $n = m_1 + \ldots + m_k$, the order of L. For $n = 1$, the statement is proved by Lemma 2.4. Let $n > 1$ and assume that the statement is true for any differential operator of order $n-1$. Consider L a differential operator of order n which has the factorization (2.81). So L can be written in the form

$$L = \left(\frac{d}{dx} - \lambda_1 I\right)\left(\frac{d}{dx} - \lambda_1 I\right)^{m_1 - 1} \ldots \left(\frac{d}{dx} - \lambda_k I\right)^{m_k}.$$

We denote by L_1 the linear differential operator of $(n-1)$th order

$$L_1 = \left(\frac{d}{dx} - \lambda_1 I\right)^{m_1 - 1} \ldots \left(\frac{d}{dx} - \lambda_k I\right)^{m_k}.$$

By Lemma 2.4, the equation

$$\left(\frac{d}{dx} - \lambda_1 I\right)[y] = P(x)\exp(\lambda x), \lambda \neq \lambda_1,$$

has a unique particular solution $y_{P_1}(x) = Q_1(x)\exp(\lambda x)$, where $Q_1(x)$ is a polynomial of the same degree as $P(x)$.

On the other hand, by the mathematical induction hypothesis, the $(n-1)$th-order equation

$$L_1[y] = Q_1(x)\exp(\lambda x)$$

has a unique particular solution of the form $y_P(x) = Q(x)\exp(\lambda x)$, where $Q(x)$ is a polynomial with complex coefficients and $\deg Q(x) = \deg Q_1(x) = \deg P(x)$. It

2.5 Nonhomogeneous Linear Equations with Constant Coefficients

is easy to see that $y_P(x) = Q(x)\exp(\lambda x)$ is the unique solution of this form for the equation $L[y] = P(x)\exp(\lambda x)$. □

Now, let us consider the case when the complex number λ in the free term of the Eq. (2.78) is a root of the characteristic Eq. (2.79).

Lemma 2.5 *Let*

$$\left(\frac{d}{dx} - \lambda_0 I\right)[y] = P(x)\exp(\lambda_0 x), \qquad (2.83)$$

be a linear first-order differential equation over $\mathbb{C}$ *($P(x) \in \mathbb{C}[x]$, $\lambda_0 \in \mathbb{C}$). Then there exists a unique polynomial $Q(x) \in \mathbb{C}[x]$ such that*

$$y_P(x) = xQ(x)\exp(\lambda_0 x)$$

is a particular solution of the Eq. (2.83). Moreover, the degree of the polynomial $Q(x)$ is equal to the degree of $P(x)$.

Proof Let $P(x) = b_0 + b_1 x + \ldots + b_k x^k$, $b_k \neq 0$, be a polynomial of degree k. Let $Q(x)$ be the sought polynomial. We introduce $y_P(x) = xQ(x)\exp(\lambda_0 x)$ in (2.83) and find the equality:

$$Q(x) + xQ'(x) = P(x).$$

It follows that $\deg P(x) = \deg Q(x)$, so $Q(x) = c_0 + c_1 x + \ldots + c_k x^k$. By equating the coefficients of the corresponding powers of x we get:

$$c_0 = \frac{b_0}{1},\ c_1 = \frac{b_1}{2},\ \ldots,\ c_k = \frac{b_k}{k+1}.$$

□

Corollary 2.6 *Let*

$$\left(\frac{d}{dx} - \lambda_0 I\right)^m [y] = P(x)\exp(\lambda_0 x), \qquad (2.84)$$

be a linear differential equation of mth order over $\mathbb{C}$ *($P(x) \in \mathbb{C}[x]$, $\lambda_0 \in \mathbb{C}$). Then there exists a unique polynomial $Q(x) \in \mathbb{C}[x]$ such that $y_P(x) = x^m Q(x)\exp(\lambda_0 x)$ is a particular solution of the Eq. (2.84). Moreover, $\deg Q(x) = \deg P(x)$.*

Proof For $m = 1$, the statement is proved by Lemma 2.5. Assume that $m > 1$ and that the statement is true for $m - 1$. We want to prove it for m. We denote by L_1 the linear differential operator

$$L_1 = \left(\frac{d}{dx} - \lambda_0 I\right)^{m-1}$$

By Lemma 2.5, there exists a unique polynomial $Q_1(x) \in \mathbb{C}[x]$ such that

$$\left(\frac{d}{dx} - \lambda_0 I\right)[xQ_1(x)\exp(\lambda_0 x)] = P(x)\exp(\lambda_0 x).$$

The degree of the polynomial $Q_1(x)$ is the same as the degree of $P(x)$.

By the induction hypothesis, there exists a unique polynomial $Q(x)$ such that

$$L_1\left[x^m Q(x)\exp(\lambda_0 x)\right] = xQ_1(x)\exp(\lambda_0 x)$$

and we have: $\deg Q(x) = \deg Q_1(x) = \deg P(x)$. If we apply the differential operator $\frac{d}{dx} - \lambda_0 I$ to the above equation, we find that the function $y_P = x^m Q(x)\exp(\lambda_0 x)$ is a solution of the Eq. (2.84) and it can be readily proved that it is the unique solution of this form: If we suppose that $z_P = x^m R(x)\exp(\lambda_0 x)$ is another solution of (2.84), then $y_P - z_P = x^m(Q(x) - R(x))\exp(\lambda_0 x)$ must be a solution of the homogeneous equation,

$$\left(\frac{d}{dx} - \lambda_0 I\right)^m [y] = 0 \tag{2.85}$$

which is impossible, because (as shown in Sect. 2.4) any solution of (2.85) should be of the form $T(x)\exp(\lambda_0 x)$, with $T(x)$ a polynomial of degree *less than m*. □

Theorem 2.9 *Consider the differential operator* (2.81),

$$L = \left(\frac{d}{dx} - \lambda_1 I\right)^{m_1} \cdots \left(\frac{d}{dx} - \lambda_k I\right)^{m_k}.$$

The differential equation

$$L[y] = P(x)\exp(\lambda_1 x) \tag{2.86}$$

(where $P(x)$ is a polynomial with complex coefficients) has a unique solution of the form

$$y_P(x) = x^{m_1} Q(x)\exp(\lambda_1 x), \tag{2.87}$$

where $Q(x)$ is a polynomial with complex coefficients. Moreover, $Q(x)$ has the same degree as $P(x)$.

Proof If $k = 1$, we obtain the statement of Corollary 2.6 for $m = m_1$. Assume that $k > 1$ and denote

$$L_2 = \left(\frac{d}{dx} - \lambda_2 I\right)^{m_2} \cdots \left(\frac{d}{dx} - \lambda_k I\right)^{m_k}.$$

2.5 Nonhomogeneous Linear Equations with Constant Coefficients

Since λ_1 is distinct from $\lambda_2, \ldots, \lambda_k$, we can apply Corollary 2.5 and find the polynomial $Q_1(x) \in \mathbb{C}[x]$ (of the same degree as $P(x)$) such that

$$L_2[Q_1(x)\exp(\lambda_1 x)] = P(x)\exp(\lambda_1 x).$$

By applying Corollary 2.6 (for $m = m_1$ and $P(x) = Q_1(x)$), we obtain that there exists the polynomial $Q(x)$ (and $\deg Q(x) = \deg Q_1(x) = \deg P(x)$) such that

$$\left(\frac{d}{dx} - \lambda_1 I\right)^{m_1} \left[x^{m_1} Q(x)\exp(\lambda_1 x)\right] = Q_1(x)\exp(\lambda_1 x).$$

Since the ring $\mathbb{R}\left[\frac{d}{dx}\right]$ is commutative, we can write

$$L = \left(\frac{d}{dx} - \lambda_1 I\right)^{m_1} L_2 = L_2 \left(\frac{d}{dx} - \lambda_1 I\right)^{m_1},$$

so we find:

$$L[x^{m_1} Q(x)\exp(\lambda_1 x)] = L_2[Q_1(x)\exp(\lambda_1 x)] = P(x)\exp(\lambda_1 x),$$

so the function $y_P = x^{m_1} Q(x)\exp(\lambda_1 x)$ is a solution of the Eq. (2.86) and it can be readily proved that it is the unique solution of this form. $\square$

Theorem 2.9 and all the previous results of this section provide some *complex* solutions of the non-homogeneous linear differential equation

$$L[y] = y^{(n)} + a_1 y^{(n-1)} + \ldots + a_{n-1} y' + a_n y = P(x)\exp(\lambda x), \ x \in \mathbb{R}, \quad (2.88)$$

where the coefficients of $a_1, \ldots, a_n$ are *real* numbers and the free term, $f(x) = P(x)\exp(\lambda x)$, may be a real or a complex function. But we usually have to solve equations where the free term is a *real* function (of the form (2.76)) and we need to find *real* solutions.

Corollary 2.7 *Consider the Eq. (2.88) where $P(x)$ is a polynomial with real coefficients, λ is a real number and $\lambda_1, \ldots, \lambda_k$ are the roots (real or complex) of the characteristic Eq. (2.79), with multiplicities $m_1, \ldots, m_k$.*

(i) *If $\lambda \neq \lambda_j$, for any $j = 1, \ldots, k$, then the Eq. (2.88) has a unique particular solution of the form:*

$$y_P(x) = Q(x)\exp(\lambda x). \quad (2.89)$$

(ii) If $\lambda = \lambda_j$, then the Eq. (2.88) has a unique particular solution of the following form (we denoted $m = m_j$):

$$y_P(x) = x^m Q(x) \exp(\lambda x). \qquad (2.90)$$

In both formulas, (2.89) and (2.90), $Q(x)$ is a polynomial with real coefficients and $\deg Q(x) = \deg P(x)$.

Proof By Corollary 2.5 (for case (i)) and Theorem 2.9 (for case (ii)), we can say that there exists a unique solution of the form (2.89) (in the first case) or (2.90) (in the second case), where $Q(x)$ is a polynomial with *complex* coefficients. But, since the real part of this (unique) polynomial should also verify the same equation (because the free term is a *real* function) it follows that $Q(x)$ is a polynomial with *real* coefficients. □

Corollary 2.8 *Consider the linear differential equation*

$$L[y] = \exp(\alpha x) \left[P_1(x) \cos \beta x + P_2(x) \sin \beta x \right], \qquad (2.91)$$

where L is the differential operator from Eq. (2.88), $\alpha, \beta \in \mathbb{R}$ and $P_1(x), P_2(x)$ are polynomials with real coefficients.

(i) *If $\lambda = \alpha + i\beta$ is not a root of the characteristic Eq. (2.79), then the Eq. (2.91) has a unique particular solution of the form:*

$$y_P(x) = \exp(\alpha x) \left[Q_1(x) \cos \beta x + Q_2(x) \sin \beta x \right]. \qquad (2.92)$$

(ii) *If $\lambda = \alpha + i\beta$ is a root (of multiplicity m) of the characteristic Eq. (2.79) then the Eq. (2.88) has a unique particular solution of the form:*

$$y_P(x) = x^m \exp(\alpha x) \left[Q_1(x) \cos \beta x + Q_2(x) \sin \beta x \right], \qquad (2.93)$$

In both formulas, (2.92) and (2.93), $Q_1(x)$ and $Q_2(x)$ are polynomials with real coefficients and

$$\max\{\deg Q_1(x), \deg Q_2(x)\} = \max\{\deg P_1(x), \deg P_2(x)\}.$$

Proof We use the formulas (2.77) in (2.91) and obtain:

$$L[y] = P_1^*(x) \exp(\lambda x) + P_2^*(x) \exp(\overline{\lambda} x), \qquad (2.94)$$

where $P_1^*(x)$ and $P_2^*(x)$ are complex polynomials, conjugate to each other:

$$P_1^*(x) = \frac{1}{2} (P_1(x) - i P_2(x)), \quad P_2^*(x) = \frac{1}{2} (P_1(x) + i P_2(x)).$$

2.5 Nonhomogeneous Linear Equations with Constant Coefficients

Now we use the superposition principle and split Eq. (2.94) into two equations:

$$L[y] = P_1^*(x) \exp(\lambda x) \tag{2.95}$$

and

$$L[y] = P_2^*(x) \exp(\overline{\lambda} x) \tag{2.96}$$

By Corollary 2.5 and Theorem 2.9, there exists a unique polynomial (with complex coefficients), $Q_1^*(x)$, such that the (complex) function

$$y_{1p}^*(x) = x^m Q_1^*(x) \exp(\lambda x)$$

(where $m = 0$ in case (i)) is a particular solution of the Eq. (2.95) and there exists a unique polynomial, $Q_2^*(x)$, such that the following function is a particular solution of the Eq. (2.96).

$$y_{2p}^*(x) = x^m Q_2^*(x) \exp(\overline{\lambda} x).$$

The sum of these functions, $y_P = y_{1p}^* + y_{2p}^* =$

$$= x^m \exp(\alpha x) \left[(Q_1^*(x) + Q_2^*(x)) \cos \beta x + i(Q_1^*(x) - Q_2^*(x)) \sin \beta x \right],$$

is a solution of the Eq. (2.94) (which is equivalent to the initial Eq. (2.91)). Since the free terms of Eqs. (2.95) and (2.96) are conjugate to each other, it follows that $\overline{y_{1p}^*}(x) = x^m \overline{Q_1^*}(x) \exp(\overline{\lambda} x)$ is a solution of Eq. (2.96), of the same form as y_{2p}^* so, by the uniqueness of the polynomial Q_2^*, we have $\overline{Q_1^*} = Q_2^*$. Therefore, the polynomials $Q_1(x)$ and $Q_2(x)$ defined by

$$Q_1(x) = Q_1^*(x) + Q_2^*(x), \quad Q_2(x) = i \left(Q_1^*(x) - Q_2^*(x) \right)$$

are polynomials with *real* coefficients. We also notice that

$$\max\{\deg Q_1(x), \deg Q_2(x)\} = \deg Q_1^*(x) = \deg Q_2^*(x) =$$

$$= \deg P_1^*(x) = \deg P_2^*(x) = \max\{\deg Q_1(x), \deg Q_2(x)\}.$$

$\square$

Example 2.16 Let us find the general solution of the equation

$$y''' - 2y'' + y' = 4 \exp x + \sin x + 3. \tag{2.97}$$

We know from Theorem 2.6 that the general solution of our equation is a sum between the general solution of the associated homogeneous equation $y''' - 2y'' +$

$y' = 0$ and a particular solution of the non-homogeneous initial equation. The characteristic equation is

$$r^3 - 2r^2 + r = 0$$

and it has the following solution: $\lambda_1 = \lambda_2 = 1$ and $\lambda_3 = 0$. Following the result of Theorem 2.8 we see that the general solution of $y''' - 2y'' + y' = 0$ is:

$$y_{GH}(x) = C_1 \exp x + C_2 x \exp x + C_3, \tag{2.98}$$

where C_1, C_2, C_3 are real arbitrary constants.

To find a particular solution y_P of the initial equation, we notice first that the free term $f(x) = 4 \exp x + \sin x + 3$ is a sum of three functions of the particular forms indicated in (2.76), so we shall search for three particular solutions y_{1p}, y_{2p} and y_{3p} for the following three non-homogeneous equations (see Remark 2.5):

$$y''' - 2y'' + y' = 4 \exp x, \tag{2.99}$$

$$y''' - 2y'' + y' = \sin x, \tag{2.100}$$

$$y''' - 2y'' + y' = 3. \tag{2.101}$$

Using Corollaries 2.7 and 2.8, we search for $y_{1p}(x) = Ax^2 \exp x$, $y_{2p}(x) = B \cos x + C \sin x$ and $y_{3p}(x) = Dx$. We have:

$$y'_{1p}(x) = (Ax^2 + 2Ax) \exp x,$$

$$y''_{1p}(x) = (Ax^2 + 4Ax + 2A) \exp x,$$

$$y'''_{1p}(x) = (Ax^2 + 6Ax + 6A) \exp x.$$

We introduce these expressions in (2.99) and obtain: $A = 2$.

For the second equation we proceed in the same way:

$$y'_{2p}(x) = -B \sin x + C \cos x,$$

$$y''_{2p}(x) = -B \cos x - C \sin x,$$

$$y'''_{2p}(x) = B \sin x - C \cos x.$$

By replacing in (2.100) we find that $B = 0$ and $C = \frac{1}{2}$.

2.5 Nonhomogeneous Linear Equations with Constant Coefficients

Finally, $y_{3P}(x) = Dx$ must be a solution of (2.101), so we find $D = 3$. Thus, a particular solution for the Eq. (2.97) is

$$y_P(x) = 2x^2 \exp x + \frac{1}{2} \sin x + 3.$$

Hence, the general solution of the initial non-homogeneous equation is:

$$y(x) = C_1 \exp x + C_2 x \exp x + C_3 + 2x^2 \exp x + \frac{1}{2} \sin x + 3.$$

Example 2.17 **Undamped forced oscillations. Resonance** (see also [3], 3.9). Consider the equation of undamped elastic vibrations under the action of an external periodic force (of period $T = \frac{2\pi}{\beta}$, where $\beta > 0$ is the frequency of this periodic force):

$$y'' + \omega^2 y = a \cos \beta t, \ \omega, \beta > 0, \ t \geq 0 \ \text{(time)} \quad (2.102)$$

The general solution of the associated homogeneous equation $y'' + \omega^2 y = 0$ is:

$$y_{GH}(t) = C_1 \cos \omega t + C_2 \sin \omega t, \quad (2.103)$$

because $\pm i\omega$ are the solutions of the characteristic equation $r^2 + \omega^2 = 0$.

Case 1) If $\beta \neq \omega$, then we search for a particular solution for (2.102) of the following simple form:

$$y_P(t) = A \cos \beta t + B \sin \beta t. \quad (2.104)$$

By forcing this function to be a solution of (2.102) we get: $A = 0$ and $B = \frac{a}{\omega^2 - \beta^2}$. So the general solution in this case is:

$$y(t) = C_1 \cos \omega t + C_2 \sin \omega t + \frac{a}{\omega^2 - \beta^2} \sin \beta t,$$

which is a *bounded* function. This means that the elastic material which vibrates will not be destroyed in time because of increasing of the amplitudes.

Case 2) If $\beta = \omega$, then we have to search for a particular solution of the form:

$$y_P(t) = Mt \cos \beta t + Nt \sin \beta t.$$

By replacing in (2.102), we find: $N = 0$ and $M = -\frac{a}{2\omega}$. Hence, the general solution in this case is:

$$y(t) = C_1 \cos \omega t + C_2 \sin \omega t - \frac{a}{2\omega} t \cos \beta t,$$

which is an *unbounded* function, i.e. the amplitude of the oscillations can increase infinitely. This phenomenon is known as *resonance*, and is very important in the design of structures (such as buildings or bridges) because in this case the oscillations might become large enough to damage or even to destroy the structure.

In order to see the difference between the two cases (*with* and *without* resonance, we will solve the following Cauchy problems:

1) $y'' + y = 12 \sin 2t$, $\quad y(0) = 0$, $y'(0) = 0$ and
2) $y'' + y = 12 \sin t$, $\quad y(0) = 0$, $y'(0) = 0$

For the first problem (without resonance: $\omega \neq \beta$), the general solution is $y = C_1 \cos t + C_2 \sin t - 4 \sin 2t$.

By the initial conditions $y(0) = 0$, $y'(0) = 0$ we obtain $C_1 = 0$, $C_2 = 8$, so the solution of the Cauchy problem is:

$$y = 8 \sin t - 4 \sin 2t.$$

The graph of this (*bounded*) function is presented in Fig. 2.7.

In the second problem we have resonance, so the general solution is $y = C_1 \cos t + C_2 \sin t - 6t \cos t$. By the initial conditions $y(0) = 0$, $y'(0) = 0$ we obtain $C_1 = 0$, $C_2 = 6$, so the solution of the Cauchy problem is:

$$y = 6 \sin t - 6t \cos t.$$

The graph of this (*unbounded*) function is presented in Fig. 2.8.

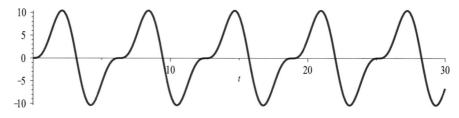

Fig. 2.7 The solution of the Cauchy problem 1)

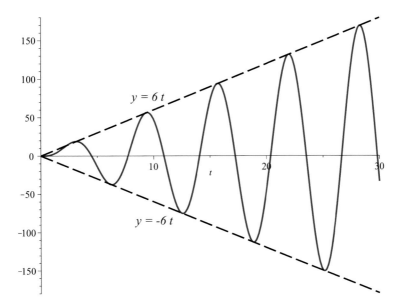

Fig. 2.8 The solution of the Cauchy problem 2)

2.6 Euler Equations

A linear differential equation (with *variable coefficients*) of the type:

$$a_0 x^n y^{(n)} + a_1 x^{n-1} y^{(n-1)} + \ldots + a_{n-1} x y' + a_n y = f(x), x \in I, \quad (2.105)$$

where $a_0, \ldots, a_n$ are real numbers, $a_0 \neq 0$, is said to be an *Euler equation* (or *Euler-Cauchy equation* [21]). The characteristic of this type of equation is that the degree k of the monomial coefficient x^k matches the order of the derivative $y^{(k)}$, for all $k = n, n-1, \ldots, 0$.

First of all, we notice that Theorem 2.2 (which states the existence and uniqueness of solution of the Cauchy problem for linear differential equations) demands the coefficient of the higher derivative to be different from zero, for any $x \in I$. So in Eq. (2.105) we have either $I = (0, \infty)$, or $I = (-\infty, 0)$. Since the second case can be easily reduced to the first one, we study only the case when $x > 0$.

We use the *change of variable* ([37, 42])

$$x = \exp(t)$$

and denote by $\widetilde{y}(t)$ the new (composite) function

$$\widetilde{y}(t) = y(\exp(t)). \quad (2.106)$$

By differentiating with respect to t the equality (2.106), we obtain:

$$\frac{d\widetilde{y}}{dt} = \frac{dy}{dx} \cdot \exp(t) = y'(\exp(t)) \cdot \exp(t), \tag{2.107}$$

so the first derivative of $y(x)$, $y'(x)$ is written

$$y'(x) = y'(\exp(t)) = \exp(-t)\frac{d\widetilde{y}}{dt}. \tag{2.108}$$

By differentiating with respect to t the last equality in (2.108), we get:

$$y''(\exp(t)) \cdot \exp(t) = -\exp(-t)\frac{d\widetilde{y}}{dt} + \exp(-t)\frac{d^2\widetilde{y}}{dt^2},$$

so the second derivative, $y''(x)$, is written

$$y''(x) = y''(\exp(t)) = \exp(-2t)\left(\frac{d^2\widetilde{y}}{dt^2} - \frac{d\widetilde{y}}{dt}\right). \tag{2.109}$$

If we denote by L_1 and L_2 the linear differential operators

$$L_1 = \frac{d}{dt}, \quad L_2 = \frac{d^2}{dt^2} - \frac{d}{dt} = \frac{d}{dt}\left(\frac{d}{dt} - I\right),$$

then the first and the second derivatives, y' and y'' can be written:

$$y'(x) = y'(\exp(t)) = \exp(-t)L_1[\widetilde{y}],$$

$$y''(x) = y''(\exp(t)) = \exp(-2t)L_2[\widetilde{y}].$$

For all $k = 1, 2, \ldots$ we denote by L_k be the linear differential operator (from the polynomial ring $\mathbb{R}\left[\frac{d}{dt}\right]$)

$$L_k = \frac{d}{dt}\left(\frac{d}{dt} - I\right)\left(\frac{d}{dt} - 2I\right)\ldots\left(\frac{d}{dt} - (k-1)I\right) \tag{2.110}$$

and we notice the following recurrence relation:

$$L_k = L_{k-1}\left(\frac{d}{dt} - I\right). \tag{2.111}$$

We prove by mathematical induction on $k = 1, 2, \ldots$ that

$$y^{(k)}(x) = \exp(-kt)L_k[\widetilde{y}]. \tag{2.112}$$

2.6 Euler Equations

For $k = 1$ the statement is already proved above. Consider $k > 1$ and suppose that

$$y^{(k-1)}(x) = y^{(k-1)}(\exp(t)) = \exp(-(k-1)t) L_{k-1}[\widetilde{y}]. \tag{2.113}$$

In order to prove (2.112), we differentiate (2.113) with respect to t:

$$\frac{d}{dt}\left[y^{(k-1)}(\exp(t))\right] = \frac{d}{dt}\left[\exp(-(k-1)t) \cdot L_{k-1}(\widetilde{y})\right],$$

so we have:

$$y^{(k)}(\exp(t)) \cdot \exp(t) = -(k-1) \exp(-(k-1)t) \cdot L_{k-1}[\widetilde{y}] +$$

$$+ \exp(-(k-1)t) \cdot \frac{d}{dt} L_{k-1}[\widetilde{y}].$$

But, as discussed in the previous section, the multiplication (composition) in the ring $\mathbb{R}\left[\frac{d}{dt}\right]$ is commutative, so

$$\frac{d}{dt} L_{k-1}[\widetilde{y}] = L_{k-1} \frac{d}{dt}[\widetilde{y}],$$

and we have:

$$y^{(k)}(\exp(t)) \cdot \exp(t) = \exp(-(k-1)t) \cdot L_{k-1}\left(\frac{d}{dt} - (k-1)I\right)[\widetilde{y}].$$

Multiplying the last equation by $\exp(-t)$ we obtain (2.112) and the proof is completed.

Thus, by using the change of variable $x = \exp(t)$, the *Euler equation* (2.105) is transformed into the following linear equation *with constant coefficients* (having the unknown function $\widetilde{y}(t)$):

$$a_0 L_n[\widetilde{y}] + a_1 L_{n-1}[\widetilde{y}] + \ldots + a_{n-1} L_1[\widetilde{y}] + a_n \widetilde{y} = f(\exp(t)). \tag{2.114}$$

Let $P_1(r), P_2(r), \ldots$ be the polynomials from $\mathbb{R}[r]$ corresponding to the differential polynomials $L_1, L_2, \ldots$ from $\mathbb{R}\left[\frac{d}{dt}\right]$:

$$P_k(r) = r(r-1)\ldots(r-k+1), \tag{2.115}$$

for all $k = 1, 2, \ldots$.

Now, if we denote by $b_0, b_1, \ldots, b_n$ the coefficients of the polynomial

$$a_0 P_n(r) + a_1 P_{n-1}(r) + \ldots + a_{n-1} P_1(r) + a_n = b_0 r^n + b_1 r^{n-1} + \ldots + b_n,$$

we can see that the differential Eq. (2.114) can be written as

$$b_0 \tilde{y}^{(n)} + b_1 \tilde{y}^{(n-1)} + \ldots + b_{n-1} \tilde{y}' + b_n \tilde{y} = f(\exp(t)). \tag{2.116}$$

Example 2.18 Let us find the function $y = y(x)$ defined on $I = (0, \infty)$, such that

$$x^3 y''' - x^2 y'' + 2xy' - 2y = \ln x, \tag{2.117}$$

and $y(1) = y'(1) = 0$ and $y''(1) = 2$.

We use the substitution $x = \exp(t)$ and denote $\tilde{y}(t) = y(\exp(t))$. As we have seen above, the derivatives y', y'' and y''' can be written using the derivatives of $\tilde{y}$ with respect to t as:

$$y' = \exp(-t) L_1[\tilde{y}] = \exp(-t) \frac{d\tilde{y}}{dt},$$

$$y'' = \exp(-2t) L_2[\tilde{y}] = \exp(-2t) \left(\frac{d^2 \tilde{y}}{dt^2} - \frac{d\tilde{y}}{dt} \right),$$

$$y''' = \exp(-3t) L_3[\tilde{y}] = \exp(-3t) \left(\frac{d^3 \tilde{y}}{dt^3} - 3 \frac{d^2 \tilde{y}}{dt^2} + 2 \frac{d\tilde{y}}{dt} \right).$$

By replacing in the initial equation we obtain the following equation with constant coefficients:

$$\frac{d^3 \tilde{y}}{dt^3} - 4 \frac{d^2 \tilde{y}}{dt^2} + 5 \frac{d\tilde{y}}{dt} - 2\tilde{y} = \ln(\exp(t)) = t. \tag{2.118}$$

The characteristic equation is

$$r^3 - 4r^2 + 5r - 2 = 0,$$

with the roots $r_1 = r_2 = 1$ and $r_3 = 2$. Thus, the general solution of the associated homogeneous equation is:

$$\tilde{y}_{GH}(t) = C_1 \exp(t) + C_2 t \exp(t) + C_3 \exp(2t).$$

Since the free term is $g(t) = t$, we can search for a particular solution of the non-homogeneous equation in t of the form: $\tilde{y}_P(t) = At + B$. By replacing in Eq. (2.118), we obtain that $A = -\frac{1}{2}$ and $B = -\frac{5}{4}$ and so the general solution of the Eq. (2.118) is

$$\tilde{y}(t) = C_1 \exp(t) + C_2 t \exp(t) + C_3 \exp(2t) - \frac{1}{2} t - \frac{5}{4}.$$

Hence the general solution of the initial (Euler) equation is

$$y(x) = C_1 x + C_2 x \ln(x) + C_3 x^2 - \frac{1}{2} \ln x - \frac{5}{4}.$$

From the initial conditions, we get $C_1 = -1$, $C_2 = -3$ and $C_3 = \frac{9}{4}$.

2.7 Exercises

1. Prove that for any two arbitrary real numbers C_1, C_2, the function

   $$y(x) = C_1 \cos 2x + C_2 \sin 2x$$

   is a solution for the second-order linear differential equation $y'' + 4y = 0$.
2. Prove that for any two arbitrary real numbers C_1, C_2, the function $y(x) = C_1 \exp 2x + C_2 \exp(-2x)$ is a solution for the second-order linear differential equation $y'' - 4y = 0$.
3. Find the linear ODEs with the following general solutions:
   (a) $y = C_1 \exp(3x) + C_2 \exp(-5x)$;
   (b) $y = C_1 \cos 4x + C_2 \sin 4x$;
   (c) $y = C_1 + C_2 x + C_3 \exp x$.
4. Solve the following homogeneous LDE with constant coefficients:
   (a) $y'' + y' - 2y = 0$;
   (b) $y'' - 6y' + 9y = 0$;
   (c) $y'' + 2y' + 10y = 0$;
   (d) $y''' - y' = 0$;
   (e) $y''' + y' = 0$;
   (f) $y''' + y = 0$;
   (g) $y^{(4)} - y = 0$;
   (h) $y^{(4)} - 5y'' + 4y = 0$;
   (i) $y^{(4)} + 5y'' + 4y = 0$;
   (j) $y^{(4)} - 2y'' + y = 0$;
   (k) $y^{(4)} + 2y'' + y = 0$;
   (l) $y^{(5)} + 3y^{(4)} + 3y''' + y'' = 0$.
5. Prove that the following sequences of functions are linearly independent and find the homogeneous linear differential equations which have these sequences as fundamental system of solutions (basis of solutions):

   (a) $\{\exp x, \exp(-x)\}$;
   (b) $\{\exp x, x \exp x, \exp(-x)\}$;
   (c) $\{1, x, x^2, \exp(2x), \sin x, \cos x\}$.

6. Solve the following non-homogeneous LDE with constant coefficients:
   (a) $y'' - 4y = 3x \exp x$;
   (b) $y'' - 4y = 4 \exp 2x$;
   (c) $y'' + 9y = \sin x$;
   (d) $y'' + 9y = \cos 3x$;
   (e) $y'' + 9y = \sin x + \cos 3x$;
   (f) $y'' + 2y' + 5y = 5x^2 - x$;
   (g) $y'' + 2y' + 2y = 8 \exp x \cdot \sin x$;
   (h) $y'' - 5y' = 5x^2 + \exp(5x)$;
   (i) $y'' - 2y' + y = \sin x + \exp x + \exp(-x)$;
   (j) $y'' - 3y' + 2y = 10x \cos x$;
   (k) $y^{(4)} - 2y''' + y'' = x + \exp 2x$;
   (l) $y''' - 4y' = 3\exp(-x)$, $y(0) = y'(0) = y''(0) = 1$;
   (m) $y'' - 4y' + 4y = \exp x$, $y(0) = y'(0) = 0$;
   (n) $y^{(4)} + y'' = 2\cos x$, $y(0) = -2$, $y'(0) = 1$; $y''(0) = y'''(0) = 0$.

7. Use Lagrange's method of variation of constants to solve the following LDEs:
   (a) $y'' + 9y = \frac{1}{\sin 3x}$;
   (b) $y'' - 2y' + y = \frac{\exp x}{\sqrt{1-x^2}}$;
   (c) $y'' + 4y = 2 \tan 2x$;
   (d) $y'' + 3y' + 2y = \frac{1}{\exp x + 1}$;
   (e) $y'' - 2y' + y = \frac{\exp x}{x}$, $y(1) = 0$, $y'(1) = e$

8. Solve the following Euler equations:
   (a) $x^2 y'' - 2xy' + 2y = \frac{\ln x}{x}$, $x > 0$.
   (b) $x^2 y'' - 2xy' + 2y = x$, $x < 0$, $y(-1) = -1$, $y'(-1) = 0$;
   (c) $x^3 y''' - x^2 y'' + 2xy' - 2y = \frac{12}{x^2}$, $x > 0$; , $y(1) = y'(1) = y''(1) = 0$;
   (d) $x^3 y''' + xy' - y = x^2$, $x < 0$;
   (e) $x^2 y'' - xy' + y = \frac{\ln x}{x} + \frac{x}{\ln x}$, $x > 1$.

9. Find an equation of the curve $y = y(x)$ which passes through $M(0, 1)$ and whose slope at $P(x, y)$ is equal to xy.

10. Find a function $f(x)$ such that $f'(x) = f(x)[1 - f(x)]$ and $f(0) = 1$. What can you say about its maximum domain of definition?

11. A sphere with radius 1 m has a temperature of 15°C. It lies inside a concentric sphere with radius 2 m and a temperature of 25°C. The temperature $u(r)$ at the distance r from the center of the spheres satisfies the following differential equation: $u'' + \frac{2}{r} u' = 0$. Find $u(1.5)$, the temperature at the distance of 1.5 m from the center.

12. Stewart [38] The gravitational force acting on a body of mass m located at the distance y above the earth surface is $F = -\frac{mgR^2}{(y+R)^2}$, where R is the earth radius and g is the gravitational acceleration (the sign "−" indicates that the vector $\vec{F}$ is directed to the negative sense of Oy axis). Since $F = m\frac{dv}{dt}$, and

2.7 Exercises

the velocity is $v = \frac{dy}{dt}$, it follows that $y = y(t)$ is the solution of the ODE-2: $m\frac{d^2y}{dt^2} + \frac{mgR^2}{(y+R)^2} = 0$. We assume that a rocket is launched vertically upward with an initial velocity v_0 and the maximum height it reaches is h. Prove that $v_0 = \sqrt{\frac{2gRh}{R+h}}$ and compute the *escape velocity* $v_e = \lim\limits_{h \to \infty} v_0$. (Hint: $v = v(y(t)) \Rightarrow \frac{dv}{dt} = \frac{dv}{dy} \cdot \frac{dy}{dt} = v\frac{dv}{dy}$, so take $v = v(y)$, etc.)

13. A flexible homogeneous cable hanging under its own weight, is suspended between two fixed points $M_1(-a, h)$ and $M_2(a, h)$, having the shape $y = y(x)$. It is known that the function $y(x)$ satisfies the following ODE-2: $y'' = k\sqrt{1 + y'^2}$, where $k > 0$ is a constant which depends on the elastic properties of the object. Find $y(x)$ and the length of the cable (Hint: put $z = y'$).

14. Stewart [38] Find $y = y(x)$ such that $[y(x)]^2 = 2 + \int\limits_0^x \left\{ [y(t)]^2 + [y'(t)]^2 \right\} dt$

 (Hint: differentiate both sides with respect to x.)

15. Stewart [38] Let xOy be a Cartesian coordinate system and let $C_0(x_0, 0)$ be a fixed point on the Ox-axis. A cat is at C_0 when she first sees a mouse which is at the origin. The mouse runs with a constant speed v_m along the positive part of the Oy-axis. The cat sees the running mouse and try to change her direction such that she is running straight for the mouse with a constant speed v_c. This means that at any moving point $C(x, y(x))$ of the cat, the tangent straight line at C to the running curve $y = y(x)$ passes through the moving point $M(0, v_m t)$ (t is time) of the mouse. (a) Show that if $v_c = v_m$ and $y = y(x)$ is the function whose graph is the cat's path, then: $xy'' = \sqrt{1 + y'^2}$. (b) Does the cat ever catch the mouse? (c) Prove that if the cat's speed is twice the mouse's speed, then the cat catches the mouse. At what point on the Oy-axis this happens? (d) If the cat's speed is a half of the mouse speed, find their positions when they are closest (Hint: $z = y'$).

Chapter 3
Systems of Differential Equations

This chapter is dedicated to systems of differential equations. We state and prove the existence and uniqueness theorem for the Cauchy problem in this case. Particular attention is paid to linear systems of differential equations and to the relation between a linear system of n equations and an n-th order linear equation. Autonomous systems are also studied, and, since first-order partial differential equations are closely related to autonomous systems, we present them in the last section of this chapter.

3.1 Introduction

In the three-dimensional space $\mathbb{R}^3$ let $M(x(t), y(t), z(t))$, $t \geq 0$, be the position of a particle of mass m, moving along the oriented curve $\gamma : \mathbf{r}(t) = x(t)\mathbf{i} + y(t)\mathbf{j} + z(t)\mathbf{k}$, where $\mathbf{r}(t)$ is the position vector of M at time t and $\mathbf{i} = (1, 0, 0)$, $\mathbf{j} = (0, 1, 0)$, $\mathbf{k} = (0, 0, 1)$ (See Fig. 3.1). Assume that the force field producing this moving is

$$\mathbf{F}(t, x, y, z, x', y', z') = f(t, x, y, z, x', y', z')\mathbf{i} + g(t, x, y, z, x', y', z')\mathbf{j} +$$
$$+ h(t, x, y, z, x', y', z')\mathbf{k},$$

where x, y, z are the space coordinates of M as functions of time t and x', y', z' are the coordinates of the velocity at the same time t. By Newton's second law of motion,

$$\mathbf{F} = m\mathbf{a},$$

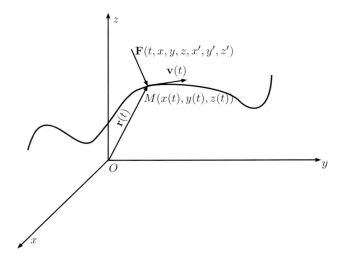

Fig. 3.1 Particle moving in the force field **F**

where $\mathbf{a}(t) = x''(t)\mathbf{i} + y''(t)\mathbf{j} + z''(t)\mathbf{k}$ is the acceleration at time t, we obtain the following set of differential equations:

$$\begin{cases} x'' = \frac{1}{m} f(t, x, y, z, x', y', z') \\ y'' = \frac{1}{m} g(t, x, y, z, x', y', z') \\ z'' = \frac{1}{m} h(t, x, y, z, x', y', z'). \end{cases} \quad (3.1)$$

To find the position of the particle at each moment t means to find the functions $x(t), y(t), z(t)$, that satisfy the system of equations (3.1), knowing the *initial position* (x_0, y_0, z_0) and the *initial velocity*, $\mathbf{v}_0 = v_{x0}\mathbf{i} + v_{y0}\mathbf{j} + v_{z0}\mathbf{k}$. This means to solve the *Cauchy problem* for the system (3.1) with the *initial conditions*:

$$\begin{array}{l} x(0) = x_0, \, y(0) = y_0, \, z(0) = z_0, \\ x'(0) = v_{x0}, \, y'(0) = v_{y0}, \, z'(0) = v_{z0}. \end{array} \quad (3.2)$$

By introducing the new functions $u = x'$, $v = y'$ and $w = z'$, the system (3.1) (of *second-order* equations) becomes:

$$\begin{cases} x' = u \\ y' = v \\ z' = w \\ u' = \frac{1}{m} f(t, x, y, z, u, v, w) \\ v' = \frac{1}{m} g(t, x, y, z, u, v, w) \\ w' = \frac{1}{m} h(t, x, y, z, u, v, w) \end{cases} \quad (3.3)$$

3.1 Introduction

This system of differential equations is said to be a *first-order system of differential equations* written in a *normal form*. A set of six differentiable functions $x(t), y(t), z(t), u(t), v(t), w(t)$ which satisfy the equations of the system (3.3) is called a *solution* of this system. The collection of all solutions of this system is called the *general solution* of the system. Systems of differential equations are used as mathematical models for many phenomena in various domains like Physics, Chemistry, Economy, etc. As we saw above, systems of higher-order equations can be transformed into systems of first-order equations (increasing the number of equations).

A *first-order system of n differential equations*, SDE-n (in a *normal form*), is a set of n differential equations of the form:

$$\begin{cases} y'_1(x) = f_1(x, y_1(x), y_2(x), \ldots, y_n(x)) \\ y'_2(x) = f_2(x, y_1(x), y_2(x), \ldots, y_n(x)) \\ \quad \vdots \\ y'_n(x) = f_n(x, y_1(x), y_2(x), \ldots, y_n(x)) \end{cases}, x \in I, \quad (3.4)$$

where $f_1, \ldots, f_n$ are continuous functions of $n + 1$ variables, defined on an open subset $D \subset \mathbb{R}^{n+1}$, $I \subset \mathbb{R}$ is a nonempty interval and $y_1, \ldots, y_n$ are the unknowns of the system: continuously differentiable functions of the free variable $x \in I$. A solution of the system (3.4) is a set of n functions of class C^1, $y_1, \ldots, y_n : I \to \mathbb{R}$, such that $f_i(x, y_1(x), \ldots, y_n(x)) \in D$ for all $x \in I$ and $i = 1, \ldots, n$, and such that equalities (3.4) hold, for any $x \in I$.

If we denote by $\mathbf{Y}(x) = (y_1(x), \ldots, y_n(x))$ the n-dimensional vector of unknown functions, by $\mathbf{Y}'(x) = (y'_1(x), \ldots, y'_n(x))$, the vector of their derivatives and by $\mathbf{F} = (f_1, \ldots, f_n)$, the vector-valued function of components $f_1, \ldots, f_n$, then the system (3.4) can be written in a more compact form, as:

$$\mathbf{Y}'(x) = \mathbf{F}(x, \mathbf{Y}(x)), \ x \in I. \quad (3.5)$$

A *Cauchy problem* for the system of differential Eq. (3.4) is the following problem: given $x_0 \in I$ and n real numbers $y_1^0, \ldots, y_n^0 \in \mathbb{R}$ such that $(x_0, y_1^0, \ldots, y_n^0) \in D$, find a solution $y_1 = y_1(x), \ldots, y_n = y_n(x)$ of the system such that the following *initial conditions* are satisfied:

$$y_1(x_0) = y_1^0, \ldots, y_n(x_0) = y_n^0. \quad (3.6)$$

If for each $i = 1, 2, \ldots, n$ one can define a family of functions $y_i(x) = y_i(x; C_1, \ldots, C_n)$ where $x \in I$ and $(C_1, \ldots, C_n) \in \Omega \subset \mathbb{R}^n$, such that the functions $y_1 = y_1(x), \ldots, y_n = y_n(x)$ form a solution of (3.4) for any $(C_1, \ldots, C_n) \in \Omega$, then we say that this set of families of functions is a *general solution* of the SDE (3.4). By giving particular values to $C_1, C_2, \ldots, C_n$, we obtain a particular solution of the system. Any other solution which cannot be obtained in this way from the general solution is called a *singular solution* of the SDE. The points $x_0 \in I$ for which the Cauchy problem has more than one solution are called *singular points* of the SDE-n.

By the same method we used in Theorem 1.3, *Picard's method of successive approximations*, we can prove the following theorem which states the existence and uniqueness of solution of the Cauchy problem for *systems* of differential equations.

Theorem 3.1 *Existence and Uniqueness Theorem for Systems of Equations. If $f_i(x, y_1, \ldots, y_n)$, $i = 1, \ldots, n$ are continuous functions defined on the $(n + 1)$-dimensional rectangle*

$$R = [x_0 - a, x_0 + a] \times [y_1^0 - b_1, y_1^0 + b_1] \times \ldots \times [y_n^0 - b_n, y_n^0 + b_n],$$

which possess bounded partial derivatives with respect to $y_1, \ldots, y_n$, then there is some interval $I = [x_0 - h, x_0 + h]$ ($0 < h \leq a$) in which there exists a unique solution $\mathbf{Y} = \mathbf{Y}(x) = (y_1(x), \ldots, y_n(x))$ of the system of differential Eq. (3.5) such that the initial conditions (3.6) are satisfied.

Proof Existence. As we saw in the proof of Theorem 1.3, the Cauchy problem (3.5)–(3.6) is equivalent to the following system of integral equations:

$$y_i(x) = y_i^0 + \int_{x_0}^{x} f_i(t, y_1(t), \ldots, y_n(t)) \, dt, \quad i = 1, \ldots, n. \tag{3.7}$$

Since the functions f_i and $\frac{\partial f_i}{\partial y_j}$ are bounded on R for all $i, j = 1, \ldots, n$ ($f_1, \ldots, f_n$ are continuous on the closed rectangle R), there exist the positive constants M and L such that

$$|f_i(x, y_1, \ldots, y_n)| \leq M, \tag{3.8}$$

$$\left|\frac{\partial f_i}{\partial j}(x, y_1, \ldots, y_n)\right| \leq L, \tag{3.9}$$

for all $(x, y_1, \ldots, y_n) \in R$, $i = 1, \ldots, n$, $j = 1, \ldots, n$.

We prove that, for any two points in the $(n + 1)$-dimensional rectangle R, $(x, y_1, \ldots, y_n)$ and $(x, z_1, \ldots, z_n)$, the following inequalities hold, for all $i = 1, \ldots, n$:

$$|f_i(x, y_1, \ldots, y_n) - f_i(x, z_1, \ldots, z_n)| \leq L \sum_{j=1}^{n} |y_j - z_j| \tag{3.10}$$

Let $i = 1, \ldots, n$. We define the function $\varphi : [0, 1] \to \mathbb{R}$,

$$\varphi(t) = f_i(x, z_1 + t(y_1 - z_1), \ldots, z_n + t(y_n - z_n)).$$

3.1 Introduction

By the Mean Value Theorem ([38], 4.2), there exists $\tau \in (0, 1)$ such that

$$\varphi'(\tau) = \varphi(1) - \varphi(0) = f_i(x, y_1, \ldots, y_n) - f_i(x, z_1, \ldots, z_n).$$

But

$$\varphi'(\tau) = \frac{\partial f_i}{\partial x} \cdot 0 + \frac{\partial f_i}{\partial y_1} \cdot (y_1 - z_1) + \ldots + \frac{\partial f_i}{\partial y_n} \cdot (y_n - z_n),$$

hence, by (3.9), the inequality (3.10) follows.

Let $h = \min\left\{a, \frac{b_1}{M}, \ldots, \frac{b_n}{M}\right\}$. For all $i = 1, \ldots, n$, we construct the following sequences of functions defined on the interval $I = [x_0 - h, x_0 + h]$:

$$y_i^{[0]}(x) = y_i^0, \ \forall x \in I, \ i = 1, \ldots, n$$

and, for $k = 1, 2, \ldots,$

$$y_i^{[k]}(x) = y_i^0 + \int_{x_0}^{x} f_i(t, y_1^{[k-1]}(t), \ldots, y_n^{[k-1]}(t)) \, dt. \tag{3.11}$$

Since

$$|y_i^{[k]}(x) - y_i^0| \le \left|\int_{x_0}^{x} \left|f_i(t, y_1^{[k-1]}(t), \ldots, y_n^{[k-1]}(t))\right| dt\right| \le M|x - x_0|, \tag{3.12}$$

it follows that $y_i^{[k]}(x) \in [y_i^0 - b_i, y_i^0 + b_i]$, for any $x \in I, k \ge 1$ and $i = 1, \ldots, n$.

By the definition of the sequences (3.11) we have, for any $i = 1, \ldots, n$:

$$\left|y_i^{[k+1]}(x) - y_i^{[k]}(x)\right| \le$$

$$\le \left|\int_{x_0}^{x} \left|f_i(t, y_1^{[k]}(t), \ldots, y_n^{[k]}(t)) - f_i(t, y_1^{[k-1]}(t), \ldots, y_n^{[k-1]}(t))\right| dt\right|$$

and, from the inequality (3.10), we obtain:

$$\left|y_i^{[k+1]}(x) - y_i^{[k]}(x)\right| \le L \sum_{j=1}^{n} \left|\int_{x_0}^{x} \left|y_j^{[k]}(t) - y_j^{[k-1]}(t)\right| dt\right|, \tag{3.13}$$

for any $x \in I, i = 1, \ldots, n$ and $k = 1, 2, \ldots$.

From (3.12) we can write (for $k = 1$):

$$\left|y_i^{[1]}(x) - y_i^{[0]}\right| \le M|x - x_0|,$$

for any $x \in I$ and $i = 1, \ldots, n$. Using (3.13), the following inequality can be easily proved by mathematical induction on $k \geq 0$:

$$\left| y_i^{[k+1]}(x) - y_i^{[k]}(x) \right| \leq \frac{M(nL)^k |x - x_0|^{k+1}}{(k+1)!}, \tag{3.14}$$

for any $x \in I$, $i = 1, \ldots, n$ and $k = 0, 1, \ldots$. It follows that

$$\left| y_i^{[k+1]}(x) - y_i^{[k]}(x) \right| \leq \frac{M(nL)^k h^{k+1}}{(k+1)!}, \tag{3.15}$$

for any $x \in I$, $i = 1, \ldots, n$ and $k = 0, 1, \ldots$. Hence, by *Weierstrass M-test* (see Theorem 4.1), since the series $\sum_{k=0}^{\infty} \frac{M(nL)^k h^{k+1}}{(k+1)!}$ converges, we obtain that, for any $i = 1, \ldots, n$, the series

$$y_i^{[0]}(x) + \sum_{n=0}^{\infty} \left(y_i^{[k+1]}(x) - y_i^{[k]}(x) \right) \tag{3.16}$$

uniformly converges on the interval I to a continuous function, $y_i(x)$. But the partial sums of the series (3.16) are the terms of the sequence $\{y_i^{[k]}\}_{k \geq 0}$:

$$y_i^{[0]}(x) + \sum_{j=0}^{k-1} \left(y_i^{[j+1]}(x) - y_i^{[j]}(x) \right) = y_i^{[k]}(x),$$

so, for every $i = 1, \ldots, n$, the sequence $\{y_i^{[k]}\}_{k \geq 0}$ constructed by (3.11) uniformly converges to $y_i(x)$, that is:

$$\lim_{k \to \infty} \sup_{x \in I} \left| y_i^{[k]}(x) - y_i(x) \right| = \lim_{k \to \infty} \left\| y_i^{[k]} - y_i \right\| = 0. \tag{3.17}$$

Since $y_i^{[k]}(x) \in [y_i^0 - b_i, y_i^0 + b_i]$, for any $x \in I$, $k \geq 1$ and $i = 1, \ldots, n$, it follows that $y_i(x) \in [y_i^0 - b_i, y_i^0 + b_i]$, for any $x \in I$, $i = 1, \ldots, n$. It can be easily proved (as in the proof of Theorem 1.3) that the functions $y_1, \ldots, y_n$ satisfy the integral Eq. (3.7), hence $\mathbf{Y} = (y_1, \ldots, y_n)$ is a solution of our Cauchy problem.

Uniqueness. Let $\mathbf{Z} = (z_1, \ldots, z_n) : I \to \mathbb{R}^n$ be another solution of the Cauchy problem (3.5)–(3.6). Then the functions $z_1, \ldots, z_n$ must also satisfy the integral Eq. (3.7):

$$z_i(x) = y_i^0 + \int_{x_0}^{x} f_i(t, z_1(t), \ldots, z_n(t)) \, dt, \quad i = 1, \ldots, n. \tag{3.18}$$

3.1 Introduction

It follows that:

$$\left| z_i(x) - y_i^0 \right| \leq \left| \int_{x_0}^{x} |f_i(t, z_1(t), \ldots, z_n(t))| \, dt \right| \leq M|x - x_0|. \tag{3.19}$$

On the other hand, by (3.11), (3.18) and (3.10), we have that:

$$|z_i(x) - y_i^{[k]}(x)| \leq \left| \int_{x_0}^{x} \left| f(t, z_1(t), \ldots, z_n(t)) - f(t, y_1^{[k-1]}(t), \ldots, y_n^{[k-1]}(t)) \right| dt \right| \leq \tag{3.20}$$

$$\leq L \sum_{j=1}^{n} \left| \int_{x_0}^{x} \left| z_j(t) - y_j^{[k-1]}(t) \right| dt \right|,$$

for all $x \in I$, $i = 1, \ldots, n$ and $k = 1, 2, \ldots$. In the same manner as in the proof of Theorem 1.3, it can be proved by mathematical induction on $k \geq 0$ that

$$|z_i(x) - y_i^{[k]}(x)| \leq M(nL)^k \frac{|x - x_0|^{k+1}}{(k+1)!}.$$

It follows that, for all $x \in I$, $i = 1, \ldots, n$ and $k = 1, 2, \ldots$, the following inequalities hold:

$$|z_i(x) - y_i^{[k]}(x)| \leq M(nL)^k \frac{h^{k+1}}{(k+1)!}.$$

Therefore,

$$\lim_{k \to \infty} y_i^{[k]}(x) = z_i(x),$$

which means that $y_i = z_i$, for all $i = 1, \ldots, n$ and the uniqueness of the solution is proved. □

Remark 3.1 As we remarked in Chap. 1, the continuity of the functions f_i, $i = 1, \ldots, n$, is sufficient to ensure the *existence* of solution. The condition regarding the existence and the boundedness of the partial derivatives guarantees the *uniqueness* of the solution. However, a weaker condition can be used instead. Thus, for the uniqueness of the solution it is sufficient that the functions $f_i(x, y_1, \ldots, y_n)$, $i = 1, \ldots, n$, satisfy the Lipschitz condition (2.22) w.r.t. $y_1, \ldots, y_n$.

Example 3.1 Find the solution of the following Cauchy problem:

$$\begin{cases} y' = z \\ z' = -y \end{cases} \quad \begin{cases} y(0) = 0 \\ z(0) = 1 \end{cases} \tag{3.21}$$

Using the *elimination method* (described in Sects. 3.2 and 3.4.4), we can "eliminate" z from the last equation by replacing it with y' (from the first equation). We obtain that $y'' = -y$, so we have to solve a second-order linear homogeneous equation in the unknown function y:

$$y'' + y = 0.$$

The general solution of this equation is $y = C_1 \cos x + C_2 \sin x$, so, since $z = y'$, the *general solution* of the system is

$$\begin{cases} y = C_1 \cos x + C_2 \sin x \\ z = C_2 \cos x - C_1 \sin x \end{cases}.$$

By the initial conditions $y(0) = 0$ and $z(0) = 1$ we obtain that $C_1 = 0$ and $C_2 = 1$, hence the solution of the Cauchy problem (3.21) is:

$$y(x) = \sin x, \quad z(x) = \cos x.$$

Now, let us apply the *method of successive approximations* described in the proof of Theorem 3.1. Thus, we construct the sequences of functions $\{y_k\}$ and $\{z_k\}$, $k = 0, 1, \ldots$, by the recurrence relation (3.11):

$$\begin{cases} y_0(x) = 0 \\ z_0(x) = 1 \end{cases} \quad \begin{cases} y_k(x) = \int_0^x z_{k-1}(t)\, dt \\ z_k(x) = 1 - \int_0^x y_{k-1}(t)\, dt \end{cases} \quad k = 1, 2, \ldots.$$

We compute $y_k(x)$, $z_k(x)$ for $k = 1, 2, \ldots$:

$$\begin{cases} y_1(x) = \int_0^x z_0(t)\, dt = \int_0^x 1\, dt = x \\ z_1(x) = 1 - \int_0^x y_0(t)\, dt = 1 - \int_0^x 0\, dt = 1 \end{cases}$$

$$\begin{cases} y_2(x) = \int_0^x z_1(t)\, dt = \int_0^x 1\, dt = x \\ z_2(x) = 1 - \int_0^x y_1(t)\, dt = 1 - \int_0^x t\, dt = 1 - \frac{x^2}{2} \end{cases}$$

$$\begin{cases} y_3(x) = \int_0^x z_2(t)\, dt = \int_0^x \left(1 - \dfrac{t^2}{2}\right) dt = x - \dfrac{x^3}{3!} \\ z_3(x) = 1 - \int_0^x y_2(t)\, dt = 1 - \int_0^x t\, dt = 1 - \dfrac{x^2}{2!} \end{cases}$$

$$\begin{cases} y_4(x) = \int_0^x z_3(t)\, dt = \int_0^x \left(1 - \dfrac{t^2}{2}\right) dt = x - \dfrac{x^3}{3!} \\ z_4(x) = 1 - \int_0^x y_3(t)\, dt = 1 - \int_0^x \left(t - \dfrac{t^3}{3!}\right) dt = 1 - \dfrac{x^2}{2!} + \dfrac{x^4}{4!} \end{cases}$$

Using mathematical induction it is easy to prove that, for any $k = 1, 2, \ldots$,

$$y_{2k-1}(x) = y_{2k}(x) = x - \dfrac{x^3}{3!} + \dfrac{x^5}{5!} - \ldots + (-1)^{k-1} \dfrac{x^{2k-1}}{(2k-1)!},$$

$$z_{2k}(x) = z_{2k+1}(x) = 1 - \dfrac{x^2}{2!} + \dfrac{x^4}{4!} - \ldots + (-1)^k \dfrac{x^{2k}}{(2k)!},$$

which means that the sequences of functions $\{y_k\}$ and $\{z_k\}$ are formed by partial sums of the series

$$\sin x = \sum_{k=0}^{\infty} (-1)^k \dfrac{x^{2k+1}}{(2k+1)!}, \quad \text{and} \quad \cos x = \sum_{k=0}^{\infty} (-1)^k \dfrac{x^{2k}}{(2k)!},$$

respectively. So, the sequences constructed by the method of successive approximations uniformly converge to the same solution found by the elimination method,

$$y(x) = \sin x \quad \text{and} \quad z(x) = \cos x.$$

3.2 First-Order Systems and Differential Equations of Order n

In Example 3.1 we used the so called *elimination method* which involves eliminating all but one of the unknown functions (let they be $y_2, \ldots, y_n$) in order to obtain a single higher order equation for the remaining unknown function, y_1.

Conversely, let

$$y^{(n)} = f(x, y, y', \ldots, y^{(n-1)}) \tag{3.22}$$

be an ordinary differential equation of order n (ODE-n) in the *normal form* (i.e. the highest derivative $y^{(n)}$ of the unknown function y is expressed as a function of x, y and the lower derivatives of y, y', y'', ..., $y^{(n-1)}$). By denoting $y = y_1$, $y' = y_2$, $y'' = y_3$, ..., $y^{(n-1)} = y_n$, the set of solutions of Eq. (3.22) is in one-to-one correspondence with the set of solutions of the following system of n first-order differential equations (SDE-n)

$$\begin{cases} y'_1 = y_2 \\ y'_2 = y_3 \\ y'_3 = y_4 \\ \vdots \\ y'_n = f(x, y_1, y_2, \ldots, y_n) \end{cases}.$$

We present the elimination method for a system of two first-order differential equations (SDE-2):

$$\begin{cases} y'_1 = f_1(x, y_1, y_2) \\ y'_2 = f_2(x, y_1, y_2) \end{cases}, \quad y_1(x_0) = y_1^0, \quad y_2(x_0) = y_2^0. \tag{3.23}$$

If $\frac{\partial f_1}{\partial y_2} = 0$, i.e if f_1 does not depend on y_2, then we can solve the first equation as an ODE-1 in the unknown function y_1 and find $y_1 = y_1(x, C_1)$. We replace this function y_1 in the second equation and find an ODE-1 in the unknown function y_2 only. By solving it we obtain $y_2 = y_2(x, C_1, C_2)$.

If $\frac{\partial f_1}{\partial y_2}(x_0) \neq 0$, assuming that $\frac{\partial f_1}{\partial y_2}$ is continuous, then there exists $h > 0$ s.t. $\frac{\partial f_1}{\partial y_2}(x) \neq 0$ for any $x \in (x_0 - h, x_0 + h)$. By the implicit function theorem (see [40], Theorem 17.8.1, or [28], Theorem 81) from the first equation we can write y_2 as a function of x, y_1 and y'_1 i.e.

$$y_2(x) = \varphi(x, y_1(x), y'_1(x)), \tag{3.24}$$

for any $x \in (x_0 - h, x_0 + h)$. Since

$$y'_2(x) = \frac{d}{dx}\left[\varphi(x, y_1(x), y'_1(x))\right] = \frac{\partial \varphi}{\partial x}(x, y_1(x), y'_1(x)) +$$

$$+ \frac{\partial \varphi}{\partial y_1}(x, y_1(x), y'_1(x)) \cdot y'_1(x) + \frac{\partial \varphi}{\partial y'_1}(x, y_1(x), y'_1(x)) \cdot y''_1(x),$$

by replacing in the second equation of (3.23), we obtain an ODE-2 in the unknown function y_1. We solve it and find $y_1 = y_1(x, C_1, C_2)$. Coming back in (3.24) we finally get $y_2 = y_2(x, C_1, C_2)$. The numerical values of the constants C_1, C_2 can be determined from the initial conditions.

3.2 First-Order Systems and Differential Equations of Order n

Example 3.2 Let us solve the Cauchy problem:

$$\begin{cases} y_1' = y_1 + y_2 - \exp(x) \\ y_2' = y_1 - y_2 + \exp(x) \end{cases}, \quad y_1(0) = 0, \ y_2(0) = 0, \ x \in \mathbb{R}_+.$$

A mechanical interpretation of the problem above would be to find the plane trajectory of a particle

$$\gamma : \mathbf{r}(t) = x(t)\mathbf{i} + y(t)\mathbf{j}, \ t \geq 0,$$

starting from the origin at time $t = 0$ and whose velocity is given by

$$\mathbf{r}'(t) = \big[x(t) + y(t) - \exp(t)\big]\mathbf{i} + \big[x(t) - y(t) + \exp(t)\big]\mathbf{j}.$$

From the first equation we get

$$y_2 = y_1' - y_1 + \exp(x). \tag{3.25}$$

So

$$y_2' = y_1'' - y_1' + \exp(x)$$

and, after introducing in the second equation these expression of y_2' and y_2, we find:

$$y_1'' - y_1 = -\exp(x),$$

which is an ODE-2 with constant coefficients. It is easy to find its general solution:

$$y_1(x) = C_1 \exp(x) + C_2 \exp(-x) - \frac{1}{2}x \exp(x)$$

and from (3.25) we obtain the second function:

$$y_2(x) = -2C_2 \exp(-x) + \frac{1}{2}\exp(x).$$

Now, the initial conditions imply:

$$\begin{cases} C_1 + C_2 = 0 \\ -2C_2 + \frac{1}{2} = 0 \end{cases},$$

thus $C_1 = -\frac{1}{4}$, $C_2 = \frac{1}{4}$. Hence, the solution of our Cauchy problem is:

$$\begin{cases} y_1(x) = -\frac{1}{4}\exp(x) + \frac{1}{4}\exp(-x) - \frac{1}{2}x \exp(x) \\ y_2(x) = -\frac{1}{2}\exp(-x) + \frac{1}{2}\exp(x) \end{cases}.$$

For systems with more than two equations, SDE-n with $n > 2$, the elimination method is more complicated. We describe the general algorithm only for $n = 3$. Let

$$\begin{cases} y'_1 = f_1(x, y_1, y_2, y_3) \\ y'_2 = f_2(x, y_1, y_2, y_3) \\ y'_3 = f_3(x, y_1, y_2, y_3) \end{cases} \quad (3.26)$$

be a system of three first-order differential equations (SDE-3). By differentiating the first equation with respect to x and substituting y'_1, y'_2, y'_3 with their expressions from (3.26), we obtain:

$$y''_1 = \frac{\partial f_1}{\partial x}(x, y_1, y_2, y_3) + \frac{\partial f_1}{\partial y_1}(x, y_1, y_2, y_3) \cdot y'_1 + \frac{\partial f_1}{\partial y_2}(x, y_1, y_2, y_3) \cdot y'_2 +$$

$$+ \frac{\partial f_1}{\partial y_3}(x, y_1, y_2, y_3) \cdot y'_3 = \frac{\partial f_1}{\partial x}(x, y_1, y_2, y_3) + \frac{\partial f_1}{\partial y_1}(x, y_1, y_2, y_3) \cdot f_1(x, y_1, y_2, y_3) +$$

$$+ \frac{\partial f_1}{\partial y_2}(x, y_1, y_2, y_3) \cdot f_2(x, y_1, y_2, y_3) + \frac{\partial f_1}{\partial y_3}(x, y_1, y_2, y_3) \cdot f_3(x, y_1, y_2, y_3).$$

We denote by $F_2(x, y_1, y_2, y_3)$ the expression above, so

$$y''_1 = F_2(x, y_1, y_2, y_3). \quad (3.27)$$

By differentiating this last equality with respect to x, we get, as above:

$$y'''_1 = F_3(x, y_1, y_2, y_3).$$

Hence the following system of implicit functions (see [40], 17.8, or [28], Theorem 83) is obtained:

$$\begin{cases} y'_1 = f_1(x, y_1, y_2, y_3) \\ y''_1 = F_2(x, y_1, y_2, y_3) \\ y'''_1 = F_3(x, y_1, y_2, y_3) \end{cases}. \quad (3.28)$$

Suppose that we have to solve a Cauchy problem for the system (3.26) with the initial conditions

$$y_1(x_0) = y_1^0, \quad y_2(x_0) = y_2^0, \quad y_3(x_0) = y_3^0$$

If the following condition

$$\frac{D(f_1, F_2)}{D(y_2, y_3)} = \begin{vmatrix} \frac{\partial f_1}{\partial y_2} & \frac{\partial f_1}{\partial y_3} \\ \frac{\partial F_2}{\partial y_2} & \frac{\partial F_2}{\partial y_3} \end{vmatrix} \neq 0$$

3.2 First-Order Systems and Differential Equations of Order n

is satisfied in a neighborhood of the point $M_0(x_0, y_1^0, y_2^0, y_3^0)$ then one can write

$$y_2(x) = \varphi_2(x, y_1, y_1', y_1''), \quad y_3(x) = \varphi_3(x, y_1, y_1', y_1''). \tag{3.29}$$

By substituting these expressions of $y_2(x)$ and $y_3(x)$ in the last equality of (3.28) we find

$$y_1''' = F_3(x, y_1, \varphi_2(x, y_1, y_1', y_1''), \varphi_3(x, y_1, y_1', y_1'')),$$

which is an ODE of order 3 in the unknown function y_1. After solving this equation we obtain the general solution:

$$y_1 = y_1(x, C_1, C_2, C_3). \tag{3.30}$$

Now, we substitute this last expression of the function y_1 in (3.29) and find $y_2 = y_2(x, C_1, C_2, C_3)$ and $y_3 = y_3(x, C_1, C_2, C_3)$, i.e. the general solution of the system (3.26).

Example 3.3 Use the elimination method to find the general solution of the system:

$$\begin{cases} y_1' = y_2 \\ y_2' = y_3 \\ y_3' = y_1 \end{cases} \tag{3.31}$$

We differentiate twice the first equation with respect to x:

$$y_1'' = y_2' \Rightarrow y_1'' = y_3$$

and then

$$y_1''' = y_3' \Rightarrow y_1''' = y_1.$$

Thus, we have obtained an ODE-3 in y_1. Its general solution is:

$$y_1(x) = C_1 \exp(x) + \exp\left(-\frac{1}{2}x\right)\left[C_2 \cos\left(\frac{\sqrt{3}}{2}x\right) + C_3 \sin\left(\frac{\sqrt{3}}{2}x\right)\right].$$

From the first and the second equations of (3.31) we obtain y_2, y_3 as functions of x, C_1, C_2, C_3. Thus we can find the general solution of the system.

Remark 3.2 Not always a SDE-n can be solved by using the elimination method. For instance, the system $\begin{cases} y_1' = y_1 \\ y_2' = y_2 \end{cases}$ cannot be solved by the above elimination method. Fortunately, in this case, the system can be broken into 2 independent first order differential equations which can be solved separately: $y_1(x) = C_1 \exp(x)$, $y_2(x) = C_2 \exp(x)$.

In Sect. 3.4.4 we present the general background for the elimination method applied to the first-order systems of *linear* differential equations with constant coefficients.

3.3 Linear Systems of Differential Equations

Solving systems of differential equations in the general form is a difficult task. However, there is a particular class of systems which can be systematically studied. We consider them in this section.

A function $E(x_1, \ldots, x_k; y_1, \ldots, y_n)$ is said to be *linear with respect to the variables* $y_1, \ldots, y_n$ if it can be written as:

$$E(x_1, \ldots, x_k; y_1, \ldots, y_n) = a_1(x_1, \ldots, x_k)y_1 + \ldots + a_n(x_1, \ldots, x_k)y_n + g(x_1, \ldots, x_k),$$

where $a_1, \ldots, a_n$ and g are functions of the independent variables $x_1, \ldots, x_k$.

In the case of the system of differential Eq. (3.4), if all the functions $f_i(x, y_1, \ldots, y_n)$ are linear with respect to $y_1, \ldots, y_n$, we say that the system is a *system of n first-order linear differential equations* (SLDE-n). Thus, a *linear system* has the following (normal) form:

$$\begin{cases} y_1'(x) = a_{11}(x)y_1 + \ldots + a_{1n}(x)y_n + g_1(x) \\ y_2'(x) = a_{21}(x)y_1 + \ldots + a_{2n}(x)y_n + g_2(x) \\ \quad \vdots \\ y_n'(x) = a_{n1}(x)y_1 + \ldots + a_{nn}(x)y_n + g_n(x) \end{cases}, x \in I, \qquad (3.32)$$

where we assume that $a_{ij}(x)$ and $g_i(x)$, $i = 1, \ldots, n$, $j = 1, \ldots, n$ are continuous functions on the interval I.

If $g_i(x) = 0$, for any $x \in I$, $i = 1, \ldots, n$, then we say that the system (3.32) is a *homogeneous* system of linear differential equations (HSLDE-n).

The system (3.32) can also be written in the matrix form

$$\mathbf{Y}' = \mathbf{A}\mathbf{Y} + \mathbf{G}, \qquad (3.33)$$

where

$$\mathbf{Y} = \mathbf{Y}(x) = \begin{pmatrix} y_1 \\ y_2 \\ \vdots \\ y_n \end{pmatrix} \qquad (3.34)$$

3.3 Linear Systems of Differential Equations

is the column vector of the unknown functions $y_1, \ldots, y_n$, (functions of class C^1 on I), $\mathbf{Y}' = \mathbf{Y}'(x)$ is the column vector of the derivatives, $y'_1, \ldots, y'_n$,

$$\mathbf{Y}' = \mathbf{Y}'(x) = \begin{pmatrix} y'_1 \\ y'_2 \\ \vdots \\ y'_n \end{pmatrix},$$

$$A = A(x) = \begin{pmatrix} a_{11} & a_{12} & \cdots & a_{1n} \\ a_{21} & a_{22} & \cdots & a_{2n} \\ \vdots & \vdots & \cdots & \vdots \\ a_{n1} & a_{n1} & \cdots & a_{nn} \end{pmatrix} \quad (3.35)$$

is an $n \times n$ matrix whose entries, $a_{ij} = a_{ij}(x)$, $i, j = 1, \ldots, n$, are continuous functions and

$$\mathbf{G} = \mathbf{G}(x) = \begin{pmatrix} g_1 \\ g_2 \\ \vdots \\ g_n \end{pmatrix} \quad (3.36)$$

is the column vector of the continuous functions $g_i(x)$, $i = 1, \ldots, n$ (the free terms of the equations).

Since the functions $f_i(x, y_1, \ldots, y_n) = a_{i1}(x)y_1 + \ldots + a_{in}(x)y_n + g_i(x)$ are continuous and their partial derivatives w.r.t. y_j, $\frac{\partial f_i}{\partial y_j} = a_{ij}(x)$, are also continuous (hence bounded on a closed rectangle R), the linear system (3.32) satisfies the conditions of Theorem 3.1 and the next theorem follows.

Theorem 3.2 (Existence and Uniqueness Theorem for Linear Systems of Equations) *If the functions $a_{ij}(x)$ and $g_i(x)$, $i = 1, \ldots, n$, $j = 1, \ldots, n$ are continuous on the interval I then, for any $x_0 \in I$, and $(y_1^0, \ldots, y_n^0) \in \mathbb{R}^n$, the linear system (3.32) has a unique solution $(y_1(x), \ldots, y_n(x))$ such that $y_i(x_0) = y_i^0$, for all $i = 1, \ldots, n$.*

Consider now the homogeneous system attached to the system (3.32),

$$\mathbf{Y}' = A\mathbf{Y}. \quad (3.37)$$

We denote by W_0 and W_1 the vector spaces

$$W_0 = \{\mathbf{Y} = (y_1(x), \ldots, y_n(x))^t \mid y_i : I \to \mathbb{R}, \; y_i \in C^0(I), \; i = 1, 2, \ldots, n\},$$

$$W_1 = \{\mathbf{Y} = (y_1(x), \ldots, y_n(x))^t \mid y_i : I \to \mathbb{R}, \; y_i \in C^1(I), \; i = 1, 2, \ldots, n\}.$$

Let $L : W_1 \to W_0$ be the differential operator $L = \dfrac{d}{dx} - A$:

$$L[\mathbf{Y}] \stackrel{def}{=} \frac{d\mathbf{Y}}{dx} - A\mathbf{Y} = \mathbf{Y}' - A\mathbf{Y} \qquad (3.38)$$

and let S be the set of all solutions of the homogeneous system (3.37):

$$S = \{\mathbf{Y} \in W_1 : \mathbf{Y}' = A\mathbf{Y}\}$$

Proposition 3.1 *The operator L is a linear operator and S is a real vector subspace of W_1, namely the kernel of L.*

Proof Let $\mathbf{Y}, \mathbf{Z} \in W_1$ be two column vectors of functions and $\alpha, \beta \in \mathbb{R}$. Since

$$L[\alpha\mathbf{Y} + \beta\mathbf{Z}] = (\alpha\mathbf{Y} + \beta\mathbf{Z})' - A(\alpha\mathbf{Y} + \beta\mathbf{Z}) =$$

$$= \alpha\left(\mathbf{Y}' - A\mathbf{Y}\right) + \beta\left(\mathbf{Z}' - A\mathbf{Z}\right) = \alpha L[\mathbf{Y}] + \beta L[\mathbf{Z}],$$

L is a linear operator. The kernel of L,

$$Ker L = \{\mathbf{Y} \in W_1 : L[\mathbf{Y}] = \mathbf{0}\},$$

is a vector subspace in W_1 and we can see that $Ker L = S$, the set of all solutions of the homogeneous system $\mathbf{Y}' = A\mathbf{Y}$. $\square$

Proposition 3.2 *The dimension of the subspace S is exactly n, i.e. there exists a basis $\{\mathbf{Y}_1, \ldots, \mathbf{Y}_n\}$ of solutions for S, which is called a **fundamental system of solutions** for the system $\mathbf{Y}' = A\mathbf{Y}$.*

Proof We fix $x_0 \in I$. For any solution of the homogeneous system, $\mathbf{Y} = (y_1, \ldots, y_n)^t \in S$, we denote by $T(\mathbf{Y}) \in \mathbb{R}^n$ the n-dimensional vector

$$T(\mathbf{Y}) = (y_1(x_0), y_2(x_0), \ldots, y_n(x_0))^t.$$

It can be readily seen that the function $T : S \to \mathbb{R}^n$ is a *linear operator*, because $T(\alpha\mathbf{Y} + \beta\mathbf{Z}) = \alpha T(\mathbf{Y}) + \beta T(\mathbf{Z})$. By Theorem 3.2, T is a bijective function, hence the vector spaces S and $\mathbb{R}^n$ are isomorphic. It follows that $\dim S = \dim \mathbb{R}^n = n$. $\square$

Example 3.4 Find a fundamental system of solutions for the homogeneous system

$$\begin{cases} y_1' = y_1 + 3y_2 \\ y_2' = y_1 - y_2 \end{cases}. \qquad (3.39)$$

We differentiate the first equation and obtain: $y_1'' = y_1' + 3y_2'$. By using the second equation one can write $y_1'' = y_1' + 3y_1 - 3y_2$. From the first equation, we have

3.3 Linear Systems of Differential Equations

$3y_2 = y_1' - y_1$, so we obtain

$$y_1'' - 4y_1 = 0,$$

which is a second order linear homogeneous equation with constant coefficients. The characteristic equation is $r^2 - 4 = 0$, so $r_1 = 2$ and $r_2 = -2$. Thus, the general solution is $y_1(x) = C_1 \exp(2x) + C_2 \exp(-2x)$. Now, from the equality $y_2 = \frac{1}{3}(y_1' - y_1)$, we finally get $y_2(x) = \frac{1}{3}C_1 \exp(2x) - C_2 \exp(-2x)$. Hence

$$\begin{pmatrix} y_1 \\ y_2 \end{pmatrix} = C_1 \begin{pmatrix} \exp(2x) \\ \frac{1}{3}\exp(2x) \end{pmatrix} + C_2 \begin{pmatrix} \exp(-2x) \\ -\exp(-2x) \end{pmatrix},$$

where C_1, C_2 are arbitrary real numbers, is the general solution of the system (3.39). It is easy to see that the vectors

$$\mathbf{Y}_1 = \begin{pmatrix} \exp(2x) \\ \frac{1}{3}\exp(2x) \end{pmatrix}, \quad \mathbf{Y}_2 = \begin{pmatrix} \exp(-2x) \\ -\exp(-2x) \end{pmatrix}$$

are linearly independent, hence they form a *fundamental system of solutions* for the homogeneous system (3.39).

Indeed, if we suppose that α_1, α_2 are real numbers such that $\alpha_1 \mathbf{Y}_1(x) + \alpha_2 \mathbf{Y}_2(x) = \mathbf{0}$, for any $x \in \mathbb{R}$, it follows that α_1, α_2 satisfy the equalities:

$$\begin{cases} \alpha_1 \exp(2x) + \alpha_2 \exp(-2x) = 0 \\ \frac{1}{3}\alpha_1 \exp(2x) - \alpha_2 \exp(-2x) = 0. \end{cases}$$

But, for a fixed x, this is a homogeneous system of algebraic equation, whose determinant is nonzero,

$$\begin{vmatrix} \exp(2x) & \exp(-2x) \\ \frac{1}{3}\exp(2x) & -\exp(-2x) \end{vmatrix} = -\frac{4}{3} \neq 0,$$

so the unique solution is $(\alpha_1, \alpha_2) = (0, 0)$, which means that the vectors $\mathbf{Y}_1, \mathbf{Y}_2$ are linearly independent.

In general, to decide if the vector functions $\mathbf{Y}_1, \ldots, \mathbf{Y}_n \in W_0$ (or W_1),

$$\mathbf{Y}_1 = \begin{pmatrix} y_{11} \\ \vdots \\ y_{n1} \end{pmatrix}, \ldots, \mathbf{Y}_n = \begin{pmatrix} y_{1n} \\ \vdots \\ y_{nn} \end{pmatrix}$$

are linearly independent, we need to compute their **Wronskian**.

Definition 3.1 Let I be a real interval and let $\mathbf{Y}_1(x) = \begin{pmatrix} y_{11}(x) \\ \vdots \\ y_{n1}(x) \end{pmatrix}, \ldots, \mathbf{Y}_n(x) = \begin{pmatrix} y_{1n}(x) \\ \vdots \\ y_{nn}(x) \end{pmatrix}$, $x \in I$, be n column vectors of functions. Then the matrix formed by these columns is called the **Wronski matrix** of $\mathbf{Y}_1, \ldots, \mathbf{Y}_n$ and its determinant,

$$W[\mathbf{Y}_1, \ldots, \mathbf{Y}_n](x) = \begin{vmatrix} y_{11}(x) & \cdots & y_{1n}(x) \\ \vdots & \vdots & \vdots \\ y_{n1}(x) & \cdots & y_{nn}(x) \end{vmatrix} \tag{3.40}$$

is called the **Wronskian** of $\mathbf{Y}_1, \ldots, \mathbf{Y}_n$.

Theorem 3.3 Let $\mathbf{Y}_1(x), \ldots, \mathbf{Y}_n(x)$ be solutions of the homogeneous system $\mathbf{Y}' = A\mathbf{Y}$. Then $\mathbf{Y}_1(x), \ldots, \mathbf{Y}_n(x)$ are linearly independent if and only if there exists a point $x_0 \in I$ such that $W[\mathbf{Y}_1, \ldots, \mathbf{Y}_n](x_0) \neq 0$. In particular, if the Wronskian of $\mathbf{Y}_1(x), \ldots, \mathbf{Y}_n(x)$ is nonzero at a point x_0, then it is nonzero at any other point $x \in I$.

Proof

a) Assume that $W[\mathbf{Y}_1, \ldots, \mathbf{Y}_n](x_0) \neq 0$ for a point $x_0 \in I$. We want to prove that $\mathbf{Y}_1(x), \ldots, \mathbf{Y}_n(x)$ are linearly independent. Suppose that they are linearly dependent. Then, there exist $\alpha_1, \ldots, \alpha_n$, n real numbers, not all zero, such that

$$\alpha_1 \mathbf{Y}_1(x) + \ldots + \alpha_n \mathbf{Y}_n(x) = \mathbf{0} = (0, 0, \ldots, 0)^t$$

for any $x \in I$, in particular for x_0. This means that

$$\alpha_1 \mathbf{Y}_1(x_0) + \ldots + \alpha_n \mathbf{Y}_n(x_0) = \mathbf{0},$$

i.e. the column vectors of the Wronski matrix are linearly dependent and so $W[\mathbf{Y}_1, \ldots, \mathbf{Y}_n](x_0) = 0$, a contradiction. This contradiction appears because we supposed that $\mathbf{Y}_1(x), \ldots, \mathbf{Y}_n(x)$ are linearly dependent. Thus, they are linearly independent.

b) Assume now that $\mathbf{Y}_1(x), \ldots, \mathbf{Y}_n(x)$ are linearly independent. Suppose that there exists a point $x_0 \in I$, such that $W[\mathbf{Y}_1, \ldots, \mathbf{Y}_n](x_0) = 0$. This means that the column vectors $\mathbf{Y}_1(x_0), \ldots, \mathbf{Y}_n(x_0)$ are linearly dependent, i.e. there exist $\beta_1, \ldots, \beta_n$, n real numbers, not all zero, such that

$$\beta_1 \mathbf{Y}_1(x_0) + \ldots + \beta_n \mathbf{Y}_n(x_0) = \mathbf{0}.$$

Let
$$Y(x) = \beta_1 Y_1(x) + \ldots + \beta_n Y_n(x), \quad x \in I$$

be a new solution of the system $Y' = AY$. Since $Y(x_0) = 0$ and since the zero-solution $Z(x) = 0$, for any $x \in I$, satisfies the same Cauchy condition, by Theorem 3.2 we obtain that $Y = Z$, i.e.

$$Y(x) = \beta_1 Y_1(x) + \ldots + \beta_n Y_n(x) = 0,$$

for any $x \in I$. This means that $Y_1(x), \ldots, Y_n(x)$ are linearly dependent, a contradiction! Thus, for any $x \in I$, $W[Y_1, \ldots, Y_n](x) \neq 0$. The last statement easily follows from a) and b).

□

Remark 3.3 For a) we did not use the fact that $Y_1(x), \ldots, Y_n(x)$ are solutions of the homogeneous system $Y' = AY$. Thus, if we have a sequence $Z_1(x), \ldots, Z_n(x)$ of n column vectors of arbitrary functions such that $W[Z_1, \ldots, Z_n](x_0) \neq 0$ for a point $x_0 \in I$, then these column vectors of functions are linearly independent. The reciprocal of this statement is not true in general (if they are not solutions of a homogeneous system $Y' = AY$). For instance, $Z_1(x) = \begin{pmatrix} 1 \\ 0 \end{pmatrix}$ and $Z_2(x) = \begin{pmatrix} x \\ 0 \end{pmatrix}$, $x \in (a, b)$ are linearly independent as column vectors of functions. Indeed,

$$\alpha_1 Z_1(x) + \alpha_2 Z_2(x) = 0, \quad x \in (a, b),$$

is equivalent to the equality $\alpha_1 + \alpha_2 x = 0$ for any $x \in (a, b)$. If $\alpha_2 \neq 0$, the polynomial $\alpha_1 + \alpha_2 x = 0$ of degree 1 has only one solution, so we must have $\alpha_1 = \alpha_2 = 0$, i.e. $Z_1(x)$ and $Z_2(x)$ are linearly independent. However, $W[Z_1, Z_2](x) = 0$, for any $x \in (a, b)$. A strange conclusion is that $Z_1(x) = \begin{pmatrix} 1 \\ 0 \end{pmatrix}$, $Z_2(x) = \begin{pmatrix} x \\ 0 \end{pmatrix}$ cannot be solutions of the same SLDE-2, which can be directly checked.

Proposition 3.3 (Liouville's Formula) *Let $Y_1, \ldots, Y_n$ be a set of n solutions for the homogeneous system $Y' = AY$. Then, for any $x \in I$,*

$$W[Y_1, \ldots, Y_n](x) = W[Y_1, \ldots, Y_n](x_0) \cdot \exp\left(\int_{x_0}^{x} \left(\sum_{i=1}^{n} a_{ii}(x)\right)\right), \quad (3.41)$$

Proof If $Y_1, \ldots, Y_n$ are linearly dependent, their Wronskian is equal to 0 at any $x \in I$ and (3.41) follows immediately.

If they are linearly independent, by Theorem 3.3 their Wronskian is nonzero at any $x \in I$. We present here the proof of the formula (3.41) for $n = 2$. It is not difficult to see that the proof can be extended to the general case, $n \geq 2$.

A determinant of functions can be differentiated by using the following formula:

$$\begin{vmatrix} y_{11}(x) & y_{12}(x) \\ y_{21}(x) & y_{22}(x) \end{vmatrix}' = \begin{vmatrix} y'_{11}(x) & y'_{12}(x) \\ y_{21}(x) & y_{22}(x) \end{vmatrix} + \begin{vmatrix} y_{11}(x) & y_{12}(x) \\ y'_{21}(x) & y'_{22}(x) \end{vmatrix}. \tag{3.42}$$

Now, we use the fact that $\mathbf{Y}_1(x) = \begin{pmatrix} y_{11}(x) \\ y_{21}(x) \end{pmatrix}$, $\mathbf{Y}_2(x) = \begin{pmatrix} y_{12}(x) \\ y_{22}(x) \end{pmatrix}$ are solutions of the system $\mathbf{Y}' = A\mathbf{Y}$, so we have:

$$\begin{vmatrix} y'_{11} & y'_{12} \\ y_{21} & y_{22} \end{vmatrix} = \begin{vmatrix} a_{11}y_{11} + a_{12}y_{21} & a_{11}y_{12} + a_{12}y_{22} \\ y_{21} & y_{22} \end{vmatrix} = a_{11}W[\mathbf{Y}_1, \mathbf{Y}_2]$$

and

$$\begin{vmatrix} y_{11} & y_{12} \\ y'_{21} & y'_{22} \end{vmatrix} = \begin{vmatrix} y_{11} & y_{12} \\ a_{21}y_{11} + a_{22}y_{21} & a_{21}y_{12} + a_{22}y_{22} \end{vmatrix} = a_{22}W[\mathbf{Y}_1, \mathbf{Y}_2].$$

Hence,

$$\frac{d}{dx}(W[\mathbf{Y}_1, \mathbf{Y}_2]) = (a_{11} + a_{22})W[\mathbf{Y}_1, \mathbf{Y}_2].$$

Since $W[\mathbf{Y}_1, \mathbf{Y}_2](x) \neq 0$, for any $x \in I$, we can write

$$\frac{W[\mathbf{Y}_1, \mathbf{Y}_2]'(x)}{W[\mathbf{Y}_1, \mathbf{Y}_2](x)} = [a_{11}(x) + a_{22}(x)],$$

and, by integrating from x_0 to x on both sides, we get:

$$\ln \frac{W[\mathbf{Y}_1, \mathbf{Y}_2](x)}{W[\mathbf{Y}_1, \mathbf{Y}_2](x_0)} = \int_{x_0}^{x} [a_{11}(x) + a_{22}(x)]dx,$$

or

$$W[\mathbf{Y}_1, \mathbf{Y}_2](x) = W[\mathbf{Y}_1, \mathbf{Y}_2](x_0) \exp\left(\int_{x_0}^{x} [a_{11}(x) + a_{22}(x)]dx\right),$$

which is exactly Liouville's formula for the particular case $n = 2$. □

In general, it is not easy to find a fundamental system of solutions for a homogeneous system $\mathbf{Y}' = A\mathbf{Y}$, where A is a matrix of arbitrary functions. In the next section we describe an algorithm for finding a fundamental system of solutions in the case when all the coefficients a_{ij} are constants.

But, if we know a fundamental system of solutions for the homogeneous system $\mathbf{Y}' = A\mathbf{Y}$, theoretically we can find the general solution for the non-homogeneous system (3.33): $\mathbf{Y}' = A\mathbf{Y} + \mathbf{G}$ as well.

3.3 Linear Systems of Differential Equations

Theorem 3.4 *Let $\{Y_1, \ldots, Y_n\}$ be a fundamental system of solutions (a basis of the vector space S) for the homogeneous system $Y' = AY$ and let Y_P be a particular solution of the non-homogeneous system $Y' = AY + G$. Then, the general solution for the non-homogeneous system is*

$$Y = C_1 Y_1 + \ldots + C_n Y_n + Y_P, \tag{3.43}$$

where $C_1, \ldots, C_n$ are arbitrary constants.

Proof It is sufficient to prove that, if Y is an arbitrary solution of the non-homogeneous system, then $Y - Y_P$ is a solution of the homogeneous system. Indeed, since $Y' = AY + G$ and $Y'_P = AY_P + G$ we have:

$$(Y - Y_P)' = Y' - Y'_P = A(Y - Y_P),$$

so $Y - Y_P \in S$. Since $\{Y_1, \ldots, Y_n\}$ is a basis of S, there exist n uniquely determined numbers $C_1, \ldots, C_n$ such that

$$Y - Y_P = C_1 Y_1 + \ldots + C_n Y_n,$$

and the formula (3.43) follows. □

Thus, in order to solve the non-homogeneous system $Y' = AY + G$ we need two things: (1) a fundamental system of solutions $\{Y_1, \ldots, Y_n\}$ for the associated homogeneous system $Y' = AY$, and (2) a particular solution Y_P of the non-homogeneous system. $Y' = AY + G$. The next theorem states that, if we know a fundamental system of solutions (or a basis of S) $\{Y_1, \ldots, Y_n\}$, we can use it to find a particular solution Y_P of the non-homogeneous system, $Y' = AY + G$.

Theorem 3.5 (Lagrange's Method of Variation of Constants) *Let $\{Y_1, \ldots, Y_n\}$ be a fundamental system of solutions for $Y' = AY$ ($Y_i(x) = (y_{1i}(x), \ldots, y_{ni}(x))^t$, $x \in I$, $i = 1, 2, \ldots, n$) and let $\varphi_1(x), \ldots, \varphi_n(x)$ be n continuously differentiable functions defined on the interval I, such that*

$$\varphi'_1(x) \cdot Y_1(x) + \varphi'_2(x) \cdot Y_2(x) + \ldots + \varphi'_n(x) \cdot Y_n(x) = G. \tag{3.44}$$

Then the vector of functions

$$Y_P = \varphi_1(x) \cdot Y_1(x) + \ldots + \varphi_n(x) \cdot Y_n(x) \tag{3.45}$$

is a particular solution of the non-homogeneous system $Y' = AY + G$.

Proof The basic rules for derivatives can be extended to the derivatives of vector functions. The derivative of such a vector function is the vector of the derivatives of its corresponding components. Let us compute the derivative of the vector function defined by (3.45):

$$\mathbf{Y}'_P = \varphi'_1(x) \cdot \mathbf{Y}_1(x) + \ldots + \varphi'_n(x) \cdot \mathbf{Y}_n(x) +$$
$$+ \varphi_1(x) \cdot \mathbf{Y}'_1(x) + \ldots + \varphi_n(x) \cdot \mathbf{Y}'_n(x) =$$
$$= \mathbf{G} + \varphi_1(x) A \mathbf{Y}_1 + \ldots + \varphi_n(x) A \mathbf{Y}_n =$$
$$= \mathbf{G} + A [\varphi_1(x) \mathbf{Y}_1 + \ldots + \varphi_n(x) \mathbf{Y}_n] = A \mathbf{Y}_P + \mathbf{G}.$$

So $\mathbf{Y}_P$ as constructed in (3.45) is a particular solution of the non-homogeneous system $\mathbf{Y}' = A\mathbf{Y} + \mathbf{G}$. □

The method of finding a particular solution of the non-homogeneous system described in the proof of Theorem 3.5 is called *Lagrange's method of variation of constants* because the formula (3.45) comes from the general solution of the homogeneous system, $\mathbf{Y} = C_1 \mathbf{Y}_1 + \ldots + C_n \mathbf{Y}_n$, by replacing the constants $C_1, C_2, \ldots, C_n$ with functions.

Example 3.5 Find the general solution of the non-homogeneous system of linear differential equations:

$$\begin{cases} y'_1 = \qquad y_3 \quad +1 \\ y'_2 = \quad y_2 + \quad x \\ y'_3 = y_1 \qquad + \exp(2x) \end{cases} \tag{3.46}$$

by using the Lagrange method. Then, find the solution $(y_1(x), y_2(x), y_3(x))$ which satisfies: $y_1(0) = 0$, $y_2(0) = 0$, $y_3(0) = 0$.
First of all let us find a basis of solutions for the associated homogeneous system:

$$\begin{cases} y'_1 = y_3 \\ y'_2 = y_2 \\ y'_3 = y_1 \end{cases}. \tag{3.47}$$

The matrix A in this case is:

$$A = \begin{pmatrix} 0 & 0 & 1 \\ 0 & 1 & 0 \\ 1 & 0 & 0 \end{pmatrix}$$

and the system (3.47) can be written in the matrix form:

$$\mathbf{Y}' = A\mathbf{Y}, \tag{3.48}$$

3.3 Linear Systems of Differential Equations

where $\mathbf{Y} = \begin{pmatrix} y_1 \\ y_2 \\ y_3 \end{pmatrix}$. Let us search for a solution of the form

$$\mathbf{Y}(x) = \begin{pmatrix} c_1 \\ c_2 \\ c_3 \end{pmatrix} \exp(\lambda x), \tag{3.49}$$

where $\begin{pmatrix} c_1 \\ c_2 \\ c_3 \end{pmatrix}$ is a numerical vector which will be determined. Since $\mathbf{Y}(x)$ must verify the Eq. (3.48), we have:

$$\lambda \begin{pmatrix} c_1 \\ c_2 \\ c_3 \end{pmatrix} \exp(\lambda x) = A \begin{pmatrix} c_1 \\ c_2 \\ c_3 \end{pmatrix} \exp(\lambda x),$$

or, equivalently,

$$(A - \lambda I_3) \begin{pmatrix} c_1 \\ c_2 \\ c_3 \end{pmatrix} = \begin{pmatrix} 0 \\ 0 \\ 0 \end{pmatrix}, \tag{3.50}$$

where I_3 is the 3×3 identity matrix. In order to have a nontrivial solution $\mathbf{Y}(x)$, we need a nontrivial column vector $(c_1, c_2, c_3)^t$, i.e. λ must be an eigenvalue for the matrix A and $(c_1, c_2, c_3)^t$, a corresponding eigenvector of it. It is easy to find the distinct eigenvalues and their corresponding eigenspaces: $\lambda_1 = \lambda_2 = 1$ with $V_1 = \{(\alpha, \beta, \alpha) : \alpha, \beta \in \mathbb{R}\}$ and $\lambda_3 = -1$ with $V_{-1} = \{(\gamma, 0, -\gamma) : \gamma \in \mathbb{R}\}$. Take now three eigenvectors which are linearly independent (a basis of eigenvectors in $\mathbb{R}^3$):

$$\mathbf{v}_1 = \begin{pmatrix} 1 \\ 0 \\ 1 \end{pmatrix}, \quad \mathbf{v}_2 = \begin{pmatrix} 0 \\ 1 \\ 0 \end{pmatrix}, \quad \mathbf{v}_3 = \begin{pmatrix} 1 \\ 0 \\ -1 \end{pmatrix}$$

and construct the vector functions

$$\mathbf{Y}_1 = \mathbf{v}_1 \exp(x), \quad \mathbf{Y}_2 = \mathbf{v}_2 \exp(x), \quad \mathbf{Y}_3 = \mathbf{v}_3 \exp(-x).$$

Since $W[\mathbf{Y}_1, \mathbf{Y}_2, \mathbf{Y}_3](x) = -2\exp(x) \neq 0$ for any $x \in \mathbb{R}$, we can see that $\{\mathbf{Y}_1, \mathbf{Y}_2, \mathbf{Y}_3\}$ is a basis of solutions for the homogeneous system $\mathbf{Y}' = A\mathbf{Y}$.

It remains now to find a particular solution for the initial system (3.46). We use the Lagrange method described above. In our case, the relation (3.44) is written:

$$\begin{pmatrix} \exp(x) & 0 & \exp(-x) \\ 0 & \exp(x) & 0 \\ \exp(x) & 0 & -\exp(-x) \end{pmatrix} \begin{pmatrix} \varphi_1' \\ \varphi_2' \\ \varphi_3' \end{pmatrix} = \begin{pmatrix} 1 \\ x \\ \exp(2x) \end{pmatrix} \qquad (3.51)$$

This is an algebraic linear system in the unknowns φ_1', φ_2', φ_3'. We solve it and then, after a simple integration, we find: $\varphi_1 = \frac{1}{2}\left[\exp(x) - \exp(-x)\right]$, $\varphi_2 = -x\exp(-x) - \exp(-x)$, $\varphi_3 = \frac{1}{2}\left[\exp(x) - \frac{1}{3}\exp(3x)\right]$. Thus, from formula (3.45), we get:

$$\mathbf{Y}_P = \frac{1}{2}\mathbf{v}_1\left[\exp(2x) - 1\right] - \mathbf{v}_2(x+1) + \frac{1}{2}\mathbf{v}_3\left[1 - \frac{1}{3}\exp(2x)\right] =$$

$$= \begin{pmatrix} \frac{1}{3}\exp(2x) \\ -x - 1 \\ \frac{2}{3}\exp(2x) - 1 \end{pmatrix}.$$

Hence, the general solution of the initial non-homogeneous system (3.46) is:

$$\mathbf{Y} = C_1\mathbf{v}_1 \exp(x) + C_2\mathbf{v}_2 \exp(x) + C_3\mathbf{v}_3 \exp(-x) + \mathbf{Y}_P =$$

$$= \begin{pmatrix} C_1 \exp(x) + C_3 \exp(-x) + \frac{1}{3}\exp(2x) \\ C_2 \exp(x) - x - 1 \\ C_1 \exp(x) - C_3 \exp(-x) + \frac{2}{3}\exp(2x) - 1 \end{pmatrix},$$

where C_1, C_2, C_3 are arbitrary real constants. We can also write this general solution in a non-vector form:

$$\begin{cases} y_1(x) = C_1 \exp(x) + C_3 \exp(-x) + \frac{1}{3}\exp(2x) \\ y_2(x) = C_2 \exp(x) - x - 1 \\ y_3(x) = C_1 \exp(x) - C_3 \exp(-x) + \frac{2}{3}\exp(2x) - 1. \end{cases} \qquad (3.52)$$

To solve the Cauchy problem, we write (3.52) for $x = 0$ and solve a linear algebraic system in the unknowns C_1, C_2, C_3:

$$\begin{cases} C_1 + C_3 + \frac{1}{3} = 0 \\ C_2 - 1 = 0 \\ C_1 - C_3 + \frac{2}{3} - 1 = 0. \end{cases}$$

Thus, $C_1 = 0$, $C_2 = 1$ and $C_3 = -\frac{1}{3}$. Coming back in (3.52) we finally find the solution of the Cauchy problem:

$$\begin{cases} y_1(x) = -\frac{1}{3}\exp(-x) + \frac{1}{3}\exp(2x) \\ y_2(x) = \exp(x) - x - 1 \\ y_3(x) = \frac{1}{3}\exp(-x) + \frac{2}{3}\exp(2x) - 1. \end{cases}$$

By this example, in addition to illustrating Lagrange's method for finding a particular solution, $\mathbf{Y}_P$, we also tried to introduce the so-called "algebraic method" for solving homogeneous systems of equations. This method works for systems with *constant coefficients* and will be presented in detail in the next section.

3.4 Linear Systems with Constant Coefficients

When all the coefficients a_{ij} of the system (3.32) are real constants, we say that it is a *linear system with constant coefficients*:

$$\begin{cases} y_1'(x) = a_{11}y_1(x) + \ldots + a_{1n}y_n(x) + g_1(x) \\ y_2'(x) = a_{21}y_1(x) + \ldots + a_{2n}y_n(x) + g_2(x) \\ \vdots \\ y_n'(x) = a_{n1}y_1(x) + \ldots + a_{nn}y_n(x) + g_n(x) \end{cases}, \quad (3.53)$$

which can also be written in the matrix form

$$\mathbf{Y}' = A\mathbf{Y} + \mathbf{G}, \quad (3.54)$$

where $A = (a_{ij})$ is an $n \times n$ real matrix and $\mathbf{G} = \mathbf{G}(x)$ is a vector of (continuous) functions.

3.4.1 The Homogeneous Case (the Algebraic Method)

Let

$$\mathbf{Y}' = A\mathbf{Y} \quad (3.55)$$

be a *homogeneous* system of linear differential equations with constant coefficients. In order to find the general solution of this system it is sufficient to find a fundamental system of solutions $\{\mathbf{Y}_1, \ldots, \mathbf{Y}_n\}$ (a basis of the vector space of solutions S). As in Example 3.5, we search for solutions of the type

$$\mathbf{Y} = \mathbf{v}\exp(\lambda x) = \begin{pmatrix} c_1 \\ c_2 \\ \vdots \\ c_n \end{pmatrix} \exp(\lambda x), \tag{3.56}$$

where $\mathbf{v} = (c_1, \ldots, c_n)^t$ is a (column) vector of real or complex numbers, not all zero (the trivial solution $\mathbf{Y} = \mathbf{0}$ is of no interest) and λ is also a real or complex number. Since

$$\mathbf{Y}' = \mathbf{v}\lambda \exp(\lambda x),$$

by replacing in (3.56) we obtain:

$$\mathbf{v}\lambda \exp(\lambda x) = A\mathbf{v}\exp(\lambda x).$$

After simplifying the equality with $\exp(\lambda x)$ and rearranging we find:

$$A\mathbf{v} - \lambda\mathbf{v} = \mathbf{0}.$$

Since $\mathbf{v} = I_n\mathbf{v}$, we obtain that (3.56) is a solution of the homogeneous system (3.55) if and only if

$$(A - \lambda I_n)\mathbf{v} = \mathbf{0}. \tag{3.57}$$

The matrix Eq. (3.57) is equivalent to the following *homogeneous* system of linear algebraic equations (in the unknowns $c_1, \ldots, c_n$):

$$\begin{cases} (a_{11} - \lambda)c_1 + a_{12}c_2 + \ldots + a_{1n}c_n = 0 \\ a_{21}c_1 + (a_{22} - \lambda)c_2 + \ldots + a_{2n}c_n = 0 \\ \quad\quad\quad\quad\quad\quad \vdots \\ a_{n1}c_1 + a_{n2}c_2 + \ldots + (a_{nn} - \lambda)c_n = 0 \end{cases} \tag{3.58}$$

The linear system (3.58) has nontrivial solutions $\mathbf{v} = (c_1, \ldots, c_n)^t \neq \mathbf{0}$ if and only if its determinant is equal to 0, which means that λ must verify the equation:

$$\det(A - \lambda I_n) = 0 \tag{3.59}$$

The polynomial Eq. (3.59) is called the *characteristic equation* of the matrix A and its solutions are the *eigenvalues* of the matrix A. A solution $\mathbf{v} \neq \mathbf{0}$ of the system (3.57) corresponding to an eigenvalue λ is an *eigenvector* of A (see [35] Ch.4 or [18] Ch.6).

3.4 Linear Systems with Constant Coefficients

The characteristic polynomial $P(\lambda) = \det(A - \lambda I_n)$ is a polynomial of degree n, so it has exactly n roots (real or complex, distinct or repeated), $\lambda_1, \lambda_2, \ldots, \lambda_n$.

Proposition 3.4 *If there exist n **linearly independent** eigenvectors of A, $\mathbf{v}_1, \mathbf{v}_2, \ldots, \mathbf{v}_n$, ($\mathbf{v}_i = (c_{1i}, \ldots, c_{ni})^t$, $i = 1, \ldots, n$) corresponding to the eigenvalues $\lambda_1, \lambda_2, \ldots, \lambda_n$ respectively, then the vector functions*

$$\mathbf{Y}_i = \exp(\lambda_i x)\mathbf{v}_i = \exp(\lambda_i x)\begin{pmatrix} c_{1i} \\ c_{2i} \\ \vdots \\ c_{ni} \end{pmatrix}, \quad i = 1, \ldots, n \tag{3.60}$$

are linearly independent solutions of the homogeneous system (3.55).

Proof It is sufficient to prove that the Wronskian $W[\mathbf{Y}_1, \ldots, \mathbf{Y}_n](x)$ is nonzero at least at one point $x_0 \in \mathbb{R}$. Indeed,

$$W[\mathbf{Y}_1, \ldots, \mathbf{Y}_n](x) = \begin{vmatrix} c_{11} & c_{12} & \cdots & c_{1n} \\ c_{21} & c_{22} & \cdots & c_{2n} \\ \vdots & \vdots & \vdots & \vdots \\ c_{n1} & c_{n2} & \cdots & c_{nn} \end{vmatrix} \exp(\lambda_1 x + \ldots + \lambda_n x) \neq 0$$

since the columns of the determinant are linearly independent vectors. □

Remark 3.4 If all the eigenvalues $\lambda_1, \ldots, \lambda_n$ are *real* numbers, then the eigenvectors $\mathbf{v}_1, \ldots, \mathbf{v}_n$ are also real, so (3.60) is a *fundamental system of solutions* of (3.55).

Remark 3.5 Suppose that not all the eigenvalues are real numbers. Let $\lambda_1 = \alpha + i\beta$ be a complex root of the characteristic Eq. (3.59). Then $\lambda_2 = \overline{\lambda_1} = \alpha - i\beta$ is also an eigenvalue of A. Moreover, if

$$\mathbf{v}_1 = \begin{pmatrix} u_1 + i w_1 \\ \vdots \\ u_n + i w_n \end{pmatrix} = \begin{pmatrix} u_1 \\ \vdots \\ u_n \end{pmatrix} + i \begin{pmatrix} w_1 \\ \vdots \\ w_n \end{pmatrix} = \mathbf{u} + i\mathbf{w}$$

is a (complex) eigenvector corresponding to λ_1, i.e.

$$(A - \lambda_1 I_n)\mathbf{v}_1 = \mathbf{0},$$

then $(A - \overline{\lambda_1} I_n)\overline{\mathbf{v}_1} = \mathbf{0}$, hence $\mathbf{v}_2 = \overline{\mathbf{v}_1} = \mathbf{u} - i\mathbf{w}$ is an eigenvector corresponding to λ_2. Thus, the *complex* (conjugate) vector functions

$$\mathbf{Y}_1 = \exp((\alpha + i\beta)x)\mathbf{v}_1 = \exp(\alpha x)(\cos \beta x + i \sin \beta x)(\mathbf{u} + i\mathbf{w})$$

and

$$\mathbf{Y}_2 = \overline{\mathbf{Y}_1} = \exp((\alpha - i\beta)x)\overline{\mathbf{v}_1} = \exp(\alpha x)(\cos \beta x - i \sin \beta x)(\mathbf{u} - i\mathbf{w})$$

are linearly independent solutions of the system (3.55). Their real part, $\mathbf{y}_1 = \text{Re}\mathbf{Y}_1$, and imaginary part, $\mathbf{y}_2 = \text{Im}\mathbf{Y}_1$,

$$\begin{aligned}\mathbf{y}_1 &= \exp(\alpha x)(\mathbf{u}\cos\beta x - \mathbf{w}\sin\beta x), \\ \mathbf{y}_2 &= \exp(\alpha x)(\mathbf{w}\cos\beta x + \mathbf{u}\sin\beta x),\end{aligned} \qquad (3.61)$$

are *real-valued* solutions of (3.55), also linearly independent. Therefore, in order to obtain a fundamental system of *real-valued* solutions, we replace the complex functions $\mathbf{Y}_1$ and $\mathbf{Y}_2$ with the real functions $\mathbf{y}_1$ and $\mathbf{y}_2$ defined by (3.61).

Recall that an $n \times n$ matrix which has n linearly independent eigenvectors is said to be *diagonalizable*. Let

$$\mathbf{v}_1 = \begin{pmatrix} c_{11} \\ c_{21} \\ \vdots \\ c_{n1} \end{pmatrix}, \ldots, \mathbf{v}_n = \begin{pmatrix} c_{1n} \\ c_{2n} \\ \vdots \\ c_{nn} \end{pmatrix}$$

be n linearly independent eigenvectors of A corresponding to the eigenvalues $\lambda_1, \ldots, \lambda_n$ (real or complex numbers, distinct or repeated) and let C be the matrix formed with the columns $\mathbf{v}_1, \ldots, \mathbf{v}_n$,

$$C = \begin{pmatrix} c_{11} & c_{12} & \cdots & c_{1n} \\ c_{21} & c_{22} & \cdots & c_{2n} \\ \vdots & \vdots & \vdots & \vdots \\ c_{n1} & c_{n2} & \cdots & c_{nn} \end{pmatrix}. \qquad (3.62)$$

Then C is invertible and $C^{-1}AC$ is a diagonal matrix (having the eigenvalues $\lambda_1, \ldots, \lambda_n$ on the main diagonal:

$$C^{-1}AC = \begin{pmatrix} \lambda_1 & 0 & \cdots & 0 \\ 0 & \lambda_2 & 0 \cdots & 0 \\ \vdots & \vdots & \ddots & \vdots \\ 0 & 0 & \cdots & \lambda_n \end{pmatrix} \stackrel{def}{=} \text{diag}(\lambda_1, \ldots, \lambda_n). \qquad (3.63)$$

We have shown above how to find a fundamental system of solutions of the system $\mathbf{Y}' = A\mathbf{Y}$ if A is a diagonalizable matrix. But how can we decide if a matrix is diagonalizable or not?

3.4 Linear Systems with Constant Coefficients

If all the eigenvalues are distinct numbers (real or complex), then A is diagonalizable, since *eigenvectors corresponding to different eigenvalues are linearly independent*.

Now, suppose that the characteristic polynomial of A has multiple roots,

$$P(\lambda) = \det(A - \lambda I_n) = (\lambda - \lambda_1)^{m_1} \ldots (\lambda - \lambda_r)^{m_r},$$

where $\lambda_1, \ldots, \lambda_r$ are the distinct roots of multiplicities $m_1, \ldots, m_r$ ($m_1 + \ldots + m_r = n$). For each eigenvalue λ_i, $i = 1, \ldots, r$, we denote by V_{λ_i} the set of all the corresponding eigenvectors:

$$V_{\lambda_i} = \{\mathbf{v} : (A - \lambda_i I_n)\mathbf{v} = \mathbf{0}\}.$$

As can be easily seen, V_{λ_i} is a vector subspace (the *eigenspace* of λ_i). The dimension of the eigenspace V_{λ_i} (the "geometric multiplicity" of λ_i) is always less than or equal to m_i (the "algebraic multiplicity"). If they are equal for any eigenvalue of A,

$$\dim V_{\lambda_i} = m_i \tag{3.64}$$

then we can find m_i linearly independent vectors corresponding to λ_i for any $i = 1, \ldots, r$. Since *eigenvectors corresponding to different eigenvalues are linearly independent* and $m_1 + \ldots + m_r = n$, we can find n linearly independent eigenvectors of A, so the matrix is diagonalizable.

Example 3.6 Solve the following Cauchy problem:

$$\frac{dx_1}{dt} = \frac{dx_2}{dt} = \frac{dx_3}{dt} = \frac{dx_4}{dt} = x_1 + x_2 + x_3 + x_4, \tag{3.65}$$

$x_1(0) = x_2(0) = x_3(0) = 0$, $x_4(0) = 4$.
Here $x_i = x_i(t)$, $i = 1, \ldots, 4$ and the matrix of the linear homogeneous system with constant coefficients is:

$$A = \begin{pmatrix} 1 & 1 & 1 & 1 \\ 1 & 1 & 1 & 1 \\ 1 & 1 & 1 & 1 \\ 1 & 1 & 1 & 1 \end{pmatrix}.$$

The characteristic equation is

$$\det(A - \lambda I_4) = \begin{vmatrix} 1-\lambda & 1 & 1 & 1 \\ 1 & 1-\lambda & 1 & 1 \\ 1 & 1 & 1-\lambda & 1 \\ 1 & 1 & 1 & 1-\lambda \end{vmatrix} = \lambda^4 - 4\lambda^3 = 0.$$

The eigenvalues and the corresponding eigenspaces of A are:

$$\lambda_1 = \lambda_2 = \lambda_3 = 0, \quad V_0 = \{(\alpha, \beta, \gamma, -\alpha - \beta - \gamma)^t : \alpha, \beta, \gamma \in \mathbb{R}\} \quad (3.66)$$

and

$$\lambda_4 = 4, \quad V_4 = \{(\delta, \delta, \delta, \delta)^t : \delta \in \mathbb{R}\}. \quad (3.67)$$

The vectors $\mathbf{v}_1 = (1, 0, 0, -1)^t$, $\mathbf{v}_2 = (0, 1, 0 - 1)^t$ and $\mathbf{v}_3 = (0, 0, 1 - 1)^t$ form a basis of the space V_0, while $\{\mathbf{v}_4 = (1, 1, 1, 1)^t\}$ is a basis of the space V_4. By Proposition 3.4 and Remark 3.4, a fundamental system of solutions for the system (3.65) is:

$$\mathbf{X}_1 = \begin{pmatrix} 1 \\ 0 \\ 0 \\ -1 \end{pmatrix}, \quad \mathbf{X}_2 = \begin{pmatrix} 0 \\ 1 \\ 0 \\ -1 \end{pmatrix}, \quad \mathbf{X}_3 = \begin{pmatrix} 0 \\ 0 \\ 1 \\ -1 \end{pmatrix}, \quad \mathbf{X}_4 = \begin{pmatrix} 1 \\ 1 \\ 1 \\ 1 \end{pmatrix} \exp(4t).$$

So, any solution of the system is of the form:

$$\mathbf{X}(t) = \begin{pmatrix} x_1(t) \\ x_2(t) \\ x_3(t) \\ x_4(t) \end{pmatrix} = C_1 \begin{pmatrix} 1 \\ 0 \\ 0 \\ -1 \end{pmatrix} + C_2 \begin{pmatrix} 0 \\ 1 \\ 0 \\ -1 \end{pmatrix} +$$

$$+ C_3 \begin{pmatrix} 0 \\ 0 \\ 1 \\ -1 \end{pmatrix} + C_4 \begin{pmatrix} 1 \\ 1 \\ 1 \\ 1 \end{pmatrix} \exp(4t),$$

or, in the expanded (non-vector) form:

$$\begin{cases} x_1(t) = C_1 + \quad\quad\quad\quad\quad\quad C_4 \exp(4t) \\ x_2(t) = \quad\quad C_2 + \quad\quad\quad\quad C_4 \exp(4t) \\ x_3(t) = \quad\quad\quad\quad C_3 + C_4 \exp(4t) \\ x_4(t) = -C_1 - C_2 - C_3 + C_4 \exp(4t) \end{cases}, t \in \mathbb{R} \quad (3.68)$$

where $C_1, \ldots, C_4$ are arbitrary real numbers. From the initial conditions we obtain a linear algebraic system in the variables C_1, C_2, C_3 and C_4:

$$\begin{cases} C_1 + \quad\quad\quad\quad\quad\quad C_4 = 0 \\ \quad\quad C_2 + \quad\quad\quad\quad C_4 = 0 \\ \quad\quad\quad\quad C_3 + C_4 = 0 \\ -C_1 - C_2 - C_3 + C_4 = 4 \end{cases},$$

3.4 Linear Systems with Constant Coefficients

so $C_1 = -1$, $C_2 = -1$, $C_3 = -1$, $C_4 = 1$. By using this values in (3.68), we find the solution of our Cauchy problem:

$$\begin{cases} x_1(t) = -1 + \exp(4t) \\ x_2(t) = -1 + \exp(4t) \\ x_3(t) = -1 + \exp(4t) \\ x_4(t) = 3 + \exp(4t) \end{cases}, t \in \mathbb{R}.$$

Example 3.7 Find the general solution of the system:

$$\begin{cases} y'_1 = y_1 & -y_3 \\ y'_2 = -y_1 + 2y_2 \\ y'_3 = & -2y_2 + y_3 \end{cases} \quad (3.69)$$

The matrix of the system is:

$$A = \begin{pmatrix} 1 & 0 & -1 \\ -1 & 2 & 0 \\ 0 & -2 & 1 \end{pmatrix}$$

and the characteristic equation:

$$\det(A - \lambda I_3) = \begin{vmatrix} 1-\lambda & 0 & -1 \\ -1 & 2-\lambda & 0 \\ 0 & -2 & 1-\lambda \end{vmatrix} = -\lambda^3 + 4\lambda^2 - 5\lambda = 0$$

The matrix A has the following eigenvalues:

$$\lambda_1 = 0, \ \lambda_2 = 2+i, \ \lambda_3 = 2-i.$$

The corresponding eigenspaces are:

$$V_0 = \left\{ \begin{pmatrix} 2\alpha \\ \alpha \\ 2\alpha \end{pmatrix} : \alpha \in \mathbb{R} \right\} = \left\{ \alpha \begin{pmatrix} 2 \\ 1 \\ 2 \end{pmatrix} : \alpha \in \mathbb{R} \right\}$$

with the basis

$$\mathbf{v}_1 = \begin{pmatrix} 2 \\ 1 \\ 2 \end{pmatrix}$$

and

$$V_{2+i} = \left\{ \begin{pmatrix} 2\beta \\ i\beta \\ (1+i)\beta \end{pmatrix} : \beta \in \mathbb{R} \right\} = \left\{ \beta \begin{pmatrix} 1 \\ i \\ 1+i \end{pmatrix} : \beta \in \mathbb{R} \right\}$$

with the basis

$$\mathbf{v}_2 = \begin{pmatrix} 1 \\ i \\ 1+i \end{pmatrix} = \begin{pmatrix} 1 \\ 0 \\ 1 \end{pmatrix} + i \begin{pmatrix} 0 \\ 1 \\ 1 \end{pmatrix} = \mathbf{u} + i\mathbf{v}.$$

We do not need to compute V_{1-i}, because we know that $V_{1-i} = \overline{V_{1+i}}$, hence the *complex-valued* vector functions

$$\begin{aligned} \mathbf{Y}_1 &= \exp(0x)\mathbf{v}_1 = \mathbf{v}_1 \\ \mathbf{Y}_2 &= \exp((1+i)x)\mathbf{v}_2 = \exp(x)(\cos x + i \sin x)(\mathbf{u} + i\mathbf{w}) \\ \overline{\mathbf{Y}_2} &= \exp((1-i)x)\mathbf{v}_2 = \exp(x)(\cos x - i \sin x)(\mathbf{u} - i\mathbf{w}) \end{aligned}$$

are linearly independent solutions of the system (3.69). By Remark 3.5, the complex functions $\mathbf{Y}_2, \overline{\mathbf{Y}_2}$ can be replaced by their real part and imaginary part, so the *real-valued* functions

$$\mathbf{Y}_1 = \mathbf{v}_1 = \begin{pmatrix} 2 \\ 1 \\ 2 \end{pmatrix}$$

$$\mathbf{y}_2 = \exp(x)(\mathbf{u}\cos x - \mathbf{w}\sin x) = \exp(x)\cos x \begin{pmatrix} 1 \\ 0 \\ 1 \end{pmatrix} - \exp(x)\sin x \begin{pmatrix} 0 \\ 1 \\ 1 \end{pmatrix}$$

$$\mathbf{y}_3 = \exp(x)(\mathbf{w}\cos x + \mathbf{u}\sin x) = \exp(x)\cos x \begin{pmatrix} 0 \\ 1 \\ 1 \end{pmatrix} + \exp(x)\sin x \begin{pmatrix} 1 \\ 0 \\ 1 \end{pmatrix}$$

form a fundamental system of solutions of the system (3.7). The general solution is

$$Y(x) = C_1 \mathbf{Y}_1 + C_2 \mathbf{y}_2 + C_3 \mathbf{y}_3,$$

or, equivalently,

$$\begin{cases} y_1(x) = 2C_1 + C_2 \exp(x)\cos x + C_3 \exp(x)\sin x \\ y_2(x) = C_1 - C_2 \exp(x)\sin x + C_3 \exp(x)\cos x \\ y_3(x) = 2C_1 + C_2 \exp(x)(\cos x - \sin x) + C_3 \exp(x)(\cos x + \sin x). \end{cases}$$

3.4 Linear Systems with Constant Coefficients

Proposition 3.4 can be applied only for diagonalizable matrices. Although not any matrix is diagonalizable, any matrix has a *Jordan canonical form*—a block diagonal matrix formed by *Jordan blocks* (see [35], 6.1). A Jordan block of order $k \geq 1$ is a square upper triangular matrix of the form:

$$k=1: (\lambda); \quad k=2: \begin{pmatrix} \lambda & 1 \\ 0 & \lambda \end{pmatrix}; \quad k=3: \begin{pmatrix} \lambda & 1 & 0 \\ 0 & \lambda & 1 \\ 0 & 0 & \lambda \end{pmatrix};$$

$$k=4: \begin{pmatrix} \lambda & 1 & 0 & 0 \\ 0 & \lambda & 1 & 0 \\ 0 & 0 & \lambda & 1 \\ 0 & 0 & 0 & \lambda \end{pmatrix} \ldots \lambda \in \mathbb{C}$$

Theorem 3.6 (Jordan Canonical Form) *For any square matrix $A \in \mathcal{M}_n(\mathbb{C})$, there exists an invertible matrix, $C \in \mathcal{M}_n(\mathbb{C})$, such that*

$$C^{-1}AC = J,$$

where J is a block diagonal matrix:

$$J = \begin{pmatrix} J_{(1)} & & & \\ & J_{(2)} & & \\ & & \ddots & \\ & & & J_{(p)} \end{pmatrix}, \qquad (3.70)$$

$J_{(1)}, \ldots, J_{(p)}$ are Jordan blocks of order $k_1, \ldots, k_p$, respectively, having the eigenvalues of A, $\lambda_1, \ldots, \lambda_p$, on the diagonal (not necessarily distinct):

$$J_{(i)} = \begin{pmatrix} \lambda_i & 1 & & \\ & \lambda_i & \ddots & \\ & & \ddots & 1 \\ & & & \lambda_i \end{pmatrix}.$$

The block diagonal matrix J is called the *Jordan canonical form* of the matrix A. The set of Jordan blocks $J_{(1)}, \ldots, J_{(p)}$ is uniquely determined, although their succession in the matrix J may differ.

Remark 3.6 If A is a diagonalizable matrix, its Jordan canonical form is the diagonal matrix (made of n Jordan blocks of order 1):

$$J = \text{diag}(\lambda_1, \lambda_2, \ldots, \lambda_n),$$

where $\lambda_1, \lambda_1, \ldots, \lambda_n$ are the eigenvalues of A (not necessarily distinct).

If we know the canonical form of A, i.e. if we find the matrices J and C such that $C^{-1}AC = J$, then the homogeneous system $\mathbf{Y}' = A\mathbf{Y}$ can be written $\mathbf{Y}' = CJC^{-1}\mathbf{Y}$. We denote $\mathbf{Z} = C^{-1}\mathbf{Y}$, so $\mathbf{Y} = C\mathbf{Z}$ and $\mathbf{Y}' = C\mathbf{Z}'$, hence the following system is obtained:

$$\mathbf{Z}' = J\mathbf{Z}. \tag{3.71}$$

Since J is a matrix of the form (3.70), the system (3.71) is composed of k *independent* systems, each system corresponding to a Jordan block J_i, $i = 1, \ldots, p$. For example, the system corresponding to the matrix

$$J = \begin{pmatrix} \lambda_1 & 1 & 0 & 0 & 0 & 0 \\ 0 & \lambda_1 & 0 & 0 & 0 & 0 \\ 0 & 0 & \lambda_2 & 0 & 0 & 0 \\ 0 & 0 & 0 & \lambda_3 & 1 & 0 \\ 0 & 0 & 0 & 0 & \lambda_3 & 1 \\ 0 & 0 & 0 & 0 & 0 & \lambda_3 \end{pmatrix}$$

which is formed by the Jordan blocks:

$$J_1 = \begin{pmatrix} \lambda_1 & 1 \\ 0 & \lambda_1 \end{pmatrix}, \quad J_2 = (\lambda_2), \text{ and } J_3 = \begin{pmatrix} \lambda_3 & 1 & 0 \\ 0 & \lambda_3 & 1 \\ 0 & 0 & \lambda_3 \end{pmatrix},$$

is composed of: a system of two equations (in the unknowns z_1 and z_2),

$$\begin{cases} z_1' = \lambda_1 z_1 + z_2 \\ z_2' = \lambda_1 z_2, \end{cases}$$

a single equation in the unknown z_3,

$$z_3' = \lambda_2 z_3,$$

and a system of three equations (in the unknowns z_4, z_5 and z_6:

$$\begin{cases} z_4' = \lambda_3 z_4 + z_5 \\ z_5' = \lambda_3 z_5 + z_6 \\ z_6' = \lambda_3 z_6. \end{cases}$$

Each system can be solved separately, and we notice that they can be easily solved, being *triangular systems*. Thus, for $k = 1$, the solution of the single equation

$$z_1' = \lambda z_1$$

3.4 Linear Systems with Constant Coefficients

is

$$z_1 = C_1 \exp(\lambda_1 x).$$

For $k = 2$, the system

$$\begin{cases} z_1' = \lambda z_1 + z_2 \\ z_2' = \lambda z_2 \end{cases} \quad (3.72)$$

can be solved as follows: from the last equation we have $z_2 = C_2 \exp(\lambda x)$, by replacing in the first one, we obtain:

$$z_1' = \lambda z_1 + C_2 \exp(\lambda x),$$

so the general solution of the system (3.72) is

$$\begin{cases} z_1 = (C_1 + C_2 x) \exp(\lambda x) \\ z_2 = C_2 \exp(\lambda x) \end{cases} \quad (3.73)$$

For $k = 3$, the system

$$\begin{cases} z_1' = \lambda z_1 + z_2 \\ z_2' = \lambda z_2 + z_3 \\ z_3' = \lambda z_3 \end{cases} \quad (3.74)$$

can be solved in the same way (starting from the last equation and replacing in the previous equation the solution found) and the general solution is:

$$\begin{cases} z_1 = \left(C_1 + C_2 x + C_3 \frac{x^2}{2}\right) \exp(\lambda x). \\ z_2 = (C_2 + C_3 x) \exp(\lambda x) \\ z_3 = C_3 \exp(\lambda x) \end{cases} \quad (3.75)$$

By mathematical induction it can be proved that the general solution of a system corresponding to a Jordan block of order k is:

$$\begin{cases} z_1 = \left(C_1 + C_2 x + C_3 \frac{x^2}{2!} + \ldots + C_k \frac{x^{k-1}}{(k-1)!}\right) \exp(\lambda x) \\ z_2 = \left(C_2 + C_3 x + \ldots + C_k \frac{x^{k-2}}{(k-2)!}\right) \exp(\lambda x) \\ \vdots \\ z_k = C_k \exp(\lambda x). \end{cases} \quad (3.76)$$

Suppose that we have solved the system (3.71) and have found the general solution

$$\mathbf{Z} = \begin{pmatrix} z_1(x) \\ \vdots \\ z_n(x) \end{pmatrix} = C_1 \mathbf{Z}_1 + \ldots + C_n \mathbf{Z}_n.$$

Since the initial vector of unknown functions is $\mathbf{Y} = C\mathbf{Z}$, and $\mathbf{Z}_1 \ldots, \mathbf{Z}_n$ is a fundamental system of solutions for the system $\mathbf{Z}' = J\mathbf{Z}$, we obtain that a fundamental system of solutions for the initial system $\mathbf{Y}' = A\mathbf{Y}$ is

$$\mathbf{Y}_1 = C\mathbf{Z}_1, \ldots, \mathbf{Y}_n = C\mathbf{Z}_n.$$

Example 3.8 Find the general solution of the system:

$$\begin{cases} y_1' = y_1 - y_2 \\ y_2' = y_1 + 3y_2. \end{cases} \quad (3.77)$$

The matrix of the system is:

$$A = \begin{pmatrix} 1 & -1 \\ 1 & 3 \end{pmatrix}$$

and the characteristic equation:

$$\det(A - \lambda I_2) = \begin{vmatrix} 1-\lambda & -1 \\ 1 & 3-\lambda \end{vmatrix} = (\lambda - 2)^2 = 0.$$

So $\lambda_1 = \lambda_2 = 2$: the matrix A has the eigenvalue $\lambda = 2$ with (algebraic) multiplicity 2. The geometric multiplicity of this eigenvalue is 1, since the dimension of the eigenspace V_2 is 1:

$$V_2 = \left\{ \alpha \begin{pmatrix} -1 \\ 1 \end{pmatrix} : \alpha \in \mathbb{R} \right\}.$$

The matrix A is not diagonalizable, but it has the canonical Jordan form

$$J = \begin{pmatrix} 2 & 1 \\ 0 & 2 \end{pmatrix}.$$

Let us find the invertible matrix

$$C = \begin{pmatrix} c_{11} & c_{12} \\ c_{21} & c_{22} \end{pmatrix}$$

3.4 Linear Systems with Constant Coefficients

such that $C^{-1}AC = J$. We denote by $\mathbf{v}_1$ and $\mathbf{v}_2$ the column of the matrix C. By the equation

$$AC = CJ$$

we can write:

$$A\mathbf{v}_1 = 2\mathbf{v}_1, \quad A\mathbf{v}_2 = \mathbf{v}_1 + 2\mathbf{v}_2,$$

so $\mathbf{v}_1$ is an eigenvalue of A, we can take $\mathbf{v}_1 = (-1, 1)^t$, and $\mathbf{v}_2 = (a, b)^t$ must satisfy

$$\begin{pmatrix} -1 & -1 \\ 1 & 1 \end{pmatrix} \begin{pmatrix} a \\ b \end{pmatrix} = \mathbf{v}_1 = \begin{pmatrix} -1 \\ 1 \end{pmatrix}.$$

It follows that $a + b = 1$, so we can take $\mathbf{v}_2 = (1, 0)^t$ and we obtained the matrix

$$C = \begin{pmatrix} -1 & 1 \\ 1 & 0 \end{pmatrix}.$$

We denote $\mathbf{Z} = C^{-1}\mathbf{Y}$ and the system (3.77) is equivalent to

$$\mathbf{Z}' = J\mathbf{Z},$$

which can be written as:

$$\begin{cases} z_1' = 2z_1 + z_2 \\ z_2' = \quad\quad 2z_2. \end{cases} \tag{3.78}$$

By (3.73), the general solution of the system (3.78) is

$$\mathbf{Z} = C_1 \exp(2x) \begin{pmatrix} 1 \\ 0 \end{pmatrix} + C_2 \exp(2x) \begin{pmatrix} x \\ 1 \end{pmatrix} = C_1 \mathbf{Z}_1 + C_2 \mathbf{Z}_2.$$

Since $\mathbf{Y} = C\mathbf{Z}$, we obtain that a fundamental system of solutions of (3.77) is formed by the functions:

$$\mathbf{Y}_1 = C\mathbf{Z}_1 = \exp(2x) \begin{pmatrix} -1 & 1 \\ 1 & 0 \end{pmatrix} \begin{pmatrix} 1 \\ 0 \end{pmatrix} = \exp(2x) \begin{pmatrix} -1 \\ 1 \end{pmatrix},$$

$$\mathbf{Y}_2 = C\mathbf{Z}_2 = \exp(2x) \begin{pmatrix} -1 & 1 \\ 1 & 0 \end{pmatrix} \begin{pmatrix} x \\ 1 \end{pmatrix} = \exp(2x) \begin{pmatrix} 1 - x \\ x \end{pmatrix},$$

so the general solution is

$$\mathbf{Y} = C_1 \exp(2x) \begin{pmatrix} -1 \\ 1 \end{pmatrix} + C_2 \exp(2x) \begin{pmatrix} 1-x \\ x \end{pmatrix}.$$

If a system is not homogeneous, $\mathbf{Y}' = A\mathbf{Y} + \mathbf{G}$, $\mathbf{G} \neq \mathbf{0}$, we can apply the Lagrange method of variation of constants, after we succeeded to find a fundamental system of solutions for the associated homogeneous system $\mathbf{Y}' = A\mathbf{Y}$. But this method usually requires a large amount of computations. We have shown in Sect. 2.5 that if the free term of an nth-order linear differential equation with constant coefficients has the form $f(x) = P(x)\exp(\lambda x)$, then, by applying the *method of undetermined coefficients*, we can search for a particular solution of the same type,

$$y_P = x^m Q(x) \exp(\lambda x), \tag{3.79}$$

where $m = 0$ if λ is not a root of the characteristic equation and $m =$ the multiplicity of λ if it is a root of the characteristic equation (the case with *resonance*); $Q(x)$ is a polynomial of the same degree as $P(x)$ and its coefficients can be determined by replacing (3.79) in the non-homogeneous equation. In the next subsection we show how this method can be applied for non-homogeneous systems of linear equations with constant coefficients.

3.4.2 The Non-Homogeneous Case (the Method of Undetermined Coefficients)

Let

$$\mathbf{Y}' = A\mathbf{Y} + \mathbf{G}, \tag{3.80}$$

be a linear system of n differential equations with constant coefficients, $A \in \mathcal{M}_n(\mathbb{R})$.

Proposition 3.5 (Superposition Principle) *Let $\mathbf{G}_j(x)$, $j = 1, \ldots, k$ be k (column) vectors of continuous functions and let $\mathbf{Y}_{j,P}$ be a particular solution for the non-homogeneous system $\mathbf{Y}' = A\mathbf{Y} + \mathbf{G}_j$, $j = 1, \ldots, k$. Then $\mathbf{Y}_P = \sum_{j=1}^{k} \mathbf{Y}_{j,P}$ is a particular solution for the system*

$$\mathbf{Y}' = A\mathbf{Y} + \sum_{j=1}^{k} \mathbf{G}_j.$$

3.4 Linear Systems with Constant Coefficients

Proof Since $\mathbf{Y}'_{j,P} = A\mathbf{Y}_{j,P} + \mathbf{G}_j$, one has that
$\sum_{j=1}^{k} \mathbf{Y}'_{j,P} = A\left(\sum_{j=1}^{k} \mathbf{Y}_{j,P}\right) + \sum_{j=1}^{k} \mathbf{G}_j$, i.e. $\mathbf{Y}'_P = A\mathbf{Y}_P + \sum_{j=1}^{k} \mathbf{G}_j$. □

Let $\mathbf{P}(x)$ be a column vector of n polynomial functions (with real coefficients):

$$\mathbf{P}(x) = \begin{pmatrix} P_1(x) \\ P_2(x) \\ \vdots \\ P_n(x) \end{pmatrix}. \tag{3.81}$$

The *degree* of $\mathbf{P}(x)$ is the maximum of $\deg P_i(x)$, $i = 1, 2, \ldots, n$.

We suppose that the free term of the system (3.80) is either a vector of functions of the form $\mathbf{G}(x) = \mathbf{P}(x) \exp(\alpha x)$, or a sum of such vector functions (we can use Proposition 3.5 in this case), so we consider linear systems of the form:

$$\mathbf{Y}'(x) = A\mathbf{Y}(x) + \mathbf{P}(x)\exp(\alpha x), \tag{3.82}$$

where $A \in \mathcal{M}_n(\mathbb{R})$, $\alpha \in \mathbb{R}$ and $\mathbf{P}(x)$ is a vector of polynomials (3.81).

3.4.2.1 The Diagonalizable Case

Suppose that A is a diagonalizable matrix, with real eigenvalues (a similar study can be performed for complex eigenvalues) and let C be a diagonalization matrix, i.e. an invertible matrix such that $D = C^{-1}AC = \text{diag}(\lambda_1, \ldots, \lambda_n)$, where $\lambda_1, \ldots, \lambda_n$ are the eigenvalues of A. The columns of the matrix C are the eigenvectors $\mathbf{v}_1, \ldots, \mathbf{v}_n$, corresponding to the eigenvalues $\lambda_1, \ldots, \lambda_n$, respectively. Let $\mathbf{Z}(x) = (z_1(x), \ldots, z_n(x))^t$ be the (column) vector function such that

$$\mathbf{Y}(x) = C\mathbf{Z}(x). \tag{3.83}$$

By replacing $\mathbf{Y}$ with this expression in (3.82) and using the equality $D = C^{-1}AC$, we get:

$$\mathbf{Z}'(x) = D\mathbf{Z}(x) + C^{-1}\mathbf{P}(x)\exp(\alpha x). \tag{3.84}$$

We denote $\mathbf{P}^*(x) = C^{-1}\mathbf{P}(x)$, a vector of polynomial functions of the same degree as $\mathbf{P}(x)$: $\deg \mathbf{P}^*(x) = \deg \mathbf{P}(x)$. Let $\mathbf{P}^*(x) = (P_1^*(x), \ldots, P_n^*(x))^t$. The system (3.84) is composed of n first-order linear equations which can be *independently* solved:

$$z_i'(x) = \lambda_i z_i(x) + P_i^*(x)\exp(\alpha x), \quad i = 1, \ldots, n. \tag{3.85}$$

If α is not equal to any eigenvalue λ_i, $i = 1, \ldots, n$, we can find a particular solution $\mathbf{Z}_p$ for the non-homogeneous system (3.84)

$$\mathbf{Z}_P(x) = \mathbf{Q}^*(x) \exp(\alpha x), \tag{3.86}$$

where $\mathbf{Q}^*(x) = \left(Q_1^*(x), \ldots, Q_n^*(x)\right)^t$, $\deg Q_i^*(x) = \deg P_i^*(x)$ for any $i = 1, \ldots, n$. Coming back to the vector variable $\mathbf{Y}$, we obtain a particular solution $\mathbf{Y}_P(x) = C\mathbf{Q}^*(x) \exp(\alpha x)$ for the initial system (3.82), so we can search for a particular solution of the form:

$$\mathbf{Y}_P(x) = \mathbf{Q}(x) \exp(\alpha x), \tag{3.87}$$

where $\mathbf{Q}(x)$ is a vector of polynomial functions of the same degree as $\mathbf{P}(x)$.

Now, let us study the case when α is equal to an eigenvalue of A: we suppose that $\lambda_1 = \lambda_2 = \ldots = \lambda_r = \alpha$ and $\alpha \neq \lambda_j$ for any $j = r+1, \ldots, n$. Thus, for the first r equations of the system (3.85) we have resonance, so we have to search for particular solutions of the form: $z_{i,P}(x) = xQ_i^*(x) \exp(\lambda_1 x)$, $i = 1, \ldots, r$, while for the other $n - r$ equations we find particular solutions of the form $z_{j,P}(x) = Q_j^*(x) \exp(\lambda_1 x)$, $j = r+1, \ldots, n$. Hence, by using (3.83) we obtain a particular solution for the initial system (3.82) of the form:

$$\mathbf{Y}_P(x) = \mathbf{v}_1 z_1(x) + \mathbf{v}_2 z_2(x) + \ldots + \mathbf{v}_n z_n(x),$$

so

$$\begin{aligned}\mathbf{Y}_P(x) = \mathbf{v}_1 x Q_1^*(x) \exp(\alpha x) + \ldots + \mathbf{v}_r x Q_r^*(x) \exp(\alpha x) + \\ + \mathbf{v}_{r+1} Q_{r+1}^*(x) \exp(\alpha x) + \ldots + \mathbf{v}_n Q_n^*(x) \exp(\alpha x)\end{aligned}, \tag{3.88}$$

where $\deg Q_i^*(x) = \deg P_i(x)$, $i = 1, \ldots, n$.

The next theorem summarizes the discussion above.

Theorem 3.7 *Let* $\mathbf{Y}'(x) = A\mathbf{Y}(x) + \mathbf{P}(x) \exp(\alpha x)$, $x \in \mathbb{R}$, *be a linear non-homogeneous system with constant coefficients (where* $A \in M_n(\mathbb{R})$ *is a diagonalizable matrix,* $\alpha \in \mathbb{R}$ *and* $\mathbf{P}(x) = (P_1(x), \ldots, P_n(x))^t$ *a vector of polynomial functions with real coefficients).*

Then we can find a particular solution of the form

$$\mathbf{Y}_P(x) = \mathbf{Q}(x) \exp(\alpha x), \tag{3.89}$$

where $\mathbf{Q}(x)$ *is a vector of polynomial functions,* $\deg \mathbf{Q}(x) = \deg \mathbf{P}(x)$ *if* α *is not an eigenvalue of the matrix* A *and* $\deg \mathbf{Q}(x) = \deg \mathbf{P}(x) + 1$ *if* α *is an eigenvalue of* A.

We remark that the formula (3.89) is difficult to be applied when $\deg \mathbf{Q}(x) > 0$. In such cases, it is better to diagonalize the system, solve each equation separately to find $\mathbf{Z}_P(x) = (z_1(x), \ldots, z_n(x))^t$ and, finally, calculate $\mathbf{Y}_P(x) = C\mathbf{Z}_P(x)$ (see the examples below).

3.4 Linear Systems with Constant Coefficients

Example 3.9 Find the general solution of the system:

$$\begin{cases} y_1' = 4y_1 + 2y_2 + 4\exp(2x) \\ y_2' = 2y_1 + y_2 + \exp(2x) \end{cases} \qquad (3.90)$$

The matrix of the system is $A = \begin{pmatrix} 4 & 2 \\ 2 & 1 \end{pmatrix}$. Its eigenvalues and eigenspaces are: $\lambda_1 = 0$, $V_0 = \{(\alpha, -2\alpha)^t : \alpha \in \mathbb{R}\}$ and $\lambda_2 = 5$, $V_5 = \{(2\beta, \beta)^t : \beta \in \mathbb{R}\}$. Thus, we can take the linearly independent eigenvectors: $\mathbf{v}_1 = (1, -2)^t$, $\mathbf{v}_2 = (2, 1)^t$. Hence the general solution of the homogeneous system attached is

$$\mathbf{Y}_{GH} = C_1 \begin{pmatrix} 1 \\ -2 \end{pmatrix} + C_2 \begin{pmatrix} 2 \\ 1 \end{pmatrix} \exp(5x).$$

Since the free term,

$$\mathbf{G}(x) = \begin{pmatrix} 4\exp(2x) \\ \exp(2x) \end{pmatrix},$$

can be written $\mathbf{G}(x) = \mathbf{P}(x)\exp(\alpha x)$, with $\alpha = 2$ (so it is not an eigenvalue of A) and $\deg \mathbf{P}(x) = 0$, we can search for a particular solution of the form:

$$\mathbf{Y}_P = \begin{pmatrix} a \\ b \end{pmatrix} \exp(2x) = \begin{pmatrix} y_{1,P}(x) \\ y_{2,P}(x) \end{pmatrix},$$

where a and b are some real coefficients which will be determined by replacing $y_{1,P}$ and $y_{2,P}$ in the initial system:

$$\begin{cases} 2a\exp(2x) = 4a\exp(2x) + 2b\exp(2x) + 4\exp(2x) \\ 2b\exp(2x) = 2a\exp(2x) + b\exp(2x) + \exp(2x). \end{cases}$$

We simplify by $\exp(2x)$ and obtain:

$$\begin{cases} a + b = -2 \\ 2a - b = -1 \end{cases} \implies a = b = -1$$

hence the general solution of the initial system is

$$\mathbf{Y} = \mathbf{Y}_{GH} + \mathbf{Y}_P = C_1 \begin{pmatrix} 1 \\ -2 \end{pmatrix} + C_2 \begin{pmatrix} 2 \\ 1 \end{pmatrix} \exp(5x) + \begin{pmatrix} -1 \\ -1 \end{pmatrix} \exp(2x).$$

Example 3.10 Find the general solution of the system:

$$\begin{cases} y_1' = 4y_1 + 2y_2 + 15x^2 \\ y_2' = 2y_1 + y_2 \end{cases}. \tag{3.91}$$

Notice that the matrix of the system is the same as in Example 3.9, so the general solution of the homogeneous system is the same.

Searching for a particular solution of the non-homogeneous system, we notice that the free term,

$$\mathbf{G}(x) = \begin{pmatrix} 15x^2 \\ 0 \end{pmatrix},$$

can be written $\mathbf{G}(x) = \mathbf{P}(x)\exp(\alpha x)$, with $\alpha = 0$ (which is an eigenvalue of A) and $\deg \mathbf{P}(x) = 2$, we should search for a particular solution of the form:

$$\mathbf{Y}_P = \begin{pmatrix} a_1 x^3 + a_2 x^2 + a_3 x + a_4 \\ b_1 x^3 + b_2 x^2 + b_3 x + b_4 \end{pmatrix} = \begin{pmatrix} y_{1,P}(x) \\ y_{2,P}(x) \end{pmatrix},$$

where $a_i, b_i, i = 1, \ldots, 4$ are some real coefficients which must be determined by replacing $y_{1,P}$ and $y_{2,P}$ in the initial system. But in this way we obtain a system of 8 equations, which is not very easy to solve. It seems that the diagonalization of the system would be a better choice (than trying to apply directly the formula (3.89)): Let C be the (invertible) matrix whose columns are the eigenvectors $\mathbf{v}_1$ and $\mathbf{v}_2$:

$$C = \begin{pmatrix} 1 & -2 \\ 2 & 1 \end{pmatrix}.$$

We multiply by C^{-1} the system $\mathbf{Y}' = A\mathbf{Y} + \mathbf{G}(x)$ and denote $\mathbf{Z} = C^{-1}\mathbf{Y}$, so the following system is obtained:

$$\mathbf{Z}' = D\mathbf{Z} + C^{-1}\mathbf{G}(x), \tag{3.92}$$

where D is the diagonal matrix formed by the eigenvalues of A,

$$D = \begin{pmatrix} 0 & 0 \\ 0 & 5 \end{pmatrix}$$

and

$$C^{-1}\mathbf{G}(x) = \frac{1}{5}\begin{pmatrix} 1 & -2 \\ 2 & 1 \end{pmatrix}\begin{pmatrix} 15x^2 \\ 0 \end{pmatrix} = \begin{pmatrix} 3x^2 \\ 6x^2 \end{pmatrix}.$$

3.4 Linear Systems with Constant Coefficients

Thus, the system (3.92) is formed by the following independent equations:

$$\begin{cases} z_1' = 3x^2 \\ z_2' = 5z_2 + 6x^2 \end{cases}$$

with the particular solutions

$$\begin{cases} z_{1,P} = x^3 \\ z_{2,P} = -\frac{6}{5}x^2 - \frac{12}{25}x - \frac{12}{125}. \end{cases}$$

Hence a particular solution of the system (3.91) is

$$\mathbf{Y}_P = C\mathbf{Z}_P = \begin{pmatrix} x^3 - \frac{12}{5}x^2 - \frac{24}{25}x - \frac{24}{125} \\ -2x^3 - \frac{6}{5}x^2 - \frac{12}{25}x - \frac{12}{125}. \end{pmatrix}.$$

Example 3.11 Find the general solution of the system:

$$\begin{cases} y_1' = y_1 + y_2 + y_3 + 1 \\ y_2' = y_1 + y_2 + y_3 + 2 \\ y_3' = y_1 + y_2 + y_3 - 12 \end{cases} \quad (3.93)$$

The matrix of the system is $A = \begin{pmatrix} 1 & 1 & 1 \\ 1 & 1 & 1 \\ 1 & 1 & 1 \end{pmatrix}$. Its eigenvalues and eigenspaces are: $\lambda_1 = \lambda_2 = 0$, $V_0 = \{(\alpha, \beta, -\alpha - \beta)^t : \alpha, \beta \in \mathbb{R}\}$ and $\lambda_3 = 3$, $V_3 = \{(\gamma, \gamma, \gamma)^t : \gamma \in \mathbb{R}\}$. Thus, we can take the linearly independent eigenvectors: $\mathbf{v}_1 = (1, 0, -1)^t$, $\mathbf{v}_2 = (0, 1, -1)^t$, $\mathbf{v}_3 = (1, 1, 1)^t$. Since

$$\mathbf{G}(x) = \begin{pmatrix} 1 \\ 2 \\ -12 \end{pmatrix} = \begin{pmatrix} 1 \\ 2 \\ -12 \end{pmatrix} \exp(0x)$$

and $\alpha = 0$ is an eigenvalue of A with multiplicity $r = 2$, the formula (3.88) becomes, in this case:

$$\mathbf{Y}_P(x) = \begin{pmatrix} y_1(x) \\ y_2(x) \\ y_3(x) \end{pmatrix} = \begin{pmatrix} 1 \\ 0 \\ -1 \end{pmatrix} x\alpha_1 + \begin{pmatrix} 0 \\ 1 \\ -1 \end{pmatrix} x\alpha_2 + \begin{pmatrix} 1 \\ 1 \\ 1 \end{pmatrix} \alpha_3.$$

(the polynomials $Q_i^*(x)$, $i = 1, 2, 3$ are polynomials of degree zero, i.e. constants $\alpha_1, \alpha_2, \alpha_3$.) Thus, a particular solution has the form:

$$\begin{cases} y_{1,p}(x) = \alpha_1 x + \alpha_3 \\ y_{2,p}(x) = \alpha_2 x + \alpha_3 \\ y_{3,p}(x) = -(\alpha_1 + \alpha_2)x + \alpha_3 \end{cases}.$$

We introduce these expressions of $y_1(x)$, $y_2(x)$, $y_3(x)$ in (3.93) and find:

$$\begin{cases} \alpha_1 = 3\alpha_3 + 1 \\ \alpha_2 = 3\alpha_3 + 2 \\ -(\alpha_1 + \alpha_2) = 3\alpha_3 - 12 \end{cases},$$

which is a linear algebraic system in $\alpha_1, \alpha_2, \alpha_3$. We solve it and find $\alpha_1 = 4$, $\alpha_2 = 5$ and $\alpha_3 = 1$. Thus, a particular solution is:

$$\begin{cases} y_{1,p}(x) = 4x + 1 \\ y_{2,p}(x) = 5x + 1 \\ y_{3,p}(x) = -9x + 1 \end{cases}.$$

By Proposition 3.4, we can write a fundamental system of solutions for the associated homogeneous system:

$$\mathbf{Y}_1 = \begin{pmatrix} 1 \\ 0 \\ -1 \end{pmatrix}, \mathbf{Y}_2 = \begin{pmatrix} 0 \\ 1 \\ -1 \end{pmatrix}, \mathbf{Y}_3 = \begin{pmatrix} 1 \\ 1 \\ 1 \end{pmatrix} \exp(3x).$$

So the general solution of the initial non-homogeneous system is:

$$\mathbf{Y} = C_1 \begin{pmatrix} 1 \\ 0 \\ -1 \end{pmatrix} + C_2 \begin{pmatrix} 0 \\ 1 \\ -1 \end{pmatrix} + C_3 \begin{pmatrix} 1 \\ 1 \\ 1 \end{pmatrix} \exp(3x) + \begin{pmatrix} 4x + 1 \\ 5x + 1 \\ -9x + 1 \end{pmatrix},$$

or,

$$\begin{cases} y_1(x) = C_1 + C_3 \exp(3x) + 4x + 1 \\ y_2(x) = C_2 + C_3 \exp(3x) + 5x + 1 \\ y_3(x) = -C_1 - C_2 + C_3 \exp(3x) - 9x + 1, \end{cases}$$

where C_1, C_2, C_3 are arbitrary real numbers.

3.4 Linear Systems with Constant Coefficients

3.4.2.2 The Non-Diagonalizable Case

We assume now that the matrix A of the system (3.82) cannot be diagonalized, but all its eigenvalues are real numbers. As we have seen in Sect. 3.4.1 (Theorem 3.6), there exists an invertible matrix $C \in \mathcal{M}(\mathbb{R})$ such that $C^{-1}AC = J$, where J is a block diagonal matrix formed by *Jordan blocks*. Recall that a Jordan block of order k is a $k \times k$ matrix of the form

$$J_k = \begin{pmatrix} \lambda & 1 & 0 & \ldots & 0 \\ 0 & \lambda & 1 & \ldots & 0 \\ \vdots & \vdots & \ddots & \ddots & \vdots \\ 0 & 0 & \ldots & \lambda & 1 \\ 0 & 0 & \ldots & 0 & \lambda \end{pmatrix}. \quad (3.94)$$

Using the Jordan canonical form of the matrix A, the non-homogeneous system (3.82) can be written

$$\mathbf{Y}' = CJC^{-1}\mathbf{Y} + \mathbf{P}(x)\exp(\alpha x),$$

and, by denoting $\mathbf{Z} = C^{-1}\mathbf{Y}$ and $\mathbf{P}^*(x) = C^{-1}\mathbf{P}(x)$, we obtain:

$$\mathbf{Z}' = J\mathbf{Z} + \mathbf{P}^*(x)\exp(\alpha x) \quad (3.95)$$

The system (3.95) is composed of *independent* systems, each system corresponding to a Jordan block. In Sect. 3.4.1 we have seen the way we can find the general solution of a homogeneous system whose matrix is a Jordan block. The next theorem constructs (using the method of undetermined coefficients) a particular solution of a non-homogenous system whose matrix is a Jordan block and the free term is of the form $\mathbf{P}(x)\exp(\alpha x)$.

Theorem 3.8 *Consider the non-homogeneous system (of k equations)*

$$\mathbf{Z}' = J_k\mathbf{Z} + \mathbf{P}(x)\exp(\alpha x), \quad (3.96)$$

where J_k is the Jordan block of order k, (3.94), λ and α are real numbers and $\mathbf{P}(x) = (P_1(x), \ldots, P_k(x))^t$ is a vector of polynomial functions of degree $h = \max\limits_{i=1,k} \deg P_i(x)$.

(i) *If $\alpha \neq \lambda$ then the system (3.96) has a particular solution of the form*

$$\mathbf{Z}_P = \mathbf{Q}(x)\exp(\alpha x), \quad (3.97)$$

where $\mathbf{Q}(x)$ is a vector of polynomials of degree h.

(ii) If $\alpha = \lambda$ then the system (3.96) has a particular solution of the form

$$\mathbf{Z}_P = x\mathbf{Q}(x)\exp(\alpha x), \qquad (3.98)$$

where $\mathbf{Q}(x)$ is a vector of polynomials of degree $h + k - 1$.

Proof We first assume that $\alpha \neq \lambda$. The system (3.96) can be written:

$$\begin{cases} z_1'(x) = \lambda z_1(x) + z_2(x) + P_1(x)\exp(\alpha x) \\ \vdots \\ z_{k-1}'(x) = \lambda z_{k-1}(x) + z_k(x) + P_{k-1}(x)\exp(\alpha x) \\ z_k'(x) = \lambda z_k(x) + P_k(x)\exp(\alpha x) \end{cases}. \qquad (3.99)$$

For the last equation of the system (3.99) we can find a particular solution of the form: $z_{k,P}(x) = Q_k(x)\exp(\alpha x)$, where $Q_k(x)$ is a polynomial of degree at most h. By introducing $z_{k,P}(x)$ in the $(k-1)$-th equation we obtain a first order equation in z_{k-1} with the free term $(Q_k(x) + P_{k-1}(x))\exp(\alpha x)$ (notice that $\deg(Q_k(x) + P_{k-1}(x)) \leq h$), so we can find a particular solution of the form $z_{k-1,P}(x) = Q_{k-1}(x)\exp(\alpha x)$, where $Q_{k-1}(x)$ is a polynomial of degree at most h. We continue in this way up to the first equation for which we can find a particular solution of the form $z_{1,P}(x) = Q_1(x)\exp(\alpha x)$, where $Q_1(x)$ is a polynomial of degree at most h and the first case of the theorem is proved.

Consider now the case when $\alpha = \lambda$. The system (3.96) is written:

$$\begin{cases} z_1'(x) = \lambda z_1(x) + z_2(x) + P_1(x)\exp(\lambda x) \\ \vdots \\ z_{k-1}'(x) = \lambda z_{k-1}(x) + z_k(x) + P_{k-1}(x)\exp(\lambda x) \\ z_k'(x) = \lambda z_k(x) + P_k(x)\exp(\lambda x) \end{cases}. \qquad (3.100)$$

In the last equation we have resonance, so we can find a particular solution of the form: $z_{k,P}(x) = xQ_k(x)\exp(\lambda x)$, where $Q_k(x)$ is a polynomial of degree at most h. By introducing $z_{k,P}(x)$ in the $(k-1)$-th equation we also obtain an equation with resonance, since the free term is $(xQ_k(x) + P_{k-1}(x))\exp(\lambda x)$ (with $\deg(Q_k(x) + P_{k-1}(x)) \leq h + 1$). We can find a particular solution of the form $z_{k-1,P}(x) = xQ_{k-1}(x)\exp(\lambda x)$, where $Q_{k-1}(x)$ is a polynomial of degree at most $h+1$. We continue in this way up to the first equation, which also has resonance, so we can find a particular solution of the form $z_{1,P}(x) = xQ_1(x)\exp(\alpha x)$, where $Q_1(x)$ is a polynomial of degree at most $h + k - 1$. Thus, the second case of the theorem is proved. □

Example 3.12 Solve the non-homogeneous system:

$$\begin{pmatrix} y_1' \\ y_2' \\ y_3' \end{pmatrix} = \begin{pmatrix} 2 & 1 & 0 \\ 0 & 2 & 1 \\ 0 & 0 & 2 \end{pmatrix} \begin{pmatrix} y_1 \\ y_2 \\ y_3 \end{pmatrix} + \begin{pmatrix} x\exp(3x) \\ \exp(2x) \\ \exp(3x) \end{pmatrix}. \qquad (3.101)$$

3.4 Linear Systems with Constant Coefficients

We notice that the matrix of the system,

$$A = \begin{pmatrix} 2 & 1 & 0 \\ 0 & 2 & 1 \\ 0 & 0 & 2 \end{pmatrix},$$

is a Jordan block of order $k = 3$ (with $\lambda = 2$), hence, by (3.75), the general solution of the homogeneous system $\mathbf{Y}' = A\mathbf{Y}$ is:

$$\begin{cases} y_{1,H} = \left(C_1 + C_2 x + C_3 \frac{x^2}{2}\right) \exp(2x). \\ y_{2,H} = (C_2 + C_3 x) \exp(2x) \\ y_{3,H} = C_3 \exp(2x) \end{cases} \quad (3.102)$$

The free term can be written as

$$\mathbf{G}(x) = \mathbf{G}_1(x) + \mathbf{G}_2(x) = \begin{pmatrix} x \\ 0 \\ 1 \end{pmatrix} \exp(3x) + \begin{pmatrix} 0 \\ 1 \\ 0 \end{pmatrix} \exp(2x).$$

By the *superposition principle* (Proposition 3.5), if $\mathbf{Y}_{P_1}$ is a particular solution of the system $\mathbf{Y}' = A\mathbf{Y} + \mathbf{G}_1(x)$ and $\mathbf{Y}_{P_2}$, a particular solution of the system $\mathbf{Y}' = A\mathbf{Y} + \mathbf{G}_2(x)$, then

$$\mathbf{Y}_P = \mathbf{Y}_{P_1} + \mathbf{Y}_{P_2}$$

is a particular solution of the initial system (with the free term $\mathbf{G}_1(x) + \mathbf{G}_2(x)$).

Let us find a particular solution of the system

$$\begin{cases} y_1' = 2y_1 + y_2 + x \exp(3x) \\ y_2' = 2y_2 + y_3 \\ y_3' = 2y_3 + \exp(3x) \end{cases}$$

We search for $y_{3,P} = a \exp 3x$; from the last equation, $a = 1$. By replacing in the second equation we get $y_{2,P} = y_{3,P} = \exp 3x$. Finally, the first equation becomes $y_1' = 2y_1 + (2x + 1)\exp(3x)$, so we search for $y_{1,P} = (ax + b)\exp 3x$. We find $y_{1,P} = (x + 1)\exp(3x)$.

Now, we search for a particular solution of the system

$$\begin{cases} y_1' = 2y_1 + y_2 \\ y_2' = 2y_2 + y_3 + \exp(2x) \\ y_3' = 2y_3 \end{cases}$$

The last equation has a particular solution $\tilde{y}_{3,P} = 0$. For the second equation (with resonance) we search for a particular solution $\tilde{y}_{2,P} = ax \exp(2x)$; we get $\tilde{y}_{2,P} =$

$x\exp(2x)$. Finally, the first equation becomes $y_1' = 2y_1 + x\exp(2x)$, so we search for $\tilde{y}_{1,P} = x(ax+b)\exp 2x$. We find $\tilde{y}_{1,P} = \frac{1}{2}x^2\exp(2x)$.

The general solution of the system (3.101) is: $\mathbf{Y} = \mathbf{Y}_{GH} + \mathbf{Y}_P =$

$$= C_1 \begin{pmatrix} 1 \\ 0 \\ 0 \end{pmatrix} \exp(2x) + C_2 \begin{pmatrix} x \\ 1 \\ 0 \end{pmatrix} \exp(2x) + C_3 \begin{pmatrix} \frac{x^2}{2} \\ x \\ 1 \end{pmatrix} \exp(2x) +$$

$$+ \begin{pmatrix} x+1 \\ 1 \\ 1 \end{pmatrix} \exp(3x) + \begin{pmatrix} \frac{x^2}{2} \\ x \\ 0 \end{pmatrix} \exp(2x).$$

3.4.3 Matrix Exponential and Linear Systems with Constant Coefficients

3.4.3.1 Fundamental Matrix

We have already seen what is a vector-valued function (or vector function),

$$F(x) = \begin{pmatrix} f_1(x) \\ f_2(x) \\ \vdots \\ f_n(x) \end{pmatrix}, \quad x \in I,$$

where $I \subset \mathbb{R}$ is a real interval (we may have $I = \mathbb{R}$ either). We can generalize this notion to define a *matrix-valued function* (or *matrix function*) as a matrix in which each entry is a function of x:

$$A(x) = \begin{pmatrix} a_{11}(x) & a_{12}(x) & \ldots & a_{1n}(x) \\ a_{21}(x) & a_{22}(x) & \ldots & a_{2n}(x) \\ \vdots & \vdots & & \vdots \\ a_{m1}(x) & a_{m2}(x) & \ldots & a_{mn}(x) \end{pmatrix}, \quad x \in I.$$

The matrix function $A(x)$ is said to be *continuous* (or *differentiable*) at a point (or on the interval I) if all the functions $a_{ij}(x)$, $i = 1, \ldots, m$, $j = 1, \ldots, n$ are continuous (or differentiable). The derivative of a differentiable matrix function

3.4 Linear Systems with Constant Coefficients

$A(x)$ is defined as

$$A'(x) = \begin{pmatrix} a'_{11}(x) & a'_{12}(x) & \ldots & a'_{1n}(x) \\ a'_{21}(x) & a'_{22}(x) & \ldots & a'_{2n}(x) \\ \vdots & \vdots & & \vdots \\ a'_{m1}(x) & a'_{m2}(x) & \ldots & a'_{mn}(x) \end{pmatrix}.$$

The following rules of differentiation can be easily proved by term-wise application of the differentiation rules for real-valued functions:

$$[A(x) + B(x)]' = A'(x) + B'(x),$$

for any $(m \times n)$ differentiable matrices A and B.

$$[A(x)B(x)]' = A'(x)B(x) + A(x)B'(x),$$

for any differentiable matrices A $(m \times n)$ and B $(n \times k)$.

$$[cA(x)]' = cA'(x),$$

for any differentiable matrix A and c a real constant.

$$[Af(x)]' = Af'(x),$$

for any constant matrix A and $f(x)$ a real-valued, differentiable function.

Consider the linear homogeneous system of differential equations with constant coefficients:

$$\mathbf{Y}' = A\mathbf{Y}, \quad x \in \mathbb{R}, \qquad (3.103)$$

where A is a constant $n \times n$ matrix, $\mathbf{Y} = \mathbf{Y}(x)$ is a (column) vector of differentiable functions, $\mathbf{Y} = (y_1(x), \ldots, y_n(x))^t$. Let $\mathbf{Y}_1, \ldots, \mathbf{Y}_n$ be n linearly independent solutions of (3.103) (*a fundamental system of solutions*),

$$\mathbf{Y}_1 = \begin{pmatrix} y_{11}(x) \\ y_{21}(x) \\ \vdots \\ y_{n1}(x) \end{pmatrix}, \mathbf{Y}_2 = \begin{pmatrix} y_{12}(x) \\ y_{22}(x) \\ \vdots \\ y_{n2}(x) \end{pmatrix}, \ldots, \mathbf{Y}_n = \begin{pmatrix} y_{1n}(x) \\ y_{2n}(x) \\ \vdots \\ y_{nn}(x) \end{pmatrix}.$$

The matrix function formed with the columns $\mathbf{Y}_1, \mathbf{Y}_2, \ldots, \mathbf{Y}_n$,

$$\boldsymbol{\Phi}(x) = \begin{pmatrix} y_{11}(x) & y_{12}(x) & \cdots & y_{1n}(x) \\ y_{21}(x) & y_{22}(x) & \cdots & y_{2n}(x) \\ \vdots & \vdots & & \vdots \\ y_{n1}(x) & y_{n2}(x) & \cdots & y_{nn}(x) \end{pmatrix} \tag{3.104}$$

is called a *fundamental matrix* for the system (3.103). It is a differentiable matrix function (since all its components y_{ij} are differentiable functions) and it verifies, for any $x \in \mathbb{R}$, the equality:

$$\boldsymbol{\Phi}'(x) = A\boldsymbol{\Phi}(x). \tag{3.105}$$

We also remark that $\boldsymbol{\Phi}(x)$ is a nonsingular matrix, because its determinant is the Wronskian of the linearly independent solutions $\mathbf{Y}_1, \ldots, \mathbf{Y}_n$ and, by Theorem 3.3, it is nonzero at any $x \in \mathbb{R}$.

The general solution of the system (3.103),

$$\mathbf{Y} = c_1 \mathbf{Y}_1 + c_2 \mathbf{Y}_2 + \ldots + c_n \mathbf{Y}_n,$$

can be written using the fundamental matrix as

$$\mathbf{Y} = \boldsymbol{\Phi}(x)\mathbf{c}, \tag{3.106}$$

where $\mathbf{c} = (c_1, \ldots, c_n)^t$ is a column vector of real arbitrary constants.

Consider the Cauchy problem

$$\begin{cases} \mathbf{Y}' = A\mathbf{Y}, \\ \mathbf{Y}(x_0) = \mathbf{Y}_0. \end{cases} \tag{3.107}$$

where $\mathbf{Y}_0 \in \mathbb{R}^n$ is an arbitrary column vector. By (3.106) we have

$$\mathbf{Y}_0 = \mathbf{Y}(x_0) = \boldsymbol{\Phi}(x_0)\mathbf{c},$$

hence

$$\mathbf{c} = \boldsymbol{\Phi}(x_0)^{-1}\mathbf{Y}_0$$

and the solution of the Cauchy problem (3.107) is

$$\mathbf{Y} = \boldsymbol{\Phi}(x)\boldsymbol{\Phi}(x_0)^{-1}\mathbf{Y}_0. \tag{3.108}$$

We consider, now, the non-homogeneous system

$$\mathbf{Y}' = A\mathbf{Y} + \mathbf{G}(x), \quad x \in \mathbb{R}, \tag{3.109}$$

3.4 Linear Systems with Constant Coefficients

where $A \in \mathcal{M}_n(\mathbb{R})$ is a constant matrix, and $\mathbf{G}(x)$ is a vector function.

Let $\mathbf{\Phi}(x)$ be a fundamental matrix of the homogeneous system attached, $\mathbf{Y}' = A\mathbf{Y}$. By Theorem 3.5 we can find a particular solution for the non-homogeneous system (3.109) of the form:

$$\mathbf{Y}_P = \mathbf{\Phi}(x)\mathbf{u}(x), \qquad (3.110)$$

where $\mathbf{u}(x) = (u_1(x), \ldots, u_n(x))^t$ is a differentiable vector function. By the differentiation rules discussed above, we obtain that

$$\mathbf{Y}'_P = \mathbf{\Phi}'(x)\mathbf{u}(x) + \mathbf{\Phi}(x)\mathbf{u}'(x).$$

Since $\mathbf{\Phi}'(x) = A\mathbf{\Phi}(x)$, by replacing (3.110) in the system (3.109) we obtain that

$$\mathbf{\Phi}(x)\mathbf{u}'(x) = \mathbf{G}(x),$$

hence

$$\mathbf{u}'(x) = \mathbf{\Phi}(x)^{-1}\mathbf{G}(x)$$

so we can write

$$\mathbf{u}(x) = \int_{x_0}^{x} \mathbf{\Phi}(t)^{-1}\mathbf{G}(t)\, dt,$$

where the integral applied to a vector function is a vector of integrals applied to each component of the vector function.

The general solution of the system (3.109) can be written

$$\mathbf{Y} = \mathbf{Y}_{GH} + \mathbf{Y}_P = \mathbf{\Phi}(x)\mathbf{c} + \mathbf{\Phi}(x)\int_{x_0}^{x} \mathbf{\Phi}(t)^{-1}\mathbf{G}(t)\, dt \qquad (3.111)$$

and the particular solution that satisfies the initial conditions $\mathbf{Y}(x_0) = \mathbf{Y}_0$ is

$$\mathbf{Y} = \mathbf{\Phi}(x)\mathbf{\Phi}(x_0)^{-1}\mathbf{Y}_0 + \mathbf{\Phi}(x)\int_{x_0}^{x} \mathbf{\Phi}(t)^{-1}\mathbf{G}(t)\, dt. \qquad (3.112)$$

3.4.3.2 Matrix Exponential

We have seen that exponential functions play a very important role in the solution of differential equations and systems of equations. For example, the general solution of the equation $y' = \alpha y$ is $y = k\exp(\alpha x)$, while $\mathbf{Y} = \mathbf{v}\exp(\lambda x)$ (with λ an eigenvalue of A and $\mathbf{v}$ a corresponding eigenvector) is a solution of the linear homogeneous system (3.103).

Recall that, for any real (or complex) number, the power series $\sum_{k=0}^{\infty} \frac{x^k}{k!}$ is convergent and its sum is the exponential function $\exp(x)$, i.e:

$$\exp(x) = 1 + \frac{x}{1!} + \frac{x^2}{2!} + \ldots + \frac{x^k}{k!} + \ldots. \tag{3.113}$$

The question that arises, inspired by this representation of $\exp(x)$, is whether it would be possible to define, for a square matrix A, the *matrix exponential* $\exp(A)$ as the sum of the following matrix power series:

$$\exp(A) = I_n + \frac{1}{1!}A + \frac{1}{2!}A^2 + \ldots + \frac{1}{k!}A^k + \ldots, \tag{3.114}$$

where I_n is the identity matrix.

We say that a matrix series $\sum_{k=0}^{\infty} A_k$ is convergent ($A_k = (a_{ij}^{[k]})_{i,j=1,\ldots,n}$) if all the series $\sum_{k=0}^{\infty} a_{ij}^{[k]}$, $i, j = 1, \ldots, n$ are convergent. In this case, the sum of the matrix series is the matrix of all the sums of the real-number series:

$$\sum_{k=0}^{\infty} A_k = \left(\sum_{k=0}^{\infty} a_{ij}^{[k]} \right)_{i,j=1,\ldots,n}.$$

Proposition 3.6 *For any square matrix $A \in M_n(\mathbb{R})$, the matrix series $\sum_{k=0}^{\infty} \frac{1}{k!} A^k$ is convergent. The sum of the series is the **matrix exponential** denoted by $\exp(A)$.*

Proof Let $A = (a_{i,j})$, $A^k = (a_{i,j}^{[k]})$ and $M = \max_{i,j=1,\ldots,n} |a_{i,j}|$. By mathematical induction on $k \geq 1$ it can be easily proved that

$$\left| a_{i,j}^{[k]} \right| \leq M^k n^{k-1}, \tag{3.115}$$

for any $i, j = 1 \ldots, n$. Since the series $\sum_{k=0}^{\infty} \frac{1}{k!} M^k n^{k-1}$ is convergent (it converges to $\frac{1}{n} \exp(Mn)$), by the Comparison test for series (see [39], Corollary 7.3.2, or [28], Theorem 24) it follows that all the series $\sum_{k=0}^{\infty} \frac{1}{k!} a_{i,j}^{[k]}$ are absolutely convergent (hence convergent) and the proposition is proved. □

3.4 Linear Systems with Constant Coefficients

Example 3.13 Calculate the exponential of the diagonal matrix

$$D = \begin{pmatrix} \lambda_1 & 0 & \cdots & 0 \\ 0 & \lambda_2 & \cdots & 0 \\ \vdots & \vdots & \ddots & \vdots \\ 0 & 0 & \cdots & \lambda_n \end{pmatrix}.$$

The powers of D are also diagonal matrices:

$$D^k = \begin{pmatrix} \lambda_1^k & 0 & \cdots & 0 \\ 0 & \lambda_2^k & \cdots & 0 \\ \vdots & \vdots & \ddots & \vdots \\ 0 & 0 & \cdots & \lambda_n^k \end{pmatrix}.$$

By applying (3.114), the exponential of D is the matrix:

$$\exp(D) = I_n + \frac{1}{1!}D + \frac{1}{2!}D^2 + \ldots =$$

$$= \begin{pmatrix} 1 + \frac{1}{1!}\lambda_1 + \frac{1}{2!}\lambda_1^2 + \ldots & 0 & \cdots & 0 \\ 0 & 1 + \frac{1}{1!}\lambda_2 + \frac{1}{2!}\lambda_2^2 + \ldots & \cdots & 0 \\ \vdots & \vdots & & \vdots \\ 0 & 0 & \cdots & 1 + \frac{1}{1!}\lambda_n + \frac{1}{2!}\lambda_n^2 + \ldots \end{pmatrix},$$

so

$$\exp(D) = \begin{pmatrix} \exp(\lambda_1) & 0 & \cdots & 0 \\ 0 & \exp(\lambda_2) & \cdots & 0 \\ \vdots & \vdots & \ddots & \vdots \\ 0 & 0 & \cdots & \exp(\lambda_n) \end{pmatrix}. \qquad (3.116)$$

Example 3.14 A *nilpotent* matrix N is a matrix with a vanishing power: $N^k = \mathbf{0}$ for some k. For such matrices, the exponential series (3.114) has only a finite number of nonzero terms. Let us calculate the exponential of the nilpotent matrix

$$N = \begin{pmatrix} 0 & x & 0 & 0 \\ 0 & 0 & x & 0 \\ 0 & 0 & 0 & x \\ 0 & 0 & 0 & 0 \end{pmatrix}.$$

The nonzero powers of N are:

$$N^2 = \begin{pmatrix} 0 & 0 & x^2 & 0 \\ 0 & 0 & 0 & x^2 \\ 0 & 0 & 0 & 0 \\ 0 & 0 & 0 & 0 \end{pmatrix}, \quad N^3 = \begin{pmatrix} 0 & 0 & 0 & x^3 \\ 0 & 0 & 0 & 0 \\ 0 & 0 & 0 & 0 \\ 0 & 0 & 0 & 0 \end{pmatrix}.$$

We can see that $N^k = \mathbf{0}$ for any $k \geq 4$, so the series (3.114) has only 4 nonzero terms:

$$\exp(N) = I_4 + \frac{1}{1!}N + \frac{1}{2!}N^2 + \frac{1}{3!}N^3 =$$

$$= \begin{pmatrix} 1 & \frac{1}{1!}x & \frac{1}{2!}x^2 & \frac{1}{3!}x^3 \\ 0 & 1 & \frac{1}{1!}x & \frac{1}{2!}x^2 \\ 0 & 0 & 1 & \frac{1}{1!}x \\ 0 & 0 & 0 & 1 \end{pmatrix}.$$

If $\sum_{k=0}^{\infty} A_k$ and $\sum_{k=0}^{\infty} B_k$ are two convergent series, their product is defined as the series $\sum_{k=0}^{\infty} C_k$, where

$$C_k = A_0 B_k + A_1 B_{k-1} + \ldots + A_{k-1} B_1 + A_k B_0. \tag{3.117}$$

If $\sum_{k=0}^{\infty} C_k$ is convergent, then its sum is equal to the product of the sums of the series $\sum_{k=0}^{\infty} A_k$ and $\sum_{k=0}^{\infty} B_k$.

It can be proved that, for any matrices $A, B \in \mathcal{M}(\mathbb{R})$, the product of the series $\exp(A)$ and $\exp(B)$ is a convergent series. Moreover, if the matrices commute, then the "law of exponents" holds, as shown in the proposition bellow:

Proposition 3.7 *If $A, B \in \mathcal{M}_n(\mathbb{R})$ such that $AB = BA$ then*

$$\exp(A)\exp(B) = \exp(A + B). \tag{3.118}$$

*The exponential of a matrix is always an **invertible** matrix and*

$$(\exp(A))^{-1} = \exp(-A). \tag{3.119}$$

3.4 Linear Systems with Constant Coefficients

Proof Since $AB = BA$, we have the following binomial formula:

$$(A+B)^k = \sum_{j=0}^{k} \binom{k}{j} A^j B^{k-j} \qquad (3.120)$$

By the formulas (3.114) and (3.117) we have:

$$\exp(A)\exp(B) = \left(\sum_{k=0}^{\infty} \frac{A^k}{k!}\right)\left(\sum_{k=0}^{\infty} \frac{B^k}{k!}\right) = \sum_{k=0}^{\infty} C_k,$$

where

$$C_k = \sum_{j=0}^{k} \frac{A^j}{j!} \cdot \frac{B^{k-j}}{(k-j)!} = \frac{1}{k!}\sum_{j=0}^{k} \binom{k}{j} A^j B^{k-j},$$

for any $k = 0, 1, \ldots$. Hence, by (3.120), we obtain:

$$\exp(A)\exp(B) = \sum_{k=0}^{\infty} \frac{1}{k!}(A+B)^k = \exp(A+B).$$

The matrices A and $-A$ commute, so $\exp(A)\exp(-A) = \exp(\mathbf{0}) = I_n$ and the last statement follows. □

Let us write the formula (3.114) for the matrix function Ax, where $A \in \mathcal{M}_n(\mathbb{R})$ is a constant matrix and $x \in \mathbb{R}$ is a scalar variable:

$$\exp(Ax) = I_n + \frac{x}{1!}A + \frac{x^2}{2!}A^2 + \ldots + \frac{x^k}{k!}A^k + \ldots = \sum_{k=0}^{\infty} \frac{x^k}{k!}A^k. \qquad (3.121)$$

While the entries of the matrix $\exp(A)$ are the (convergent) real-number series $\sum_{k=0}^{\infty} \frac{1}{k!}a_{i,j}^{[k]}$, the entries of the *matrix function* $\exp(Ax)$ are the *power series*

$$\sum_{k=0}^{\infty} \frac{1}{k!}a_{i,j}^{[k]}x^k. \qquad (3.122)$$

By the inequality (3.115) we obtain that, for any $i, j = 1, \ldots, n$,

$$\lim_{k \to \infty} \sqrt[k]{\left|\frac{a_{i,j}^{[k]}}{k!}\right|} = 0,$$

hence all the power series (3.122) have the radius of convergence ∞ (they are uniformly convergent on $\mathbb{R}$). As a consequence, we can apply for each power series (3.122) the term-by term differentiation and obtain:

$$[\exp(Ax)]' = \frac{1}{1!}A + \frac{2x}{2!}A^2 + \ldots + \frac{kx^{k-1}}{k!}A^k + \ldots =$$

$$= A\left(I_n + \frac{x}{1!}A + \ldots + \frac{x^{k-1}}{(k-1)!}A^{k-1} + \ldots\right),$$

hence

$$[\exp(Ax)]' = A\exp(Ax). \tag{3.123}$$

By this equation it follows that the columns of the matrix function $\mathbf{\Psi}(x) = \exp(Ax)$ are solutions of the homogeneous system (3.103). They are linearly independent, since $\exp(Ax)$ is an invertible matrix, so $\mathbf{\Psi}(x) = \exp(Ax)$ is a *fundamental matrix* for the system $\mathbf{Y}' = A\mathbf{Y}$. The general solution of this homogeneous system is

$$\mathbf{Y} = \exp(Ax)\mathbf{c}. \tag{3.124}$$

If we need to find the particular solution that satisfies the initial conditions $\mathbf{Y}(x_0) = \mathbf{Y}_0$, we obtain

$$\mathbf{Y}_0 = \exp(Ax_0)\mathbf{c},$$

so $\mathbf{c} = \exp(Ax_0)^{-1} = \exp(-Ax_0)$ and the next theorem follows.

Theorem 3.9 *For any matrix* $A \in \mathcal{M}_n(\mathbb{R})$, $x_0 \in \mathbb{R}$ *and* $\mathbf{Y}_0 \in \mathbb{R}^n$, *the unique solution of the Cauchy problem*

$$\begin{cases} \mathbf{Y}' = A\mathbf{Y}, \\ \mathbf{Y}(x_0) = \mathbf{Y}_0. \end{cases}$$

is

$$\mathbf{Y} = \exp(A(x - x_0))\mathbf{Y}_0. \tag{3.125}$$

Moreover, using the fundamental matrix $\mathbf{\Psi}(x) = \exp(Ax)$ in the formula (3.111), we obtain the general solution of the non-homogeneous system $\mathbf{Y}' = A\mathbf{Y} + \mathbf{G}$:

$$\mathbf{Y} = \exp(Ax)\mathbf{c} + \int_{x_0}^{x} \exp(A(x-t))\mathbf{G}(t)\,dt. \tag{3.126}$$

3.4 Linear Systems with Constant Coefficients

Theorem 3.10 *Let $\mathbf{G}(x)$ be a vector of real functions, $A \in \mathcal{M}_n(\mathbb{R})$, $x_0 \in \mathbb{R}$ and $\mathbf{Y}_0 \in \mathbb{R}^n$. Then the unique solution of the Cauchy problem*

$$\begin{cases} \mathbf{Y}' = A\mathbf{Y} + \mathbf{G}, \\ \mathbf{Y}(x_0) = \mathbf{Y}_0. \end{cases}$$

is

$$\mathbf{Y} = \exp(A(x - x_0))\mathbf{Y}_0 + \int_{x_0}^{x} \exp(A(x - t))\mathbf{G}(t)\, dt. \tag{3.127}$$

3.4.3.3 The Exponential of a Diagonalizable Matrix

Suppose that $A \in \mathcal{M}_n(\mathbb{R})$ is a diagonalizable matrix. Let $\lambda_1, \ldots, \lambda_n$ be its eigenvalues and $\mathbf{v}_1, \ldots, \mathbf{v}_n$ corresponding (linearly independent) eigenvectors. We denote by C the (nonsingular) matrix formed by the columns $\mathbf{v}_1, \ldots, \mathbf{v}_n$ and by D the diagonal matrix $D = \text{diag}(\lambda_1, \ldots, \lambda_n)$. Since $A = CDC^{-1}$, it follows that any power of A can be written

$$A^k = CDC^{-1}CDC^{-1} \ldots CDC^{-1} = CD^kC^{-1},$$

hence the expression of $\exp(Ax)$ is

$$\exp(Ax) = \sum_{k=0}^{\infty} \frac{1}{k!} C(Dx)^k C^{-1} = C \left(\sum_{k=0}^{\infty} \frac{1}{k!} (Dx)^k \right) C^{-1} \Rightarrow$$

$$\exp(Ax) = C \exp(Dx) C^{-1}, \tag{3.128}$$

where, by (3.116),

$$\exp(Dx) = \text{diag}(\exp(\lambda_1 x), \ldots, \exp(\lambda_n x)).$$

Example 3.15 Find the solution of the system:

$$\begin{cases} y_1' = 3y_2 - 4y_3 \\ y_2' = - y_3 \\ y_3' = -2y_1 + y_2 \end{cases} \tag{3.129}$$

that satisfies the initial conditions

$$y_1(0) = 12, \quad y_2(0) = 5, \quad y_3(0) = 3.$$

The matrix of the system is:

$$A = \begin{pmatrix} 0 & 3 & -4 \\ 0 & 0 & -1 \\ -2 & 1 & 0 \end{pmatrix}$$

and the characteristic equation:

$$\det(A - \lambda I_3) = \begin{vmatrix} -\lambda & 3 & -4 \\ 0 & -\lambda & -1 \\ -2 & 1 & -\lambda \end{vmatrix} = -\lambda^3 + 7\lambda + 6 = 0$$

The matrix A has the following eigenvalues:

$$\lambda_1 = -1, \ \lambda_2 = -2, \ \lambda_3 = 3.$$

The corresponding eigenspaces are:

$$V_{-1} = \left\{ \begin{pmatrix} \alpha \\ \alpha \\ \alpha \end{pmatrix} : \alpha \in \mathbb{R} \right\} = \left\{ \alpha \begin{pmatrix} 1 \\ 1 \\ 1 \end{pmatrix} : \alpha \in \mathbb{R} \right\}$$

with the basis

$$\mathbf{v}_1 = \begin{pmatrix} 1 \\ 1 \\ 1 \end{pmatrix},$$

$$V_{-2} = \left\{ \begin{pmatrix} 5\beta \\ 2\beta \\ 4\beta \end{pmatrix} : \beta \in \mathbb{R} \right\} = \left\{ \beta \begin{pmatrix} 5 \\ 2 \\ 4 \end{pmatrix} : \beta \in \mathbb{R} \right\}$$

with the basis

$$\mathbf{v}_2 = \begin{pmatrix} 5 \\ 2 \\ 4 \end{pmatrix},$$

$$V_3 = \left\{ \begin{pmatrix} 5\gamma \\ \gamma \\ -3\gamma \end{pmatrix} : \gamma \in \mathbb{R} \right\} = \left\{ \gamma \begin{pmatrix} 5 \\ 1 \\ -3 \end{pmatrix} : \gamma \in \mathbb{R} \right\}$$

with the basis

$$\mathbf{v}_3 = \begin{pmatrix} 5 \\ 1 \\ -3 \end{pmatrix}.$$

We denote by C the matrix of the eigenvectors $\mathbf{v}_1, \mathbf{v}_2, \mathbf{v}_3$:

$$C = \begin{pmatrix} 1 & 5 & 5 \\ 1 & 2 & 1 \\ 1 & 4 & -3 \end{pmatrix}.$$

The fundamental matrix is

$$\exp(Ax) = C \cdot \text{diag}(\exp(-x), \exp(-2x), \exp(3x)) \cdot C^{-1}.$$

By Theorem 3.9, the solution of the Cauchy problem is

$$\mathbf{Y} = C \cdot \text{diag}(\exp(-x), \exp(-2x), \exp(3x)) \cdot C^{-1} \mathbf{Y}_0,$$

where $\mathbf{Y}_0 = (12, 5, 3)^t$. The vector $\mathbf{c} = C^{-1} \mathbf{Y}_0$ is the solution of the linear algebraic system

$$\begin{cases} c_1 + 5c_2 + 5c_3 = 12 \\ c_1 + 2c_2 + c_3 = 5 \\ c_1 + 4c_2 - 3c_3 = 3 \end{cases}.$$

Hence $\mathbf{c} = (2, 1, 1)^t$ and the solution of the Cauchy problem is:

$$\mathbf{Y} = \begin{pmatrix} \exp(-x) & 5\exp(-2x) & 5\exp(3x) \\ \exp(-x) & 2\exp(-2x) & \exp(3x) \\ \exp(-x) & 4\exp(-2x) & -3\exp(3x) \end{pmatrix} \begin{pmatrix} 2 \\ 1 \\ 1 \end{pmatrix} =$$

$$= \begin{pmatrix} 2\exp(-x) + 5\exp(-2x) + 5\exp(3x) \\ 2\exp(-x) + 2\exp(-2x) + \exp(3x) \\ 2\exp(-x) + 4\exp(-2x) - 3\exp(3x) \end{pmatrix}.$$

3.4.3.4 The Exponential of a Nondiagonalizable Matrix

Consider now the case when the matrix A is not diagonalizable. As we discussed in Sect. 3.4.1 (Theorem 3.6) any square matrix has a *Jordan canonical form*: there exists an invertible matrix C such that

$$C^{-1}AC = J,$$

where J is a block diagonal matrix:

$$J = \begin{pmatrix} J_{(1)} & & & \\ & J_{(2)} & & \\ & & \ddots & \\ & & & J_{(p)} \end{pmatrix} \qquad (3.130)$$

$J_{(1)}, \ldots, J_{(p)}$ are Jordan blocks of order $k_1, \ldots, k_p$, respectively, having the eigenvalues of A, $\lambda_1, \ldots, \lambda_p$ on the diagonal (not necessarily distinct):

$$J_{(i)} = \begin{pmatrix} \lambda_i & 1 & 0 & 0 & \cdots & 0 \\ 0 & \lambda_i & 1 & 0 & \cdots & 0 \\ 0 & 0 & \lambda_i & 1 & & \vdots \\ \vdots & \vdots & \vdots & \ddots & \ddots & 0 \\ 0 & 0 & 0 & & \lambda_i & 1 \\ 0 & 0 & 0 & \cdots & 0 & \lambda_i \end{pmatrix}.$$

As can be readily seen, a relation similar to (3.128) can be written:

$$\exp(Ax) = C \exp(Jx) C^{-1}, \qquad (3.131)$$

where

$$\exp(Jx) = \begin{pmatrix} \exp(J_{(1)}x) & & & \\ & \exp(J_{(2)}x) & & \\ & & \ddots & \\ & & & \exp(J_{(p)}x) \end{pmatrix}.$$

We compute $\exp(J_k x)$ where $J_k \in \mathcal{M}_k(\mathbb{R})$ is a Jordan block of order $k \geq 2$ (for $k = 1$, we have a scalar: $\exp(J_1 x) = \exp(\lambda x)$):

$$J_k = \begin{pmatrix} \lambda & 1 & 0 & 0 & \cdots & 0 \\ 0 & \lambda & 1 & 0 & \cdots & 0 \\ 0 & 0 & \lambda & 1 & & \vdots \\ \vdots & \vdots & \vdots & \ddots & \ddots & 0 \\ 0 & 0 & 0 & & \lambda & 1 \\ 0 & 0 & 0 & \cdots & 0 & \lambda \end{pmatrix}.$$

3.4 Linear Systems with Constant Coefficients

We denote by N the matrix

$$N = J_k - \lambda I_k = \begin{pmatrix} 0 & 1 & 0 & 0 & \cdots & 0 \\ 0 & 0 & 1 & 0 & \cdots & 0 \\ 0 & 0 & 0 & 1 & & \vdots \\ \vdots & \vdots & \vdots & \ddots & \ddots & 0 \\ 0 & 0 & 0 & & 0 & 1 \\ 0 & 0 & 0 & \cdots & 0 & 0 \end{pmatrix}.$$

We have:

$$\exp(J_k x) = \exp(\lambda x I_k + Nx) = \exp(\lambda x I_k) \exp(Nx) =$$
$$= \exp(\lambda x) \exp(Nx).$$

But, as can be easily noticed, N is a *nilpotent matrix* since $N^j = \mathbf{0}$ for any $j \geq k$ (see Example 3.14). Hence the exponential series $\exp(Nx)$ has only a finite number of nonzero terms:

$$\exp(Nx) = \sum_{j=0}^{k-1} \frac{1}{j!} N^j x^j.$$

It follows that

$$\exp(J_k x) = \exp(\lambda x) \begin{pmatrix} 1 & \frac{1}{1!}x & \frac{1}{2!}x^2 & \frac{1}{3!}x^3 & \cdots & \cdots & \frac{1}{(k-1)!}x^{k-1} \\ 0 & 1 & \frac{1}{1!}x & \frac{1}{2!}x^2 & \cdots & \cdots & \frac{1}{(k-2)!}x^{k-2} \\ 0 & 0 & 1 & \frac{1}{1!}x & \cdots & \cdots & \frac{1}{(k-3)!}x^{k-3} \\ 0 & 0 & 0 & 1 & \cdots & \cdots & \frac{1}{(k-4)!}x^{k-4} \\ \vdots & \vdots & \vdots & \vdots & \ddots & & \vdots \\ 0 & 0 & 0 & 0 & \cdots & 1 & \frac{1}{1!}x \\ 0 & 0 & 0 & 0 & \cdots & 0 & 1 \end{pmatrix}. \quad (3.132)$$

Example 3.16 Find the general solution of the non-homogeneous system:

$$\begin{cases} y_1' = 4y_1 - y_2 + y_3 \\ y_2' = y_1 + 3y_2 - y_3 + \exp(2x), \quad x \in \mathbb{R}. \\ y_3' = y_2 + y_3 \end{cases} \quad (3.133)$$

We denote $\mathbf{Y} = (y_1, y_2, y_3)^t$, $\mathbf{G} = (0, \exp(2x), 0)^t$,

$$A = \begin{pmatrix} 4 & -1 & 1 \\ 1 & 3 & -1 \\ 0 & 1 & 1 \end{pmatrix},$$

and (3.133) becomes:

$$\mathbf{Y}' = A\mathbf{Y} + \mathbf{G}. \tag{3.134}$$

The general solution of this system is given by the formula (3.126). So we need only to compute $\exp(Ax)$ and $\exp(A(x - t))$.

The characteristic equation of A is:

$$P_A(\lambda) = \det(A - \lambda I_3) = (\lambda - 3)^2(2 - \lambda) = 0.$$

So the eigenvalues of the matrix A are:

$$\lambda_1 = \lambda_2 = 3, \quad \lambda_3 = 2.$$

The corresponding eigenspaces are:

$$V_{\lambda_1} = \{(\alpha, 2\alpha, \alpha)^t : \alpha \in \mathbb{R}\},$$

$$V_{\lambda_3} = \{(0, \beta, \beta)^t : \beta \in \mathbb{R}\}.$$

Since $\dim V_{\lambda_1} = 1 < 2 = m_1$ (m_1 = the multiplicity of $\lambda_1 = 3$), the matrix A is not diagonalizable. Let us find a nonsingular matrix

$$C = \begin{pmatrix} c_{11} & c_{12} & c_{13} \\ c_{21} & c_{22} & c_{23} \\ c_{31} & c_{32} & c_{33} \end{pmatrix}$$

such that $C^{-1}AC = J$, where J is the Jordan canonical form of A:

$$J = \begin{pmatrix} \lambda_1 & 1 & 0 \\ 0 & \lambda_1 & 0 \\ 0 & 0 & \lambda_3 \end{pmatrix} = \begin{pmatrix} 3 & 1 & 0 \\ 0 & 3 & 0 \\ 0 & 0 & 2 \end{pmatrix}.$$

We have:

$$AC = CJ \tag{3.135}$$

3.4 Linear Systems with Constant Coefficients

and, if we denote by v_1, v_2 and v_3 the columns of the matrix C, equality (3.135) becomes:

$$\begin{cases} Av_1 = \lambda_1 v_1 \\ Av_2 = \lambda_1 v_2 + v_1 \\ Av_3 = \lambda_3 v_3. \end{cases}$$

Thus, v_1 is an eigenvector corresponding to $\lambda_1 = 3$, we take $v_1 = (1, 2, 1)^t$, and v_3 is an eigenvector corresponding to $\lambda_3 = 2$: $v_3 = (0, 1, 1)^t$. To find $v_2 = (x_1, x_2, x_3)^t$, we use the second equation of the system above:

$$(A - \lambda_1 I_3)v_2 = v_1$$

which can be written as:

$$\begin{cases} x_1 - x_2 + x_3 = 1 \\ x_1 \quad\quad - x_3 = 2 \\ \quad\quad x_2 - 2x_3 = 1 \end{cases}.$$

The set of solutions for this system is

$$\{(\gamma + 2, 2\gamma + 1, \gamma)^t : \gamma \in \mathbb{R}\},$$

so we can take $v_2 = (2, 1, 0)^t$ and the matrix C is:

$$C = \begin{pmatrix} 1 & 2 & 0 \\ 2 & 1 & 1 \\ 1 & 0 & 1 \end{pmatrix}.$$

The Jordan canonical matrix J is composed of two Jordan blocks (of order 2 and 1, respectively):

$$J_{(1)} = \begin{pmatrix} 3 & 1 \\ 0 & 3 \end{pmatrix}, \quad J_{(2)} = (2).$$

By applying the formula (3.132) for $k = 2$ (and $\lambda = 3$) we get:

$$\exp(J_{(1)}x) = \exp(3x) \begin{pmatrix} 1 & x \\ 0 & 1 \end{pmatrix} = \begin{pmatrix} \exp(3x) & x \exp(3x) \\ 0 & \exp(3x) \end{pmatrix},$$

hence

$$\exp(Jx) = \begin{pmatrix} \exp(3x) & x \exp(3x) & 0 \\ 0 & \exp(3x) & 0 \\ 0 & 0 & \exp(2x) \end{pmatrix}.$$

So

$$\exp(Ax) = C \begin{pmatrix} \exp(3x) & x\exp(3x) & 0 \\ 0 & \exp(3x) & 0 \\ 0 & 0 & \exp(2x) \end{pmatrix} C^{-1} =$$

$$= \begin{pmatrix} 1 & 2 & 0 \\ 2 & 1 & 1 \\ 1 & 0 & 1 \end{pmatrix} \begin{pmatrix} \exp(3x) & x\exp(3x) & 0 \\ 0 & \exp(3x) & 0 \\ 0 & 0 & \exp(2x) \end{pmatrix} \begin{pmatrix} -1 & 2 & -2 \\ 1 & -1 & 1 \\ 1 & -2 & 3 \end{pmatrix} =$$

$$= \begin{pmatrix} (x+1)e^{3x} & -xe^{3x} & xe^{3x} \\ (2x-1)e^{3x} + e^{2x} & (-2x+3)e^{3x} - 2e^{2x} & (2x-3)e^{3x} + 3e^{2x} \\ (x-1)e^{3x} + e^{2x} & (-x+2)e^{3x} - 2e^{2x} & (x-2)e^{3x} + 3e^{2x} \end{pmatrix}. \quad (3.136)$$

By replacing x with $x-t$, we get:

$$\exp(A(x-t)) \cdot \mathbf{G}(t) = \begin{pmatrix} (-x+t)e^{3x-t} \\ (-2x+3+2t)e^{3x-t} - 2e^{2x} \\ (-x+2+t)e^{3x-t} - 2e^{2x} \end{pmatrix}.$$

We integrate each component of this vector function with respect to t on the interval $[0, x]$ and obtain:

$$\mathbf{Y}_P = \int_0^x \exp((x-t)A) \cdot \mathbf{G}(t)\, dt = \begin{pmatrix} (-x+1)e^{3x} - e^{2x} \\ (-2x+5)e^{3x} - (5+2x)e^{2x} \\ (-x+3)e^{3x} - (3+2x)e^{2x} \end{pmatrix}. \quad (3.137)$$

By applying the formula (3.126) with $\mathbf{c} = (c_1, c_2, c_3)^t$ a vector of arbitrary constants, the general solution of the system (3.133)) is:

$$\mathbf{Y} = \mathbf{Y}_{GH} + \mathbf{Y}_P = \exp(Ax)\mathbf{c} + \mathbf{Y}_P.$$

3.4.4 Elimination Method for Linear Systems with Constant Coefficients

In this section we give a general theoretical background to the elimination method for linear systems of differential equations with constant coefficients. The method works not only for first-order equations, but also for higher order linear equations.

3.4 Linear Systems with Constant Coefficients

For the beginning, we consider a linear first-order system with constant coefficients:

$$\begin{cases} y'_1(x) - a_{11}y_1(x) - \ldots - a_{1n}y_1(x) = g_1(x) \\ y'_2(x) - a_{21}y_1(x) - \ldots - a_{2n}y_n(x) = g_2(x) \\ \vdots \\ y'_n(x) - a_{n1}y_1(x) - \ldots - a_{nn}y_1(x) = g_n(x) \end{cases}, \quad (3.138)$$

where $x \in [a, b]$ and $a_{ij} \in \mathbb{R}$ for any $i, j = 1, 2, \ldots, n$.

We denote by $p = \frac{d}{dx}$, the usual differential operator, by I, the identity operator and write the differential equations above using the linear operators $L_{ii}(p) = p - a_{ii}\mathrm{I}$, and $L_{ij}(p) = -a_{ij}\mathrm{I}$ for $i \neq j$ and $i, j = 1, 2, \ldots, n$:

$$\begin{pmatrix} L_{11}(p) & L_{12}(p) & \cdots & L_{1n}(p) \\ L_{21}(p) & L_{22}(p) & \cdots & L_{2n}(p) \\ \vdots & \vdots & \ddots & \vdots \\ L_{n1}(p) & L_{n1}(p) & \cdots & L_{nn}(p) \end{pmatrix} \begin{pmatrix} y_1 \\ y_2 \\ \vdots \\ y_n \end{pmatrix} = \begin{pmatrix} g_1 \\ g_2 \\ \vdots \\ g_n \end{pmatrix} \quad (3.139)$$

The system (3.139) can represent *any* linear system of differential equations with constant coefficients (even of higher-order) using *polynomial differential operators* $L_{ij}(p)$. As we have seen in Sect. 2.4, these polynomial operators can be added or multiplied (and the multiplication is *commutative*) and they form a *commutative ring*, denoted by $\mathbb{R}[p]$ or $\mathbb{R}\left[\frac{d}{dx}\right]$. Let us multiply, for example, the operators $L_{11}(p)$ and $L_{22}(p)$ defined above:

$$L_{11}(p) \cdot L_{22}(p) = (p - a_{11}\mathrm{I})(p - a_{22}\mathrm{I}) = p^2 - (a_{11} + a_{22})p + a_{11}a_{22}\mathrm{I},$$

where $p^2 = \frac{d^2}{dx^2}$.

We denote by $D(p) \in \mathbb{R}[p]$ the determinant of the matrix on the left and by $A_{ij}(p)$ the *cofactor* of the element $L_{ij}(p)$, $i, j = 1, 2, \ldots, n$. Recall that $A_{ij}(p) = (-1)^{i+j}M_{ij}(p)$, where the *minor* $M_{ij}(p)$ is the determinant of the matrix obtained by removing from our matrix the ith row and the jth column.

We know (see [35], 1.51 and 1.52) that:

$$\sum_{k=1}^{n} A_{ki}(p)L_{ki}(p) = D(p), \quad (3.140)$$

for any $i = 1, 2, \ldots, n$ and

$$\sum_{k=1}^{n} A_{ki}(p)L_{kj}(p) = 0, \quad (3.141)$$

for any $i \neq j$, $i = 1, 2, \ldots, n$.

We denote by $L = \left(L_{ij}(p)\right)_{i,j=\overline{1,n}}$ the initial operator matrix of the system and by $L^* = \left(A_{ij}(p)\right)^t_{i,j=\overline{1,n}}$, the transpose of the cofactor matrix of L (the adjugate matrix). Let $\mathbf{G} = (g_1, g_2, \ldots, g_n)^t$ be the column vector functions on the right side of the equality (3.139). Thus (3.139) becomes:

$$\begin{pmatrix} D(p) & 0 & \cdots & 0 \\ 0 & D(p) & \cdots & 0 \\ \vdots & \vdots & \ddots & \vdots \\ 0 & 0 & \cdots & D(p) \end{pmatrix} \begin{pmatrix} y_1 \\ y_2 \\ \vdots \\ y_n \end{pmatrix} = L^*\mathbf{G}, \tag{3.142}$$

i.e. $y_i(x)$ is a solution of the linear differential equations of order n:

$$D(p)y_i = \sum_{k=1}^{n} A_{ki}(p)g_k, \tag{3.143}$$

$i = 1, 2, \ldots, n$.

We have obtained n linear differential equations of order n which can be solved independently. The general solution of each equation, $y_i(x, C_1^i, C_2^i, \ldots, C_n^i)$, $i = 1, \ldots, n$, depends on n arbitrary parameters, so we have n^2 parameters in total. But they are not free because $y_1, y_2, \ldots, y_n$ have to satisfy the initial system (3.138) for any $x \in [a, b]$. Thus, at the end, the solution of the system, $y_1, y_2, \ldots, y_n$ will depend on n free parameters, say $\alpha_1, \alpha_2, \ldots, \alpha_n$.

Example 3.17 Let us use the method described above to find the general solution of the system:

$$\begin{cases} y_1' = y_1 + y_2 + y_3 \\ y_2' = y_1 + y_2 + y_3 + 2x \\ y_3' = y_1 + y_2 + y_3 + 4x \end{cases} \tag{3.144}$$

The system can be written in the form (3.139) as:

$$\begin{pmatrix} p-1 & -1 & -1 \\ -1 & p-1 & -1 \\ -1 & -1 & p-1 \end{pmatrix} \begin{pmatrix} y_1 \\ y_2 \\ y_3 \end{pmatrix} = \begin{pmatrix} 0 \\ 2x \\ 4x \end{pmatrix}. \tag{3.145}$$

The determinant of the matrix is $D(p) = p^3 - 3p^2$ and the free term in Eq. (3.142) is:

$$L^*\mathbf{G} = \begin{pmatrix} p^2 - 2p & p & p \\ p & p^2 - 2p & p \\ p & p & p^2 - 2p \end{pmatrix} \begin{pmatrix} 0 \\ 2x \\ 4x \end{pmatrix} = \begin{pmatrix} 6 \\ 0 \\ -6 \end{pmatrix}.$$

3.4 Linear Systems with Constant Coefficients

Thus, we obtain the following differential equations of order 3:

$$\begin{cases} (p^3 - 3p^2)(y_1) = 6, \\ (p^3 - 3p^2)(y_2) = 0, \\ (p^3 - 3p^2)(y_3) = -6 \end{cases} \Leftrightarrow \begin{cases} y_1''' - 3y_1'' = 6 \\ y_2''' - 3y_2'' = 0 \\ y_3''' - 3y_3'' = -6 \end{cases}$$

with the general solutions:

$$\begin{aligned} y_1(x) &= C_1 + C_2 x + C_3 \exp(3x) - x^2 \\ y_2(x) &= D_1 + D_2 x + D_3 \exp(3x) \\ y_3(x) &= E_1 + E_2 x + E_3 \exp(3x) + x^2 \end{aligned},$$

where C_1, C_2, C_3, D_1, D_2, D_3, E_1, E_2, E_3 are real parameters, but only three of them are free. From the first equation of the system (3.144) we get:

$$\begin{cases} C_1 + D_1 + E_1 = C_2 \\ C_2 + D_2 + E_2 = -2 \\ C_3 + D_3 + E_3 = 3C_3 \end{cases}.$$

The second equation implies:

$$\begin{cases} C_1 + D_1 + E_1 = D_2 \\ C_2 + D_2 + E_2 = -2 \\ C_3 + D_3 + E_3 = 3D_3 \end{cases}.$$

The last equation gives us:

$$\begin{cases} C_1 + D_1 + E_1 = E_2 \\ C_2 + D_2 + E_2 = -2 \\ C_3 + D_3 + E_3 = 3E_3. \end{cases}$$

Finally we find: $C_3 = D_3 = E_3 = \alpha$, $C_2 = D_2 = E_2 = -\frac{2}{3}$, $C_1 = \beta$, $D_1 = \gamma$, $E_1 = -\frac{2}{3} - \beta - \gamma$. Thus, the general solution of the system (3.144) is:

$$\begin{cases} y_1(x) = \alpha \exp(3x) - x^2 - \frac{2}{3}x + \beta \\ y_2(x) = \alpha \exp(3x) - \frac{2}{3}x + \gamma \\ y_3(x) = \alpha \exp(3x) + x^2 - \frac{2}{3}(x+1) - \beta - \gamma, \end{cases}$$

where α, β, γ are arbitrary real numbers.

In the next example we show how the method can be used to find the general solution for higher order linear systems of differential equations.

Example 3.18 Find the general solution for the following system:

$$\begin{cases} y_1'' + y_2' = \exp x \\ y_1'' - y_1 + y_2'' - y_2 = -x \end{cases}, \quad x \in \mathbb{R}. \tag{3.146}$$

We write the system using the differential operator $p = \frac{d}{dx}$ and the identity operator I:

$$\begin{cases} p^2 y_1 + p y_2 = \exp x \\ (p^2 - \mathrm{I}) y_1 + (p^2 - \mathrm{I}) y_2 = -x \end{cases},$$

or, equivalently,

$$\begin{pmatrix} p^2 & p \\ p^2 - \mathrm{I} & p^2 - \mathrm{I} \end{pmatrix} \begin{pmatrix} y_1 \\ y_2 \end{pmatrix} = \begin{pmatrix} \exp x \\ -x \end{pmatrix}.$$

The determinant of the matrix on the left is: $D(p) = p(p - \mathrm{I})^2(p + \mathrm{I})$, and the adjugate matrix is:

$$L^* = \begin{pmatrix} p^2 - \mathrm{I} & -p \\ -p^2 + \mathrm{I} & p^2 \end{pmatrix}.$$

So

$$\begin{pmatrix} D(p) & 0 \\ 0 & D(p) \end{pmatrix} \begin{pmatrix} y_1 \\ y_2 \end{pmatrix} = \begin{pmatrix} p^2 - \mathrm{I} & -p \\ -p^2 + \mathrm{I} & p^2 \end{pmatrix} \begin{pmatrix} \exp x \\ -x \end{pmatrix} = \begin{pmatrix} 1 \\ 0 \end{pmatrix},$$

and the differential equations for y_1, y_2 are:

$$\begin{cases} D(p) y_1 = 1 \\ D(p) y_2 = 0 \end{cases}. \tag{3.147}$$

Consider the first equation: $p(p - \mathrm{I})^2 (p + \mathrm{I}) y_1 = 1$. The general solution is:

$$y_1(x) = C_1 \exp x + C_2 x \exp x + C_3 \exp(-x) + C_4 + y_{1,P}(x),$$

where $y_{1,P}(x) = C_5 x$, is a particular solution of our non-homogeneous equation. Since

$$p(p - \mathrm{I})^2 (p + \mathrm{I}) y_1 = y_1^{(4)} - y_1''' - y_1'' + y_1',$$

3.4 Linear Systems with Constant Coefficients

the solution $y_{1,P}(x) = C_5 x$ has to satisfy the differential equation:

$$y_1^{(4)} - y_1''' - y_1'' + y_1' = 1.$$

We find $C_5 = 1$, so

$$y_1(x) = C_1 \exp x + C_2 x \exp x + C_3 \exp(-x) + C_4 + x \tag{3.148}$$

The second equation in (3.147) is homogeneous,

$$p(p-1)^2(p+1)y_2 = 0,$$

and its general solution is:

$$y_2(x) = D_1 \exp x + D_2 x \exp x + D_3 \exp(-x) + D_4. \tag{3.149}$$

Since our initial system is equivalent (has the same set of solutions) with a first-order linear system of four differential equations, only four of the parameters C_i, D_i, $i = 1, \ldots, 4$ are free.

The solutions $y_1(x)$, $y_2(x)$ from (3.148) and (3.149) respectively have to satisfy the initial system (3.146). We calculate the derivatives:

$$\begin{aligned}
y_1' &= (C_1 + C_2) \exp x + C_2 x \exp x - C_3 \exp(-x) + 1, \\
y_1'' &= (C_1 + 2C_2) \exp x + C_2 x \exp x + C_3 \exp(-x), \\
y_2' &= (D_1 + D_2) \exp x + D_2 x \exp x - D_3 \exp(-x), \\
y_2'' &= (D_1 + 2D_2) \exp x + D_2 x \exp x + D_3 \exp(-x),
\end{aligned}$$

and replace in the initial system (3.146):

$$\begin{cases} (C_1 + 2C_2 + D_1 + D_2)e^x + (C_2 + D_2)xe^x + (C_3 - D_3)e^{-x} = e^x \\ 2(C_2 + D_2)e^x - (C_4 + D_4) - x = -x. \end{cases}$$

Hence we obtain:

$$\begin{cases} C_1 + 2C_2 + D_1 + D_2 = 1, \\ C_2 + D_2 = 0, \\ C_3 - D_3 = 0 \\ C_4 + D_4 = 0. \end{cases}$$

Thus

$$D_1 = 1 - C_1 - C_2, \quad D_2 = -C_2, \quad D_3 = C_3, \quad D_4 = -C_4,$$

and the general solution of the system (3.146) is:

$$\begin{pmatrix} y_1 \\ y_2 \end{pmatrix} = C_1 \exp(x) \begin{pmatrix} 1 \\ -1 \end{pmatrix} + C_2 \exp(x) \begin{pmatrix} x \\ -x-1 \end{pmatrix} +$$

$$+ C_3 \exp(-x) \begin{pmatrix} 1 \\ 1 \end{pmatrix} + C_4 \begin{pmatrix} 1 \\ -1 \end{pmatrix} + \begin{pmatrix} x \\ \exp x \end{pmatrix}.$$

3.5 Autonomous Systems of Differential Equations

A system of differential equations is said to be an *autonomous system* if the equations does not explicitly depend on the independent variable. When the independent variable is time, autonomous systems are also called time-invariant systems. Such systems are the mathematical expression of many laws of Physics (since these laws remain unchanged in time).

Consider the following first-order autonomous system of differential equations:

$$\begin{cases} y_1' = f_1(y_1, y_2, \ldots, y_n) \\ y_2' = f_2(y_1, y_2, \ldots, y_n) \\ \quad \vdots \\ y_n' = f_n(y_1, y_2, \ldots, y_n) \end{cases}, x \in I, \qquad (3.150)$$

where I is a real interval and $f_1, \ldots, f_n$ are continuously differentiable functions defined on a domain $D \subset \mathbb{R}^n$.

Using the notations $\mathbf{Y} = (y_1, \ldots, y_n)$ and $\mathbf{F} = (f_1, \ldots, f_n)$, the system (3.150) can be also written in the compact form:

$$\mathbf{Y}' = \mathbf{F}(\mathbf{Y}). \qquad (3.151)$$

Definition 3.2 (Pontryagin [27], 23) A *first integral* for the autonomous system (3.150) is a non-constant, continuously differentiable function of n variables, $\varphi(z_1, \ldots, z_n)$ with the property that, for any solution of the system, $(y_1, \ldots, y_n)$, the expression $\varphi(y_1(x), \ldots, y_n(x))$ is constant with respect to x, i.e.

$$\frac{d}{dx} [\varphi(x, y_1(x), \ldots, y_n(x))] = 0, \qquad (3.152)$$

for any $x \in I$.

Theorem 3.11 *A continuously differentiable, non-constant function $\varphi(z_1, \ldots, z_n)$ defined on $D \subset \mathbb{R}^n$ is a first integral of the system (3.150) if and only if the following*

3.5 Autonomous Systems of Differential Equations

equality holds, for any $\mathbf{z} = (z_1, \ldots, z_n) \in D$

$$f_1(\mathbf{z}) \cdot \frac{\partial \varphi}{\partial z_1}(\mathbf{z}) + f_2(\mathbf{z}) \cdot \frac{\partial \varphi}{\partial z_2}(\mathbf{z}) + \ldots + f_n(\mathbf{z}) \cdot \frac{\partial \varphi}{\partial z_n}(\mathbf{z}) = 0. \qquad (3.153)$$

Proof The relation (3.152) can be also written as:

$$\frac{\partial \varphi}{\partial z_1}(y_1(x), \ldots, y_n(x)) \cdot y_1'(x) + \ldots + \frac{\partial \varphi}{\partial z_n}(y_1(x), \ldots, y_n(x)) \cdot y_n'(x) = 0$$

hence φ is a first integral of the system (3.150) if and only if, for any solution of the system, $\mathbf{Y} = (y_1, \ldots, y_n)$, we have:

$$\frac{\partial \varphi}{\partial z_1}(\mathbf{Y}(x)) \cdot f_1(\mathbf{Y}(x)) + \ldots + \frac{\partial \varphi}{\partial z_n}(\mathbf{Y}(x)) \cdot f_n(\mathbf{Y}(x)) = 0, \qquad (3.154)$$

for all $x \in I$.

Let φ be a first integral of the system (3.150), $\mathbf{z} = (z_1, \ldots, z_n) \in D$ and x_0 be a point in the interval I. By Theorem 3.1, there exists a unique solution $(y_1, \ldots, y_n)$ of the system such that $y_1(x_0) = z_1, \ldots, y_n(x_0) = z_n$. We write the equality (3.154) for $x = x_0$ and the equality (3.153) follows.

Conversely, suppose that (3.153) holds for any $\mathbf{z} = (z_1, \ldots, z_n) \in D$. Let $\mathbf{Y} = (y_1, \ldots, y_n)$ be a solution of the system (3.150). For any $x \in I$, if we write the equality (3.153) for $z_1 = y_1(x), \ldots, z_n = y_n(x)$ we obtain (3.154), so φ is a first integral of the system. □

Example 3.19 (The Law of Conservation of Energy for Conservative Fields) We recall that a $3D$-field of forces

$$\mathbf{F}(x, y, z) = f_1(x, y, z)\mathbf{i} + f_2(x, y, z)\mathbf{j} + f_3(x, y, z)\mathbf{k},$$

defined on a domain $\Omega \subset \mathbb{R}^3$ is said to be a *conservative field* if there exists a scalar function $U : \Omega \to \mathbb{R}$ whose gradient,

$$\operatorname{grad} U = \nabla U = \frac{\partial U}{\partial x}\mathbf{i} + \frac{\partial U}{\partial y}\mathbf{j} + \frac{\partial U}{\partial z}\mathbf{k},$$

is $\nabla U = -\mathbf{F}$ or, equivalently,

$$\frac{\partial U}{\partial x} = -f_1, \quad \frac{\partial U}{\partial y} = -f_2, \quad \frac{\partial U}{\partial z} = -f_3. \qquad (3.155)$$

The function $U(x, y, z)$ is called the potential energy of $\mathbf{F}$ at the point $P(x, y, z)$. If $\mathbf{v} = x'(t)\mathbf{i} + y'(t)\mathbf{j} + z'(t)\mathbf{k}$ is the velocity of a moving point $M(x(t), y(t), z(t))$ at time t, in the field $\mathbf{F}$, then the kinetic energy is

$$T = \frac{m}{2}\|\mathbf{v}\|^2 = \frac{m}{2}\left(x'^2 + y'^2 + z'^2\right).$$

Let us prove that the total energy $E = T + U$ is constant with respect to t, i.e. it is a first integral of the following dynamic system of differential equations:

$$\begin{cases} x''(t) = \frac{1}{m} f_1(x(t), y(t), z(t)) \\ y''(t) = \frac{1}{m} f_2(x(t), y(t), z(t)) \\ z''(t) = \frac{1}{m} f_3(x(t), y(t), z(t)) \end{cases} \quad (3.156)$$

Indeed,

$$\frac{dE}{dt} = \frac{dT}{dt} + \frac{dU}{dt} = m(x'x'' + y'y'' + z'z'') + \frac{\partial U}{\partial x}x' + \frac{\partial U}{\partial y}y' + \frac{\partial U}{\partial z}z'.$$

From (3.155) and (3.156) we have:

$$mx'' = f_1 = -\frac{\partial U}{\partial x}, \quad my'' = f_2 = -\frac{\partial U}{\partial y}, \quad mz'' = f_3 = -\frac{\partial U}{\partial z},$$

so we obtain that

$$\frac{dE}{dt} = 0,$$

i.e. the total energy E is a constant. This result is the mathematical expression of the *law of conservation of energy*.

Remark 3.7 It is easy to see that, if $\varphi_1, \ldots, \varphi_k$ are first integrals for the system (3.4) and if $G(x_1, \ldots, x_k)$ is a non-constant function of k variables of class C^1, then the function

$$\varphi(z_1, \ldots, z_n) = G\left(\varphi_1(z_1, \ldots, z_n), \ldots, \varphi_k(z_1, \ldots, z_n)\right)$$

is also a first integral for our system.

A set $\{\varphi_1, \ldots, \varphi_k\}$, $2 \le k \le n$, of first integrals is said to be *independent* if no one of them can be written as a function of the other $k - 1$, i.e if the functions $\varphi_1, \ldots, \varphi_k$ are functionally independent (see [28], 11.3). We know that $\{\varphi_1, \ldots, \varphi_k\}$, $\varphi_i = \varphi_i(x, z_1, \ldots, z_n)$, $i = 1, \ldots, k$, $k \le n$, are functionally

3.5 Autonomous Systems of Differential Equations

independent if and only if the Jacobian matrix

$$\begin{pmatrix} \frac{\partial \varphi_1}{\partial z_1} & \frac{\partial \varphi_1}{\partial z_2} & \cdots & \frac{\partial \varphi_1}{\partial z_n} \\ \frac{\partial \varphi_2}{\partial z_1} & \frac{\partial \varphi_2}{\partial z_2} & \cdots & \frac{\partial \varphi_2}{\partial z_n} \\ \vdots & \vdots & & \vdots \\ \frac{\partial \varphi_k}{\partial z_1} & \frac{\partial \varphi_k}{\partial z_2} & \cdots & \frac{\partial \varphi_k}{\partial z_n} \end{pmatrix} \tag{3.157}$$

has the rank equal to k.

A point $\mathbf{z} \in D$ is said to be a *critical point* of the system (3.150) if $f_i(\mathbf{z}) = 0$, $i = 1, \ldots, n$.

The next theorem states that the maximum number of *independent* first integrals of the system (3.150) is $n - 1$ (or, equivalently, any n first integrals of the system are functionally dependent). At the same time, there always exist $n - 1$ independent first integrals in a subset of D which does not contain any *critical point*.

Theorem 3.12 (Arnold [2], p. 127) *If $f_1, \ldots, f_n$ are continuously differentiable functions such that*

$$(f_1(\mathbf{z}), \ldots, f_n(\mathbf{z})) \neq (0, \ldots, 0), \ \forall \mathbf{z} \in D,$$

then the system (3.150) has $n - 1$ independent first integrals, $\varphi_1, \ldots, \varphi_{n-1}$, and any first integral of the system is a function of $\varphi_1, \ldots, \varphi_{n-1}$.

Proof Let $\varphi_1, \ldots, \varphi_{n-1}$ be $n - 1$ independent first integrals of the system (3.150) (see [27], p. 184 for the proof of the existence of $n - 1$ independent first integrals). Let φ_n be another first integral of the system. By Theorem 3.11 we have:

$$\begin{cases} \frac{\partial \varphi_1}{\partial z_1}(\mathbf{z}) \cdot f_1(\mathbf{z}) + \ldots + \frac{\partial \varphi_1}{\partial z_n}(\mathbf{z}) \cdot f_n(\mathbf{z}) = 0 \\ \vdots \\ \frac{\partial \varphi_n}{\partial z_1}(\mathbf{z}) \cdot f_1(\mathbf{z}) + \ldots + \frac{\partial \varphi_n}{\partial z_n}(\mathbf{z}) \cdot f_n(\mathbf{z}) = 0. \end{cases} \tag{3.158}$$

Since $(f_1(\mathbf{z}), \ldots, f_n(\mathbf{z})) \neq (0, \ldots, 0)$, the homogeneous system of (algebraic) Eq. (3.105) must have the determinant equal to 0:

$$\frac{D(\varphi_1 \ldots, \varphi_n)}{D(z_1, \ldots, z_n)} = \begin{vmatrix} \frac{\partial \varphi_1}{\partial z_1}(\mathbf{z}) & \cdots & \frac{\partial \varphi_1}{\partial z_n}(\mathbf{z}) \\ \vdots & & \vdots \\ \frac{\partial \varphi_n}{\partial z_1}(\mathbf{z}) & \cdots & \frac{\partial \varphi_n}{\partial z_n}(\mathbf{z}) \end{vmatrix} = 0$$

It follows that the rank of the Jacobian matrix is less than n, hence $\varphi_1, \ldots, \varphi_n$ are functionally dependent. By the implicit function theorem ([28], Theorem 83), since

$\varphi_1, \ldots, \varphi_{n-1}$ are independent, we obtain that the first integral φ_n can be written as a function of $\varphi_1, \ldots, \varphi_{n-1}$. $\square$

The general solution of an autonomous system having the independent first integrals $\varphi_1, \ldots, \varphi_{n-1}$, is represented by the intersection of the hypersurfaces

$$\begin{cases} \varphi_1(y_1, \ldots, y_n) = C_1 \\ \vdots \\ \varphi_{n-1}(y_1, \ldots, y_n) = C_{n-1} \end{cases} \tag{3.159}$$

Notice that we do not find the *explicit solution* $y_1 = y_1(x), \ldots, y_2 = y_2(x)$, but the *trajectory* in the *phase space* (3.159).

Thus, to solve an autonomous system means to find $n - 1$ independent first integrals. By using for derivatives the Leibniz notation $y'_i = \dfrac{dy_i}{dx}$, the system (3.150) can be written as

$$\begin{cases} \dfrac{dy_1}{dx} = f_1 = f_1(y_1, \ldots, y_n) \\ \vdots \\ \dfrac{dy_n}{dx} = f_n = f_n(y_1, \ldots, y_n) \end{cases}$$

and the following *symmetric form* of the system is obtained:

$$\frac{dy_1}{f_1} = \frac{dy_2}{f_2} = \ldots = \frac{dy_n}{f_n}. \tag{3.160}$$

In this notation, some pairs of ratios may be directly integrated (by separation of variables) to find a first integral. Another way to find a first integral is the *integrable combinations method*, based on an elementary property of a sequence of equal fractions. If

$$\frac{a_1}{b_1} = \frac{a_2}{b_2} = \ldots = \frac{a_n}{b_n}, \tag{3.161}$$

then, for any $\lambda_1, \ldots, \lambda_n \in \mathbb{R}$, we have:

$$\frac{a_1}{b_1} = \frac{a_2}{b_2} = \ldots = \frac{a_n}{b_n} = \frac{\sum_{i=1}^n \lambda_i a_i}{\sum_{i=1}^n \lambda_i b_i}. \tag{3.162}$$

The integrable combinations method consists of finding n functions $\lambda_1 = \lambda_1(y_1, \ldots, y_n), \lambda_2 = \lambda_2(y_1, \ldots, y_n), \ldots, \lambda_n = \lambda_n(y_1, \ldots, y_n)$ such that

$$\lambda_1 f_1 + \ldots + \lambda_n f_n = 0 \tag{3.163}$$

3.5 Autonomous Systems of Differential Equations

and the differential form $\omega = \lambda_1 dy_1 + \ldots + \lambda_n dy_n$ is *exact*, i.e. there exists a differentiable function $U(y_1, \ldots, y_n)$ such that

$$dU = \lambda_1 dy_1 + \ldots + \lambda_n dy_n. \tag{3.164}$$

By applying (3.162) we obtain:

$$\frac{dy_1}{f_1} = \frac{dy_2}{f_2} = \ldots = \frac{dy_n}{f_n} = \frac{\lambda_1 dy_1 + \ldots + \lambda_n dy_n}{\lambda_1 f_1 + \ldots + \lambda_n f_n} = \frac{dU}{0}.$$

Since all these fractions are finite, we have to conclude that $dU = 0$, or, equivalently, for any solution $(y_1(x), \ldots, y_n(x))$ of the autonomous system (3.160), we have: $U(y_1(x), \ldots, y_n(x)) = C$, a constant. Hence, the function $U(y_1, \ldots, y_n)$ is a first integral of the autonomous system.

Consider the autonomous system

$$\begin{cases} x' = f_1(x, y, z) \\ y' = f_2(x, y, z) \\ z' = f_3(x, y, z), \end{cases} \tag{3.165}$$

where the independent variable is time t and $(x(t), y(t), z(t))$ is the position of a particle at time t. Suppose that we have found two independent first integrals of the system, $\varphi_1(x, y, z)$ and $\varphi_2(x, y, z)$. The general solution of the system is:

$$\begin{cases} \varphi_1(x, y, z) = C_1 \\ \varphi_2(x, y, z) = C_2. \end{cases} \tag{3.166}$$

If we know that the trajectory of the particle passes through a given point (x_0, y_0, z_0), then we can calculate the constants C_1 and C_2,

$$C_1 = \varphi_1(x_0, y_0, z_0), \quad C_2 = \varphi_2(x_0, y_0, z_0),$$

and find the unique solution of the Cauchy problem: the curve represented as the intersection of the surfaces (3.166).

Example 3.20 Solve the following Cauchy problem:

$$\begin{cases} y_1' = y_1 y_3 \\ y_2' = y_2 y_3 \\ y_3' = \left(y_1^2 + y_2^2\right) y_3 \end{cases}, \quad y_1(0) = y_2(0) = y_3(0) = 1, \; x \in \mathbb{R}.$$

It is an autonomous system, so we can write it in the symmetric form:

$$\frac{dy_1}{y_1 y_3} = \frac{dy_2}{y_2 y_3} = \frac{dy_3}{(y_1^2 + y_2^2) y_3},$$

$y_3 \neq 0$ and $y_1^2 + y_2^2 \neq 0$. From the first equality we find that $\frac{dy_1}{y_1} = \frac{dy_2}{y_2}$, or $\int \frac{dy_1}{y_1} = \int \frac{dy_2}{y_2}$, which implies that $y_2 = C_1 y_1$. Hence $\varphi_1(y_1, y_2, y_3) = \frac{y_2}{y_1}$ is a first integral of the system. Now, for $\lambda_1(y_1, y_2, y_3) = y_1$, $\lambda_2(y_1, y_2, y_3) = y_2$ and $\lambda_3(y_1, y_2, y_3) = -1$, one obtains

$$\omega = y_1 dy_1 + y_2 dy_2 - dy_3 = 0.$$

Hence $U(y_1, y_2, y_3) = \frac{y_1^2}{2} + \frac{y_2^2}{2} - y_3 = C_2$, so $\varphi_2(y_1, y_2, y_3) = y_1^2 + y_2^2 - 2y_3$ is another first integral of our system. Since φ_2 contains the variable y_3 and φ_1 does not contain it, φ_2 cannot be a function of φ_1. Thus $\{\varphi_1, \varphi_2\}$ are functionally independent. Hence, the general solution of the system can be described as the following family of trajectories:

$$\begin{cases} y_2 = C_1 y_1 \\ y_1^2 + y_2^2 - 2y_3 = C_2 \end{cases},$$

i.e. a family of parabolas in the phases space of the variables y_1, y_2, y_3. The initial conditions imply that $C_1 = 1$ and $C_2 = 0$, so the solution is the parabola which results as the intersection between the plane $y_2 = y_1$ and the paraboloid $y_1^2 + y_2^2 - 2y_3 = 0$.

Sometimes it is useful to use a first integral in order to find another one.

Example 3.21 Let us find two independent first integrals for the autonomous system (written in the symmetric form):

$$\frac{dy_1}{y_1} = \frac{dy_2}{y_2} = \frac{dy_3}{(y_1 + y_2)y_3}.$$

From the first equality we find $y_2 = C_1 y_1$. Let us take now the equality:

$$\frac{dy_1}{y_1} = \frac{dy_3}{(y_1 + y_2)y_3}$$

and substitute here y_2 with $C_1 y_1$. We obtain:

$$\frac{dy_1}{y_1} = \frac{dy_3}{y_1 y_3 (1 + C_1)},$$

or

$$\frac{dy_1}{1} = \frac{dy_3}{y_3 (1 + C_1)}.$$

3.5 Autonomous Systems of Differential Equations

Hence $y_1 = \frac{1}{1+C_1} \ln|y_3| + \ln C_2$, or $C_2 y_3^{\frac{1}{1+C_1}} = \exp(y_1)$. Now, instead of C_1 we put $\frac{y_2}{y_1}$. So $\exp(y_1) = C_2 y_3^{\frac{y_1}{y_1+y_2}}$. Hence, the first integrals are $\varphi_1(y_1, y_2, y_3) = \frac{y_2}{y_1}$ and $\varphi_2(y_1, y_2, y_3) = y_3^{-\frac{y_1}{y_1+y_2}} \exp(y_1)$. It is clear that φ_1 and φ_2 are functionally independent because the second function depends on the variable y_3, while the first does not.

Note that we could also use the integrable combinations method in order to find the second first integral. For $\lambda_1(y_1, y_2, y_3) = 1$, $\lambda_2(y_1, y_2, y_3) = 1$ and $\lambda_3(y_1, y_2, y_3) = -\frac{1}{y_3}$, one obtains

$$\omega = dy_1 + dy_2 - \frac{dy_3}{y_3} = 0.$$

Hence $U(y_1, y_2, y_3) = y_1 + y_2 - \ln|y_3| = \ln C_2$, so $\varphi_3(y_1, y_2, y_3) = \frac{\exp(y_1+y_2)}{y_3}$ is another first integral of our system (independent of φ_1).

Example 3.22 Solve the following autonomous system

$$\frac{dx}{y-z} = \frac{dy}{z-x} = \frac{dz}{x-y} \tag{3.167}$$

by finding two independent first integrals and the corresponding trajectory for the initial conditions $x(0) = 1$, $y(0) = 2$ and $z(0) = 3$.

It is easy to see that

$$\frac{dx}{y-z} = \frac{dy}{z-x} = \frac{dz}{x-y} = \frac{d(x+y+z)}{0},$$

so the first integral is $\varphi_1(x, y, z) = x + y + z = C_1$; which generates a family of parallel planes. Now, one also can write

$$\frac{2x\,dx}{2x(y-z)} = \frac{2y\,dy}{2y(z-x)} = \frac{2z\,dz}{2z(x-y)} = \frac{d(x^2+y^2+z^2)}{0}.$$

Thus, another first integral is $\varphi_2(x, y, z) = x^2 + y^2 + z^2 = C_2$, which generates a family of spheres. Let us find the domain where these two first integrals are independent. For this, let us compute the corresponding functional Jacobian matrix

$$\begin{pmatrix} \frac{\partial \varphi_1}{\partial x} & \frac{\partial \varphi_1}{\partial y} & \frac{\partial \varphi_1}{\partial z} \\ \frac{\partial \varphi_2}{\partial x} & \frac{\partial \varphi_2}{\partial y} & \frac{\partial \varphi_2}{\partial z} \end{pmatrix} = \begin{pmatrix} 1 & 1 & 1 \\ 2x & 2y & 2z \end{pmatrix}.$$

We can see that its rank is 2 if and only if the point (x, y, z) is not on the line $x = y = z$. Thus, if we consider an arbitrary domain D which does not intersect

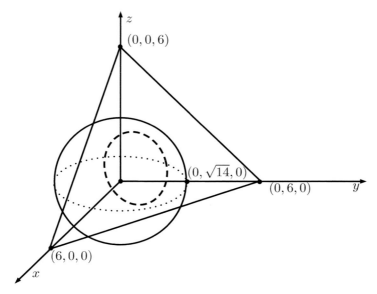

Fig. 3.2 The circle (C) of intersection between the sphere $x^2 + y^2 + z^2 = 14$ and the plane $x + y + z = 6$

this line, on such a domain, φ_1 and φ_2 are independent. The point $(1, 2, 3)$ which appears in the initial conditions is not on this line, so, in a neighborhood D of this point which does not intersect the line $x = y = z$, the first integrals are independent and we can solve the Cauchy problem. Namely, for $x = 1$, $y = 2$, $z = 3$, we get $C_1 = 6$ and $C_2 = 14$. Thus, the trajectory is the circle (see Fig. 3.2):

$$(C): \begin{cases} x + y + z = 6 \\ x^2 + y^2 + z^2 = 14 \end{cases}.$$

3.6 First-Order Partial Differential Equations

The last section of this chapter is devoted to *first-order partial differential equations*. The reason why we included this section in the chapter about systems of *ordinary* differential equations is the close relation between the linear/quasilinear partial differential equations and the autonomous systems presented in Sect. 3.5. Theorem 3.11 makes this close relation obvious.

3.6 First-Order Partial Differential Equations

3.6.1 Linear Homogeneous First-Order PDE

We begin with the following example:

Example 3.23 Consider a domain $D \subset \mathbb{R}^2$ (an open connected set) in the xOy plane and $\mathbf{F}(x, y) = P(x, y)\mathbf{i} + Q(x, y)\mathbf{j}$, a continuous vector field defined on D such that $\mathbf{F}(x, y) \neq \mathbf{0}$, for any $(x, y) \in D$. We want to find all the smooth surfaces $(S) : z = u(x, y)$ (*smooth* means that $u \in C^1(D)$) such that, for any point of (S), $M_0(x_0, y_0, u(x_0, y_0))$, the tangent plane to the surface at M_0 is parallel to the direction of the vector $\mathbf{F}(x_0, y_0) = P(x_0, y_0)\mathbf{i} + Q(x_0, y_0)\mathbf{j}$. Since the implicit equation of the surface (S) is $z - u(x, y) = 0$, the normal vector to the tangent plane at M_0 is $\mathbf{n}(x_0, y_0) = -\frac{\partial u}{\partial x}(x_0, y_0)\mathbf{i} - \frac{\partial u}{\partial y}(x_0, y_0)\mathbf{j} + \mathbf{k}$. Thus, the required condition becomes:

$$P(x_0, y_0)\frac{\partial u}{\partial x}(x_0, y_0) + Q(x_0, y_0)\frac{\partial u}{\partial y}(x_0, y_0) = 0 \qquad (3.168)$$

for any point $M_0(x_0, y_0, z_0) \in (S)$ (see Fig. 3.3). Hence, we want to find all functions $u = u(x, y)$ of class C^1, defined on D such that

$$P(x, y)\frac{\partial u}{\partial x}(x, y) + Q(x, y)\frac{\partial u}{\partial y}(x, y) = 0, \qquad (3.169)$$

The Eq. (3.169) is a *first order partial differential equation* (it contains the first order partial derivatives $\frac{\partial u}{\partial x}$ and $\frac{\partial u}{\partial y}$ of an unknown function $u(x, y)$). The equation is linear (with respect to $\frac{\partial u}{\partial x}$ and $\frac{\partial u}{\partial y}$) and homogeneous (the free term is zero).

Fig. 3.3 The surface (S) in Example 3.23

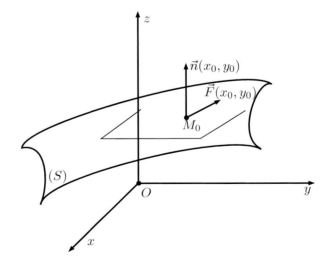

The general form of a *linear homogeneous first-order partial differential equation* is:

$$f_1(x_1,\ldots,x_n)\frac{\partial u}{\partial x_1}(x_1,\ldots,x_n)+\ldots+f_n(x_1,\ldots,x_n)\frac{\partial u}{\partial x_n}(x_1,\ldots,x_n)=0, \tag{3.170}$$

where $u=u(x_1,\ldots,x_n)$ is the unknown function, of class C^1 on a fixed domain $D\subset\mathbb{R}^n$, $x_1,\ldots,x_n$ are free variables such that $(x_1,\ldots,x_n)\in D$ and $f_1,\ldots,f_n$ are functions of class C^1 defined on D, such that $\sum_{i=1}^{n}f_i^2(\mathbf{x})\neq 0$ at any point $\mathbf{x}=(x_1,\ldots,x_n)\in D$.

We associate to the partial differential Eq. (3.170) the autonomous system in the unknown functions $x_1,\ldots,x_n$, where $x_1=x_1(t),\ldots,x_n=x_n(t)$, $t\in I$, (I is a real interval) considered in the phase space $\mathbb{R}^n$ (see (3.160)):

$$\frac{dx_1}{f_1}=\frac{dx_2}{f_2}=\ldots=\frac{dx_n}{f_n}. \tag{3.171}$$

The solutions of this associated system are called the *characteristic curves* of the partial differential Eq. (3.170) (see [9]).

The next theorem, which gives the general solution of Eq. (3.170) is a direct result of Theorems 3.11 and 3.12.

Theorem 3.13 *A non-constant function $u(x_1,\ldots,x_n)$ is a solution of the linear homogeneous Eq. (3.170) if and only if it is a first integral of the autonomous system (3.171). Moreover, if $\varphi_1,\ldots,\varphi_{n-1}$ are independent first integrals of the autonomous system (3.171), then the general solution of Eq. (3.170) is*

$$u(x_1,\ldots,x_n)=F(\varphi_1(x_1,\ldots,x_n),\ldots,\varphi_{n-1}(x_1,\ldots,x_n)), \tag{3.172}$$

where $F(z_1,\ldots,z_{n-1})$ is an arbitrary function of class C^1.

Example 3.24 Find the general solution of the Eq. (3.169) from Example 3.23 in the particular case when $P(x,y)=x$ and $Q(x,y)=y$.

We have to find the general solution of the partial differential equation

$$x\frac{\partial u}{\partial x}+y\frac{\partial u}{\partial y}=0. \tag{3.173}$$

Its associated system is:

$$\frac{dx}{x}=\frac{dy}{y}.$$

By a simple integration we find that $\frac{x}{y}=C_1$, (the characteristic curves of the partial differential Eq. (3.173) are the straight lines $\frac{x}{y}=C_1$). Hence $\varphi(x,y)=\frac{x}{y}$ is a

3.6 First-Order Partial Differential Equations

first integral of the system. Since $n = 2$ in this case, the maximum number of independent first integrals is 1. So, the general solution of the Eq. (3.173) is:

$$u(x, y) = F\left(\frac{x}{y}\right), \quad (3.174)$$

where F is an arbitrary function of class C^1. Let us verify that the function $u(x, y)$ defined by (3.174) satisfies Eq. (3.173):

$$\frac{\partial u}{\partial x}(x, y) = F'\left(\frac{x}{y}\right) \cdot \frac{1}{y}, \quad \frac{\partial u}{\partial y}(x, y) = F'\left(\frac{x}{y}\right) \cdot \left(-\frac{x}{y^2}\right),$$

hence

$$x \frac{\partial u}{\partial x} + y \frac{\partial u}{\partial y} = F'\left(\frac{x}{y}\right) \cdot \frac{x}{y} - F'\left(\frac{x}{y}\right) \cdot \frac{x}{y} = 0.$$

Equation (3.174) defines a family of surfaces $z = u(x, y)$, representing the general solution of the partial differential Eq. (3.173). If we have to find a particular surface in this family knowing the curve of intersection with the plane $y = a$, $z = g(x)$, we say that we have to solve a *Cauchy problem*. For instance, a Cauchy problem for Eq. (3.173) is to find the particular solution $u(x, y)$ that verifies the initial condition $u(x, 1) = 4x$ (find the surface $z = u(x, y) = F\left(\frac{x}{y}\right)$ that contains the straight line $y = 1, z = 4x$).

By (3.174) and the initial condition we obtain that $F(x) = 4x$, hence, the solution of the Cauchy problem is $u(x, y) = \frac{4x}{y}$. The surface defined by this equation, $z = \frac{4x}{y}$, or, equivalently, $4x = yz$, is a hyperbolic paraboloid.

A *Cauchy problem* for the linear homogeneous partial differential Eq. (3.170) is the problem of finding a solution $u(x_1, \ldots, x_n)$ which satisfies an initial condition of the type:

$$u(x_1, \ldots, x_{i-1}, a, x_{i+1}, \ldots, x_n) = g(x_1, \ldots, x_{i-1}, x_{i+1}, \ldots, x_n),$$

where g is a given function of class C^1 of $n - 1$ variables and a is a fixed real number. Here i is an arbitrary index between 1 and n. For simplicity, we assume that $i = n$ and write the initial condition:

$$u(x_1, \ldots, x_{n-1}, a) = g(x_1, \ldots, x_{n-1}), \quad (3.175)$$

How can the Cauchy problem be solved ? First of all, we find the general solution of (3.170),

$$u(x_1, \ldots, x_n) = F(\varphi_1(x_1, \ldots, x_n), \ldots, \varphi_{n-1}(x_1, \ldots, x_n)), \quad (3.176)$$

where $\varphi_1, \ldots, \varphi_{n-1}$ are $n-1$ independent first integrals of the autonomous system (3.171) and F is an arbitrary function of class C^1. Using the initial condition (3.175) in the general solution (3.176) we obtain:

$$g(x_1, \ldots, x_{n-1}) = F(\varphi_1(x_1, \ldots, x_{n-1}, a), \ldots, \varphi_{n-1}(x_1, \ldots, x_{n-1}, a)). \tag{3.177}$$

In order to find the expression of F, we denote

$$\begin{cases} \varphi_1(x_1, \ldots, x_{n-1}, a) = z_1 \\ \varphi_2(x_1, \ldots, x_{n-1}, a) = z_2 \\ \vdots \\ \varphi_{n-1}(x_1, \ldots, x_{n-1}, a) = z_{n-1} \end{cases}. \tag{3.178}$$

Since $\varphi_1, \ldots, \varphi_{n-1}$ are independent, by the implicit function theorem ([28], Theorem 83), the system (3.178) can be solved for $x_1, \ldots, x_{n-1}$: we can write $x_1, \ldots, x_{n-1}$ as continuously differentiable functions of $z_1, \ldots, z_{n-1}$:

$$\begin{cases} x_1 = h_1(z_1, \ldots, z_{n-1}) \\ x_2 = h_2(z_1, \ldots, z_{n-1}) \\ \vdots \\ x_{n-1} = h_{n-1}(z_1, \ldots, z_{n-1}). \end{cases} \tag{3.179}$$

Using (3.179) in (3.177), we find the expression of the function F:

$$F(z_1, \ldots, z_{n-1}) = g(h_1(z_1, \ldots, z_{n-1}), \ldots, h_{n-1}(z_1, \ldots, z_{n-1}))$$

and finally, replacing this expression of F in (3.176), we obtain the unique solution of the Cauchy problem:

$$u(\mathbf{x}) = g(h_1(\varphi_1(\mathbf{x}), \ldots, \varphi_{n-1}(\mathbf{x})), \ldots, h_{n-1}(\varphi_1(\mathbf{x}), \ldots, \varphi_{n-1}(\mathbf{x}))),$$

where $\mathbf{x} = (x_1, \ldots, x_n)$.

Example 3.25 Let us find a function $u(x, y, z)$ such that

$$u(x, y, 0) = x^2 + y^2$$

and

$$y \frac{\partial u}{\partial x} + x \frac{\partial u}{\partial y} + 2xy \frac{\partial u}{\partial z} = 0.$$

3.6 First-Order Partial Differential Equations

The associated system is:

$$\frac{dx}{y} = \frac{dy}{x} = \frac{dz}{2xy}.$$

From the first equality we find the first integral $\varphi_1(x, y, z) = x^2 - y^2 = C_1$.

Now, let us multiply by x the numerator and the denominator of the first fraction, by y the numerator and the denominator of the second, by -1, the numerator and the denominator of the last one and make the sum of numerators and denominators, so we get:

$$\frac{x\,dx}{xy} = \frac{y\,dy}{yx} = \frac{-dz}{-2xy} = \frac{d\left(\frac{x^2}{2} + \frac{y^2}{2} - z\right)}{0}.$$

Hence, $x^2 + y^2 - 2z = C_2$. Thus we have just obtained two independent first integrals $\varphi_1(x, y, z) = x^2 - y^2$ and $\varphi_2(x, y, z) = x^2 + y^2 - 2z$. They are independent because it is not possible to express one of them as a function of the other (one contains the variable z and the other not). Thus the general solution of our equation is:

$$u(x, y, z) = F(x^2 - y^2, x^2 + y^2 - 2z). \qquad (3.180)$$

Now, let us use the initial condition $u(x, y, 0) = x^2 + y^2$:

$$x^2 + y^2 = F(x^2 - y^2, x^2 + y^2). \qquad (3.181)$$

To find the expression of F, we denote $z_1 = x^2 - y^2$ and $z_2 = x^2 + y^2$ and express x^2 and y^2 as functions of z_1 and z_2. We get: $x^2 = \frac{z_1 + z_2}{2}$ and $y^2 = \frac{z_2 - z_1}{2}$. Hence, from (3.181) we obtain $F(z_1, z_2) = z_2$ and so, from (3.180), the solution of the Cauchy problem is:

$$u(x, y, z) = x^2 + y^2 - 2z.$$

3.6.2 Quasilinear First-Order Partial Differential Equations

A *quasilinear first-order partial differential equation* is an equation of the form:

$$f_1(x_1, \ldots, x_n, u)\frac{\partial u}{\partial x_1}(x_1, \ldots, x_n) + \ldots + f_n(x_1, \ldots, x_n, u)\frac{\partial u}{\partial x_n}(x_1, \ldots, x_n) =$$
$$= f_{n+1}(x_1, \ldots, x_n, u), \qquad (3.182)$$

where $f_1, \ldots, f_{n+1}$ are functions of class C^1 on a domain $D^* \in \mathbb{R}^{n+1}$, which cannot be simultaneously equal to zero at any point of D^*.

We search for the solution of the Eq. (3.182) $u = u(x_1, \ldots, x_n)$ *implicitly defined by an equation of the form*

$$\Phi(x_1, \ldots, x_n, u) = 0, \tag{3.183}$$

where $\Phi(z_1, \ldots, z_n, z_{n+1})$ is a function of class C^1 such that $\dfrac{\partial \Phi}{\partial z_{n+1}} \neq 0$ on D^*.

By the implicit function theorem, there exists a unique function of class C^1, $u = u(x_1, \ldots, x_n)$, such that

$$\Phi(x_1, \ldots, x_n, u(x_1, \ldots, x_n)) = 0, \text{ for all } (x_1, \ldots, x_n) \in D. \tag{3.184}$$

We differentiate (3.184) with respect to x_i, $i = 1, 2, \ldots, n$ and denote $\mathbf{x} = (x_1, \ldots, x_n)$. Hence we obtain:

$$\frac{\partial \Phi}{\partial z_i}(\mathbf{x}, u(\mathbf{x})) + \frac{\partial \Phi}{\partial z_{n+1}}(\mathbf{x}, u(\mathbf{x})) \cdot \frac{\partial u}{\partial x_i}(\mathbf{x}) = 0,$$

so,

$$\frac{\partial u}{\partial x_i}(\mathbf{x}) = -\frac{\dfrac{\partial \Phi}{\partial z_i}(\mathbf{x}, u(\mathbf{x}))}{\dfrac{\partial \Phi}{\partial z_{n+1}}(\mathbf{x}, u(\mathbf{x}))} \tag{3.185}$$

for any $i = 1, 2, \ldots, n$. We substitute these expressions of $\dfrac{\partial u}{\partial x_i}$ in (3.182) and replace the variables x_i by z_i, $i = 1, 2, \ldots, n$ and $u(x_1, \ldots, x_n)$ by z_{n+1}. Thus, we obtain the following homogeneous partial differential equation in the unknown function $\Phi(z_1, \ldots, z_{n+1})$:

$$f_1(z_1, \ldots, z_{n+1})\frac{\partial \Phi}{\partial z_1}(z_1, \ldots, z_{n+1}) + \ldots + f_n(z_1, \ldots, z_{n+1})\frac{\partial \Phi}{\partial z_n}(z_1, \ldots, z_{n+1}) +$$

$$+ f_{n+1}(z_1, \ldots, z_{n+1})\frac{\partial \Phi}{\partial z_{n+1}}(z_1, \ldots, z_{n+1}) = 0, \tag{3.186}$$

We write the autonomous system associated to this homogeneous linear equation,

$$\frac{dz_1}{f_1} = \frac{dz_2}{f_2} = \ldots = \frac{dz_{n+1}}{f_{n+1}},$$

3.6 First-Order Partial Differential Equations

find n first integrals of it, $\varphi_1, \ldots, \varphi_n$, and then we can write the general solution of (3.182) in the implicit form

$$F(\varphi_1(x_1, \ldots, x_n, u), \ldots, \varphi_n(x_1, \ldots, x_n, u)) = 0, \quad (3.187)$$

where F is an arbitrary function of class C^1.

Example 3.26 Consider the following problem related to Example 3.24:

Find all the surfaces of equation $S : z = u(x, y)$ such that the tangent plane at any point $M_0(x_0, y_0, z_0)$ ($z_0 = u(x_0, y_0)$) is parallel to the direction of the vector $\mathbf{F}(x_0, y_0, z_0) = x_0\mathbf{i} + y_0\mathbf{j} + z_0\mathbf{k}$ (where $\mathbf{F}(x, y, z) = x\mathbf{i} + y\mathbf{j} + z\mathbf{k}$, is a vector field defined on a domain $D \subset \mathbb{R}^3$ which contains the surface S).

Since the implicit equation of the surface S is $u(x, y) - z = 0$, the normal vector to S at M_0 is $\mathbf{n} = \frac{\partial u}{\partial x}(x_0, y_0)\mathbf{i} + \frac{\partial u}{\partial y}(x_0, y_0)\mathbf{j} - \mathbf{k}$. Thus, the required condition becomes:

$$x_0 \frac{\partial u}{\partial x}(x_0, y_0) + y_0 \frac{\partial u}{\partial y}(x_0, y_0) - z_0 = 0,$$

or

$$x_0 \frac{\partial u}{\partial x}(x_0, y_0) + y_0 \frac{\partial u}{\partial y}(x_0, y_0) = u(x_0, y_0).$$

Thus, we have to solve the quasilinear equation (to find its *integral surfaces*)

$$x \frac{\partial u}{\partial x}(x, y) + y \frac{\partial u}{\partial y}(x, y) = u(x, y). \quad (3.188)$$

We search for an implicit solution $\Phi(x, y, u(x, y)) = 0$. From the above remarks, the function $\Phi(x, y, z)$ is a solution of the linear *homogeneous* equation:

$$x \frac{\partial \Phi}{\partial x} + y \frac{\partial \Phi}{\partial y} + z \frac{\partial \Phi}{\partial z} = 0. \quad (3.189)$$

Its associated system is:

$$\frac{dx}{x} = \frac{dy}{y} = \frac{dz}{z}.$$

From the first equality we find $\frac{x}{y} = C_1$, and from the second, $\frac{z}{y} = C_2$ (the characteristic curves are the straight lines $x = C_1 y, z = C_2 y$). Thus, we have found the independent first integrals $\varphi_1(x, y, z) = \frac{x}{y}$ and $\varphi_2(x, y, z) = \frac{z}{y}$, so $\Phi(x, y, z) = F(\frac{x}{y}, \frac{z}{y})$, where F is an arbitrary function of class C^1 is the general solution of the linear homogeneous Eq. (3.189). Consequently, the general solution of (3.188) is:

$$F\left(\frac{x}{y}, \frac{z}{y}\right) = 0. \qquad (3.190)$$

The surfaces defined by (3.190) are the *integral surfaces* of the partial differential Eq. (3.188).

Now, let us solve a *Cauchy problem* for the Eq. (3.188):
Find the integral surface that contains the circle

$$y = 1, \quad x^2 + z^2 = 4. \qquad (3.191)$$

We need to find the particular function $F(w_1, w_2)$ such that the initial condition (3.191) is satisfied. For this, we eliminate x, y, z from the equations:

$$w_1 = \frac{x}{y}, \quad w_2 = \frac{z}{y}, \quad y = 1, \quad x^2 + z^2 = 4$$

and find a relation between w_1 and w_2.

Thus, we have $w_1 = x$ and $w_2 = z$ and, from the last equation, we obtain:

$$w_1^2 + w_2^2 - 4 = 0$$

i.e. $F(w_1, w_2) = w_1^2 + w_2^2 - 4$. So the solution of the Cauchy problem is the surface defined by the equation $\frac{x^2}{y^2} + \frac{z^2}{y^2} - 4 = 0$, or $x^2 + z^2 = 4y^2$ (the surface is a cone along the y-axis).

3.7 Exercises

The unknown functions are $x = x(t)$, $y = y(t)$, $z = z(t)$ and sometimes the derivatives are denoted by $\dot{x} = x'$, $\dot{y} = y'$, $\dot{z} = z'$.

1. Solve the following systems of linear differential equations by two methods: the elimination method and the algebraic method. Solve the Cauchy problem when initial conditions are given.

    (a) $\begin{cases} x' = x + y \\ y' = -2x - 2y \end{cases}$.

    (b) $\begin{cases} x' - x - y = 0 \\ y' = 4y - 2x \end{cases}$, $x(0) = 0$, $y(0) = -1$.

    (c) $\begin{cases} \frac{dx}{dt} - 3x + y = 0 \\ \frac{dy}{dt} - 10x + 4y = 2 \end{cases}$, $x(0) = 1$, $y(0) = 6$.

3.7 Exercises

(d) $\begin{cases} \dot{x} + 2x + 4y = 1 + 4t \\ \dot{y} + x - y = \frac{3}{2}t^2 \end{cases}$.

(e) $\begin{cases} x' = y + z \\ y' = x + z \\ z' = x + y \end{cases}$.

2. Solve the following systems of linear differential equations by using the algebraic method.

(a) $\begin{cases} x' = -x + y + z \\ y' = x - y + z \\ z' = x + y - z \end{cases}$.

(b) $\begin{cases} x' = 2x - z \\ y' = -x + 2y \\ z' = -x + 2z \end{cases}$.

(c) $\begin{cases} x' = x + y - 2z + \exp(3t) \\ y' = x + z \\ z' = -2x + y + z - \exp(3t) \end{cases}$.

(d) $\begin{cases} \dot{x} = x + y + z + 3 \\ \dot{y} = x + y + z - \exp t \\ \dot{z} = x + y + z - \exp t \end{cases}$, $x(0) = 1, y(0) = 0, z(0) = -1$.

3. Apply the Lagrange method of variation of constants to solve the following non-homogeneous systems of linear differential equations:

(a) $\begin{cases} \dot{x} = -2x + y - \exp(2t) \\ \dot{y} = -3x + 2y + 3\exp(2t) \end{cases}$.

(b) $\begin{cases} \dot{x} = 2y + \cos t + 2\sin t \\ \dot{y} = x - y - \cos t + \sin t \end{cases}$.

(c) $\begin{cases} \dot{x} = -4x - 2y + \frac{2}{\exp t - 1} \\ \dot{y} = 6x + 3y - \frac{3}{\exp t - 1} \end{cases}$.

4. Solve the following systems of linear differential equations by using the elimination method. Solve the Cauchy problem when initial conditions are given.

(a) $\begin{cases} x' = -2x - y + \sin t \\ y' = 4x + 2y + \cos t \end{cases}$.

(b) $\begin{cases} \dot{x} = x - y + \frac{1}{\cos t} \\ \dot{y} = 2x - y \end{cases}$.

(c) $\begin{cases} \dot{x} = y + \tan^2 t - 1 \\ \dot{y} = -x + \tan t \end{cases}$.

(d) $\begin{cases} x' + y - z = 0 \\ y' - z = 0 \\ z' + x - z = 0 \end{cases}$, $x(0) = y(0) = 1$, $z(0) = 0$.

(e) $\begin{cases} x' - 2x - y + 2z = 2 - t \\ y' + x = 1 \\ z' - x - y + z = 1 - t \end{cases}$.

(f) $\dot{\mathbf{Y}} = \begin{pmatrix} 3 & 1 & 0 \\ 0 & 3 & 1 \\ 0 & 0 & 3 \end{pmatrix} \mathbf{Y} + \begin{pmatrix} 0 \\ 0 \\ \exp(3t) \end{pmatrix}$, $\mathbf{Y} = \begin{pmatrix} x(t) \\ y(t) \\ z(t) \end{pmatrix}$, $\mathbf{Y}(0) = \mathbf{0}$.

(g) $\dot{\mathbf{Y}} = \begin{pmatrix} -1 & -2 & 2 \\ 0 & -1 & -1 \\ 0 & 0 & -1 \end{pmatrix} \mathbf{Y} + \begin{pmatrix} 2\exp(-t) \\ 1 \\ 1 \end{pmatrix}$, $\mathbf{Y}(0) = \begin{pmatrix} 1 \\ 1 \\ 1 \end{pmatrix}$.

(h) $\begin{cases} \ddot{x} + \dot{x} + \dot{y} - 2y = 0 \\ \dot{x} - \dot{y} + x = 0 \end{cases}$.

(i) $\begin{cases} \ddot{x} + 3\ddot{y} - x = 0 \\ \dot{x} + 3\dot{y} - 2y = 0 \end{cases}$.

(j) $\begin{cases} \ddot{x} = 2y + \exp t \\ \ddot{y} = -2x \end{cases}$.

(k) $\begin{cases} \ddot{x} = 3x + 4y + \sin t \\ \ddot{y} = -x - y \end{cases}$.

(l) $\begin{cases} t\dot{x} = x + y \\ t\dot{y} = 3x - y \end{cases}$, $x(1) = 0$, $y(1) = 1$.

(m) $\begin{cases} t\dot{x} = 3x - 2y \\ t\dot{y} = 2x - y + \ln t \end{cases}$, $x(1) = 1$, $y(1) = 1$.

5. Use the method of integrable combinations to solve the following autonomous systems (give the solution in the phase space, i.e. as an intersection of surfaces or hypersurfaces):

(a) $\begin{cases} \dot{x} = x + 2y \\ \dot{y} = \frac{x^2}{y} + 2x \end{cases}$;

(b) $\begin{cases} \dot{x} = -\frac{1}{y} \\ \dot{y} = -\frac{1}{x} \end{cases}$;

(c) $\begin{cases} \dot{x} = x^2 + y^2 \\ \dot{y} = 2xy \end{cases}$;

(d) $\frac{dx}{x+y} = \frac{dy}{-2x+4y}$;

(e) $\frac{dx}{z} = \frac{dy}{xz} = \frac{dz}{y}$;

(f) $\frac{dx}{x(y^2+z^2)} = \frac{dy}{y(x^2+z^2)} = \frac{dz}{z(x^2-y^2)}$;

(g) $\frac{dx}{x+y^2+z^2} = \frac{dy}{y} = \frac{dz}{z}$;

(h) $\frac{dx}{x(z-y)} = \frac{dy}{y(y-x)} = \frac{dz}{y^2-xz}$;

(i) $\frac{dx}{xz} = \frac{dy}{yz} = \frac{dz}{xy\sqrt{z^2+1}}$;

(j) $\frac{dx}{y+z} = \frac{dy}{x+z} = \frac{dz}{x+y}$;

(k) $\frac{dx}{z} = \frac{dy}{u} = \frac{dz}{x} = \frac{du}{y}$.

6. Find the general solution of the following linear homogeneous partial differential equations:

   (a) $(x + 2y)\frac{\partial z}{\partial x} - y\frac{\partial z}{\partial y} = 0$;

   (b) $x\frac{\partial u}{\partial x} + y\frac{\partial u}{\partial y} + (x + y)\frac{\partial u}{\partial z} = 0$;

   (c) $yz\frac{\partial u}{\partial x} + xz\frac{\partial u}{\partial y} + xy\frac{\partial u}{\partial z} = 0$;

   (d) $(y^2 + z^2 - x^2)\frac{\partial u}{\partial x} - 2xy\frac{\partial u}{\partial y} - 2xz\frac{\partial u}{\partial z} = 0$.

7. Solve the following Cauchy problems for the partial differential equations:

   (a) $y\frac{\partial z}{\partial x} - x\frac{\partial z}{\partial y} = 0$, $z(x, 1) = x^2$.

   (b) $(1 + x^2)\frac{\partial z}{\partial x} + xy\frac{\partial z}{\partial y}$, $z(0, y) = y^2$.

   (c) $\frac{\partial z}{\partial x} + (2\exp x - y)\frac{\partial z}{\partial y} = 0$, $z = y$ if $x = 0$.

   (d) $xz\frac{\partial u}{\partial x} + yz\frac{\partial u}{\partial y} + xy\frac{\partial u}{\partial z} = 0$, $u(x, y, 2) = xy$.

   (e) $(1, 1, 2) \cdot \nabla u = 0$, $u(1, y, z) = yz$;

   (f) $(x, y, xy) \cdot \nabla u = 0$, $u(x, y, 0) = x^2 + y^2$.

8. Find a surface $u(x, y, z) = 0$, if $\mathbf{r} \cdot \nabla u = 0$ and $u = x + z$ if $y = 1$.

9. Find the surface which passes through the line $y = x$, $z = 1$ and which is orthogonal to the family of spheres: $x^2 + y^2 + z^2 = Cx$.

10. Find the general solution of the following quasilinear partial differential equations:

    (a) $y\frac{\partial z}{\partial x} + x\frac{\partial z}{\partial y} = x - 2y$;

    (b) $xy\frac{\partial z}{\partial x} - y^2\frac{\partial z}{\partial y} = x^2 + 1$;

    (c) $\sin^2 x \frac{\partial z}{\partial x} + \tan z \frac{\partial z}{\partial y} = \cos^2 z$.

    (d) $x\frac{\partial u}{\partial x} + (z + u)\frac{\partial u}{\partial y} + (y + u)\frac{\partial u}{\partial z} = y + z$;

11. Solve the following Cauchy problems for quasilinear partial differential equations:

    (a) $(z, -xy) \cdot \nabla z = 2xz$, $x + y = 2$, $yz = 1$, where $z = z(x, y)$.

    (b) $(x, -y) \cdot \nabla z = z^2(x - 3y)$, $x = 1$, $yz = -1$, where $z = z(x, y)$.

    (c) $(x, y) \cdot \nabla z = z - xy$, $x = 2$, $z = y^2 + 1$, where $z = z(x, y)$.

12. Find the surface of equation $z = z(x, y)$, which passes through the circle $z = 0$, $x^2 + y^2 = y$ and such that $(2yz, xz) \cdot \nabla z = -xy$.

Chapter 4
Fourier Series

In Chap. 4 we discuss the expansion of periodic functions into infinite trigonometric series (introduced by Joseph Fourier in the early 1800s). Fourier series will be used in Chap. 7 to solve partial differential equations such as the wave equation, the heat equation, or the Laplace equation. General orthogonal systems of functions are also introduced.

4.1 Introduction: Periodic, Piecewise Smooth Functions

Trying to describe the heat diffusion in solid bodies, the French mathematician Joseph Fourier (1768–1830) proposed the revolutionary idea that any periodic function $f(x)$ could be written as an infinite sum of harmonics of the form:

$$A_n \sin(n\omega x + \alpha_n) = A_n \sin \alpha_n \cdot \cos n\omega x + A_n \cos \alpha_n \cdot \sin n\omega x,$$

where $\omega = 2\pi/T$ is the first fundamental frequency ($T > 0$ is the period of $f(x)$) and $n = 0, 1, \ldots$. If we denote $a_0 = 2A_0 \sin \alpha_0$, $a_n = A_n \sin \alpha_n$, and $b_n = A_n \cos \alpha_n$ for $n = 1, 2, \ldots$, we get:

$$f(x) = \frac{a_0}{2} + \sum_{n=1}^{\infty} (a_n \cos n\omega x + b_n \sin n\omega x), \quad x \in \mathbb{R}. \qquad (4.1)$$

Fourier proposed this ingenious and strange idea in 1807, in a paper submitted to the Academy of Sciences in Paris. Although the paper was rejected for lack of rigor, Fourier's ideas conquered the world of scientists of that time, leading to important advances in mathematics, science and engineering. Later, other mathematicians, like Peter Gustav Lejeune Dirichlet and Bernhard Riemann expressed Fourier's results in a more rigorous way.

It is difficult to accept that a continuous (or discontinuous) function can be completely described by a *countable* system of functions

$$\{1, \cos \omega x, \sin \omega x, \cos 2\omega x, \sin 2\omega x, \ldots\}.$$

The formula (4.1) raises some difficult questions:

(Q1) What is the kind of convergence of the series on the right-hand of the equality (4.1)?
(Q2) How to calculate the sequences of numbers $\{a_n\}$, $\{b_n\}$?
(Q3) What is the most general class of functions which admit such a representation (4.1)?
(Q4) What happens when f is not a periodic function, i.e. for $f : [-T/2, T/2] \to \mathbb{R}$ and $T \to \infty$?

In this chapter (and also in Chap. 5 for (Q4)) we present some basic results on Fourier series, trying to answer (at least partially) the above questions. More detailed proofs and more general results can be found in [4, 41] and [43].

4.1.1 Periodic Functions

Definition 4.1 A real or complex-valued function $f : D \subset \mathbb{R} \to \mathbb{R}(\mathbb{C})$ is said to be a *periodic function* if there exists a positive constant $T > 0$ such that

$$f(x + T) = f(x), \text{ for all } x \in D. \tag{4.2}$$

The positive constant T is called a *period* of the function f and we notice that it is not unique: if T is a period, then any number of the form nT, where n is a positive integer, is also a period of the function f, since

$$f(x) = f(x + T) = f(x + 2T) = f(x + 3T) = \ldots, \text{ for all } x \in D.$$

Moreover, if T_1 and T_2 are two periods of the function f, then $T_1 \pm T_2$ is also a period of the function. The least period $T_0 > 0$ (if it exists) is called the *fundamental period* (or the basic period) of the function f. We often say "the" period of a function to mean its fundamental period. If there exists the fundamental period T_0, then any other period of the function must be of the form nT_0. (Otherwise, if T were a period of the function such that $nT_0 < T < (n+1)T_0$, then the positive number $T - nT_0$ would be also a period, smaller than the principal period T_0, a contradiction.)

The most familiar periodic functions are the trigonometric functions $\sin x$, $\cos x$, $\tan x$, etc. Notice that $\tan x$ is an example of periodic function which is not defined on the whole axis: in this case $D = \mathbb{R} - \{\frac{\pi}{2} + n\pi : n \in \mathbb{Z}\}$. Obviously, the sum,

4.1 Introduction: Periodic, Piecewise Smooth Functions

the difference, the product, or the quotient of two functions of period T is also a function of period T. We remark that the fundamental period of the functions $\sin x$ and $\cos x$ is 2π, while the fundamental period of $\tan x = \frac{\sin x}{\cos x}$ is π.

Remark 4.1 Not all the periodic functions have a principal period. A trivial example is the constant function $f(x) = c$ which has no fundamental period because any positive real number is a period of it. Another example is the *Dirichlet function*,

$$d(x) = \begin{cases} 1, & x \in \mathbb{Q} \\ 0, & x \notin \mathbb{Q} \end{cases} \quad (4.3)$$

for which any positive rational number is a period. Notice that $d(x)$ is discontinuous everywhere and all the discontinuities are of the second kind.

We can prove that a non-constant periodic *continuous* function $f : \mathbb{R} \to \mathbb{R}$ has always a principal period T_0.

Indeed, if such a T_0 did not exist, then a sequence of periods decreasing to 0 could be found:

$$T_1 > T_2 > \ldots > T_n > \ldots, \quad T_n \to 0.$$

Let x be a fixed number such that $f(x) \neq f(0)$ and $\varepsilon > 0$ be a positive number such that $\varepsilon < |f(x) - f(0)|$. Since f is continuous at 0, there exists a positive number $\delta > 0$ such that $|f(z) - f(0)| < \varepsilon$, for any $z \in (-\delta, \delta)$. Since $T_n \to 0$, we have $T_n < \delta$ for a great enough integer n. Let $k = \left[\frac{x}{T_n}\right]$ be the greatest integer smaller than or equal to $\frac{x}{T_n}$. Thus, we have:

$$0 \leq x - kT_n < T_n < \delta,$$

hence $z = x - kT_n \in (-\delta, \delta)$. But, since T_n is a period of f, it follows that $f(x - kT_n) = f(x)$, hence $|f(x - kT_n) - f(0)| > \varepsilon$, a contradiction. Thus, f must have a least period T_0.

Finally, we remark that the same result can be proved for *piecewise continuous* periodic functions.

The next proposition states that, for an *integrable* function of period T, the integral on any interval of length T has the same value, so we can compute the integral on the interval of length T centered at 0, to possibly use the oddness or the evenness of the function (see Fig. 4.1).

Proposition 4.1 *If f is a periodic function of a period T, integrable on any finite interval, then*

$$\int_\alpha^{\alpha+T} f(x)\,dx = \int_{-\frac{T}{2}}^{\frac{T}{2}} f(x)\,dx, \quad (4.4)$$

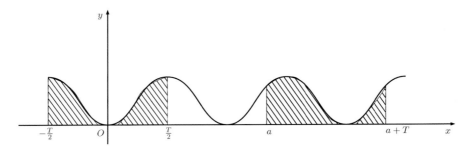

Fig. 4.1 Illustration of equality (4.4) for a periodic function

for any real number $\alpha \in \mathbb{R}$.

In particular, if f is an odd function ($f(-x) = -f(x)$ for all x), then

$$\int_\alpha^{\alpha+T} f(x)\,dx = 0 \qquad (4.5)$$

and if f is an even function ($f(-x) = f(x)$ for all x), then

$$\int_\alpha^{\alpha+T} f(x)\,dx = 2\int_0^{\frac{T}{2}} f(x)\,dx \qquad (4.6)$$

Proof To prove the formula (4.4) we write:

$$\int_{-\frac{T}{2}}^{\frac{T}{2}} f(x)\,dx = \int_{-\frac{T}{2}}^{\alpha} f(x)\,dx + \int_\alpha^{\alpha+T} f(x)\,dx + \int_{\alpha+T}^{\frac{T}{2}} f(x)\,dx. \qquad (4.7)$$

Making the change of variable $x = t + T$ in the last integral, we obtain:

$$\int_{\alpha+T}^{\frac{T}{2}} f(x)\,dx = \int_\alpha^{-\frac{T}{2}} f(t)dt = -\int_{-\frac{T}{2}}^{a} f(t)dt,$$

hence, by replacing in (4.7), the formula (4.4) follows.

To prove (4.5) and (4.6), we write

$$\int_a^{a+T} f(x)\,dx = \int_{-\frac{T}{2}}^{\frac{T}{2}} f(x)\,dx = \int_{-\frac{T}{2}}^{0} f(x)\,dx + \int_0^{\frac{T}{2}} f(x)\,dx$$

4.1 Introduction: Periodic, Piecewise Smooth Functions

and make the change of variable $t = -x$ in the first integral. We obtain:

$$\int_{-\frac{T}{2}}^{\frac{T}{2}} f(x)\,dx = \int_0^{\frac{T}{2}} f(-t)\,dt + \int_0^{\frac{T}{2}} f(x)\,dx.$$

If the function is odd, then

$$\int_0^{\frac{T}{2}} f(-t)\,dt = -\int_0^{\frac{T}{2}} f(t)\,dt$$

and so the sum of the integrals will be 0. If the function is even, then

$$\int_0^{\frac{T}{2}} f(-t)\,dt = \int_0^{\frac{T}{2}} f(t)\,dt$$

and (4.6) follows. □

4.1.2 Piecewise Continuous and Piecewise Smooth Functions

Definition 4.2 We say that a function f has one-sided limits at the point c if there exist the limits

$$\lim_{x \nearrow c} f(x) = f(c_-) \quad \text{and} \quad \lim_{x \searrow c} f(x) = f(c_+).$$

A function f is said to be *piecewise continuous* on the interval $[a, b]$ if it is continuous everywhere on $[a, b]$, possibly except for a finite number of points $a \leq x_1 < x_2 < \ldots < x_n \leq b$ where the function does have *finite one-sided limits*. If the left-hand and the right-hand limits of f at x_k are different, then we say that x_k is a point of discontinuity of the first kind ("jump discontinuity"). If $x_k = a$ (or $x_k = b$) then only the right-hand limit $f(a_+)$ (the left-hand limit $f(b_-)$) must exist and be finite. *The function f does not even need to be defined at the points* $x_1, \ldots, x_n$.

A function is said to be piecewise continuous on $\mathbb{R}$ if it is piecewise continuous on any bounded interval $[a, b]$.

Note that a *piecewise continuous* function is *integrable* on $[a, b]$, since it is integrable on each interval $[a, x_1], [x_1, x_2], \ldots, [x_n, b]$ and we can write

$$\int_a^b f(x)\,dx = \int_a^{x_1} f(x)\,dx + \int_{x_1}^{x_2} f(x)\,dx + \ldots + \int_{x_n}^b f(x)\,dx.$$

Definition 4.3 A function f with at most a finite number of discontinuities in $[a, b]$, $a \leq x_1 < x_2 < \ldots < x_n \leq b$, is said to be *absolutely integrable* on $[a, b]$ if each integral (which may be *improper*)

$$\int_{x_i}^{x_{i+1}} |f(x)|\, dx, \quad i = 0, \ldots, n$$

exists and is finite (for simplicity, we denoted $x_0 = a$ and $x_{n+1} = b$).

Remark 4.2 A piecewise continuous function is also absolutely integrable (since $|f|$ is a piecewise continuous function). But the converse is not true: the function

$$f : [-1, 1] \setminus \{0\}, \quad f(x) = \frac{1}{\sqrt[3]{x}}$$

is absolutely integrable on $[-1, 1]$, $\int_{-1}^{1} |f(x)|\, dx = 3 < \infty$, but it is not piecewise continuous on $[-1, 1]$ since the one-sided limits at 0 are

$$f(0_-) = \lim_{x \nearrow 0} f(x) = -\infty, \quad f(0_+) = \lim_{x \searrow 0} f(x) = \infty.$$

Definition 4.4 Let $f(x)$ be a function which has finite one-sided limits at the point x_0, $f(x_{0-})$ and $f(x_{0+})$. We say that f has a left-hand derivative at x_0 if the limit below exists and is finite. We denote it by $f'_l(x_0)$:

$$f'_l(x_0) = \lim_{x \nearrow x_0} \frac{f(x) - f(x_{0-})}{x - x_0}.$$

We say that f has a right-hand derivative at x_0 if the limit below exists and is finite. We denote it by $f'_r(x_0)$:

$$f'_r(x_0) = \lim_{x \searrow x_0} \frac{f(x) - f(x_{0+})}{x - x_0}.$$

The existence of a left-hand derivative of the function at $x = x_0$ is equivalent to the existence of the tangent at this point to the curve $y = f_-(x)$, equal to $f(x)$ for $x < x_0$ and equal to $f(x_{0-})$ for $x = x_0$. In the same way, if the right-hand derivative of f at $x = x_0$ exists, there exists the tangent at this point to the curve $y = f_+(x)$, equal to $f(x)$ for $x > x_0$ and equal to $f(x_{0+})$ for $x = x_0$.

If the function is continuous at x_0 and has (finite) one-sided derivatives at x_0, there are two possibilities: either $f'_l(x_0) = f'_r(x_0)$ and so the function is differentiable at x_0, or $f'_l(x_0) \neq f'_r(x_0)$, hence the curve $y = f(x)$ has a "corner" at the point x_0 (see Fig. 4.2a).

4.1 Introduction: Periodic, Piecewise Smooth Functions

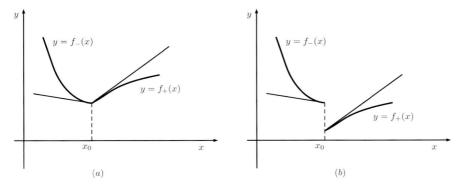

Fig. 4.2 Examples of piecewise smooth functions that are: continuous, but not differentiable (**a**); discontinuous (**b**)

If the function has a "jump discontinuity" at x_0 (see Fig. 4.2b), then f cannot be differentiable at this point, even if the one-sided derivatives are equal: for example, the function

$$f(x) = \begin{cases} x - 1 & \text{if } x < 0 \\ x + 1 & \text{if } x \geq 0 \end{cases}$$

is not differentiable at 0, although $f'_l(0) = f'_r(0) = 1$.

Definition 4.5 A function $f(x)$ is said to be *piecewise smooth* on the interval $[a, b]$ if it is piecewise continuous on $[a, b]$ and has a derivative f' which is also piecewise continuous on $[a, b]$. The function is said to be piecewise smooth on $\mathbb{R}$ if it is piecewise smooth on any bounded interval $[a, b]$.

Proposition 4.2 *If the function f is piecewise smooth on $[a, b]$, then it has one-sided derivatives at every point $x_0 \in [a, b]$ and*

$$f'_l(x_0) = \lim_{x \nearrow x_0} f'(x) = f'(x_0-), \quad f'_r(x_0) = \lim_{x \searrow x_0} f'(x) = f'(x_0+).$$

Proof Let x_0 be a point in the interval $[a, b]$. Since f is piecewise smooth, it is piecewise continuous, so it has one-sided limits at the point x_0, $f(x_0-)$ and $f(x_0+)$. By l'Hospital rule,

$$\lim_{x \nearrow x_0} \frac{f(x) - f(x_0-)}{x - x_0} \overset{\frac{0}{0}}{=} \lim_{x \nearrow x_0} f'(x) = f'(x_0-),$$

$$\lim_{x \searrow x_0} \frac{f(x) - f(x_0+)}{x - x_0} \overset{\frac{0}{0}}{=} \lim_{x \searrow x_0} f'(x) = f'(x_0+),$$

□

Note that the reciprocal statement is not true: a function can fail to be piecewise smooth, even if it has (finite) one-sided derivatives. For example, the function

$$f(x) = \begin{cases} x^2 \sin \frac{1}{x} & \text{if } x \neq 0 \\ 0 & \text{if } x = 0 \end{cases}$$

is continuous and has one-sided derivatives at $x_0 = 0$,

$$f'_l(0) = f'_r(0) = \lim_{x \to 0} \frac{f(x) - f(0)}{x - 0} = \lim_{x \to 0} x \sin \frac{1}{x} = 0,$$

but the derivative $f'(x) = 2x \sin \frac{1}{x} - \cos \frac{1}{x}$, $x \neq 0$ is not piecewise continuous (it does not have limit at 0).

For any interval $I = [a, b]$, we denote by $PC(I)$ the set of all the functions piecewise continuous on I and by $PS(I)$ the set of all the functions piecewise smooth on the interval I. Obviously, if f and g are two piecewise continuous / smooth functions on I and α, β are two real numbers, then the function $\alpha f + \beta g$ (defined everywhere on the interval I, except a finite set where either f or g are discontinuous) is also a piecewise continuous / smooth function. Thus, the set $PC(I)$ is a real vector space and $PS(I) \subset PC(I)$ is a vector subspace of it.

Moreover, we can extend these spaces to the complex vector spaces, by extending Definitions 4.2 and 4.5 for complex-valued functions: thus, we say that the function $f = u + iv$ is piecewise continuous / smooth on the interval I if its real part u and imaginary part v are both piecewise continuous / smooth on I.

4.2 Fourier Series Expansions

Now, let us take a closer look at the series of functions in the formula (4.1),

$$\frac{a_0}{2} + \sum_{k=1}^{\infty} (a_k \cos k\omega x + b_k \sin k\omega x). \tag{4.8}$$

Since all the terms of the series (4.8) are periodic functions of period $T = \frac{2\pi}{\omega}$, the partial sum,

$$s_n(x) = \frac{a_0}{2} + \sum_{k=1}^{n} (a_k \cos k\omega x + b_k \sin k\omega x)$$

is also periodic. The function $s_n(x)$ is called a *trigonometric polynomial* of order n and period T If the *trigonometric series* (4.8) is convergent, then its sum, $f(x)$ is

4.2 Fourier Series Expansions

also a periodic function of period T,

$$f(x) = \frac{a_0}{2} + \sum_{k=1}^{\infty} (a_k \cos k\omega x + b_k \sin k\omega x). \tag{4.9}$$

Using the change of variable $t = \omega x$ and denoting $\varphi(t) = f\left(\frac{x}{\omega}\right)$, we find

$$\varphi(t) = \frac{a_0}{2} + \sum_{k=1}^{\infty} (a_k \cos kt + b_k \sin kt), \tag{4.10}$$

a series of period 2π. Since the relations (4.9) and (4.10) are equivalent, we can study only the functions of period 2π.

4.2.1 Series of Functions

First, let us recall some basic facts on the convergence of sequences and series of functions (see also Chap. 11, Sect. 11.1.2).

A sequence of functions $\{f_n(x)\}$ is said to be pointwise convergent to $f(x)$ on the set A if $\lim_{n \to \infty} f_n(x) = f(x)$, for all $x \in A$.

A series of functions,

$$f_0(x) + f_1(x) + \ldots + f_k(x) + \ldots = \sum_{k=0}^{\infty} f_k(x). \tag{4.11}$$

is said to be pointwise convergent on the set A if, for any $x \in A$, the sequence of partial sums,

$$s_n(x) = \sum_{k=0}^{n} f_k(x)$$

is convergent. The function $f(x) = \lim_{n \to \infty} s_n(x)$, $x \in A$, is said to be the sum of the series. We write:

$$f(x) = \sum_{k=0}^{\infty} f_k(x).$$

But the pointwise convergence does not guarantee the transfer of some important properties: for instance, if the terms of the series, the functions $f_k(x)$, are continuous (or integrable), the sum of the series $f(x)$ is not necessarily a continuous (or integrable) function. Take, for instance, the series

$$\sum_{k=1}^{\infty} (x^{k-1} - x^k), \quad x \in [0, 1].$$

Since $s_n = 1 - x^n$, we obtain that the sum of the series is

$$\sum_{k=1}^{\infty} (x^{k-1} - x^k) = \lim_{n \to \infty} s_n(x) = \begin{cases} 1, & \text{if } x \in [0, 1) \\ 0, & \text{if } x = 1 \end{cases},$$

a discontinuous function (although the functions $x^{k-1} - x^k$ are continuous).

For this reason, a new type of convergence (stronger than the pointwise convergence) was defined: A sequence of functions $\{f_n(x)\}$ is said to be uniformly convergent to $f(x)$ on the set A if, for any $\varepsilon > 0$, there exists a positive integer N such that

$$|f_n(x) - f(x)| \leq \varepsilon, \text{ for all } n \geq N \text{ and } x \in A.$$

The series (4.11) is said to be *uniformly convergent* if the sequence of partial sums, $s_n(x)$ converges uniformly.

Obviously, the uniform convergence implies the pointwise convergence (but the converse is not true).

Theorem 4.1 (Weierstrass M-Test) *If $\{M_n\}_{n \geq 0}$ is a sequence of positive numbers such that $|f_n(x)| \leq M_n$ for all $x \in A$ and the series $\sum_{n=0}^{\infty} M_n$ is convergent, then the series $\sum_{n=0}^{\infty} f_n(x)$ converges uniformly on A.*

We also say in this case that the series converges absolutely uniformly, since the moduli series, $\sum_{n=0}^{\infty} |f_n(x)|$ is uniformly convergent.

Theorem 4.2 *Consider the series $\sum_{n=0}^{\infty} f_n(x)$, uniformly convergent on $[a, b]$.*

(i) If the functions $f_n(x)$, $n = 0, 1, \ldots$ are continuous, then the sum of the series,

$$f(x) = \sum_{n=0}^{\infty} f_n(x)$$

is also continuous on $[a, b]$;

(ii) If the functions $f_n(x)$, $n = 0, 1, \ldots$ are integrable on $[a, b]$, then $f(x)$ is also integrable on $[a, b]$ and the series can be integrated term by term, i.e.

$$\int_a^b f(x)\, dx = \sum_{n=0}^{\infty} \int_a^b f_n(x)\, dx. \tag{4.12}$$

(iii) If the functions $f_n(x)$, $n = 0, 1, \ldots$ are differentiable on $[a, b]$, and the series of derivatives, $\sum_{n=0}^{\infty} f_n'(x)$ uniformly converges on $[a, b]$, then $f(x)$ is also differentiable on $[a, b]$ and the series can be differentiated term by term, i.e.

$$f'(x)b = \sum_{n=0}^{\infty} f_n'(x). \tag{4.13}$$

Thus, by (4.12) and (4.13), the familiar results "the integral of the sum equals the sum of the integrals" and "the derivative of the sum equals the sum of the derivatives" hold also for *infinite sums*, provided the *uniform* convergence of the series (note that, for differentiation, the uniform convergence of the derivatives series is needed too).

4.2.2 A Basic Trigonometric System

Lemma 4.1 *Consider the following system of trigonometric functions (having the common period 2π):*

$$\mathcal{T} = \{1, \cos x, \sin x, \cos 2x, \sin 2x, \ldots, \cos nx, \sin nx, \ldots\}. \tag{4.14}$$

Then, for any functions $\phi, \psi \in \mathcal{T}$, we have:

$$\int_{-\pi}^{\pi} \phi(x)\psi(x)\, dx = 0 \quad \text{if } \phi \neq \psi \tag{4.15}$$

and, for any $\phi \neq 1$,

$$\int_{-\pi}^{\pi} \phi^2(x)\, dx = \pi. \tag{4.16}$$

In Sect. 4.3 we shall define the inner product and the orthogonality of two functions and we shall see that Eq. (4.15) states that any two functions in $\mathcal{T}$, $\phi \neq \psi$ are *orthogonal* on $[-\pi, \pi]$ (in fact, by Proposition 4.1, on any interval of length 2π).

Proof If $\phi(x) = 1$ and $\psi(x) = \cos nx$ or $\psi(x) = \sin nx$, then (4.15) follows immediately:

$$\int_{-\pi}^{\pi} \cos nx\, dx = \left.\frac{\sin nx}{n}\right|_{-\pi}^{\pi} = 0, \tag{4.17}$$

$$\int_{-\pi}^{\pi} \sin nx\, dx = -\left.\frac{\cos nx}{n}\right|_{-\pi}^{\pi} = 0. \tag{4.18}$$

If $\phi(x) = \cos kx$ and $\psi(x) = \cos nx$ (or $\phi(x) = \sin kx$ and $\psi(x) = \sin nx$) with $k \neq n$, then we use the trigonometric formulas:

$$\cos\alpha \cos\beta = \frac{1}{2}(\cos(\alpha - \beta) + \cos(\alpha + \beta)),$$

$$\sin\alpha \sin\beta = \frac{1}{2}(\cos(\alpha - \beta) - \cos(\alpha + \beta)).$$

We have:

$$\int_{-\pi}^{\pi} \cos kx \cos nx\, dx = \frac{1}{2}\left[\int_{-\pi}^{\pi} \cos(k-n)x\, dx + \int_{-\pi}^{\pi} \cos(k+n)x\, dx\right],$$

$$\int_{-\pi}^{\pi} \sin kx \sin nx\, dx = \frac{1}{2}\left[\int_{-\pi}^{\pi} \cos(k-n)x\, dx - \int_{-\pi}^{\pi} \cos(k+n)x\, dx\right],$$

hence, by the formula (4.17), we obtain:

$$\int_{-\pi}^{\pi} \cos kx \cos nx\, dx = 0, \tag{4.19}$$

$$\int_{-\pi}^{\pi} \sin kx \sin nx\, dx = 0. \tag{4.20}$$

If $\phi(x) = \sin kx$ and $\psi(x) = \cos nx$, we use the formula

$$\sin\alpha \cos\beta = \frac{1}{2}(\sin(\alpha - \beta) + \sin(\alpha + \beta)).$$

We have:

$$\int_{-\pi}^{\pi} \sin kx \cos nx\, dx = \frac{1}{2}\left[\int_{-\pi}^{\pi} \sin(k-n)x\, dx + \int_{-\pi}^{\pi} \sin(k+n)x\, dx\right],$$

4.2 Fourier Series Expansions

and by the formula (4.18) we obtain:

$$\int_{-\pi}^{\pi} \sin kx \cos nx \, dx = 0, \tag{4.21}$$

To prove (4.16), we use the trigonometric formulas:

$$\cos^2 \alpha = \frac{1 + \cos 2\alpha}{2}, \quad \sin^2 \alpha = \frac{1 - \cos 2\alpha}{2}.$$

We have:

$$\int_{-\pi}^{\pi} 1 \, dx = 2\pi, \tag{4.22}$$

$$\int_{-\pi}^{\pi} \cos^2 nx \, dx = \frac{1}{2} \left[\int_{-\pi}^{\pi} dx + \int_{-\pi}^{\pi} \cos 2nx \, dx \right] = \pi, \tag{4.23}$$

$$\int_{-\pi}^{\pi} \sin^2 nx \, dx = \frac{1}{2} \left[\int_{-\pi}^{\pi} dx - \int_{-\pi}^{\pi} \cos 2nx \, dx \right] = \pi. \tag{4.24}$$

□

4.2.3 Fourier Coefficients

Suppose that the function $f(x)$ of period 2π is the sum of the following trigonometric series

$$f(x) = \frac{a_0}{2} + \sum_{k=1}^{\infty} (a_k \cos kx + b_k \sin kx). \tag{4.25}$$

To determine the coefficients $a_0, a_k, b_k, k = 1, 2, \ldots$, we assume that the functions $f(x), f(x) \cos nx, f(x) \sin nx$ are integrable on $[-\pi, \pi]$ (or any interval of length 2π) and the series (4.25) (as well as the series obtained by multiplying it with $\cos nx$ or $\sin nx$) can be integrated term by term. Thus, one can write:

$$\int_{-\pi}^{\pi} f(x) \, dx = \frac{a_0}{2} \int_{-\pi}^{\pi} dx + \sum_{k=1}^{\infty} \left(a_k \int_{-\pi}^{\pi} \cos kx \, dx + b_k \int_{-\pi}^{\pi} \sin kx \, dx \right)$$

By the formulas (4.17)–(4.18), it follows that

$$\int_{-\pi}^{\pi} f(x)\,dx = a_0 \pi. \qquad (4.26)$$

Furthermore, we multiply (4.25) by $\cos nx$ and integrate on $[-\pi, \pi]$:

$$\int_{-\pi}^{\pi} f(x) \cos nx\, dx = \frac{a_0}{2} \int_{-\pi}^{\pi} \cos nx\, dx +$$

$$+ \sum_{k=1}^{\infty} \left(a_k \int_{-\pi}^{\pi} \cos kx \cos nx\, dx + b_k \int_{-\pi}^{\pi} \sin kx \cos nx\, dx \right).$$

By (4.17) the first integral is 0 and, by (4.19) and (4.21) all the integrals in the sum also vanish, except one, corresponding to coefficient a_n:

$$\int_{-\pi}^{\pi} f(x) \cos nx\, dx = a_n \int_{-\pi}^{\pi} \cos^2 nx\, dx = a_n \pi. \qquad (4.27)$$

In the same way, multiplying (4.25) by $\sin nx$ and integrating on $[-\pi, \pi]$, we obtain

$$\int_{-\pi}^{\pi} f(x) \sin nx\, dx = \frac{a_0}{2} \int_{-\pi}^{\pi} \sin nx\, dx +$$

$$+ \sum_{k=1}^{\infty} \left(a_k \int_{-\pi}^{\pi} \cos kx \sin nx\, dx + b_k \int_{-\pi}^{\pi} \sin kx \sin nx\, dx \right)$$

and, since all the integrals in the right-hand side vanish, except the one corresponding to b_n, we get:

$$\int_{-\pi}^{\pi} f(x) \sin nx\, dx = b_n \int_{-\pi}^{\pi} \sin^2 nx\, dx = b_n \pi. \qquad (4.28)$$

Thus, by the formulas (4.26)–(4.28) we find that the coefficients of the series (4.25) have the following expressions:

$$a_n = \frac{1}{\pi} \int_{-\pi}^{\pi} f(x) \cos nx\, dx, \quad n = 0, 1, 2, \ldots, \qquad (4.29)$$

$$b_n = \frac{1}{\pi} \int_{-\pi}^{\pi} f(x) \sin nx\, dx, \quad n = 1, 2, \ldots. \qquad (4.30)$$

The coefficients given by the formulas (4.29) and (4.30) are called the *Fourier coefficients* of the function $f(x)$ and the trigonometric series (4.25) having these coefficients is said to be the *Fourier series* of the function $f(x)$ (or the expansion of $f(x)$ in Fourier series). We write:

$$f(x) \sim \frac{a_0}{2} + \sum_{n=1}^{\infty} (a_n \cos nx + b_n \sin nx), \quad (4.31)$$

where the sign "$\sim$" means that the series above *corresponds* to the function $f(x)$. It can be replaced by the sign "$=$" if we succeed to prove that the series is (pointwise) convergent and its sum is equal to $f(x)$. In Sect. 4.4, *sufficient* conditions for the convergence of the Fourier series of a function $f(x)$ are given. We shall see that, even if the Fourier series of a function is convergent everywhere, the sum of the series may be different from $f(x)$.

4.3 Orthogonal Systems of Functions

4.3.1 Inner Product

Definition 4.6 Let $I = [a, b]$ be a bounded, closed interval. For any two functions $f, g \in PC(I)$, the product $f \cdot g$ is also a piecewise continuous, hence integrable function. Therefore, we can define the *inner product* of the functions f and g as the real number denoted by $\langle f, g \rangle$,

$$\langle f, g \rangle = \int_a^b f(x) g(x) \, dx. \quad (4.32)$$

Using the definition (4.32), it can be easily shown that the inner product has the following properties:

$$\langle f, g \rangle = \langle g, f \rangle \quad (4.33)$$

$$\langle f, g + h \rangle = \langle f, g \rangle + \langle f, h \rangle \quad (4.34)$$

$$\langle \alpha f, g \rangle = \alpha \langle f, g \rangle \quad (4.35)$$

$$\langle f, f \rangle \geq 0 \quad (4.36)$$

for any functions $f, g, h \in PC(I)$ and $\alpha \in \mathbb{R}$.

Remark 4.3 The exact definition of an *inner product* requires that

$$\langle f, f \rangle > 0 \text{ for any } f \neq \mathbf{0}, \quad (4.37)$$

which is not true in our case, since $\langle f, f \rangle = 0$ for any function $f \in PC(I)$ such that $f(x) \neq 0$ for $x \in S$, a finite set of points from I and $f(x) = 0$ for $x \in I - S$. To fix this problem, we define on $PC(I)$ the following relation: we say that $f, g \in PC(I)$

are *equal almost everywhere*, $f \sim g$ if $f(x) = g(x)$ for any $x \in I$, possibly except for a finite number of points. Obviously, this is an equivalence relation, so it provides a partition of $PC(I)$ into disjoint equivalence classes. We denote the set of equivalence classes by $\widehat{PC}(I) = PC(I)/\sim$ (the quotient space). We prove that the condition (4.37) is fulfilled on this space, so we have a true inner product. Thus, we have to prove that if f is a piecewise continuous function such that $f \not\sim \mathbf{0}$, then

$$\int_a^b f^2(x)\,dx > 0.$$

Since $f \not\sim \mathbf{0}$ is piecewise continuous function, there exists a point of continuity such that $f(x_0) \neq 0$. It follows that there exist two positive numbers $\varepsilon, \delta > 0$ such that $f^2(x) > \varepsilon$, for any $x \in (x_0 - \delta, x_0 + \delta)$. We can write:

$$\int_a^b f^2(x)\,dx = \int_a^{x_0-\delta} f^2(x)\,dx + \int_{x_0-\delta}^{x_0+\delta} f^2(x)\,dx + \int_{x_0+\delta}^b f^2(x)\,dx \geq \varepsilon \cdot 2\delta > 0$$

hence (4.37) is fulfilled in the space $\widehat{PC}(I)$.

Definition 4.7 The *norm* of a function $f \in PC(I)$ is the nonnegative number $\|f\|$ defined by:

$$\|f\| = \sqrt{\langle f, f \rangle} = \sqrt{\int_a^b f^2(x)\,dx}. \tag{4.38}$$

Proposition 4.3 (Cauchy-Schwarz Inequality) *For any functions $f, g \in PC(I)$ we have:*

$$|\langle f, g \rangle| \leq \|f\| \cdot \|g\|. \tag{4.39}$$

Proof Let α be an arbitrary real number. Using the properties (4.34)–(4.36) we can write:

$$0 \leq \langle \alpha f + g, \alpha f + g \rangle = \alpha^2 \langle f, f \rangle + 2\alpha \langle f, g \rangle + \langle g, g \rangle.$$

Hence the polynomial function $\varphi(\alpha)$ defined by

$$\varphi(\alpha) = \|f\|^2 \alpha^2 + 2 \langle f, g \rangle \alpha + \|g\|^2$$

4.3 Orthogonal Systems of Functions

has only nonnegative values: $\varphi(\alpha) \geq 0$, for all $\alpha \in \mathbb{R}$. It follows that its discriminant must satisfy the inequality:

$$\langle f, g \rangle^2 - |f|^2 |g|^2 \leq 0,$$

which is equivalent to (4.38). □

As a direct consequence of Cauchy-Schwarz inequality, the **triangle inequality**:

$$\|f + g\| \leq \|f\| + \|g\|, \text{ for any } f, g \in PC(I)$$

can be proved. Thus, we can write:

$$\|f + g\|^2 = \langle f + g, f + g \rangle = \|f\|^2 + 2 \langle f, g \rangle + \|g\|^2 \leq$$

$$\leq \|f\|^2 + 2\|f\| \cdot \|g\| + \|g\|^2 = (\|f\| + \|g\|)^2$$

and the inequality follows. So the function $\|\ \| : PC(I) \to \mathbb{R}_+$ defined above has the following properties:

$$\|f + g\| \leq \|f\| + \|g\| \qquad (4.40)$$

$$\|\alpha f\| = |\alpha| \|f\| \qquad (4.41)$$

for any functions $f, g \in PC(I)$ and $\alpha \in \mathbb{R}$.

As shown in Remark 4.3, if we consider the function $\|\ \|$ on the quotient space $\widehat{PC}(I)$, it also satisfies:

$$\|f\| = 0 \Leftrightarrow f \sim \mathbf{0}, \qquad (4.42)$$

so $\widehat{PC}(I)$ is a normed vector space.

Definition 4.8 The functions $f, g \in PC(I)$ are said to be *orthogonal* on the interval $[a, b]$ if

$$\langle f, g \rangle = 0.$$

A set of functions $\{\phi_0, \phi_1, \ldots, \phi_n, \ldots\}$ is said to be an *orthogonal system* if $\|\phi_j\| \neq 0$ for any $j = 0, 1, \ldots$ and

$$\langle \phi_i, \phi_j \rangle = 0 \text{ if } i \neq j.$$

If, in addition, $\|\phi_j\| = 1$ for all $j = 0, 1, \ldots$, then the above set of functions is said to be an *orthonormal system*. In this case, we can write:

$$\langle \phi_i, \phi_j \rangle = \begin{cases} 0, & \text{if } i \neq j \\ 1, & \text{if } i = j \end{cases} = \delta_{i,j} \quad \text{(Kronecker's symbol)}.$$

Remark 4.4 If $\{\psi_0, \psi_1, \ldots, \psi_n, \ldots\}$ is an *orthogonal* system, then the functions

$$\phi_k(x) = \frac{\psi_k(x)}{\|\psi_k\|}, \quad k = 0, 1, \ldots$$

form an *orthonormal* system.

By Lemma 4.1, the basic trigonometric system

$$\mathcal{T} = \{1, \cos x, \sin x, \cos 2x, \sin 2x, \ldots\} \tag{4.43}$$

is orthogonal on the interval $[-\pi, \pi]$ (in fact, it is orthogonal on any interval of length 2π). Since

$$\|1\| = \sqrt{\int_{-\pi}^{\pi} dx} = \sqrt{2\pi}, \quad \|\cos nx\| = \sqrt{\int_{-\pi}^{\pi} \cos^2 nx \, dx} = \sqrt{\pi},$$

$$\|\sin nx\| = \sqrt{\int_{-\pi}^{\pi} \sin^2 nx \, dx} = \sqrt{\pi}, \quad n = 1, 2, \ldots,$$

the following system is an *orthonormal* system of functions:

$$\mathcal{S} = \left\{ \frac{1}{\sqrt{2\pi}}, \frac{\cos x}{\sqrt{\pi}}, \frac{\sin x}{\sqrt{\pi}}, \frac{\cos 2x}{\sqrt{\pi}}, \frac{\sin 2x}{\sqrt{\pi}}, \ldots \right\}. \tag{4.44}$$

4.3.2 Best Approximation in the Mean: Bessel's Inequality

Let f be a piecewise continuous function on the interval $[-\pi, \pi]$ and s_n, $n = 1, 2, \ldots$, denote the partial sums of its Fourier series:

$$s_n(x) = \frac{a_0}{2} + \sum_{k=1}^{n} (a_k \cos kx + b_k \sin kx), \tag{4.45}$$

4.3 Orthogonal Systems of Functions

where the Fourier coefficients a_k, b_k are given by the formulas (4.29) and (4.30). We consider in this section the problem of approximating the function $f(x)$ by the partial sums $s_n(x)$. An important result (which will be used in the next section for proving the convergence of the Fourier series) will be obtained:

Lemma 4.2 *If the function f is piecewise continuous on the interval $[-\pi, \pi]$, and $a_0, a_n, b_n, n = 1, 2, \ldots$ are its Fourier coefficients (defined by the formulas (4.29)–(4.30)), then*

$$\lim_{n \to \infty} a_n = 0, \quad \lim_{n \to \infty} b_n = 0. \tag{4.46}$$

The proof of this result will be given later; for the moment, we extend our discussion to a more general framework. Consider an *orthogonal system* of functions in $PC(I)$, $I = [a, b]$:

$$\{\phi_0(x), \phi_1(x), \ldots, \phi_n(x), \ldots\}, \tag{4.47}$$

and suppose that the function $f \in PC(I)$ can be represented as the sum of the series:

$$f(x) = \sum_{k=0}^{\infty} c_k \phi_k(x), \tag{4.48}$$

where $c_0, c_1, \ldots$ are real constants. To find the expression of these coefficients, we assume that all the series

$$f(x)\phi_n(x) = \sum_{k=0}^{\infty} c_k \phi_k(x) \phi_n(x)$$

can be integrated term by term on the interval $[a, b]$ and we find:

$$\int_a^b f(x)\phi_n(x)\, dx = \sum_{k=0}^{\infty} c_k \int_a^b \phi_k(x)\phi_n(x)\, dx,$$

which can be written as

$$\langle f, \phi_n \rangle = \sum_{k=0}^{\infty} c_k \langle \phi_k, \phi_n \rangle = c_n \|\phi_n\|^2.$$

The series (4.48) with the coefficients c_n given by the formula

$$c_n = \frac{\langle f, \phi_n \rangle}{\|\phi_n\|^2} = \frac{1}{\|\phi_n\|^2} \int_a^b f(x)\phi_n(x)\, dx, \quad n = 0, 1 \ldots, \tag{4.49}$$

is called the *generalized Fourier series* of the function f, with respect to the orthogonal system (4.47). The constants c_n are called the *Fourier coefficients* of f with respect to the system (4.47).

Let f be a function piecewise continuous on $[a, b]$ and let $\sigma_n(x)$ be a linear combination of the first $n + 1$ terms of the orthogonal system (4.47),

$$\sigma_n(x) = \gamma_0 \phi_0(x) + \gamma_1 \phi_1(x) + \ldots + \gamma_n \phi_n(x),$$

where $\gamma_0, \ldots, \gamma_n \in \mathbb{R}$. The nonnegative number δ_n defined by

$$\delta_n = \|f - \sigma_n\|^2 = \int_a^b (f(x) - \sigma_n(x))^2 \, dx \qquad (4.50)$$

represents the *mean square error* in approximating the function f by the linear combination σ_n. We search for the coefficients $\gamma_0, \gamma_1, \ldots, \gamma_n$ which minimize δ_n. We can write:

$$\delta_n = \|f - \sigma_n\|^2 = \langle f - \sigma_n, f - \sigma_n \rangle =$$
$$= \|f\|^2 - 2 \langle f, \sigma_n \rangle + \|\sigma_n\|^2.$$

Since

$$\|\sigma_n\|^2 = \langle \sigma_n, \sigma_n \rangle = \sum_{j,k=0}^{n} \gamma_j \gamma_k \langle \phi_j, \phi_k \rangle = \sum_{k=0}^{n} \gamma_k^2 \|\phi_k\|^2,$$

$$\langle f, \sigma_n \rangle = \sum_{k=0}^{n} \gamma_k \langle f, \phi_k \rangle = \sum_{k=0}^{n} \gamma_k c_k \|\phi_k\|^2,$$

we obtain

$$\delta_n = \|f\|^2 - 2 \sum_{k=0}^{n} \gamma_k c_k \|\phi_k\|^2 + \sum_{k=0}^{n} \gamma_k^2 \|\phi_k\|^2 =$$
$$= \|f\|^2 + \sum_{k=0}^{n} (\gamma_k - c_k)^2 \|\phi_k\|^2 - \sum_{k=0}^{n} c_k^2 \|\phi_k\|^2,$$

so δ_n attains its minimum value when the coefficients of σ_n are the Fourier coefficients: $\gamma_k = c_k$, $k = 0, 1, \ldots, n$. Thus, we have proved that

$$s_n(x) = c_0 \phi_0(x) + c_1 \phi_1(x) + \ldots + c_n \phi_n(x) \qquad (4.51)$$

4.3 Orthogonal Systems of Functions

is the *best approximation in the mean* of $f(x)$ and the minimum value of δ_n is

$$\Delta_n = \|f\|^2 - \sum_{k=0}^{n} c_k^2 \|\phi_k\|^2. \tag{4.52}$$

Since $\Delta_n \geq 0$ for any $n = 0, 1, \ldots$, we obtain the *Bessel's inequality*:

$$\sum_{k=0}^{\infty} c_k^2 \|\phi_k\|^2 \leq \|f\|^2. \tag{4.53}$$

When the interval is $I = [-\pi, \pi]$ and the orthogonal system is (4.43), the Bessel's inequality is written:

$$\frac{1}{2}a_0^2 + \sum_{k=1}^{\infty} \left(a_k^2 + b_k^2\right) \leq \frac{1}{\pi} \int_{-\pi}^{\pi} f^2(x)\, dx, \tag{4.54}$$

where a_n, b_n are the Fourier coefficients defined by (4.29) and (4.30) and the next proposition follows immediately.

Proposition 4.4 *If f is a piecewise continuous function on the interval $[-\pi, \pi]$, then the series of the squares of its Fourier coefficients,*

$$\frac{1}{2}a_0^2 + \sum_{k=1}^{\infty} \left(a_k^2 + b_k^2\right) \tag{4.55}$$

is convergent.

Lemma 4.2 follows as a direct consequence of Proposition 4.4.

As a matter of fact, we shall see in Sect. 4.6 that the sum of this series (4.55) is $\frac{1}{\pi} \int_{-\pi}^{\pi} f^2(x)\, dx$, which means that the inequality (4.54) is actually an identity (*Parseval's identity*).

An orthogonal system (4.47) is said to be *complete* if, for any $f \in PC(I)$, Bessel's inequality (4.53) becomes equality:

$$\sum_{k=0}^{\infty} c_k^2 \|\phi_k\|^2 = \|f\|^2 = \int_a^b f^2(x)\, dx. \tag{4.56}$$

Formula (4.56) is known as *Parseval's identity* and, as mentioned above, it holds for the "classic" Fourier coefficients a_n, b_n defined by the formulas (4.29) and (4.30) (it will be proved in Sect. 4.6 that the orthogonal system (4.43) is complete). Thus, the inequality (4.54) is actually an equality:

$$\frac{1}{2}a_0^2 + \sum_{k=1}^{\infty}\left(a_k^2 + b_k^2\right) = \frac{1}{\pi}\int_{-\pi}^{\pi} f^2(x)\,dx. \qquad (4.57)$$

If instead of the interval $I = [-\pi, \pi]$ we have an interval of length $2l$, $I = [-l, l]$, then, with the coefficients a_n, b_n defined by the formulas (4.80) and (4.81), Parseval's identity is written:

$$\frac{1}{2}a_0^2 + \sum_{k=1}^{\infty}\left(a_k^2 + b_k^2\right) = \frac{1}{l}\int_{-l}^{l} f^2(x)\,dx. \qquad (4.58)$$

4.4 The Convergence of Fourier Series

By Lemma 4.2 we know that, for any function f piecewise continuous on the interval $[-\pi, \pi]$,

$$\lim_{k\to\infty}\int_{-\pi}^{\pi} f(x)\cos kx\,dx = 0 \;,\quad \lim_{k\to\infty}\int_{-\pi}^{\pi} f(x)\sin kx\,dx = 0. \qquad (4.59)$$

Let f be a function piecewise continuous on the interval $[0, \pi]$. We extend it to an even function $\bar{f}$ on the interval $[-\pi, \pi]$, $\bar{f}(x) = f(x)$, if $x \in [0, \pi]$ and $\bar{f}(x) = f(-x)$ if $x \in [-\pi, 0)$ (see Fig. 4.3). We find that

$$\int_0^{\pi} f(x)\cos kx\,dx = \frac{1}{2}\int_{-\pi}^{\pi} \bar{f}(x)\cos kx\,dx \longrightarrow 0 \text{ as } k \to \infty.$$

In the same way, by extending f to an odd function on the interval $[-\pi, \pi]$, $\bar{f}(x) = f(x)$, if $x \in [0, \pi]$ and $\bar{f}(x) = -f(-x)$ if $x \in [-\pi, 0)$, since the function $\sin kx$ is also odd and the product of two odd functions is an even function, we obtain that

$$\int_0^{\pi} f(x)\sin kx\,dx = \frac{1}{2}\int_{-\pi}^{\pi} \bar{f}(x)\sin kx\,dx \longrightarrow 0 \text{ as } k \to \infty.$$

and the following lemma is proved:

Lemma 4.3 *If f is a piecewise continuous function on the interval $[0, \pi]$, then*

$$\lim_{k\to\infty}\int_0^{\pi} f(x)\cos kx\,dx = 0,\quad \lim_{k\to\infty}\int_0^{\pi} f(x)\sin kx\,dx = 0. \qquad (4.60)$$

4.4 The Convergence of Fourier Series

Lemma 4.4 *If f is a piecewise continuous function on the interval $[0, \pi]$, then*

$$\lim_{k \to \infty} \int_0^\pi f(x) \sin\left(k + \tfrac{1}{2}\right) x \, dx = 0. \tag{4.61}$$

Proof We can write

$$\sin\left(k + \tfrac{1}{2}\right) x = \sin kx \cos \frac{x}{2} + \cos kx \sin \frac{x}{2}$$

and apply Lemma 4.3 to the piecewise continuous functions $f(x) \cos \frac{x}{2}$ and $f(x) \sin \frac{x}{2}$. □

We note that the above lemmas are particular cases of the *Riemann-Lebesgue Lemma* (see [41], p. 70).

Lemma 4.5 (Riemann-Lebesgue Lemma) *If the function $f(x)$ is absolutely integrable on $[a, b]$, then the following limits are 0 ($\lambda \in \mathbb{R}$):*

$$\lim_{\lambda \to \infty} \int_a^b f(x) \sin \lambda x \, dx = \lim_{\lambda \to \infty} \int_a^b f(x) \cos \lambda x \, dx = 0.$$

Now, we consider the following sequence of functions, known as the *Dirichlet kernel*:

$$D_n(x) = \frac{1}{2} + \cos x + \cos 2x + \ldots + \cos nx. \tag{4.62}$$

Obviously, for any $n = 1, 2, \ldots$, $D_n(x)$ is a continuous, periodic function (with period 2π). We also note that it is an even function and

$$\int_0^\pi D_n(x) \, dx = \frac{\pi}{2}. \tag{4.63}$$

Multiplying the equality (4.62) by $2 \sin \frac{x}{2}$ we obtain:

$$2 D_n(x) \sin \frac{x}{2} = \sin \frac{x}{2} + 2 \cos x \sin \frac{x}{2} + 2 \cos 2x \sin \frac{x}{2} + \ldots + 2 \cos nx \sin \frac{x}{2} =$$

$$= \sin \frac{x}{2} + \sin \frac{3x}{2} - \sin \frac{x}{2} + \sin \frac{5x}{2} - \sin \frac{3x}{2} + \ldots +$$

$$+ \sin \frac{(2n+1)x}{2} - \sin \frac{(2n-1)x}{2},$$

hence
$$D_n(x) = \frac{\sin\left(n + \frac{1}{2}\right)x}{2\sin\frac{x}{2}}. \tag{4.64}$$

Lemma 4.6 *Let f be a piecewise continuous function on the interval $[0, \pi]$ such that the right-hand derivative $f_r'(0)$ exists and is finite. Then*

$$\lim_{n \to \infty} \int_0^\pi f(x) D_n(x)\, dx = \frac{\pi}{2} f(0_+). \tag{4.65}$$

Proof By (4.63) and (4.64), the Eq. (4.65) is equivalent to

$$\lim_{n \to \infty} \int_0^\pi (f(x) - f(0_+)) \cdot \frac{\sin\left(n + \frac{1}{2}\right)x}{2\sin\frac{x}{2}}\, dx = 0,$$

or

$$\lim_{n \to \infty} \int_0^\pi \frac{f(x) - f(0_+)}{2\sin\frac{x}{2}} \cdot \sin\left(n + \frac{1}{2}\right)x\, dx = 0. \tag{4.66}$$

The function

$$g(x) = \frac{f(x) - f(0_+)}{2\sin\frac{x}{2}}$$

is piecewise continuous on $[0, \pi]$: although the denominator vanishes at the point $x = 0$, the existence of $f_r'(0)$ ensures the existence of the finite right-hand limit $g(0_+)$:

$$\lim_{x \searrow 0} g(x) = \lim_{x \searrow 0} \frac{f(x) - f(0_+)}{x - 0} \cdot \frac{\frac{x}{2}}{\sin\frac{x}{2}} = f_r'(0).$$

Thus, we can apply Lemma 4.4 for the function $g(x)$ and (4.66) follows. □

Theorem 4.3 (Dirichlet's Theorem) *If $f(x)$ is a piecewise continuous, periodic function with period 2π, then its Fourier series*

$$S_f(x) = \frac{a_0}{2} + \sum_{n=1}^\infty (a_n \cos nx + b_n \sin nx), \tag{4.67}$$

4.4 The Convergence of Fourier Series

where

$$a_n = \frac{1}{\pi} \int_{-\pi}^{\pi} f(x) \cos nx \, dx, \quad n = 0, 1, \ldots \tag{4.68}$$

and

$$b_n = \frac{1}{\pi} \int_{-\pi}^{\pi} f(x) \sin nx \, dx, \quad n = 1, 2, \ldots, \tag{4.69}$$

converges to the arithmetic mean of the one-sided limits of f,

$$\frac{f(x_+) + f(x_-)}{2} = \frac{a_0}{2} + \sum_{n=1}^{\infty} (a_n \cos nx + b_n \sin nx) \tag{4.70}$$

at each point x where both of the one-sided derivatives $f'_r(x)$ and $f'_l(x)$ exist. If such a point x is a point of continuity for the function f, then

$$f(x) = \frac{a_0}{2} + \sum_{n=1}^{\infty} (a_n \cos nx + b_n \sin nx). \tag{4.71}$$

Proof By (4.68) and (4.69), the partial sums of the series (4.67) can be written

$$s_n(x) = \frac{1}{2\pi} \int_{-\pi}^{\pi} f(t) dt + \frac{1}{\pi} \sum_{k=1}^{n} \int_{-\pi}^{\pi} f(t)[\cos kt \cos kx + \sin kt \sin kx] dt =$$

$$= \frac{1}{\pi} \int_{-\pi}^{\pi} f(t) \left[\frac{1}{2} + \sum_{k=1}^{n} \cos k(t-x) \right] dt =$$

$$= \frac{1}{\pi} \int_{-\pi}^{\pi} f(t) D_n(t-x) dt.$$

Using the change of variable $t - x = y$, we obtain

$$s_n(x) = \frac{1}{\pi} \int_{-\pi-x}^{\pi-x} f(x+y) D_n(y) \, dy.$$

For a fixed x, the function $\varphi(y) = f(x+y)D_n(y)$ is periodic with period 2π, so its definite integral on any interval of length 2π has the same value. Hence

$$S_n(x) = \frac{1}{\pi} \int_{-\pi}^{\pi} f(x+y)D_n(y)\,dy = \qquad (4.72)$$

$$= \frac{1}{\pi} \int_{-\pi}^{0} f(x+y)D_n(y)\,dy + \frac{1}{\pi} \int_{0}^{\pi} f(x+y)D_n(y)\,dy =$$

$$= \frac{1}{\pi} \int_{0}^{\pi} f(x-y)D_n(y)\,dy + \frac{1}{\pi} \int_{0}^{\pi} f(x+y)D_n(y)\,dy,$$

because the function $D_n(y)$ is even. For x fixed, we denote $g(y) = f(x-y)$ and $h(y) = f(x+y)$. Obviously, g and h are piecewise continuous and they have a right-hand derivative at 0,

$$g'_r(0) = f'_l(x), \quad h'_r(0) = f'_r(x).$$

By Lemma 4.6, since $g(0_+) = f(x_-)$ and $h(0_+) = f(x_+)$, we obtain

$$\lim_{n\to\infty} s_n = \frac{1}{\pi} \cdot \frac{\pi}{2} f(x_-) + \frac{1}{\pi} \cdot \frac{\pi}{2} f(x_+) = \frac{f(x_-) + f(x_+)}{2}$$

and the proof is complete. $\qquad\square$

A wide range of functions encountered in practical problems are *piecewise smooth* functions. By Proposition 4.2, we know that any piecewise smooth function has finite one-sided derivatives at any point, so the next corollary follows.

Corollary 4.1 *If $f(x)$ is a piecewise smooth function of period 2π and $S_f(x)$ is the corresponding Fourier series (4.67) with the coefficients a_n, b_n defined by (4.68) and (4.69), then $S_f(x)$ is convergent everywhere and the relation (4.70) holds for any $x \in \mathbb{R}$.*

Dirichlet's Theorem (and its Corollary 4.1) consider only the *simple* (pointwise) convergence of the Fourier series. Anyway, if the function f has "jump" discontinuities, then it is clear that the Fourier series cannot converge *uniformly* since the sum of a uniformly convergent series of continuous functions must be also continuous (see Theorem 4.2). The next theorem ([41], p 81) gives *sufficient* conditions for the uniform convergence of the Fourier series S_f (as discussed above, the continuity of the function f is also a *necessary* condition).

4.4 The Convergence of Fourier Series

Theorem 4.4 *If $f : \mathbb{R} \to \mathbb{R}$ is a continuous, piecewise smooth function of period 2π, then its Fourier series,*

$$\frac{a_0}{2} + \sum_{n=1}^{\infty}(a_n \cos nx + b_n \sin nx), \tag{4.73}$$

with the coefficients a_n, b_n defined by (4.68) and (4.69), converges to $f(x)$ absolutely and uniformly on $\mathbb{R}$.

Proof Since the function f is continuous, we can apply the integration by parts for the integrals (4.68) and (4.69) defining the Fourier coefficients. Note that, since the derivative f' is piecewise continuous, the functions $f'(x) \cos nx$ and $f'(x) \sin nx$ are integrable on $[-\pi, \pi]$. We find:

$$a_n = \frac{1}{\pi} \int_{-\pi}^{\pi} f(x) \cos nx\, dx = \frac{1}{n\pi} f(x) \sin nx \bigg|_{-\pi}^{\pi} - \frac{1}{n\pi} \int_{-\pi}^{\pi} f'(x) \sin nx\, dx,$$

$$b_n = \frac{1}{\pi} \int_{-\pi}^{\pi} f(x) \sin nx\, dx = -\frac{1}{n\pi} f(x) \cos nx \bigg|_{-\pi}^{\pi} + \frac{1}{n\pi} \int_{-\pi}^{\pi} f'(x) \cos nx\, dx.$$

The first terms vanish in both formulas (the last one, due to the periodicity of f). We denote by α_n and β_n the Fourier coefficients of the function f' and obtain:

$$a_n = -\frac{\beta_n}{n}, \quad b_n = \frac{\alpha_n}{n}, \quad n = 1, 2, \ldots. \tag{4.74}$$

Since the function f' is piecewise continuous, it follows from Proposition 4.4 that the series $\sum_{n=1}^{\infty}(\alpha_n^2 + \beta_n^2)$ is convergent.

By the inequalities

$$\left(|\alpha_n| - \frac{1}{n}\right)^2 = \alpha_n^2 - \frac{2|\alpha_n|}{n} + \frac{1}{n^2} \geq 0,$$

$$\left(|\beta_n| - \frac{1}{n}\right)^2 = \beta_n^2 - \frac{2|\beta_n|}{n} + \frac{1}{n^2} \geq 0,$$

and using (4.74), we obtain:

$$|a_n| + |b_n| = \frac{|\beta_n|}{n} + \frac{|\alpha_n|}{n} \leq \frac{1}{2}\left(\beta_n^2 + \alpha_n^2\right) + \frac{1}{n^2},$$

for all $n = 1, 2, \ldots$. The series $\sum_{n=1}^{\infty} \frac{1}{2}\left(\alpha_n^2 + \beta_n^2\right)$ and $\sum_{n=1}^{\infty} \frac{1}{n^2}$ are convergent, so we obtain that the series

$$\sum_{n=1}^{\infty} (|a_n| + |b_n|) \tag{4.75}$$

is also convergent. Now, we apply the Weierstrass M-test (Theorem 4.1) for the series (4.73): since

$$|(a_n \cos nx + b_n \sin nx)| \leq |a_n \cos nx| + |b_n \sin nx| \leq |a_n| + |b_n|$$

for all $x \in \mathbb{R}$, $n = 1, 2, \ldots$ and since the series (4.75) converges, it follows that the Fourier series (4.73) is absolutely and uniformly convergent. □

Remark 4.5 The conditions in Theorems 4.3 and 4.4 are just *sufficient* conditions and are not claimed to be *necessary* conditions. More general conditions can be set: thus, in Theorem 4.3, it is not necessary for f to be *piecewise continuous*, it suffices to be *absolutely integrable* ([41], p. 77). In Theorem 4.4 also, is not necessary for f to be *piecewise smooth* (which means that f' is piecewise continuous), it is sufficient that f' is *absolutely integrable*. Moreover, if the function f is not continuous everywhere, but it is continuous on an interval $[a, b]$ (of length less than 2π) and has an absolutely integrable derivative on $[a, b]$ then the Fourier series converges absolutely and uniformly to $f(x)$ on every interval $[a + \delta, b - \delta]$, $\delta > 0$ ([41], pp. 84–86).

Remark 4.6 Fourier series for functions defined on an interval of length 2π. If the function $f(x)$ is defined only on the interval $[-\pi, \pi]$ (possibly except for a finite number of points), and it is piecewise smooth on this interval, then it can be *extended by periodicity* to a piecewise smooth function of period 2π, $\bar{f}$ s.t $\bar{f}(x) = f(x)$ in $(-\pi, \pi)$, (recall that a piecewise smooth function may have some points at which it is not defined; so, if $f(-\pi) \neq f(\pi)$, we simply do not define the function $\bar{f}$ at the points $k\pi$, $k \in \mathbb{Z}$). The Fourier series of the function $f(x)$, $S_f(x)$ converges everywhere and and the formulas (4.70)–(4.71) holds for any $x \in (-\pi, \pi)$. For $x = \pm\pi$ we have:

$$S_f(\pi) = S_f(-\pi) = \frac{1}{2}(f(\pi_-) + f(-\pi_+)).$$

The function f can actually be defined on any interval of length 2π, $[a, a + 2\pi]$, since the integral of a function of period T on any interval of length T has the same value (Proposition 4.1).

If the function is continuous and piecewise smooth on $[a, a + 2\pi]$ and $f(a) = f(a + 2\pi)$, then its Fourier series converges absolutely and uniformly on $\mathbb{R}$. But

4.4 The Convergence of Fourier Series

even if $f(a) \neq f(a + 2\pi)$, the Fourier series is still uniformly convergent to f on every interval $[a + \delta, a + 2\pi - \delta], \delta > 0$.

Remark 4.7 Fourier series for functions of an arbitrary period $T = 2l$.

If the function $f(x)$ is a piecewise smooth function of period $T = 2l$, then the function $\varphi(t)$ defined by

$$\varphi(t) = f\left(\frac{lt}{\pi}\right) \tag{4.76}$$

is piecewise smooth and has the period 2π. Thus,

$$\varphi(t) \sim \frac{a_0}{2} + \sum_{n=1}^{\infty} (a_n \cos nt + b_n \sin nt) \tag{4.77}$$

and, using (4.76), we can write the Fourier series (of period T) corresponding to $f(x)$ as:

$$f(x) = \varphi\left(\frac{\pi x}{l}\right) \sim \frac{a_0}{2} + \sum_{n=1}^{\infty} \left(a_n \cos \frac{n\pi x}{l} + b_n \sin \frac{n\pi x}{l}\right) = S_f(x), \tag{4.78}$$

where $S_f(x) = f(x)$ at any point x where f is continuous and

$$S_f(x) = \frac{f(x_-) + f(x_+)}{2} \tag{4.79}$$

at any point of discontinuity.

The coefficients a_n, b_n of the Fourier series (4.77) are given by the formulas:

$$a_n = \frac{1}{\pi} \int_{-\pi}^{\pi} f\left(\frac{lt}{\pi}\right) \cos nt \, dt, \quad n = 0, 1, 2, \ldots,$$

$$b_n = \frac{1}{\pi} \int_{-\pi}^{\pi} f\left(\frac{lt}{\pi}\right) \sin nt \, dt, \quad n = 1, 2, \ldots,$$

and using in both integrals the substitution $x = \frac{lt}{\pi}$ we find:

$$a_n = \frac{1}{l} \int_{-l}^{l} f(x) \cos \frac{n\pi x}{l} \, dx, \quad n = 0, 1, 2, \ldots, \tag{4.80}$$

$$b_n = \frac{1}{l} \int_{-l}^{l} f(x) \sin \frac{n\pi x}{l} \, dx, \quad n = 1, 2, \ldots. \tag{4.81}$$

As we have noted in Remark 4.6, one can write the Fourier series for a function which is defined only on an interval of length $T = 2l$, since it can be extended by periodicity to the whole axis of real numbers. Let $[a, b]$ be an interval of length $T = b - a$. If we denote by $\omega = \frac{2\pi}{T}$, then we can write the Fourier coefficients (4.80) and (4.81) in the form:

$$a_n = \frac{2}{T} \int_a^b f(x) \cos n\omega x \, dx, \quad n = 0, 1, 2, \ldots, \tag{4.82}$$

$$b_n = \frac{2}{T} \int_a^b f(x) \sin n\omega x \, dx, \quad n = 1, 2, \ldots. \tag{4.83}$$

Remark 4.8 Fourier series of odd/even functions.

If f is an *even* function then $f(x) \sin \frac{n\pi x}{l}$ is odd (the product between an odd function and an even function is odd). Consequently, $b_n = 0$, $n = 1, 2, \ldots$ and the Fourier series expansion of f is a series of cosines:

$$f(x) \sim \frac{a_0}{2} + \sum_{n=1}^{\infty} a_n \cos \frac{n\pi x}{l}, \tag{4.84}$$

where

$$a_n = \frac{2}{l} \int_0^l f(x) \cos \frac{n\pi x}{l} \, dx, \quad n = 0, 1, \ldots \tag{4.85}$$

because $f(x) \cos \frac{n\pi x}{l}$ is even (being the product of two even functions).

Similarly, if f is an *odd* function then the product $f(x) \cos \frac{n\pi x}{l}$ is odd, so $a_n = 0$, $n = 0, 1, \ldots$ and the Fourier series expansion of f is a series of sines:

$$f(x) \sim \sum_{n=1}^{\infty} b_n \sin \frac{n\pi x}{l}, \tag{4.86}$$

where

$$b_n = \frac{2}{l} \int_0^l f(x) \sin \frac{n\pi x}{l} \, dx, \quad n = 1, 2, \ldots \tag{4.87}$$

because $f(x) \sin \frac{n\pi x}{l}$ is even (being the product of two odd functions).

These formulas reflect a kind of an hereditary property of f, namely, if f is even, then its Fourier expansion contains only cosines, while in the Fourier expansion of an odd function appear only sin-functions.

Remark 4.9 A problem which arises frequently is that of making an expansion in *sine series* or in *cosine series* of a piecewise smooth function $f(x)$ defined on

4.4 The Convergence of Fourier Series

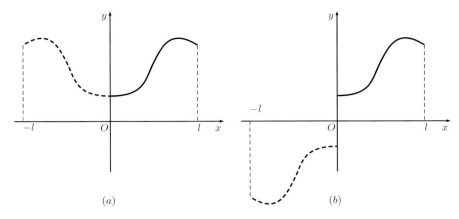

Fig. 4.3 The extension of a function defined on $[0, l]$ to an even function (**a**); to an odd function (**b**)

an interval $[0, l]$. If we are interested to expand the function in a Fourier series of cosines on $[0, l]$, then we can extend it to an even function (see Fig. 4.3a), $\bar{f} : [-l, l] \to \mathbb{R}$,

$$\bar{f}(x) = \begin{cases} f(x), & \text{for } x \in [0, l] \\ f(-x), & \text{for } x \in [-l, 0) \end{cases}$$

and we can apply the formulas (4.84), (4.85) to $\bar{f}(x)$. Since $\bar{f}(x) = f(x), \forall x \in [0, l]$, the formula (4.84) holds for any $x \in [0, l]$ (so there is no need to write explicitly the even extension, $\bar{f}(x)$).

In the same way, if we have to expand the function $f(x)$ in a Fourier series of sines on the interval $[0, l]$, then we extend it to an odd function (see Fig. 4.3b), $\bar{f} : [-l, l] \to \mathbb{R}$,

$$\bar{f}(x) = \begin{cases} f(x), & \text{for } x \in [0, l] \\ -f(-x), & \text{for } x \in [-l, 0) \end{cases}$$

and we can apply the formulas (4.86), (4.87) to the function $\bar{f}(x)$. Since $\bar{f}(x) = f(x), \forall x \in [0, l]$, the formula (4.84) holds for any $x \in (0, l)$.

Example 4.1 Periodic rectangular wave [21]. Let us find the Fourier series corresponding to the function $f(x)$ which is defined on the fundamental interval $(-\pi, \pi)$ as follows:

$$f(x) = \begin{cases} -\pi/4, & \text{if } x \in (-\pi, 0) \\ \pi/4, & \text{if } x \in (0, \pi) \end{cases} \qquad (4.88)$$

First of all, we notice that the function is odd, so, by Remark 4.8, it has a Fourier expansion in series of sines:

$$f(x) \sim \sum_{n=1}^{\infty} b_n \sin nx,$$

where, for $n = 1, 2, \ldots,$

$$b_n = \frac{2}{\pi} \int_0^{\pi} f(x) \sin nx \, dx = \frac{1}{2} \int_0^{\pi} \sin nx \, dx = -\left.\frac{\cos nx}{2n}\right|_0^{\pi} =$$

$$= -\frac{(-1)^n - 1}{2n} = \begin{cases} 0, & \text{if } n = 2k \\ \frac{1}{2k+1}, & \text{if } n = 2k+1 \end{cases}.$$

Hence we obtain that

$$f(x) = \sum_{k=0}^{\infty} \frac{\sin(2k+1)x}{2k+1} = \sin x + \frac{\sin 3x}{3} + \frac{\sin 5x}{5} + \ldots, \tag{4.89}$$

for any $x \in (-\pi, \pi)$, $x \neq 0$. The point $x = 0$ is a point of discontinuity and it is obvious that the sum of the series (4.89) at this point equals 0, which is the arithmetic mean of the one-sided limits $-\frac{\pi}{4}$ and $\frac{\pi}{4}$.

The graph of the function $f(x)$ (extended by periodicity to the whole x-axis) and the graphs of the partial sums

$$s_1(x) = \sin x, \quad s_2(x) = \sin x + \frac{1}{3} \sin 3x,$$

$$s_6(x) = \sin x + \frac{1}{3} \sin 3x + \frac{1}{5} \sin 5x + \frac{1}{7} \sin 7x + \frac{1}{9} \sin 9x + \frac{1}{11} \sin 11x,$$

are represented in Fig. 4.4.

We notice also that, for $x = \frac{\pi}{2}$ the Eq. (4.89) becomes

$$f\left(\frac{\pi}{2}\right) = \sum_{k=0}^{\infty} \frac{1}{2k+1} \sin \frac{(2k+1)\pi}{2} = \sum_{k=0}^{\infty} \frac{(-1)^k}{2k+1},$$

or, equivalently,

$$\frac{\pi}{4} = 1 - \frac{1}{3} + \frac{1}{5} - \frac{1}{7} + \ldots,$$

a famous result obtained by Leibniz in 1673 (using geometrical methods).

4.4 The Convergence of Fourier Series

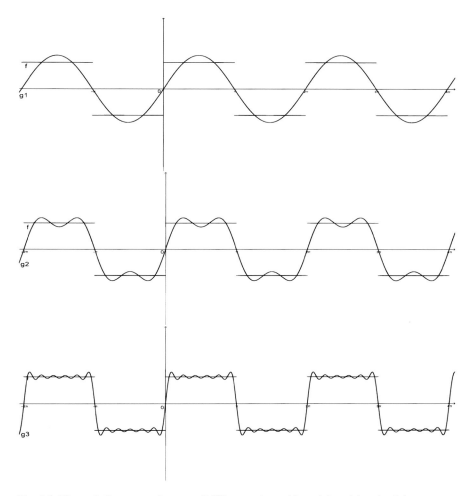

Fig. 4.4 The periodic rectangular wave (4.88) approximated by $s_1(x)$, $s_2(x)$ and $s_6(x)$

Remark 4.10 Gibbs phenomenon. We remark in Fig. 4.4 that, near the points of discontinuity $k\pi, k \in \mathbb{Z}$, the difference between $f(x)$ and the partial sum $s_n(x)$ has some extreme values and we might expect these extrema to decrease to zero as n increases, but this is not true. By adding more and more terms, the extrema are only shifted closer to the discontinuity points, while the maximum amplitude of the oscillations does not fade away as n approaches infinity, but rather seems to tend to a constant value. This strange behavior of the Fourier series in the neighborhood of the discontinuity points is known as the *Gibbs phenomenon*.

The partial sum of the Fourier series (4.89),

$$s_n(x) = \sin x + \frac{\sin 3x}{3} + \ldots + \frac{\sin(2n-1)x}{2n-1},$$

has the derivative

$$s'_n(x) = \cos x + \cos 3x + \ldots + \cos(2n-1)x.$$

We multiply by $2 \sin x$ this equality and obtain

$$2 \sin x \cdot s'_n(x) = \sin 2x + \sin 4x - \sin 2x + \ldots + \sin 2nx - \sin(2n-2)x,$$

hence

$$s'_n(x) = \frac{\sin 2nx}{2 \sin x}$$

and we notice that s_n has $2n - 1$ local extrema in the interval $(0, \pi)$:
n local maxima at the points $\frac{\pi}{2n}, \frac{3\pi}{2n}, \ldots, \frac{(2n-1)\pi}{2n}$ and
$n - 1$ local minima at the points $\frac{\pi}{n}, \frac{2\pi}{n}, \ldots, \frac{(n-1)\pi}{n}$.
We calculate the value of s_n at the point $\frac{\pi}{2n}$:

$$s_n\left(\frac{\pi}{2n}\right) = \sin \frac{\pi}{2n} + \frac{1}{3} \sin \frac{3\pi}{2n} + \ldots + \frac{1}{2n-1} \sin \frac{(2n-1)\pi}{2n} =$$

$$= \frac{1}{2}\left[\frac{\sin \frac{\pi}{2n}}{\frac{\pi}{2n}}\left(\frac{\pi}{n} - 0\right) + \frac{\sin \frac{3\pi}{2n}}{\frac{3\pi}{2n}}\left(\frac{2\pi}{n} - \frac{\pi}{n}\right) + \ldots + \frac{\sin \frac{(2n-1)\pi}{2n}}{\frac{(2n-1)\pi}{2n}}\left(\pi - \frac{(n-1)\pi}{n}\right)\right].$$

Notice that the expression in square brackets is a Riemann sum approximation to the integral $\int_0^\pi \frac{\sin x}{x} dx$. Since the function $\frac{\sin x}{x}$ is continuous, it follows that

$$\lim_{n \to \infty} s_n(x) = \frac{1}{2} \int_0^\pi \frac{\sin x}{x}$$

and so

$$\lim_{n \to \infty} \left| f\left(\frac{\pi}{2n}\right) - s_n\left(\frac{\pi}{2n}\right) \right| = \frac{1}{2} \int_0^\pi \frac{\sin x}{x} - \frac{\pi}{4} = \frac{\pi}{2} \cdot 0.08948987\ldots.$$

It can be proved that, for any piecewise continuous function which has a piecewise continuous derivative, the maximum value of $|f(x) - s_n(x)|$ in a neighborhood of a point of discontinuity x_0 tends to

$$|f(x_{0+}) - f(x_{0-})| \cdot 0.08948987 \ldots .$$

4.5 Differentiation and Integration of the Fourier Series

Theorem 4.5 *If $f(x)$ is a continuous, piecewise smooth function of period 2π, then the Fourier series of the derivative $f'(x)$ can be obtained from the Fourier series of the function $f(x)$ using term by term differentiation ([41], p. 129).*

Proof Since f is continuous everywhere, we can write:

$$f(x) = \frac{a_0}{2} + \sum_{n=1}^{\infty}(a_n \cos nx + b_n \sin nx). \tag{4.90}$$

The derivative, f', is not continuous everywhere, but it is piecewise continuous. We denote by α_n and β_n its Fourier coefficients. First, we notice that

$$\alpha_0 = \frac{1}{\pi} \int_{-\pi}^{\pi} f'(x)\,dx = f(\pi) - f(-\pi) = 0.$$

For $n = 1, 2, \ldots$, using integration by parts, we obtain:

$$\alpha_n = \frac{1}{\pi} \int_{-\pi}^{\pi} f'(x) \cos nx\,dx = \frac{1}{\pi} f(x) \cos nx \Big|_{-\pi}^{\pi} + \frac{n}{\pi} \int_{-\pi}^{\pi} f(x) \sin nx\,dx = nb_n,$$

$$\beta_n = \frac{1}{\pi} \int_{-\pi}^{\pi} f'(x) \sin nx\,dx = \frac{1}{\pi} f(x) \sin nx \Big|_{-\pi}^{\pi} - \frac{n}{\pi} \int_{-\pi}^{\pi} f(x) \cos nx\,dx = -na_n.$$

Thus,

$$f'(x) \sim \sum_{n=1}^{\infty}(nb_n \cos nx - na_n \sin nx),$$

and this is exactly the series obtained from (4.90) by term by term differentiation. □

Theorem 4.6 *If $f(x)$ is a piecewise continuous function of period 2π with the Fourier series*

$$f(x) \sim \frac{a_0}{2} + \sum_{n=1}^{\infty}(a_n \cos nx + b_n \sin nx), \tag{4.91}$$

then

$$\int_a^b f(x)\,dx = \frac{a_0}{2}(b-a) + \sum_{n=1}^{\infty} \int_a^b (a_n \cos nx + b_n \sin nx)\,dx, \qquad (4.92)$$

for any interval $[a,b]$ and

$$\int_0^x f(t)\,dt = \sum_{n=1}^{\infty} \frac{b_n}{n} + \sum_{n=1}^{\infty} \frac{-b_n \cos nx + (a_n + (-1)^{n+1} a_0)\sin nx}{n}, \qquad (4.93)$$

for any $x \in (-\pi, \pi)$ ([41], p. 125).

Proof The function $F : \mathbb{R} \to \mathbb{R}$,

$$F(x) = \int_0^x \left(f(t) - \frac{a_0}{2} \right) dt$$

is continuous, with a piecewise continuous derivative ($F'(x) = f(x) - \frac{a_0}{2}$) and it is periodic of period 2π:

$$F(x + 2\pi) = \int_0^x \left(f(t) - \frac{a_0}{2} \right) dt + \int_x^{x+2\pi} \left(f(t) - \frac{a_0}{2} \right) dt =$$

$$= F(x) + \int_x^{x+2\pi} f(t)\,dt - a_0 \pi = F(x),$$

for any $x \in \mathbb{R}$. So, by Dirichlet's Theorem, we can write:

$$F(x) = \frac{A_0}{2} + \sum_{n=1}^{\infty} (A_n \cos nx + B_n \sin nx),$$

where

$$A_n = \frac{1}{\pi} \int_{-\pi}^{\pi} F(x) \cos nx\,dx =$$

$$= \frac{1}{\pi} F(x) \frac{\sin nx}{n} \bigg|_{-\pi}^{\pi} - \frac{1}{n\pi} \int_{-\pi}^{\pi} \left(f(x) - \frac{a_0}{2} \right) \sin nx\,dx = -\frac{b_n}{n}$$

and

$$B_n = \frac{1}{\pi} \int_{-\pi}^{\pi} F(x) \sin nx \, dx =$$

$$= -\frac{1}{\pi} F(x) \frac{\cos nx}{n} \Big|_{-\pi}^{\pi} + \frac{1}{n\pi} \int_{-\pi}^{\pi} \left(f(x) - \frac{a_0}{2} \right) \cos nx \, dx = \frac{a_n}{n}.$$

So it follows that

$$F(x) = \frac{A_0}{2} + \sum_{n=1}^{\infty} \frac{a_n \sin nx - b_n \cos nx}{n}$$

and we obtain

$$\int_0^x f(t) dt = \frac{A_0}{2} + \frac{a_0 x}{2} + \sum_{n=1}^{\infty} \frac{a_n \sin nx - b_n \cos nx}{n}. \qquad (4.94)$$

Since

$$\int_a^b f(t) dt = \int_0^b f(t) dt - \int_0^a f(t) dt$$

the equality (4.92) follows.

If we set $x = 0$ in (4.94) we obtain that

$$\frac{A_0}{2} = \sum_{n=1}^{\infty} \frac{b_n}{n}. \qquad (4.95)$$

The expansion in Fourier series of the function $g(x) = \frac{x}{2}$ is

$$\frac{x}{2} = \sum_{n=1}^{\infty} (-1)^{n+1} \frac{\sin nx}{n}. \qquad (4.96)$$

By replacing (4.95) and (4.96) in Eq. (4.94), the equality (4.93) follows. □

4.6 The Convergence in the Mean: Complete Systems

Definition 4.9 Let $f, f_1, f_2, \ldots$ be piecewise continuous functions on the interval $I = [a, b]$. The sequence $\{f_n\}$ is said to be *convergent in the mean* to f if

$\lim_{n\to\infty} \|f_n - f\| = 0$ or, equivalently, if

$$\lim_{n\to\infty} \int_a^b (f_n(x) - f(x))^2 \, dx = 0.$$

The series $\sum_{n=1}^{\infty} f_n$ is said to be *convergent in the mean* to f if

$$\lim_{n\to\infty} \left\| \sum_{k=1}^n f_k - f \right\| = 0.$$

Lemma 4.7 *If the sequence of functions $\{f_n\}$ uniformly converges to f on the interval $[a, b]$, then $\{f_n\}$ is also convergent in the mean to the function f.*

Proof Since $\{f_n\}$ uniformly converges to f on $[a, b]$, it follows that, for any $\varepsilon > 0$, there exists $N \in \mathbb{N}$ such that

$$|f_n(x) - f(x)| < \frac{\varepsilon}{\sqrt{b-a}}, \quad \text{for all } x \in [a, b], n \geq N.$$

We have:

$$\|f_n - f\|^2 = \int_a^b (f_n(x) - f(x))^2 \, dx \leq \frac{\varepsilon^2}{b-a} \cdot (b-a) = \varepsilon^2, \ \forall n \geq N$$

hence $\lim_{n\to\infty} \|f_n - f\| = 0$ and the proof is completed. □

Note that the reciprocal statement is not true: the sequence of functions $f_n(x) = x^n$ is not uniformly convergent on $[0, 1]$, since it is pointwise convergent to the *discontinuous* function $f(x) = 0$, if $x \in [0, 1)$, $f(1) = 1$. Nevertheless, it is convergent in the mean to the function $g(x) = 0$, $x \in [0, 1]$:

$$\int_0^1 (f_n(x) - g(x))^2 \, dx = \int_0^1 x^{2n} \, dx = \frac{1}{2n+1} \longrightarrow 0.$$

Consider the orthogonal system of functions in the space of piecewise continuous functions, $PC(I)$:

$$\{\phi_0(x), \phi_1(x), \ldots, \phi_n(x), \ldots\}. \tag{4.97}$$

4.6 The Convergence in the Mean: Complete Systems

For every $f \in PC(I)$, we denote by c_n, $n = 0, 1, \ldots$, the Fourier coefficients of the function f with respect to the system (4.97),

$$c_n = \frac{\langle f, \phi_n \rangle}{\|\phi_n\|^2} = \frac{1}{\|\phi_n\|^2} \int_a^b f(x)\phi_n(x)\,dx. \tag{4.98}$$

We proved in Sect. 4.3 the Bessel's inequality:

$$\sum_{k=0}^{\infty} c_k^2 \|\phi_k\|^2 \leq \|f\|^2. \tag{4.99}$$

Definition 4.10 The orthogonal system (4.97) is said to be *complete* if, for any $f \in PC(I)$, the following equality holds (instead of Bessel's inequality (4.99)):

$$\sum_{k=0}^{\infty} c_k^2 \|\phi_k\|^2 = \|f\|^2 = \int_a^b f^2(x)\,dx. \tag{4.100}$$

The equality (4.100) is known as Parseval's identity.

Remark 4.11 If the orthogonal system (4.97) is complete, and $f \in PC(I)$ is orthogonal on every function of the system, then $c_k = 0$ for all $k = 0, 1, \ldots$. By Definition 4.10 it follows that $\|f\| = 0$, hence, from (4.42), we obtain that $f \sim \mathbf{0}$ ($f(x) = 0$ on the interval $[a, b]$, possibly except for a finite number of points: we say that $f(x) = 0$ *almost everywhere* (a.e.)).

Theorem 4.7 *The orthogonal system (4.97) is complete if and only if the Fourier series of any piecewise continuous function $f \in PC(I)$,*

$$\sum_{k=0}^{\infty} c_k \phi_k(x), \tag{4.101}$$

converges in the mean to f.

Proof Let

$$s_n(x) = \sum_{k=0}^{n} c_k \phi_k(x), \quad n = 0, 1, \ldots, \tag{4.102}$$

be the partial sums of the Fourier series (4.101). By the orthogonality of the system (4.97) we have that

$$\|s_n - f\|^2 = \left\langle \sum_{k=0}^{n} c_k \phi_k - f, \sum_{k=0}^{n} c_k \phi_k - f \right\rangle =$$

$$= \sum_{k=0}^{n} c_k^2 \|\phi_k(x)\|^2 - 2 \sum_{k=0}^{n} c_k \langle \phi_k, f \rangle + \|f\|^2$$

and, by (4.98), we get

$$\|s_n - f\|^2 = \|f\|^2 - \sum_{k=0}^{n} c_k^2 \|\phi_k(x)\|^2.$$

So $\lim_{n \to \infty} \|s_n - f\| = 0$ if and only if the equality (4.100) holds (see also Theorem 11.15). □

Lemma 4.8 *For any function f piecewise continuous on the interval $I = [a, b]$ and any $\varepsilon > 0$, there exists a continuous function $F : [a, b] \to \mathbb{R}$ such that $F(a) = F(b)$ and*

$$\|f - F\| \leq \varepsilon.$$

Proof Let

$$a < x_1 < x_2 < \ldots < x_p < b$$

be the discontinuity points of the function f and

$$M = \sup_{x \in [a,b]} |f(x)|.$$

Let $\delta > 0$ be a positive number small enough such that the intervals $[a, a + \delta]$, $[x_1 - \delta, x_1 + \delta]$, ..., $[x_p - \delta, x_p + \delta]$ and $[b - \delta, b]$ do not overlap. We define the function F as follows:

$F(x) = \frac{1}{\delta} \left(\frac{1}{2}(f(a_+) + f(b_-))(a + \delta - x) + f(a + \delta)(x - a) \right)$, if $x \in [a, a+\delta]$,
$F(x) = \frac{1}{2\delta} \left(f(x_k - \delta)(x_k + \delta - x) + f(x_k + \delta)(x - x_k + \delta) \right)$,
if $x \in [x_k - \delta, x_k + \delta]$, $k = 1, \ldots, p$,
$F(x) = \frac{1}{\delta} \left(\frac{1}{2}(f(a_+) + f(b_-))(x - b + \delta) + f(b - \delta)(b - x) \right)$, if $x \in [b-\delta, b]$,

4.6 The Convergence in the Mean: Complete Systems

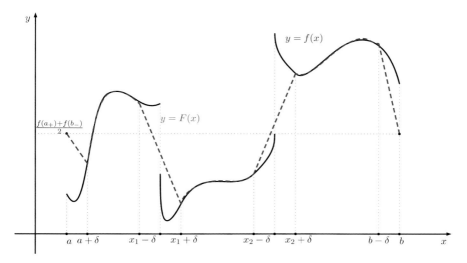

Fig. 4.5 A piecewise continuous function $f(x)$ and a continuous function $F(x)$ s.t. $F(a) = F(b)$ and $\|f - F\| \leq \varepsilon$

and $F(x) = f(x)$ otherwise (see Fig. 4.5). The function F is continuous, $F(a) = F(b)$ and $|f(x) - F(x)| \leq 2M$ for all $x \in [a, b]$. Thus, we have:

$$\|f - F\|^2 = \int_a^{a+\delta} (f(x) - F(x))^2 \, dx + \int_{b-\delta}^b (f(x) - F(x))^2 \, dx +$$

$$+ \sum_{k=1}^p \int_{x_k-\delta}^{x_k+\delta} (f(x) - F(x))^2 \, dx \leq 8M^2 \delta (p+1).$$

Thus, in order to have $\|f - F\| \leq \varepsilon$, it suffices to take

$$\delta \leq \frac{\varepsilon^2}{8M^2(p+1)}$$

hence the lemma is proved. □

We say that $f : [a, b] \to \mathbb{R}$ is a *piecewise linear function* on $[a, b]$ if there exists a partition of $[a, b]$, $\Delta : a = x_0 < x_1 < \ldots < x_n = b$ such that f is a linear function on each subinterval (x_{i-1}, x_i), $i = 1, 2, \ldots, n$. It is clear that any piecewise linear function is a piecewise smooth function.

Lemma 4.9 *Any continuous function $F : [a, b] \to \mathbb{R}$ can be well approximated by a continuous piecewise linear function, i.e. for any $\varepsilon > 0$, there exists a partition $\Delta : a = x_0 < x_1 < \ldots < x_n = b$ of the interval $[a, b]$ and a piecewise linear* function $\Phi : [a, b] \to \mathbb{R}$ such that

$$|F(x) - \Phi(x)| \leq \varepsilon,$$

for all $x \in [a, b]$.

Proof The function F is continuous on the interval $[a, b]$, so it is also *uniformly continuous* on $[a, b]$ (see Proposition 11.18 from Chap. 11). This means that, for any $\varepsilon > 0$, there exists $\delta > 0$ such that

$$|F(x) - F(x')| < \varepsilon, \text{ for any } x, x' \text{ with } |x - x'| < \delta.$$

Let $\Delta : a = x_0 < x_1 < \ldots < x_n = b$ be an equidistant partition of $[a, b]$, i.e. $x_i - x_{i-1} = \frac{b-a}{n}$ for any $i = 1, 2, \ldots, n$. The number of points is chosen such that $\frac{b-a}{n} \leq \delta$, hence

$$|F(x) - F(x')| < \varepsilon, \text{ for any } x, x' \in [x_{i-1}, x_i], \ i = 1, \ldots, n. \tag{4.103}$$

We define the piecewise linear function Φ as follows:

$$\Phi(x) = F(x_{i-1}) \cdot \frac{x_i - x}{x_i - x_{i-1}} + F(x_i) \cdot \frac{x - x_{i-1}}{x_i - x_{i-1}},$$

for all $x \in [x_{i-1}, x_i]$, $i = 1, \ldots n$. Obviously, Φ is continuous on $[a, b]$ (see Fig. 4.6). We evaluate the distance between $F(x)$ and $\Phi(x)$ on each interval interval

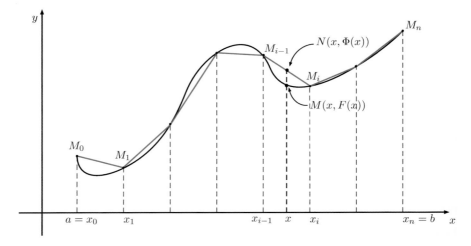

Fig. 4.6 Approximation of a continuous function $F(x)$ by a piecewise linear function $\Phi(x)$

4.6 The Convergence in the Mean: Complete Systems

$[x_{i-1}, x_i]$:

$$|F(x) - \Phi(x)| \le$$

$$\le |F(x) - F(x_{i-1})| \cdot \frac{x_i - x}{x_i - x_{i-1}} + |F(x) - F(x_i)| \cdot \frac{x - x_{i-1}}{x_i - x_{i-1}} < \varepsilon,$$

where the last inequality follows from (4.103). □

Corollary 4.2 *For any continuous function $F : [a, b] \to \mathbb{R}$, and any $\varepsilon > 0$, there exists a function $\Phi : [a, b] \to \mathbb{R}$, continuous and piecewise smooth such that*

$$\|F - \Phi\| \le \varepsilon.$$

Proof By Lemma 4.9, there exists a continuous function, piecewise linear (hence piecewise smooth) such that

$$|F(x) - \Phi(x)| < \frac{\varepsilon}{\sqrt{b-a}},$$

for all $x \in [a, b]$. It follows that:

$$\|F - \Phi\|^2 = \int_a^b |F(x) - \Phi(x)|^2 \, dx \le \frac{\varepsilon^2}{b-a} \cdot (b-a) = \varepsilon^2.$$

□

Consider now the following orthogonal system on the interval $I = [-l, l]$:

$$\mathcal{T} = \left\{1, \cos\frac{\pi x}{l}, \sin\frac{\pi x}{l}, \ldots, \cos\frac{n\pi x}{l}, \sin\frac{n\pi x}{l}, \ldots\right\}. \tag{4.104}$$

Theorem 4.8 *The orthogonal system (4.104) is complete in $PC(I)$.*

Proof By Theorem 4.7, it is sufficient to prove that, for any piecewise continuous function $f(x)$ and $\varepsilon > 0$, there exists $N \in \mathbb{N}$ such that

$$\|s_n - f\| < \varepsilon, \text{ for all } n \ge N,$$

where $s_n(x)$ is the partial sum of the Fourier series of the function f,

$$s_n(x) = \frac{a_0}{2} + \sum_{k=1}^n \left(a_k \cos\frac{k\pi x}{l} + b_k \sin\frac{k\pi x}{l}\right).$$

By Lemma 4.8 we know that there exists a function $F(x)$, continuous on $[-l, l]$, such that $F(-l) = F(l)$ and

$$\|f - F\| < \frac{\varepsilon}{3}.$$

On the other hand, by Corollary 4.2, there exists a function $\Phi(x)$, continuous and piecewise smooth on $[-l, l]$ such that

$$\|F - \Phi\| < \frac{\varepsilon}{3}.$$

We also remark (from the proof of Lemma 4.9) that $\Phi(-l) = F(-l) = F(l) = \Phi(l)$.

Finally, since $\Phi(-l) = \Phi(l)$, the function $\Phi(x)$ can be extended by periodicity to a continuous and piecewise smooth function and, from Theorem 4.4 and Remark 4.7, its Fourier series is uniformly convergent on $[-l, l]$. But, as follows from Lemma 4.7, a uniformly convergent series must also converge in the mean, so there exists a natural number N such that

$$\|\Phi - \sigma_n\| < \frac{\varepsilon}{3}, \quad \forall n \geq N,$$

where

$$\sigma_n(x) = \frac{\alpha_0}{2} + \sum_{k=1}^{n} \left(\alpha_k \cos \frac{k\pi x}{l} + \beta_k \sin \frac{k\pi x}{l} \right)$$

is the partial sum of the Fourier series of the function $\Phi(x)$.

By the triangle inequality (4.40), we obtain:

$$\|f - \sigma_n\| \leq \|f - F\| + \|F - \Phi\| + \|\Phi - \sigma_n\| <$$

$$< \frac{\varepsilon}{3} + \frac{\varepsilon}{3} + \frac{\varepsilon}{3} = \varepsilon,$$

for any $n \geq N$.

But we have proved in Sect. 4.3.2 that the partial sum $s_n(x)$ of the Fourier series represents the *best approximation in the mean* of the function f by a trigonometric polynomial of order n, hence

$$\|f - s_n\| \leq \|f - \sigma_n\| \leq \varepsilon,$$

for any $n \geq N$ and the theorem is proved. □

4.7 Examples of Fourier Expansions

Remark 4.12 From Theorems 4.7 and 4.8 we can conclude that the Fourier series of any piecewise continuous function $f \in PC(I)$ is *convergent in the mean* to f, even if it may be not piecewise convergent at some points.

Remark 4.13 Hilbert space. Hilbert basis.

A *Hilbert space* H is an inner product space that is also a complete metric space with respect to the distance function induced by the inner product (for a detailed presentation, see Sects. 11.1.5 and 11.1.7 in Chap. 11). This means that if $\{f_n\} \subset H$ is a Cauchy sequence (for any $\varepsilon > 0$ there is a number $N \in \mathbb{N}$ such that $\|f_n - f_m\| < \varepsilon$ if $m, n \geq N$), then it converges to a limit in H (there exists $f \in H$ such that $\lim_{n \to \infty} \|f_n - f\| = 0$.)

The inner product space $\widehat{PC}(I)$ introduced in Sect. 4.3 is not a Hilbert space (it is not complete), but, according to Theorem 11.11, it can be completed to a Hilbert space, which will be denoted by $\overline{PC}(I)$.

A complete orthonormal system in a Hilbert space is said to be a *Hilbert basis*. Thus, the *complete orthonormal system*

$$S = \left\{ \frac{1}{\sqrt{2l}}, \frac{1}{\sqrt{l}} \cos \frac{\pi x}{l}, \frac{1}{\sqrt{l}} \sin \frac{\pi x}{l}, \ldots, \frac{1}{\sqrt{l}} \cos \frac{n \pi x}{l}, \frac{1}{\sqrt{l}} \sin \frac{n \pi x}{l}, \ldots \right\}. \tag{4.105}$$

is a Hilbert basis in the space $\overline{PC}(I)$.

4.7 Examples of Fourier Expansions

Example 4.2 Let us find the Fourier expansion of $f : [-1, 1] \to \mathbb{R}$, $f(x) = 3x$ (Fig. 4.7).

Since f is an odd function, we have $a_k = 0$ for $k = 0, 1, \ldots$ and the coefficients b_k for $k = 1, 2, \ldots$ are given by the formula (4.87):

$$b_k = 2 \int_0^1 3x \sin k\pi x \, dx = 6 \int_0^1 x \left(-\frac{\cos k\pi x}{k\pi} \right)' dx =$$

$$= 6 \left\{ \left[x \left(-\frac{\cos k\pi x}{k\pi} \right) \right]_0^1 + \frac{1}{k\pi} \int_0^1 \cos k\pi x \, dx \right\} =$$

$$= 6 \left[\frac{1}{k\pi} (-1)^{k+1} + \frac{1}{k^2 \pi^2} \sin k\pi x \Big|_0^1 \right] = \frac{6}{k\pi} (-1)^{k+1}.$$

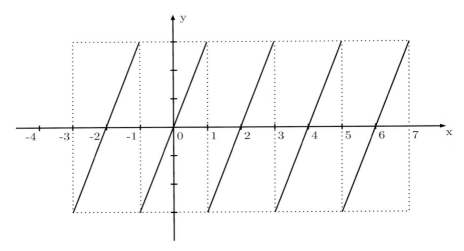

Fig. 4.7 The periodic extension of the function $f(x) = 3x$, $x \in (-1, 1)$ to the whole real axis

So the Fourier expansion of f on $(-1, 1)$ is:

$$f(x) = \frac{6}{\pi} \sum_{k=1}^{\infty} \frac{(-1)^{k+1}}{k} \sin k\pi x = S_f(x). \tag{4.106}$$

At the point $x = 1$, we have $S_f(1) = 0$, but $f(1) = 3$. Let $\widetilde{f} : \mathbb{R} \to \mathbb{R}$ be the extension by periodicity of f to the whole $\mathbb{R}$. Since $\widetilde{f}$ is not continuous at $x = 1$, but it has side limits, by the formula (4.79) we obtain:

$$S_{\widetilde{f}}(1) = \frac{1}{2}\left[\widetilde{f}(1-0) + \widetilde{f}(1+0)\right] = \frac{1}{2}[3 - 3] = 0.$$

We can use the equality (4.106) to find the values of some important numerical series. For instance, if $x = \frac{1}{2}$ in (4.106), we have:

$$\frac{3}{2} = \frac{6}{\pi} \sum_{k=1}^{\infty} \frac{(-1)^{k+1}}{k} \sin k\pi \frac{1}{2} = \frac{6}{\pi} \sum_{p=0}^{\infty} \frac{(-1)^p}{2p+1},$$

or

$$\sum_{p=0}^{\infty} \frac{(-1)^p}{2p+1} = \frac{\pi}{4},$$

4.7 Examples of Fourier Expansions

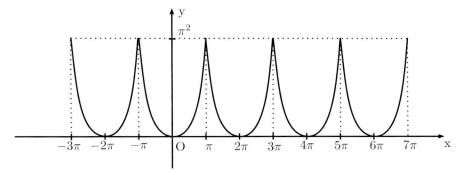

Fig. 4.8 The periodic extension of the function $f(x) = x^2$, $x \in [-\pi, \pi]$, to the whole real axis

and we can find an approximation of π by rational numbers. We also can apply the Parseval's identity (4.58) in our case:

$$\frac{36}{\pi^2} \sum_{k=1}^{\infty} \frac{1}{k^2} = 2 \int_0^1 9x^2 \, dx = 6,$$

so we find the Euler identity:

$$\sum_{k=1}^{\infty} \frac{1}{k^2} = \frac{\pi^2}{6}. \tag{4.107}$$

Example 4.3 Let us find the Fourier expansion of the function $f(x) = x^2$ defined on the interval $[-\pi, \pi]$ (Fig. 4.8).

Let $\widetilde{f} : \mathbb{R} \to \mathbb{R}$ be the expansion by periodicity of f.
For instance,

$$\widetilde{f}\left(\frac{7\pi}{2}\right) = \widetilde{f}(4\pi - \frac{\pi}{2}) = f\left(-\frac{\pi}{2}\right) = f\left(\frac{\pi}{2}\right) = \frac{\pi^2}{4}.$$

Since our function is even, we have $b_k = 0$ for $k = 1, 2, \ldots$ and the coefficients a_k for $k = 0, 1, \ldots$ are given by the formula (4.85):

$$a_0 = \frac{1}{\pi} \int_{-\pi}^{\pi} f(x) \, dx = \frac{2}{\pi} \int_0^{\pi} x^2 \, dx = \frac{2\pi^2}{3}$$

and, for $k = 1, 2, \ldots$ we get:

$$a_k = \frac{2}{\pi} \int_0^\pi x^2 \cos kx \, dx = \frac{2}{\pi} \int_0^\pi x^2 \left(\frac{\sin kx}{k}\right)' dx =$$

$$= \frac{2}{\pi} \left[x^2 \frac{\sin kx}{k} \Big|_0^\pi - \frac{2}{k} \int_0^\pi x \sin kx \, dx \right] =$$

$$= \frac{4}{k\pi} \int_0^\pi x \left(\frac{\cos kx}{k}\right)' dx = \frac{4}{k\pi} \left[x \frac{\cos kx}{k} \Big|_0^\pi \right] = \frac{4}{k^2}(-1)^k.$$

So, the Fourier expansion of $\tilde{f}$ defined on the entire $\mathbb{R}$ is:

$$\tilde{f}(x) = S_{\tilde{f}}(x) = \frac{\pi^2}{3} + 4 \sum_{k=1}^\infty \frac{(-1)^k}{k^2} \cos kx.$$

We can write the equality $\tilde{f}(x) = S_{\tilde{f}}(x)$ for any $x \in \mathbb{R}$, because $\tilde{f}$ is a continuous function. In particular, if $x \in [-\pi, \pi]$, one can write:

$$x^2 = \frac{\pi^2}{3} + 4 \sum_{k=1}^\infty \frac{(-1)^k}{k^2} \cos kx. \qquad (4.108)$$

We can again compute the values of some particular numerical series by using this last equality. Let us firstly make $x = 0$ and find

$$0 = \frac{\pi^2}{3} + 4 \sum_{k=1}^\infty \frac{(-1)^k}{k^2},$$

or

$$\sum_{k=1}^\infty \frac{(-1)^{k-1}}{k^2} = \frac{\pi^2}{12},$$

i.e.

$$\frac{\pi^2}{12} = 1 - \frac{1}{2^2} + \frac{1}{3^2} - \frac{1}{4^2} + \ldots$$

Since $\tilde{f}$ is continuous at π, if we take $x = \pi$ in the formula (4.108) we obtain

$$\pi^2 = \frac{\pi^2}{3} + 4 \sum_{k=1}^\infty \frac{1}{k^2},$$

4.7 Examples of Fourier Expansions

or

$$\sum_{k=1}^{\infty} \frac{1}{k^2} = 1 + \frac{1}{2^2} + \frac{1}{3^2} + \frac{1}{4^2} + \ldots = \frac{\pi^2}{6}.$$

Example 4.4 Let us find the Fourier expansions: a) only in sines, and b) only in cosines, for $f(x) = l - x$, $x \in [0, l]$, where l is a positive real number.

a) We have to extend our function to an odd function (see Remark 4.9), $\bar{f} : [-l, l] \to \mathbb{R}$,

$$\bar{f} = \begin{cases} f(x) = l - x, & x \in [0, l] \\ -f(-x) = -l - x, & x \in [-l, 0) \end{cases}.$$

In fact, we do not need in the following the expression of $\bar{f}$. Since it is odd, we have $a_k = 0$, $k = 0, 1, \ldots$ and the coefficients b_k, $k = 1, 2, \ldots$ are given by the formula (4.87):

$$b_k = \frac{2}{l} \int_0^l f(x) \sin \frac{k\pi}{l} x \, dx = -\frac{2}{l} \int_0^l (l - x) \left(\frac{\cos \frac{k\pi}{l} x}{\frac{k\pi}{l}} \right)' dx =$$

$$= -\frac{2}{l} \left[(l - x) \frac{\cos \frac{k\pi}{l} x}{\frac{k\pi}{l}} \bigg|_0^l + \frac{l}{k\pi} \int_0^l \cos \frac{k\pi}{l} x \, dx \right] = \frac{2}{l} \left[\frac{l^2}{k\pi} \right] = \frac{2l}{\pi} \frac{1}{k}.$$

So, the expansion of f in a Fourier series of sines (see formula (4.86)) is

$$l - x = \frac{2l}{\pi} \sum_{k=1}^{\infty} \frac{1}{k} \sin \frac{k\pi}{l} x,$$

for $x \in (0, l)$.

b) This time we extend f to an even function $\bar{f} : [-l, l] \to \mathbb{R}$,

$$\bar{f} = \begin{cases} f(x) = l - x, & x \in [0, l] \\ f(-x) = l + x, & x \in [-l, 0) \end{cases}.$$

Since the function is even, we have $b_k = 0$, $k = 1, 2, \ldots$ and the coefficients a_k, $k = 0, 1, \ldots$ are given by the formula (4.85):

$$a_0 = \frac{2}{l} \int_0^l (l - x) \, dx = l.$$

$$a_k = \frac{2}{l} \int_0^l (l - x) \left(\frac{\sin \frac{k\pi}{l} x}{\frac{k\pi}{l}} \right)' dx =$$

$$= \frac{2}{l} \left[(l - x) \frac{\sin \frac{k\pi}{l} x}{\frac{k\pi}{l}} \bigg|_0^l + \frac{1}{k\pi} \int_0^l \sin \frac{k\pi}{l} x \, dx \right] =$$

$$= -\frac{2l}{k^2 \pi^2} \cos \frac{k\pi}{l} x \bigg|_0^l = -\frac{2l}{k^2 \pi^2} \left[(-1)^k - 1 \right] =$$

$$= \begin{cases} 0, & \text{if } k = 2p \\ \frac{4l}{k^2 \pi^2}, & \text{if } k = 2p + 1, \end{cases}$$

where $p = 0, 1, \ldots$. Thus,

$$l - x = \frac{l}{2} + \frac{4l}{\pi^2} \sum_{p=0}^{\infty} \frac{1}{(2p + 1)^2} \cos \frac{k\pi}{l} x.$$

Here, if we make $x = 0$, then, by the formula (4.79) we get

$$\frac{l}{2} = \frac{4l}{\pi^2} \sum_{p=0}^{\infty} \frac{1}{(2p + 1)^2},$$

or

$$\frac{\pi^2}{8} = \sum_{p=0}^{\infty} \frac{1}{(2p + 1)^2}.$$

By applying the Parseval's formula (4.58) we obtain:

$$\frac{l^2}{2} + \frac{16 l^2}{\pi^4} \sum_{p=0}^{\infty} \frac{1}{(2p + 1)^4} = \frac{2}{l} \int_0^l (l - x)^2 \, dx = \frac{2}{l} \cdot \frac{l^3}{3}.$$

Hence,
$$\sum_{p=0}^{\infty} \frac{1}{(2p+1)^4} = \frac{\pi^4}{96}.$$

4.8 The Complex form of the Fourier Series

Let $f(x)$ be a piecewise smooth function of period T and denote by $\omega = \frac{2\pi}{T}$. By (4.78), the Fourier series of the function f can be written as

$$S_f(x) = \frac{a_0}{2} + \sum_{k=1}^{\infty} (a_k \cos k\omega x + b_k \sin k\omega x), \qquad (4.109)$$

where the coefficients a_k, b_k are given by the integrals in the formulas (4.80)–(4.81). Since the integral of a function of period T on any interval of length T has the same value, one can write:

$$a_k = \frac{2}{T} \int_0^T f(x) \cos k\omega x, \text{ for } k = 0, 1, 2, \ldots \qquad (4.110)$$

and

$$b_k = \frac{2}{T} \int_0^T f(x) \sin k\omega x, \text{ for } k = 1, 2, \ldots. \qquad (4.111)$$

Recall that a complex-valued function of a real variable is a function $\varphi(x)$ defined on a real interval (a, b) (a, b may be $-\infty$ or ∞ respectively) with complex values, $\varphi : (a, b) \to \mathbb{C}$. It can be represented as a couple of two real-valued functions: $\varphi(x) = u(x) + iv(x)$, $x \in (a, b)$. The function u is called the real part of φ and v is called the imaginary part of φ. We know that $f' = u' + iv'$ and $\int_c^d f(x)\, dx = \int_c^d u(x)\, dx + i \int_c^d f(x)\, dx$.

Using the Euler's formula (2.56) we can write:

$$\exp(ik\omega x) = \cos k\omega x + i \sin k\omega x,$$

$$\exp(-ik\omega x) = \cos k\omega x - i \sin k\omega x,$$

so we find:

$$\cos k\omega x = \frac{\exp(ik\omega x) + \exp(-ik\omega x)}{2},$$

$$\sin k\omega x = \frac{\exp(ik\omega x) - \exp(-ik\omega x)}{2i}.$$

We replace in (4.109) and obtain:

$$S_f(x) = \frac{a_0}{2} + \sum_{k=1}^{\infty} \left(a_k \frac{\exp(ik\omega x) + \exp(-ik\omega x)}{2} + b_k \frac{\exp(ik\omega x) - \exp(-ik\omega x)}{2i} \right)$$

$$= \frac{a_0}{2} + \sum_{k=1}^{\infty} \left(\frac{a_k - ib_k}{2} \exp(ik\omega x) + \frac{a_k + ib_k}{2} \exp(-ik\omega x) \right).$$

Let us denote $c_0 = \frac{a_0}{2}$, $c_k = \frac{a_k - ib_k}{2}$ and $c_{-k} = \frac{a_k + ib_k}{2}$ for $k = 1, 2, \ldots$. Thus,

$$S_f(x) = \sum_{k \in \mathbb{Z}} c_k \exp(ik\omega x), \qquad (4.112)$$

is the *complex form of the Fourier series* (4.109). Since

$$c_k = \frac{a_k - ib_k}{2} = \frac{1}{T} \int_0^T f(x) \left(\cos k\omega x - i \sin k\omega x \right) dx =$$

$$= \frac{1}{T} \int_0^T f(x) \exp(-ik\omega x) \, dx$$

and since

$$c_{-k} = \frac{a_k + ib_k}{2} = \frac{1}{T} \int_0^T f(x) \exp ik\omega x \, dx, \qquad (4.113)$$

one can put everything into a unique formula:

$$c_k = \frac{1}{T} \int_0^T f(x) \exp(-ik\omega x) \, dx \qquad (4.114)$$

for any $k = 0, \pm 1, \pm 2, \ldots$. These numbers are called the *complex Fourier coefficients* of f.

4.8 The Complex form of the Fourier Series

Remark 4.14 From (4.113) we get:

$$a_k + ib_k = \frac{2}{T} \int_0^T f(x) \exp(ik\omega x) \, dx,$$

$k = 0, 1, \ldots$. This formula is very useful in the computation of the Fourier coefficients a_k, b_k. For instance, if $T = 2\pi$, i.e. $\omega = 1$, then we have:

$$a_k + ib_k = \frac{1}{\pi} \int_0^{2\pi} f(x) \exp(ikx) \, dx. \tag{4.115}$$

If $f(x) = R(\sin x, \cos x)$, a rational function of $\sin x, \cos x$, then one can make the change of variable $z = \exp(ix)$ and find:

$$a_k + ib_k = \frac{1}{\pi i} \int_{|z|=1} R\left(\frac{z - z^{-1}}{2i}, \frac{z + z^{-1}}{2}\right) \cdot z^{k-1} \, dz. \tag{4.116}$$

This last integral can be easily computed by the calculus of residue method (see Sect. 11.2 in Chap. 11).

Let $I = [0, T]$ and denote by $\mathcal{PC}(I)$ be the *complex* vector space of all the piecewise continuous functions on I. Obviously, any such function can be extended by periodicity to a piecewise continuous function on $\mathbb{R}$.

For any two functions $f, g \in \mathcal{PC}(I)$, we define the *inner product* of f and g as the *complex* number denoted by $\langle f, g \rangle$,

$$\langle f, g \rangle = \int_0^T f(x) \overline{g(x)} \, dx. \tag{4.117}$$

As can be readily seen, this inner product is a natural extension of the inner product defined in Sect. 4.3 (on the real vector space $PC(I)$) to the complex vector space $\mathcal{PC}(I)$: if f and g are real-valued functions then $\overline{g(x)} = g(x)$, so (4.117) becomes (4.32). It satisfies the properties (4.34)–(4.36), while (4.33) is slightly different:

$$\langle g, f \rangle = \overline{\langle f, g \rangle} \tag{4.118}$$

The functions $f, g \in \mathcal{PC}(I)$ are said to be *orthogonal* if $\langle f, g \rangle = 0$

It is easy to see that for any $k = 0, \pm 1, \pm 2, \ldots$, the functions $\exp(ik\omega x)$ are periodic of period T (recall that $\omega = \frac{2\pi}{T}$). Moreover, the sequence

$$\{\exp(ik\omega x)\}_{k \in \mathbb{Z}}$$

is *orthogonal* relative to the inner product defined by (4.117).

Let f be a piecewise continuous function with finite one-sided derivatives on $[0, T]$. By the Dirichlet's Theorem 4.3 we can write, for every point x at which f is continuous,

$$f(x) = \sum_{k=-\infty}^{\infty} c_k \exp(ik\omega x), \qquad (4.119)$$

where

$$c_k = \frac{1}{T} \langle f(x), \exp(ik\omega x) \rangle, \quad for \ k = 0, \pm 1, \pm 2, \ldots. \qquad (4.120)$$

If f is discontinuous at x, then

$$\frac{f(x_-) + f(x_+)}{2} = \sum_{k=-\infty}^{\infty} c_k \exp(ik\omega x). \qquad (4.121)$$

Example 4.5 Find the complex form of the Fourier expansion for the periodic function $f : \mathbb{R} \to \mathbb{R}$ of period $T = 4$ which has the following expression on the interval $[-2, 2]$:

$$f(x) = \begin{cases} -1, & x \in [-2, 0] \\ 1, & x \in (0, 2]. \end{cases}$$

The complex Fourier coefficients are:

$$c_k = \frac{1}{4} \int_{-2}^{2} f(x) \exp\left(-ik\frac{\pi}{2}x\right) dx = -\frac{1}{4} \int_{-2}^{0} \exp\left(-ik\frac{\pi}{2}x\right) dx +$$

$$+ \frac{1}{4} \int_{0}^{2} \exp\left(-ik\frac{\pi}{2}x\right) dx = \frac{1}{2k\pi i} \exp\left(-ik\frac{\pi}{2}x\right) \bigg|_{-2}^{0} -$$

$$- \frac{1}{2k\pi i} \exp\left(-ik\frac{\pi}{2}x\right) \bigg|_{0}^{2} = \frac{1}{2k\pi i} \left[1 - \exp(ik\pi) - \exp(-ik\pi) + 1 \right] =$$

$$= -\frac{i}{k\pi} [1 - \cos k\pi] = \begin{cases} -\frac{2i}{k\pi}, & \text{if } k = 2p + 1 \\ 0, & \text{if } k = 2p. \end{cases}$$

Hence

$$f(x) = -\frac{2i}{\pi} \sum_{p=-\infty}^{\infty} \frac{1}{2p+1} \exp\left[i(2p+1)\frac{\pi}{2}x\right].$$

4.8 The Complex form of the Fourier Series

Prove that the series above is identical to

$$f(x) = \frac{4}{\pi} \sum_{p=0}^{\infty} \frac{1}{2p+1} \sin\left[(2p+1)\frac{\pi}{2}x\right].$$

For $x = 0$ we get $0 \neq f(0) = -1$ Why?

Example 4.6 Let us find the Fourier expansion for the function:

$$f(x) = \frac{1}{\sin x + \cos x + 2}, \ x \in [0, 2\pi].$$

The function f is a continuous periodic function of period $T = 2\pi$ and frequency $\omega = 1$. Thus,

$$f(x) = \frac{a_0}{2} + \sum_{k=1}^{\infty}(a_k \cos kx + b_k \sin kx),$$

where

$$a_k = \frac{1}{\pi} \int_0^{2\pi} f(x) \cos kx \, dx, \ \text{for } k = 0, 1, \ldots$$

and

$$b_k = \frac{1}{\pi} \int_0^{2\pi} f(x) \sin kx \, dx, \ \text{for } k = 1, 2, \ldots$$

Since the computation of the above integrals is not possible by usual elementary substitutions, we shall apply the calculus of residues from complex analysis (see Sect. 11.2 in Chap. 11) to compute a_k, $k = 0, 1, \ldots$ and b_k, $k = 1, 2, \ldots$. By the formula (4.115) we find:

$$a_k + ib_k = \frac{1}{\pi} \int_0^{2\pi} f(x) \exp(ikx) \, dx$$

Hence, using the change of variable $z = \exp(ix)$, we obtain:

$$a_k + ib_k = \frac{2}{\pi} \int_{|z|=1} \frac{z^k}{(1+i)z^2 + 4iz + i - 1} \, dz, \ k = 0, 1, \ldots$$

The roots of the equation $(1+i)z^2 + 4iz + i - 1 = 0$ are:

$$z_1 = \frac{(1+i)(-2+\sqrt{2})}{2}, \ z_2 = \frac{(1+i)(-2-\sqrt{2})}{2}.$$

Since $|z_1| < 1$ and $|z_2| > 1$, we have a pole of order 1 at z_1 inside the circle of radius 1. Using the residue theorem (Theorem 11.33, Chap. 11), we obtain:

$$a_k + ib_k = \frac{2}{\pi} \cdot 2\pi i \cdot \text{Res}\left(\frac{z^k}{(1+i)z^2 + 4iz + i - 1}, z_1\right)$$

$$= 4i \lim_{z \to z_1} \frac{z^k(z - z_1)}{(1+i)(z - z_1)(z - z_2)} = \frac{4i \cdot z_1^k}{(1+i)(z_1 - z_2)} = \sqrt{2} \cdot z_1^k.$$

Since z_1 can be written

$$z_1 = (1 - \sqrt{2}) \exp\left(i\frac{\pi}{4}\right),$$

we obtain:

$$a_k + ib_k = \sqrt{2}(1 - \sqrt{2})^k \exp\left(i\frac{k\pi}{4}\right) = \sqrt{2}(1 - \sqrt{2})^k \left(\cos\frac{k\pi}{4} + i\sin\frac{k\pi}{4}\right).$$

Hence, for $k = 0, 1, \ldots$,

$$a_k = \sqrt{2}(1 - \sqrt{2})^k \cos\frac{k\pi}{4}, \quad b_k = \sqrt{2}(1 - \sqrt{2})^k \sin\frac{k\pi}{4}.$$

4.9 Exercises

1. (a) Find the Fourier expansion $S_f(x)$ for the function

$$f(x) = \begin{cases} 0, & \text{if } -\pi \leq x < 0 \\ 4x, & \text{if } 0 \leq x < \pi \end{cases}.$$

(b) Use (a) to compute $\sum_{n=1}^{\infty} \frac{1}{(2n-1)^2}$ and $\sum_{n=1}^{\infty} \frac{1}{(2n-1)^4}$.

(c) Let $\tilde{f}$ be the unique periodic function of period $T = 2\pi$ which extends f to the entire $\mathbb{R}$. Find the set of points in $\mathbb{R}$ at which $\tilde{f}$ is not continuous. Prove that $S_{\tilde{f}}(x)$ is not uniformly convergent on $\mathbb{R}$.

4.9 Exercises

2. Find the sin and the cosine Fourier expansions for $f(x) = 4x$, $x \in [0, \pi)$. Compute $\sum_{n=1}^{\infty} \frac{(-1)^{n+1}}{2n-1}$.

3. Consider the function $f(x) = \frac{\pi^2 - 3x^2}{12}$, $x \in [-\pi, \pi]$.

   (a) Find its Fourier expansion.
   (b) Compute $\sum_{k=1}^{\infty} (-1)^{k-1} \frac{1}{k^2}$ and $\sum_{k=1}^{\infty} \frac{1}{k^4}$.
   (c) Find n such that the approximation $f(x) \approx S_n(x)$, where $S_n(x)$ is the trigonometric polynomial of order n (and $T = 2\pi$), has an error less than or equal to $1/10^2$ for any $x \in [-\pi, \pi]$.

4. Consider the function $f(x) = 3 - x$, $x \in [-2, 2)$.

   (a) Draw the graph of the function $\widetilde{f} : \mathbb{R} \to \mathbb{R}$, the extension of f by periodicity ($T = 4$) and find the points at which the equality $\widetilde{f}(x) = S_{\widetilde{f}}(x)$ fails.
   (b) Find its Fourier expansion.
   (c) Compute $\sum_{n=1}^{\infty} \frac{1}{n^2}$ using the Fourier expansion.
   (d) Find its complex representation.

5. (a) Find the periodic ($T = 2$) even function $\widetilde{f} : \mathbb{R} \to \mathbb{R}$ such that $\widetilde{f}(x) = x$ for any $x \in [2, 3]$.
   (b) Write its Fourier expansion and find the set of points x at which $\widetilde{f}(x) \neq S_{\widetilde{f}}(x)$.
   (c) Compute $\sum_{n=0}^{\infty} \frac{1}{(2n+1)^2}$ and $\sum_{n=0}^{\infty} \frac{1}{(2n+1)^4}$.

6. Let $f : \mathbb{R} \to \mathbb{R}$ be a periodic function of period $T = 3$ such that $f(x) = (x - 3)(x - 6)$ for any $x \in [3, 6]$.

   (a) Find its Fourier expansion.
   (b) Compute $\sum_{k=1}^{\infty} \frac{1}{k^4}$.

7. Find (a) the cosine and (b) the sine Fourier expansion for $f(x) = \sin \frac{x}{2}$, $x \in [0, \pi]$.

   (c) Compute:
   $$\sum_{k=1}^{\infty} \frac{1}{4k^2 - 1}, \quad \sum_{k=1}^{\infty} \frac{1}{(4k^2 - 1)^2}, \quad \sum_{m=0}^{\infty} (-1)^m \frac{2m+1}{16m^2 + 16m + 3}.$$

8. Find the Fourier expansion for $f(x) = \cos \frac{x}{3}$, $x \in [-3\pi, 3\pi]$.

9. Use the approximation $\ln(1 + x) \approx x - \frac{x^2}{2}$, $x \in [0, 1]$ to find an approximation of the Fourier expansion of $g(x) = \ln(x + 1)$, $x \in [0, 1]$. Find its complex representation.

Chapter 5
Fourier Transform

In Chap. 4 we have seen that Fourier series are used to represent periodic functions (or functions defined on a finite interval, which can be extended by periodicity to the entire axis). In this chapter, we introduce the Fourier transform, which generalizes this idea to the integral representation of non-periodic functions defined on the whole set of real numbers. We present the properties and some applications of the Fourier transform. Finally, we introduce the discrete Fourier transform which can be used when a function is given only in terms of values at a finite number of points.

5.1 Improper Integrals

The Riemann integral $\int_a^b f(x)dx$ is defined on a *bounded* interval $[a, b]$ and the *boundedness* of the function $f(x)$ is also a necessary condition for integrability. But the notion of integral can be extended to the case of unbounded intervals or unbounded functions. These integrals are called *improper integrals*. We consider in what follows functions of a real variable $x \in \mathbb{R}$ with real or complex values $f(x) \in \mathbb{R}$ or $f(x) \in \mathbb{C}$. Recall that, if $f(x) = u(x) + iv(x)$, then $\int_a^b f(x)dx = \int_a^b u(x)dx + i \int_a^b v(x)dx$.

Definition 5.1 Let $f(x)$ be a function which is piecewise continuous (see Definition 4.2) on any interval $[a, t] \subset [a, b)$, where

$$b = \infty \text{ or } b \in \mathbb{R} \text{ but } \lim_{x \to b} |f(x)| = \infty. \tag{5.1}$$

If there exists the (finite) limit

$$\lim_{t \nearrow b} \int_a^t f(x)\,dx, \tag{5.2}$$

© The Author(s), under exclusive license to Springer Nature Switzerland AG 2022
S. A. Popescu, M. Jianu, *Advanced Mathematics for Engineers and Physicists*,
https://doi.org/10.1007/978-3-031-21502-5_5

then the *improper* integral

$$\int_a^b f(x)\,dx \tag{5.3}$$

is said to be convergent and

$$\int_a^b f(x)\,dx = \lim_{t \nearrow b} \int_a^t f(x)\,dx.$$

We also say in this case that the function f is *integrable* on the interval $[a, b)$. Otherwise, if the limit (5.2) does not exist or it is infinite, the integral (5.3) is said to be divergent.

In the same way, if $f(x)$ is a function piecewise continuous on any interval $[t, b] \subset (a, b]$, where

$$a = -\infty \text{ or } a \in \mathbb{R} \text{ but } \lim_{x \to a} |f(x)| = \infty, \tag{5.4}$$

and there exists the (finite) limit $\lim_{t \searrow a} \int_t^b f(x)\,dx$, then the *improper* integral (5.3) is said to be convergent, the function f is said to be *integrable* on the interval $(a, b]$ and

$$\int_a^b f(x)\,dx = \lim_{t \searrow a} \int_t^b f(x)\,dx.$$

Otherwise, the integral (5.3) is said to be divergent.

If the function $f(x)$ is piecewise continuous on every interval $[t_1, t_2] \subset (a, b)$, if (5.1) and (5.4) hold and there exists $c \in (a, b)$ such that the improper integrals $\int_a^c f(x)\,dx$ and $\int_c^b f(x)\,dx$ are both convergent, then the (improper) integral (5.3) is convergent, the function f is said to be integrable on the interval (a, b) and

$$\int_a^b f(x)\,dx = \int_a^c f(x)\,dx + \int_c^b f(x)\,dx.$$

We say that a function $f(x)$ has a *singularity* at the point c if at least one of the one-sided limits $f(c_+)$ and $f(c_-)$ is infinite. In Definition 5.1 we assumed that the function $f(x)$ has no singularity *inside* the interval (a, b). Now, we extend the notion of integrability for functions with one or more singularities inside the interval (a, b).

Definition 5.2 A function f is said to be *continuous almost everywhere* (a.e.) on the interval I if it is continuous at every point of I, possibly except for a subset

5.1 Improper Integrals

$\Omega \subset I$ such that, for any interval $[\alpha, \beta]$, the intersection $\Omega \cap [\alpha, \beta]$ is either the empty set, or a finite set of real numbers. *The function f does not even need to be defined at the points of discontinuity.*

For a given interval I (bounded or not), we denote by $\mathcal{C}(I)$ the set of all the functions continuous almost everywhere on I.

Recall that a *piecewise continuous* function has *finite* one-sided limits at every point. So we remark that a function which is piecewise continuous is also continuous almost everywhere on I, but the converse is not true since a function continuous a.e. on I may have infinite one-sided limits at the discontinuity points.

Definition 5.3 Let $f(x)$ be a function continuous a.e., which has the following singularities in the interval (a, b): $a < c_1 < c_2 < \ldots < c_n < b$. If the function is integrable (in the sense of Definition 5.1) on every interval $(a, c_1), (c_1, c_2), \ldots, (c_n, b)$, then we say that the function is integrable on the interval (a, b), the improper integral $\int_a^b f(x)\,dx$ is convergent and

$$\int_a^b f(x)\,dx = \int_a^{c_1} f(x)\,dx + \int_{c_1}^{c_2} f(x)\,dx + \ldots + \int_{c_n}^b f(x)\,dx.$$

If the function $f(x)$ is integrable on $[a, b]$ for any $b > a$ and there exists the finite limit $\lim_{b \to \infty} \int_a^b f(x)\,dx$, then the function $f(x)$ is said to be integrable on $[a, \infty)$, the improper integral $\int_a^\infty f(x)\,dx$ is convergent and

$$\int_a^\infty f(x)\,dx = \lim_{b \to \infty} \int_a^b f(x)\,dx. \tag{5.5}$$

Similarly, if the function $f(x)$ is integrable on $[a, b]$ for any $a < b$ and there exists the finite limit $\lim_{a \to -\infty} \int_a^b f(x)\,dx$, then the function $f(x)$ is said to be integrable on $(-\infty, b]$, the improper integral $\int_{-\infty}^b f(x)\,dx$ is convergent and

$$\int_{-\infty}^b f(x)\,dx = \lim_{a \to -\infty} \int_a^b f(x)\,dx. \tag{5.6}$$

If there exists a point c such that the function $f(x)$ is integrable on both intervals $(-\infty, c]$ and $[c, \infty)$, then we say that $f(x)$ is integrable on $\mathbb{R}$, the improper integral $\int_{-\infty}^\infty f(x)\,dx$ is convergent and

$$\int_{-\infty}^\infty f(x)\,dx = \int_{-\infty}^c f(x)\,dx + \int_c^\infty f(x)\,dx.$$

Moreover, it can be easily proved that in this case the integrals $\int_{-\infty}^{\xi} f(x)\,dx$ and $\int_{\xi}^{\infty} f(x)\,dx$ are both convergent for any $\xi \in \mathbb{R}$ and

$$\int_{-\infty}^{\infty} f(x)\,dx = \int_{-\infty}^{\xi} f(x)\,dx + \int_{\xi}^{\infty} f(x)\,dx.$$

To conclude, we say that a function is *integrable* on an interval I if it is continuous a.e. on I and the integral $\int_I f(x)\,dx$ is convergent.

Remark 5.1 If f is integrable on $\mathbb{R}$ then

$$\int_{-\infty}^{\infty} f(x)\,dx = \lim_{l \to \infty} \int_{-l}^{l} f(x)\,dx. \tag{5.7}$$

Indeed,

$$\int_{-\infty}^{\infty} f(x)\,dx = \int_{-\infty}^{c} f(x)\,dx + \int_{c}^{\infty} f(x)\,dx =$$

$$= \lim_{l \to \infty} \int_{-l}^{c} f(x)\,dx + \lim_{l \to \infty} \int_{c}^{l} f(x)\,dx = \lim_{l \to \infty} \int_{-l}^{l} f(x)\,dx.$$

The next theorem gives an important test of convergence for improper integrals of real-valued functions with non-negative values.

Theorem 5.1 (Comparison Test for Improper Integrals) *Let $f(x)$ and $g(x)$ be two functions continuous almost everywhere on the interval I such that*

$$0 \le f(x) \le g(x)$$

for any $x \in I$. If the integral $\int_I g(x)\,dx$ is convergent, then the integral $\int_I f(x)\,dx$ is also convergent.

Example 5.1 Let us prove that the improper integral

$$\int_0^{\infty} x^{t-1} \exp(-x)\,dx \tag{5.8}$$

is convergent for any $t > 0$.

5.1 Improper Integrals

Suppose, first, that $t \in (0, 1)$. Then $\lim_{x \to 0} x^{t-1} \exp(-x) = \infty$, so we write the initial integral as

$$\int_0^\infty x^{t-1} \exp(-x)\, dx = \int_0^1 x^{t-1} \exp(-x)\, dx + \int_1^\infty x^{t-1} \exp(-x)\, dx$$

and prove that both improper integrals are convergent.

For the first one, we notice that $x^{t-1} \exp(-x) \leq x^{t-1}$ for every $x \in [0, 1]$ and the integral $\int_0^1 x^{t-1}\, dx = \left.\dfrac{x^t}{t}\right|_0^1 = \dfrac{1}{t}$ is convergent. Hence, by Theorem 5.1 we obtain that the integral $\int_0^1 x^{t-1} \exp(-x)\, dx$ converges.

For the second integral, we use the inequality $x^{t-1} \exp(-x) \leq \exp(-x)$, which holds for every $x \geq 1$. Since the integral

$$\int_1^\infty \exp(-x)\, dx = -\exp(-x)\Big|_1^\infty = e^{-1}$$

is convergent, by applying again the comparison test for improper integrals, we find the convergence of the integral $\int_1^\infty x^{t-1} \exp(-x)\, dx$. It follows that the integral (5.8) is convergent for $t \in (0, 1)$.

Consider now the second case, $t \geq 1$. Since $\lim_{x \to \infty} x^{t-1} \exp\left(-\dfrac{x}{2}\right) = 0$, there exists $\lambda > 0$ such that $x^{t-1} \exp\left(-\dfrac{x}{2}\right) < 1$, for every $x \geq \lambda$. Thus, we obtain that

$$x^{t-1} \exp(-x) < \exp\left(-\dfrac{x}{2}\right), \quad \forall x \geq \lambda.$$

Since the integral $\int_\lambda^\infty \exp\left(-\dfrac{x}{2}\right) dx = -2 \exp\left(-\dfrac{x}{2}\right)\Big|_\lambda^\infty = 2 \exp\left(-\dfrac{\lambda}{2}\right)$ is convergent, and the function $x^{t-1} \exp(-x)$ does not have a singularity at 0 if $t \geq 1$, it follows that the integral

$$\int_0^\infty x^{t-1} \exp(-x)\, dx = \int_0^\lambda x^{t-1} \exp(-x)\, dx + \int_\lambda^\infty x^{t-1} \exp(-x)\, dx$$

is convergent.

The function $\Gamma : (0, \infty) \to \mathbb{R}$, $\Gamma(t) = \int_0^\infty x^{t-1} \exp(-x)\, dx$ is called the *gamma function* and is also known as the Euler's integral of the second kind. Some of its properties and applications will be presented in Example 5.11.

Definition 5.4 We say that a function $f \in C(I)$ is *absolutely integrable* on the interval I if the function $|f(x)|$ is integrable on I. We denote by $R_1(I)$ the set of the functions which are absolutely integrable on I.

Since

$$\left| \int_I f(x)\,dx \right| \le \int_I |f(x)|\,dx,$$

any absolutely integrable function is also integrable, but the converse is not true, as can be seen in Example 5.2.

Definition 5.5 We say that a function $f \in C(I)$ is *square integrable* on the interval I if the function $|f(x)|^2$ is integrable on I. We denote by $R_2(I)$ the set of functions square integrable on I.

It can be easily proved that $C(I)$ is a vector space and $R_1(I)$ and $R_2(I)$ are its vector subspaces.

Let $f, g \in R_1(I)$. Since $|f(x) + g(x)| \le |f(x)| + |g(x)|$ for any $x \in I$, by Theorem 5.1 we obtain that $f + g$ is absolutely integrable on I.

Suppose that $f, g \in R_2(I)$. Since $|f(x) + g(x)|^2 \le 2\left(|f(x)|^2 + |g(x)|^2\right)$ for any $x \in I$, by Theorem 5.1 we get that $f + g$ is square integrable on I.

The function $f(x) = \dfrac{1}{\sqrt{x}}$ is absolutely integrable on $I = (0, 1]$,

$$\int_0^1 \frac{dx}{\sqrt{x}} = 2\sqrt{x}\Big|_0^1 = 2,$$

but it is not square integrable on I since

$$\int_0^1 \frac{dx}{x} = \ln x\Big|_0^1 = \infty.$$

On the other hand, the function $f(x) = \dfrac{1}{x}$ is square integrable on $[1, \infty)$,

$$\int_1^\infty \frac{dx}{x^2} = -\frac{1}{x}\Big|_1^\infty = 1,$$

but it is not absolutely integrable on I since

$$\int_1^\infty \frac{dx}{x} = \ln x\Big|_1^\infty = \infty.$$

But it can be proved that $R_2(I) \subset R_1(I)$ if I is a *bounded* interval. Let f be a square integrable function on the bounded interval I. Since $2|f(x)| \le |f(x)|^2 + 1$ for any

$x \in I$ and the constant function $g(x) = 1$ is integrable on the bounded interval I, it follows that $f(x)$ is absolutely integrable on I.

Example 5.2 (Dirichlet Integral) The improper integral $\int_0^\infty \frac{\sin x}{x} dx$ is convergent but not absolutely convergent. The value of the integral is

$$\int_0^\infty \frac{\sin x}{x} dx = \frac{\pi}{2}. \tag{5.9}$$

We can write

$$\int_0^\infty \frac{\sin x}{x} dx = \int_0^1 \frac{\sin x}{x} dx + \int_1^\infty \frac{\sin x}{x} dx$$

and we notice that the first integral is a definite, not an improper integral since the function $f(x) = \frac{\sin x}{x}$ defined for $x \neq 0$ can be extended to a continuous function by taking $f(0) = 1$. So we need to prove only the convergence of the second integral. Applying integration by parts, we get:

$$\int_1^\infty \frac{\sin x}{x} dx = -\frac{\cos x}{x}\Big|_1^\infty - \int_1^\infty \frac{\cos x}{x^2} dx =$$

$$= \cos 1 - \int_1^\infty \frac{\cos x}{x^2} dx$$

Since $\left|\frac{\cos x}{x^2}\right| \leq \frac{1}{x^2}$ for all $x \in [1, \infty)$ and the integral $\int_1^\infty \frac{dx}{x^2}$ is convergent, it follows that the integral $\int_1^\infty \frac{\cos x}{x^2} dx$ is also convergent.

Now let us prove that the integral $\int_1^\infty \left|\frac{\sin x}{x}\right| dx$ is divergent. Since, for any $x \in \mathbb{R}$,

$$|\sin x| \geq \sin^2 x = \frac{1 - \cos 2x}{2},$$

it follows that

$$\int_1^\infty \frac{|\sin x|}{x} dx \geq \frac{1}{2} \int_1^\infty \frac{dx}{x} - \frac{1}{2} \int_1^\infty \frac{\cos 2x}{x} dx.$$

Similarly as above, using integration by parts, it can be proved that the last integral is convergent. But the integral $\int_1^\infty \frac{dx}{x}$ diverges, so it follows that the integral $\int_1^\infty \frac{|\sin x|}{x} dx$ also diverges.

To prove (5.9) we use the Dirichlet kernel (introduced in Sect. 4.4):

$$D_n(t) = \frac{\sin\left(n + \frac{1}{2}\right)t}{2\sin\frac{t}{2}}. \qquad (5.10)$$

Since the integral $\int_0^\infty \frac{\sin x}{x} dx$ is convergent, we can write:

$$\int_0^\infty \frac{\sin x}{x} dx = \lim_{l \to \infty} \int_0^l \frac{\sin x}{x} dx = \lim_{n \to \infty} \int_0^{\left(n+\frac{1}{2}\right)\pi} \frac{\sin x}{x} dx =$$

$$= \lim_{n \to \infty} \int_0^\pi \frac{\sin\left(n + \frac{1}{2}\right)t}{t} dt = \lim_{n \to \infty} \int_0^\pi D_n(t) \cdot \frac{2\sin\frac{t}{2}}{t} dt$$

$$= \lim_{n \to \infty} \int_0^\pi D_n(t) \cdot g(t) dt,$$

where the function $g(t) = \frac{2\sin\frac{t}{2}}{t}$ for $t \neq 0$ and $g(0) = 1$ is continuous and has a (finite) right-hand derivative at 0 (which can be calculated by l'Hospital rule):

$$g'_r(0) = \lim_{t \searrow 0} \frac{g(t) - g(0)}{t - 0} = \lim_{t \searrow 0} \frac{g(t) - 1}{t} =$$

$$= \lim_{t \searrow 0} \frac{2\sin\frac{t}{2} - t}{t^2} \stackrel{l'H}{=} \lim_{t \searrow 0} \frac{\cos\frac{t}{2} - 1}{2t} \stackrel{l'H}{=} \lim_{t \searrow 0} \left(-\frac{1}{4}\sin\frac{t}{2}\right) = 0$$

Hence, by Lemma 4.6, we obtain that

$$\lim_{n \to \infty} \int_0^\pi D_n(t) \cdot g(t) dt = \frac{\pi}{2} g(0_+) = \frac{\pi}{2}$$

and the Eq. (5.9) is proved. We notice that, since the function $\frac{\sin x}{x}$ is even, one can write:

$$\int_{-\infty}^\infty \frac{\sin x}{x} dx = \pi.$$

5.1 Improper Integrals

Example 5.3 The improper integral $I = \int_{-\infty}^{\infty} \exp(-x^2)\, dx$ is known as the *Gaussian integral* (or the *Euler-Poisson integral*). We prove that it is convergent and its value is

$$\int_{-\infty}^{\infty} \exp(-x^2)\, dx = \sqrt{\pi}. \tag{5.11}$$

Since the function $\exp(-x^2)$ is even, it suffices to prove that the integral $\int_0^{\infty} \exp(-x^2)\, dx$ is convergent. We can write:

$$\int_0^{\infty} \exp(-x^2)\, dx = \int_0^1 \exp(-x^2)\, dx + \int_1^{\infty} \exp(-x^2)\, dx.$$

The first integral is a definite integral; the convergence of the second one follows by Theorem 5.1 since $\exp(-x^2) \leq \exp(-x)$ for any $x \geq 1$ and

$$\int_1^{\infty} \exp(-x)\, dx = -\exp(-x)\Big|_1^{\infty} = e^{-1} < \infty.$$

So the integral I is convergent and

$$\int_{-\infty}^{\infty} \exp(-x^2)\, dx = 2\int_0^{\infty} \exp(-x^2)\, dx.$$

Following Poisson's idea, we compute I^2:

$$I^2 = \int_{-\infty}^{\infty} \exp(-x^2)\, dx \cdot \int_{-\infty}^{\infty} \exp(-y^2)\, dy = \int_{-\infty}^{\infty}\int_{-\infty}^{\infty} \exp(-x^2 - y^2)\, dx dy.$$

Using the polar coordinates

$$\begin{cases} x = \rho \cos\theta \\ y = \rho \sin\theta \end{cases}, \quad \rho \in [0, \infty), \theta \in [0, 2\pi]$$

we obtain:

$$I^2 = \int_0^{\infty} \rho \exp(-\rho^2)\, d\rho \cdot \int_0^{2\pi} d\theta = -\frac{1}{2}\exp(-\rho^2)\Big|_0^{\infty} \cdot 2\pi = \pi,$$

and (5.11) follows.

5.2 The Fourier Integral Formula

Trying to find a mathematical description for the heat transfer phenomenon, J. B. Fourier discovered what we call today the "Fourier transform". It is a $\mathbb{C}$-linear mapping $\mathcal{F}$ which associates to a function $f(x)$ another function $F(\xi) = \mathcal{F}\{f(x)\}(\xi)$. This mapping has an "inverse", $\mathcal{F}^{-1}$. The Fourier Transform $\mathcal{F}$ is used to convert some difficult situations into simpler ones, solve them there and, by the inverse mapping $\mathcal{F}^{-1}$, come back with the obtained information.

The Fourier transform can be viewed as a "generalization" of the Fourier series. We present in the following the intuitive way used by Fourier himself to construct this generalization called today the *Fourier integral formula*. The function $g(\xi)$ that appears here is a particular case of the general Fourier transform that will be defined later.

Recall that a *piecewise smooth* function is a piecewise continuous function whose derivative is also piecewise continuous.

Let $f : \mathbb{R} \to \mathbb{R}$ be a function which is piecewise smooth on every interval $[-l, l]$ ($l > 0$). We know (see Remark 4.7) that f can be expanded as a Fourier series on $[-l, l]$: at any point of continuity we can write

$$f(x) = \frac{a_0}{2} + \sum_{n=1}^{\infty} a_n \cos \frac{n\pi x}{l} + b_n \sin \frac{n\pi x}{l}, \qquad (5.12)$$

where

$$\begin{cases} a_n = \dfrac{1}{l} \displaystyle\int_{-l}^{l} f(t) \cos \dfrac{n\pi t}{l}\, dt, \ n = 0, 1, 2, \ldots \\ b_n = \dfrac{1}{l} \displaystyle\int_{-l}^{l} f(t) \sin \dfrac{n\pi t}{l}\, dt, \ n = 1, 2, \ldots \end{cases} \qquad (5.13)$$

(At discontinuity points we should write $\frac{1}{2}(f(x_+) + f(x_-))$ instead of $f(x)$ in (5.12)). We introduce these expressions of a_n, b_n in (5.12) and obtain:

$$f(x) = \frac{1}{2l} \int_{-l}^{l} f(t)\, dt + \sum_{n=1}^{\infty} \frac{1}{l} \int_{-l}^{l} f(t) \cos \frac{n\pi (x-t)}{l}\, dt. \qquad (5.14)$$

Now, let as assume that the function $f(x)$ is *absolutely integrable* on $\mathbb{R}$, i.e.

$$\int_{-\infty}^{\infty} |f(t)|\, dt < \infty.$$

5.2 The Fourier Integral Formula

Since

$$\left|\frac{1}{2l}\int_{-l}^{l} f(t)\,dt\right| \leq \frac{1}{2l}\int_{-l}^{l} |f(t)|\,dt \leq \frac{1}{2l}\int_{-\infty}^{\infty} |f(t)|\,dt \to 0$$

when $l \to \infty$, we obtain that

$$f(x) = \lim_{l\to\infty} \sum_{n=1}^{\infty} \frac{1}{l}\int_{-l}^{l} f(t)\cos\frac{n\pi(x-t)}{l}\,dt. \qquad (5.15)$$

To compute the limit, we consider the following partition of the interval $[0, \infty)$:

$$0 = \xi_0 < \xi_1 = \frac{\pi}{l} < \xi_2 = \frac{2\pi}{l} < \ldots < \xi_n = \frac{n\pi}{l} < \ldots.$$

Thus, the sum under the limit in (5.15) becomes:

$$\frac{1}{\pi}\sum_{n=1}^{\infty}(\xi_n - \xi_{n-1})\int_{-l}^{l} f(t)\cos\xi_n(x-t)\,dt.$$

Since $l \to \infty$, we may consider the sum (note that the improper integral is absolutely convergent):

$$\frac{1}{\pi}\sum_{n=1}^{\infty}(\xi_n - \xi_{n-1})\int_{-\infty}^{\infty} f(t)\cos\xi_n(x-t)\,dt,$$

which is like a Riemann sum for the function

$$g(\xi) = \frac{1}{\pi}\int_{-\infty}^{\infty} f(t)\cos\xi(x-t)\,dt,$$

defined on $[0, \infty)$. For $l \to \infty$ we have $\xi_n - \xi_{n-1} = \frac{\pi}{l} \to 0$, so we expect that the limit in (5.15) is equal to $\int_0^{\infty} g(\xi)\,d\xi$ and we obtain the *Fourier integral formula*:

$$f(x) = \frac{1}{\pi}\int_0^{\infty}\left[\int_{-\infty}^{\infty} f(t)\cos\xi(x-t)\,dt\right]d\xi \qquad (5.16)$$

We can use the formula for the cosine of a difference to write:

$$f(x) = \int_0^{\infty}\left[\left(\frac{1}{\pi}\int_{-\infty}^{\infty} f(t)\cos\xi t\,dt\right)\cos\xi x + \left(\frac{1}{\pi}\int_{-\infty}^{\infty} f(t)\sin\xi t\,dt\right)\sin\xi x\right]d\xi.$$

Finally we obtain:

$$f(x) = \int_0^\infty [a(\xi)\cos\xi x + b(\xi)\sin\xi x]\,d\xi, \tag{5.17}$$

where

$$a(\xi) = \frac{1}{\pi}\int_{-\infty}^\infty f(t)\cos\xi t\,dt, \quad b(\xi) = \frac{1}{\pi}\int_{-\infty}^\infty f(t)\sin\xi t\,dt. \tag{5.18}$$

Note that the formula (5.17) holds if the function is continuous at x. Otherwise, if x is a point of discontinuity, then we need to replace $f(x)$ with the arithmetic mean of the one-sided limits.

However, what we presented above is just an intuitive argument. The rigorous proof of the Fourier integral formula will be given later. First of all, we need some basic results regarding improper integrals with parameters (for more details and proofs see [41], Ch. 7, Sec. 6).

Definition 5.6 Consider the following improper integral depending on the parameter λ:

$$\mathcal{I}(\lambda) = \int_a^\infty F(x,\lambda)\,dx. \tag{5.19}$$

Suppose that the integral $\mathcal{I}(\lambda)$ is convergent for any $\lambda \in I$ (where I is a real interval). We say that it is *uniformly convergent* on I if for any $\varepsilon > 0$, there exists $L \geq a$ such that

$$\left|\int_l^\infty F(x,\lambda)\,dx\right| \leq \varepsilon, \tag{5.20}$$

for all $l \geq L$ and $\lambda \in I$.

Theorem 5.2 *If the function $F(x,\lambda)$ is continuous with respect to $\lambda \in [\alpha,\beta]$ and integrable with respect to x on the interval (a,∞) and if the integral (5.19) is uniformly convergent on the interval $[\alpha,\beta]$, then $\mathcal{I}(\lambda)$ is continuous on I and*

$$\int_\alpha^\beta \left[\int_a^\infty F(x,\lambda)\,dx\right]d\lambda = \int_a^\infty \left[\int_\alpha^\beta F(x,\lambda)\,d\lambda\right]dx. \tag{5.21}$$

Theorem 5.3 (Leibniz' Formula) *Let $F(x,\lambda)$ be a continuous function on $(a,\infty) \times I$ that has a continuous partial derivative $\frac{\partial F}{\partial \lambda}$. If the improper integrals*

$$\int_a^\infty F(x,\lambda)\,dx \quad \text{and} \quad \int_a^\infty \frac{\partial F}{\partial \lambda}(x,\lambda)\,dx \tag{5.22}$$

5.2 The Fourier Integral Formula

are convergent for any $\lambda \in I$ and the last one is uniformly convergent on $I = (\alpha, \beta)$, then

$$\frac{d}{d\lambda} \int_a^\infty F(x, \lambda)\, dx = \int_a^\infty \frac{\partial F}{\partial \lambda}(x, \lambda)\, dx \tag{5.23}$$

for all $\lambda \in I$. The theorem holds true if instead of the interval $[a, \infty)$ we have $(-\infty, a]$ or $(-\infty, \infty)$.

Theorem 5.4 (Weierstrass M-test for Improper Integrals) *Consider the function $F(x, \lambda)$ which is continuous with respect to $\lambda \in I$ and integrable with respect to x on the interval $[a, \infty)$. If $f(x)$ is a function absolutely integrable on $[a, \infty)$ such that*

$$|F(x, \lambda)| \leq |f(x)| \tag{5.24}$$

for all $x \in [a, \infty)$ and $\lambda \in I$ then the integral

$$\int_a^\infty F(x, \lambda)\, dx$$

is uniformly convergent on I.

The following Lemma shows that the Riemann-Lebesgue Lemma 4.5 holds not only for functions integrable on bounded intervals $[a, b]$, but also for functions integrable on unbounded intervals of the form $[a, \infty)$, $(-\infty, a]$ or $(-\infty, \infty)$.

Lemma 5.1 *If the function $f(x)$ is absolutely integrable on $[a, \infty)$, then the following limits are 0 ($\lambda \in \mathbb{R}$):*

$$\lim_{\lambda \to \infty} \int_a^\infty f(x) \sin \lambda x\, dx = \lim_{\lambda \to \infty} \int_a^\infty f(x) \cos \lambda x\, dx = 0.$$

Proof Since $|f(x) \sin \lambda x| \leq |f(x)|$ for all $x \geq a$ and $\lambda \in \mathbb{R}$, and the function $f(x)$ is absolutely integrable on $[a, \infty)$, it follows (from Theorem 5.4) that the integral $\int_a^\infty f(x) \sin \lambda x\, dx$ uniformly converges on $\mathbb{R}$. Therefore, for any $\varepsilon > 0$, there is a $b > a$ sufficiently large such that

$$\left| \int_b^\infty f(x) \sin \lambda x\, dx \right| \leq \frac{\varepsilon}{2}$$

for any $\lambda \in \mathbb{R}$.

Moreover, by Lemma 4.5, there exists L such that

$$\left| \int_a^b f(x) \sin \lambda x\, dx \right| \leq \frac{\varepsilon}{2}$$

for any $\lambda \geq L$. It follows that

$$\left| \int_a^\infty f(x) \sin \lambda x \, dx \right| \leq \varepsilon$$

for any $\lambda \geq L$.

Obviously, the same reasoning can be applied to prove that the second limit is 0, using $\cos \lambda x$ instead of $\sin \lambda x$.

On the other hand, if $f(x)$ is a function absolutely integrable on $(-\infty, a]$, then $g(x) = f(-x)$ is absolutely integrable on the interval $[-a, \infty)$, hence:

$$\lim_{\lambda \to \infty} \int_{-\infty}^a f(x) \sin \lambda x \, dx = \lim_{\lambda \to \infty} \int_{-\infty}^a f(x) \cos \lambda x \, dx = 0.$$

□

Lemma 5.2 *Let f be a function absolutely integrable on $\mathbb{R}$ and let $x \in \mathbb{R}$ be a point at which f has finite one-sided limits and derivatives. Then:*

$$\lim_{\lambda \to \infty} \frac{1}{\pi} \int_{-\infty}^\infty f(x+t) \frac{\sin \lambda t}{t} \, dt = \frac{f(x_+) + f(x_-)}{2}. \tag{5.25}$$

Proof Since $|\sin x| \leq |x|$ for any $x \in \mathbb{R}$ and the function $f(x)$ is absolutely integrable on $\mathbb{R}$, the improper integral in (5.25) is convergent, for any $\lambda \in \mathbb{R}$. We will prove that

$$\lim_{\lambda \to \infty} \int_0^\infty f(x+t) \frac{\sin \lambda t}{t} \, dt = \frac{\pi}{2} f(x_+). \tag{5.26}$$

The function f is integrable, so it is continuous a.e. on $\mathbb{R}$. Since f has a finite right-hand limit at x, $f(x_+)$, there exists $\delta > 0$ such that the function $g(t)$ defined by $g(t) = f(x+t)$ for $t > 0$ and $g(0) = f(x_+)$, is continuous on $[0, \delta]$. We can write:

$$\lim_{\lambda \to \infty} \int_0^\infty f(x+t) \frac{\sin \lambda t}{t} \, dt = \tag{5.27}$$

$$= \lim_{\lambda \to \infty} \int_0^\delta f(x+t) \frac{\sin \lambda t}{t} \, dt + \lim_{\lambda \to \infty} \int_\delta^\infty \frac{f(x+t)}{t} \sin \lambda t \, dt$$

As $\dfrac{|f(x+t)|}{t} \leq \dfrac{1}{\delta} |f(x+t)|$ for all $t \geq \delta$, the function $\dfrac{f(x+t)}{t}$ is absolutely integrable on $[\delta, \infty)$ and by Lemma 5.1 it follows that

$$\lim_{\lambda \to \infty} \int_\delta^\infty \frac{f(x+t)}{t} \sin \lambda t \, dt = 0. \tag{5.28}$$

5.2 The Fourier Integral Formula

Since f has a finite right-hand derivative at x, $f'_r(x) = \lim_{t \searrow 0} \dfrac{f(x+t) - f(x_+)}{t}$, the function $h(t)$ defined by

$$h(t) = \begin{cases} \dfrac{f(x+t) - f(x_+)}{t}, & \text{if } t > 0, \\ f'_r(x), & \text{if } t = 0 \end{cases}$$

is continuous on $[0, \delta]$ and by the Riemann-Lebesgue Lemma 4.5 we obtain that

$$\lim_{\lambda \to \infty} \int_0^\delta \frac{f(x+t) - f(x_+)}{t} \sin \lambda t \, dt = 0. \tag{5.29}$$

Using (5.28) and (5.29) in (5.27), we obtain

$$\lim_{\lambda \to \infty} \int_0^\infty f(x+t) \frac{\sin \lambda t}{t} dt = f(x_+) \lim_{\lambda \to \infty} \int_0^\delta \frac{\sin \lambda t}{t} dt =$$

$$= f(x_+) \lim_{\lambda \to \infty} \int_0^{\lambda \delta} \frac{\sin u}{u} du = f(x_+) \int_0^\infty \frac{\sin u}{u} du = f(x_+) \frac{\pi}{2}$$

(see Example 5.2). In the same way we obtain

$$\lim_{\lambda \to \infty} \int_{-\infty}^0 f(x+t) \cdot \frac{\sin \lambda t}{t} dt = \frac{\pi}{2} f(x_-) \tag{5.30}$$

and the Lemma follows by adding (5.26) and (5.30). □

Now we can give a rigorous proof of the *Fourier integral formula* (5.16).

Theorem 5.5 (Fourier Integral Theorem) *Let $f(x)$ be a function absolutely integrable on $\mathbb{R}$. Then, at every point x where f has finite one-sided limits and derivatives, we have:*

$$\frac{f(x_+) + f(x_-)}{2} = \frac{1}{\pi} \int_0^\infty \int_{-\infty}^\infty f(t) \cos \xi(x-t) \, dt \, d\xi. \tag{5.31}$$

This formula can be also written in the exponential form:

$$\frac{f(x_+) + f(x_-)}{2} = \frac{1}{2\pi} \int_{-\infty}^\infty \int_{-\infty}^\infty f(t) \exp(i\xi(x-t)) \, dt \, d\xi. \tag{5.32}$$

Proof Let $\mathcal{I}(\xi)$ and $\mathcal{J}(\xi)$ denote the improper integrals (depending on the parameter $\xi \in \mathbb{R}$):

$$\mathcal{I}(\xi) = \int_{-\infty}^{\infty} f(t) \cos \xi(x-t)\, dt, \quad \mathcal{J}(\xi) = \int_{-\infty}^{\infty} f(t) \sin \xi(x-t)\, dt.$$

Since

$$|f(t) \cos \xi(x-t)| \le |f(t)|, \quad |f(t) \sin \xi(x-t)| \le |f(t)|$$

for any $t, \xi \in \mathbb{R}$ and the function $f(t)$ is absolutely integrable on $\mathbb{R}$, by Weierstrass M-test for improper integrals (Theorem 5.4) it follows that $\mathcal{I}(\xi)$ and $\mathcal{J}(\xi)$ are uniformly convergent on $\mathbb{R}$. By Theorem 5.2 we obtain that

$$\int_0^\lambda \int_{-\infty}^{\infty} f(t) \cos \xi(x-t)\, dt\, d\xi = \int_{-\infty}^{\infty} \int_0^\lambda f(t) \cos \xi(x-t)\, d\xi\, dt =$$

$$= \int_{-\infty}^{\infty} f(t) \cdot \frac{\sin \lambda(t-x)}{t-x}\, dt = \int_{-\infty}^{\infty} f(x+u) \cdot \frac{\sin \lambda u}{u}\, du.$$

Hence

$$\frac{1}{\pi} \int_0^{\infty} \int_{-\infty}^{\infty} f(t) \cos \xi(x-t)\, dt\, d\xi = \lim_{\lambda \to \infty} \frac{1}{\pi} \int_0^\lambda \int_{-\infty}^{\infty} f(t) \cos \xi(x-t)\, dt\, d\xi =$$

$$= \lim_{\lambda \to \infty} \frac{1}{\pi} \int_{-\infty}^{\infty} f(x+u) \cdot \frac{\sin \lambda u}{u}\, du,$$

and, by Lemma 5.2, the formula (5.31) follows.

To prove the second formula, we apply Theorem 5.2 for the improper integral $\mathcal{J}(\xi)$:

$$\int_{-\lambda}^{\lambda} \int_{-\infty}^{\infty} f(t) \sin \xi(x-t)\, dt\, d\xi = \int_{-\infty}^{\infty} f(t) \int_{-\lambda}^{\lambda} \sin \xi(x-t)\, d\xi\, dt = 0,$$

because the function $\sin \xi(t-x)$ is odd (as a function of ξ).

On the other hand, since the function $\cos \xi(x-t)$ is even (considered as a function of ξ), we have:

$$\int_{-\lambda}^{\lambda} \int_{-\infty}^{\infty} f(t) \cos \xi(x-t)\, dt\, d\xi = 2 \int_{-\infty}^{\infty} f(t) \int_0^\lambda \cos \xi(x-t)\, d\xi\, dt.$$

So

$$\frac{1}{2\pi} \int_{-\infty}^{\infty} \int_{-\infty}^{\infty} f(t) \exp(i\xi(x-t)) \, dt \, d\xi =$$

$$= \lim_{\lambda \to \infty} \frac{1}{2\pi} \left[\int_{-\lambda}^{\lambda} \int_{-\infty}^{\infty} f(t) \cos \xi(x-t) \, dt \, d\xi + i \int_{-\lambda}^{\lambda} \int_{-\infty}^{\infty} f(t) \sin \xi(x-t) \, dt \, d\xi \right] =$$

$$= \lim_{\lambda \to \infty} \frac{1}{\pi} \int_{0}^{\lambda} \int_{-\infty}^{\infty} f(t) \cos \xi(x-t) \, dt \, d\xi = \frac{1}{\pi} \int_{0}^{\infty} \int_{-\infty}^{\infty} f(t) \cos \xi(x-t) \, dt \, d\xi$$

and, by (5.31), the formula (5.32) follows. □

Remark 5.2 Note that, if the (absolutely integrable) function f is continuous at x (and has finite one-sided derivatives at x), then the formula (5.32) becomes:

$$f(x) = \frac{1}{2\pi} \int_{-\infty}^{\infty} \int_{-\infty}^{\infty} f(t) \exp(i\xi(x-t)) \, dt \, d\xi, \tag{5.33}$$

which can be also written as:

$$f(x) = \frac{1}{\sqrt{2\pi}} \int_{-\infty}^{\infty} \left[\frac{1}{\sqrt{2\pi}} \int_{-\infty}^{\infty} f(t) \exp(-i\xi t) \, dt \right] \exp(i\xi x) \, d\xi, \tag{5.34}$$

5.3 The Fourier Transform

Now we begin a systematic study of the Fourier transform, defined by the expression in the brackets from (5.34).

Definition 5.7 Let $f(x)$ be a function absolutely integrable on $\mathbb{R}$. The *Fourier transform* of the function f is the function $F(\xi) = \mathcal{F}\{f(x)\}$, $F : \mathbb{R} \to \mathbb{C}$,

$$F(\xi) = \frac{1}{\sqrt{2\pi}} \int_{-\infty}^{\infty} f(x) \exp(-i\xi x) \, dx. \tag{5.35}$$

Remark 5.3 Since $|f(x) \exp(-i\xi x)| = |f(x)|$ for any x and ξ and the function $f(x)$ is absolutely integrable on $\mathbb{R}$, by Weierstrass M-test for improper integrals (Theorem 5.4) it follows that the complex improper integral (5.35) is *uniformly convergent* on $\mathbb{R}$. Moreover, by Theorem 5.2, the function $F(\xi)$ is continuous on $\mathbb{R}$. and, by Lemma 5.1, we have:

$$\lim_{|\xi| \to \infty} F(\xi) = 0 \tag{5.36}$$

Remark 5.4 By the *Fourier Integral Theorem* (see Remark 5.2) we can also define the *inverse Fourier transform* of the function $F(\xi)$ (absolutely integrable on $\mathbb{R}$) as the function $f(x) = \mathcal{F}^{-1}\{F(\xi)\}(x)$,

$$f(x) = \frac{1}{\sqrt{2\pi}} \int_{-\infty}^{\infty} F(\xi) \exp(i\xi x) \, d\xi, \tag{5.37}$$

The formula (5.35) which defines the Fourier transform can be also written as:

$$F(\xi) = \frac{1}{\sqrt{2\pi}} \int_{-\infty}^{\infty} f(x) \cos \xi x \, dx - \frac{i}{\sqrt{2\pi}} \int_{-\infty}^{\infty} f(x) \sin \xi x \, dx. \tag{5.38}$$

Now, if the function $f(x)$ is even, then we get:

$$F(\xi) = \mathcal{F}\{f(x)\}(\xi) = \sqrt{\frac{2}{\pi}} \int_0^{\infty} f(x) \cos \xi x \, dx.$$

We denote $F(\xi) = F_c(\xi)$ (the Fourier cosine transform). The function $F(\xi)$ is also even and we have:

$$f(x) = \sqrt{\frac{2}{\pi}} \int_0^{\infty} F_c(\xi) \cos \xi x \, d\xi. \tag{5.39}$$

If $f(x)$ is odd, then

$$F(\xi) = \mathcal{F}\{f(x)\}(\xi) = -i\sqrt{\frac{2}{\pi}} \int_0^{\infty} f(x) \sin \xi x \, dx$$

In this case, we denote $F_s(\xi) = iF(\xi)$ (the Fourier sine transform). It is also an odd function and we have:

$$f(x) = \sqrt{\frac{2}{\pi}} \int_0^{\infty} F_s(\xi) \sin \xi x \, d\xi. \tag{5.40}$$

Any function $f(x)$ defined on the interval $(0, \infty)$ which is absolutely integrable on this interval can be extend to an even function $f_e(x)$ or to an odd function $f_o(x)$, where f_e and f_o are absolutely integrable on the whole axis (for $x < 0$ we take $f_e(x) = f(-x)$ and $f_o(x) = -f(-x)$, respectively). For such a function we can define the Fourier cosine transform and the Fourier sine transform:

Definition 5.8 Let $f(x)$ be a function absolutely integrable on the interval $(0, \infty)$. The function $F_c(\xi) = \mathcal{F}_c\{f(x)\}(\xi)$ defined by

$$F_c(\xi) = \sqrt{\frac{2}{\pi}} \int_0^{\infty} f(x) \cos \xi x \, dx, \tag{5.41}$$

5.3 The Fourier Transform

is called the **Fourier cosine transform** of $f(x)$ and the function $F_s(\xi) = \mathcal{F}_s\{f(x)\}(\xi)$ defined by

$$\mathcal{F}_s\{f(x)\} = F_s(\xi) = \sqrt{\frac{2}{\pi}} \int_0^\infty f(x) \sin \xi x \, dx, \qquad (5.42)$$

is called the **Fourier sine transform** of $f(x)$.

Remark 5.5 By the formulas (5.39) and (5.40) we can see that

$$\mathcal{F}_c^{-1} = \mathcal{F}_c \text{ and } \mathcal{F}_s^{-1} = \mathcal{F}_s.$$

Theorem 5.6 (Linearity of the Fourier Transform) *For any functions f and g, absolutely integrable on $\mathbb{R}$, and for any constants $a, b \in \mathbb{C}$, the Fourier transform of the function $af + bg$ exists and*

$$\mathcal{F}\{af(x) + bg(x)\} = a\mathcal{F}\{f(x)\} + b\mathcal{F}\{g(x)\}. \qquad (5.43)$$

Proof Since integration is a linear operation, from (5.35) we have:

$$\mathcal{F}\{af(x) + bg(x)\} = \frac{1}{\sqrt{2\pi}} \int_{-\infty}^\infty (af(x) + bg(x)) \exp(-i\xi x) \, dx =$$

$$= \frac{a}{\sqrt{2\pi}} \int_{-\infty}^\infty f(x) \exp(-i\xi x) \, dx + \frac{b}{\sqrt{2\pi}} \int_{-\infty}^\infty g(x) \exp(-i\xi x) \, dx =$$

$$= a\mathcal{F}\{f(x)\} + b\mathcal{F}\{g(x)\}.$$

□

Theorem 5.7 (Shifting Theorem) *Let $f(x)$ be a function absolutely integrable on $\mathbb{R}$ and $a \in \mathbb{R}$. Then the function $g(x) = f(x + a)$ is also absolutely integrable on $\mathbb{R}$ and its Fourier transform is*

$$G(\xi) = \mathcal{F}\{f(x + a)\}(\xi) = F(\xi) \exp(ia\xi), \qquad (5.44)$$

where $F(\xi) = \mathcal{F}\{f(x)\}(\xi)$ is the Fourier transform of f.

Proof It is clear that the function $g(x)$ is continuous almost everywhere on $\mathbb{R}$ and

$$\int_{-\infty}^\infty |g(x)| \, dx = \int_{-\infty}^\infty |f(x + a)| \, dx \stackrel{x+a=t}{=} \int_{-\infty}^\infty |f(t)| \, dt < \infty.$$

Now we compute

$$G(\xi) = \frac{1}{\sqrt{2\pi}} \int_{-\infty}^{\infty} f(x+a) \exp(-i\xi x) dx \stackrel{x+a=t}{=}$$

$$= \frac{1}{\sqrt{2\pi}} \exp(i\xi a) \int_{-\infty}^{\infty} f(t) \exp(-i\xi t) dt = F(\xi) \exp(i\xi a).$$

□

Theorem 5.8 (Shifting the Image) *Let $f(x)$ be a function absolutely integrable on $\mathbb{R}$ and $a \in \mathbb{R}$. Then the function $g(x) = f(x)\exp(iax)$ is also absolutely integrable on $\mathbb{R}$ and its Fourier transform is*

$$G(\xi) = \mathcal{F}\{f(x)\exp(iax)\}(\xi) = F(\xi - a), \tag{5.45}$$

where $F(\xi) = \mathcal{F}\{f(x)\}(\xi)$ is the Fourier transform of f.

Proof The function $g(x)$ is continuous almost everywhere on $\mathbb{R}$ and

$$\int_{-\infty}^{\infty} |g(x)| \, dx = \int_{-\infty}^{\infty} |f(x)| \, dx < \infty,$$

so g is absolutely integrable on $\mathbb{R}$ and we have:

$$G(\xi) = \frac{1}{\sqrt{2\pi}} \int_{-\infty}^{\infty} f(x) \exp(-i(\xi - a)x) \, dx = F(\xi - a),$$

where $F(\xi) = \mathcal{F}\{f(x)\}(\xi)$ is the Fourier transform of f. □

Theorem 5.9 (Dilation Theorem) *Let $f(x)$ be a function absolutely integrable on $\mathbb{R}$ and $b \in \mathbb{R}$, $b \neq 0$. Then the function $g(x) = f(bx)$ is also absolutely integrable on $\mathbb{R}$ and its Fourier transform is*

$$G(\xi) = \mathcal{F}\{f(bx)\}(\xi) = \frac{1}{|b|} F\left(\frac{\xi}{b}\right). \tag{5.46}$$

where $F(\xi) = \mathcal{F}\{f(x)\}(\xi)$ is the Fourier transform of f.
In particular, $\mathcal{F}\{f(-x)\}(\xi) = F(-\xi)$ for all $\xi \in \mathbb{R}$,

Proof The function $g(x)$ is continuous almost everywhere on $\mathbb{R}$ and

$$\int_{-\infty}^{\infty} |g(x)| \, dx = \int_{-\infty}^{\infty} |f(bx)| \, dx \stackrel{bx=t}{=} \frac{1}{|b|} \int_{-\infty}^{\infty} |f(x)| \, dx < \infty.$$

Let us compute $G(\xi)$, the Fourier transform of the function $g(x)$.

$$G(\xi) = \frac{1}{\sqrt{2\pi}} \int_{-\infty}^{\infty} f(bx) \exp(-i\xi x) \, dx \stackrel{bx=t}{=}$$

$$= \frac{1}{|b|} \frac{1}{\sqrt{2\pi}} \int_{-\infty}^{\infty} f(t) \exp\left(-i\frac{\xi}{b}t\right) dt = \frac{1}{|b|} F\left(\frac{\xi}{b}\right).$$

$\square$

Theorem 5.10 (Fourier Transform of the Derivative) *Let $f(x)$ be a continuous function which is differentiable almost everywhere on $\mathbb{R}$ such that f and f' are both absolutely integrable on $\mathbb{R}$. Then*

$$\mathcal{F}\{f'(x)\}(\xi) = i\xi F(\xi), \quad \xi \in \mathbb{R}, \tag{5.47}$$

where $F(\xi) = \mathcal{F}\{f(x)\}(\xi)$ is the Fourier transform of f.

If $f(x)$ is a function of class $C^{n-1}(\mathbb{R})$ such that there exists $f^{(n)}$ almost everywhere on $\mathbb{R}$ and $f, f', f'', \ldots, f^{(n)}$ are absolutely integrable on $\mathbb{R}$, then

$$\mathcal{F}\{f^{(n)}(x)\}(\xi) = (i\xi)^n F(\xi), \quad \xi \in \mathbb{R}, \tag{5.48}$$

Proof We can write:

$$\mathcal{F}\{f'(x)\}(\xi) = \frac{1}{\sqrt{2\pi}} \int_{-\infty}^{\infty} f'(x) \exp(-i\xi x) \, dx =$$

$$= \frac{1}{\sqrt{2\pi}} \lim_{N \to \infty} \int_{-N}^{N} f'(x) \exp(-i\xi x) \, dx =$$

$$= \frac{1}{\sqrt{2\pi}} \lim_{N \to \infty} \left[f(x) \exp(-i\xi x) \Big|_{-N}^{N} + i\xi \int_{-N}^{N} f(x) \exp(-i\xi x) \, dx \right] = i\xi F(\xi),$$

because $\lim_{|x| \to \infty} f(x) = 0$ (f is absolutely integrable on $\mathbb{R}$).

The formula (5.48) follows by repeated application of (5.47):

$$\mathcal{F}\{f^{(n)}(x)\}(\xi) = i\xi \mathcal{F}\{f^{(n-1)}(x)\}(\xi) = (i\xi)^2 \mathcal{F}\{f^{(n-2)}(x)\}(\xi) =$$

$$= \ldots = (i\xi)^{n-1} \mathcal{F}\{f'(x)\}(\xi) = (i\xi)^n \mathcal{F}\{f(x)\}(\xi).$$

$\square$

Theorem 5.11 (Derivative of the Image) Let $f(x)$ be a function absolutely integrable on $\mathbb{R}$ such that $\int_{-\infty}^{\infty} |xf(x)|\, dx < \infty$. Then the Fourier transform of the function $f(x)$, $F(\xi) = \mathcal{F}\{f(x)\}(\xi)$, is differentiable and

$$F'(\xi) = -i\mathcal{F}\{xf(x)\}(\xi), \ \xi \in \mathbb{R}. \tag{5.49}$$

Moreover, if the function $x^n f(x)$ is absolutely integrable on $\mathbb{R}$, then $F(\xi)$ is differentiable n times and

$$F^{(n)}(\xi) = (-i)^n \mathcal{F}\{x^n f(x)\}(\xi), \ \xi \in \mathbb{R}. \tag{5.50}$$

Proof We denote by $G(\xi) = \mathcal{F}\{xf(x)\}(\xi)$ the Fourier transform of the function $xf(x)$. Then, By Weierstrass M-test (Theorem 5.4), the improper integrals

$$F(\xi) = \frac{1}{\sqrt{2\pi}} \int_{-\infty}^{\infty} f(x) \exp(-i\xi x)\, dx$$

and

$$G(\xi) = \frac{1}{\sqrt{2\pi}} \int_{-\infty}^{\infty} xf(x) \exp(-i\xi x)\, dx$$

are uniformly convergent for $\xi \in \mathbb{R}$. Since

$$\frac{d}{d\xi}\left(f(x) \exp(-i\xi x)\right) = -ixf(x) \exp(-i\xi x),$$

by Theorem 5.3 we obtain (5.49).

The formula (5.50) follows by repeated application of (5.49):

$$\mathcal{F}\{x^n f(x)\}(\xi) = \mathcal{F}\{x \cdot x^{n-1} f(x)\}(\xi) = i\mathcal{F}\{x^{n-1} f(x)\}(\xi) =$$

$$= i^2 \mathcal{F}\{x^{n-1} f(x)\}(\xi) = \ldots = i^{n-1}\mathcal{F}\{xf(x)\}(\xi) = i^n \mathcal{F}\{f(x)\}(\xi). \qquad \square$$

Definition 5.9 The **convolution** of the functions $f(x)$ and $g(x)$ is a new function denoted by $(f * g)(x)$, defined as

$$(f * g)(x) = \int_{-\infty}^{\infty} f(t)g(x-t)\, dt = \int_{-\infty}^{\infty} f(x-t)g(t)\, dt. \tag{5.51}$$

5.3 The Fourier Transform

Theorem 5.12 (The Convolution Theorem for the Fourier Transform) *If the functions $f(x)$ and $g(x)$ are piecewise continuous and absolutely integrable on $\mathbb{R}$ and their Fourier transforms are $F(\xi) = \mathcal{F}\{f(x)\}$ and $G(\xi) = \mathcal{F}\{g(x)\}$ respectively, then:*

$$\mathcal{F}\{(f * g)(x)\}(\xi) = \sqrt{2\pi}\, F(\xi)G(\xi), \quad \xi \in \mathbb{R} \tag{5.52}$$

and, conversely,

$$(f * g)(x) = \int_{-\infty}^{\infty} F(\xi) G(\xi) \exp(i\xi x)\, d\xi \quad x \in \mathbb{R}. \tag{5.53}$$

Proof By definition,

$$\mathcal{F}\{(f * g)(x)\}(\xi) = \frac{1}{\sqrt{2\pi}} \int_{-\infty}^{\infty} \left(\int_{-\infty}^{\infty} f(t) g(x-t)\, dt \right) \exp(-i\xi x)\, dx.$$

We change the order of integration (see Theorem 5.2) and obtain:

$$\mathcal{F}\{(f * g)(x)\}(\xi) = \frac{1}{\sqrt{2\pi}} \int_{-\infty}^{\infty} \left(\int_{-\infty}^{\infty} f(t) g(x-t) \exp(-i\xi x)\, dx \right) dt =$$

$$= \frac{1}{\sqrt{2\pi}} \int_{-\infty}^{\infty} f(t) \exp(-i\xi t)\, dt \int_{-\infty}^{\infty} g(x-t) \exp(-i\xi(x-t))\, dx \quad \overset{x-t=u}{=}$$

$$= \frac{1}{\sqrt{2\pi}} \int_{-\infty}^{\infty} f(t) \exp(-i\xi t)\, dt \cdot \frac{1}{\sqrt{2\pi}} \int_{-\infty}^{\infty} g(u) \exp(-i\xi u)\, du \cdot \sqrt{2\pi} =$$

$$= \sqrt{2\pi}\, F(\xi) G(\xi).$$

The formula (5.53) is obtained by taking the inverse Fourier transform on both sides of (5.52). □

Theorem 5.13 (The Parseval's Formula for the Fourier Transform) *If the functions $f(x)$ and $g(x)$ are absolutely and square integrable on $\mathbb{R}$ and their Fourier transforms are $F(\xi) = \mathcal{F}\{f(x)\}$ and $G(\xi) = \mathcal{F}\{g(x)\}$, respectively, then*

$$\int_{-\infty}^{\infty} f(x) \overline{g(x)}\, dx = \int_{-\infty}^{\infty} F(\xi) \overline{G(\xi)}\, d\xi. \tag{5.54}$$

In particular, if $f = g$, then

$$\int_{-\infty}^{\infty} |f(x)|^2\, dx = \int_{-\infty}^{\infty} |F(\xi)|^2\, d\xi. \tag{5.55}$$

If $f(t)$ describes a signal f relative to time t, then $\mathcal{F}\{f\}(\xi)$ describes the evolution of the signal in terms of frequencies ξ. The equality (5.55) says that the "power" of the signal as a function of time is equal to the "power" of it relative to its frequency.

Proof First we prove that, if $G(\xi)$ is the Fourier transform of the function $g(x)$,

$$G(\xi) = \frac{1}{\sqrt{2\pi}} \int_{-\infty}^{\infty} g(t) \exp(-i\xi t) \, dt,$$

then its conjugate, $\overline{G(\xi)}$, is the Fourier transform of the function $\overline{g(-x)}$:

$$\overline{G(\xi)} = \frac{1}{\sqrt{2\pi}} \int_{-\infty}^{\infty} \overline{g(t)} \exp(i\xi t) \, dt \stackrel{u=-t}{=}$$

$$= \frac{1}{\sqrt{2\pi}} \int_{-\infty}^{\infty} \overline{g(-u)} \exp(-i\xi u) \, du = \mathcal{F}\{h(x)\}.$$

where $h(x) = \overline{g(-x)}$, $x \in \mathbb{R}$.

Now, we use the convolution theorem (Theorem 5.12) and write the Eq. (5.53) for $x = 0$ and for the functions f and h:

$$(f * h)(0) = \int_{-\infty}^{\infty} F(\xi)\overline{G(\xi)} \, d\xi. \tag{5.56}$$

On the other hand, by the definition of the convolution (5.51) we have:

$$(f * h)(0) = \int_{-\infty}^{\infty} f(t) h(-t) \, dt = \int_{-\infty}^{\infty} f(t) \overline{g(t)} \, dt. \tag{5.57}$$

Hence, by Eqs. (5.56) and (5.57), the theorem follows. □

In the space of the continuous (complex-valued) functions which are absolutely and square integrable on $\mathbb{R}$ we can define the inner product of two functions f and g as

$$\langle f, g \rangle = \int_{-\infty}^{\infty} f(x) \overline{g(x)} \, dx \tag{5.58}$$

and the corresponding norm,

$$\|f\| = \sqrt{\langle f, f \rangle} = \sqrt{\int_{-\infty}^{\infty} |f(x)|^2 \, dx}. \tag{5.59}$$

5.3 The Fourier Transform

It is easy to prove that (5.58) satisfies the properties of an inner product (on a *complex* vector space):

$$\langle g, f \rangle = \overline{\langle f, g \rangle} \tag{5.60}$$

$$\langle f, g + h \rangle = \langle f, g \rangle + \langle f, h \rangle \tag{5.61}$$

$$\langle \alpha f, g \rangle = \alpha \langle f, g \rangle \tag{5.62}$$

$$\langle f, f \rangle \geq 0, \text{ and } \langle f, f \rangle = 0 \Leftrightarrow f = \mathbf{0}. \tag{5.63}$$

In an inner product space we have the **Cauchy-Schwarz inequality**:

$$|\langle f, g \rangle| \leq \|f\| \cdot \|g\|. \tag{5.64}$$

Using the inner product and the norm defined above, the Parseval's Theorem can be written:

$$\langle f, g \rangle = \langle \mathcal{F}\{f\}, \mathcal{F}\{g\} \rangle \tag{5.65}$$

and

$$\|f\| = \|\mathcal{F}\{f\}\|, \tag{5.66}$$

which means that the Fourier transform is an isometry (it preserves the inner product and the norm—see Sect. 11.1 in Chap. 11).

Proposition 5.1 (Two-Sided Exponential Function) *For any $a > 0$, the function $f(x) = \exp(-a|x|)$, $x \in \mathbb{R}$ is continuous and absolutely integrable on $\mathbb{R}$. Let $F(\xi) = \mathcal{F}\{f(x)\}(\xi)$ be the Fourier transform of the function $f(x)$. Then*

$$F(\xi) = \sqrt{\frac{2}{\pi}} \cdot \frac{a}{a^2 + \xi^2}, \xi \in \mathbb{R}. \tag{5.67}$$

Proof We notice that f is even, so

$$\int_{-\infty}^{\infty} |\exp(-a|x|)| \, dx = 2 \int_0^{\infty} \exp(-ax) \, dx = -\frac{2}{a} \exp(-ax) \Big|_0^{\infty} = \frac{2}{a},$$

hence f is integrable on $\mathbb{R}$. We compute its Fourier transform:

$$F(\xi) = \frac{1}{\sqrt{2\pi}} \int_{-\infty}^{\infty} \exp(-a|x|) \exp(-i\xi x) \, dx =$$

$$= \frac{1}{\sqrt{2\pi}} \left[\int_{-\infty}^{0} \exp((-i\xi + a)x) \, dx + \int_0^{\infty} \exp((-i\xi - a)x) \, dx \right] =$$

$$= \frac{1}{\sqrt{2\pi}} \left[\frac{\exp((a-i\xi)x)}{a-i\xi} \bigg|_{-\infty}^{0} + \frac{\exp((-a-i\xi)x)}{-a-i\xi} \bigg|_{0}^{\infty} \right] =$$

$$= \frac{1}{\sqrt{2\pi}} \left[\frac{1}{a-i\xi} + \frac{1}{a+i\xi} \right] = \sqrt{\frac{2}{\pi}} \cdot \frac{a}{a^2+\xi^2}.$$

□

Proposition 5.2 (Rectangular Pulse Function) *Let $a > 0$ and $\chi_a : \mathbb{R} \to \mathbb{R}$ be the rectangular pulse function:*

$$\chi_a(x) = \begin{cases} 1, & -a \leq x \leq a, \\ 0, & \text{otherwise} \end{cases}.$$

If $F(\xi) = \mathcal{F}\{\chi_a(x)\}(\xi)$ is the Fourier transform of χ_a, then

$$F(\xi) = a\sqrt{\frac{2}{\pi}} \operatorname{sinc}(a\xi), \tag{5.68}$$

where

$$\operatorname{sinc}(x) = \begin{cases} \dfrac{\sin x}{x}, & x \neq 0, \\ 1, & x = 0 \end{cases} \tag{5.69}$$

*is the **sinc function** ("sine cardinal" is the full name of this function which arises frequently in signal processing).*

Proof Since $\int_{-\infty}^{\infty} |\chi_a(x)|\, dx = \int_{-a}^{a} dx = 2a$, the function χ_a is absolutely integrable on $\mathbb{R}$. We compute its Fourier transform:

$$F(\xi) = \frac{1}{\sqrt{2\pi}} \int_{-\infty}^{\infty} \chi_a(x) \exp(-i\xi x)\, dx = \frac{1}{\sqrt{2\pi}} \int_{-a}^{a} \exp(-i\xi x)\, dx.$$

Therefore, for $\xi \neq 0$ we obtain:

$$F(\xi) = \frac{\exp(-i\xi x)}{-i\xi\sqrt{2\pi}} \bigg|_{-a}^{a} = -\frac{1}{i\xi\sqrt{2\pi}} \left[\exp(-i\xi a) - \exp(i\xi a) \right] =$$

$$= a\sqrt{\frac{2}{\pi}} \cdot \frac{\sin \xi a}{\xi a}.$$

If $\xi = 0$, $F(0) = \dfrac{1}{\sqrt{2\pi}} \int_{-a}^{a} dx = a\sqrt{\dfrac{2}{\pi}}$ and (5.68) is proved. □

5.3 The Fourier Transform

The functions which are "piecewise defined" can be expressed using the *unit step function* (or *Heaviside step function*):

$$\eta(x) = \begin{cases} 1, & x \geq 0, \\ 0, & \text{otherwise} \end{cases}. \tag{5.70}$$

For instance,

$$\eta(x) f(x) = \begin{cases} f(x), & x \geq 0, \\ 0, & \text{otherwise} \end{cases},$$

or

$$\eta(-x) f(x) = \begin{cases} f(x), & x \leq 0, \\ 0, & \text{otherwise} \end{cases}.$$

Proposition 5.3 (One-Sided Exponential Function) *Let* $f(x) = \eta(x) \exp(-\alpha x)$, *where* $\alpha = a + ib$, $a > 0$, *be the one-sided exponential function and let* $F(\xi) = \mathcal{F}\{f(x)\}(\xi)$ *be the Fourier transform of* $f(x)$. *Then*

$$F(\xi) = \frac{1}{\sqrt{2\pi}} \cdot \frac{1}{\alpha + i\xi}, \quad \xi \in \mathbb{R}. \tag{5.71}$$

Proof Since

$$f(x) = \begin{cases} \exp(-\alpha x), & x \geq 0, \\ 0, & \text{otherwise} \end{cases},$$

we can see that f is absolutely integrable on $\mathbb{R}$, because

$$\int_{-\infty}^{\infty} |f(x)| \, dx = \int_0^{\infty} \exp(-ax) \, dx = -\frac{1}{a} \exp(-ax) \Big|_0^{\infty} = \frac{1}{a} < \infty.$$

Now, the Fourier transform of $f(x)$ is:

$$F(\xi) = \frac{1}{\sqrt{2\pi}} \int_{-\infty}^{\infty} \eta(x) \exp(-\alpha x) \exp(-i\xi x) \, dx =$$

$$= \frac{1}{\sqrt{2\pi}} \int_0^{\infty} \exp((-\alpha - i\xi)x) \, dx = -\frac{1}{\sqrt{2\pi}} \cdot \frac{1}{\alpha + i\xi} \exp(-\alpha - i\xi)x \Big|_0^{\infty} =$$

$$= \frac{1}{\sqrt{2\pi}} \cdot \frac{1}{\alpha + i\xi},$$

because

$$|\exp(-\alpha - i\xi)x| = \exp(-ax) \longrightarrow 0 \text{ as } x \to \infty.$$

□

Proposition 5.4 *Let $\eta(x)$ be the Heaviside step function, $k \in \{1, 2, \ldots\}$ and $\alpha = a + ib$, $a, b \in \mathbb{R}$, $a > 0$. If $\mathcal{F}^{-1}\{F(\xi)\}$ is the inverse Fourier transform of the function $F(\xi)$, then*

$$\mathcal{F}^{-1}\left\{\frac{1}{(i\xi + \alpha)^k}\right\}(x) = \frac{\sqrt{2\pi}}{(k-1)!}\eta(x)x^{k-1}\exp(-\alpha x). \tag{5.72}$$

Proof Let $f(x) = \sqrt{2\pi}\,\eta(x)\exp(-\alpha x)$. By Proposition 5.3, the Fourier transform of the function $f(x)$ is

$$F(\xi) = \mathcal{F}\{f(x)\}(\xi) = \frac{1}{i\xi + \alpha},$$

so the formula (5.72) for $k = 1$ follows.

Moreover, since $|x^n f(x)| = \sqrt{2\pi}\,|x^n \exp(-ax)|$ and the integral

$$\int_0^\infty x^n \exp(-ax)\,dx \stackrel{ax=y}{=} \frac{1}{a^{n+1}}\int_0^\infty y^{n+1-1}\exp(-y)\,dy = \frac{1}{a^{n+1}}\Gamma(n+1)$$

is convergent for any $a > 0$ and $n \in \mathbb{N}$ (see Example 5.1 which introduces the gamma function), it follows that the function $x^n f(x)$ is absolutely integrable on $\mathbb{R}$ and so, by Theorem 5.11, we can write:

$$F^{(k-1)}(\xi) = (-i)^{k-1}\mathcal{F}\left\{x^{k-1}f(x)\right\}$$

for any $k \geq 2$. But, as can be readily seen,

$$F^{(k-1)}(\xi) = \left(\frac{1}{i\xi + \alpha}\right)^{(k-1)} = \frac{(-i)^{k-1}(k-1)!}{(i\xi + \alpha)^k},$$

so we obtain that

$$\frac{(k-1)!}{(i\xi + \alpha)^k} = \mathcal{F}\left\{x^{k-1}f(x)\right\}$$

and the formula (5.72) follows. □

5.3 The Fourier Transform

Proposition 5.5 *Let $a > 0$ and $f(x) = \dfrac{a}{x^2 + a^2}$, $x \in \mathbb{R}$. Then f is continuous and absolutely integrable on $\mathbb{R}$ and its Fourier transform is:*

$$F(\xi) = \sqrt{\frac{\pi}{2}} \exp(-a|\xi|). \tag{5.73}$$

Proof The function f is integrable on $\mathbb{R}$ because

$$\int_{-\infty}^{\infty} \frac{a}{x^2 + a^2} \, dx = 2a \cdot \frac{1}{a} \tan^{-1} \frac{x}{a} \Big|_0^{\infty} = \pi < \infty.$$

We calculate its Fourier transform. We know from Proposition 5.5 that

$$\mathcal{F}\{\exp(-a|x|)\}(\xi) = \sqrt{\frac{2}{\pi}} \cdot \frac{a}{\xi^2 + a^2}$$

It follows that

$$\sqrt{\frac{\pi}{2}} \exp(-a|u|) = \mathcal{F}^{-1}\left\{\frac{a}{\xi^2 + a^2}\right\}(u) = \frac{1}{\sqrt{2\pi}} \int_{-\infty}^{\infty} \frac{a}{\xi^2 + a^2} \exp(i\xi u) \, d\xi$$

$$\stackrel{\xi = -t}{=} \frac{1}{\sqrt{2\pi}} \int_{-\infty}^{\infty} \frac{a}{t^2 + a^2} \exp(-itu) \, dt = \mathcal{F}\left\{\frac{a}{t^2 + a^2}\right\}(u)$$

and the proposition is proved. □

Proposition 5.6 *Let $a > 0$ and $f(x) = x \exp(-a|x|)$, $x \in \mathbb{R}$. Then f is continuous and absolutely integrable on $\mathbb{R}$ and its Fourier transform is:*

$$F(\xi) = -2\sqrt{\frac{2}{\pi}} \cdot \frac{ia\xi}{(\xi^2 + a^2)^2}. \tag{5.74}$$

Proof We prove that the function f is absolutely integrable on $\mathbb{R}$:

$$\int_{-\infty}^{\infty} |f(x)| \, dx = 2 \int_0^{\infty} x \exp(-ax) \, dx = -\frac{2}{a} \int_0^{\infty} x \left[\exp(-ax)\right]' \, dx \stackrel{\text{by parts}}{=}$$

$$= -\frac{2}{a} \left[x \exp(-ax) \Big|_0^{\infty} - \int_0^{\infty} \exp(-ax) \, dx \right] =$$

$$= -\frac{2}{a} \left[0 + \frac{1}{a} \exp(-ax) \Big|_0^{\infty} \right] = \frac{2}{a^2} < \infty.$$

Now we compute the Fourier transform, $F(\xi)$:

$$F(\xi) = \frac{1}{\sqrt{2\pi}} \int_{-\infty}^{\infty} x \exp(-a|x| - i\xi x)\, dx =$$

$$= \frac{1}{\sqrt{2\pi}} \left[\int_{-\infty}^{0} x \exp[(a - i\xi)x]\, dx + \int_{0}^{\infty} x \exp[-(a + i\xi)x]\, dx \right].$$

We use integration by parts to compute the first integral:

$$\int_{-\infty}^{0} x \exp[(a - i\xi)x]\, dx = \frac{x \exp[(a - i\xi)x]}{a - i\xi} \bigg|_{-\infty}^{0} - \frac{1}{a - i\xi} \int_{-\infty}^{0} \exp[(a - i\xi)x]\, dx =$$

$$= -\frac{1}{(a - i\xi)^2} \exp[(a - i\xi)x] \bigg|_{-\infty}^{0} = -\frac{1}{(a - i\xi)^2}.$$

In the same way we calculate the second integral:

$$\int_{0}^{\infty} x \exp[-(a+i\xi)x]\, dx = -\frac{x \exp[-(a+i\xi)x]}{a + i\xi} \bigg|_{0}^{\infty} + \frac{1}{a + i\xi} \int_{0}^{\infty} \exp[-(a+i\xi)x]\, dx =$$

$$= -\frac{1}{(a + i\xi)^2} \exp[(a - i\xi)x] \bigg|_{0}^{\infty} = \frac{1}{(a + i\xi)^2}.$$

Hence

$$F(\xi) = \frac{1}{\sqrt{2\pi}} \left[\frac{1}{(a + i\xi)^2} - \frac{1}{(a - i\xi)^2} \right] = \frac{1}{\sqrt{2\pi}} \cdot \frac{-4ia\xi}{(a^2 + \xi^2)^2}$$

and (5.74) follows. □

Proposition 5.7 *The Fourier transform of the triangular function*

$$f(x) = \begin{cases} 1 - |x|, & |x| \leq 1, \\ 0, & |x| > 1, \end{cases}$$

is:

$$F(\xi) = \frac{1}{\sqrt{2\pi}} \operatorname{sinc}^2 \frac{\xi}{2}, \qquad (5.75)$$

where $\operatorname{sinc}(x)$ *is the sine cardinal function defined by* (5.69).

5.3 The Fourier Transform

Proof It is obvious that f is absolutely integrable on $\mathbb{R}$. Since f is an even function, we have:

$$F(\xi) = \frac{1}{\sqrt{2\pi}} \int_{-1}^{1} (1 - |x|) \exp(-i\xi x) \, dx = \sqrt{\frac{2}{\pi}} \int_{0}^{1} (1 - x) \cos \xi x \, dx =$$

$$= \sqrt{\frac{2}{\pi}} \left((1 - x) \frac{\sin \xi x}{\xi} \bigg|_0^1 + \frac{1}{\xi} \int_0^1 \sin \xi x \, dx \right) = \sqrt{\frac{2}{\pi}} \cdot \frac{-\cos \xi x}{\xi^2} \bigg|_0^1 =$$

$$= \sqrt{\frac{2}{\pi}} \cdot \frac{1 - \cos \xi}{\xi^2} = \sqrt{\frac{2}{\pi}} \cdot \frac{2 \sin^2 \frac{\xi}{2}}{\xi^2} = \frac{1}{\sqrt{2\pi}} \operatorname{sinc}^2 \frac{\xi}{2},$$

for any $\xi \neq 0$. Now, for $\xi = 0$ we have:

$$F(0) = \frac{1}{\sqrt{2\pi}} \int_{-1}^{1} (1 - |x|) \, dx = \frac{2}{\sqrt{2\pi}} \int_0^1 (1 - x) \, dx = \frac{1}{\sqrt{2\pi}}.$$

The same result could be obtained using the continuity of $F(\xi)$ at 0 and the limit $\lim_{t \to 0} \frac{\sin t}{t} = 1$:

$$\lim_{\xi \to 0} \frac{1}{\sqrt{2\pi}} \left(\frac{\sin \frac{\xi}{2}}{\frac{\xi}{2}} \right)^2 = \frac{1}{\sqrt{2\pi}},$$

□

Example 5.4 Let $a > 0$ and

$$f(x) = \begin{cases} \cos ax, & |x| \leq \frac{\pi}{2a}, \\ 0 & |x| > \frac{\pi}{2a}. \end{cases}$$

Then f is absolutely integrable on $\mathbb{R}$ and its Fourier transform is:

$$F(\xi) = \begin{cases} \sqrt{\frac{2}{\pi}} \frac{a}{a^2 - \xi^2} \cos \frac{\pi \xi}{2a}, & \xi \neq \pm a \\ \frac{\sqrt{\pi}}{2\sqrt{2a}}, & \xi = \pm a \end{cases}. \tag{5.76}$$

The function f is absolutely integrable on $\mathbb{R}$, since

$$\int_{-\infty}^{\infty} |f(x)| \, dx = \int_{-\frac{\pi}{2a}}^{\frac{\pi}{2a}} |\cos ax| \, dx < \infty.$$

We compute its Fourier transform, $F(\xi)$. Since f is even, we have:

$$F(\xi) = \frac{1}{\sqrt{2\pi}} \int_{-\frac{\pi}{2a}}^{\frac{\pi}{2a}} \cos ax \exp(-i\xi x)\, dx = \frac{2}{\sqrt{2\pi}} \int_{0}^{\frac{\pi}{2a}} \cos ax \cos \xi x\, dx =$$

$$= \frac{1}{\sqrt{2\pi}} \int_{0}^{\frac{\pi}{2a}} (\cos(a-\xi)x + \cos(a+\xi)x)\, dx =$$

$$= \frac{1}{\sqrt{2\pi}} \left(\frac{\sin(a-\xi)x}{a-\xi} \bigg|_{0}^{\frac{\pi}{2a}} + \frac{\sin(a+\xi)x}{a+\xi} \bigg|_{0}^{\frac{\pi}{2a}} \right) =$$

$$= \frac{1}{\sqrt{2\pi}} \cos\frac{\pi \xi}{2a} \left(\frac{1}{a-\xi} + \frac{1}{a+\xi} \right) = \sqrt{\frac{2}{\pi}} \frac{a}{a^2 - \xi^2} \cos\frac{\pi \xi}{2a},$$

if $\xi \neq \pm a$. For $\xi = \pm a$ we can write:

$$F(\pm a) = \frac{2}{\sqrt{2\pi}} \int_{0}^{\frac{\pi}{2a}} \cos^2 ax\, dx = \frac{1}{\sqrt{2\pi}} \int_{0}^{\frac{\pi}{2a}} (1 + \cos 2ax)\, dx = \frac{\sqrt{\pi}}{2\sqrt{2a}}.$$

Notice that he same result can be obtained using the continuity of $F(\xi)$:

$$F(\pm a) = \lim_{\xi \to \pm a} \sqrt{\frac{2}{\pi}} \frac{a \cos \frac{\pi \xi}{2a}}{a^2 - \xi^2} \stackrel{l'H}{=} \sqrt{\frac{2}{\pi}} \frac{\pi}{2} \lim_{\xi \to \pm a} \frac{\sin \frac{\pi}{2a}\xi}{2\xi} = \frac{\sqrt{\pi}}{2\sqrt{2a}}.$$

Example 5.5 Let $a > 0$ and

$$f(x) = \begin{cases} \sin ax, & |x| \leq \frac{\pi}{a}, \\ 0 & |x| > \frac{\pi}{a}. \end{cases}$$

Then f is absolutely integrable on $\mathbb{R}$ and its Fourier transform is:

$$F(\xi) = \begin{cases} \sqrt{\frac{2}{\pi}} \frac{ai}{\xi^2 - a^2} \sin \frac{\pi \xi}{a}, & \xi \neq \pm a \\ \mp \frac{i\sqrt{\pi}}{a\sqrt{2}} & \xi = \pm a \end{cases}. \tag{5.77}$$

5.3 The Fourier Transform

Obviously, the function f is absolutely integrable on $\mathbb{R}$, and, since it is odd, we have:

$$F(\xi) = \frac{1}{\sqrt{2\pi}} \int_{-\frac{\pi}{a}}^{\frac{\pi}{a}} \sin ax \exp(-i\xi x)\, dx = \frac{-2i}{\sqrt{2\pi}} \int_0^{\frac{\pi}{a}} \sin ax \sin \xi x\, dx =$$

$$= \frac{i}{\sqrt{2\pi}} \int_0^{\frac{\pi}{a}} (\cos(a+\xi)x - \cos(a-\xi)x)\, dx =$$

$$= \frac{i}{\sqrt{2\pi}} \left(\frac{\sin(a+\xi)x}{a+\xi} \Big|_0^{\frac{\pi}{a}} - \frac{\sin(a-\xi)x}{a-\xi} \Big|_0^{\frac{\pi}{a}} \right) =$$

$$= \frac{-i}{\sqrt{2\pi}} \sin \frac{\pi \xi}{a} \left(\frac{1}{a+\xi} + \frac{1}{a-\xi} \right) = \sqrt{\frac{2}{\pi}} \frac{ai}{\xi^2 - a^2} \sin \frac{\pi \xi}{a},$$

if $\xi \neq \pm a$. For $\xi = \pm a$ we have:

$$F(a) = \frac{-2i}{\sqrt{2\pi}} \int_0^{\frac{\pi}{a}} \sin^2 ax\, dx = \frac{-i}{\sqrt{2\pi}} \int_0^{\frac{\pi}{a}} (1 - \cos 2ax)\, dx = \frac{-i\sqrt{\pi}}{a\sqrt{2}}$$

and

$$F(-a) = \frac{2i}{\sqrt{2\pi}} \int_0^{\frac{\pi}{a}} \sin^2 ax\, dx = \frac{i}{\sqrt{2\pi}} \int_0^{\frac{\pi}{a}} (1 - \cos 2ax)\, dx = \frac{i\sqrt{\pi}}{a\sqrt{2}}.$$

We let the reader to prove these results using the continuity of $F(\xi)$.

Example 5.6 Let $\alpha > 0$ and $f(x) = \exp(-\alpha x^2)$. We compute the Fourier transform $F(\xi) = \mathcal{F}\{f(x)\}(\xi)$.

Since

$$\left| \exp(-\alpha x^2) \exp(-i\xi x) \right| = \exp(-\alpha x^2),$$

and the integral

$$\int_{-\infty}^{\infty} \exp(-\alpha x^2)\, dx \stackrel{\sqrt{\alpha}x=u}{=} \frac{1}{\sqrt{\alpha}} \int_{-\infty}^{\infty} \exp(-u^2)\, du = \frac{1}{\sqrt{\alpha}} \cdot \sqrt{\pi} = \sqrt{\frac{\pi}{\alpha}} \quad (5.78)$$

is convergent (see the Gaussian integral in Example 5.3), by Weierstrass M-test (Theorem 5.4) we obtain that the integral

$$F(\xi) = \frac{1}{\sqrt{2\pi}} \int_{-\infty}^{\infty} \exp(-\alpha x^2) \exp(-i\xi x)\, dx$$

is uniformly convergent for $\xi \in \mathbb{R}$.

In the same way, since

$$\left|-ix \exp(-\alpha x^2) \exp(-i\xi x)\right| = |x| \exp(-\alpha x^2),$$

and the integral

$$\int_{-\infty}^{\infty} |x| \exp(-\alpha x^2) \, dx = 2 \int_0^{\infty} x \exp(-\alpha x^2) \, dx \stackrel{x^2 = u}{=} \int_0^{\infty} \exp(-\alpha u) \, du = \frac{1}{\alpha}$$

is convergent, by Weierstrass M-test we obtain that the integral

$$\frac{1}{\sqrt{2\pi}} \int_{-\infty}^{\infty} \frac{\partial}{\partial \xi} \left[\exp(-\alpha x^2) \exp(-i\xi x)\right] dx = -\frac{i}{\sqrt{2\pi}} \int_{-\infty}^{\infty} x \exp(-\alpha x^2) \exp(-i\xi x) dx$$

is uniformly convergent for $\xi \in \mathbb{R}$ and, by Theorem 5.3, we can write:

$$F'(\xi) = -\frac{i}{\sqrt{2\pi}} \int_{-\infty}^{\infty} x \exp(-\alpha x^2) \exp(-i\xi x) dx.$$

We use the integration by parts formula and get:

$$F'(\xi) = \frac{i}{2\alpha \sqrt{2\pi}} \int_{-\infty}^{\infty} \exp(-i\xi x) \left[\exp(-\alpha x^2)\right]' dx =$$

$$= \frac{i}{2\alpha \sqrt{2\pi}} \left[\exp(-i\xi x) \exp(-\alpha x^2) \Big|_{-\infty}^{\infty} + i\xi \int_{-\infty}^{\infty} \exp(-i\xi x) \exp(-\alpha x^2) \, dx \right].$$

Since $\alpha > 0$, we have

$$\left|\exp(-i\xi x) \exp(-\alpha x^2)\right| = \exp(-\alpha x^2) \longrightarrow 0 \text{ as } x \to \pm\infty$$

and so we get:

$$F'(\xi) = -\frac{\xi}{2\alpha} F(\xi). \tag{5.79}$$

But (5.79) is a first-order differential equation with separable variables. Its general solution is

$$F(\xi) = C_1 \exp\left(-\frac{\xi^2}{4\alpha}\right).$$

5.3 The Fourier Transform

Since

$$F(0) = C_1 = \frac{1}{\sqrt{2\pi}} \int_{-\infty}^{\infty} \exp(-\alpha x^2)\, dx,$$

by (5.78) we obtain that $C_1 = \frac{1}{\sqrt{2\pi}} \cdot \frac{\sqrt{\pi}}{\sqrt{\alpha}} = \frac{1}{\sqrt{2\alpha}}$, so

$$\mathcal{F}\left\{\exp(-\alpha x^2)\right\}(\xi) = F(\xi) = \frac{1}{\sqrt{2\alpha}} \exp\left(-\frac{\xi^2}{4\alpha}\right). \tag{5.80}$$

We remark that for $\alpha = 1/2$ we get:

$$\mathcal{F}\left\{\exp\left(-\frac{x^2}{2}\right)\right\}(\xi) = \exp\left(-\frac{\xi^2}{2}\right), \tag{5.81}$$

so $f(x) = \exp\left(-\frac{x^2}{2}\right)$ is like a "fixed point" of the Fourier transform $\mathcal{F}$.

Example 5.7 Let $\alpha > 0$ and $f(x) = x\exp(-\alpha x^2)$. We compute the Fourier transform $F(\xi) = \mathcal{F}\{f(x)\}(\xi)$ using Theorem 5.10.
Let $g(x) = \exp(-\alpha x^2)$. Since $g'(x) = -2\alpha x \exp(-\alpha x^2) = -2\alpha f(x)$ we have:

$$\mathcal{F}\{f(x)\}(\xi) = -\frac{1}{2\alpha} \mathcal{F}\{g'(x)\}(\xi) = -\frac{i\xi}{2\alpha} \mathcal{F}\{g(x)\}(\xi)$$

and by (5.80) we obtain

$$\mathcal{F}\{f(x)\}(\xi) = -\frac{i\xi}{2\alpha\sqrt{2\alpha}} \exp\left(-\frac{\xi^2}{4\alpha}\right).$$

Notice that the same result could be obtained using Theorem 5.11.

Example 5.8 Let $\alpha > 0$ and $f(x) = x\exp(-a|x|)$. We compute the Fourier transform $F(\xi) = \mathcal{F}\{f(x)\}(\xi)$ using Theorem 5.11.
Let $g(x) = \exp(-a|x|)$ and let $G(\xi) = \sqrt{\frac{2}{\pi}} \cdot \frac{a}{a^2+\xi^2}$ be the Fourier transform of $g(x)$ (see Proposition 5.1). Since $f(x) = xg(x)$, we have:

$$\mathcal{F}\{f(x)\}(\xi) = iG'(\xi) = \sqrt{\frac{2}{\pi}} \cdot \frac{-2ia\xi}{(a^2+\xi^2)^2}.$$

Example 5.9 (An Integral Equation) Let us find a function $y(x)$ defined on the whole real line, piecewise smooth and absolutely integrable, such that:

$$\int_{-\infty}^{\infty} y(x) \exp(-i\xi x)\, dx = 2\pi \exp(-|\xi|). \tag{5.82}$$

It is easy to see that this equation can also be written as:

$$\mathcal{F}\{y(x)\}(\xi) = \sqrt{2\pi} \exp(-|\xi|).$$

We use the *inverse Fourier transform*, and get:

$$y(x) = \sqrt{2\pi}\,\mathcal{F}^{-1}\{\exp(-|\xi|)\} = \int_{-\infty}^{\infty} \exp(-|\xi|) \exp(i\xi x)\, d\xi =$$

$$= \int_{-\infty}^{0} \exp[\xi(1+ix)]\, d\xi + \int_{0}^{\infty} \exp[\xi(-1+ix)]\, d\xi =$$

$$= \left.\frac{\exp[\xi(1+ix)]}{1+ix}\right|_{\xi=-\infty}^{\xi=0} + \left.\frac{\exp[\xi(-1+ix)]}{-1+ix}\right|_{\xi=0}^{\xi=\infty} =$$

$$= \frac{1}{1+ix} - \frac{1}{-1+ix} = \frac{2}{1+x^2}.$$

So, the solution of the integral equation is $y(x) = \dfrac{2}{1+x^2}$.

We summarize in Table 5.1 the most important results regarding the Fourier transform of some basic functions ($a > 0$).

5.4 Solving Linear Differential Equations

In Chap. 7 we use the Fourier transform to find a solution of the heat flow equation in an infinite rod. Here we show how to use the Fourier transform for solving linear differential equations (LDE) with constant coefficients. The idea is to apply the linear operator $\mathcal{F}$ to a LDE with constant coefficients, solve the algebraic equation obtained in this way, then come back through $\mathcal{F}^{-1} = U$ and find a particular solution of our initial LDE. We formally perform all these operations and, at the end, we have to check if this formal solution is indeed a particular solution of the LDE. In the following we use some ideas from the succinct and elegant presentation of Hsu [19].

5.4 Solving Linear Differential Equations

Table 5.1 Basic Fourier transforms

$f(x)$	$F(\xi) = \mathcal{F}\{f(x)\}(\xi)$
$f(x) = \exp(-a\|x\|)$	$F(\xi) = \sqrt{\dfrac{2}{\pi}} \cdot \dfrac{a}{a^2 + \xi^2}$
$f(x) = \dfrac{1}{a^2 + x^2}$	$F(\xi) = \sqrt{\dfrac{\pi}{2}} \cdot \dfrac{\exp(-a\|\xi\|)}{a}$
$f(x) = x \exp(-a\|x\|)$	$F(\xi) = \sqrt{\dfrac{2}{\pi}} \cdot \dfrac{-2a\xi i}{(a^2 + \xi^2)^2}$
$f(x) = \exp(-ax^2)$	$F(\xi) = \dfrac{1}{\sqrt{2a}} \exp\left(-\dfrac{\xi^2}{4a}\right)$
$f(x) = \begin{cases} 1, & \|x\| \leq a, \\ 0, & \|x\| > a \end{cases}$	$F(\xi) = \sqrt{\dfrac{2}{\pi}} \cdot \dfrac{\sin a\xi}{\xi}$
$f(x) = \begin{cases} 1 - \|x\|, & \|x\| \leq 1, \\ 0, & \|x\| > 1 \end{cases}$	$F(\xi) = 2\sqrt{\dfrac{2}{\pi}} \left(\dfrac{\sin \frac{\xi}{2}}{\xi}\right)^2$
$f(x) = \begin{cases} \cos ax, & \|x\| \leq \frac{\pi}{2a}, \\ 0, & \|x\| > \frac{\pi}{2a} \end{cases}$	$F(\xi) = \sqrt{\dfrac{2}{\pi}} \cdot \dfrac{a}{a^2 - \xi^2} \cos \dfrac{\pi \xi}{2a}$
$f(x) = \begin{cases} \sin ax, & \|x\| \leq \frac{\pi}{a}, \\ 0, & \|x\| > \frac{\pi}{a} \end{cases}$	$F(\xi) = \sqrt{\dfrac{2}{\pi}} \cdot \dfrac{ai}{\xi^2 - a^2} \sin \dfrac{\pi \xi}{a}$

Let

$$c_0 y^{(n)}(x) + c_1 y^{(n-1)}(x) + \ldots + c_{n-1} y'(x) + c_n y(x) = f(x), \quad (5.83)$$

$c_0, c_1, \ldots, c_n \in \mathbb{C}$, $c_0 \neq 0$ and $f \in R_1(\mathbb{R})$, be such a LDE. Now, we apply the $\mathbb{C}$-linear operator $\mathcal{F}$ to the equality (5.83), denote by $\widehat{y}$ the Fourier transform of the function y and, by Theorem 5.10, we get:

$$c_0 (i\xi)^n \widehat{y}(\xi) + c_1 (i\xi)^{n-1} \widehat{y}(\xi) + \ldots + c_{n-1}(i\xi)\widehat{y}(\xi) + c_n \widehat{y}(\xi) = \widehat{f}(\xi). \quad (5.84)$$

We consider the polynomial

$$P(X) = c_0 X^n + c_1 X^{n-1} + \ldots + c_{n-1} X + c_n \in \mathbb{C}[X].$$

If $P(X)$ has no root of the form $i\xi$, then, by Eq. (5.84), one can write:

$$\widehat{y}(\xi) = \frac{\widehat{f}(\xi)}{P(i\xi)}.$$

Suppose that $g(x)$ is the inverse Fourier transform of the function $\dfrac{1}{P(i\xi)}$. So

$$\widehat{g}(x) = \mathcal{F}\left\{\frac{1}{P(i\xi)}\right\}.$$

and, using Theorem 5.12, we can write:

$$\widehat{y} = \widehat{g} \cdot \widehat{f} = \frac{1}{\sqrt{2\pi}} \mathcal{F}\{g * f\}.$$

It follows that

$$y(x) = \frac{1}{\sqrt{2\pi}} (g * f)(x) =$$

$$= \frac{1}{\sqrt{2\pi}} \int_{-\infty}^{\infty} g(t) f(x-t) \, dt = \frac{1}{\sqrt{2\pi}} \int_{-\infty}^{\infty} g(x-t) f(t) \, dt.$$

To find the function $g(x) = \mathcal{F}^{-1}\left\{\frac{1}{P(i\xi)}\right\}$ we have to write the partial fraction expansion of the function $\frac{1}{P(X)}$. Assume that $c_0 = 1$ and the polynomial $P(X)$ can be factorized as

$$P(X) = (X - \alpha_1)^{m_1} \ldots (X - \alpha_p)^{m_p}.$$

Then,

$$\frac{1}{P(X)} = \sum_{j=1}^{p} \sum_{k=1}^{m_j} \frac{A_{j,k}}{(X - \alpha_j)^k},$$

where $A_{j,k}, \alpha_j \in \mathbb{C}$ and $m_1, \ldots, m_p$ are positive integers.

Now, we use Proposition 5.4 to find the inverse Fourier transform of each fraction $\frac{A_{j,k}}{(i\xi - \alpha_j)^k}$: If $\text{Re}(\alpha_j) < 0$ then

$$\mathcal{F}^{-1}\left\{\frac{1}{(i\xi - \alpha_j)^k}\right\}(x) = \frac{\sqrt{2\pi}}{(k-1)!} \eta(x) x^{k-1} \exp(\alpha_j x).$$

If $\text{Re}(\alpha_j) > 0$ then

$$\mathcal{F}^{-1}\left\{\frac{1}{(i\xi - \alpha_j)^k}\right\}(x) = \mathcal{F}^{-1}\left\{\frac{(-1)^k}{(-i\xi + \alpha_j)^k}\right\}(x).$$

By Theorem 5.9, $\mathcal{F}^{-1}\{F(-\xi)\}(x) = f(-x)$, so

$$\mathcal{F}^{-1}\left\{\frac{1}{(i\xi - \alpha_j)^k}\right\}(x) = -\frac{\sqrt{2\pi}}{(k-1)!} \eta(-x) x^{k-1} \exp(\alpha_j x).$$

5.4 Solving Linear Differential Equations

Example 5.10 Use the Fourier transform to find a particular solution of the following LDE with constant coefficients:

$$y^{(4)} - 8y'' + 16y = \exp(-x^2), \quad x \in \mathbb{R} \tag{5.85}$$

and then to write its general solution. It is clear that $f(x) = \exp(-x^2)$ is absolutely integrable on $\mathbb{R}$ (see Example 5.3).

In this case $P(X) = X^4 - 8X^2 + 16 = (X^2 - 4)^2 = (X - 2)^2(X + 2)^2$ and

$$\frac{1}{P(X)} = -\frac{1}{32}\frac{1}{X-2} + \frac{1}{16}\frac{1}{(X-2)^2} + \frac{1}{32}\frac{1}{X+2} + \frac{1}{16}\frac{1}{(X+2)^2}.$$

So

$$\frac{1}{P(i\xi)} = -\frac{1}{32}\frac{1}{i\xi - 2} + \frac{1}{16}\frac{1}{(i\xi - 2)^2} + \frac{1}{32}\frac{1}{i\xi + 2} + \frac{1}{16}\frac{1}{(i\xi + 2)^2}$$

and

$$\mathcal{F}^{-1}\left\{\frac{1}{P(i\xi)}\right\}(x) = -\frac{1}{32}\mathcal{F}^{-1}\left\{\frac{1}{i\xi - 2}\right\}(x) + \frac{1}{16}\mathcal{F}^{-1}\left\{\frac{1}{(i\xi - 2)^2}\right\}(x) +$$

$$+ \frac{1}{32}\mathcal{F}^{-1}\left\{\frac{1}{i\xi + 2}\right\}(x) + \frac{1}{16}\mathcal{F}^{-1}\left\{\frac{1}{(i\xi + 2)^2}\right\}(x).$$

We find:

$$g(x) = \mathcal{F}^{-1}\left\{\frac{1}{P(i\xi)}\right\}(x) = \frac{\sqrt{2\pi}}{32}\eta(-x)\exp 2x - \frac{\sqrt{2\pi}}{16}\eta(-x)x\exp 2x +$$

$$+ \frac{\sqrt{2\pi}}{32}\eta(x)\exp(-2x) + \frac{\sqrt{2\pi}}{16}\eta(x)x\exp(-2x) =$$

$$= \frac{\sqrt{2\pi}}{32}\left[(1 - 2x)\eta(-x)\exp 2x + (1 + 2x)\eta(x)\exp(-2x)\right].$$

Thus,

$$g(x) = \begin{cases} \frac{\sqrt{2\pi}}{32}(1 - 2x)\exp 2x, & x \leq 0, \\ \frac{\sqrt{2\pi}}{32}(1 + 2x)\exp(-2x), & x > 0. \end{cases}$$

A particular solution $y_P(x)$ of the LDE with constant coefficients (5.85) is

$$y_P(x) = \frac{1}{\sqrt{2\pi}} \int_{-\infty}^{\infty} g(t) f(x-t) \, dt =$$

$$= \frac{1}{32} \int_{-\infty}^{0} (1-2t) \exp 2t \exp(-(x-t)^2) \, dt +$$

$$\frac{1}{32} \int_{0}^{\infty} (1+2t) \exp(-2t) \exp(-(x-t)^2) \, dt =$$

$$= \frac{1}{32} \int_{0}^{\infty} (1+2t) \exp(-2t) \left[\exp(-(x+t)^2) + \exp(-(x-t)^2) \right] dt.$$

Hence

$$y_P(x) = \frac{1}{16} \exp(-x^2) \int_{0}^{\infty} (1+2t) \exp(-2t - t^2) \cosh(2xt) \, dt.$$

The general solution of the LDL (5.85) is:

$$y(x) = y_P(x) + (C_1 + C_2 x) \exp 2x + (C_3 + C_4 x) \exp(-2x),$$

where C_i, $i = 1, 2, 3, 4$, are arbitrary real or complex constants.

5.5 Moments Theorems

Let f be a function continuous almost everywhere on $\mathbb{R}$ such that $f(x) \geq 0$ for any $x \in \mathbb{R}$ and $0 < \int_{-\infty}^{\infty} f(x) \, dx < \infty$.

The *n-th moment* of f is the real number

$$\mu_f^{(n)} = \int_{-\infty}^{\infty} x^n f(x) \, dx, \tag{5.86}$$

provided that the integral absolutely converges.

For $n = 0$, $\mu_f^{(0)}$ is the area of the region bonded by the graph of $f(x)$ and the x-axis. The real number

$$x_{G,f} = \frac{\mu_f^{(1)}}{\mu_f^{(0)}} = \frac{\int_{-\infty}^{\infty} x f(x) \, dx}{\int_{-\infty}^{\infty} f(x) \, dx} \tag{5.87}$$

is called the *centroid* of f [17].

5.5 Moments Theorems

If the function f is the density function of the infinite rod represented by the x-axis, then the centroid is exactly the x-coordinate of the center of mass of this infinite rod.

In order to measure how the mass of the rod is spread around its center of mass $x_{G,f}$, we introduce the *variance* of f, based on the second moment of the function about its centroid (the second *central moment*):

$$\sigma_f^2 = \frac{1}{\mu_f^{(0)}} \int_{-\infty}^{\infty} (t - x_{G,f})^2 f(t)\, dt. \tag{5.88}$$

It can be easily proved that

$$\sigma_f^2 = \frac{\mu_f^{(2)}}{\mu_f^{(0)}} - \left(\frac{\mu_f^{(1)}}{\mu_f^{(0)}}\right)^2. \tag{5.89}$$

Lemma 5.3 *With the above notation and hypotheses, for any $\alpha, \beta \in \mathbb{R}^*$, one has:*

$$\mu_{\alpha f + \beta g}^{(n)} = \alpha \mu_f^{(n)} + \beta \mu_g^{(n)}, \tag{5.90}$$

$$x_{G,\alpha f} = x_{G,f}, \tag{5.91}$$

$$\sigma_{\alpha f}^2 = \sigma_f^2. \tag{5.92}$$

Moreover, if $h(x) = \dfrac{f(x + x_{G,f})}{\mu_f^{(0)}}$, *then:* $x_{G,h} = 0$, $\mu_h^{(0)} = 1$, $\sigma_h^2 = \mu_h^{(2)} = \sigma_f^2$.

Proof The formula (5.90) easily follows from the definition of the n-th moment (5.86). The formulas (5.91) and (5.92) are direct consequences of the equality $\mu_{\alpha f}^{(n)} = \alpha \mu_f^{(n)}$.

Now, let us prove the last statements. We have:

$$\mu_h^{(0)} = \int_{-\infty}^{\infty} h(x)\, dx = \frac{1}{\mu_f^{(0)}} \int_{-\infty}^{\infty} f(x + x_{G,f})\, dx = \frac{1}{\mu_f^{(0)}} \int_{-\infty}^{\infty} f(t)\, dt = 1,$$

$$x_{G,h} = \frac{1}{\mu_f^{(0)}} \int_{-\infty}^{\infty} x f(x + x_{G,f})\, dx = \frac{1}{\mu_f^{(0)}} \int_{-\infty}^{\infty} (t - x_{G,f}) f(t)\, dt =$$

$$= \frac{\mu_f^{(1)}}{\mu_f^{(0)}} - x_{G,f} = 0,$$

$$\sigma_h^2 = \frac{1}{\mu_h^{(0)}} \int_{-\infty}^{\infty} (x - x_{G,h})^2 h(x)\, dx = \int_{-\infty}^{\infty} x^2 h(x)\, dx =$$

$$= \frac{1}{\mu_f^{(0)}} \int_{-\infty}^{\infty} x^2 f(x + x_{G,f})\, dx = \frac{1}{\mu_f^{(0)}} \int_{-\infty}^{\infty} (t - x_{G,f})^2 f(t)\, dx = \sigma_f^2.$$

$\square$

Theorem 5.14 *Let f be an absolutely integrable function and let $F(\xi) = \mathcal{F}\{f(x)\}(\xi)$ be its Fourier transform. We assume that there exist the moments of order n, $\mu_f^{(n)}$ and $\mu_F^{(n)}$, and that both functions f and F have continuous n-th derivatives in a neighborhood $(-\varepsilon, \varepsilon)$ of 0. Then,*

$$\mu_f^{(n)} = i^n \sqrt{2\pi} F^{(n)}(0), \tag{5.93}$$

$$\mu_F^{(n)} = (-i)^n \sqrt{2\pi} f^{(n)}(0), \tag{5.94}$$

Proof For $n = 0$, we use the definition of the Fourier transform (5.35) and the definition of the inverse Fourier transform (5.37). Since

$$F(\xi) = \frac{1}{\sqrt{2\pi}} \int_{-\infty}^{\infty} f(x) \exp(-i\xi x) \, dx,$$

for $\xi = 0$ we obtain that $F(0)\sqrt{2\pi} = \mu_f^{(0)}$. In the same way, since

$$f(x) = \frac{1}{\sqrt{2\pi}} \int_{-\infty}^{\infty} F(\xi) \exp(i\xi x) \, d\xi,$$

for $x = 0$ we find: $f(0)\sqrt{2\pi} = \mu_F^{(0)}$.

If $n \geq 1$, by Theorem 5.11 (Eq. (5.50)) we have:

$$F^{(n)}(\xi) = \frac{(-i)^n}{\sqrt{2\pi}} \int_{-\infty}^{\infty} x^n f(x) \exp(-i\xi x) \, dx.$$

By taking $\xi = 0$ we obtain (5.93).

On the other hand, by Theorem 5.10 (Eq. (5.48)) we have:

$$f^{(n)}(x) = \mathcal{F}^{-1}[(i\xi)^n F(\xi)] = \frac{i^n}{\sqrt{2\pi}} \int_{-\infty}^{\infty} \xi^n F(\xi) \exp(i\xi x) \, d\xi.$$

By taking $x = 0$ we obtain (5.94). □

Example 5.11 Let $f(x) = \exp(-x^2)$. Since

$$\int_{-\infty}^{\infty} \exp(-x^2) \, dx = \sqrt{\pi} \tag{5.95}$$

(the Gaussian integral—see Example 5.3), $f(x)$ is absolutely integrable on $\mathbb{R}$. Moreover, since for any $n \in \mathbb{N}$,

$$\lim_{x \to \infty} x^2 \cdot x^n \exp(-x^2) = \lim_{x \to \infty} \frac{x^{n+2}}{\exp(x^2)} = 0,$$

5.5 Moments Theorems

the integral $\int_{-\infty}^{\infty} x^n \exp(-x^2)\,dx$ is absolutely convergent, so there exists the n-th moment of $f(x)$. Let us compute $\mu_f^{(n)}$ for $n = 1, 2, \ldots$.

If n is odd, the n-th moment of f is 0: $\mu_f^{(2k+1)} = 0$ (because the function $x^{2k+1} \exp(-x^2)$ is odd).

If n is even, then $\mu_f^{(2k)} = 2 \int_0^{\infty} x^{2k} \exp(-x^2)\,dx$ and we shall use the *gamma function* to compute this integral.

The *gamma function* is defined by the improper integral

$$\Gamma(x) = \int_0^{\infty} t^{x-1} \exp(-t)\,dt, \tag{5.96}$$

which is convergent for any $x > 0$, as proved above (see Example 5.1). We list below the most used properties of the gamma function:

$$\Gamma(x+1) = x\Gamma(x), \quad \forall x > 0 \tag{5.97}$$

$$\Gamma(n) = (n-1)!, \quad \forall n \in \mathbb{N}^* \tag{5.98}$$

$$\Gamma(1/2) = \sqrt{\pi}. \tag{5.99}$$

We use integration by parts to prove (5.97):

$$\Gamma(x+1) = \int_0^{\infty} t^x \exp(-t)\,dt = -t^x \exp(-t)\Big|_0^{\infty} + x \int_0^{\infty} t^{x-1} \exp(-t)\,dt = x\Gamma(x).$$

Since $\Gamma(1) = \int_0^{\infty} \exp(-t)\,dt = -\exp(-t)\Big|_0^{\infty} = 1$, from (5.97) we find:

$$\Gamma(n) = (n-1)(n-2)\ldots 2 \cdot 1 \cdot \Gamma(1) = (n-1)!$$

To prove (5.99), we use the Gaussian integral:

$$\sqrt{\pi} = \int_{-\infty}^{\infty} \exp(-x^2)\,dx = 2 \int_0^{\infty} \exp(-x^2)\,dx \quad \stackrel{x^2=t}{=}$$

$$= \int_0^{\infty} t^{-1/2} \exp(-t)\,dt = \int_0^{\infty} t^{1/2-1} \exp(-t)\,dt = \Gamma(1/2).$$

Now, let us calculate the even moments of f:

$$\mu_f^{(2k)} = 2\int_0^\infty x^{2k}\exp(-x^2)\,dx \stackrel{x=\sqrt{t}}{=} \int_0^\infty t^{k-\frac{1}{2}}\exp(-t)\,dt \stackrel{(5.96)}{=}$$

$$= \Gamma\left(k+\frac{1}{2}\right) \stackrel{(5.97)}{=} \left(k-\frac{1}{2}\right)\left(k-\frac{3}{2}\right)\cdots\frac{1}{2}\Gamma\left(\frac{1}{2}\right).$$

Since $\Gamma\left(\frac{1}{2}\right) = \sqrt{\pi}$, we obtain

$$\mu_f^{(2k)} = \frac{1\cdot 3\cdot 5\cdot\ldots\cdot(2k-1)}{2^k}\sqrt{\pi},\ k=1,2,\ldots. \tag{5.100}$$

For $k=0$, by (5.95) we have $\mu_f^{(0)} = \sqrt{\pi}$ (the area bounded by the graph of f—the famous Gauss' bell—and the x-axis).

Using (5.100) for $k=1$ we find $\mu_f^{(2)} = \frac{\sqrt{\pi}}{2}$, so the variance of f is

$$\sigma_f^2 = \frac{\mu_f^{(2)}}{\sqrt{\pi}} = \frac{1}{2}. \tag{5.101}$$

We can also apply Theorem 5.14 to compute the moments of f. Thus, from (5.80), the Fourier transform of the function $f(x) = \exp(-x^2)$ is

$$F(\xi) = \frac{1}{\sqrt{2}}\exp\left(-\frac{\xi^2}{4}\right),$$

so we have:

$$F'(\xi) = \frac{1}{\sqrt{2}}\exp\left(-\frac{\xi^2}{4}\right)\left(-\frac{\xi}{2}\right)$$

$$F''(\xi) = \frac{1}{\sqrt{2}}\exp\left(-\frac{\xi^2}{4}\right)\left(-\frac{1}{2}+\frac{\xi^2}{4}\right)$$

$$F'''(\xi) = \frac{1}{\sqrt{2}}\exp\left(-\frac{\xi^2}{4}\right)\left(\frac{3\xi}{4}-\frac{\xi^3}{8}\right)$$

$$F^{(4)}(\xi) = \frac{1}{\sqrt{2}}\exp\left(-\frac{\xi^2}{4}\right)\left(\frac{3}{4}-\frac{3\xi^2}{4}+\frac{\xi^4}{16}\right),\ldots$$

Hence we obtain by applying the formula (5.93) the same results as above:

$$\mu_f^{(1)} = \mu_f^{(3)} = 0,\ \mu_f^{(2)} = \frac{\sqrt{\pi}}{2},\ \mu_f^{(4)} = \frac{3\sqrt{\pi}}{4},\ldots.$$

5.5 Moments Theorems

Theorem 5.15 *Let $f, g : \mathbb{R} \to \mathbb{R}$ be two integrable functions with nonnegative values such that $\mu_f^{(0)}, \mu_g^{(0)} \neq 0$ and there exist $\mu_f^{(1)}$ and $\mu_g^{(1)}$.*

(a) *Then, the moments $\mu_{f*g}^{(0)}$ and $\mu_{f*g}^{(1)}$ exist and we have:*

$$\frac{\mu_{f*g}^{(1)}}{\mu_{f*g}^{(0)}} = \frac{\mu_f^{(1)}}{\mu_f^{(0)}} + \frac{\mu_g^{(1)}}{\mu_g^{(0)}}. \tag{5.102}$$

(b) *If we also assume that $\mu_f^{(2)}$ and $\mu_g^{(2)}$ exist, then*

$$\sigma_{f*g}^2 = \sigma_f^2 + \sigma_g^2. \tag{5.103}$$

Proof

(a) We compute the moments of the convolution $f * g$:

$$\mu_{f*g}^{(0)} = \int_{-\infty}^{\infty} \int_{-\infty}^{\infty} f(t) g(x - t) \, dt \, dx \stackrel{x-t=u}{=}$$

$$= \int_{-\infty}^{\infty} \int_{-\infty}^{\infty} f(t) g(u) \, dt \, du = \mu_f^{(0)} \cdot \mu_g^{(0)},$$

and

$$\mu_{f*g}^{(1)} = \int_{-\infty}^{\infty} x \int_{-\infty}^{\infty} f(t) g(x - t) \, dt \, dx \stackrel{x=u+t}{=} \int_{-\infty}^{\infty} \int_{-\infty}^{\infty} (u+t) f(t) g(u) \, dt \, du =$$

$$= \int_{-\infty}^{\infty} u g(u) \, du \cdot \int_{-\infty}^{\infty} f(t) \, dt + \int_{-\infty}^{\infty} g(u) \, du \cdot \int_{-\infty}^{\infty} t f(t) \, dt =$$

$$= \mu_g^{(1)} \cdot \mu_f^{(0)} + \mu_g^{(0)} \cdot \mu_f^{(1)},$$

so the equality (5.102) follows.

(b) Let us compute the second moment of $f * g$:

$$\mu_{f*g}^{(2)} = \int_{-\infty}^{\infty} x^2 \int_{-\infty}^{\infty} f(t) g(x - t) \, dt \, dx \stackrel{x=t+u}{=}$$

$$= \int_{-\infty}^{\infty} \int_{-\infty}^{\infty} (t+u)^2 f(t) g(u) \, dt \, du =$$

$$= \int_{-\infty}^{\infty} t^2 f(t) \, dt \int_{-\infty}^{\infty} g(u) \, du + \int_{-\infty}^{\infty} f(t) \, dt \int_{-\infty}^{\infty} u^2 g(u) \, du +$$

$$+ 2 \int_{-\infty}^{\infty} t f(t) \, dt \int_{-\infty}^{\infty} u g(u) \, du =$$

$$= \mu_f^{(2)} \mu_g^{(0)} + \mu_f^{(0)} \mu_g^{(2)} + 2 \mu_f^{(1)} \mu_g^{(1)}.$$

The variance of $f * g$ is

$$\sigma^2_{f*g} = \frac{\mu^{(2)}_{f*g}}{\mu^{(0)}_{f*g}} - \left[\frac{\mu^{(1)}_{f*g}}{\mu^{(0)}_{f*g}}\right]^2 = \frac{\mu^{(2)}_{f*g}}{\mu^{(0)}_f \mu^{(0)}_g} - \left[\frac{\mu^{(1)}_f}{\mu^{(0)}_f} + \frac{\mu^{(1)}_g}{\mu^{(0)}_g}\right]^2 =$$

$$= \frac{\mu^{(2)}_f}{\mu^{(0)}_f} + \frac{\mu^{(2)}_g}{\mu^{(0)}_g} + 2\frac{\mu^{(1)}_f}{\mu^{(0)}_f} \cdot \frac{\mu^{(1)}_g}{\mu^{(0)}_g} - \left[\frac{\mu^{(1)}_f}{\mu^{(0)}_f} + \frac{\mu^{(1)}_g}{\mu^{(0)}_g}\right]^2 =$$

$$= \frac{\mu^{(2)}_f}{\mu^{(0)}_f} - \left[\frac{\mu^{(1)}_f}{\mu^{(0)}_f}\right]^2 + \frac{\mu^{(2)}_g}{\mu^{(0)}_g} - \left[\frac{\mu^{(1)}_g}{\mu^{(0)}_g}\right]^2 = \sigma^2_f + \sigma^2_g$$

and the theorem is proved.

$\square$

Theorem 5.16 (Uncertainty Principle) *Let* $f : \mathbb{R} \to \mathbb{C}$ *be a continuously differentiable function such that* $f(x)$, $xf(x)$ *and* $f'(x)$ *are absolutely and square integrable on* $\mathbb{R}$. *Let* $F(\xi) = \mathcal{F}\{f(x)\}$ *be the Fourier transform of the function* $f(x)$. *Then*

$$\sigma^2_{|f|^2} \cdot \sigma^2_{|F|^2} \geq \frac{1}{4}. \tag{5.104}$$

Proof Since the functions $f(x)$ and $xf(x)$ are absolutely and square integrable on $\mathbb{R}$, there exist the moments $\mu^{(n)}_{|f|^2}$, for $n = 0, 1, 2$. We notice that

$$\mu^{(0)}_{|f|^2} = \int_{-\infty}^{\infty} |f(x)|^2 dx = \|f\|^2,$$

$$\mu^{(2)}_{|f|^2} = \int_{-\infty}^{\infty} x^2 |f(x)|^2 dx = \|xf\|^2.$$

By Theorem 5.10, we have

$$\mathcal{F}\{f'(x)\} = i\xi F(\xi),$$

hence, from Parseval's formula (5.66), we obtain that there exist also the moments $\mu^{(n)}_{|F|^2}$, for $n = 0, 1, 2$ and

$$\mu^{(0)}_{|F|^2} = \int_{-\infty}^{\infty} |F(\xi)|^2 d\xi = \|F\|^2 = \|f\|^2 = \mu^{(0)}_{|f|^2}$$

$$\mu^{(2)}_{|F|^2} = \int_{-\infty}^{\infty} \xi^2 |F(\xi)|^2 d\xi = \|\xi F\|^2 = \|f'\|^2.$$

5.5 Moments Theorems

Let α and β be the centroids of the functions $|f|^2$ and $|F|^2$ respectively:

$$\alpha = \frac{1}{\|f\|^2} \int_{-\infty}^{\infty} x|f(x)|^2 dx, \quad \beta = \frac{1}{\|F\|^2} \int_{-\infty}^{\infty} \xi |F(\xi)|^2 d\xi.$$

Consider the function $g : \mathbb{R} \to \mathbb{C}$,

$$g(x) = f(x + \alpha) \exp(-i\beta x).$$

Then, by Theorems 5.7 and 5.8, it follows that the function $G : \mathbb{R} \to \mathbb{C}$,

$$G(\xi) = F(\xi + \beta) \exp[i\alpha(\xi + \beta)]$$

is the Fourier transform of the function $g(x)$.

We notice that $|g(x)| = |f(x + \alpha)|$ and $|G(\xi)| = |F(\xi + \beta)|$, so we have:

$$\|g\| = \|f\| = \|F\| = \|G\|. \tag{5.105}$$

By the formula (5.88) one can write:

$$\sigma^2_{|f|^2} = \frac{1}{\|f\|^2} \int_{-\infty}^{\infty} (t - \alpha)^2 |f(t)|^2 dt \stackrel{t-\alpha=x}{=}$$

$$= \frac{1}{\|f\|^2} \int_{-\infty}^{\infty} x^2 |f(x+\alpha)|^2 dx = \frac{1}{\|g\|^2} \int_{-\infty}^{\infty} x^2 |g(x)|^2 dx = \frac{\|xg\|^2}{\|g\|^2}$$

and

$$\sigma^2_{|F|^2} = \frac{1}{\|F\|^2} \int_{-\infty}^{\infty} (\tau - \beta)^2 |F(\tau)|^2 d\tau = \frac{1}{\|F\|^2} \int_{-\infty}^{\infty} \xi^2 |F(\xi + \beta)|^2 d\xi =$$

$$= \frac{1}{\|g\|^2} \int_{-\infty}^{\infty} \xi^2 |G(\xi)|^2 d\xi = \frac{\|\xi G\|^2}{\|g\|^2} = \frac{\|g'\|^2}{\|g\|^2},$$

where the last equality follows by Parseval's formula (5.66), since

$$\mathcal{F}\{g'(x)\} = i\xi G(\xi).$$

Now, from Cauchy-Schwarz inequality (5.64) we can write:

$$2\|xg\| \cdot \|g'\| \geq |\langle xg, g'\rangle| + |\langle g', xg\rangle| \geq |\langle xg, g'\rangle + \langle g', xg\rangle|.$$

Since

$$\langle xg, g'\rangle + \langle g', xg\rangle = \int_{-\infty}^{\infty} x\left(g(x)\cdot \overline{g(x)}' + g'(x)\cdot \overline{g(x)}\right) dx =$$

$$= \int_{-\infty}^{\infty} x\left(g(x)\cdot \overline{g(x)}\right)' dx = x|g(x)|^2\Big|_{-\infty}^{\infty} - \int_{-\infty}^{\infty} |g(x)|^2 dx = -\|g\|^2,$$

we obtain that

$$\|xg\| \cdot \|g'\| \geq \frac{1}{2}\|g\|^2,$$

hence

$$\sigma^2_{|f|^2} \cdot \sigma^2_{|F|^2} = \frac{\|xg\|^2}{\|g\|^2} \cdot \frac{\|g'\|^2}{\|g\|^2} \geq \frac{1}{4}$$

and the theorem is proved. □

Remark 5.6 Theorem 5.16 has an interesting interpretation in the signal theory. As long as the frequency function $F(\xi)$ is concentrated around its centroid $x_{G,F}$, i.e. the values of $F(\xi)$ are very small on $\mathbb{R}$ except for a narrow interval $(x_{G,F}-\varepsilon, x_{G,F}+\varepsilon)$, the signal $f(t)$ is longer in time, i.e. $f(t)$ cannot be concentrated in a narrow interval $(x_{G,f} - \eta, x_{G,f} + \eta)$, $\eta > 0$. Thus, if one has to make a good approximation of $f(t)$ on a narrow interval, then the band of high frequencies ξ must be increased.

Example 5.12 (Gaussian Signal) Let $f(x) = \exp(-ax^2)$, $a > 0$ be the Gaussian "signal". We know (see Example 5.6) that

$$\mathcal{F}\{\exp(-ax^2)\}(\xi) = F(\xi) = \frac{1}{\sqrt{2a}}\exp\left(-\frac{\xi^2}{4a}\right).$$

Since $\int_{-\infty}^{\infty} \exp(-x^2)\, dx = \sqrt{\pi}$, we obtain that

$$\mu^{(0)}_{|f|^2} = \mu^{(0)}_{|F|^2} = \int_{-\infty}^{\infty} \exp(-2ax^2)\, dx = \frac{1}{\sqrt{2a}} \int_{-\infty}^{\infty} \exp(-t^2)\, dt = \sqrt{\frac{\pi}{2a}}.$$

Obviously, $\mu^{(1)}_{|f|^2} = \mu^{(1)}_{|F|^2} = 0$ and we compute the moments of order 2 using the formula (5.100) for $k = 1$:

$$\mu^{(2)}_{|f|^2} = \int_{-\infty}^{\infty} x^2 \exp(-2ax^2)\, dx = \frac{1}{2a\sqrt{2a}} \int_{-\infty}^{\infty} t^2 \exp(-t^2)\, dt = \frac{\sqrt{\pi}}{4a\sqrt{2a}}$$

and

$$\mu^{(2)}_{|F|^2} = \frac{1}{2a} \int_{-\infty}^{\infty} \xi^2 \exp\left(-\frac{\xi^2}{2a}\right) dx = \sqrt{2a} \int_{-\infty}^{\infty} t^2 \exp(-t^2)\, dt = \frac{\sqrt{\pi a}}{\sqrt{2}}.$$

Hence

$$\sigma^2_{|f|^2} = \frac{1}{4a}, \quad \sigma^2_{|F|^2} = a$$

and the inequality (5.104) becomes equality:

$$\sigma^2_{|f|^2} \cdot \sigma^2_{|F|^2} = \frac{1}{4}.$$

The Gaussian signal has minimum uncertainty and it is the only function for which the inequality (5.104) becomes equality.

5.6 Sampling Theorem

Problem Let $f(t)$ be a continuous-time signal (a piecewise continuous function, absolutely integrable on $\mathbb{R}$). Would it be possible to completely determine the signal $f(t)$ knowing its values (or *samples*) at a discrete set of points, equally-spaced in time, $f(nh)$, $n \in \mathbb{Z}$? The answer is "yes" under certain conditions: the signal must be band-limited (there exists $b > 0$ such that its Fourier transform $F(\xi)$ equals zero for $|\xi| > b$) and the samples must be taken with a sampling interval small enough, depending on the bandwidth, $h \leq \frac{\pi}{b}$. The mathematical expression of this result is called the *Sampling Theorem*. It was progressively improved by Whittacker— its initiator (1915), Nyquist (1928), Kotelnikov (1933) and extensively applied by Shannon (1948) in the communication technology. This is why this fundamental result is also known as "Whittaker–Kotelnikov-Shannon Theorem" or "Nyquist-Shannon Theorem". The theorem provides a formula for reconstructing the original signal $f(t)$ from the samples $f(nh)$, $n \in \mathbb{Z}$, using the sine cardinal function, $\text{sinc}(x) = \frac{\sin x}{x}$ if $x \neq 0$ and $\text{sinc}(0) = 1$.

Theorem 5.17 (Sampling Theorem) *Let $f : \mathbb{R} \to \mathbb{R}$ be a piecewise continuous function, absolutely integrable on $\mathbb{R}$, and let $F(\xi) = \mathcal{F}\{f(t)\}(\xi)$ be its Fourier transform. Suppose that there exist $b > 0$ such that $F(\xi) = 0$ for all ξ with $|\xi| > b$ and denote $h = \frac{\pi}{b}$. Then, for any $t \in \mathbb{R}$,*

$$f(t) = \sum_{n \in \mathbb{Z}} f(nh) \cdot \text{sinc}(b(t - nh)). \tag{5.106}$$

Proof Since F is a continuous function (see Remark 5.3), for any $\xi \in (-b, b)$ one can write:

$$F(\xi) = \sum_{k \in \mathbb{Z}} c_k \exp(ikh\xi)$$

(see the formula (4.119) in Sect. 4.8), where

$$c_k = \frac{1}{2b} \int_{-b}^{b} F(\xi) \exp(-ikh\xi)\, d\xi = \frac{1}{2b} \int_{-\infty}^{\infty} F(\xi) \exp(-ikh\xi)\, d\xi \stackrel{(5.37)}{=} \frac{\sqrt{2\pi}}{2b} f(-kh).$$

Thus,

$$F(\xi) = \frac{h}{\sqrt{2\pi}} \sum_{k \in \mathbb{Z}} f(-kh) \exp(ikh\xi) \stackrel{k=-n}{=} \frac{h}{\sqrt{2\pi}} \sum_{n \in \mathbb{Z}} f(nh) \exp(-inh\xi).$$

Finally, from (5.37), for $t \neq nh$, we get:

$$f(t) = \frac{1}{\sqrt{2\pi}} \int_{-b}^{b} F(\xi) \exp(it\xi)\, d\xi = \frac{h}{2\pi} \sum_{n \in \mathbb{Z}} f(nh) \int_{-b}^{b} \exp(i\xi(t-nh))\, d\xi =$$

$$= \frac{1}{2b} \sum_{n \in \mathbb{Z}} f(nh) \left[\frac{\exp(ib(t-nh)) - \exp(-ib(t-nh))}{i(t-nh)} \right] =$$

$$= \frac{1}{2b} \sum_{n \in \mathbb{Z}} f(nh) \cdot \frac{2 \sin b(t-nh)}{t - nh} = \sum_{n \in \mathbb{Z}} f(nh) \cdot \text{sinc}(b(t-nh)).$$

$\square$

The *Sampling Theorem* establishes sufficient conditions and states a formula for perfectly reconstructing a continuous function from a discrete sequence of samples, acting like a bridge between continuous-time signals and discrete-time signals.

5.7 Discrete Fourier Transform

Consider the set of "finite signals" of length N,

$$\mathbb{C}^N = \{x = (x_0, x_1, \ldots, x_{N-1})^t : x_i \in \mathbb{C}, i = 0, 1, \ldots, N-1\}.$$

5.7 Discrete Fourier Transform

In the complex vector space $\mathbb{C}^N$ we define the *inner product* of two vectors $x = (x_0, x_1, \ldots, x_{N-1})^t$ and $y = (y_0, y_1, \ldots, y_{N-1})^t$ as:

$$\langle x, y \rangle = \sum_{j=0}^{N-1} x_j \bar{y}_j = x^t \bar{y} \tag{5.107}$$

and the *norm* of a vector as

$$\|x\| = \sqrt{\langle x, x \rangle} = \sqrt{\sum_{j=0}^{N-1} |x_j|^2}. \tag{5.108}$$

The space $\mathbb{C}^N$ with the inner product defined above is a Hilbert space (see Example 11.15 in Chap. 11).

Recall that two vectors $x, y \in \mathbb{C}^N$ are said to be *orthogonal* if $\langle x, y \rangle = 0$. A set of N nonzero vectors $V = \{v_0, v_1, \ldots, v_{N-1}\}$ such that $\langle v_p, v_q \rangle = 0$ for any $p \neq q$ is an *orthogonal basis* of the N-dimensional space $\mathbb{C}^N$. If, in addition, $\|v_p\| = 1$ for all $p = 0, 1, \ldots, N-1$, then V is said to be an *orthonormal basis* of $\mathbb{C}^N$.

Let $v = \exp\left(-i\frac{2\pi}{N}\right) = \exp(-i\lambda)$ (where $\lambda = \frac{2\pi}{N}$) and consider in the space $\mathbb{C}^N$ the following vectors:

$$\begin{aligned}
v_0 &= (1, 1, 1, \ldots, 1)^t, \\
v_1 &= (1, v, v^2, \ldots, v^{N-1})^t, \\
v_2 &= (1, v^2, v^4, \ldots, v^{2(N-1)})^t, \\
v_3 &= (1, v^3, v^6, \ldots, v^{3(N-1)})^t, \\
&\vdots \\
v_{N-1} &= (1, v^{N-1}, v^{2(N-1)}, \ldots, v^{(N-1)(N-1)})^t.
\end{aligned} \tag{5.109}$$

We prove that $V = \{v_0, v_1, \ldots, v_{N-1}\}$ is an orthogonal basis of $\mathbb{C}^N$. Indeed,

$$\langle v_p, v_q \rangle = \sum_{k=0}^{N-1} v^{kp} \cdot \bar{v}^{kq} = \sum_{l=0}^{N-1} [\exp(\lambda(q-p)i)]^k = \begin{cases} N, & p = q, \\ 0, & p \neq q \end{cases},$$

because

$$\sum_{l=0}^{N-1} [\exp(\lambda(q-p)i)]^k = \frac{\alpha^N - 1}{\alpha - 1},$$

where $\alpha = \exp(q-p)\lambda i$, $\alpha \neq 1$ if $q \neq p$, and $\alpha^N = \exp(q-p)2\pi i = 1$.

So V is an orthogonal basis. Since, for every $p = 0, 1, \ldots, N-1$,
$$\|v_p\| = \sqrt{N}, \ p = 0, 1, \ldots, N-1,$$
we obtain that
$$V^* = \left\{ \frac{1}{\sqrt{N}} v_0, \frac{1}{\sqrt{N}} v_1, \ldots, \frac{1}{\sqrt{N}} v_{N-1} \right\}$$
is an orthonormal basis of $\mathbb{C}^N$.

Consider the matrix
$$W = \frac{1}{\sqrt{N}} \begin{pmatrix} 1 & 1 & 1 & \cdots & 1 \\ 1 & v & v^2 & \cdots & v^{N-1} \\ 1 & v^2 & v^4 & \cdots & v^{2(N-1)} \\ \vdots & \vdots & \vdots & & \vdots \\ 1 & v^{N-1} & v^{2(N-1)} & \cdots & v^{(N-1)(N-1)} \end{pmatrix} \tag{5.110}$$

As can be readily seen, the matrix W is *symmetric* ($W = W^t$), and *unitary* ($W^t \cdot \overline{W} = I_N$), where I_N is the identity matrix. It follows that
$$\overline{W} = W^{-1}. \tag{5.111}$$

Definition 5.10 The **discrete Fourier transform** is the linear mapping $\mathcal{F} : \mathbb{C}^N \to \mathbb{C}^N$ defined by the symmetric unitary matrix (5.110). For any (column) vector $f \in \mathbb{C}^N$, the vector
$$F = \mathcal{F}(f) = Wf \tag{5.112}$$
is said to be the **discrete Fourier transform** (DFT) of f. Thus, if $f = (f_0, f_1, \ldots, f_{N-1})^t$ and $F = \mathcal{F}(f) = (F_0, F_1, \ldots, F_{N-1})^t$, then
$$F_n = \frac{1}{\sqrt{N}} \sum_{n=0}^{N-1} f_k v^{nk} = \frac{1}{\sqrt{N}} \sum_{n=0}^{N-1} f_k \exp\left(-i \frac{2\pi nk}{N}\right), \tag{5.113}$$
for all $n = 0, 1, \ldots, N-1$.

By (5.111) we can define the **inverse discrete Fourier transform** of a (column) vector $F \in \mathbb{C}^N$ as
$$f = \mathcal{F}^{-1}(F) = \overline{W} F, \tag{5.114}$$

5.7 Discrete Fourier Transform

i.e., for all $n = 0, 1, \ldots, N-1$, we have:

$$f_n = \frac{1}{\sqrt{N}} \sum_{n=0}^{N-1} F_k \overline{v}^{nk} = \frac{1}{\sqrt{N}} \sum_{n=0}^{N-1} F_k \exp\left(i\frac{2\pi nk}{N}\right). \tag{5.115}$$

Theorem 5.18 (The Parseval's Formula for DFT) *For any $f, g \in \mathbb{C}^N$, we have:*

$$\langle f, g \rangle = \langle \mathcal{F}(f), \mathcal{F}(g) \rangle. \tag{5.116}$$

In particular, when $f = g$ we obtain:

$$\|f\| = \|\mathcal{F}(f)\|. \tag{5.117}$$

So $\mathcal{F}$ is an isometry of the Hilbert space $\mathbb{C}^N$.

Proof By (5.107), $\langle f, g \rangle = f^t \overline{g}$. Since $\mathcal{F}(f) = Wf$ and $\mathcal{F}(g) = Wg$, one can write:

$$\langle \mathcal{F}(f), \mathcal{F}(g) \rangle = \langle Wf, Wg \rangle = (Wf)^t \overline{Wg} = f^t \underbrace{W^t \overline{W}}_{I_N} \overline{g} = \langle f, g \rangle$$

because W is an unitary matrix. □

Remark 5.7 Let $\varphi(t)$, $\varphi : \mathbb{R} \to \mathbb{R}$ be a continuous "signal", (φ is a continuous function, absolutely integrable on $\mathbb{R}$) and let $\Phi = \mathcal{F}\{\varphi\}$ be its Fourier transform:

$$\Phi(\xi) = \frac{1}{\sqrt{2\pi}} \int_{-\infty}^{\infty} \varphi(t) \exp(-i\xi t)\, dt.$$

Suppose that $\varphi(t)$ has significant values in a finite interval of time $[0, T]$. In this interval, we consider the moments $t_k = \frac{kT}{N}$, $k = 0, 1, \ldots, N-1$ and the vector $f = (f_0, f_1, \ldots, f_{N-1})^t$, where $f_k = \varphi(t_k)$, $k = 0, 1, \ldots, N-1$. The Fourier transform $\Phi(\xi)$ can be approximated as:

$$\Phi(\xi) = \frac{1}{\sqrt{2\pi}} \int_0^T \varphi(t) \exp(-i\xi t)\, dt \approx \frac{1}{\sqrt{2\pi}} \cdot \frac{T}{N} \sum_{k=0}^{N-1} f_k \exp(-i\xi t_k).$$

Using this formula, we evaluate $\Phi(\xi_n)$, where $\xi_n = \frac{2\pi}{T} n$, $n = 0, 1, \ldots, N-1$:

$$\Phi(\xi_n) \approx \frac{1}{\sqrt{2\pi}} \cdot \frac{T}{N} \sum_{k=0}^{N-1} f_k \exp\left(-i\frac{2\pi nk}{N}\right) = \frac{T}{\sqrt{2\pi N}} \cdot F_n,$$

where $F = \mathcal{F}(f) = (F_0, F_1, \ldots, F_{N-1})^t$ is the discrete Fourier transform of the vector f.

5.8 Exercises

1. Find the Fourier transform of the following functions:

   (a) $f(x) = \begin{cases} 3, & |x| < 2 \\ 1, & x = \pm 2 \\ 0, & |x| > 2 \end{cases}$;

   (b) $f(x) = \begin{cases} 4, & a < x < b \\ 0, & x \notin [a, b] \end{cases}$;

   (c) $f(x) = \exp(-2|x|)$;

   (d) $f(x) = \dfrac{x}{(x^2 + a^2)^2}$;

   (e) $f(x) = \begin{cases} \cos x, & |x| \leq \pi \\ 0, & |x| > \pi \end{cases}$.

2. Apply the Fourier inverse cosine transform to find the even function $f(x)$ knowing that $\displaystyle\int_0^\infty f(t)\cos(\xi t)\,dt = \dfrac{1}{1+\xi^2}$.

3. Apply the Parseval's formula (5.66) to show that

$$\int_{-\infty}^{\infty} \frac{\sin^2 \xi}{\xi^2}\,d\xi = \pi.$$

4. Given $a > 0$, find the Fourier transform of the function

$$f(x) = \begin{cases} x, & |x| < a \\ 0, & |x| > a \end{cases}$$

by two methods: using the definition formula (5.35) and then, using Theorem 5.11.

5. Let $a > 0$ and $b \in \mathbb{R}$. Compute the Fourier transform of the function

$$f(x) = \begin{cases} \exp(iax), & |x| < b \\ 0, & |x| > b \end{cases}$$

by two methods: using the definition formula (5.35) and then, using Theorem 5.8.

6. Consider the rectangular pulse function

$$f_a(x) = \begin{cases} 1, & |x| < a \\ 0, & |x| > a \end{cases},$$

5.8 Exercises

where $a > 0$. Prove that $f_a * f_a = (2a - |x|) f_{2a}$.

Using this result and the convolution theorem (Theorem 5.12), compute the Fourier transform of the function

$$g_a(x) = \begin{cases} |x|, & |x| < a \\ 0, & |x| > a \end{cases}.$$

Verify the result using the definition formula (5.35).

7. Consider $N = 4$ measurements (sample values). Write the matrix W in this case.

   (a) Given the sample values $f = (f_0, f_1, f_2, f_3)^t$, find $F = \mathcal{F}(f)$, the discrete Fourier transform of the signal f.
   (b) Given the frequency vector $F = (F_0, F_1, F_2, F_3)^t$, recreate the signal, $f = \mathcal{F}^{-1}(F)$.

Chapter 6
Laplace Transform

In this chapter we present another important mathematical tool—the Laplace transform. As the Fourier transform, the Laplace transform simplifies the solution of linear differential equations by transforming them into algebraic equations. It is also applied for solving partial differential equations—in Chap. 7 we use the Laplace transform for the solution of the finite vibrating string equation. By using the Heaviside step function or the Dirac delta function, the Laplace transform can be applied in problems where the free term has some discontinuities or represents short impulses.

6.1 Introduction

Let I be a real interval (bounded or not), $A \subset \mathbb{C}$ be a *connected* set (any two points of A can be connected by a continuous path $\gamma \subset A$) and $K : A \times I \to \mathbb{C}$, $K(p, t)$, be a continuous function (with respect to both p and t) which is differentiable with respect to the complex variable p. We also assume that $K(p, t) \neq 0$ on $A \times I$. Given $f(t)$, a real or complex-valued function of a real variable $t \in I$, the complex-valued function $F(p)$ (of the complex variable p) defined by

$$F(p) = \int_I f(t) K(p, t) \, dt, \qquad (6.1)$$

is said to be the K–transform of the function f. We denote $F(p) = \mathcal{T}_K\{f\}(p)$. The function $K(p, t)$ is called the *kernel* of the transform $\mathcal{T}_K$ (which is a $\mathbb{C}$-linear mapping). Usually, the integral (6.1) is an improper integral (with a complex parameter p). The set of all $p \in A$ for which this integral is convergent is called the *region of convergence* of f.

Since the simple convergence of the integral $\mathcal{T}_K\{f\}(p)$ is not sufficient to ensure the continuity, differentiability and integrability properties of the complex function

$F(p)$, we need to assume this integral is absolutely and uniformly convergent, i.e. the integral $\int_I |f(t)K(p,t)|\,dt$ uniformly converges. The set of all $p \in A$ for which this last integral is uniformly convergent is called the *region of uniform convergence* of f (with respect to the kernel $K(p,t)$).

It appears that L. Euler was the first who used integral transforms of the type

$$y(x) = \int_a^b \phi(s)\exp(sx)\,ds$$

in order to solve linear differential equations of order two ([34]).

In Mathematics there are several integral transforms (or transformations) with different kernels $K(p,t)$. Two of them are of the greatest interest to Engineering and Physics: the Fourier transform (presented in Chap. 5) and the Laplace transform.

Recall that the Fourier transform of a function $f(t)$ which is absolutely integrable on $\mathbb{R}$ ($f \in R_1(\mathbb{R})$) is the function

$$F(\xi) = \frac{1}{\sqrt{2\pi}} \int_{-\infty}^{\infty} f(t)\exp(-i\xi t)\,dt,$$

so in this case the kernel is $K(p,t) = \frac{1}{\sqrt{2\pi}}\exp(-pt)$, where $p = 0 + i\xi$, $\xi \in \mathbb{R}$ and $I = (-\infty, \infty) = \mathbb{R}$. As we have seen in Chap. 5, the Fourier transform changes the differentiation, integration, or convolution operations in the space $R_1(\mathbb{R})$ into algebraic scalar multiplication, scalar division, or the usual multiplication in the space of the functions $F(\xi)$. This basic feature of $\mathcal{F}$ is very useful in solving differential equations and systems or in some problems of probability and statistics. But the space $R_1(\mathbb{R})$, is relatively "poor" (it does not contain, for instance, polynomial functions). Euler, Laplace and other mathematicians used the kernel function $K(p,t) = \exp(-pt)$ with $\operatorname{Re} p > 0$ (not $\operatorname{Re} p = 0$ as in the Fourier transform) to enlarge the class of functions $f(t)$ for which the integral (6.1) is absolutely and uniformly convergent. Laplace used this kernel $K(p,t) = \exp(-pt)$ in his work on probability theory "Théorie analytique des Probabilités"(1812).

Later, in 1920, Bernstein used the term "Laplace transformation" for the linear mapping:

$$\mathcal{L}\{f(t)\}(p) = F(p) = \int_0^\infty f(t)\exp(-pt)\,dt,$$

where $p = s + i\sigma \in \mathbb{C}$.

Definition 6.1 A real or complex-valued function $f : \mathbb{R} \to \mathbb{C}$ of a real variable t is said to be an **original (function)** if it satisfies the following conditions:

(i) $f(t) = 0$ if $t < 0$.

(ii) $f(t)$ is a piecewise continuous function, i.e. f is continuous on any interval $[a, b]$, possibly except for a finite number of points in $[a, b]$, at which f has finite one-sided limits.

(iii) $f(t)$ is a **function of exponential order**, i.e. there exist $s \in \mathbb{R}$ and $M > 0$ such that

$$|f(t)| \leq M \exp(st), \ \forall t \in \mathbb{R}. \tag{6.2}$$

It is easy to see that, if (6.2) holds for $s = s_1$ then it holds for any $s > s_1$. The greatest lower bound of the set

$$\{s \in \mathbb{R} : \exists M > 0 \text{ such that } |f(t)| \leq M \exp(st), \forall t \in \mathbb{R}\}$$

is denoted by s_0 and it is called the **abscissa of convergence**.

In general, the inequality (6.2) does not hold for $s = s_0$, but, for any $\varepsilon > 0$ one has

$$|f(t)| \leq M_\varepsilon \exp((s_0 + \varepsilon)t)$$

for all $t \geq 0$. For instance, the function $f(t) = t$, $t \geq 0$, and $f(t) = 0$, for $t < 0$, is an original function with $s_0 = 0$, but $|f(t)| = t$ cannot be less than or equal to a fixed number $M > 0$. However, since $\lim\limits_{t \to \infty} t \exp(-\varepsilon t) = 0$ for any $\varepsilon > 0$, there exists $M > 0$ such that $t \leq M \exp(\varepsilon t)$ for any $t \geq 0$.

Example 6.1 **Examples of original functions**

1. The simplest original function is the *unit step function (Heaviside function)*:

$$\eta(t) = \begin{cases} 1, & \text{if } t \geq 0 \\ 0, & \text{if } t < 0 \end{cases}.$$

It is an original function with $s_0 = 0$.

If some function $f(t)$ satisfies only the conditions (ii) and (iii) from Definition 6.1, then the function $f^*(t) = \eta(t) f(t)$ is an original function (since it satisfies all the three conditions from the definition). We shall consider in the following that all functions which satisfies (ii) and (iii) are multiplied by the unit step function. For example, we shall identify $f(t) = \sin t$ with $f^*(t) = \eta(t) \sin t$ etc.

2. The trigonometric functions $f(t) = \sin t$ and $g(t) = \cos t$ are original functions with $s_0 = 0$: Obviously, $|\sin t| \leq \exp(0t) = 1$ and $|\cos t| \leq \exp(0t) = 1$, for any $t \geq 0$. On the other hand, the abscissa of convergence cannot be negative, since $\sin t$ and $\cos t$ are periodic functions (and, for $s < 0$, $\exp st \to 0$ as $t \to \infty$).

3. $f(t) = t^n$ is an original function with $s_0 = 0$, for any positive integer n (as we noted above, we consider the function as $f(t) = \eta(t) t^n$).

Moreover, for any $a \geq 0$, the function $f(t) = t^a$ is an original function with $s_0 = 0$: for any $s > 0$, since the function $t^a \exp(st)$ is continuous on $[0, \infty)$ and

$$\lim_{t \to \infty} t^a \exp(-st) = 0,$$

there exist $M > 0$ such that $t^a \exp(-st) < M$ for all $t \geq 0$.

4. The exponential function $f(t) = \exp(\xi t)$, $\xi = \alpha + i\beta \in \mathbb{C}$, is an original function with $s_0 = \operatorname{Re} \xi = \alpha$:

$$f(t) = \exp((\alpha + i\beta)t) = \exp(\alpha t)(\cos \beta t + i \sin \beta t) \Rightarrow |f(t)| = \exp(\alpha t).$$

Not all the exponential functions are original functions. For example, the function $f(t) = \exp(t^2)$ does not satisfy the condition iii) from Definition 6.1, since $\lim_{t \to \infty} \exp(t^2) \exp(-st) = \infty$ for any $s \in \mathbb{R}$.

We denote by $\mathcal{O}$ the set of all the original functions. It can be proved that $\mathcal{O}$ is a complex vector space.

Proposition 6.1 *If f and g are original functions with the abscissa of convergence s_1 and s_2, respectively, then:*

(a) $f + g$ *is an original function with the abscissa of convergence less than or equal to* $\max\{s_1, s_2\}$;
(b) αf *is an original function with the abscissa of convergence equal to s_1 if $\alpha \neq 0$ and $-\infty$ if $\alpha = 0$;*
(c) $f \cdot g$ *is an original function with the abscissa of convergence less than or equal to* $s_1 + s_2$.

Proof

(a) For any $s > \max\{s_1, s_2\}$, there exist $M_1, M_2 > 0$ such that

$$|f(t) + g(t)| \leq |f(t)| + |g(t)| < (M_1 + M_2) \exp(st) \text{ for all } t \in \mathbb{R}.$$

(b) If $\alpha \neq 0$, the functions αf and f have the same abscissa of convergence. If $\alpha = 0$, then $\alpha f \equiv 0$ and the null function has the abscissa of convergence $-\infty$.
(c) For any $s = s_1 + s_2 + \varepsilon$, $\varepsilon > 0$, there exists $M_1, M_2 > 0$ such that

$$|f(t)g(t)| = |f(t)||g(t)| < M_1 \exp((s_1 + \varepsilon/2)t) \cdot M_2 \exp((s_2 + \varepsilon/2)t)$$

$$= M_1 M_2 \exp((s_1 + s_2 + \varepsilon)t) \text{ for all } t \in \mathbb{R}.$$

□

Remark 6.1 We have seen, in Example 6.1 that t^n is an original function with $s_0 = 0$. By Proposition 6.1 (c) we obtain that, if $f(t)$ is an original function with the

6.1 Introduction

abscissa of convergence s_0, then $t^n f(t)$ is also an original function with the abscissa of convergence less than or equal to s_0.

Definition 6.2 Let $f(t)$ be an original function. The *Laplace transform* of the function f, denoted by $\mathcal{L}\{f\}$, is a new complex-valued function $F(p)$ of a complex variable $p = s + i\sigma$, $s, \sigma \in \mathbb{R}$, defined by the improper integral

$$F(p) = \int_0^\infty f(t) \exp(-pt)\, dt. \tag{6.3}$$

Thus, the Laplace transform carries an original function $f(t)$ into another function $F(p)$ called *image function*. We shall write $F(p) = \mathcal{L}\{f(t)\}(p)$ and $f(t) = \mathcal{L}^{-1}\{F(p)\}$ (the inverse Laplace transform, $\mathcal{L}^{-1}$, will be defined in Sect. 6.3).

Theorem 6.1 *Let $f(t)$ be an original function with the abscissa of convergence s_0. Then the improper integral (6.3) converges absolutely and uniformly in the right half-plane*

$$D_a = \{p \in \mathbb{C} : \operatorname{Re} p \geq a\},$$

for any $a > s_0$. As a consequence, the Laplace transform of the function $f(t)$, $F(p) = \mathcal{L}\{f(t)\}(p)$, is a continuous function defined on the domain

$$D_{s_0} = \{p \in \mathbb{C} : \operatorname{Re} p > s_0\}.$$

and

$$\lim_{\operatorname{Re} p \to \infty} F(p) = 0.$$

Moreover, $F(p)$ is an analytic function on D_{s_0} and

$$F^{(k)}(p) = (-1)^k \mathcal{L}\{t^k f(t)\},\ k = 1, 2, \ldots, \tag{6.4}$$

or, equivalently,

$$\mathcal{L}\{t^k f(t)\}(p) = (-1)^k F^{(k)}(p)\ k = 1, 2, \ldots. \tag{6.5}$$

Proof Let $a > s_1 > s_0$. There exists $M > 0$ such that

$$|f(t)| \leq M \exp(s_1 t) < M \exp(at)$$

for any $t \geq 0$. For any complex number $p = s + i\sigma$ with $s \geq a$, we have:

$$|f(t)\exp(-pt)| = |f(t)|\exp(-st) \leq M\exp(-(a-s_1)t)$$

for all $t \geq 0$. Since the function $\exp(-(a-s_1)t)$ is integrable on $[0, \infty)$,

$$\int_0^\infty \exp(-(a-s_1)t)\,dt = \frac{1}{a-s_1},$$

by Weierstrass M-test for improper integrals (Theorem 5.4) it follows that the integral (6.2) is absolutely and uniformly convergent on the domain D_a. Consequently, by Theorem 5.2 it follows that the function $F(p)$ is continuous on the domain D_{s_0}. Moreover, since

$$|F(p)| = \left|\int_0^\infty f(t)\exp(-pt)\,dt\right| \leq M\int_0^\infty \exp(-(s-s_1)t)\,dt = \frac{M}{s-s_1},$$

it follows that $F(p) \to 0$, when $\operatorname{Re} p \to \infty$.

Now, the partial derivative with respect to p of the function $f(t)\exp(-pt)$ is

$$\frac{\partial}{\partial p}(f(t)\exp(-pt)) = -tf(t)\exp(-pt).$$

Since $tf(t)$ is an original function with the abscissa of convergence $\leq s_0$ (see Remark 6.1), it follows that the integral $\int_0^\infty tf(t)\exp(-pt)\,dt$ is uniformly convergent on D_a, for any $a > s_0$. By Theorem 5.3 we obtain that the integral (6.3) can be differentiated with respect to p and we get:

$$F'(p) = -\int_0^\infty tf(t)\exp(-pt)\,dt$$

for any $p \in \mathbb{C}$ with $\operatorname{Re} p > s_0$. The formula (6.4) follows by mathematical induction on $k \geq 1$. □

Remark 6.2 There exist functions $f(t)$ which are not piecewise continuous, but the Laplace transform (6.2) is still convergent. We denote by $\overline{\mathcal{O}}$ the set of functions satisfying the conditions bellow (together with the condition (i) from Definition 6.1):

(ii') $f(t)$ is absolutely integrable on $[0, T]$ for any $T > 0$;
(iii') There exist $s \in \mathbb{R}$, $M > 0$ and $T \geq 0$ such that

$$|f(t)| \leq M\exp(st) \; \forall t \geq T. \tag{6.6}$$

6.1 Introduction

The set $\overline{\mathcal{O}}$ is also a complex vector space and $\overline{\mathcal{O}} \supset \mathcal{O}$. Using the Weierstrass M-test, as in the proof of Theorem 6.1, it can be proved that the integral (6.2) converges absolutely and uniformly, for any function $f \in \overline{\mathcal{O}}$.

For example, the function

$$f(t) = \begin{cases} t^a, & \text{for } t > 0 \\ 0, & \text{for } t \leq 0 \end{cases}$$

with $a \in (-1, 0)$ is not piecewise continuous (because $\lim_{t \searrow 0} t^a = \infty$), so $f \notin \mathcal{O}$, but $f \in \overline{\mathcal{O}}$ and it has a Laplace transform $F(p)$, defined for every $p \in \mathbb{C}$ with $\operatorname{Re} p > s_0 = 0$.

In Sect. 6.2 we present some basic properties of the Laplace transform of original functions. It can be proved that all these results hold also for functions in in the space $\overline{\mathcal{O}}$.

Example 6.2 **The Laplace transform of the Heaviside step function** $\eta(t)$ is

$$F(p) = \mathcal{L}\{\eta(t)\}(p) = \frac{1}{p}, \quad \operatorname{Re} p > 0. \tag{6.7}$$

We have:

$$\mathcal{L}\{\eta(t)\}(p) = \int_0^\infty \exp(-pt)\, dt = \left.\frac{\exp(-pt)}{-p}\right|_0^\infty.$$

Let $p = s + i\sigma$. Since $s = \operatorname{Re} > 0$, it follows that

$$\lim_{t \to \infty} |\exp(-pt)| = \lim_{t \to \infty} \exp(-st) = 0$$

and the equality (6.7) follows.

Example 6.3 The Laplace transform of function $f(t) = t^n$ is

$$F(p) = \mathcal{L}\{t^n\}(p) = \frac{n!}{p^{n+1}}, \quad \operatorname{Re} p > 0. \tag{6.8}$$

The original function considered is $f^*(t) = \eta(t) t^n$. By Theorem 6.1 (Eq. (6.5)) and using (6.7) one can write:

$$F(p) = \mathcal{L}\{t^n \eta(t)\}(p) = (-1)^n \left(\frac{1}{p}\right)^{(n)} = \frac{n!}{p^{n+1}}.$$

We can show a more general result: as discussed in Remark 6.2, the Laplace transform of the function $g(t) = t^a$ exists for any $a > -1$ and

$$G(p) = \mathcal{L}\{t^a\}(p) = \frac{\Gamma(a+1)}{p^{a+1}}, \quad \text{Re } p > 0, \qquad (6.9)$$

where

$$\Gamma(\alpha) = \int_0^\infty x^{\alpha-1} \exp(-x) dx, \quad \alpha > 0$$

is the gamma function (see (5.96)–(5.99)), which extends the factorial function to non-integer numbers (recall that $\Gamma(n+1) = n!$ for $n = 0, 1, \ldots$).

Indeed, by the definition formula (6.3), we have:

$$G(s) = \mathcal{L}\{t^a\}(s) = \int_0^\infty t^a \exp(-st)\, dt \stackrel{st=u}{=}$$

$$= \frac{1}{s^{a+1}} \int_0^\infty u^a \exp(-u)\, du = \frac{1}{s^{a+1}} \Gamma(a+1),$$

for any $s > 0$. By the Identity Theorem (see Theorem 11.32 in Chap. 11), since $G(p)$ is analytic on the domain Re $p > 0$, the formula we have found for $G(s)$, $s > 0$ can be extended to the whole domain, hence the Eq. (6.9) follows.

Finally, let us write the formula (6.9) for $a = -\frac{1}{2}$:

$$\mathcal{L}\left\{\frac{1}{\sqrt{t}}\right\}(p) = \frac{\Gamma(1/2)}{p^{1/2}} = \frac{\sqrt{\pi}}{\sqrt{p}}, \qquad (6.10)$$

since $\Gamma(1/2) = \sqrt{\pi}$.

6.2 Properties of the Laplace Transform

First of all, from Definition 6.2 we can easily prove that $\mathcal{L}$ is a linear operator.

Theorem 6.2 (Linearity) *If $f(t)$ and $g(t)$ are original functions and $\alpha, \beta \in \mathbb{C}$, then*

$$\mathcal{L}\{\alpha f(t) + \beta g(t)\} = \alpha \mathcal{L}\{f(t)\} + \beta \mathcal{L}\{g(t)\}.$$

for any $\alpha, \beta \in \mathbb{C}$ and $f(t), g(t) \in \mathcal{O}$.

Proof Suppose that $f(t)$ has the abscissa of convergence s_1 and the Laplace transform $F(p) = \mathcal{L}\{f(t)\}$, and $g(t)$ has the abscissa of convergence s_2 and the

6.2 Properties of the Laplace Transform

Laplace transform $G(p) = \mathcal{L}\{g(t)\}$. Then,

$$\mathcal{L}\{\alpha f(t) + \beta g(t)\} = \int_0^\infty (\alpha f(t) + \beta g(t)) \exp(-pt)\, dt =$$

$$= \alpha \int_0^\infty f(t) \exp(-pt)\, dt + \beta \int_0^\infty g(t) \exp(-pt)\, dt = \alpha F(p) + \beta G(p),$$

for any $p \in \mathbb{C}$ with $\operatorname{Re} p > \max\{s_1, s_2\}$. □

Proposition 6.2 *For any complex number $\xi \in \mathbb{C}$, $\xi = \alpha + i\beta$, $\alpha, \beta \in \mathbb{R}$, the Laplace transform of the function $f(t) = \exp(\xi t)$ is*

$$\mathcal{L}\{\exp(\xi t)\}(p) = \frac{1}{p - \xi}, \quad \operatorname{Re} p > \alpha = \operatorname{Re} \xi. \tag{6.11}$$

Proof Actually, as we have seen above, the original function considered is $f(t) = \eta(t) \exp(\xi t)$, so the formula (6.7) from Example 6.2 is a particular case ($\xi = 0$) of (6.11). We can write:

$$\mathcal{L}\{\exp(\xi t)\}(p) = \int_0^\infty \exp((\xi - p)t)\, dt = \left.\frac{\exp((\xi - p)t)}{\xi - p}\right|_0^\infty = \frac{1}{p - \xi},$$

because $\lim_{t \to \infty} \exp((\xi - p)t) = 0$ since $\operatorname{Re}(\xi - p) < 0$. □

Using the linearity of the Laplace transform and the Eq. (6.8), the next propositions follows.

Proposition 6.3 *Let $P(t) = b_0 + b_1 t + \ldots + b_n t^n$ be a polynomial with complex coefficients. Then,*

$$\mathcal{L}\{P(t)\}(p) = \sum_{k=0}^n \frac{b_k \cdot k!}{p^{k+1}}, \quad \operatorname{Re} p > 0. \tag{6.12}$$

Proposition 6.4 *For any $\omega \in \mathbb{R}$, the following Laplace transforms are defined for any $p \in \mathbb{C}$ with $\operatorname{Re} p > 0$:*

$$\mathcal{L}\{\cos \omega t\}(p) = \frac{p}{p^2 + \omega^2}, \tag{6.13}$$

$$\mathcal{L}\{\sin \omega t\}(p) = \frac{\omega}{p^2 + \omega^2}. \tag{6.14}$$

The Laplace transforms of the hyperbolic functions $\cosh \omega t$, $\sinh \omega t$ are defined for any $p \in \mathbb{C}$ such that $\operatorname{Re} p > |\omega|$:

$$\mathcal{L}\{\cosh \omega t\}(p) = \frac{p}{p^2 - \omega^2}, \tag{6.15}$$

$$\mathcal{L}\{\sinh \omega t\}(p) = \frac{\omega}{p^2 - \omega^2}. \tag{6.16}$$

Proof Using the Euler's formula (2.56) one can write:

$$\cos \omega t = \frac{\exp(i\omega t) + \exp(-i\omega t)}{2}, \quad \sin \omega t = \frac{\exp(i\omega t) - \exp(-i\omega t)}{2i}. \tag{6.17}$$

By the linearity of the Laplace transform and Proposition 6.2 we have:

$$\mathcal{L}\{\cos \omega t\} = \frac{1}{2}\left(\mathcal{L}\{\exp(i\omega t)\} + \mathcal{L}\{\exp(-i\omega t)\}\right) =$$

$$\stackrel{(6.11)}{=} \frac{1}{2}\left(\frac{1}{p - i\omega} + \frac{1}{p + i\omega}\right) = \frac{p}{p^2 + \omega^2}$$

and

$$\mathcal{L}\{\sin \omega t\} = \frac{1}{2i}\left(\mathcal{L}\{\exp(i\omega t)\} - \mathcal{L}\{\exp(-i\omega t)\}\right) =$$

$$\stackrel{(6.11)}{=} \frac{1}{2i}\left(\frac{1}{p - i\omega} - \frac{1}{p + i\omega}\right) = \frac{\omega}{p^2 + \omega^2}.$$

The formulas (6.15) and (6.16) can be proved in the same manner. □

Theorem 6.3 (Similarity) *Let $\mathcal{L}\{f(t)\}(p) = F(p)$, be the Laplace transform of the original function $f(t)$, defined on the half-plane $\operatorname{Re} p > s_0$, where s_0 is the abscissa of convergence of f. Then, for any $\alpha > 0$, we have:*

$$\mathcal{L}\{f(\alpha t)\}(p) = \frac{1}{\alpha} F\left(\frac{p}{\alpha}\right), \quad \operatorname{Re} p > \alpha s_0. \tag{6.18}$$

6.2 Properties of the Laplace Transform

Proof

$$\mathcal{L}\{f(\alpha t)\}(p) = \int_0^\infty f(\alpha t)\exp(-pt)\,dt \overset{\alpha t=u}{=}$$

$$= \frac{1}{\alpha}\int_0^\infty f(u)\exp\left(-\frac{p}{\alpha}u\right) du = \frac{1}{\alpha}F\left(\frac{p}{\alpha}\right).$$

$\square$

Theorem 6.4 (Delay) *Let $f(t)$ be an original function with the abscissa of convergence s_0 and let $\mathcal{L}\{f(t)\}(p) = F(p)$ be the Laplace transform of f, defined on the half-plane $\mathrm{Re}\, p > s_0$. Then, for any $\tau > 0$ (delay), we have (see Figs. 6.1 and 6.2):*

$$\mathcal{L}\{f(t-\tau)\}(p) = \exp(-\tau p)F(p), \quad \mathrm{Re}\, p > s_0. \tag{6.19}$$

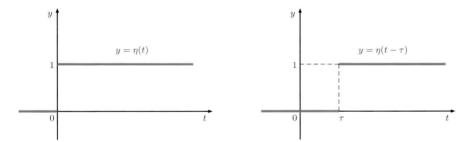

Fig. 6.1 The Heaviside step function $\eta(t)$ and $\eta(t-\tau)$

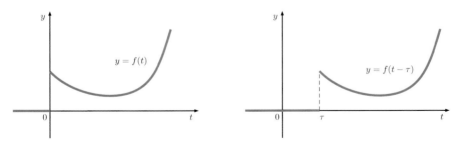

Fig. 6.2 Original function $f(t)$ and the function $f(t-\tau)$

Proof Since $f(t-\tau) = 0$ if $t < \tau$, one can write:

$$\mathcal{L}\{f(t-\tau)\}(p) = \int_\tau^\infty f(t-\tau)\exp(-pt)\,dt \stackrel{u=t-\tau}{=}$$

$$= \int_0^\infty f(u)\exp(-p(u+\tau))\,du = \exp(-\tau p)F(p).$$

□

Example 6.4 Use Theorem 6.4 to compute the Laplace transforms of the functions $\cos(t-\alpha)$ and $\sin(t-\alpha)$, where $\alpha > 0$.

Since $\mathcal{L}\{\cos t\}(p) = \frac{p}{p^2+1}$ and $\mathcal{L}\{\sin t\}(p) = \frac{1}{p^2+1}$ (by the formulas (6.13) and (6.14)), we obtain:

$$\mathcal{L}\{\cos(t-\alpha)\}(p) = \exp(-\alpha p)\frac{p}{p^2+1}, \tag{6.20}$$

and

$$\mathcal{L}\{\sin(t-\alpha)\}(p) = \exp(-\alpha p)\frac{1}{p^2+1}. \tag{6.21}$$

Example 6.5 ([22]) Find the Laplace transform of the function in Fig. 6.3a:

$$f(t) = \begin{cases} 1, & \text{if } t \in [0,1] \\ -1, & \text{if } t \in [1,2] \\ 0, & \text{otherwise.} \end{cases}$$

We notice that a function as presented in Fig. 6.3b,

$$\varphi_{a,b}(t) = \begin{cases} 1, & \text{if } t \in [a,b] \\ 0, & \text{otherwise} \end{cases}$$

can be written using the Heaviside function $\eta(t)$ as

$$\varphi_{a,b}(t) = \eta(t-a) - \eta(t-b).$$

Hence

$$f(t) = \varphi_{0,1} - \varphi_{1,2} = \eta(t) - \eta(t-1) - [\eta(t-1) - \eta(t-2)]$$

$$= \eta(t) - 2\eta(t-1) + \eta(t-2).$$

6.2 Properties of the Laplace Transform

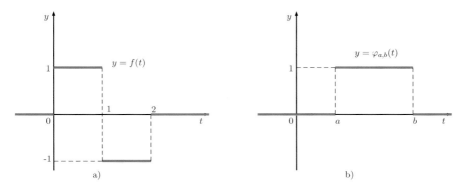

Fig. 6.3 The functions (**a**) $f(t)$ and (**b**) $\varphi_{a,b}(t)$

Since $\mathcal{L}\{\eta(t)\} = \dfrac{1}{p}$, by Theorem 6.4 (and using the linearity) we get:

$$\mathcal{L}\{f(t)\} = \frac{1}{p} - 2\frac{\exp(-p)}{p} + \frac{\exp(-2p)}{p}.$$

Theorem 6.5 (Displacement) *Let $f(t)$ be an original function with the abscissa of convergence s_0 and $\mathcal{L}\{f(t)\}(p) = F(p)$. Then, for any complex number p_0,*

$$\mathcal{L}\{\exp(p_0 t) f(t)\}(p) = F(p - p_0) \qquad (6.22)$$

for any $p \in \mathbb{C}$ such that $\operatorname{Re} p > \operatorname{Re} p_0 + s_0$.

Proof

$$\mathcal{L}\{\exp(p_0 t) f(t)\}(p) = \int_0^\infty f(t) \exp(p_0 t) \exp(-pt)\, dt =$$

$$\int_0^\infty f(t) \exp(-(p - p_0)t)\, dt = F(p - p_0),$$

for any $p \in \mathbb{C}$ such that $\operatorname{Re}(p - p_0) > s_0$. □

Using Theorem 6.5 and Proposition 6.4

$$\mathcal{L}\{\exp(\lambda t) \cos \omega t\}(p) = \frac{p - \lambda}{(p - \lambda)^2 + \omega^2}. \qquad (6.23)$$

and

$$\mathcal{L}\{\exp(\lambda t)\sin\omega t\}(p) = \frac{\omega}{(p-\lambda)^2+\omega^2} \quad (6.24)$$

Example 6.6 Find an original function $f(t)$ such that its Laplace transform is

$$F(p) = \mathcal{L}\{f(t)\} = \frac{p+3}{p^2-2p+5}.$$

We can write:

$$F(p) = \frac{p-1+4}{(p-1)^2+2^2} = \frac{p-1}{(p-1)^2+2^2} + 2 \cdot \frac{2}{(p-1)^2+2^2} \stackrel{(6.23),(6.24)}{=}$$

$$= \mathcal{L}\{\exp(t)\cos(2t)\} + 2\cdot\mathcal{L}\{\exp(t)\sin(2t)\} = \mathcal{L}\{\exp(t)[\cos(2t)+2\sin(2t)]\},$$

so the sought original function is $f(t) = \eta(t)\exp(t)[\cos(2t)+2\sin(2t)]$. We also say that $f(t)$ is the inverse transform of the function $F(p)$. (see Sect. 6.3).

By Theorem 6.1 we know that the Laplace transform $F(p) = \mathcal{L}\{f(t)\}$ is an analytic function, and the derivative $F'(p)$ is the Laplace transform of the original function $f(t)$ multiplied by $(-t)$ (see the formula (6.4)). The next theorem gives the Laplace transform of the derivatives of the original function $f(t)$.

Theorem 6.6 (Differentiation of the Original) *Assume that the function $f(t)$ as well as its derivatives $f'(t),\ldots,f^{(n)}(t)$ are original functions with abscissas of convergence $s_0, s_1, \ldots, s_n$. We denote by $s^* = \max\{s_0, \ldots, s_n\}$ and let $\mathcal{L}\{f(t)\}(p) = F(p)$ be the Laplace transform of $f(t)$. Then, for p in the right half-plane* $\operatorname{Re} p > s^*$,

$$\mathcal{L}\{f'(t)\}(p) = pF(p) - f(0) \quad (6.25)$$

and

$$\mathcal{L}\{f^{(n)}(t)\}(p) = p^n F(p) - p^{n-1}f(0) - p^{n-2}f'(0) - \ldots - pf^{(n-2)}(0) - f^{(n-1)}(0), \quad (6.26)$$

where $f^{(k)}(0)$, $k = 0, 1, \ldots, n-1$ stands for the right-hand limit at 0:

$$f^{(k)}(0) = \lim_{t\searrow 0} f^{(k)}(t).$$

6.2 Properties of the Laplace Transform

Proof It is sufficient to prove formula (6.25), because we can easily use mathematical induction to prove the other formula (6.26).

$$\mathcal{L}\{f'(t)\}(p) = \int_0^\infty f'(t)\exp(-pt)\,dt \overset{\text{by parts}}{=} f(t)\exp(-pt)\Big|_0^\infty +$$

$$+ p\int_0^\infty f(t)\exp(-pt)\,dt = -f(0) + pF(p).$$

□

This theorem is very important because it establishes that the Laplace transform can be used to convert a differential equation into an algebraic equation.

For instance, the differential equation $y'(t) + y(t) = 1$, $t \geq 0$ with the initial condition $y(0) = 0$, becomes, using the Laplace transform:

$$\mathcal{L}\{y'(t)\} + \mathcal{L}\{y(t)\} = \mathcal{L}\{\eta(t)\}.$$

We denote $\mathcal{L}\{y(t)\}(p) = F(p)$. By Theorem 6.6 (using the initial condition $y(0) = 0$), the image of the differential equation is $pF(p) + F(p) = \frac{1}{p}$. So $F(p) = \frac{1}{p(p+1)} = \frac{1}{p} - \frac{1}{p+1}$.

From formulas (6.7) and (6.11) we know that $\frac{1}{p} = \mathcal{L}\{\eta(t)\}(p)$ and $\frac{1}{p+1} = \mathcal{L}\{\eta(t)\exp(-t)\}(p)$ so, by using the linearity of $\mathcal{L}$, one can write:

$$\mathcal{L}\{y(t)\} = \mathcal{L}\{\eta(t)\} - \mathcal{L}\{\eta(t)\exp(-t)\} = \mathcal{L}\{\eta(t)(1-\exp(-t))\}$$

We will prove in Sect. 6.3 (Theorem 6.12) that two continuous functions having the same Laplace transform are equal. Since $y(t)$ is continuous, it follows that $y(t) = 1 - \exp(-t)$ for all $t \geq 0$.

Remark 6.3 Let $f(t)$ be an original function and $\mathcal{L}\{f(t)\}(p) = F(p)$. From Theorem 6.1 we know that $\lim_{\text{Re } p \to \infty} F(p) = 0$. If $f'(t)$ is also an original function, it follows by (6.25) that

$$\lim_{\text{Re } p \to \infty} pF(p) = f(0).$$

The following result is a direct consequence of Theorem 6.6 and of the formula (6.5). It is useful for solving linear differential equations with polynomial coefficients.

Corollary 6.1 *Assume that $f(t), f'(t), \ldots, f^{(n)}(t) \in \mathcal{O}$ and let*

$$G(p) = \mathcal{L}\{f^{(n)}(t)\}(p) \stackrel{(6.26)}{=}$$
$$= p^n F(p) - p^{n-1} f(0) - p^{n-2} f'(0) - \ldots - f^{(n-1)}(0),$$

where $F(p) = \mathcal{L}\{f(t)\}(p)$. Then

$$\mathcal{L}\{t^k f^{(n)}(t)\}(p) \stackrel{(6.5)}{=} (-1)^k \frac{d^k}{dp^k} G(p) = \qquad (6.27)$$

$$= (-1)^k \left[\frac{d^k}{dp^k} [p^n F(p)] - \sum_{j=1}^{n-k} \frac{(n-j)!}{(n-k-j)!} p^{n-k-j} f^{(j-1)}(0) \right].$$

Theorem 6.7 (Integration of the Original) *Let $f(t)$ be an original function with the abscissa of convergence s_0 and $\mathcal{L}\{f(t)\}(p) = F(p)$. Then $g(t) = \int_0^t f(u)\,du$ is also an original function with the abscissa of convergence $s_1 = \max\{0, s_0\}$ and*

$$\mathcal{L}\{g(t)\}(p) = \frac{F(p)}{p}. \qquad (6.28)$$

Proof Let $s > s_0$ be a positive number. There exists $M > 0$ such that $|f(u)| \le M \exp(su)$ for all $u \ge 0$, so

$$\left| \int_0^t f(u)\,du \right| \le \int_0^t |f(u)|\,du \le M \int_0^t \exp(su)\,du < \frac{M}{s} \exp(st),$$

for all $t \ge 0$, so $g(t)$ is an original function.

Denote by $G(p) = \mathcal{L}\{g(t)\}$, $\operatorname{Re} p > s_1$. From Theorem 6.6 we can write:

$$\mathcal{L}\{g'(t)\}(p) = pG(p) - g(0) = pG(p).$$

Since $g'(t) = f(t)$, we have $pG(p) = F(p)$, so $G(p) = \dfrac{F(p)}{p}$. □

Example 6.7 Let us apply Theorem 6.7 to compute $\mathcal{L}\{\operatorname{erf}(t)\}$, where

$$\operatorname{erf}(t) = \frac{2}{\sqrt{\pi}} \int_0^t \exp(-u^2)\,du$$

is the famous Gauss error function which appears frequently in Probability theory and Statistics. We know that $\operatorname{erf}(\infty) = 1$ (see Example 5.3). Since $f(t) =$

6.2 Properties of the Laplace Transform

$\eta(t)\exp(-t^2) \in \mathcal{O}$ has the abscissa of convergence $s_0 = -\infty$ (for any $s \in \mathbb{R}$, there exists $M > 0$ such that $\exp(-t^2) \le M\exp(st)$ for all $t \ge 0$), the Laplace transform $F(p) = \mathcal{L}\{f(t)\}(p)$ is an analytic function on $\mathbb{C}$ (see Theorem 6.1). From Theorem 6.7 we deduce that

$$\mathcal{L}\{\mathrm{erf}(t)\}(p) = \frac{2}{\sqrt{\pi}} \frac{F(p)}{p}, \quad \mathrm{Re}\, p > 0.$$

We have to find the function $F(p)$. Since $F(p)$ is analytic on $\mathbb{C}$, it is sufficient to find a formula for $F(s)$, where $s \in \mathbb{R}$. Indeed, two analytic functions which coincide on a subset of $\mathbb{C}$ that has at least one limit point, coincide on their common definition domain (see the Identity Theorem 11.32 in Chap. 11).

$$F(s) = \int_0^\infty \exp(-t^2)\exp(-st)\,dt = \int_0^\infty \exp(-(t^2+st))\,dt =$$

$$= \int_0^\infty \exp\left(-\left(t+\frac{s}{2}\right)^2 + \frac{s^2}{4}\right) dt = \exp\left(\frac{s^2}{4}\right)\int_0^\infty \exp\left[-\left(t+\frac{s}{2}\right)^2\right]dt =$$

$$\stackrel{u=t+\frac{s}{2}}{=} \exp\left(\frac{s^2}{4}\right)\int_{s/2}^\infty \exp\left(-u^2\right)du =$$

$$= \exp\left(\frac{s^2}{4}\right)\left[\int_0^\infty \exp\left(-u^2\right)du - \int_0^{s/2} \exp\left(-u^2\right)du\right] =$$

$$= \exp\left(\frac{s^2}{4}\right)\frac{\sqrt{\pi}}{2}\left[1 - \mathrm{erf}\left(\frac{s}{2}\right)\right].$$

Since the function $\exp\left(\frac{p^2}{4}\right)[1 - \mathrm{erf}(\frac{p}{2})]$ is analytic for $p \in \mathbb{C}$, we obtain:

$$\mathcal{L}\left\{\exp(-t^2)\right\}(p) = \exp\left(\frac{p^2}{4}\right)\frac{\sqrt{\pi}}{2}\left[1 - \mathrm{erf}\left(\frac{p}{2}\right)\right]. \tag{6.29}$$

Hence, by Theorem 6.7, we find:

$$\mathcal{L}\{\mathrm{erf}(t)\}(p) = \frac{1}{p}\exp\left(\frac{p^2}{4}\right)\left[1 - \mathrm{erf}\left(\frac{p}{2}\right)\right], \quad \mathrm{Re}\, p > 0. \tag{6.30}$$

Theorem 6.8 (Integration of the Image) *Let $f(t)$ be an original function with the abscissa of convergence s_0 and $\mathcal{L}\{f(t)\}(p) = F(p)$. If there exists the (finite) limit*

$\lim\limits_{t\searrow 0} \dfrac{f(t)}{t}$, then $\dfrac{f(t)}{t}$ is also an original function and

$$\int_s^\infty F(x)dx = \mathcal{L}\left\{\dfrac{f(t)}{t}\right\}(s), \quad \forall s > s_0. \tag{6.31}$$

Proof By the hypothesis, the function $\dfrac{f(t)}{t}$ is an original function (with the abscissa of convergence less than or equal to s_0). We denote by $G(p) = \mathcal{L}\left\{\dfrac{f(t)}{t}\right\}(p)$ its Laplace transform.

For any $x > s_0$ we have

$$F(x) = \int_0^\infty f(t)\exp(-xt)\,dt.$$

By integrating this relation on the interval $[s, w]$ ($w > s > s_0$) we obtain:

$$\int_s^w F(x)dx = \int_s^w \left(\int_0^\infty f(t)\exp(-xt)\,dt\right)dx.$$

From Theorem 6.1 the integral $\int_0^\infty f(t)\exp(-xt)\,dt$ is uniformly convergent on $[s, w]$, so we can reverse the order of integration:

$$\int_s^w F(x)dx = \int_0^\infty \left(\int_s^w f(t)\exp(-xt)\,dx\right)dt =$$

$$= \int_0^\infty f(t)\left(\int_s^w \exp(-xt)\,dx\right)dt = \int_0^\infty f(t)\left(\dfrac{\exp(-st) - \exp(-wt)}{t}\right)dt$$

$$= G(s) - G(w)$$

and, since $\lim\limits_{w\to\infty} G(w) = 0$ (see Theorem 6.1), we obtain that

$$\int_s^\infty F(x)dx = \lim_{w\to\infty}\int_s^w F(x)dx = G(s).$$

$\square$

Example 6.8 Find the Laplace transform of the function $\dfrac{\sin t}{t}$.

6.2 Properties of the Laplace Transform

First of all, we notice that $\lim_{t \searrow 0} \frac{\sin t}{t} = 1$, so we can apply Theorem 6.8. Since

$$\mathcal{L}\{\sin t\}(p) = \frac{1}{p^2 + 1}, \quad \text{Re } p > s_0,$$

for any $s > 0$, one can write:

$$\mathcal{L}\left\{\frac{\sin t}{t}\right\}(s) = \int_s^\infty \frac{dx}{x^2 + 1} = \arctan s \Big|_s^\infty,$$

hence

$$\mathcal{L}\left\{\frac{\sin t}{t}\right\}(s) = \frac{\pi}{2} - \arctan s = \arctan \frac{1}{s}, \quad s > 0.$$

Using the Identity Theorem 11.32, we obtain

$$\mathcal{L}\left\{\frac{\sin t}{t}\right\}(p) = \arctan \frac{1}{p}, \quad \text{Re } p > 0. \tag{6.32}$$

Example 6.9 Find the Laplace transform of the function $\frac{\sinh \omega t}{t}$.

The function $\sinh \omega t$ is an original function with the abscissa of convergence $s_0 = |\omega|$. Since $\lim_{t \searrow 0} \frac{\sinh \omega t}{t} = \omega$, we can apply Theorem 6.8. By (6.16),

$$\mathcal{L}\{\sinh \omega t\}(p) = \frac{\omega}{p^2 - \omega^2}, \quad \text{Re } p > |\omega|,$$

so, for any $s > |\omega|$, we have:

$$\mathcal{L}\left\{\frac{\sinh \omega t}{t}\right\}(s) = \int_s^\infty \frac{\omega dx}{x^2 - \omega^2} = \frac{1}{2} \ln \frac{x - \omega}{x + \omega} \Big|_s^\infty.$$

Hence

$$\mathcal{L}\left\{\frac{\sinh \omega t}{t}\right\}(s) = \frac{1}{2} \ln \frac{s + \omega}{s - \omega}, \quad s > |\omega|. \tag{6.33}$$

By Proposition 6.3, the Laplace transform of a polynomial $f(t) = \sum_{n=0}^{N} a_n t^n$ is defined for Re $p > 0$ and has the expression:

$$\mathcal{L}\left\{\sum_{n=0}^{N} a_n t^n\right\}(p) = \sum_{n=0}^{N} a_n \cdot \frac{n!}{p^{n+1}}.$$

A question that arises naturally is, if the original function $f(t)$ is the sum of a *power series*, $f(t) = \sum_{n=0}^{\infty} a_n t^n$, could we conclude that

$$\mathcal{L}\left\{\sum_{n=0}^{\infty} a_n t^n\right\}(p) = \sum_{n=0}^{\infty} a_n \cdot \frac{n!}{p^{n+1}}?$$

The answer is "no", as shown in the example below. But Theorem 6.9 gives sufficient conditions for the above statement to be true.

Example 6.10 Since $\exp(t) = \sum_{n=0}^{\infty} \frac{t^n}{n!}$, for any $t \in \mathbb{R}$, it follows that

$$\exp(-t^2) = \sum_{n=0}^{\infty} \frac{(-1)^n t^{2n}}{n!}, \quad t \in \mathbb{R}.$$

We know that $f(t) = \exp(-t^2)$ is an original function, so there exists the Laplace transform $\mathcal{L}\{f(t)\}$ (and its expression is given by Eq. (6.29)). But if we take the Laplace transform of each term of the series above, we obtain a divergent series, as shown below.

$$\sum_{n=0}^{\infty} \frac{(-1)^n}{n!} \mathcal{L}\{t^{2n}\}(p) = \sum_{n=0}^{\infty} \frac{(-1)^n (2n)!}{n!} p^{-2n-1}$$

and, by ratio test, the series diverges for all p:

$$\lim_{n \to \infty} \left|\frac{u_{n+1}}{u_n}\right| = \lim_{n \to \infty} \frac{2(2n+1)}{|p|^2} = \infty.$$

6.2 Properties of the Laplace Transform

Theorem 6.9 ([34], Theorem 1.18) *If* $f(t) = \sum_{n=0}^{\infty} a_n t^n$ *is convergent for* $t \geq 0$ *and there exist the positive constants* $K, \alpha > 0$ *such that*

$$|a_n| \leq K \cdot \frac{\alpha^n}{n!}, \text{ for } n = 0, 1, \ldots, \quad (6.34)$$

then $f(t)$ *is an original function and its Laplace transform satisfies the equation:*

$$\mathcal{L}\left\{\sum_{n=0}^{\infty} a_n t^n\right\}(p) = \sum_{n=0}^{\infty} a_n \mathcal{L}\{t^n\}(p) = \sum_{n=0}^{\infty} a_n \frac{n!}{p^{n+1}}. \quad (6.35)$$

for any $p \in \mathbb{C}$ *with* $\operatorname{Re} p > \alpha$.

Proof By Eq. (6.34) and using the Weierstrass M-test for series of functions (Theorem 4.1), we find that the series $\sum_{n=0}^{\infty} a_n t^n$ uniformly converges on any interval $[0, R]$, so the function $f(t)$ is continuous on $[0, \infty)$. On the other hand, since

$$|f(t)| \leq \sum_{n=0}^{\infty} |a_n| t^n \leq K \sum_{n=0}^{\infty} \frac{(\alpha t)^n}{n!} = K \exp(\alpha t), \ \forall t \geq 0,$$

the function $f(t)$ is of exponential order α, so it is an original function.

We have to prove that the difference

$$\mathcal{L}\{f(t)\} - \sum_{n=0}^{N} a_n \mathcal{L}\{t^n\} = \mathcal{L}\left\{f(t) - \sum_{n=0}^{N} a_n t^n\right\} = \mathcal{L}\left\{\sum_{n=N+1}^{\infty} a_n t^n\right\}$$

converges to 0 as $N \to \infty$. For all $p \in \mathbb{C}$ with $\operatorname{Re} p = s > \alpha$ we have:

$$\left|\mathcal{L}\left\{\sum_{n=N+1}^{\infty} a_n t^n\right\}(p)\right| \leq \int_0^{\infty} |\exp(-pt)| \sum_{n=N+1}^{\infty} |a_n| t^n \, dt \stackrel{(6.34)}{\leq}$$

$$\leq K \int_0^{\infty} \exp(-st) \sum_{n=N+1}^{\infty} \frac{(\alpha t)^n}{n!} \, dt =$$

$$= K\left(\int_0^\infty \exp((\alpha-s)t)\,dt - \sum_{n=0}^N \frac{\alpha^n}{n!}\int_0^\infty t^n \exp(-st)\,dt\right)$$

$$= K\left(\frac{1}{s-\alpha} - \frac{1}{s}\sum_{n=0}^N \frac{\alpha^n}{s^n}\right) \xrightarrow[N\to\infty]{} 0$$

since the geometric series of ratio $\frac{\alpha}{s} < 1$, $\frac{1}{s}\sum_{n=0}^N \left(\frac{\alpha}{s}\right)^n$ converges to $\frac{1}{s-\alpha}$. So we have

$$\mathcal{L}\{f(t)\}(p) = \lim_{N\to\infty} \sum_{n=0}^N a_n \mathcal{L}\{t^n\}(p)$$

for any $p \in \mathbb{C}$ with $\operatorname{Re} p > \alpha$ and the Eq. (6.35) is proved. □

We remark that the coefficients of the series in Example 6.10 ($a_{2n} = \frac{(-1)^n}{n!}$, $a_{2n+1} = 0$) do not satisfy the hypothesis of Theorem 6.9: There are no constants $K, \alpha > 0$ such that $\frac{1}{n!} \leq K\frac{\alpha^{2n}}{(2n)!}$ for all $n = 0, 1, \ldots$. An example of function which satisfies the conditions of Theorem 6.9 is the function $\frac{\sin t}{t}$ from Example 6.8. We compute again its Laplace transform, using the power series expansion this time.

Example 6.11 ([34], Example 1.19) Consider the power series representation of the function $f(t) = \frac{\sin t}{t}$:

$$\frac{\sin t}{t} = \sum_{n=0}^\infty \frac{(-1)^n}{(2n+1)!} t^{2n}.$$

Since $|a_{2n}| < \frac{1}{(2n)!}$ and $|a_{2n+1}| = 0 < \frac{1}{(2n+1)!}$, the inequality (6.34) is verified for $K = \alpha = 1$. By Theorem 6.9 we can write:

$$\mathcal{L}\left\{\frac{\sin t}{t}\right\}(p) = \sum_{n=0}^\infty \frac{(-1)^n}{(2n+1)!} \mathcal{L}\{t^{2n}\}(p) =$$

$$= \sum_{n=0}^\infty \frac{(-1)^n}{(2n+1)!} \frac{(2n)!}{p^{2n+1}} = \sum_{n=0}^\infty \frac{(-1)^n}{2n+1} p^{-2n-1}.$$

But we know that

$$\arctan(x) = \int_0^x \frac{ds}{s^2+1} = \int_0^x (-1)^n s^{2n} ds = \sum_{n=0}^{\infty} \frac{(-1)^n x^{2n+1}}{2n+1},$$

for any $x \in [-1, 1]$. It follows that

$$\mathcal{L}\left\{\frac{\sin t}{t}\right\}(s) = \arctan \frac{1}{s}$$

for any $s \geq 1$ and, by Identity Theorem 11.32, since the Laplace transform $\mathcal{L}\left\{\frac{\sin t}{t}\right\}(p)$ is analytic on the domain $\operatorname{Re} p > 0$, we obtain that

$$\mathcal{L}\left\{\frac{\sin t}{t}\right\}(p) = \arctan \frac{1}{p},$$

for any p with $\operatorname{Re} p > 0$.

Theorem 6.10 (Laplace Transform of a Periodic Function) *Let $f(t)$ be piecewise continuous function which is periodic of period $T > 0$. Then $f^*(t) = f(t)\eta(t)$ is an original function with the abscissa of convergence $s_0 = 0$ and*

$$F(p) = \mathcal{L}\{f(t)\}(p) = \frac{1}{1 - \exp(-pT)} \int_0^T f(t) \exp(-pt) \, dt, \qquad (6.36)$$

for any $p \in \mathbb{C}$, $\operatorname{Re} p > 0$.

Proof Indeed, since the function is piecewise continuous and periodic, it follows that it is bounded: there exists $M > 0$ such that $|f(t)| \leq M \exp(0 \cdot t)$ for all $t \geq 0$, so $s_0 = 0$. Moreover,

$$\int_{nT}^{(n+1)T} f(t) \exp(-pt) \, dt \stackrel{u=t-nT}{=} \exp(-pnT) \int_0^T f(u) \exp(-pu) \, du.$$

So

$$F(p) = \int_0^{\infty} f(t) \exp(-pt) \, dt = \sum_{n=0}^{\infty} \int_{nT}^{(n+1)T} f(t) \exp(-pt) \, dt =$$

$$= \left[\sum_{n=0}^{\infty} \exp^n(-pT)\right] \int_0^T f(u) \exp(-pu) \, dt.$$

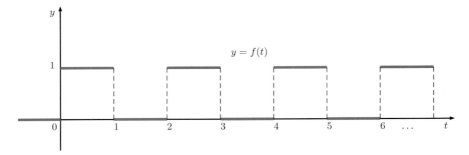

Fig. 6.4 The square wave function $f(t)$

Since $|\exp(-pT)| < 1$ for any p with $\operatorname{Re} p > 0$, the series $\sum_{n=0}^{\infty} \exp^n(-pT)$ is convergent and

$$\sum_{n=0}^{\infty} \exp^n(-pT) = \frac{1}{1-\exp(-pT)},$$

so the formula (6.36) follows. □

Example 6.12 Let us find the Laplace transform of the *square-wave function* $f(t) = 1$, if $t \in [2k, 2k+1]$, $f(t) = 0$, if $t \in (2k+1, 2k+2)$ for any $k = 0, 1, \ldots$ and $f(t) = 0$, if $t < 0$ (a well-known function in signal theory) (Fig. 6.4).

It is clear that f is a periodic function of period $T = 2$. By the formula (6.36) we find:

$$\mathcal{L}\{f(t)\}(p) = \frac{1}{1-\exp(-2p)} \int_0^2 f(t) \exp(-pt)\, dt =$$

$$= \frac{1}{1-\exp(-2p)} \int_0^1 \exp(-pt)\, dt =$$

$$= \frac{1}{1-\exp(-2p)} \frac{1}{p} [1 - \exp(-p)] = \frac{1}{p(1+\exp(-p))}.$$

As we have seen in Chap. 5 (see Definition 5.9), the **convolution** of the functions $f(x)$ and $g(x)$ is a new function denoted by $(f * g)(x)$, defined as

$$(f * g)(t) = \int_{-\infty}^{\infty} f(u) g(t-u)\, du \qquad (6.37)$$

(if the integral is convergent). If $f(t)$ and $g(t)$ are original functions, then the Eq. (6.37) becomes:

$$(f * g)(t) = \int_0^t f(u) g(t-u)\, du. \tag{6.38}$$

Theorem 6.11 (Multiplication Theorem) *If $f(t)$ and $g(t)$ are two original functions, then $f * g$ is also an original function and*

$$\mathcal{L}\{(f * g)(t)\} = \mathcal{L}\{f(t)\} \cdot \mathcal{L}\{g(t)\}. \tag{6.39}$$

(convolution formula for the Laplace transform).

Proof Let s_1, s_2 be the abscissas of convergence for the functions f and g, respectively, and let $s > s' > s_* = \max\{s_1, s_2\}$. We know that there exist the positive numbers M_1, M_2 such that $|f(t)| \leq M_1 \exp(s't)$ and $|g(t)| \leq M_2 \exp(s't)$, for all $t \geq 0$, so we have:

$$|(f * g)(t)| \leq \int_0^t |f(u) g(t-u)|\, du \leq M_1 M_2 \exp(s't) \cdot t.$$

Hence, there exists $M > 0$ such that $|(f * g)(t)| \leq M \exp(st)$, $\forall t \geq 0$.

We have proved that $f * g$ is an original function. Now, let $F(p) = \mathcal{L}\{f(t)\}(p)$ and $G(p) = \mathcal{L}\{g(t)\}$. The Laplace transform of the convolution $f * g$ is defined for any p with $\mathrm{Re}\, p > s_*$ as

$$\mathcal{L}\{f * g\}(p) = \int_0^\infty \exp(-pt) \int_0^t f(u) g(t-u)\, du\, dt =$$

$$= \int_0^\infty \int_0^\infty \exp(-pt) f(u) g(t-u)\, du\, dt.$$

Since we can change the order of integration (by Theorem 5.2), we obtain:

$$\mathcal{L}\{f * g\}(p) = \int_0^\infty f(u) \int_0^\infty \exp(-pt) g(t-u)\, dt\, du \stackrel{t-u=v}{=}$$

$$= \int_0^\infty \exp(-pu) f(u)\, du \int_0^\infty \exp(-pv) g(v)\, dv = F(p) \cdot G(p).$$

□

Absolute convergence was necessarily when we changed the order of integration. In the particular case, when $f, g \in \mathcal{O}$ (so they are defined at $t = 0$), for all p with $\mathrm{Re}\, p > s_3 = \max\{s_1, s_2\}$ the integrals $\mathcal{L}\{f(t)\}(p)$ and $\mathcal{L}\{g(t)\}(p)$ are absolutely and uniformly convergent (see Theorem 6.1).

Example 6.13 Compute (by two methods!) the Laplace transform of the function $h(t)$ defined by

$$h(t) = \int_0^t u \sin(t-u)\, du.$$

We consider the original functions $f(t) = t$ and $g(t) = \sin t$. We notice that $h = f * g$, so, by the Multiplication Theorem 6.11, we can write:

$$\mathcal{L}\{h(t)\}(p) = \mathcal{L}\{f(t)\}(p) \cdot \mathcal{L}\{g(t)\}(p) = \frac{1}{p^2(p^2+1)}.$$

The second method is to find the expression of $h(t)$ (using integration by parts):

$$h(t) = \int_0^t u(\cos(t-u))'\, du = u\cos(t-u)\Big|_0^t - \int_0^t \cos(t-u)\, du =$$

$$= t + \sin(t-u)\Big|_0^t = t - \sin t.$$

Hence

$$\mathcal{L}\{h(t)\}(p) = \mathcal{L}\{t\}(p) - \mathcal{L}\{\sin t\}(p) = \frac{1}{p^2} - \frac{1}{p^2+1} = \frac{1}{p^2(p^2+1)}.$$

Now we shall use Theorem 6.11 to prove a basic identity for the beta function, $B(a, b)$, and the gamma function, $\Gamma(a)$ (also known as the Euler integral of the first kind and of the second kind, respectively).

Example 6.14 The beta function is defined by the integral

$$B(a,b) = \int_0^1 u^{a-1}(1-u)^{b-1}\, a, b > 0, \tag{6.40}$$

and the gamma function is defined as (see (5.96))

$$\Gamma(a) = \int_0^\infty t^{a-1} \exp(-t)\, dt,\ a > 0. \tag{6.41}$$

We want to prove that

$$B(a,b) = \frac{\Gamma(a)\Gamma(b)}{\Gamma(a+b)}. \tag{6.42}$$

6.2 Properties of the Laplace Transform

Let us denote $f(t) = t^{a-1}$, $g(t) = t^{b-1}$, $t \geq 0$, $a, b > 0$. We see that when $a, b < 1$, these functions are not defined at $t = 0$, so f, g are not necessarily in $\mathcal{O}$. But, as discussed in Remark 6.2, they are in $\overline{\mathcal{O}}$ and Theorem 6.11 can be applied.

The Laplace transforms of the functions $f(t)$ and $g(t)$ are given by the formula formula (6.9) (see Example 6.3):

$$\mathcal{L}\{f(t)\}(p) = \frac{\Gamma(a)}{p^a}, \quad \text{Re } p > 0,$$

and

$$\mathcal{L}\{g(t)\}(p) = \frac{\Gamma(b)}{p^b}, \quad \text{Re } p > 0.$$

By the convolution formula (6.39) we have, for Re $p > 0$:

$$\mathcal{L}\{f(t)\}(p) \cdot \mathcal{L}\{g(t)\}(p) = \mathcal{L}\{(f * g)(t)\}(p),$$

where

$$(f * g)(t) = \int_0^t \tau^{a-1}(t-\tau)^{b-1} d\tau \stackrel{\tau=tu}{=} t^{a+b-1} B(a, b).$$

Hence

$$\frac{\Gamma(a)}{p^a} \cdot \frac{\Gamma(b)}{p^b} = B(a, b)\mathcal{L}\{t^{a+b-1}\} = B(a, b)\frac{\Gamma(a+b)}{p^{a+b}}.$$

Finally we get:

$$\Gamma(a)\Gamma(b) = B(a, b)\Gamma(a+b),$$

and the formula (6.42) follows.

Example 6.15 Find a function $f \in \overline{\mathcal{O}}$ such that of $\mathcal{L}\{f\}(p) = \frac{1}{\sqrt{p(p-a)}}$, where $a > 0$. Here $\sqrt{p}$ is considered to be the principal branch of the two-valued complex function $\sqrt{p}$, i.e. if $p = |p|\exp(i\theta)$, where $\theta \in [0, 2\pi)$, then $\sqrt{p} = \sqrt{|p|}\exp(i\theta/2)$.

Using Eq. (6.10) from Example 6.3 one can write:

$$\mathcal{L}\left\{\frac{1}{\sqrt{\pi t}}\right\} = \frac{1}{\sqrt{p}}.$$

By Eq. (6.11) we get:

$$\mathcal{L}\{\exp(at)\}(p) = \frac{1}{p-a}, \qquad \operatorname{Re} p > \alpha = \operatorname{Re} a$$

Let us denote $f(t) = \frac{1}{\sqrt{\pi t}}$ and $g(t) = \exp(at)$. We apply the convolution formula (6.39) and find:

$$\frac{1}{\sqrt{p}(p-a)} = \mathcal{L}\{(f*g)(t)\} =$$

$$\int_0^t \frac{1}{\sqrt{\tau}} \exp(a(t-\tau))\, d\tau$$

$$= \frac{\exp(at)}{\sqrt{\pi}} \int_0^t \frac{1}{\sqrt{\tau}} \exp(-a\tau)\, d\tau \stackrel{u=\sqrt{a\tau}}{=} \frac{2\exp(at)}{\sqrt{\pi a}} \int_0^{\sqrt{at}} \exp(-u^2)\, du =$$

$$= \frac{\exp(at)}{\sqrt{a}} \operatorname{erf}\left(\sqrt{at}\right),$$

where $\operatorname{erf}(x) = \frac{2}{\sqrt{\pi}} \int_0^x \exp(-t^2)\, dt$ is the **Gauss error function**. Hence

$$\mathcal{L}\left\{\frac{\exp(at)}{\sqrt{a}} \operatorname{erf}\left(\sqrt{at}\right)\right\} = \frac{1}{\sqrt{p}(p-a)}.$$

As we will see in the next section, this means that the *inverse Laplace transform* of the function $F(p) = \frac{1}{\sqrt{p}(p-a)}$ is the function

$$\varphi(t) = \mathcal{L}^{-1}\left\{\frac{1}{\sqrt{p}(p-a)}\right\} = \frac{\exp(at)}{\sqrt{a}} \operatorname{erf}\left(\sqrt{at}\right).$$

6.3 Inverse Laplace Transform

In this section we discuss the following problem: Given a (complex) function $F(p)$, how can we find a function $f(t)$ such that $F(p)$ is the Laplace transform (the image) of the function $f(t)$? If such a function $f(t)$ exists, then it is said to be the *inverse Laplace transform* of $F(p)$, we write $f(t) = \mathcal{L}^{-1}\{F(p)\}(t)$. We are interested to know what conditions must be fulfilled by the function $F(p)$ to ensure the existence

6.3 Inverse Laplace Transform

of the inverse Laplace transform $f(t)$ and to see if this inverse transform is unique or not.

Theorem 6.12 (Inversion Theorem for the Laplace Transform) *Let $f(t)$ be an original function with the abscissa of convergence s_0 and let $F(p) = \mathcal{L}\{f(t)\}(p)$ be its Laplace transform, defined for Re $p > s_0$. Then, for any $s > s_0$, the following inversion formula (also known as the **Mellin formula**) holds:*

$$f(t) = \frac{1}{2\pi} \int_{-\infty}^{\infty} F(s+i\xi) \exp((s+i\xi)t)\, d\xi = \quad (6.43)$$

$$= \frac{1}{2\pi i} \int_{s-i\infty}^{s+i\infty} F(p) \exp(pt)\, dp, \quad (6.44)$$

at any point $t > 0$ where the function f is continuous. If t is a discontinuity point for f, then the Mellin formula is written:

$$\frac{f(t_+) + f(t_-)}{2} = \quad (6.45)$$

$$= \frac{1}{2\pi} \int_{-\infty}^{\infty} F(s+i\xi) \exp((s+i\xi)t)\, d\xi = \frac{1}{2\pi i} \int_{s-i\infty}^{s+i\infty} F(p) \exp(pt)\, dp.$$

Proof The integration in (6.44) is along the straight line Re $p = s = \text{const} > s_0$, so $dp = i\,d\xi$ and the equivalence of the formulas (6.43) and (6.44) is evident.

We apply the Fourier Integral Theorem 5.5 to the function $\varphi(t) = f(t) \exp(-st)$, which is absolutely integrable on $\mathbb{R}$, for any $s > s_0$:

$$\frac{\varphi(t_+) + \varphi(t_-)}{2} = \frac{1}{2\pi} \int_{-\infty}^{\infty} \int_{-\infty}^{\infty} \varphi(u) \exp(i\xi(t-u))\, du\, d\xi =$$

$$= \frac{1}{2\pi} \int_{-\infty}^{\infty} \int_{-\infty}^{\infty} f(u) \exp(-su) \exp(-i\xi u) \exp(i\xi t)\, du\, d\xi =$$

$$= \frac{1}{2\pi} \int_{-\infty}^{\infty} \exp(i\xi t) \int_{0}^{\infty} f(u) \exp(-(s+i\xi)u)\, du\, d\xi.$$

Hence,

$$\frac{f(t_+) + f(t_-)}{2} \cdot \exp(-st) = \frac{1}{2\pi} \int_{-\infty}^{\infty} \exp(i\xi t) F(s+i\xi u)\, d\xi$$

and the Eq. (6.45) follows by multiplication with $\exp(st)$. Obviously, if f is continuous at t, then $f(t_+) = f(t_-) = f(t)$ and so the formulas (6.43) and (6.44) follows. □

The next theorem gives sufficient conditions for a complex function $F(p)$ to have an inverse Laplace transform, $f(t)$ (for the proof, see [25], 79, Theorem 4).

Theorem 6.13 *Let $F(p)$ be an analytic complex function defined on a domain $D_{s_0} = \{p \in \mathbb{C} : \operatorname{Re} p > s_0\}$ such that:*

(1) *$F(p)$ converges to 0 as $|p| \to \infty$ in the half-plane $\operatorname{Re} p \geq s > s_0$, uniformly in $\arg p$ and*

(2) *the integral $\int_{s-i\infty}^{s+i\infty} F(p)\,dp$ is absolutely convergent.*

Then $F(p)$ is the Laplace transform of the function $f(t)$ defined by the Mellin formula:

$$f(t) = \frac{1}{2\pi i} \int_{s-i\infty}^{s+i\infty} F(p) \exp(pt)\,dp.$$

For instance, the function $F(p) = \frac{p+1}{p}$ does not have an inverse Laplace transform, because $\lim\limits_{|p|\to\infty} F(p) = 1 \neq 0$.

Now let us consider the problem of the uniqueness of the inverse Laplace transform.

We say that two functions are *equal almost everywhere* (a.e.) if they have different values only on a set A such that $A \cap [a, b]$ is finite, for any interval $[a, b]$. Obviously, two original functions which are equal almost everywhere have the same Laplace transform, hence the inverse Laplace transform is not unique. As a matter of fact, if $\mathcal{L}\{f(t)\} = F(p)$, then we write $\mathcal{L}^{-1}\{F(p)\}(t) = f(t)$ but we understand that $\mathcal{L}^{-1}\{F(p)\}(t)$ is the set of all the functions of the form $g(t) = f(t) + n(t)$, $n(t) \in \ker \mathcal{L}$, where $\ker \mathcal{L}$ is the *kernel* of the linear operator $\mathcal{L}$,

$$\ker \mathcal{L} = \left\{n(t) \in \overline{\mathcal{O}} : \mathcal{L}\{n(t)\} \equiv 0\right\}.$$

By the Inversion Theorem 6.12, if the Laplace transform of an original function $f(t)$ is identical 0, $F(p) \equiv 0$, then $f(t) = 0$ at any point of continuity. Since $f(t)$ is piecewise continuous, it follows that $f(t) = 0$ almost everywhere. Hence,

$$\ker \mathcal{L} = \left\{n(t) \in \overline{\mathcal{O}} : n(t) = 0 \text{ a.e.}\right\}$$

so we can state the next theorem, which establishes the sense of the uniqueness of the inverse Laplace transform (see also [12], Theorem 5.1).

Theorem 6.14 (Uniqueness Theorem) *Two original functions $f(t)$ and $g(t)$ having the same Laplace transform, $\mathcal{L}\{f(t)\} = \mathcal{L}\{g(t)\}$ are equal almost everywhere. If the functions $f(t)$ and $g(t)$ are continuous, they must be equal.*

6.3 Inverse Laplace Transform

So, whenever $\mathcal{L}\{f(t)\} = F(p)$, the inverse Laplace transform of the function $F(p)$ is the set of functions

$$\mathcal{L}^{-1}\{F(p)\} = f(t) + \ker \mathcal{L},$$

but we do not use this notation: we simply write $\mathcal{L}^{-1}\{F(p)\} = f(t)$.

Theorem 6.15 ([12], Theorem 5.6.) *A Laplace transform $F(p) \not\equiv 0$ cannot be periodic.*

Proof Suppose that $F(p)$ is a periodic function, i.e. that there exists $q \in \mathbb{C}$, $q \neq 0$, $\operatorname{Re} q \geq 0$, such that $F(p) = F(p+q)$ for all p with $\operatorname{Re} p > s_0$. Thus,

$$\int_0^\infty f(t) \exp(-pt)\, dt = \int_0^\infty f(t) \exp(-(p+q)t)\, dt,$$

so it follows that

$$\int_0^\infty f(t)[1 - \exp(-qt)] \exp(-pt)\, dt = 0, \quad \operatorname{Re} p > s_0$$

i.e. $\mathcal{L}\{f(t)[1 - \exp(-qt)]\} = 0$. Thus, from Theorem 6.14, $f(t)[1 - \exp(-qt)]$ is zero almost everywhere, so $f(t) = 0$ a.e., which implies that $F(p) \equiv 0$, a contradiction. Hence $F(p)$ cannot be a periodic function. □

Remark 6.4 Since the complex function $h_a(p) = \exp(-ap)$, $a > 0$ is a periodic function (of period $q = 2\pi i/a$, for example), by Theorem 6.15 it cannot be the image of an original function. We shall see later that $h_a(p)$ is the image of the Dirac Delta function $\delta_a(t)$ through an extended Laplace transform (see Sect. 6.5).

To compute the inverse transform of a function we can use the properties of the Laplace transform presented in Sect. 6.2.

Example 6.16 Let $a < b$. Find the inverse transform of the function

$$F(p) = \log \frac{p+a}{p+b}, \quad \operatorname{Re} p > -a.$$

Here $\log z = \log |z| + i \arg z$, i.e. the principal branch of the multivalued complex logarithmic function.

By Theorem 6.1 one can write:

$$F'(p) = \frac{1}{p+a} - \frac{1}{p+b} \stackrel{(6.4)}{=} -\mathcal{L}\{tf(t)\}(p). \quad (6.4)$$

Thus,

$$\mathcal{L}^{-1}\left\{\log\frac{p+a}{p+b}\right\} = -\frac{1}{t}\left[\mathcal{L}^{-1}\left\{\frac{1}{p+a}\right\} - \mathcal{L}^{-1}\left\{\frac{1}{p+b}\right\}\right] \stackrel{(6.11)}{=} \tag{6.11}$$

$$\frac{\exp(-bt) - \exp(-at)}{t}.$$

Since $\lim_{t\searrow 0}\frac{\exp(-bt)-\exp(-at)}{t} = a - b$, we see that

$$f(t) = \eta(t)\frac{\exp(-bt) - \exp(-at)}{t} = \begin{cases} \frac{\exp(-bt) - \exp(-at)}{t}, & t > 0 \\ 0, & t < 0 \end{cases} \tag{6.46}$$

is an original function (with the abscissa of convergence $s_0 = -a$. Note that $f(0)$ can be any real number (since f needs to be only *piecewise* continuous). Anyway, here, as well as in the next examples, whenever we find the inverse Laplace transform, $f(t) = \mathcal{L}^{-1}\{F(p)\}$, we know that the original function is actually $f(t)\eta(t)$, even if we do not write it explicitly. We write

$$\mathcal{L}^{-1}\left\{\log\frac{p+a}{p+b}\right\} = \frac{\exp(-bt) - \exp(-at)}{t}, \tag{6.47}$$

but we mean that $\mathcal{L}^{-1}\left\{\log\frac{p+a}{p+b}\right\}$ is defined by (6.46).

Example 6.17 Find the inverse Laplace transform of the function $F(p) = \frac{p}{p^2+4p+6}$ which is analytical in the domain Re $p > -2$.

As we did in Example 6.6, we write the fraction $F(p)$ in the form

$$\frac{p}{p^2+4p+6} = \frac{p}{(p+2)^2+2} = \frac{p+2}{(p+2)^2+2} - \sqrt{2}\cdot\frac{\sqrt{2}}{(p+2)^2+2}$$

and use the formulas (6.23) and (6.24) to write

$$\frac{p+2}{(p+2)^2+2} = \mathcal{L}\{\exp(-2t)\cos\sqrt{2}t\},$$

$$\frac{\sqrt{2}}{(p+2)^2+2} = \mathcal{L}\{\exp(-2t)\sin\sqrt{2}t\}.$$

Thus,

$$\frac{p}{p^2+4p+6} = \mathcal{L}\{\exp(-2t)\cos\sqrt{2}t - \sqrt{2}\exp(-2t)\sin\sqrt{2}t\},$$

6.3 Inverse Laplace Transform

and we can write

$$f(t) = \mathcal{L}^{-1}\left\{\frac{p}{p^2+4p+6}\right\} = \exp(-2t)\cos\sqrt{2}t - \sqrt{2}\exp(-2t)\sin\sqrt{2}t.$$

The same method can be used for any rational function (a quotient of two polynomials) $F(p)$ such that $\lim\limits_{|p|\to\infty} F(p) = 0$ (i.e. the degree of the denominator is greater than the degree of the numerator). To find the inverse transform of a rational function, we write its partial fraction decomposition, find the inverse transform of each fraction and then use the linearity of the Laplace transform.

Example 6.18 Let

$$F(p) = \frac{1}{p^2+p+1} + \frac{3p+1}{p^2+5} + \frac{2}{(p-9)^3} + \frac{1}{3p+5}.$$

It is an analytic function for $\operatorname{Re} p > 9$. Let us compute separately the original function for each term of $F(p)$.

$$\frac{1}{p^2+p+1} = \frac{1}{\left(p+\frac{1}{2}\right)^2 + \frac{3}{4}} = \frac{2}{\sqrt{3}}\frac{\frac{\sqrt{3}}{2}}{\left(p+\frac{1}{2}\right)^2 + \frac{3}{4}} \stackrel{(6.24)}{=}$$

$$= \mathcal{L}\left\{\frac{2}{\sqrt{3}}\exp\left(-\frac{1}{2}t\right)\sin\frac{\sqrt{3}}{2}t\right\}.$$

$$\frac{3p+1}{p^2+5} = 3\frac{p}{p^2+5} + \frac{1}{\sqrt{5}}\frac{\sqrt{5}}{p^2+5} \stackrel{(6.23)}{\underset{(6.24)}{=}}$$

$$= \mathcal{L}\left\{3\cos\sqrt{5}t + \frac{1}{\sqrt{5}}\sin\sqrt{5}t\right\}.$$

By the formula (6.8) we know that $\mathcal{L}\{t^2\}(p) = \dfrac{2}{p^3}$. From Theorem 6.5 we obtain

$$\frac{2}{(p-9)^3} = \mathcal{L}\left\{\exp(9t)\cdot t^2\right\}.$$

Finally,

$$\frac{1}{3p+5} = \frac{1}{3}\cdot\frac{1}{p+\frac{5}{3}} \stackrel{(6.11)}{=} \frac{1}{3}\mathcal{L}\left\{\exp\left(-\frac{5}{3}t\right)\right\}.$$

At last we use the linearity of the Laplace transform to find $f(t) = \mathcal{L}^{-1}\{F(p)\}(t)$:

$$f(t) = \frac{2}{\sqrt{3}} \exp\left(-\frac{1}{2}t\right) \sin \frac{\sqrt{3}}{2}t + 3\cos\sqrt{5}t + \frac{1}{\sqrt{5}} \sin\sqrt{5}t +$$

$$+ \exp(9t) \cdot t^2 + \frac{1}{3} \exp\left(-\frac{5}{3}t\right).$$

If the partial fraction decomposition is too complicated, then one can apply the following theorem, based on the calculus of residues (see Sect. 11.2 in Chap. 11).

Theorem 6.16 *Let* $F(p) = \frac{P(p)}{Q(p)}$ *be a rational function such that* $P(z), Q(z) \in \mathbb{C}[z]$ *have no common factor and* $\deg P < \deg Q$. *Let* $p_1, p_2, \ldots, p_m$ *be the roots of the denominator* $Q(p)$ *(the poles of the function* $F(p)$*) and* $s_0 = \max\limits_{k=1,\ldots,m} \operatorname{Re} p_k$. *Then* $F(p)$ *is analytical in the right half-plane* $\operatorname{Re} p > s_0$ *and it satisfies the conditions from Theorem 6.13. Its inverse Laplace transform* $f(t) = \mathcal{L}^{-1}\{F(p)\}$ *is given by the formula*

$$f(t) = \mathcal{L}^{-1}\{F(p)\} = \sum_{k=1}^{m} \operatorname{Res}\left[F(p)\exp(pt), p_k\right]. \tag{6.48}$$

Proof By the Mellin formula (6.43), for any fixed $s > s_0$ we have:

$$f(t) = \frac{\exp(st)}{2\pi} \int_{-\infty}^{\infty} G(\xi) \exp(i\xi t)\, d\xi,$$

where $G(\xi) = F(s + i\xi)$ is also a rational function whose poles, $\xi_k = i(s - p_k)$, $k = 1, \ldots, m$, are in the upper half-plane ($\operatorname{Im} \xi_k > 0$, for all $k = 1, \ldots, m$).

By Theorem 11.35 we know that

$$\int_{-\infty}^{\infty} G(\xi) \exp(i\xi t)\, d\xi = 2\pi i \sum_{k=1}^{m} \operatorname{Res}\left[G(\xi)\exp(i\xi t), \xi_k\right].$$

It remains to prove that, for all $k = 1, 2, \ldots, m$,

$$\operatorname{Res}\left[F(p)\exp(pt), p_k\right] = i \cdot \exp(st) \operatorname{Res}\left[G(\xi)\exp(i\xi t), \xi_k\right] \tag{6.49}$$

6.3 Inverse Laplace Transform

If $p_k = s + i\xi_k$ is a pole of the function $F(p)$ then, for $r > 0$ sufficiently small (such that p_k is the only root of $Q(p)$ inside the circle $|p - p_k| = r$), we have:

$$\mathrm{Res}\left[F(p)\exp(pt), p_k\right] = \frac{1}{2\pi i} \int_{|p-p_k|=r} F(p)\exp(pt)\,dp \quad \stackrel{p=s+i\xi}{=}$$

$$= \frac{1}{2\pi} \int_{|\xi-\xi_k|=r} F(s+i\xi)\exp(st)\exp(i\xi t)\,d\xi$$

so the Eq. (6.49) follows and the theorem is proved. □

Recall that, if z_0 is a pole of order n of the function $\varphi(z)$, then

$$\mathrm{Res}[\varphi(z), z_0] = \frac{1}{(n-1)!} \left[(z-z_0)^n \varphi(z)\right]^{(n-1)} \bigg|_{z=z_0}. \tag{6.50}$$

This formula becomes very simple when the multiplicity of the pole z_0 is $n = 1$:

$$\mathrm{Res}[\varphi(z), z_0] = ((z-z_0)\varphi(z))\big|_{z=z_0}. \tag{6.51}$$

Example 6.19 Let us find the inverse Laplace transform of the function

$$F(p) = \frac{1}{(p-1)(p-2)(p-3)\ldots(p-10)(p-11)}.$$

To find the partial fraction decomposition of $F(p)$ seems to be a difficult way. Let us apply the formula (6.48) instead. The poles are the zeros of the denominator of $F(p)$, $p_1 = 1, p_2 = 2, \ldots, p_{11} = 11$. We use the formula (6.51) to compute the residue of the function $G(p) = F(p)\exp(pt)$ at $p_k = k$, $k = 1, 2, \ldots, 11$:

$$\mathrm{Res}\left[F(p)\exp(pt), k\right] = \frac{\exp(kt)}{(k-1)\ldots(k-k+1)(k-k-1)\ldots(k-11)} =$$

$$= \frac{\exp(kt)}{(-1)^{11-k}(k-1)!(11-k)!} = \frac{\exp t}{10!}\binom{10}{k-1}(-1)^{10-(k-1)}(\exp t)^{k-1}.$$

Thus,

$$\mathcal{L}^{-1}\{F(p)\} = \frac{\exp t}{10!} \sum_{k=0}^{10} \binom{10}{k}(-1)^{10-k}(\exp t)^k = \frac{\exp t}{10!}(\exp t - 1)^{10}.$$

Remark 6.5 If the rational function $F(p) = \frac{P(p)}{Q(p)}$ (with deg P < deg Q) has only simple poles ($Q(p)$ has only simple zeros), then the formula (6.48) becomes

$$\mathcal{L}^{-1}\{F(p)\} = \sum_{k=1}^{m} \frac{P(p_k)}{Q'(p_k)} \exp(p_k t). \qquad (6.52)$$

Indeed, we can assume that $Q(p) = (p - p_1)\ldots(p - p_m)$, where p_i, $i = 1, 2, \ldots, m$ are mutually distinct. Since

$$Q'(p) = \sum_{k=1}^{m} (p - p_1)\ldots(p - p_{k-1})(p - p_{k+1})\ldots(p - p_m),$$

for $p = p_k$ we have $Q'(p_k) = (p_k - p_1)\ldots(p_k - p_{k-1})(p_k - p_{k+1})\ldots(p_k - p_m)$ and we see that

$$\text{Res}\left[F(p)\exp(pt), p_k\right] \stackrel{(6.51)}{=} \left[(p - p_k)\frac{P(p)}{Q(p)}\exp(pt)\right]_{p=p_k} =$$

$$= \left[\frac{P(p)}{(p - p_1)\ldots(p - p_{k-1})(p - p_{k+1})\ldots(p - p_m)}\exp(pt)\right]_{p=p_k} =$$

$$= \frac{P(p_k)\exp(p_k t)}{(p_k - p_1)\ldots(p_k - p_{k-1})(p_k - p_{k+1})\ldots(p_k - p_m)} = \frac{P(p_k)}{Q'(p_k)}\exp(p_k t).$$

Example 6.20 Let us compute $\mathcal{L}^{-1}\left\{\frac{1}{p^3-1}\right\}$ using (6.52).
We can write:

$$Q(p) = p^3 - 1 = (p - 1)(p - \omega)(p - \omega^2),$$

where ω is a root of the equation $p^2 + p + 1 = 0$, say $\omega = -\frac{1}{2} + i\frac{\sqrt{3}}{2}$, so the poles of $F(p)\exp(pt)$ are $p_1 = 1$, $p_2 = \omega$ and $p_3 = \omega^2$. Since $Q'(p) = 3p^2$, we have:

$$\mathcal{L}^{-1}\left\{\frac{1}{p^3-1}\right\} = \sum_{k=1}^{3} \frac{1}{Q'(p_k)}\exp(p_k t) = \sum_{k=1}^{3} \frac{1}{3p_k^2}\exp(p_k t) =$$

$$= \frac{1}{3}\left[\exp(t) + \omega\exp(\omega t) + \omega^2\exp(\omega^2 t)\right].$$

6.3 Inverse Laplace Transform

Since $\omega^2 = -\frac{1}{2} - i\frac{\sqrt{3}}{2}$, $\exp(\omega t) = \exp\left(-\frac{1}{2}t\right)\left[\cos\left(\frac{\sqrt{3}}{2}t\right) + i\sin\left(\frac{\sqrt{3}}{2}t\right)\right]$ and $\exp(\omega^2 t) = \exp\left(-\frac{1}{2}t\right)\left[\cos\left(\frac{\sqrt{3}}{2}t\right) - i\sin\left(\frac{\sqrt{3}}{2}t\right)\right]$, we finally obtain:

$$\mathcal{L}^{-1}\left\{\frac{1}{p^3-1}\right\} = \frac{1}{3}\exp(t) - \frac{1}{3}\exp\left(-\frac{1}{2}t\right)\left[\cos\left(\frac{\sqrt{3}}{2}t\right) - \sqrt{3}\sin\left(\frac{\sqrt{3}}{2}t\right)\right].$$

Prove that the same result is obtained by using the method of partial fractions decomposition:

$$\frac{1}{p^3-1} = \frac{1}{3}\left[\frac{1}{p-1} - \frac{p+2}{p^2+p+1}\right].$$

Example 6.21 Find the inverse Laplace transform of the function

$$F(p) = \frac{1}{(p^2+a^2)^2}, \quad p \in \mathbb{C}, \ a > 0.$$

We can see that $F(p)$ is analytic on the domain $D_0 = \{p \in \mathbb{C} : \operatorname{Re} p > 0\}$. We calculate its inverse transform using Theorem 6.16. The poles of the function $F(p)$ are $z_1 = ai$ and $z_2 = -ai$, so

$$f(t) = \mathcal{L}^{-1}\{F(p)\} = \operatorname{Res}\left[\frac{\exp(pt)}{(p^2+a^2)^2}, ai\right] + \operatorname{Res}\left[\frac{\exp(pt)}{(p^2+a^2)^2}, -ai\right].$$

Both roots of the denominator are poles of order 2, so, by the formula (6.50), we have:

$$\operatorname{Res}\left[\frac{\exp(pt)}{(p^2+a^2)^2}, ai\right] = \left[\frac{\exp(pt)}{(p+ai)^2}\right]'_{p=ai} = \left(-\frac{i}{4a^3} - \frac{t}{4a^2}\right)\exp(iat),$$

$$\operatorname{Res}\left[\frac{\exp(pt)}{(p^2+a^2)^2}, -ai\right] = \left[\frac{\exp(pt)}{(p-ai)^2}\right]'_{p=-ai} = \left(\frac{i}{4a^3} - \frac{t}{4a^2}\right)\exp(-iat),$$

Hence,

$$f(t) = \frac{1}{2a^3}\sin at - \frac{t}{2a^2}\cos at. \tag{6.53}$$

Let us now use some basic formulas from Sect. 6.2 in order to verify that

$$F(p) = \mathcal{L}\{f(t)\}(p) = \frac{1}{(p^2+a^2)^2}$$

for $f(t)$ defined by (6.53). First of all we use the linearity of $\mathcal{L}$:

$$\mathcal{L}\{f(t)\} = \frac{1}{2a^3}\mathcal{L}\{\sin at\} - \frac{1}{2a^2}\mathcal{L}\{t\cos at\}. \tag{6.54}$$

We know that $\mathcal{L}\{\sin at\}(p) = \frac{a}{p^2+a^2}$ (see (6.14)) and $\mathcal{L}\{\cos at\}(p) = \frac{p}{p^2+a^2}$ (see (6.13)). From Theorem 6.1, formula (6.4), we have:

$$\mathcal{L}\{t\cos at\} = -[\mathcal{L}\{\cos at\}(p)]' = -\left[\frac{p}{p^2+a^2}\right]' = \frac{p^2-a^2}{(p^2+a^2)^2}.$$

Thus, coming back to (6.54), we finally obtain:

$$\mathcal{L}\{f(t)\}(p) = \frac{1}{2a^3}\cdot\frac{a}{p^2+a^2} - \frac{1}{2a^2}\cdot\frac{p^2-a^2}{(p^2+a^2)^2} = \frac{1}{(p^2+a^2)^2}.$$

Theorem 6.17 (Complex Convolution (see also [12], 31.14)) *Let f_1, f_2 be two original functions with the abscissas of convergence s_1 and s_2, respectively, and let $F_1(p) = \mathcal{L}\{f_1(t)\}(p)$, $F_2(p) = \mathcal{L}\{f_2(t)\}(p)$ be their Laplace transforms. Then, for all $p \in \mathbb{C}$ with $\operatorname{Re} p > s_1 + s_2$,*

$$\int_{x-i\infty}^{x+i\infty} F_1(q)F_2(p-q)\,dq = 2\pi i \int_0^\infty f_1(t)f_2(t)\exp(-pt)\,dt = \tag{6.55}$$

$$= 2\pi i \mathcal{L}\{f_1 f_2\}(p) =$$

$$= \int_{y-i\infty}^{y+i\infty} F_2(q)F_1(p-q)\,dq,$$

for any x, y such that $s_1 < x < \operatorname{Re} p - s_2$, and $s_2 < y < \operatorname{Re} p - s_1$.

Proof Let $p \in \mathbb{C}$ with $\operatorname{Re} p > s_1 + s_2$ and $x \in (s_2, \operatorname{Re} p - s_2)$. Then, for any $q = x + i\omega$, there exist $F_1(q)$ ($\operatorname{Re} q = x > s_1$ and $F_2(p-q)$ ($\operatorname{Re}(p-q) = \operatorname{Re} p - x > s_2$ and we have:

$$\int_{x-i\infty}^{x+i\infty} F_1(q)F_2(p-q)\,dq = \int_{x-i\infty}^{x+i\infty} F_1(q)\left[\int_0^\infty f_2(t)\exp(-(p-q)t)\,dt\right]dq =$$

$$= \int_0^\infty \left[f_2(t)\exp(-pt)\int_{x-i\infty}^{x+i\infty} F_1(q)\exp(qt)\,dq\right]dt \stackrel{(6.44)}{=}$$

$$= 2\pi i\int_0^\infty f_2(t)\exp(-pt)f_1(t)\,dt =$$

$$= 2\pi i\mathcal{L}\{f_1 f_2\}(p).$$

6.3 Inverse Laplace Transform

The last equality can be proved in a similar way. □

In practice the above formula (6.55) is used in the following form:

$$\mathcal{L}\{f_1 f_2\}(p) = \frac{1}{2\pi i} \int_{x-i\infty}^{x+i\infty} F_1(q) F_2(p-q) \, dq \tag{6.56}$$

for all p with $\operatorname{Re} p \geq s_1 + s_2$ and for any x, $s_1 \leq x \leq \operatorname{Re} p - s_2$.

Example 6.22 Let us calculate the Laplace transform of the original function

$$f(t) = t^n \sin(\omega t), \quad n \in \mathbb{N}, \ \omega > 0.$$

We know that $F_1(p) = \frac{n!}{p^{n+1}}$, $\operatorname{Re} p > 0$, is the Laplace transform of the function $f_1(t) = t^n$ (see formula (6.8)) and $F_2(p) = \frac{\omega}{p^2+\omega^2}$, $\operatorname{Re} p > 0$, is the Laplace transform of $f_2(t) = \sin \omega t$ (see formula (6.14)). By Theorem 6.17 one can write, for $0 < x < \operatorname{Re} p$:

$$\mathcal{L}\{t^n \sin \omega t\}(p) = \frac{n!\omega}{2\pi i} \int_{x-i\infty}^{x+i\infty} \frac{1}{q^{n+1}} \cdot \frac{1}{(p-q)^2 + \omega^2} \, dq. \tag{6.57}$$

It remains to compute the complex integral

$$\int_{x-i\infty}^{x+i\infty} \frac{1}{q^{n+1}} \cdot \frac{1}{(p-q)^2 + \omega^2} \, dq. \tag{6.58}$$

Let $C_R = \{z : |z - x| = R, \operatorname{Re}(z - x) > 0\}$ be the semicircle of radius R with the center at the point $M(x, 0)$ and consider the points $A(x, R)$ and $B(x, R)$ (the limits of the semicircle) and the closed curve $\gamma_R = [AB] \cup C_R$.

The rational function $\varphi(z) = \frac{1}{z^{n+1}} \cdot \frac{1}{(z-p)^2 + \omega^2}$ has two (simple) poles in the right half-plane $\operatorname{Re} z > x$:

$$z_1 = p + i\omega \text{ and } z_2 = p - i\omega.$$

Hence, for $R > |p - x \pm i\omega|$, one can write:

$$\int_{\gamma_R} \varphi(z) \, dz = 2\pi i \left[\operatorname{Res}\left[\varphi(z), p + i\omega\right] + \operatorname{Res}\left[\varphi(z), p - i\omega\right] \right].$$

We have:

$$\int_{\gamma_R} \varphi(z) \, dz = \int_{[AB]} \varphi(z) \, dz + \int_{C_R} \varphi(z) \, dz = \int_{x+iR}^{x-iR} \varphi(z) \, dz + \int_{C_R} \varphi(z) \, dz$$

and it can be shown that

$$\lim_{R\to\infty}\int_{C_R} \varphi(z)dz = 0$$

(see Theorem 11.5 in Sect. 11.2). It follows that

$$\int_{x-i\infty}^{x+i\infty} \frac{1}{q^{n+1}} \cdot \frac{1}{(p-q)^2+\omega^2} dq = -\lim_{R\to\infty}\int_{x+iR}^{x-iR} \varphi(z)dz =$$

$$= -2\pi i \left[\text{Res}\left[\varphi(z), p+i\omega\right] + \text{Res}\left[\varphi(z), p-i\omega\right]\right] \stackrel{(6.51)}{=}$$

$$= -2\pi i \left(\frac{1}{2i\omega(p+i\omega)^{n+1}} - \frac{1}{2i\omega(p-i\omega)^{n+1}}\right) =$$

$$= \frac{\pi}{\omega}\left(\frac{1}{(p-i\omega)^{n+1}} - \frac{1}{(p+i\omega)^{n+1}}\right).$$

By replacing in (6.57) we find:

$$\mathcal{L}\{t^n \sin \omega t\}(p) = \frac{n!}{2i}\left[\frac{1}{(p-i\omega)^{n+1}} - \frac{1}{(p+i\omega)^{n+1}}\right], \quad \text{Re } p > 0$$

The same result could be obtained by applying (6.5):

$$\mathcal{L}\{t^n \sin \omega t\} = (-1)^n \left[\frac{\omega}{p^2+\omega^2}\right]^{(n)} =$$

$$= \frac{(-1)^n}{2i}\left[\frac{1}{p-i\omega} - \frac{1}{p+i\omega}\right]^{(n)} =$$

$$= \frac{n!}{2i}\left[\frac{1}{(p-i\omega)^{n+1}} - \frac{1}{(p+i\omega)^{n+1}}\right].$$

6.4 Solving Linear Differential Equations

Let us consider a linear ordinary differential equation of order n with constant coefficients:

$$a_0 y^{(n)}(t) + a_1 y^{(n-1)}(t) + \ldots + a_{n-1} y'(t) + a_n y(t) = f(t), \tag{6.59}$$

6.4 Solving Linear Differential Equations

where $a_0, \ldots, a_n$ are real numbers and $a_0 \neq 0$. We assume that $t \geq 0$ and $f(t)$ is an original function. We shall prove in the following that the Cauchy problem with the initial conditions:

$$y(0) = y_0, \ldots, y^{(n-1)}(0) = y_{n-1} \qquad (6.60)$$

has a unique solution $y(t)$ in $\mathcal{O}$ (or in $\overline{\mathcal{O}}$—see Remark 6.2) and we shall give an algorithm to find it by using the Laplace transform $\mathcal{L}$.

We apply the linear operator $\mathcal{L}$ to the equality (6.59) and obtain:

$$a_0 \mathcal{L}\{y^{(n)}(t)\}(p) + a_1 \mathcal{L}\{y^{(n-1)}(t)\}(p) + \cdots + a_{n-1}\mathcal{L}\{y'(t)\}(p) + a_n \mathcal{L}\{y(t)\}(p) = \qquad (6.61)$$

$$= \mathcal{L}\{f(t)\}(p) \stackrel{def}{=} F(p).$$

We denote $\mathcal{L}\{y(t)\}(p) = Y(p)$. We apply Theorem 6.6 and use the initial conditions (6.60) to write the formula (6.26):

$$\mathcal{L}\{y^{(k)}(t)\} = p^k Y(p) - p^{k-1} y_0 - p^{k-2} y_1 - \ldots - p y_{k-2} - y_{k-1}, \qquad (6.62)$$

$k = n, n-1, \ldots 0$. Thus,

$$\mathcal{L}\{y^{(n)}(t)\}(p) = p^n Y(p) - p^{n-1} y_0 - p^{n-2} y_1 - \ldots - p y_{n-2} - y_{n-1}$$

$$\mathcal{L}\{y^{(n-1)}(t)\}(p) = p^{n-1} Y(p) - p^{n-2} y_0 - p^{n-3} y_1 - \ldots - p y_{n-3} - y_{n-2}$$

$$\vdots$$

$$\mathcal{L}\{y'(t)\} = pY(p) - y_0$$

$$\mathcal{L}\{y(t)\} = Y(p).$$

We multiply the first equality by a_0, the second by $a_1, \ldots,$ the last by a_n and add everything to find:

$$F(p) = Y(p)P(p) - Q(p)$$

where $P(p)$ is a polynomial of degree n ($a_0 \neq 0$) and $Q(p)$ is a polynomial of degree at most $n - 1$:

$$P(p) = a_0 p^n + a_1 p^{n-1} + \ldots + a_{n-1} p + a_n,$$

$$Q(p) = a_0 y_0 p^{n-1} + (a_0 y_1 + a_1 y_0) p^{n-2} + (a_0 y_2 + a_1 y_1 + a_2 y_0) p^{n-3} + \ldots$$

$$+ a_0 y_{n-1} + a_1 y_{n-2} + \ldots + a_{n-1} y_0.$$

Hence we obtain:

$$Y(p) = \frac{F(p) + Q(p)}{P(p)}. \tag{6.63}$$

Now, we have to find the inverse Laplace transform, $y(t) = \mathcal{L}^{-1}\{Y(p)\}$.

If $F(p)$ is a rational function (with the degree of the numerator less than the degree of the denominator), then $Y(p)$ is also a rational function (with the same property). Using partial fraction decomposition over the field $\mathbb{C}$, everything reduces to the computation of $\mathcal{L}^{-1}\left\{\frac{1}{(p-\lambda)^k}\right\}$, where λ is a complex number and k is a positive integer. By the formula (6.8) and Theorem 6.5 we obtain:

$$\mathcal{L}\left\{\exp(\lambda t) t^{k-1}\right\}(p) = \frac{(k-1)!}{(p-\lambda)^k}, \tag{6.64}$$

or, equivalently,

$$\mathcal{L}^{-1}\left\{\frac{1}{(p-\lambda)^k}\right\}(t) = \exp(\lambda t)\frac{t^{k-1}}{(k-1)!}. \tag{6.65}$$

We also notice that $F(p)$ is a rational function if and only if $f(t)$ can be written in the form

$$f(t) = \sum_{j=1}^{N} P_j(t) \exp(\lambda_j t), \tag{6.66}$$

where $\lambda_j \in \mathbb{C}$ and $P_j(t)$ are polynomials with complex coefficients.

Example 6.23 Let us solve the following Cauchy problem by using the Laplace transform:

$$y''' - 7y'' + 14y' - 8y = 2\exp(3t), \tag{6.67}$$

$$y(0) = y'(0) = 0, \qquad y''(0) = 1.$$

We apply the Laplace transform to Eq. (6.67):

$$\mathcal{L}\{y'''\} - 7\mathcal{L}\{y''\} + 14\mathcal{L}\{y'\} - 8\mathcal{L}\{y\} = 2\mathcal{L}\{\exp(3t)\} = \frac{2}{p-3}.$$

6.4 Solving Linear Differential Equations

We denote $\mathcal{L}\{y\} = Y(p)$ and use (6.62) to write the Laplace transforms of the derivatives:

$$Y(p)(p^3 - 7p^2 + 14p - 8) - 1 = \frac{2}{p-3}.$$

Hence

$$Y(p)(p-1)(p-2)(p-4) = \frac{p-1}{p-3},$$

and so

$$Y(p) = \frac{1}{(p-2)(p-3)(p-4)}.$$

It remains to calculate the inverse Laplace transform, $y(t) = \mathcal{L}^{-1}\{Y(p)\}$. The partial fraction decomposition decomposition of the rational function $Y(p)$ is:

$$Y(p) = \frac{1}{2} \cdot \frac{1}{p-2} - \frac{1}{p-3} + \frac{1}{2} \cdot \frac{1}{p-4}$$

so

$$y(t) = \frac{1}{2} \exp(2t) - \exp(3t) + \frac{1}{2} \exp(4t)$$

is the (unique) solution of our Cauchy problem.

An important feature of the Laplace transform method is that it can be used for equations where the free term $f(t)$ is a discontinuous function, as in the next example.

Example 6.24 Solve the Cauchy problem:

$$y'' + y = k\eta(t-a) = \begin{cases} 0, & t < a, \\ k, & t \geq a, \end{cases} \quad y(0) = 0, \ y'(0) = 1,$$

where k and $a > 0$ are given constants.

Let $Y(p) = \mathcal{L}\{y(t)\}$. We apply the Laplace transform to the differential equation and obtain:

$$p^2 Y(p) - 1 + Y(p) = k\,\mathcal{L}\{\eta(t-a)\} = k\exp(-ap) \cdot \frac{1}{p}.$$

It follows that

$$Y(p) = \frac{1}{p^2+1} + k\exp(-ap) \cdot \frac{1}{p(p^2+1)} =$$

$$= \frac{1}{p^2+1} + k\exp(-ap)\left(\frac{1}{p} - \frac{p}{p^2+1}\right).$$

By Theorem 6.4 we know that, if $F(p) = \mathcal{L}\{f(t)\}$, then

$$\mathcal{L}^{-1}\{\exp(-ap)F(p)\}(t) = f(t-a), \tag{6.68}$$

hence

$$y(t) = \mathcal{L}^{-1}\{Y(p)\} = \eta(t)\sin t + k\eta(t-a)(1 - \cos(t-a)) =$$

$$= \begin{cases} \sin t, & t \in [0, a), \\ \sin t + k(1 - \cos(t-a)), & t \geq a, \end{cases}$$

is the solution of our Cauchy problem.

Corollary 6.1 of Differentiation Theorem 6.6 enables us to use the Laplace transform for solving linear equations with *polynomial coefficients*, as in the next example.

Example 6.25 The following differential equation

$$ty'' + (1-t)y' + ny = 0, \tag{6.69}$$

$n = 0, 1, \ldots$ is known as **Laguerre's differential equation**. We solve it by using the Laplace transform. Let $Y = Y(p)$ be the image of $y(t)$, $t \geq 0$. By the formula (6.27) we can write:

$$\mathcal{L}\{ty''\}(p) = -p^2 Y' - 2pY + y(0),$$

$$\mathcal{L}\{(1-t)y'\}(p) = \mathcal{L}\{y'\}(p) - \mathcal{L}\{ty'\}(p) =$$

$$= pY - y(0) - [-Y - pY'] = pY' + (p+1)Y - y(0).$$

Thus, the image of the equation is

$$(p - p^2)Y' + (n+1-p)Y = 0,$$

6.4 Solving Linear Differential Equations

a first order linear differential equation in Y. We solve it (as a separable equation):

$$\frac{dY}{Y} = \left(\frac{n}{p-1} - \frac{n+1}{p}\right) dp,$$

so we obtain by integration

$$Y(p) = C\frac{(p-1)^n}{p^{n+1}},$$

where C is a constant. Now, we apply the inverse Laplace transform and find

$$y(t) = C\mathcal{L}^{-1}\left\{\frac{(p-1)^n}{p^{n+1}}\right\} = C\sum_{k=0}^{n}(-1)^k\binom{n}{k}\mathcal{L}^{-1}\left\{\frac{1}{p^{k+1}}\right\},$$

so we obtain that $y(t) = CL_n(t)$, where $L_n(t)$ is the *Laguerre polynomial* of order n,

$$L_n(t) = \sum_{k=0}^{n}\frac{(-1)^k}{k!}\binom{n}{k}t^k.$$

We prove that Laguerre polynomials satisfy Rodrigues formula (6.70)

$$L_n(t) = \frac{\exp(t)}{n!}\frac{d^n}{dt^n}\left[t^n \exp(-t)\right] \qquad (6.70)$$

for all $n = 1, 2, \ldots$.

By Eq. (6.8) and using the displacement formula (6.22) we get

$$\mathcal{L}\{t^n \exp(-t)\}(p) = \frac{n!}{(p+1)^{n+1}}.$$

Since all the derivatives of the function $t^n \exp(-t)$ (up to the order $(n-1)$) are zero at $t = 0$, by Theorem 6.6 we obtain:

$$\mathcal{L}\left\{\frac{d^n}{dt^n}\left[t^n \exp(-t)\right]\right\}(p) = \frac{p^n n!}{(p+1)^{n+1}}.$$

By applying again the displacement formula (6.22) we find

$$\mathcal{L}\left\{\frac{\exp(t)}{n!}\frac{d^n}{dt^n}\left[t^n \exp(-t)\right]\right\}(p) = \frac{(p-1)^n}{p^{n+1}} = \mathcal{L}\{L_n(t)\}$$

and the Rodrigues formula (6.70) follows.

But the Eq. (6.69) is a linear second-order equation, so it has two linearly independent solutions. We denote by $y_1 = L_n(t)$ the solution we found above and search for another solution (which is not an original function) of the form $y_2 = y_1 \cdot u$, where $u = u(t)$ is a non-constant function. We compute the derivatives of y_2 and replace them in Eq. (6.69). Taking into account that y_1 is also a solution of (6.69), we find:

$$u'' y_1 + u' \left(2y_1' + \frac{1-t}{t} y_1 \right) = 0.$$

We denote by v the derivative of the non-constant function u, $v = u'$, and obtain the following first order equation with separable variables:

$$v' + v \left(\frac{2y_1'}{y_1} + \frac{1-t}{t} \right) = 0.$$

A solution of this equation is

$$v(t) = \frac{\exp(t)}{t \cdot y_1^2} = \frac{(n!)^2}{t \exp t} \left[\frac{d^n}{dt^n} (t^n \exp(-t)) \right]^{-2},$$

which is not identical zero. So u is not a constant function. Thus,

$$y_2(t) = y_1(t) \int v(t)\,dt$$

is a new solution of (6.69), linearly independent to y_1.

We can also use the Laplace transform for solving linear systems of differential equations. First of all we apply the operator $\mathcal{L}$ to the equations of the system, then solve the algebraic linear system which is obtained and, finally, we calculate the inverse Laplace transforms to obtain the solution.

Example 6.26 Find the general solution of the following system of differential equations:

$$\begin{cases} x' = 2x + 4y \\ y' = 4x + 2y \end{cases}. \tag{6.71}$$

We denote $\mathcal{L}\{x(t)\}(p) = X(p)$, $\mathcal{L}\{y(t)\}(p) = Y(p)$ and apply $\mathcal{L}$ to both equations in (6.71):

$$\begin{cases} pX(p) - x(0) = 2X(p) + 4Y(p) \\ pY(p) - y(0) = 4X(p) + 2Y(p) \end{cases}. \tag{6.72}$$

6.4 Solving Linear Differential Equations

We denote $x(0) = C_1$ and $y(0) = C_2$, two arbitrary constants and solve the algebraic linear system (6.72) for $X(p)$ and $Y(p)$:

$$\begin{cases} X(p) = C_1 \dfrac{p-2}{(p-2)^2 - 16} + C_2 \dfrac{4}{(p-2)^2 - 16} \\ Y(p) = C_1 \dfrac{4}{(p-2)^2 - 16} + C_2 \dfrac{p-2}{(p-2)^2 - 16} \end{cases}. \qquad (6.73)$$

To find the original functions $x(t) = \mathcal{L}^{-1}\{X(p)\}$ and $y(t) = \mathcal{L}^{-1}\{Y(p)\}$, we need to find $\mathcal{L}^{-1}\left\{\dfrac{p-2}{(p-2)^2-16}\right\}$ and $\mathcal{L}^{-1}\left\{\dfrac{1}{(p-2)^2-16}\right\}$. Since the functions $\dfrac{p-2}{(p-2)^2-16}$ and $\dfrac{4}{(p-2)^2-16}$ are obtained from $\dfrac{p}{p^2-16}$ and $\dfrac{4}{p^2-16}$, respectively, by a translation of the variable p with $p_0 = 2$, we can use Theorem 6.5 and find:

$$\mathcal{L}^{-1}\left\{\dfrac{p-2}{(p-2)^2-16}\right\}(t) = \exp(2t)f(t),$$

where

$$f(t) = \mathcal{L}^{-1}\left\{\dfrac{p}{p^2-16}\right\}(t) = \cosh(4t) = \dfrac{1}{2}\left[\exp(4t) + \exp(-4t)\right],$$

and

$$\mathcal{L}^{-1}\left\{\dfrac{4}{(p-2)^2-16}\right\}(t) = \exp(2t)g(t),$$

where

$$g(t) = \mathcal{L}^{-1}\left\{\dfrac{4}{p^2-16}\right\}(t) = \sinh(4t) = \dfrac{1}{2}\left[\exp(4t) - \exp(-4t)\right].$$

Thus,

$$\mathcal{L}^{-1}\left\{\dfrac{p-2}{(p-2)^2-16}\right\}(t) = \dfrac{1}{2}\left[\exp(6t) + \exp(-2t)\right]$$

and

$$\mathcal{L}^{-1}\left\{\dfrac{4}{(p-2)^2-16}\right\}(t) = \dfrac{1}{2}\left[\exp(6t) - \exp(-2t)\right].$$

Finally, coming back to (6.73), we find:

$$x(t) = \dfrac{C_1}{2}\left[\exp(6t) + \exp(-2t)\right] + \dfrac{C_2}{2}\left[\exp(6t) - \exp(-2t)\right]$$

and

$$y(t) = \frac{C_1}{2}\left[\exp(6t) - \exp(-2t)\right] + \frac{C_2}{2}\left[\exp(6t) + \exp(-2t)\right],$$

which is the general solution of our system.

In Chap. 7 we shall apply the Laplace transform to solve the partial differential equation of the finite vibrating string.

6.5 The Dirac Delta Function

The *Dirac delta function* was introduced by the English physicist Paul Dirac (one of the founders of the quantum mechanics) for modeling physical phenomena of impulsive nature, characterized by the sudden action of a force of large magnitude (like a lightening strike, a mechanical system hit by a hammerblow, a billiard ball being struck, etc.). It is not a *function* in the ordinary sense, it is a *generalized function* or a *distribution* (the theory of distributions was created later by the French mathematician Laurent Schwartz).

Let $a \geq 0$ be a fixed real number and consider the function defined for all $t > 0, \tau > 0$:

$$f_a(t, \tau) = \begin{cases} \dfrac{1}{\tau}, & \text{if } a \leq t \leq a + \tau \\ 0, & \text{otherwise.} \end{cases} \quad (6.74)$$

For a fixed $\tau > 0$, we can think of $f_a(t, \tau)$ as representing a force of magnitude $\frac{1}{\tau}$ acting in the interval of time $a \leq t \leq a + \tau$. The *impulse* of this force (the integral of the force over the time interval) is

$$\int_0^\infty f_a(t, \tau)\,dt = \int_a^{a+\tau} \frac{1}{\tau}\,dt = 1, \quad (6.75)$$

for any $\tau > 0$. Figure 6.5 shows the graph of the function $y = f_a(t, \tau)$ for different values of τ. As the length of the time interval τ approaches 0, the area of the rectangle (the integral (6.75)) remains constant, equal to 1.

The "function" obtained by taking the limit of $f_a(t, \tau)$ as $\tau \to 0$ is called the *Dirac delta function* (or the *unit impulse function*):

$$\delta_a(t) = \begin{cases} \infty, & \text{if } t = a \\ 0, & \text{otherwise.} \end{cases} \quad (6.76)$$

Although it is not a "true" function, it proved to be a very useful instrument in problems related to impulse phenomena. However, as mentioned above, the Dirac

6.5 The Dirac Delta Function

Fig. 6.5 The functions $f_a(t, \tau)$ as $\tau \to 0$

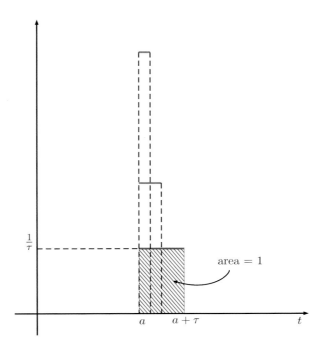

delta function has a rigorous mathematical foundation in the theory of distributions, but this is not a suitable subject for the present book. So we shall make some formal assumptions which allow us to operate on $\delta_a(t)$ as though it were an ordinary function.

One such formal assumption we make is that the limit (as $\tau \to 0$) and the integral can swap (the limit of the integral equals the integral of the limit). Thus, by taking the limit as $\tau \to 0$ in (6.75), we obtain the following property of the Dirac delta function:

$$\int_0^\infty \delta_a(t)\,dt = 1. \tag{6.77}$$

Moreover, let $g(t)$ be a continuous function and $G(t)$ be a primitive of it. Then

$$\int_0^\infty \delta_a(t)g(t)\,dt = \lim_{\tau \to 0} \frac{1}{\tau} \int_a^{a+\tau} g(t)\,dt =$$

$$= \lim_{\tau \to 0} \frac{G(a+\tau) - G(a)}{\tau} = G'(a) = g(a).$$

Thus, we have proved the so-called *sifting* property of δ_a:

$$\int_0^\infty \delta_a(t)g(t)\,dt = g(a). \tag{6.78}$$

Using the formula (6.78), one can compute the Laplace transform of the Dirac delta function $\delta_a(t)$:

$$\mathcal{L}\{\delta_a(t)\}(p) = \int_0^\infty \delta_a(t)\exp(-pt)\,dt = \exp(-ap). \tag{6.79}$$

Consequently, the inverse Laplace transform of the function $\exp(-ap)$ is

$$\mathcal{L}^{-1}\{\exp(-ap)\}(t) = \delta_a(t). \tag{6.80}$$

By Theorem 6.15, the Laplace transform cannot be a periodic function. Obviously, $F(p) = \exp(-ap)$ is a periodic function, but we should not be surprised, since $\delta_a(t)$ is not an original function. So the formulas (6.79) and (6.79) express a *formal extension* of the Laplace transform and of the inverse Laplace transform, respectively. However, they can be used to solve differential equations where the free term contains the Dirac delta function.

Example 6.27 We consider the equation of a damped mass-spring system with the mass $m = 1$, the spring constant $k = 5$ and the damping constant $c = 2$ (see Example 2.15). At the initial moment the displacement is $y(0) = 4$ and the velocity 0. Suppose that the system freely oscillates and receives a unit impulse at time $t = 6$. The Cauchy problem we have to solve is:

$$y'' + 2y' + 5y = \delta_6(t), \quad y(0) = 4, \quad y'(0) = 0. \tag{6.81}$$

We apply the Laplace transform and obtain:

$$\mathcal{L}\{y''\}(p) + 2\mathcal{L}\{y'\}(p) + 5\mathcal{L}\{y\}(p) = \mathcal{L}\{\delta_6(t)\} = \exp(-6p).$$

Denote $\mathcal{L}\{y\}(p) = Y(p)$. Since

$$\mathcal{L}\{y''\}(p) = p^2 Y(p) - py(0) - y'(0) = p^2 Y(p) - 4p,$$

$$\mathcal{L}\{y'\}(p) = pY(p) - y(0) = pY(p) - 4,$$

we obtain:

$$(p^2 + 2p + 5)Y(p) - 4p - 8 = \exp(-6p),$$

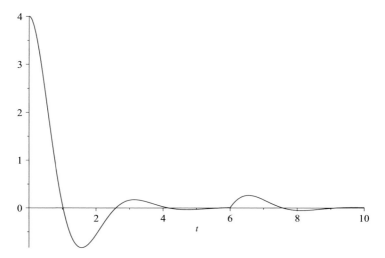

Fig. 6.6 The solution $y(t)$ of the Cauchy problem (6.81)

hence

$$Y(p) = \frac{4p + 8 + \exp(-6p)}{p^2 + 2p + 5} =$$

$$= 4 \cdot \frac{p+1}{(p+1)^2 + 4} + 2 \cdot \frac{2}{(p+1)^2 + 4} + \frac{1}{2} \cdot \exp(-6t) \cdot \frac{2}{(p+1)^2 + 4}.$$

It follows that

$$y(t) = \mathcal{L}^{-1}\{Y(p)\} = 4\exp(-t)\cos(2t)\eta(t) + 2\exp(-t)\sin(2t)\eta(t) +$$

$$+ \frac{1}{2}\exp(-(t-6))\sin(2(t-6))\eta(t-6).$$

The graph of the solution is presented in Fig. 6.6.

We present in Table 6.1 the most important properties of the Laplace transform and in Table 6.2 the Laplace transforms of some basic functions.

Table 6.1 Basic properties of the Laplace transform

No.	Formula	Reference
1.	$F(p) = \mathcal{L}\{f(t)\}(p) = \int_0^\infty f(t)\exp(-pt)\,dt$ (Definition)	(6.3)
2.	$\mathcal{L}\{\alpha f(t) + \beta g(t)\} = \alpha \mathcal{L}\{f(t)\} + \beta \mathcal{L}\{g(t)\}$ (Linearity)	Theorem 6.2
3.	$\mathcal{L}\{t^k f(t)\}(p) = (-1)^k F^{(k)}(p),\ k = 1, 2, \ldots$	(6.5)
4.	$\mathcal{L}\{f(at)\}(p) = \dfrac{1}{a} F\left(\dfrac{p}{a}\right),\ a > 0$ (Similarity)	(6.18)
5.	$\mathcal{L}\{f(t-u)\}(p) = \exp(-up) F(p),\ u > 0$ (Delay)	(6.19)
6.	$\mathcal{L}\{\exp(p_0 t) f(t)\}(p) = F(p - p_0)$ (Displacement)	(6.22)
7.	$\mathcal{L}\{f'(t)\}(p) = p F(p) - f(0)$ (Differentiation)	(6.25)
8.	$\mathcal{L}\{f^{(n)}(t)\}(p) = p^n F(p) - \sum_{j=0}^{n-1} p^{n-1-j} f^{(j)}(0)$	(6.26)
9.	$\mathcal{L}\left\{\int_0^t f(u)\,du\right\}(p) = \dfrac{F(p)}{p}$ (Integration of original)	(6.28)
10.	$\mathcal{L}\left\{\dfrac{f(t)}{t}\right\}(p) = \int_p^\infty F(q)\,dq$ (Integration of image)	(6.31)
11.	$\mathcal{L}\left\{\dfrac{f(t)}{t}\right\}(p) = \dfrac{1}{1 - \exp(-Tp)} \int_0^T f(t)\exp(-pt)\,dt$ whenever $f(t)$ is periodic of period $T > 0$	(6.36)
12.	$\mathcal{L}\{f * g\} = \mathcal{L}\{f\} \cdot \mathcal{L}\{g\} = F(p)G(p)$ (Multiplication (convolution))	(6.39)
13.	$\mathcal{L}\{f(t)g(t)\}(p) = \dfrac{1}{2\pi i} \int_{x-i\infty}^{x+i\infty} F(q)G(p-q)\,dq$ (Complex convolution)	(6.56)

6.5 The Dirac Delta Function

Table 6.2 Basic Laplace transforms

No.	$f(t)$	$F(p) = \mathcal{L}\{f(t)\}(p)$	Reference
1.	$\eta(t)$	$\dfrac{1}{p}$	(6.7)
2.	$\eta(t)t^n, n \in \mathbb{N}$	$\dfrac{n!}{p^{n+1}}$	(6.8)
3.	$\eta(t)t^a, a > -1$	$\dfrac{\Gamma(a+1)}{p^{a+1}}$	(6.9)
4.	$\eta(t)\exp(at)$	$\dfrac{1}{p-a}$	(6.11)
5.	$\eta(t)\cos\omega t$	$\dfrac{p}{p^2+\omega^2}$	(6.13)
6.	$\eta(t)\sin\omega t$	$\dfrac{\omega}{p^2+\omega^2}$	(6.14)
7.	$\eta(t)\cosh\omega t$	$\dfrac{p}{p^2-\omega^2}$	(6.15)
8.	$\eta(t)\sinh\omega t$	$\dfrac{\omega}{p^2-\omega^2}$	(6.16)
9.	$\eta(t-a)\cos(t-a)$	$\exp(-ap)\dfrac{p}{p^2+1}$	(6.20)
10.	$\eta(t-a)\sin(t-a)$	$\exp(-ap)\dfrac{1}{p^2+1}$	(6.21)
11.	$\eta(t)\exp(\lambda t)\cos\omega t$	$\dfrac{p-\lambda}{(p-\lambda)^2+\omega^2}$	(6.23)
12.	$\eta(t)\exp(\lambda t)\sin\omega t$	$\dfrac{\omega}{(p-\lambda)^2+\omega^2}$	(6.24)
13.	$\eta(t)\exp(\lambda t)\cdot t^{a-1}, a > 0$	$\dfrac{\Gamma(a)}{(p-\lambda)^a}$	(6.64)
14.	$\eta(t)\exp(-t^2)$	$\dfrac{\sqrt{\pi}}{2}\exp\left(\dfrac{p^2}{4}\right)\left[1-\text{erf}\left(\dfrac{p}{2}\right)\right]$	(6.29)
15.	$\eta(t)\,\text{erf}\,t$	$\dfrac{1}{p}\exp\left(\dfrac{p^2}{4}\right)\left[1-\text{erf}\left(\dfrac{p}{2}\right)\right]$	(6.30)
16.	$\delta_a(t)$ Dirac delta function, $a \geq 0$	$\exp(-ap)$	(6.79)

6.6 Exercises

1. Find the Laplace transforms of the following functions:

   (a) $\sin^2 3t$; (b) $\sin 2t \cos 3t$; (c) $t \cos 2t$; (d) $(2t + 3) \sin 4t$; (e) $\exp(2t) \sin t$; (f) $\exp(-2t) \cos 3t$; (g) $\frac{1}{t}(\exp(-at) - \exp(-bt))$.

   Note: All the functions are considered to be multiplied by the unit step function $\eta(t)$.

2. Find the inverse Laplace transforms of the following functions:

   (a) $F(p) = \frac{1}{p^2+3p+2}$; (b) $F(p) = \frac{p}{(p+3)^2}$; (c) $F(p) = \frac{p}{(p^2+4)^2}$, (d) $F(p) = \frac{1}{(p^2+1)^2}$; (e) $F(p) = \frac{p}{p^2-4p+5}$; (f) $F(p) = \frac{p}{p^3+8}$; (g) $F(p) = \frac{\exp(-p)}{p(p-1)}$.

3. Solve the following Cauchy problems using the Laplace transform:

   (a) $y' + y = \exp(-t)$, $y(0) = 2$;
   (b) $y'' + 4y = t$, $y(0) = 0$, $y'(0) = 2$;
   (c) $y'' + 9y = 1$, $y(0) = -1$, $y'(0) = 0$;
   (d) $y''' + y' = 4$, $y(0) = y'(0) = y''(0) = 0$;
   (e) $y'' + y = \sin t$, $y(0) = y'(0) = 1$.
   (f) $\begin{cases} x' + 2y = 0 \\ y' + 2x = 0 \end{cases}$, $x(0) = 1$, $y(0) = -2$.
   (g) $\begin{cases} x' + x - 2y = t \\ y' + 2x - y = 1 \end{cases}$, $x(0) = 2$, $y(0) = 3$.

4. Consider the equation of a mass-spring system moving in a medium where the damping is negligible. We know the mass $m = 1$ and the spring constant $k = 1$. At the moment $t = 0$ the mass is released to oscillate freely from the initial position $y(0) = 2$ (the initial velocity is $y'(0) = 0$). At time $t = 2\pi$ the mass is given a sharp blow. Find the expression of $y(t)$.

   Hint: the Cauchy problem we have to solve is:

   $$y'' + y = 4\delta_{2\pi}(t), \quad y(0) = 2, \quad y'(0) = 0.$$

Chapter 7
Second-Order Partial Differential Equations

Chapter 7 is an introduction to "The Equations of Mathematical Physics"—quasilinear second-order partial differential equations that are mostly used in Physics and Engineering. We begin by classifying these equations and introduce the characteristics method to find their canonical form. We present the solution of the one-dimensional and two-dimensional wave equation, the heat flow equation and the Dirichlet problem for Laplace's equation.

7.1 Classification: Canonical Form

Many physical phenomena (such as vibrations of strings, heat conduction, fluid flow or the behavior of electric and magnetic fields) can be described by *partial differential equations*—equations which contain some partial derivatives of an unknown function of two or more variables. The *order* of the partial differential equation is the order of the highest derivative. In Sect. 3.6 of Chap. 3 we studied the linear and quasilinear first-order partial differential equations (whose solution is closely connected to autonomous systems). This chapter is mainly devoted to *second-order* (quasi)linear partial differential equations, but it also contains a section dedicated to the fourth-order PDE which describes the vibrations of an elastic beam (Sect. 7.3). We present below some of the most used second-order partial differential equations:

$$\frac{\partial^2 u}{\partial t^2} = a^2 \frac{\partial^2 u}{\partial x^2} \qquad \text{one-dimensional wave equation} \qquad (7.1)$$

$$\frac{\partial u}{\partial t} = a^2 \frac{\partial^2 u}{\partial x^2} \qquad \text{one-dimensional heat equation} \qquad (7.2)$$

$$\frac{\partial^2 u}{\partial x^2} + \frac{\partial^2 u}{\partial y^2} = 0 \qquad \text{two-dimensional Laplace equation} \qquad (7.3)$$

$$\frac{\partial^2 u}{\partial x^2} + \frac{\partial^2 u}{\partial y^2} = f(x, y) \qquad \text{two-dimensional Poisson equation} \qquad (7.4)$$

$$\frac{\partial^2 u}{\partial t^2} = a^2 \left(\frac{\partial^2 u}{\partial x^2} + \frac{\partial^2 u}{\partial y^2} \right) \qquad \text{two-dimensional wave equation} \qquad (7.5)$$

$$\frac{\partial^2 u}{\partial x^2} + \frac{\partial^2 u}{\partial y^2} + \frac{\partial^2 u}{\partial z^2} = 0 \qquad \text{three-dimensional Laplace equation.} \qquad (7.6)$$

In all these equations, u is the unknown function, which can be a function of two variables (in most cases) or a function of three variables (in the last two equations), t is the temporal coordinate, x, y, z are spatial (Cartesian) coordinates (the *dimension of the equation* is given by the number of spatial coordinates) and a is a positive constant.

The sum of the second-order unmixed partial derivatives w.r.t. the spatial (Cartesian) variables is called the *Laplacian* of the function u and is denoted by Δu or $\nabla^2 u$ using the *Del* (or *Nabla*) operator which defines the gradient of a function:

$$\nabla = \frac{\partial}{\partial x}\mathbf{i} + \frac{\partial}{\partial y}\mathbf{j}$$

in a 2-dimensional space, or, in a 3-dimensional space,

$$\nabla = \frac{\partial}{\partial x}\mathbf{i} + \frac{\partial}{\partial y}\mathbf{j} + \frac{\partial}{\partial z}\mathbf{k}.$$

Thus, the Laplacian of a function is

$$\Delta u = \nabla^2 u = \nabla \cdot \nabla u = \frac{\partial^2 u}{\partial x^2} + \frac{\partial^2 u}{\partial y^2} \qquad (7.7)$$

for a function which depends on two spatial variables x and y, and

$$\Delta u = \nabla^2 u = \nabla \cdot \nabla u = \frac{\partial^2 u}{\partial x^2} + \frac{\partial^2 u}{\partial y^2} + \frac{\partial^2 u}{\partial z^2} \qquad (7.8)$$

for a function defined in a three-dimensional space (note that only the spatial variables are considered in the Laplacian), so the Eqs. (7.3)–(7.6) can be written as:

$$\nabla^2 u = 0, \quad \nabla^2 u = f(x, y), \quad \frac{\partial^2 u}{\partial t^2} = a^2 \nabla^2 u, \quad \text{and} \quad \nabla^2 u = 0.$$

7.1 Classification: Canonical Form

In general, a partial differential equation has a very broad class of solutions. For example, as can be readily verified, all the functions bellow are solutions of the Laplace equation (7.3):

$$u(x, y) = x^2 - y^2, \quad u(x, y) = e^x \cos y, \quad u(x, y) = \ln(x^2 + y^2).$$

As will be shown later, the (unique) solution of a PDE related to a given physical problem can be obtained by using some additional conditions (which may be "boundary conditions" or "initial conditions").

We introduce in the following the general linear and quasilinear second-order PDE in two variables. We shall see how they can be classified and reduced to a standard form called the *canonical form* or the *normal form* (see [9], [20]).

Let Ω be a plane domain (an open and connected subset of $\mathbb{R}^2$). A *linear second-order partial differential equation* (PDE-2) for an unknown function of two variables $u(x, y)$ (which is twice continuously differentiable i.e. of class C^2) is an equation of the form

$$A(x, y)\frac{\partial^2 u}{\partial x^2} + 2B(x, y)\frac{\partial^2 u}{\partial x \partial y} + C(x, y)\frac{\partial^2 u}{\partial y^2} + \qquad (7.9)$$

$$+ P(x, y)\frac{\partial u}{\partial x} + Q(x, y)\frac{\partial u}{\partial y} + R(x, y)u = f(x, y), \quad \forall (x, y) \in \Omega,$$

where $A, B, C, P, Q, R, f : \Omega \to \mathbb{R}$ are continuous functions such that $A(x, y)$, $B(x, y)$ and $C(x, y)$ are not simultaneously 0 at any point $(x, y) \in \Omega$. The factor 2 multiplying B was introduced to simplify the calculations (as we shall see in what follows). The functions $A(x, y), \ldots, R(x, y)$ are called the *coefficients* of the partial differential equation, while $f(x, y)$ is the *free term* of the equation. If $f(x, y)$ is identically 0, the PDE (7.9) is said to be *homogeneous*. Otherwise, it is said to be *non-homogeneous* (or inhomogeneous).

The general form of a linear second-order partial differential equation in n variables, $x_1, \ldots, x_n$, is:

$$\sum_{1 \le i \le j \le n} a_{i,j}(x_1, \ldots, x_n)\frac{\partial^2 u}{\partial x_i \partial x_j} + \sum_{i=1}^{n} b_i(x_1, \ldots, x_n)\frac{\partial u}{\partial x_i} + \qquad (7.10)$$

$$+ c(x_1, \ldots, x_n)u = f(x_1, \ldots, x_n).$$

We notice that all the Eqs. (7.1)–(7.6) are *linear* PDE-2 and all of them are homogeneous, except for Poisson equation (7.4) which is non-homogeneous.

To solve the Eq. (7.9) means to find all the possible solutions i.e. all the functions $u(x, y)$ of class $C^2(\Omega)$ which satisfy the equality (7.9) for any $(x, y) \in \Omega$. The set of all these functions is called the *general solution* of Eq. (7.9). As in the case of first-order partial differential equations, the general solution of a PDE-2 depends on some arbitrary functions. We can see this in the following examples.

Example 7.1 Find the general solution of the PDE-2:

$$\frac{\partial^2 u}{\partial x \partial y} = 0.$$

This is a homogeneous equation and all the coefficients are 0, except for $B(x, y) = \frac{1}{2}$. We can write:

$$\frac{\partial^2 u}{\partial x \partial y}(x, y) = \frac{\partial}{\partial x}\left(\frac{\partial u}{\partial y}(x, y)\right) = 0,$$

so the function $\frac{\partial u}{\partial y}(x, y)$ is constant with respect to x, i.e. $\frac{\partial u}{\partial y}(x, y) = h(y)$ (where $h(y)$ is an arbitrary, continuous function). If we integrate this last equality with respect to y, we get:

$$u(x, y) = G(y) + F(x),$$

where $G(y)$ is a primitive of $h(y)$ and $F(x)$ is the "constant" of integration—a constant with respect to y, but an arbitrary function of x. It is easy to check that $u(x, y) = F(x) + G(y)$ is a solution of our equation, for any functions $F(x)$, $G(y)$ of class C^2.

Example 7.2 Let us find now the general solution of the PDE-2:

$$\frac{\partial^2 u}{\partial x^2} = 0.$$

This is also a homogeneous equation and all the coefficients are 0, except for $A(x, y) = 1$. Since

$$\frac{\partial^2 u}{\partial x^2}(x, y) = \frac{\partial}{\partial x}\left(\frac{\partial u}{\partial x}(x, y)\right) = 0,$$

it follows that $\frac{\partial u}{\partial x}(x, y) = h(y)$. We integrate with respect to x and obtain

$$u(x, y) = xh(y) + g(y),$$

where $g(y)$ is the "constant" of integration (which may depend on y). This is the general solution of our equation, where $h(y)$, $g(y)$ are arbitrary functions of class C^2.

7.1 Classification: Canonical Form

Example 7.3 Find the general solution of the PDE-2:

$$\frac{\partial^2 u}{\partial y^2} = \frac{\partial u}{\partial y}.$$

This homogeneous equation has all the coefficients 0, except for $C(x, y) = 1$ and $Q(x, y) = -1$. We write the equation as:

$$\frac{\partial}{\partial y}\left(\frac{\partial u}{\partial y}(x, y) - u(x, y)\right) = 0$$

and we find that $\frac{\partial u}{\partial y}(x, y) - u(x, y)$ does not depend on y, but only on x:

$$\frac{\partial u}{\partial y}(x, y) - u(x, y) = h(x). \qquad (7.11)$$

This equation can be solved like an ordinary (linear) equation with the independent variable y (considering x as a parameter). Thus, the associated homogeneous equation is: $\frac{\partial u}{\partial y} - u = 0$, with the general solution

$$u_{GH}(x, y) = C_1(x)\exp y$$

(notice that the "constant" $C_1(x)$ may depend on the "parameter" x). Since the free term $h(x)$ does not depend on y, we search for a particular solution of the same form: $u_P(x, y) = C_2(x)$. By replacing in (7.11), we find $C_2(x) = -h(x)$. Hence, the general solution of our initial PDE-2 is

$$u(x, y) = C_1(x)\exp y + C_2(x),$$

where $C_1(x)$ and $C_2(x)$ are arbitrary functions of class C^2.

Coming back to the linear partial differential Eq. (7.9), we denote by $D\left(x, y, u(x, y), \frac{\partial u}{\partial x}(x, y), \frac{\partial u}{\partial y}(x, y)\right)$ the part of the equation containing the first-order derivatives of u, the function u itself and also the free term $(-f(x, y))$. Thus, the equation is written:

$$A(x, y)\frac{\partial^2 u}{\partial x^2} + 2B(x, y)\frac{\partial^2 u}{\partial x \partial y} + C(x, y)\frac{\partial^2 u}{\partial y^2} +$$
$$+ D\left(x, y, u, \frac{\partial u}{\partial x}, \frac{\partial u}{\partial y}\right) = 0, \ \forall (x, y) \in \Omega. \qquad (7.12)$$

The Eq. (7.12) can be considered more general than (7.9) if we drop the condition for the expression $D\left(x, y, u, \frac{\partial u}{\partial x}, \frac{\partial u}{\partial y}\right)$ to be linear in u, $\frac{\partial u}{\partial x}$, $\frac{\partial u}{\partial y}$. Thus, if we consider

D as a continuous function of five variables, then (7.12) is said to be a *quasilinear second order partial differential equation*.

For instance, the equation

$$a^2 \frac{\partial^2 u}{\partial x^2} - \frac{\partial u}{\partial t} - u \frac{\partial u}{\partial x} = 0,$$

known as the Burger's equation (with applications in fluid mechanics), is not a linear equation (because of the nonlinear term $u \frac{\partial u}{\partial x}$), but it is a *quasilinear* equation.

We denote by $L[u]$ the part of the Eq. (7.12) containing the second-order derivatives, called the *principal part* of the equation,

$$L[u] = A(x, y) \frac{\partial^2 u}{\partial x^2} + 2B(x, y) \frac{\partial^2 u}{\partial x \partial y} + C(x, y) \frac{\partial^2 u}{\partial y^2}.$$

Examples 7.1, 7.2 and 7.3 show how to find the general solution for some simple partial differential equations. But *any* quasilinear PDE-2 (or at least, its principal part $L[u]$), can be reduced to a simpler form by using a *change of variables*—a transformation of the independent variables x and y to some new variables ξ and η,

$$\begin{cases} \xi = \xi(x, y) \\ \eta = \eta(x, y), \end{cases} \quad (7.13)$$

where ξ and η are functions of the class C^2 such that one point in the domain Ω in the (x, y)-plane corresponds to one point in the corresponding domain D in the (ξ, η)-plane, and conversely (the transformation is a bijection between the two domains. The transformation is a bijection if and only if the Jacobian matrix $J(x, y)$,

$$J(x, y) = \begin{pmatrix} \frac{\partial \xi}{\partial x} & \frac{\partial \xi}{\partial y} \\ \frac{\partial \eta}{\partial x} & \frac{\partial \eta}{\partial y} \end{pmatrix}$$

is *invertible* at any point in the domain Ω: $\det J(x, y) \neq 0, \forall (x, y) \in \Omega$.

In this case, for any $(\xi, \eta) \in D$, there is a unique solution of the system (7.13), $x = x(\xi, \eta)$, $y = y(\xi, \eta)$ and the unknown function $u(x, y)$ becomes a new unknown function $\overline{u}(\xi, \eta) = u(x(\xi, \eta), y(\xi, \eta))$. Thus, coming back to $u(x, y)$, we get:

$$u(x, y) = \overline{u}(\xi, \eta) = \overline{u}(\xi(x, y), \eta(x, y)). \quad (7.14)$$

Using the chain rule to differentiate the composite function (7.14), we can express the partial derivatives of u with respect to x and y in terms of the derivatives of $\overline{u}$ with respect to ξ and η:

$$\frac{\partial u}{\partial x} = \frac{\partial \overline{u}}{\partial \xi} \frac{\partial \xi}{\partial x} + \frac{\partial \overline{u}}{\partial \eta} \frac{\partial \eta}{\partial x} \quad (7.15)$$

7.1 Classification: Canonical Form

$$\frac{\partial u}{\partial y} = \frac{\partial \bar{u}}{\partial \xi}\frac{\partial \xi}{\partial y} + \frac{\partial \bar{u}}{\partial \eta}\frac{\partial \eta}{\partial y} \qquad (7.16)$$

By (7.15) and (7.16) we can write the operator relationships:

$$\frac{\partial}{\partial x} = \frac{\partial}{\partial \xi}\frac{\partial \xi}{\partial x} + \frac{\partial}{\partial \eta}\frac{\partial \eta}{\partial x} \qquad (7.17)$$

$$\frac{\partial}{\partial y} = \frac{\partial}{\partial \xi}\frac{\partial \xi}{\partial y} + \frac{\partial}{\partial \eta}\frac{\partial \eta}{\partial y} \qquad (7.18)$$

and use them to compute the second-order derivatives $\frac{\partial^2 u}{\partial x^2}$, $\frac{\partial^2 u}{\partial y^2}$ and $\frac{\partial^2 u}{\partial x \partial y}$. Thus, starting from the definition of $\frac{\partial^2 u}{\partial x^2}$, one can write:

$$\frac{\partial^2 u}{\partial x^2} = \frac{\partial}{\partial x}\left(\frac{\partial u}{\partial x}\right) \stackrel{(7.15)}{=} \frac{\partial}{\partial x}\left(\frac{\partial \bar{u}}{\partial \xi}\frac{\partial \xi}{\partial x} + \frac{\partial \bar{u}}{\partial \eta}\frac{\partial \eta}{\partial x}\right) =$$

$$= \frac{\partial}{\partial x}\left(\frac{\partial \bar{u}}{\partial \xi}\right)\frac{\partial \xi}{\partial x} + \frac{\partial \bar{u}}{\partial \xi}\frac{\partial^2 \xi}{\partial x^2} + \frac{\partial}{\partial x}\left(\frac{\partial \bar{u}}{\partial \eta}\right)\frac{\partial \eta}{\partial x} + \frac{\partial \bar{u}}{\partial \eta}\frac{\partial^2 \eta}{\partial x^2} \stackrel{(7.17)}{=}$$

$$= \left(\frac{\partial^2 \bar{u}}{\partial \xi^2}\frac{\partial \xi}{\partial x} + \frac{\partial^2 \bar{u}}{\partial \eta \partial \xi}\frac{\partial \eta}{\partial x}\right)\frac{\partial \xi}{\partial x} + \frac{\partial \bar{u}}{\partial \xi}\frac{\partial^2 \xi}{\partial x^2} +$$

$$+ \left(\frac{\partial^2 \bar{u}}{\partial \xi \partial \eta}\frac{\partial \xi}{\partial x} + \frac{\partial^2 \bar{u}}{\partial \eta^2}\frac{\partial \eta}{\partial x}\right)\frac{\partial \eta}{\partial x} + \frac{\partial \bar{u}}{\partial \eta}\frac{\partial^2 \eta}{\partial x^2}.$$

Since $\bar{u}$ is a function of class C^2, the mixed derivatives are equal $\frac{\partial^2 \bar{u}}{\partial \xi \partial \eta} = \frac{\partial^2 \bar{u}}{\partial \eta \partial \xi}$ and we have:

$$\frac{\partial^2 u}{\partial x^2} = \frac{\partial^2 \bar{u}}{\partial \xi^2}\left(\frac{\partial \xi}{\partial x}\right)^2 + 2\frac{\partial^2 \bar{u}}{\partial \xi \partial \eta}\frac{\partial \xi}{\partial x}\frac{\partial \eta}{\partial x} + \frac{\partial^2 \bar{u}}{\partial \eta^2}\left(\frac{\partial \eta}{\partial x}\right)^2 + \frac{\partial \bar{u}}{\partial \xi}\frac{\partial^2 \xi}{\partial x^2} + \frac{\partial \bar{u}}{\partial \eta}\frac{\partial^2 \eta}{\partial x^2}. \qquad (7.19)$$

In the same way we can find the formulas for $\frac{\partial^2 u}{\partial y^2}$ and $\frac{\partial^2 u}{\partial x \partial y}$:

$$\frac{\partial^2 u}{\partial y^2} = \frac{\partial^2 \bar{u}}{\partial \xi^2}\left(\frac{\partial \xi}{\partial y}\right)^2 + 2\frac{\partial^2 \bar{u}}{\partial \xi \partial \eta}\frac{\partial \xi}{\partial y}\frac{\partial \eta}{\partial y} + \frac{\partial^2 \bar{u}}{\partial \eta^2}\left(\frac{\partial \eta}{\partial y}\right)^2 + \frac{\partial \bar{u}}{\partial \xi}\frac{\partial^2 \xi}{\partial y^2} + \frac{\partial \bar{u}}{\partial \eta}\frac{\partial^2 \eta}{\partial y^2}, \qquad (7.20)$$

$$\frac{\partial^2 u}{\partial x \partial y} = \frac{\partial^2 \bar{u}}{\partial \xi^2}\frac{\partial \xi}{\partial x}\frac{\partial \xi}{\partial y} + \frac{\partial^2 \bar{u}}{\partial \xi \partial \eta}\left(\frac{\partial \xi}{\partial x}\frac{\partial \eta}{\partial y} + \frac{\partial \xi}{\partial y}\frac{\partial \eta}{\partial x}\right) + \frac{\partial^2 \bar{u}}{\partial \eta^2}\frac{\partial \eta}{\partial x}\frac{\partial \eta}{\partial y} + \qquad (7.21)$$

$$+ \frac{\partial \bar{u}}{\partial \xi}\frac{\partial^2 \xi}{\partial x \partial y} + \frac{\partial \bar{u}}{\partial \eta}\frac{\partial^2 \eta}{\partial x \partial y}.$$

By replacing in Eq. (7.12) the partial derivatives of u with the expressions above, we obtain the following quasilinear equation in the unknown function $\bar{u} = \bar{u}(\xi, \eta)$:

$$A^*(\xi, \eta) \frac{\partial^2 \bar{u}}{\partial \xi^2} + 2B^*(\xi, \eta) \frac{\partial^2 \bar{u}}{\partial \xi \partial \eta} + C^*(\xi, \eta) \frac{\partial^2 \bar{u}}{\partial \eta^2} + \qquad (7.22)$$

$$+ D^*\left(\xi, \eta, \bar{u}, \frac{\partial \bar{u}}{\partial \xi}, \frac{\partial \bar{u}}{\partial \eta}\right) = 0,$$

where

$$A^*(\xi, \eta) = A\left(\frac{\partial \xi}{\partial x}\right)^2 + 2B \frac{\partial \xi}{\partial x} \frac{\partial \xi}{\partial y} + C\left(\frac{\partial \xi}{\partial y}\right)^2 \qquad (7.23)$$

$$B^*(\xi, \eta) = A \frac{\partial \xi}{\partial x} \frac{\partial \eta}{\partial x} + B\left(\frac{\partial \xi}{\partial x} \frac{\partial \eta}{\partial y} + \frac{\partial \xi}{\partial y} \frac{\partial \eta}{\partial x}\right) + C \frac{\partial \xi}{\partial y} \frac{\partial \eta}{\partial y} \qquad (7.24)$$

$$C^*(\xi, \eta) = A\left(\frac{\partial \eta}{\partial x}\right)^2 + 2B \frac{\partial \eta}{\partial x} \frac{\partial \eta}{\partial y} + C\left(\frac{\partial \eta}{\partial y}\right)^2. \qquad (7.25)$$

The function $\Delta(x, y) = B^2(x, y) - A(x, y)C(x, y)$ is said to be the *discriminant* of the Eq. (7.12). We shall prove that the sign of this discriminant is invariant through a change of variables.

Proposition 7.1 *With the above notation, if*

$$\Delta^*(\xi, \eta) = B^*(\xi, \eta)^2 - A^*(\xi, \eta)C^*(\xi, \eta),$$

then, for any $(x, y) \in \Omega$,

$$\Delta^*(\xi(x, y), \eta(x, y)) = \Delta(x, y) \cdot (\det J(x, y))^2. \qquad (7.26)$$

As a consequence, $\Delta^* > 0$ *if and only if* $\Delta > 0$; $\Delta^* < 0$ *if and only if* $\Delta < 0$; $\Delta^* = 0$ *if and only if* $\Delta = 0$.

Proof We denote by $M = M(x, y)$ and $M^* = M^*(\xi, \eta)$ the matrices

$$M = \begin{pmatrix} A & B \\ B & C \end{pmatrix} \quad \text{and} \quad M^* = \begin{pmatrix} A^* & B^* \\ B^* & C^* \end{pmatrix}.$$

7.1 Classification: Canonical Form

Obviously, the discriminant of the initial Eq. (7.12) is $\Delta = -\det M$, while the discriminant of the transformed Eq. (7.22) is $\Delta^* = -\det M^*$. On the other hand, the following equality can be readily verified using the formulas (7.23)–(7.25):

$$\begin{pmatrix} A^* & B^* \\ B^* & C^* \end{pmatrix} = \begin{pmatrix} \frac{\partial \xi}{\partial x} & \frac{\partial \xi}{\partial y} \\ \frac{\partial \eta}{\partial x} & \frac{\partial \eta}{\partial y} \end{pmatrix} \begin{pmatrix} A & B \\ B & C \end{pmatrix} \begin{pmatrix} \frac{\partial \xi}{\partial x} & \frac{\partial \eta}{\partial x} \\ \frac{\partial \xi}{\partial y} & \frac{\partial \eta}{\partial y} \end{pmatrix}.$$

Thus, we have

$$M^* = J M J^T,$$

so

$$\det M^* = \det M \, (\det J)^2$$

and, multiplying this equality by -1, the Eq. (7.26) follows. □

Based on this invariance of the sign of the discriminant Δ, we can give a classification of the second-order quasilinear partial differential equations.

Definition 7.1 (Classification) The PDE-2 (7.12) is said to be:

(a) of **hyperbolic type** on Ω if $\Delta(x, y) > 0$ for all $(x, y) \in \Omega$;
(b) of **parabolic type** on Ω if $\Delta(x, y) = 0$ for all $(x, y) \in \Omega$;
(c) of **elliptic type** on Ω if $\Delta(x, y) < 0$ for all $(x, y) \in \Omega$.

Remark 7.1 The same equation can be of different types on different domains: For instance, the *Tricomi equation* (see [9] p.162)

$$\frac{\partial^2 u}{\partial x^2} + x \frac{\partial^2 u}{\partial y^2} = 0 \qquad (7.27)$$

is an equation of elliptic type in the half-plane $x > 0$ and of hyperbolic type in the half-plane $x < 0$. This equation is used to approximate the motion of a fluid at supersonic speeds ($x < 0$) and at subsonic speeds ($x > 0$) respectively.

We assume in what follows that the coefficients $A(x, y)$, $B(x, y)$ and $C(x, y)$ do not change their sign on Ω. We want to find a change of variables (7.13) such that the transformed Eq. (7.22) has a simpler form than the initial Eq. (7.12). More exactly, we want to make 0 one or two of the coefficients A^*, B^* and C^*.

Suppose that $A^* = 0$ and $C^* = 0$. By (7.23) and (7.25) we have that the functions $\xi = \xi(x, y)$ and $\eta = \eta(x, y)$ must satisfy the following (nonlinear) first-order partial differential equation (in the unknown function $\varphi(x, y)$):

$$A(x, y) \left(\frac{\partial \varphi}{\partial x} \right)^2 + 2B(x, y) \frac{\partial \varphi}{\partial x} \frac{\partial \varphi}{\partial y} + C(x, y) \left(\frac{\partial \varphi}{\partial y} \right)^2 = 0. \qquad (7.28)$$

Let $\varphi(x, y)$ be a (nontrivial) solution of Eq. (7.28). The curve (included in the domain Ω) defined by the implicit equation

$$\varphi(x, y) = 0 \qquad (7.29)$$

is called a *characteristic curve* (or, simply, a *characteristic*) of the PDE (7.12).

Assume that (x_0, y_0) is a point of the characteristic curve (7.29) such that $\frac{\partial \varphi}{\partial y}(x_0, y_0) \neq 0$. It follows that $\frac{\partial \varphi}{\partial y}(x, y) \neq 0$ in a neighborhood of the point (x_0, y_0) and, by the implicit function theorem, there exists a function $y = y(x)$, defined on a neighborhood of x_0, such that

$$\varphi(x, y(x)) = 0, \qquad (7.30)$$

for any x in this neighborhood. Thus, the characteristic curve is the graph of the function $y = y(x)$. By differentiating the Eq. (7.30) with respect to x, we get

$$y'(x) = -\frac{\frac{\partial \varphi}{\partial x}(x, y(x))}{\frac{\partial \varphi}{\partial y}(x, y(x))}. \qquad (7.31)$$

Now, we can see that if we divide by $\left(\frac{\partial \varphi}{\partial y}\right)^2$ the Eq. (7.28) and use the formula (7.31), we obtain:

$$A(x, y(x))y'(x)^2 - 2B(x, y(x))y'(x) + C(x, y(x)) = 0. \qquad (7.32)$$

If $A(x, y) \neq 0$, then (7.32) can be solved for $y'(x)$ like a quadratic equation. We call it the *characteristic differential equation* associated to our initial PDE-2 (7.12).

What happens if $A(x, y) \equiv 0$? In this case, if $C(x, y) \neq 0$, by a similar reasoning as above, we write the explicit equation of the characteristic curve as $x = x(y)$. Since

$$x'(y) = -\frac{\frac{\partial \varphi}{\partial y}(x(y), y)}{\frac{\partial \varphi}{\partial x}(x(y), y)}, \qquad (7.33)$$

by the equality (7.28) we obtain the characteristic differential equation

$$A(x(y), y) - 2B(x(y), y)x'(y) + C(x(y), y)x'(y)^2 = 0. \qquad (7.34)$$

If both coefficients A and C are 0 (and compulsory $B(x, y) \neq 0$) then $\Delta(x, y) = [B(x, y)]^2 > 0$ and the Eq. (7.12) is already in the canonical form of the hyperbolic-type equation.

7.1 Classification: Canonical Form

Suppose that $\Delta(x, y) = B^2(x, y) - A(x, y)C(x, y) > 0$ and $A(x, y) \neq 0$ for all $(x, y) \in \Omega$. Then, the quadratic Eq. (7.32) has two distinct real roots. Thus, we may have either

$$y' = \frac{dy}{dx} = \frac{B(x, y) + \sqrt{\Delta(x, y)}}{A(x, y)} \quad \text{or} \quad y' = \frac{dy}{dx} = \frac{B(x, y) - \sqrt{\Delta(x, y)}}{A(x, y)}. \tag{7.35}$$

By integrating these two first-order differential equations we obtain the general solutions (in an implicit form) $\varphi_1(x, y) = C_1$ and $\varphi_2(x, y) = C_2$ respectively, where C_1, C_2 are real constants. These are the characteristic curves of the of the partial differential Eq. (7.12).

If $A(x, y) \equiv 0$ but $C(x, y) \neq 0$, then we solve for x' the quadratic Eq. (7.34) which has also two real distinct roots:

$$x' = \frac{dx}{dy} = \frac{B(x, y) + \sqrt{\Delta(x, y)}}{C(x, y)} = \frac{2B(x, y)}{C(x, y)} \quad \text{or}$$

$$x' = \frac{dx}{dy} = \frac{B(x, y) - \sqrt{\Delta(x, y)}}{C(x, y)} = 0. \tag{7.36}$$

The characteristic curves $\varphi_1(x, y) = C_1$ and $\varphi_2(x, y) = C_2$ are obtained in this case by integrating (7.36).

Let us prove that the functions $\varphi_1(x, y)$ and $\varphi_2(x, y)$ are functionally independent (i.e. their Jacobian matrix is invertible). Thus, if $A(x, y) \neq 0$,

$$\det J(x, y) = \begin{vmatrix} \frac{\partial \varphi_1}{\partial x} & \frac{\partial \varphi_1}{\partial y} \\ \frac{\partial \varphi_2}{\partial x} & \frac{\partial \varphi_2}{\partial y} \end{vmatrix} = 0 \Leftrightarrow \frac{\frac{\partial \varphi_1}{\partial x}}{\frac{\partial \varphi_1}{\partial y}} = \frac{\frac{\partial \varphi_2}{\partial x}}{\frac{\partial \varphi_2}{\partial y}}. \tag{7.37}$$

By (7.31) and (7.35) one can write:

$$\frac{\frac{\partial \varphi_1}{\partial x}}{\frac{\partial \varphi_1}{\partial y}} = -\frac{B(x, y) + \sqrt{\Delta(x, y)}}{A(x, y)} \quad \text{and} \quad \frac{\frac{\partial \varphi_2}{\partial x}}{\frac{\partial \varphi_2}{\partial y}} = -\frac{B(x, y) - \sqrt{\Delta(x, y)}}{A(x, y)}.$$

Since $\Delta(x, y) > 0$, by (7.37) it follows that $\det J(x, y) \neq 0$. (we let the reader to prove that the Jacobian is also nonzero when $A(x, y) \equiv 0$).

Thus, we can make the change of variables

$$\begin{cases} \xi = \varphi_1(x, y) \\ \eta = \varphi_2(x, y). \end{cases}$$

and, as we saw above, the new coefficients A^* and B^* will be identically 0 (while $B^*(\xi, \eta) \neq 0$ at any point (ξ, η) of the new domain), so the initial Eq. (7.12) becomes

$$2B^*(\xi, \eta) \frac{\partial^2 \bar{u}}{\partial \xi \partial \eta} + D^*\left(\xi, \eta, \bar{u}, \frac{\partial \bar{u}}{\partial \xi}, \frac{\partial \bar{u}}{\partial \eta}\right) = 0.$$

We can divide this equation by $2B^*(\xi, \eta)$ and obtain the *canonical form of hyperbolic-type equations*, (7.38), so we have proved the following theorem.

Theorem 7.1 *If $\Delta(x, y) = B^2(x, y) - A(x, y)C(x, y) > 0$, $\forall (x, y) \in \Omega$ (the Eq. (7.12) is of hyperbolic type), then the characteristic differential Eq. (7.32) (or (7.34), if $A(x, y) = 0$) has two distinct families of real solutions: $\varphi_1(x, y) = C_1$ and $\varphi_2(x, y) = C_2$ and by the change of variables*

$$\begin{cases} \xi = \varphi_1(x, y) \\ \eta = \varphi_2(x, y), \end{cases} \quad u(x, y) = \bar{u}(\xi, \eta)$$

the PDE-2 (7.12) becomes

$$\frac{\partial^2 \bar{u}}{\partial \xi \partial \eta} + \overline{D}\left(\xi, \eta, \bar{u}, \frac{\partial \bar{u}}{\partial \xi}, \frac{\partial \bar{u}}{\partial \eta}\right) = 0. \tag{7.38}$$

Consider now the case of a parabolic-type equation. Thus, we suppose that the Eq. (7.12) has the discriminant

$$\Delta(x, y) = B^2(x, y) - A(x, y)C(x, y) = 0 \quad \forall (x, y) \in \Omega$$

and $A(x, y) \neq 0$ on Ω (otherwise, if $A(x, y) \equiv 0$, then we have $B(x, y) \equiv 0$, so the Eq. (7.12) is already in the canonical form).

Since $\Delta = 0$, the characteristic Eq. (7.32) has only one real solution,

$$y' = \frac{B(x, y)}{A(x, y)},$$

and we obtain by integration the family of characteristic curves $\varphi(x, y) = C$. Assuming that $\frac{\partial \varphi}{\partial y} \neq 0$ on Ω, we can make the change of variables

$$\begin{cases} \xi = \varphi(x, y) \\ \eta = x. \end{cases}$$

Thus, the new-obtained Eq. (7.22) in the unknown function $\bar{u} = \bar{u}(\xi, \eta)$ will have the coefficient $A^* = 0$. Since

$$B^{*2} - A^*C^* = B^2 - AC = 0,$$

7.1 Classification: Canonical Form

we obtain that $B^* = 0$ for any choice of the function $\eta(x, y)$ (such that the Jacobian is different of 0); we take $\eta(x, y) = x$ for simplicity. In this case we have $C^* = A \neq 0$ and, dividing by C^* the transformed Eq. (7.22) we obtain the *canonical form of parabolic-type equations*, (7.39), as stated in the next theorem.

Theorem 7.2 *If* $\Delta(x, y) = B^2(x, y) - A(x, y)C(x, y) = 0, \forall (x, y) \in \Omega$ *(the Eq. (7.12) is of parabolic type), then, assuming $A(x, y) \neq 0$, the characteristic differential Eq. (7.32) has one family of real solutions:* $\varphi(x, y) = C$ *and, by the change of variables*

$$\begin{cases} \xi = \varphi(x, y) \\ \eta = x, \end{cases} \quad u(x, y) = \overline{u}(\xi, \eta),$$

the PDE-2 (7.12) becomes

$$\frac{\partial^2 \overline{u}}{\partial \eta^2} + \overline{D}\left(\xi, \eta, \overline{u}, \frac{\partial \overline{u}}{\partial \xi}, \frac{\partial \overline{u}}{\partial \eta}\right) = 0. \quad (7.39)$$

Finally, let us consider the case of an elliptic-type equation. Since

$$\Delta(x, y) = B^2(x, y) - A(x, y)C(x, y) < 0 \quad \forall (x, y) \in \Omega,$$

we have $A(x, y) \neq 0$ on Ω and the characteristic Eq. (7.32) has complex conjugate solutions:

$$y' = \frac{B(x, y) \pm i\sqrt{-\Delta(x, y)}}{A(x, y)}, \quad (7.40)$$

and we obtain by integration

$$\varphi_{1,2}(x, y) = \alpha(x, y) \pm i\beta(x, y) = C_{1,2}.$$

It can be proved that the functions $\alpha(x, y)$ and $\beta(x, y)$ are functionally independent. Thus,

$$\det J(x, y) = \begin{vmatrix} \frac{\partial \alpha}{\partial x} & \frac{\partial \alpha}{\partial y} \\ \frac{\partial \beta}{\partial x} & \frac{\partial \beta}{\partial y} \end{vmatrix} = 0 \Leftrightarrow \frac{\frac{\partial \alpha}{\partial x}}{\frac{\partial \alpha}{\partial y}} = \frac{\frac{\partial \beta}{\partial x}}{\frac{\partial \beta}{\partial y}}.$$

From the last equality it follows that

$$\frac{\frac{\partial \alpha}{\partial x}}{\frac{\partial \alpha}{\partial y}} = \frac{\frac{\partial \beta}{\partial x}}{\frac{\partial \beta}{\partial y}} = \frac{\frac{\partial \varphi_1}{\partial x}}{\frac{\partial \varphi_1}{\partial y}} = \frac{\frac{\partial \varphi_2}{\partial x}}{\frac{\partial \varphi_2}{\partial y}},$$

and from (7.40) we get

$$\frac{B(x, y) + i\sqrt{-\Delta(x, y)}}{A(x, y)} = \frac{B(x, y) - i\sqrt{-\Delta(x, y)}}{A(x, y)},$$

a contradiction, since $\Delta(x, y) \neq 0$. So $\det J(x, y) \neq 0$ and we can make the change of variables

$$\begin{cases} \xi = \alpha(x, y) \\ \eta = \beta(x, y). \end{cases}$$

Since the function $\varphi_1(x, y) = \alpha(x, y) + i\beta(x, y)$ satisfies the Eq. (7.28), we have:

$$A\left(\frac{\partial \alpha}{\partial x} + i\frac{\partial \beta}{\partial x}\right)^2 + 2B\left(\frac{\partial \alpha}{\partial x} + i\frac{\partial \beta}{\partial x}\right)\left(\frac{\partial \alpha}{\partial y} + i\frac{\partial \beta}{\partial y}\right) + C\left(\frac{\partial \alpha}{\partial y} + i\frac{\partial \beta}{\partial y}\right)^2 = 0,$$

so it follows that

$$A\left[\left(\frac{\partial \alpha}{\partial x}\right)^2 - \left(\frac{\partial \beta}{\partial x}\right)^2\right] + 2B\left[\frac{\partial \alpha}{\partial x}\frac{\partial \alpha}{\partial y} - \frac{\partial \beta}{\partial x}\frac{\partial \beta}{\partial y}\right] + C\left[\left(\frac{\partial \alpha}{\partial y}\right)^2 - \left(\frac{\partial \beta}{\partial y}\right)^2\right] = 0$$

and

$$2A\frac{\partial \alpha}{\partial x}\frac{\partial \beta}{\partial x} + 2B\left(\frac{\partial \alpha}{\partial x}\frac{\partial \beta}{\partial y} + \frac{\partial \alpha}{\partial y}\frac{\partial \beta}{\partial x}\right) + 2C\frac{\partial \alpha}{\partial y}\frac{\partial \beta}{\partial y} = 0.$$

Thus, the new-obtained Eq. (7.22) in the unknown function $\overline{u} = \overline{u}(\xi, \eta)$ has the coefficients $B^* = 0$ and $A^* = C^* \neq 0$. We divide by A^* and obtain the *canonical form of elliptic-type equations*, (7.39).

Theorem 7.3 *If $\Delta(x, y) = B^2(x, y) - A(x, y)C(x, y) < 0, \forall (x, y) \in \Omega$ (the Eq. (7.12) is of elliptic type), then the characteristic differential Eq. (7.32) has complex (conjugate) solutions: $\varphi_{1,2}(x, y) = \alpha(x, y) + i\beta(x, y) = C_{1,2}$. Using the change of variables*

$$\begin{cases} \xi = \alpha(x, y) \\ \eta = \beta(x, y), \end{cases} \quad u(x, y) = \overline{u}(\xi, \eta),$$

the PDE-2 (7.12) becomes

$$\frac{\partial^2 \overline{u}}{\partial \xi^2} + \frac{\partial^2 \overline{u}}{\partial \eta^2} + \overline{D}\left(\xi, \eta, \overline{u}, \frac{\partial \overline{u}}{\partial \xi}, \frac{\partial \overline{u}}{\partial \eta}\right) = 0. \tag{7.41}$$

7.1 Classification: Canonical Form

Remark 7.2 A *Cauchy problem* for the partial differential Eq. (7.12) with $A(x, y) \neq 0$ on Ω is the problem of finding a particular solution $u(x, y)$ that satisfies the initial conditions:

$$u(x_0, y) = f(y), \quad \frac{\partial u}{\partial x}(x_0, y) = g(y), \tag{7.42}$$

for any y such that $(x_0, y) \in \Omega$, where $f(y)$ is a given function of class C^2 and $g(y)$ is a given function of class C^1. We do not discuss the existence and uniqueness of the solution of the Cauchy problem. If A, B, C, D are *analytic* functions (i.e. they can be expanded into power series converging in a suitably small domain), the Cauchy-Kowalewsky Theorem states that there exists a unique analytic solution of the Cauchy problem (see [9], Chapter I).

We illustrate each type of partial differential equation by the following examples. The first example deals with a Cauchy problem for a hyperbolic-type equation.

Example 7.4 Let us solve the following PDE-2:

$$\frac{\partial^2 u}{\partial x^2} + 2 \frac{\partial^2 u}{\partial x \partial y} - 3 \frac{\partial^2 u}{\partial y^2} - 2 = 0, \tag{7.43}$$

After getting its general solution, let us find the particular solution that satisfies the initial conditions: $u(x, 0) = 0$ and $\frac{\partial u}{\partial y}(x, 0) = x + \cos x$.

First of all, we have to find an appropriate change of variables in order to obtain the canonical form of the equation. Since $\Delta = 1 + 3 = 4 > 0$, the Eq. (7.43) is of the hyperbolic type and we can apply Theorem 7.1. The associated characteristic differential Eq. (7.32) is:

$$y'^2 - 2y' - 3 = 0.$$

Its roots are $y'_1 = 3$ and $y'_2 = -1$. Thus, the characteristic curves are: $3x - y = C_1$ and $x + y = C_2$, so we shall use the change of variables $\xi = x + y$ and $\eta = 3x - y$ (we could also take $\xi = 3x - y$ and $\eta = x + y$, it does not matter!). We denote by $\bar{u}(\xi, \eta)$ the new function,

$$u(x, y) = \bar{u}(\xi, \eta) = \bar{u}(\xi(x, y), \eta(x, y)),$$

and, by using the formulas (7.15), (7.16), we obtain:

$$\frac{\partial u}{\partial x} = \frac{\partial \bar{u}}{\partial \xi} \frac{\partial \xi}{\partial x} + \frac{\partial \bar{u}}{\partial \eta} \frac{\partial \eta}{\partial x} = \frac{\partial \bar{u}}{\partial \xi} + 3 \frac{\partial \bar{u}}{\partial \eta},$$

$$\frac{\partial u}{\partial y} = \frac{\partial \bar{u}}{\partial \xi} \frac{\partial \xi}{\partial y} + \frac{\partial \bar{u}}{\partial \eta} \frac{\partial \eta}{\partial y} = \frac{\partial \bar{u}}{\partial \xi} - \frac{\partial \bar{u}}{\partial \eta}.$$

The second-order partial derivatives are given by the formulas (7.19), (7.21) and (7.20):

$$\frac{\partial^2 u}{\partial x^2} = \frac{\partial^2 \bar{u}}{\partial \xi^2} \cdot 1 + 2\frac{\partial^2 \bar{u}}{\partial \xi \partial \eta} \cdot 3 + \frac{\partial^2 \bar{u}}{\partial \eta^2} \cdot 9, \tag{7.44}$$

$$\frac{\partial^2 u}{\partial x \partial y} = \frac{\partial^2 \bar{u}}{\partial \xi^2} \cdot 1 \cdot 1 + \frac{\partial^2 \bar{u}}{\partial \xi \partial \eta}[1 \cdot (-1) + 1 \cdot 3] + \frac{\partial^2 \bar{u}}{\partial \eta^2} \cdot 3 \cdot (-1), \tag{7.45}$$

$$\frac{\partial^2 u}{\partial y^2} = \frac{\partial^2 \bar{u}}{\partial \xi^2} \cdot 1 + 2\frac{\partial^2 \bar{u}}{\partial \xi \partial \eta} \cdot 1 \cdot (-1) + \frac{\partial^2 \bar{u}}{\partial \eta^2} \cdot 1. \tag{7.46}$$

We multiply now the equality (7.45) by 2, the equality (7.46) by (-3), and add everything by columns. So, the Eq. (7.43) becomes:

$$\frac{\partial^2 \bar{u}}{\partial \xi \partial \eta} = \frac{1}{8}. \tag{7.47}$$

Note that Eq. (7.47) is in the canonical form (7.38). Now, to solve this equation, we write it as follows:

$$\frac{\partial}{\partial \xi}\left(\frac{\partial \bar{u}}{\partial \eta} - \frac{1}{8}\xi\right) = 0,$$

so $\frac{\partial \bar{u}}{\partial \eta} - \frac{1}{8}\xi$ does not depend on ξ:

$$\frac{\partial \bar{u}}{\partial \eta} - \frac{1}{8}\xi = h(\eta).$$

We integrate this equation with respect to η and obtain

$$\bar{u}(\xi, \eta) = \frac{1}{8}\xi\eta + G(\eta) + F(\xi), \tag{7.48}$$

where $G(\eta)$ is a primitive of the arbitrary function $h(\eta)$ and $F(\xi)$ is a "constant" of integration, which may depend on ξ. To find $u(x, y)$ we substitute ξ and η in (7.48) with their expressions, $\xi = x + y$ and $\eta = 3x - y$. So the general solution of the Eq. (7.43) is:

$$u(x, y) = \frac{1}{8}(x + y)(3x - y) + G(3x - y) + F(x + y), \tag{7.49}$$

where F and G are arbitrary functions of class C^2 on $\mathbb{R}$.

In order to solve the Cauchy problem, we have to find the expressions of the functions F and G such that the initial conditions $u(x, 0) = 0$ and $\frac{\partial u}{\partial y}(x, 0) =$

7.1 Classification: Canonical Form

$x + \cos x$ are fulfilled. Thus, from the first condition we have

$$0 = u(x, 0) = \frac{3}{8}x^2 + G(3x) + F(x), \tag{7.50}$$

and, since

$$\frac{\partial u}{\partial y}(x, y) = \frac{2}{8}(x - y) - G'(3x - y) + F'(x + y),$$

the second condition can be written:

$$\frac{\partial u}{\partial y}(x, 0) = x + \cos x = \frac{x}{4} - G'(3x) + F'(x). \tag{7.51}$$

Let us differentiate w.r.t. x the equality (7.50) and then consider also the last equality from (7.51) to obtain a linear algebraic system in the unknowns $F'(x)$ and $G'(3x)$:

$$\begin{cases} 3G'(3x) + F'(x) = -\frac{3}{4}x \\ G'(3x) - F'(x) = -\frac{3}{4}x - \cos x \end{cases}.$$

We solve this system and find:

$$F'(x) = \frac{3}{8}x + \frac{3}{4}\cos x, \quad G'(3x) = -\frac{3}{8}x - \frac{1}{4}\cos x.$$

If we put $3x = t$ in the expression of $G'(3x)$, we get

$$G'(t) = -\frac{t}{8} - \frac{1}{4}\cos\frac{t}{3}.$$

In order to obtain the expression of F and G, we integrate and obtain:

$$F(x) = \frac{3}{16}x^2 + \frac{3}{4}\sin x + C_1, \quad G(x) = -\frac{1}{16}x^2 - \frac{3}{4}\sin\frac{x}{3} + C_2.$$

If we come back to (7.50) with these expressions of F and G, we get:

$$0 = \frac{3}{8}x^2 - \frac{9}{16}x^2 - \frac{3}{4}\sin x + C_2 + \frac{3}{16}x^2 + \frac{3}{4}\sin x + C_1,$$

so $C_2 + C_1 = 0$. Thus, the solution of the initial Cauchy problem is:

$$u(x, y) = \frac{1}{8}(x + y)(3x - y) - \frac{1}{16}(3x - y)^2 - \frac{3}{4}\sin\frac{3x - y}{3} +$$

$$+ \frac{3}{16}(x + y)^2 + \frac{3}{4}\sin(x + y).$$

After some computations we find:

$$u(x, y) = xy - \frac{3}{4}\sin\frac{3x - y}{3} + \frac{3}{4}\sin(x + y).$$

The reader can check that this is indeed the solution of the above Cauchy problem.

Example 7.5 Let us find the canonical form and the general solution of the PDE-2:

$$x^2\frac{\partial^2 u}{\partial x^2} - 2xy\frac{\partial^2 u}{\partial x \partial y} + y^2\frac{\partial^2 u}{\partial y^2} + 2y\frac{\partial u}{\partial y} = 0, \quad x, y > 0. \tag{7.52}$$

The coefficients of the equation are: $A = x^2$, $B = -xy$ and $C = y^2$. So $\Delta = 0$, i.e the equation is of the parabolic type and we can apply Theorem 7.2. The associated differential equation is:

$$x^2 y'^2 + 2xy y' + y^2 = 0.$$

The discriminant of this quadratic equation is zero, so it has a double root, $y' = -\frac{y}{x}$. The general solution of this ODE-1 is $xy = C$. Thus, we can use the change of variables $\xi(x, y) = xy$ and $\eta(x, y) = x$ because the Jacobian does not vanish in the domain $\Omega = (0, \infty) \times (0, \infty)$:

$$\det J(x, y) = \begin{vmatrix} y & x \\ 1 & 0 \end{vmatrix} = -x \neq 0.$$

We denote by $\bar{u}(\xi, \eta)$ the new function, $u(x, y) = \bar{u}(\xi, \eta) = \bar{u}(\xi(x, y), \eta(x, y))$, and calculate the partial derivatives of $u(x, y)$:

$$\frac{\partial u}{\partial x} = \frac{\partial \bar{u}}{\partial \xi}\frac{\partial \xi}{\partial x} + \frac{\partial \bar{u}}{\partial \eta}\frac{\partial \eta}{\partial x} = \frac{\partial \bar{u}}{\partial \xi} \cdot y + \frac{\partial \bar{u}}{\partial \eta} \cdot 1, \tag{7.53}$$

$$\frac{\partial u}{\partial y} = \frac{\partial \bar{u}}{\partial \xi}\frac{\partial \xi}{\partial y} + \frac{\partial \bar{u}}{\partial \eta}\frac{\partial \eta}{\partial y} = \frac{\partial \bar{u}}{\partial \xi} \cdot x + \frac{\partial \bar{u}}{\partial \eta} \cdot 0, \tag{7.54}$$

$$\frac{\partial^2 u}{\partial x^2} = \frac{\partial^2 \bar{u}}{\partial \xi^2} \cdot y^2 + 2\frac{\partial^2 \bar{u}}{\partial \xi \partial \eta} \cdot y \cdot 1 + \frac{\partial^2 \bar{u}}{\partial \eta^2} + \frac{\partial \bar{u}}{\partial \xi} \cdot 0 + \frac{\partial \bar{u}}{\partial \eta} \cdot 0. \tag{7.55}$$

$$\frac{\partial^2 u}{\partial x \partial y} = \frac{\partial^2 \bar{u}}{\partial \xi^2} \cdot y \cdot x + \frac{\partial^2 \bar{u}}{\partial \xi \partial \eta}(y \cdot 0 + x \cdot 1) + \frac{\partial^2 \bar{u}}{\partial \eta^2} \cdot 1 \cdot 0 + \tag{7.56}$$

$$+ \frac{\partial \bar{u}}{\partial \xi} \cdot 1 + \frac{\partial \bar{u}}{\partial \eta} \cdot 0.$$

$$\frac{\partial^2 u}{\partial y^2} = \frac{\partial^2 \bar{u}}{\partial \xi^2} \cdot x^2 + 2\frac{\partial^2 \bar{u}}{\partial \xi \partial \eta} \cdot x \cdot 0 + \frac{\partial^2 \bar{u}}{\partial \eta^2} \cdot 0 + \frac{\partial \bar{u}}{\partial \xi} \cdot 0 + \frac{\partial \bar{u}}{\partial \eta} \cdot 0. \tag{7.57}$$

7.1 Classification: Canonical Form

We multiply the equality (7.54) by $2y$, (7.55) by x^2, (7.56) by $-2xy$, and (7.57) by y^2 and add them to obtain the canonical form of the parabolic Eq. (7.52):

$$\frac{\partial^2 \bar{u}}{\partial \eta^2} = 0.$$

As we have seen in Example 7.2, the general solution of this equation is

$$\bar{u}(\xi, \eta) = \eta F(\xi) + G(\xi),$$

where F and G are arbitrary functions of class C^2. Hence the general solution of the Eq. (7.52) is

$$u(x, y) = x F(xy) + G(xy).$$

Example 7.6 Let us find the canonical form of the PDE-2:

$$\frac{\partial^2 u}{\partial x^2} - 2 \frac{\partial^2 u}{\partial x \partial y} + 5 \frac{\partial^2 u}{\partial y^2} = 0. \tag{7.58}$$

Since $A = 1$, $B = -1$, $C = 5$ and $\Delta = -4 < 0$, we have an equation of elliptic type. The characteristic differential equation is:

$$y'^2 + 2y' + 5 = 0.$$

Its roots are $y' = -1 \pm 2i$ and we find by integration: $y + x - 2ix = C_1$ and $y + x + 2ix = C_2$, respectively. So we make the change of variables $\xi = x + y$ and $\eta = 2x$. We denote by $\bar{u}(\xi, \eta)$ the new function, $u(x, y) = \bar{u}(\xi, \eta) = \bar{u}(\xi(x, y), \eta(x, y))$, and calculate the partial derivatives of $u(x, y)$:

$$\frac{\partial u}{\partial x} = \frac{\partial \bar{u}}{\partial \xi} \frac{\partial \xi}{\partial x} + \frac{\partial \bar{u}}{\partial \eta} \frac{\partial \eta}{\partial x} = \frac{\partial \bar{u}}{\partial \xi} + 2 \frac{\partial \bar{u}}{\partial \eta},$$

$$\frac{\partial u}{\partial y} = \frac{\partial \bar{u}}{\partial \xi} \frac{\partial \xi}{\partial y} + \frac{\partial \bar{u}}{\partial \eta} \frac{\partial \eta}{\partial y} = \frac{\partial \bar{u}}{\partial \xi},$$

$$\frac{\partial^2 u}{\partial x^2} = \frac{\partial^2 \bar{u}}{\partial \xi^2} + 4 \frac{\partial^2 \bar{u}}{\partial \xi \partial \eta} + 4 \frac{\partial^2 \bar{u}}{\partial \eta^2}. \tag{7.59}$$

$$\frac{\partial^2 u}{\partial x \partial y} = \frac{\partial^2 \bar{u}}{\partial \xi^2} + 2 \frac{\partial^2 \bar{u}}{\partial \xi \partial \eta}. \tag{7.60}$$

$$\frac{\partial^2 u}{\partial y^2} = \frac{\partial^2 \bar{u}}{\partial \xi^2}. \tag{7.61}$$

Let us multiply the equality (7.60) by -2 and (7.61) by 5 and add them by columns. Thus, the canonical form of the Eq. (7.58) is the Laplace equation (7.3):

$$\nabla^2 \bar{u} = \frac{\partial^2 \bar{u}}{\partial \xi^2} + \frac{\partial^2 \bar{u}}{\partial \eta^2} = 0.$$

7.2 The Wave Equation

Consider a perfectly elastic string which is homogeneous (the mass per unit length is a constant, ρ) and performs small transverse motions in the vertical plane xOu: the x-axis represents the equilibrium position of the string and we assume that each point of the string moves on the vertical direction only. Let $u(x, t)$ denote the deflection at the point x and at time t (the vertical displacement of any point on the string, x, measured from the x-axis). We assume that the displacements $u(x, t)$ are small compared to the length of the string, the slope of the curve is also small, and the tension T acts tangent to the string, and its magnitude is the same at all points (see Fig. 7.1). We also suppose that the tension is so large that the action of the gravitational force, as well as any other exterior force on the string can be neglected (i.e. we have *free vibrations*). We shall prove that the function $u(x, t)$ satisfies the following relation (see also [44]):

$$\rho \frac{\partial^2 u}{\partial t^2} = T \frac{\partial^2 u}{\partial x^2}, \qquad (7.62)$$

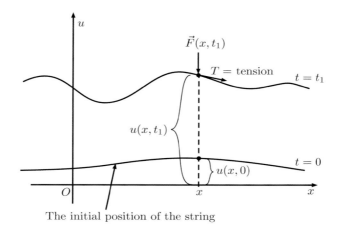

Fig. 7.1 Vibrating string

7.2 The Wave Equation

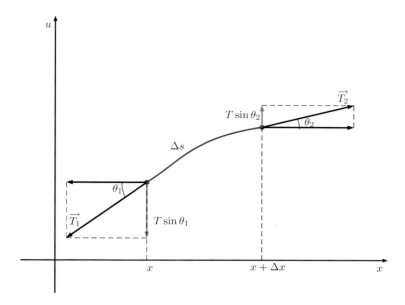

Fig. 7.2 A small portion of the string

Let us examine, at time t, a small portion of the string (of length Δs) corresponding to the interval $[x, x + \Delta x]$ (see Fig. 7.2).

Let $\vec{T_1}$ and $\vec{T_2}$ be the tensions at the endpoints. Since we assumed that the slope of the curve is small, we can approximate $\cos\theta_1 \approx \cos\theta_2 \approx 1$, so the horizontal components of the tensions $\vec{T_1}$ and $\vec{T_2}$ are equal:

$$T\cos\theta_1 = T\cos\theta_2 = T$$

(we do not have movement on the horizontal direction). By writing Newton's second law $F = ma$ on the vertical direction we obtain:

$$T\sin\theta_2 - T\sin\theta_1 = m \cdot a = \rho\Delta x \cdot \frac{\partial^2 u}{\partial t^2}. \tag{7.63}$$

Using again the approximation $\cos\theta_1 \approx \cos\theta_2 \approx 1$ one can write:

$$\sin\theta_1 \approx \tan\theta_1 = \frac{\partial u}{\partial x}(x, t) \quad \text{and} \quad \sin\theta_2 \approx \tan\theta_2 = \frac{\partial u}{\partial x}(x + \Delta x, t).$$

We replace in (7.63) and divide the equality by Δx:

$$T \frac{\frac{\partial u}{\partial x}(x+\Delta x, t) - \frac{\partial u}{\partial x}(x, t)}{\Delta x} = \rho \frac{\partial^2 u}{\partial t^2}.$$

Since $\lim\limits_{\Delta x \to 0} \frac{\frac{\partial u}{\partial x}(x+\Delta x, t) - \frac{\partial u}{\partial x}(x, t)}{\Delta x} = \frac{\partial^2 u}{\partial x^2}$, we obtain the Eq. (7.62).

We denote the positive constant T/ρ by a^2 ($a>0$) and obtain the homogeneous *vibrating string equation* (or the homogeneous *one-dimensional wave equation*):

$$\frac{\partial^2 u}{\partial t^2} = a^2 \frac{\partial^2 u}{\partial x^2}. \tag{7.64}$$

If the resultant of all the exterior forces cannot be neglected, then we have *forced vibrations*. Let $F(x, t)$ be the acceleration due to the external forces acting vertically at time t at the point $(x, u(x, t))$ (see Fig. 7.1). The partial differential equation in this case is the following non-homogeneous equation:

$$\frac{\partial^2 u}{\partial t^2} = a^2 \frac{\partial^2 u}{\partial x^2} + F(x, t). \tag{7.65}$$

If instead of a string (a one dimensional object) we consider a two dimensional object (a vibrating membrane), then one can prove that the function $u(x, y, t)$ satisfies the *two-dimensional wave equation*:

$$\frac{\partial^2 u}{\partial t^2} = a^2 \left(\frac{\partial^2 u}{\partial x^2} + \frac{\partial^2 u}{\partial y^2} \right) + F(x, y, t). \tag{7.66}$$

In the case of *free vibrations* (when $F(x, y, t) \equiv 0$), Eq. (7.66) becomes the homogeneous Eq. (7.5):

$$\frac{\partial^2 u}{\partial t^2} = a^2 \left(\frac{\partial^2 u}{\partial x^2} + \frac{\partial^2 u}{\partial y^2} \right). \tag{7.67}$$

Finally, the three-dimensional (homogeneous) wave equation is:

$$\frac{\partial^2 u}{\partial t^2} = a^2 \left(\frac{\partial^2 u}{\partial x^2} + \frac{\partial^2 u}{\partial y^2} + \frac{\partial^2 u}{\partial z^2} \right) = a^2 \nabla^2 u, \tag{7.68}$$

where $\nabla^2 u = \frac{\partial^2 u}{\partial x^2} + \frac{\partial^2 u}{\partial y^2} + \frac{\partial^2 u}{\partial z^2}$ is the *Laplacian* of the function u.

7.2 The Wave Equation

This equation is extremely important in the study of electromagnetic phenomena (governed by Maxwell's equations). If $\mathbf{E} = \mathbf{E}(x, y, z, t)$ and $\mathbf{H} = \mathbf{H}(x, y, z, t)$ represent electric and magnetic fields in empty space, then Eq. (7.68) is satisfied by each component of the vector functions $\mathbf{E} = E_1\mathbf{i} + E_2\mathbf{j} + E_3\mathbf{k}$ and $\mathbf{H} = H_1\mathbf{i} + H_2\mathbf{j} + H_3\mathbf{k}$, respectively. Thus, one can write:

$$\frac{\partial^2 \mathbf{E}}{\partial t^2} = c^2 \nabla^2 \mathbf{E} \quad \text{and} \quad \frac{\partial^2 \mathbf{H}}{\partial t^2} = c^2 \nabla^2 \mathbf{H},$$

where c is the velocity of light (see [20], p. 963).

7.2.1 Infinite Vibrating String: D'Alembert Formula

An infinite vibrating string corresponds to an infinite domain for the spatial variable ($x \in \mathbb{R}$). In the following, we show how to solve a Cauchy problem for the one dimensional homogeneous wave equation, that is, how to find a function $u(x, t)$ of class C^2 on $\Omega = \mathbb{R} \times [0, \infty)$ such that:

$$\frac{\partial^2 u}{\partial t^2}(x, t) = a^2 \frac{\partial^2 u}{\partial x^2}(x, t), \quad (x, t) \in \Omega \tag{7.69}$$

and

$$\begin{aligned} u(x, 0) &= f(x), \\ \frac{\partial u}{\partial t}(x, 0) &= g(x), \end{aligned} \quad x \in \mathbb{R}, \tag{7.70}$$

where $f(x)$ is a function of class C^2 on $\mathbb{R}$ and $g(x)$ is a function of class C^1 on $\mathbb{R}$. The relations (7.70) are the *initial conditions* of the Cauchy problem. The first function, $f(x)$ gives the position of the string at the initial moment, $t = 0$, while the second function, $g(x)$, provides the initial velocity for each point on the string, $M(x)$.

The Eq. (7.69) can be written as

$$\frac{\partial^2 u}{\partial t^2} - a^2 \frac{\partial^2 u}{\partial x^2} = 0. \tag{7.71}$$

By using the theory described in the previous section, let us find its general solution by the change of variables method (d'Alembert method). Note that in our case the free variables are t and x (instead of x and y). The coefficients are: $A = 1$, $B = 0$ and $C = -a^2$ and we have $\Delta = B^2 - AC = a^2 > 0$, so the equation is of the hyperbolic type. Its associated characteristic equation (relative to the function $x = x(t)$) is $x'^2 - a^2 = 0$, so $x - at = C_1$ and $x + at = C_2$ are its general solutions.

Thus, the corresponding change of variables is: $\xi = x - at$ and $\eta = x + at$. By applying the chain rule formulas (7.19), (7.20) we find the canonical form of this PDE-2:

$$\frac{\partial^2 \bar{u}}{\partial \xi \partial \eta} = 0. \tag{7.72}$$

As shown in Example 7.1, the general solution of the Eq. (7.72) is:

$$\bar{u}(\xi, \eta) = F(\xi) + G(\eta),$$

where F, G are two arbitrary functions of class C^2. Thus, the general solution of the Eq. (7.71) is:

$$u(x, t) = F(x - at) + G(x + at). \tag{7.73}$$

The initial conditions (7.70) of the Cauchy problem become:

$$u(x, 0) = F(x) + G(x) = f(x) \tag{7.74}$$

and

$$\frac{\partial u}{\partial t}(x, 0) = -aF'(x) + aG'(x) = g(x). \tag{7.75}$$

This last equation can also be written in the form:

$$F'(x) - G'(x) = -\frac{1}{a}g(x)$$

and, if we fix an arbitrary real number x_0 and integrate this last equality on the interval $[x_0, x]$, we obtain:

$$F(x) - G(x) = -\frac{1}{a} \int_{x_0}^{x} g(z)\, dz + C, \tag{7.76}$$

where $C = F(x_0) - G(x_0)$. The relations (7.74) and (7.76) give rise to a linear system in the unknowns $F(x)$ and $G(x)$. So we find:

$$F(x) = \frac{1}{2} f(x) - \frac{1}{2a} \int_{x_0}^{x} g(z)\, dz + \frac{C}{2},$$

$$G(x) = \frac{1}{2} f(x) + \frac{1}{2a} \int_{x_0}^{x} g(z)\, dz - \frac{C}{2}.$$

7.2 The Wave Equation

Coming back to the expression of $u(x, t)$ from (7.73), we obtain:

$$u(x, t) = \frac{1}{2} f(x - at) - \frac{1}{2a} \int_{x_0}^{x-at} g(z) \, dz + \frac{1}{2} f(x + at) + \frac{1}{2a} \int_{x_0}^{x+at} g(z) \, dz.$$

Thus, by the obvious relation: $\int_c^b - \int_c^a = \int_a^c + \int_c^b = \int_a^b$, we finally get:

$$u(x, t) = \frac{1}{2} [f(x + at) + f(x - at)] + \frac{1}{2a} \int_{x-at}^{x+at} g(z) \, dz, \qquad (7.77)$$

which is the unique solution of the Cauchy problem. Formula (7.77) is called the *d'Alembert formula* for the infinite vibrating string.

When the initial velocity $g(x) \equiv 0$, the formula (7.77) becomes

$$u(x, t) = \frac{1}{2} [f(x + at) + f(x - at)]. \qquad (7.78)$$

This relation can be interpreted as a superposition of two traveling waves, one moving to the right, $\frac{1}{2} f(x - at)$, and one moving to the left $\frac{1}{2} f(x + at)$, with the same speed, a, and having the same shape as the initial displacement $f(x)$ (as shown in the following example).

Example 7.7 Consider an infinitely long, perfectly elastic string, initially held at the three points $(-1, 0)$, $(0, 1)$ and $(1, 0)$ (outside the interval $(-1, 1)$ the string coincides with the x-axis). At time $t = 0$ the string is simultaneously released at all three points. Draw the position of the string at time $t = 1$ and $t = 2$, knowing that $a = 1$.

We have to solve the equation

$$\frac{\partial^2 u}{\partial t^2} = \frac{\partial^2 u}{\partial x^2}, \quad x \in \mathbb{R}, \; t \geq 0,$$

with the initial conditions:

$$u(x, 0) = f(x) = \begin{cases} 1 - |x|, & |x| \leq 1 \\ 0, & |x| > 1, \end{cases}$$

$$\frac{\partial u}{\partial t}(x, 0) = 0, \; x \in \mathbb{R}.$$

We remark that the function $f(x)$ is not of class C^2, but it is continuous and, even if the function $u(x, t)$ constructed by d'Alembert formula (7.77) does not satisfy the PDE (7.69) for any $(x, t) \in \Omega$, since $u \notin C^2(\Omega)$, we can still call it a solution in a *generalized sens* (such cases, with the functions $f(x)$ and $g(x)$ not differentiable, are often met in practice).

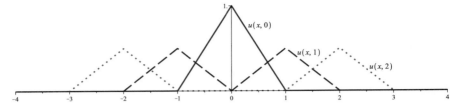

Fig. 7.3 The function $u(x, t)$ for $t = 0$, $t = 1$ and $t = 2$

Using the formula (7.78), the solution is:

$$u(x, t) = \frac{1}{2}[f(x + t) + f(x - t)].$$

The positions of the string at time $t = 0$, $t = 1$ and $t = 2$ are presented in Fig. 7.3.

Example 7.8 Let us solve the following Cauchy problem for the infinite vibrating string:

$$4\frac{\partial^2 u}{\partial t^2} = \frac{\partial^2 u}{\partial x^2}, \quad x \in \mathbb{R}, \ t \geq 0,$$

$$u(x, 0) = \sin x, \quad x \in \mathbb{R},$$

$$\frac{\partial u}{\partial t}(x, 0) = \cos^2 x, \quad x \in \mathbb{R}.$$

By d'Alembert formula (7.77) (note that $a = \frac{1}{2}$ in this case) we get:

$$u(x, t) = \frac{1}{2}\left[\sin\left(x + \tfrac{1}{2}t\right) + \sin\left(x - \tfrac{1}{2}t\right)\right] + \int_{x-t/2}^{x+t/2} \frac{1 + \cos 2z}{2}\, dz =$$

$$= \sin x \cos \frac{t}{2} + \frac{t}{2} + \frac{\sin 2z}{4}\bigg|_{x-t/2}^{x+t/2}.$$

Thus,

$$u(x, t) = \sin x \cos \frac{t}{2} + \frac{t}{2} + \frac{1}{2}\cos 2x \sin t.$$

The positions of the string at time $t = 0$, $t = \pi/2$, $t = 3\pi/4$ and $t = \pi$ are presented in Fig. 7.4. Note that for $t = \pi$, the position of the string is the straight line $u = \pi/2$.

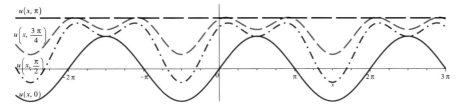

Fig. 7.4 The solution $u(x, t)$ for $t = 0, t = \pi/2, t = 3\pi/4$ and $t = \pi$

7.2.2 Finite Vibrating String: Fourier Method

The *finite vibrating string* is a homogeneous, perfectly elastic, one-dimensional object of length l, ($l > 0$), which performs small motions in a vertical plane (see Fig. 7.5). The equation of the displacement $u(x, t)$ is the same:

$$\frac{\partial^2 u}{\partial t^2} - a^2 \frac{\partial^2 u}{\partial x^2} = 0, \quad x \in (0, l); \; t > 0. \tag{7.79}$$

The initial conditions are also the same: the initial deflection of the string and the initial velocity at each point x are known:

$$u(x, 0) = f(x), \; x \in [0, l] \tag{7.80}$$

$$\frac{\partial u}{\partial t}(x, 0) = g(x), \; x \in (0, l). \tag{7.81}$$

But in the case of the *finite* string we have also another type of conditions, called *boundary conditions*. Thus, the conditions

$$\begin{matrix} u(0, t) = 0, \\ u(l, t) = 0, \end{matrix} \quad t \geq 0, \tag{7.82}$$

state that during the vibrating process the ends of the string remain fixed at the origin $O(0, 0)$ and at $L(l, 0)$ respectively. It is easy to see that the boundary conditions imply that

$$f(0) = f(l) = 0. \tag{7.83}$$

Let us remark that the Eq. (7.79) with the initial and boundary conditions is a completely different problem of that one studied in the previous section (for the *infinite* string). Because of the boundary conditions (the ends of the string remain fixed during vibrations), d'Alembert method cannot be applied for a *finite* string. This is why Fourier used two main ideas to find a solution for the finite vibrating

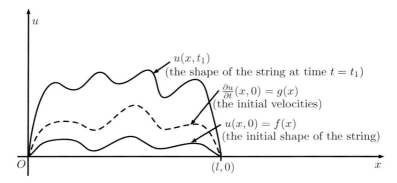

Fig. 7.5 Finite vibrating string

string problem. The first one was to the search for a solution $u(x, t)$ with "separated variables", i.e. a function of the form:

$$u(x, t) = X(x) \cdot T(t), \tag{7.84}$$

where $X(x)$ and $T(t)$ are functions of one variable such that $u(x, t)$ satisfies the Eq. (7.79), and the boundary conditions (7.82). The second idea of Fourier was to "superpose" an infinite sequence of solutions $u_n(x, t) = X_n(x) \cdot T_n(t)$, $n = 1, 2, \ldots$, with separated variables, in order to obtain a solution $u(x, t) = \sum_{n=1}^{\infty} u_n(x, t)$ of the Eq. (7.79) that satisfies not only the boundary conditions (7.82), but also the initial conditions (7.80), (7.81). This is why two distinct "steps" naturally appear in solving our problem.

The First Step (Separation of Variables) Let us search for a solution $u(x, t) = X(x) \cdot T(t)$ of the partial problem:

$$\frac{\partial^2 u}{\partial t^2} - a^2 \frac{\partial^2 u}{\partial x^2} = 0, \text{ with } u(0, t) = 0, \ u(l, t) = 0, \ t \geq 0. \tag{7.85}$$

By introducing $u(x, t) = X(x) \cdot T(t)$ in Eq. (7.79) we get:

$$X(x)T''(t) = a^2 X''(x)T(t),$$

or, equivalently,

$$\frac{X''(x)}{X(x)} = \frac{1}{a^2} \frac{T''(t)}{T(t)}.$$

7.2 The Wave Equation

On the left side we have a function of x and on the right side, a function of t, so the equality is possible only if the two functions are equal to the same constant, say λ:

$$\frac{X''(x)}{X(x)} = \frac{1}{a^2} \frac{T''(t)}{T(t)} = \lambda.$$

These equalities give rise to the following ODEs:

$$X''(x) - \lambda X(x) = 0 \tag{7.86}$$

and

$$T''(t) - a^2 \lambda T(t) = 0. \tag{7.87}$$

The boundary conditions (7.82) imply $X(0) = 0$ and $X(l) = 0$, so the function $X(x)$ is a solution of the following boundary value problem:

$$X''(x) - \lambda X(x) = 0, \quad X(0) = X(l) = 0. \tag{7.88}$$

We prove that this boundary value problem has only trivial solutions if $\lambda \geq 0$.

Case 1 $\lambda > 0$. The equation $X''(x) - \lambda X(x) = 0$ is an ODE-2 with constant coefficients. Its general solution is:

$$X(x) = C_1 \exp(\sqrt{\lambda} x) + C_2 \exp(-\sqrt{\lambda} x), \; x \in [0, l]. \tag{7.89}$$

The boundary conditions $X(0) = 0$, $X(l) = 0$ give rise to the following algebraic linear homogeneous system in the variables C_1 and C_2 respectively:

$$\begin{cases} C_1 + C_2 = 0 \\ \exp(\sqrt{\lambda} l) C_1 + \exp(-\sqrt{\lambda} l) C_2 = 0. \end{cases}$$

Its determinant is equal to $\exp(-\sqrt{\lambda} l) - \exp(\sqrt{\lambda} l)$ which cannot be zero because $\lambda, l > 0$. Thus, the solution is $C_1 = C_2 = 0$ and so $X(x) = 0$ which implies $u(x, t) = 0$ for any $x \in [0, l]$ and $t \geq 0$. But this trivial solution does not satisfy the initial conditions (7.80) and (7.81), except the trivial case when $f = g = 0$, i.e. when we have no vibration at all.

Case 2 $\lambda = 0$. In this situation, the Eq. (7.86) has the general solution:

$$X(x) = C_1 x + C_2, \; x \in [0, l].$$

The same boundary conditions $X(0) = 0$, $X(l) = 0$ imply that $C_1 = C_2 = 0$, so we obtain the trivial solution $u(x, t) = 0$. So the only possible case is $\lambda < 0$:

Case 3 $\lambda < 0$. In order to avoid to use radicals, we put $\lambda = -\mu^2$, $\mu > 0$. The general solution of our ODE-2 (7.86) is:

$$X(x) = C_1 \cos \mu x + C_2 \sin \mu x, \ x \in [0, l]. \tag{7.90}$$

The first boundary condition, $X(0) = 0$ implies $C_1 = 0$. The second boundary condition $X(l) = 0$ implies $\sin \mu l = 0$ (because $C_2 = 0$ would lead to the same trivial solution $u(x, t) = 0$). Thus the constant μ cannot be arbitrary. From $\sin \mu l = 0$ we deduce that

$$\mu = \mu_n = \frac{n\pi}{l}, \ n = 1, 2, \ldots \tag{7.91}$$

We do not take here $n < 0$ because the sign "$-$" is included in the arbitrary constant C_2 which appear in front of $\sin \mu x$.

Thus, the infinite sequence of solutions of the boundary value problem (7.88),

$$X_n(x) = \sin\left(\frac{n\pi}{l}x\right), \ n = 1, 2, \ldots. \tag{7.92}$$

are called the *eigenfunctions* of the boundary value problem. The (positive) values $-\lambda_n = \frac{n^2\pi^2}{l^2}$ are said to be the *eigenvalues* of the Eq. (7.86) with the boundary conditions $X(0) = X(l) = 0$ (see also Example 2.3). From an algebraic perspective, if we consider the linear operator $\frac{d^2}{dx^2}$ (in the space of the functions of class $C^2(0, l)$ that satisfy the boundary conditions $X(0) = X(l) = 0$), then the (negative) numbers $\lambda_n = -\frac{n^2\pi^2}{l^2}$, $n = 1, 2, \ldots$, are its *eigenvalues* and the functions (7.92) are corresponding eigenvectors.

Now, let us solve Eq. (7.87) with $\lambda = \lambda_n = -\frac{n^2\pi^2}{l^2}$. For each $n = 1, 2, \ldots$, the linear ODE-2 with constant coefficients:

$$T_n''(t) + \frac{n^2\pi^2 a^2}{l^2} T_n(t) = 0, \ t \geq 0,$$

has the general solution:

$$T_n(t) = A_n \cos\left(\frac{n\pi a}{l}t\right) + B_n \sin\left(\frac{n\pi a}{l}t\right), \ t \geq 0,$$

where A_n, B_n are arbitrary constants.

Thus, coming back to (7.84), we obtain the following sequence of solutions for the partial problem (7.85):

$$u_n(x, t) = \left[A_n \cos\left(\frac{n\pi a}{l}t\right) + B_n \sin\left(\frac{n\pi a}{l}t\right)\right] \sin\left(\frac{n\pi}{l}x\right), \ n = 1, 2, \ldots. \tag{7.93}$$

7.2 The Wave Equation

As Fourier remarked almost two hundred years ago, each of these functions $u_n(x,t)$, separately considered, does not satisfy the initial conditions (7.80) and (7.81). But these solutions can be "superposed" in order to obtain a new solution that would satisfy the initial conditions.

The Second Step (Superposition Principle) Thus, let us search for a solution of the following type:

$$u(x,t) = \sum_{n=1}^{\infty} u_n(x,t)$$
$$= \sum_{n=1}^{\infty} \left[A_n \cos\left(\frac{n\pi a}{l}t\right) + B_n \sin\left(\frac{n\pi a}{l}t\right) \right] \sin\left(\frac{n\pi}{l}x\right). \tag{7.94}$$

If this series is convergent, then $u(x,t)$ satisfies the boundary conditions (7.82) (because $u_n(x,t)$ does, for every $n = 1, 2, \ldots$). If, in addition, the series is uniformly convergent with respect to both variables x and t and the series resulted by partial differentiations are also uniformly convergent, then $u(x,t)$ is also a solution of the PDE-2 (7.79) (see Remark 7.3).

In order to find the coefficients A_n, B_n, $n = 1, 2, \ldots$, we shall force $u(x,t)$ from (7.94) to satisfy the initial conditions. Thus, from (7.80) we get:

$$u(x,0) = f(x) = \sum_{n=1}^{\infty} A_n \sin\left(\frac{n\pi}{l}x\right), \quad x \in [0, l], \tag{7.95}$$

and, since

$$\frac{\partial u}{\partial t}(x,t) = \sum_{n=1}^{\infty} \frac{n\pi a}{l} \left[-A_n \sin\left(\frac{n\pi a}{l}t\right) + B_n \cos\left(\frac{n\pi a}{l}t\right) \right] \sin\left(\frac{n\pi}{l}x\right),$$

by (7.80) we obtain:

$$\frac{\partial u}{\partial t}(x,0) = g(x) = \sum_{n=1}^{\infty} \left(B_n \frac{n\pi a}{l} \right) \sin\left(\frac{n\pi}{l}x\right), \quad x \in [0, l]. \tag{7.96}$$

The formulas (7.95) and (7.96) represent the expansion in Fourier series of sines of the function $f(x)$ and $g(x)$, respectively (see Chap. 4, Remark 4.9). Thus, by using the formula (4.87), we find the coefficients:

$$A_n = \frac{2}{l} \int_0^l f(x) \sin\left(\frac{n\pi}{l}x\right) dx \tag{7.97}$$

and

$$B_n = \frac{2}{n\pi a} \int_0^l g(x) \sin\left(\frac{n\pi}{l}x\right) dx. \tag{7.98}$$

Remark 7.3 (see [22], Theorem 30.2) If $f(x)$ is a function of class C^3 on $[0, l]$ such that

$$f(0) = f(l) = 0, \ f''(0) = f''(l) = 0$$

and $g(x)$ is a function of class C^2 on $[0, l]$ with

$$g(0) = g(l) = 0,$$

then the function $u(x, t)$ defined by (7.94), with A_n and B_n given by the formulas (7.97) and (7.98), is a function of class C^2 on $[0, l] \times [0, \infty)$ and it is a solution of the PDE-2 (7.79) that satisfies the boundary conditions (7.82) and the initial conditions (7.80), (7.81).

If the functions f and g do not meet the conditions above, but they are continuous on $[0, l]$ and satisfy $f(0) = f(l) = 0$, $g(0) = g(l) = 0$, then the series (7.94) is still uniformly convergent to the continuous function $u(x, t)$ which is said to be a *generalized solution* of the problem (see Example 7.10).

Theorem 7.4 (Krasnov et al. [22], 30.9) *The solution $u(x, t)$ of the PDE-2 (7.79) which satisfies the boundary conditions (7.82) and the initial conditions (7.80), (7.81) is unique.*

Proof Let $u_1(x, t)$ and $u_2(x, t)$ be two solutions with the properties stated in the theorem. Then the function $v(x, t) = u_1(x, t) - u_2(x, t)$ satisfies the following conditions:

$$\frac{\partial^2 v}{\partial t^2} - a^2 \frac{\partial^2 v}{\partial x^2} = 0, \ x \in (0, l), \ t \geq 0, \tag{7.99}$$

$$v(0, t) = v(l, t) = 0, \ t \geq 0, \tag{7.100}$$

$$v(x, 0) = 0, \quad \frac{\partial v}{\partial t}(x, 0) = 0, \ x \in [0, l]. \tag{7.101}$$

Let us prove that in this case $v(x, t) = 0$ on $\Omega = [0, l] \times [0, \infty)$, i.e. $u_1 = u_2$.

We consider the function

$$E(t) = \frac{1}{2} \int_0^l \left[\left(\frac{\partial v}{\partial t}(x, t)\right)^2 + a^2 \left(\frac{\partial v}{\partial x}(x, t)\right)^2 \right] dx. \tag{7.102}$$

7.2 The Wave Equation

By using Leibniz' formula for differentiating an integral with parameter (1.72), we compute its derivative with respect to t:

$$E'(t) = \int_0^l \frac{\partial v}{\partial t} \cdot \frac{\partial^2 v}{\partial t^2} dx + a^2 \int_0^l \frac{\partial v}{\partial x} \cdot \frac{\partial}{\partial x}\left(\frac{\partial v}{\partial t}\right) dx. \qquad (7.103)$$

But, integrating by parts the last integral, we get:

$$\int_0^l \frac{\partial v}{\partial x} \cdot \frac{\partial}{\partial x}\left(\frac{\partial v}{\partial t}\right) dx = \left.\frac{\partial v}{\partial x}(x,t) \cdot \frac{\partial v}{\partial t}(x,t)\right|_0^l - \int_0^l \frac{\partial^2 v}{\partial x^2} \cdot \frac{\partial v}{\partial t} dx$$

$$= -\int_0^l \frac{\partial^2 v}{\partial x^2} \cdot \frac{\partial v}{\partial t} dx$$

(the first term vanishes because $\frac{\partial v}{\partial t}(0,t) = \frac{\partial v}{\partial t}(l,t) = 0$ from the boundary conditions (7.100)). Coming back to (7.103), we finally obtain:

$$E'(t) = \int_0^l \frac{\partial v}{\partial t}\left[\frac{\partial^2 v}{\partial t^2} - a^2 \frac{\partial^2 v}{\partial x^2}\right] dx = 0,$$

because of (7.99). Thus, $E(t) = c$ (a constant) for any $t \geq 0$. However, from the initial conditions (7.101) we find $E(0) = 0$, so $E(t) = 0$ for any $t \geq 0$. Therefore,

$$\int_0^l \left[\left(\frac{\partial v}{\partial t}(x,t)\right)^2 + a^2\left(\frac{\partial v}{\partial x}(x,t)\right)^2\right] dx = 0, \qquad (7.104)$$

for any $t \geq 0$ and so

$$\left(\frac{\partial v}{\partial t}(x,t)\right)^2 + a^2\left(\frac{\partial v}{\partial x}(x,t)\right)^2 \equiv 0.$$

It follows that

$$\frac{\partial v}{\partial t}(x,t) = 0, \quad \frac{\partial v}{\partial x}(x,t) = 0, \forall (x,t) \in \Omega,$$

so $v(x,t)$ is a constant function. Taking into account anyone of the boundary conditions (7.100), or the initial condition $v(x,0) = 0$, we conclude that $v(x,t) = 0$ for any $(x,t) \in \Omega$ and the theorem is proved. □

Recall that $a^2 = T/\rho$, where T is the tension in the string and ρ is the mass of the string per unit length (both are constants). Thus, using for the partial derivatives the notations $\frac{\partial v}{\partial x} = v_x$ and $\frac{\partial v}{\partial t} = v_t$ (the velocity), the integral $E(t)$ can be written:

$$E(t) = \frac{1}{\rho}\int_0^l \left(\frac{1}{2}\rho v_t^2 + \frac{1}{2}T v_x^2\right) dx.$$

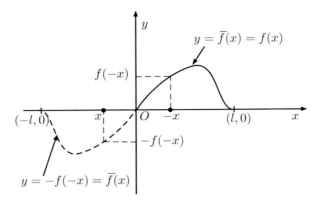

Fig. 7.6 The extension of $f(x)$ to an odd function $\overline{f}(x)$

The integral $\int_0^l \frac{1}{2}\rho v_t^2 dx$ is the kinetic energy of the string at time t, while $\int_0^l \frac{1}{2}T v_x^2 dx$ represents its potential energy, so $E(t)\rho$ is the total energy of the string. The equality $E'(t) = 0$ is the mathematical expression of the law of conservation of energy.

Remark 7.4 Suppose that the function f is continuous and piecewise smooth on the interval $[0, l]$, and $f(0) = f(l) = 0$. We extend it to an odd function on the interval $[-l, l]$ and then, extend this odd function by periodicity to the whole x-axis. Let $\overline{f}$ denote this extension, which is a continuous, odd, periodic function (of period $2l$) (see Fig. 7.6). Then,

$$\overline{f}(x) = \sum_{n=1}^\infty A_n \sin \frac{n\pi x}{l}, \quad x \in \mathbb{R},$$

where the coefficients A_n are given by the formula (7.97):

$$A_n = \frac{2}{l}\int_0^l f(x)\sin\frac{n\pi x}{l}\,dx, \quad n = 1, 2, \ldots.$$

We can see that, for any $x \in [0, l]$ and $t \geq 0$, we have:

$$\frac{\overline{f}(x+at) + \overline{f}(x-at)}{2} = \frac{1}{2}\sum_{n=1}^\infty A_n\left(\sin\frac{n\pi(x+at)}{l} + \sin\frac{n\pi(x-at)}{l}\right),$$

7.2 The Wave Equation

so

$$\frac{\overline{f}(x+at)+\overline{f}(x-at)}{2} = \sum_{n=1}^{\infty} A_n \cos\frac{n\pi at}{l} \sin\frac{n\pi x}{l}. \tag{7.105}$$

Similarly, if g is a continuous, piecewise smooth function on the interval $[0, l]$ such that $g(0) = g(l) = 0$ and $\overline{g}$ is the extension of g to an odd, periodic function of period $2l$, then, for any $x \in \mathbb{R}$,

$$\overline{g}(x) = \sum_{n=1}^{\infty} C_n \sin\frac{n\pi x}{l}, \tag{7.106}$$

where the coefficients C_n are given by the formula:

$$C_n = \frac{2}{l}\int_0^l g(x)\sin\frac{n\pi x}{l}\,dx, \quad n = 1, 2, \ldots.$$

We integrate (7.106) on the interval $[x - at, x + at]$ and obtain:

$$\int_{x-at}^{x+at} \overline{g}(z)\,dz = \sum_{n=1}^{\infty} C_n \frac{l}{n\pi}\left(-\cos\frac{n\pi z}{l}\right)\Big|_{x-at}^{x+at}$$

$$= \sum_{n=1}^{\infty} C_n \frac{2l}{n\pi} \sin\frac{n\pi at}{l} \sin\frac{n\pi x}{l}.$$

Consequently, if B_n are the coefficients defined by (7.98),

$$B_n = \frac{2}{n\pi a}\int_0^l g(x)\sin\left(\frac{n\pi}{l}x\right)dx = C_n \frac{l}{n\pi a},$$

then we can write:

$$\frac{1}{2a}\int_{x-at}^{x+at} \overline{g}(z)\,dz = \sum_{n=1}^{\infty} B_n \sin\frac{n\pi at}{l} \sin\frac{n\pi x}{l}. \tag{7.107}$$

Thus, if we consider $f(x)$ the initial shape and $g(x)$ the initial velocity of the *finite* vibrating string with fixed ends, we can see that the solution (7.94),

$$u(x,t) = \sum_{n=1}^{\infty}\left[A_n \cos\left(\frac{n\pi a}{l}t\right) + B_n \sin\left(\frac{n\pi a}{l}t\right)\right]\sin\left(\frac{n\pi}{l}x\right),$$

is identical to the d'Alembert solution (7.77) for an *infinite* vibrating string having as initial shape and initial velocity the odd, periodic functions $\overline{f}$ and $\overline{g}$, respectively:

$$u(x,t) = \frac{\overline{f}(x+at) + \overline{f}(x-at)}{2} + \frac{1}{2a}\int_{x-at}^{x+at} \overline{g}(z)\,dz. \quad (7.108)$$

The function $u(x,t)$ is periodic in t of period $\frac{2l}{a}$. Moreover if the initial velocity is 0, then

$$u(x,t) = \sum_{n=1}^{\infty} A_n \cos\left(\frac{n\pi a}{l}t\right) \sin\left(\frac{n\pi}{l}x\right)$$

and it is sufficient to find $u(x,t)$ for $t \in \left(0, \frac{l}{2a}\right)$, since $u\left(x, \frac{l}{a} \pm t\right) = -u(x,t)$ and $u\left(x, \frac{2l}{a} - t\right) = u(x,t)$. Notice that the expression of $\overline{f}(x)$ on the interval $[-l, 2l]$ is:

$$\overline{f}(x) = \begin{cases} f(x), & x \in [0, l) \\ -f(-x), & x \in [-l, 0) \\ -f(2l-x), & x \in [l, 2l). \end{cases}$$

Thus, for any $t \in \left(0, \frac{l}{2a}\right)$, one can write:

$$u(x,t) = \frac{\overline{f}(x+at) + \overline{f}(x-at)}{2}$$

$$= \begin{cases} \dfrac{f(x+at) - f(at-x)}{2}, & x \in [0, at] \\ \dfrac{f(x+at) + f(x-at)}{2}, & x \in [at, l-at] \\ \dfrac{-f(2l-x-at) + f(x-at)}{2}, & x \in [l-at, l] \end{cases} \quad (7.109)$$

Example 7.9 Let us find the solution $u(x,t)$ of the following finite vibrating string problem:

$$\begin{cases} \dfrac{\partial^2 u}{\partial t^2} - \dfrac{\partial^2 u}{\partial x^2} = 0, \ x \in [0,2], \ t \geq 0, \\ u(0,t) = u(2,t) = 0, \ t \geq 0, \\ u(x,0) = \sin \pi x, \ \dfrac{\partial u}{\partial t}(x,0) = 0, \ x \in [0,2]. \end{cases}$$

7.2 The Wave Equation

In this case $a = 1$, $l = 2$, $f(x) = \sin \pi x$ and $g(x) = 0$, $x \in [0, 2]$. First of all we compute A_n and B_n from formulas (7.97) and (7.98) respectively. Since $g(x) = 0$, we have $B_n = 0$, $n = 1, 2, \ldots$. Let us compute A_n:

$$A_n = \frac{2}{l} \int_0^l f(x) \sin\left(\frac{n\pi}{l}x\right) dx = \int_0^2 \sin \pi x \cdot \sin\left(\frac{n\pi}{2}x\right) dx =$$

$$= \frac{1}{2} \int_0^2 \left[\cos\left(\pi - \frac{n\pi}{2}\right)x - \cos\left(\pi + \frac{n\pi}{2}\right)x\right] dx =$$

$$= \frac{1}{2} \left[\frac{1}{\pi - \frac{n\pi}{2}} \sin\left(\pi - \frac{n\pi}{2}\right)x - \frac{1}{\pi + \frac{n\pi}{2}} \sin\left(\pi + \frac{n\pi}{2}\right)x\right]_0^2 = 0,$$

if $n \neq 2$. For $n = 2$ we get:

$$A_2 = \int_0^2 \sin^2(\pi x)\, dx = \frac{1}{2} \int_0^2 (1 - \cos 2\pi x)\, dx = 1 - \frac{1}{4\pi} \sin 2\pi x \bigg|_0^2 = 1.$$

Thus, the series in the formula (7.94) has only one nonzero term, so we finally obtain:

$$u(x, t) = \cos(\pi t) \cdot \sin(\pi x).$$

This formula can also be obtained by using directly the formula (7.108): since $f(x) = \sin \pi x$ is an odd, periodic function and $g \equiv 0$, we can write:

$$u(x, t) = \frac{1}{2}(\sin \pi(x + t) + \sin \pi(x - t)) = \cos(\pi t) \cdot \sin(\pi x).$$

The positions of the string at time $t = 0$, $t = 1/3$, $t = 1/2$ and $t = 1$ are presented in Fig. 7.7. Note that for $t = 1/2$, the string coincides with the x-axis.

Example 7.10 We consider now a string of length l with fixed endpoints which is initially "pinched" in the midpoint such that $u(l/2, 0) = h$ and released to vibrate

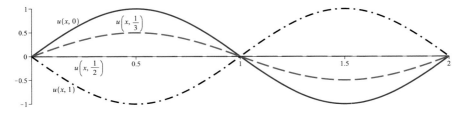

Fig. 7.7 The function $u(x, t)$ for $t = 0$, $t = 1/3$, $t = 1/2$ and $t = 1$

freely (the initial velocity is 0). Supposing that $a = 1$, we have to find the solution of the following problem:

$$\begin{cases} \dfrac{\partial^2 u}{\partial t^2} - \dfrac{\partial^2 u}{\partial x^2} = 0, \ x \in [0, l], \ t \geq 0, \\ u(0, t) = u(l, t) = 0, \ t \geq 0, \\ u(x, 0) = f(x) = \begin{cases} \dfrac{2h}{l} x, & x \in \left[0, \dfrac{l}{2}\right] \\ \dfrac{2h}{l} (l - x), & x \in \left[\dfrac{l}{2}, l\right], \end{cases} \\ \dfrac{\partial u}{\partial t}(x, 0) = 0, \ x \in [0, l]. \end{cases}$$

We remark that in this case the function $u(x, t)$ constructed using either the formula (7.94) or (7.109) is a solution in a *generalized sense* (see Remark 7.3): it describes the shape of our vibrating string at any moment t, even if it is not a function of class C^2 (since $f(x)$ is not continuously differentiable).

Since $g(x) = 0$, we have $B_n = 0$, $n = 1, 2, \ldots$. We compute A_n:

$$A_n = \frac{2}{l} \int_0^l f(x) \sin\left(\frac{n\pi}{l} x\right) dx =$$

$$= \frac{4h}{l^2} \left[\int_0^{l/2} x \sin\left(\frac{n\pi}{l} x\right) dx + \int_{l/2}^l (l - x) \sin\left(\frac{n\pi}{l} x\right) dx \right] =$$

$$= \frac{4h}{n\pi l} \left[-x \cos \frac{n\pi x}{l} \Big|_0^{l/2} + \int_0^{l/2} \cos \frac{n\pi x}{l} dx \right.$$

$$\left. -(l - x) \cos \frac{n\pi x}{l} \Big|_{l/2}^l - \int_{l/2}^l \cos \frac{n\pi x}{l} dx \right] =$$

$$= \frac{4h}{n\pi l} \left[-\frac{l}{2} \cos \frac{n\pi}{2} + \frac{l}{n\pi} \sin \frac{n\pi}{2} + \frac{l}{2} \cos \frac{n\pi}{2} + \frac{l}{n\pi} \sin \frac{n\pi}{2} \right] =$$

$$= \frac{8h}{n^2 \pi^2} \sin \frac{n\pi}{2} = \begin{cases} 0, & n = 2k \\ \dfrac{8h(-1)^k}{(2k+1)^2 \pi^2} & n = 2k + 1. \end{cases}$$

Thus, by replacing in the formula (7.94), we finally obtain:

$$u(x, t) = \sum_{k=0}^\infty \frac{8h(-1)^k}{(2k+1)^2 \pi^2} \cos \frac{(2k+1)\pi t}{l} \sin \frac{(2k+1)\pi x}{l}. \tag{7.110}$$

The shape of the string at time $t = 0$, $t = l/4$, $t = l/2$, $t = 2l/3$ and $t = l$ are presented in Fig. 7.8. We remark that for $t = (2n + 1)l/2$ the string coincides with the x-axis and $u(x, t)$ is a periodic function in t, of period $2l$.

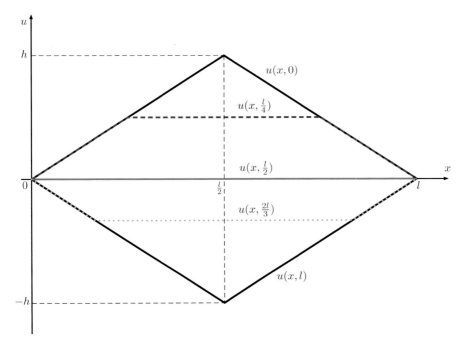

Fig. 7.8 The function $u(x,t)$ for $t = 0, t = l/4, t = l/2, t = 2l/3$

As shown in Remark 7.4, we can find a closed formula for $u(x,t)$. Thus, using (7.109), for any $t \in \left[0, \frac{l}{2}\right]$, one can write:

$$u(x,t) = \begin{cases} \frac{f(x+t)-f(t-x)}{2}, & x \in [0,t) \\ \frac{f(x+t)+f(x-t)}{2}, & x \in [t, l-t] \\ \frac{-f(2l-x-t)+f(x-t)}{2}, & x \in [l-t, l] \end{cases}$$

$$= \begin{cases} \frac{2h}{l}x, & x \in \left[0, \frac{l}{2}-t\right] \\ \frac{2h}{l}\left(\frac{l}{2}-t\right), & x \in \left[\frac{l}{2}-t, \frac{l}{2}+t\right] \\ \frac{2h}{l}(l-x), & x \in \left[\frac{l}{2}+t, l\right] \end{cases}, \quad \text{for } t \in \left[0, \frac{l}{2}\right].$$

Since $u(x, l-t) = -u(x,t)$, we have:

$$u(x,t) = \begin{cases} -\frac{2h}{l}x, & x \in \left[0, t-\frac{l}{2}\right] \\ -\frac{2h}{l}\left(t-\frac{l}{2}\right), & x \in \left[t-\frac{l}{2}, \frac{3l}{2}-t\right] \\ -\frac{2h}{l}(l-x), & x \in \left[\frac{3l}{2}-t, l\right] \end{cases}, \quad \text{for } t \in \left[\frac{l}{2}, l\right].$$

To complete the interval $[0, 2l]$, it is sufficient to notice that $u(x, 2l - t) = u(x,t)$, for any $t \in [l, 2l]$.

7.2.3 Laplace Transform Method for the Vibrating String

Consider a finite vibrating string as in (7.79), but with *nonhomogeneous boundary conditions*. Thus, the problem we have to solve is:

$$\frac{\partial^2 u}{\partial t^2} - a^2 \frac{\partial^2 u}{\partial x^2} = 0, \quad x \in (0, l), \ t \geq 0, \tag{7.111}$$

$$u(0, t) = a(t), \ u(l, t) = b(t), \ t \geq 0, \tag{7.112}$$

$$u(x, 0) = f(x), \ \frac{\partial u}{\partial t}(x, 0) = g(x), \ x \in [0, l]. \tag{7.113}$$

Here $a(t)$ and $b(t)$ are functions of class C^2 on $[0, \infty)$ and the functions $f(x), g(x)$ satisfy the same conditions as in the previous subsection.

Let $U(x, t)$ denote the Laplace transform of the function $u(x, t)$ (with respect to t):

$$\mathcal{L}\{u(x, t)\} = U(x, p) = \int_0^\infty u(x, t) \exp(-pt) dt,$$

where p is a complex parameter.

By Leibniz' formula for improper integrals (5.22), the Laplace transform of $\frac{\partial^2 u}{\partial x^2}$ is:

$$\mathcal{L}\left\{\frac{\partial^2 u}{\partial x^2}(x, t)\right\} = \int_0^\infty \frac{\partial^2 u}{\partial x^2} \exp(-pt) dt = \frac{d^2 U}{dx^2}(x, p). \tag{7.114}$$

On the other hand, by Theorem 6.6, the Laplace transform of $\frac{\partial^2 u}{\partial t^2}$ is:

$$\mathcal{L}\left\{\frac{\partial^2 u}{\partial t^2}(x, t)\right\} = p^2 U(x, p) - pu(x, 0) - \frac{\partial u}{\partial t}(x, 0),$$

and, using the initial conditions (7.113), we can write:

$$\mathcal{L}\left\{\frac{\partial^2 u}{\partial t^2}(x, t)\right\} = p^2 U(x, p) - pf(x) - g(x). \tag{7.115}$$

Let $A(p)$ and $B(p)$ be the Laplace transforms of the functions $a(t)$ and $b(t)$.

$$\mathcal{L}\{a(t)\} = A(p), \ \mathcal{L}\{b(t)\} = B(p).$$

7.2 The Wave Equation

With these notations, one can write:

$$U(0, p) = \int_0^\infty u(0, t) \exp(-pt) dt = A(p), \quad U(l, p) = B(p). \tag{7.116}$$

Thus, the "image" of the problem (7.111)–(7.113) through the Laplace transform is an ordinary differential equation (of order 2) in the unknown function $U(x, p)$ (considered as a function of x only) with two boundary conditions:

$$\frac{d^2 U}{dx^2} - \frac{p^2}{a^2} U = -\frac{p}{a^2} f(x) - \frac{1}{a^2} g(x), \tag{7.117}$$

$$U(0, p) = A(p), \quad U(l, p) = B(p). \tag{7.118}$$

We solve this equation, find $U(x, p)$ and then write $u(x, t) = \mathcal{L}^{-1}\{U(x, p)\}$, the inverse Laplace transform of $U(x, p)$ (considered as a function of the complex variable p).

Example 7.11 Let us use the Laplace transform method to solve the problem from Example 7.9:

$$\begin{cases} \dfrac{\partial^2 u}{\partial t^2} - \dfrac{\partial^2 u}{\partial x^2} = 0, \ x \in [0, 2], \ t \geq 0, \\ u(0, t) = u(2, t) = 0, \ t \geq 0, \\ u(x, 0) = \sin \pi x, \ \dfrac{\partial u}{\partial t}(x, 0) = 0, \ x \in [0, 2]. \end{cases}$$

The "image" of this problem through the Laplace transform is:

$$\frac{d^2 U}{dx^2} - p^2 U = -p \sin \pi x, \quad U(0, p) = 0, \quad U(2, p) = 0.$$

The general solution of this equation is

$$U(x, p) = C_1 \exp(px) + C_2 \exp(-px) + \frac{p}{p^2 + \pi^2} \sin \pi x,$$

where C_1, C_2 are arbitrary constants. From the boundary conditions $U(0, p) = U(2, p) = 0$ we find these constants $C_1 = C_2 = 0$, so the solution is:

$$U(x, p) = \frac{p}{p^2 + \pi^2} \sin \pi x.$$

Since

$$\mathcal{L}^{-1}\left\{\frac{p}{p^2 + \pi^2}\right\} = \cos \pi t$$

(see the formula (6.13)), the inverse transform of $U(x, p)$ is

$$u(x, t) = \cos \pi t \cdot \sin \pi x,$$

the solution of our problem.

7.2.4 Vibrations of a Rectangular Membrane: Two-Dimensional Wave Equation

Let us consider a homogeneous, perfectly elastic, rectangular membrane covering the region R in the plane (see Fig. 7.9),

$$R: \ 0 \le x \le a, \ \ 0 \le y \le b.$$

The membrane vibrates, performing small motions in the vertical plane. The displacement at the point (x, y) and time t is denoted by $u(x, y, t)$. The function u satisfies the *two-dimensional wave equation* (7.67):

$$\frac{\partial^2 u}{\partial t^2} = c^2 \nabla^2 u = c^2 \left(\frac{\partial^2 u}{\partial x^2} + \frac{\partial^2 u}{\partial y^2} \right), \ (x, y) \in R, \ t > 0, \tag{7.119}$$

($c > 0$ is a constant depending on the material properties). We suppose that the sides of the rectangle remain fixed, so we have the *boundary conditions*:

$$\begin{aligned} u(x, 0, t) = u(x, b, t) = 0, \ x \in [0, a], \ t \ge 0, \\ u(0, y, t) = u(a, y, t) = 0, \ y \in [0, b] \ t \ge 0. \end{aligned} \tag{7.120}$$

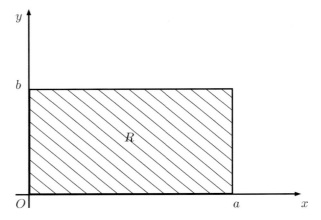

Fig. 7.9 Rectangular domain $R: \ 0 \le x \le a, \ 0 \le y \le b$

7.2 The Wave Equation

The initial deflection and the initial velocity at each point $(x, y) \in R$ are known, so we have the *initial conditions*:

$$u(x, y, 0) = f(x, y), \quad (x, y) \in R, \tag{7.121}$$

$$\frac{\partial u}{\partial t}(x, y, 0) = g(x, y), \quad (x, y) \in R, \tag{7.122}$$

where the (continuous) functions $f(x, y)$ and $g(x, y)$ are given.

We apply the Fourier method. First of all, using the *separation of variables*, we search for functions of the form

$$u(x, y, t) = X(x)Y(y)T(t)$$

that satisfy Eq. (7.119) and the boundary conditions (7.120). Thus, from (7.119) we have:

$$\frac{T''}{T} = c^2 \left(\frac{X''}{X} + \frac{Y''}{Y} \right) \tag{7.123}$$

and from the boundary conditions (7.120) it follows that

$$X(0) = X(a) = 0, \quad Y(0) = Y(b) = 0. \tag{7.124}$$

If we write (7.123) in the form

$$\frac{1}{c^2} \frac{T''}{T} - \frac{Y''}{Y} = \frac{X''}{X},$$

in the left side of the equality we have a function of y and t, and in the right side, a function of x, so the equality is possible only if both functions are equal to the same constant, say λ. In the same way it can be proved that the function $\frac{Y''}{Y}$ must be a constant, say μ, hence we can write:

$$\frac{X''}{X} = \lambda, \quad \frac{Y''}{Y} = \mu, \quad \frac{T''}{T} = c^2(\lambda + \mu).$$

These equalities are actually three linear ODEs with constant coefficients:

$$X'' - \lambda X = 0 \tag{7.125}$$

$$Y'' - \mu Y = 0 \tag{7.126}$$

$$T'' - c^2(\lambda + \mu)T = 0 \tag{7.127}$$

As we have seen in Sect. 7.2.2, the Eqs. (7.125) and (7.126) with the boundary conditions (7.124) have non-trivial solutions if and only if

$$\lambda = \lambda_m = -\left(\frac{m\pi}{a}\right)^2, \, m = 1, 2, \ldots \text{ and}$$

$$\mu = \mu_n = -\left(\frac{n\pi}{b}\right)^2, \, n = 1, 2, \ldots,$$

(7.128)

and these solutions are the eigenfunctions (multiplied by constants)

$$X_m = \sin\left(\frac{m\pi}{a}x\right), \, m = 1, 2, \ldots$$

$$Y_n = \sin\left(\frac{n\pi}{b}y\right), \, n = 1, 2, \ldots.$$

(7.129)

To find the general solution of Eq. (7.127) with λ and μ defined by (7.128), we denote

$$\omega_{m,n} = c\pi \sqrt{\frac{m^2}{a^2} + \frac{n^2}{b^2}}, \, m, n = 1, 2, \ldots. \quad (7.130)$$

Thus, for every $m, n = 1, 2, \ldots$, the Eq. (7.128) is written:

$$T'' + \omega_{m,n}^2 T = 0$$

and has the general solution:

$$T_{m,n} = A_{m,n} \cos \omega_{m,n} t + B_{m,n} \sin \omega_{m,n} t, \quad (7.131)$$

where $A_{m,n}$ and $B_{m,n}$ are arbitrary constants, $m, n = 1, 2, \ldots$.

Thus, we have obtained the following solutions of the two-dimensional wave Eq. (7.119) with the boundary conditions (7.120):

$$u_{m,n}(x, y, t) = \left(A_{m,n} \cos \omega_{m,n} t + B_{m,n} \sin \omega_{m,n} t\right) \sin\left(\frac{m\pi}{a}x\right) \sin\left(\frac{n\pi}{b}y\right).$$

The second step is to apply the *superposition principle* and search for a solution of the form

$$u(x, y, t) = \sum_{m=1}^{\infty} \sum_{n=1}^{\infty} \left(A_{m,n} \cos \omega_{m,n} t + B_{m,n} \sin \omega_{m,n} t\right) \sin\left(\frac{m\pi}{a}x\right) \sin\left(\frac{n\pi}{b}y\right).$$

(7.132)

7.2 The Wave Equation

Thus, we have to find the coefficients $A_{m,n}$, $B_{m,n}$ such that the double series (7.132) satisfies the initial conditions (7.121), (7.122) as well. From (7.121) we have:

$$u(x, y, 0) = f(x, y) = \sum_{m=1}^{\infty} \sum_{n=1}^{\infty} A_{m,n} \sin\left(\frac{m\pi}{a}x\right) \sin\left(\frac{n\pi}{b}y\right) \quad (7.133)$$

This is a *double Fourier series* of sines (see [41], Chapter 7). Since the functions

$$\phi_{m,n}(x, y) = \sin\left(\frac{m\pi}{a}x\right) \sin\left(\frac{n\pi}{b}y\right), \quad m, n = 1, 2, \ldots,$$

form an *orthogonal system*, that is,

$$\int_0^b \int_0^a \phi_{m,n}(x, y) \phi_{p,q}(x, y) \, dx \, dy = 0 \text{ for } (m, n) \neq (p, q),$$

and

$$\int_0^b \int_0^a \phi_{m,n}^2(x, y) \, dx \, dy = \int_0^a \sin^2\left(\frac{m\pi}{a}x\right) dx \int_0^b \sin^2\left(\frac{n\pi}{b}y\right) dy = \frac{a}{2} \cdot \frac{b}{2} = \frac{ab}{4},$$

the coefficients $A_{m,n}$, $m, n = 1, 2, \ldots$ are given by the formula:

$$A_{m,n} = \frac{4}{ab} \int_0^b \int_0^a f(x, y) \sin\left(\frac{m\pi}{a}x\right) \sin\left(\frac{n\pi}{b}y\right) dx \, dy. \quad (7.134)$$

On the other hand, since

$$\frac{\partial u}{\partial t} = \sum_{m=1}^{\infty} \sum_{n=1}^{\infty} \omega_{m,n} \left(-A_{m,n} \sin \omega_{m,n} t + B_{m,n} \cos \omega_{m,n} t\right) \sin\left(\frac{m\pi}{a}x\right) \sin\left(\frac{n\pi}{b}y\right),$$

from the second initial condition (7.122) we get

$$\frac{\partial u}{\partial t}(x, y, 0) = g(x, y) = \sum_{m=1}^{\infty} \sum_{n=1}^{\infty} \omega_{m,n} B_{m,n} \sin\left(\frac{m\pi}{a}x\right) \sin\left(\frac{n\pi}{b}y\right), \quad (7.135)$$

so the coefficients $B_{m,n}$, $m, n = 1, 2, \ldots$ are given by the formula:

$$B_{m,n} = \frac{4}{ab\omega_{m,n}} \int_0^b \int_0^a g(x, y) \sin\left(\frac{m\pi}{a}x\right) \sin\left(\frac{n\pi}{b}y\right) dx \, dy. \quad (7.136)$$

Example 7.12 Solve the two-dimensional wave equation

$$\frac{\partial^2 u}{\partial t^2} = 0.05\left(\frac{\partial^2 u}{\partial x^2} + \frac{\partial^2 u}{\partial y^2}\right), \quad 0 < x < 1, \ 0 < y < 1, \ t > 0,$$

with the boundary conditions

$$u(x, 0, t) = u(x, 1, t) = 0, \quad x \in [0, 1],$$
$$u(0, y, t) = u(1, y, t) = 0, \quad y \in [0, 1]$$

and the initial conditions:

$$u(x, y, 0) = \sin \pi x \sin 2\pi y,$$
$$\frac{\partial u}{\partial t}(x, y, 0) = 0, \quad x \in [0, 1], \ y \in [0, 1].$$

We have $a = b = 1$ and $c = \frac{\sqrt{5}}{10}$. Since $g(x, y) = 0$, by (7.136), all the coefficients $B_{m,n}$ are zero. We calculate the coefficients $A_{m,n}$ using the formula (7.134):

$$A_{m,n} = 4\int_0^1\int_0^1 \sin \pi x \sin 2\pi y \sin m\pi x \sin n\pi y \, dx \, dy =$$

$$= \begin{cases} 1 & \text{if } (m, n) = (1, 2) \\ 0 & \text{if } (m, n) \neq (1, 2) \end{cases}$$

Hence the series (7.132) has only one nonzero term:

$$u(x, y, t) = \cos\left(\frac{\pi}{2}t\right) \sin \pi x \sin 2\pi y.$$

Notice that, for $t = 1, 3, 5, \ldots$ the deflection is 0 at any point (x, y).

In Fig. 7.10 the initial shape of the square membrane is represented.

7.3 Vibrations of a Simply Supported Beam: Fourier Method

A *simply supported beam* is a rectangular parallelepiped homogeneous object which is perfectly elastic. Let $[OL]$ be the symmetry axis of its longest part, where $O(0, 0)$ is the origin and $L(l, 0)$.

Even if we have a 3D-object, since its behavior during the vibration process is the same for all the points in a fixed cross section $A(x)$ we can identify this cross section with its center of mass, $G(x)$, which is on the symmetry axis. Moreover,

7.3 Vibrations of a Simply Supported Beam: Fourier Method

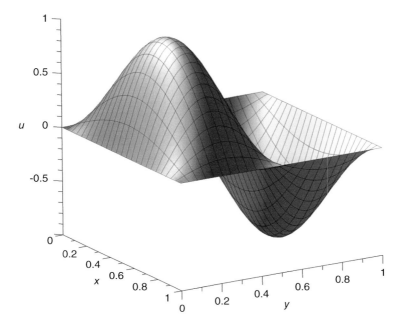

Fig. 7.10 The initial shape $u(x, y, 0) = \sin \pi x \sin 2\pi y$

the ends of the beam are (simply) supported by the rigid points O and L. If $u(x, t)$ denotes the displacement of the center of mass $G(x)$ at time t, it can be shown that small free vertical vibrations of the uniform elastic beam in Fig. 7.11 are modeled by the fourth-order PDE (see [21], p. 552):

$$\frac{\partial^2 u}{\partial t^2} + c^2 \frac{\partial^4 u}{\partial x^4} = 0, \quad x \in (0, l), \quad t \geq 0, \tag{7.137}$$

where $c^2 = \frac{EI}{\rho A}$ (E is the Young's modulus of elasticity, I is the moment of inertia of the cross section with respect to the y-axis, orthogonal on the plane of displacement (xOu), ρ is the density and A is the cross section area). We have to find the solution $u(x, t)$ which satisfies the boundary conditions

$$u(0, t) = u(l, t) = 0, \quad t \geq 0 \tag{7.138}$$

(simply supported beam—the endpoints O and L are fixed),

$$\frac{\partial^2 u}{\partial x^2}(0, t) = \frac{\partial^2 u}{\partial x^2}(l, t) = 0 \quad t \geq 0 \tag{7.139}$$

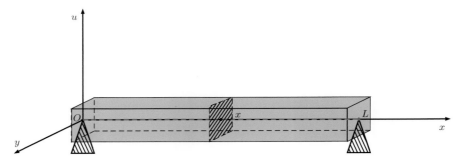

Fig. 7.11 Simply supported beam

(the curvatures or the moments at the endpoints O and L are zero), and the initial conditions

$$u(x, 0) = f(x), \quad \frac{\partial u}{\partial t}(x, 0) = g(x), \quad x \in [0, l]. \tag{7.140}$$

We let the reader find the conditions of compatibility at the endpoints for these given functions, $f(x)$ and $g(x)$.

We follow the same steps as in the study of the finite vibrating string with fixed ends. So, first of all, we use the Fourier's method of separation of variables and search for a solution of the form:

$$u(x, t) = X(x) \cdot T(t)$$

Thus, the Eq. (7.137) becomes:

$$X(x)T''(t) + c^2 X^{(4)}(x)T(t) = 0.$$

Now, we separate the functions depending on x from the functions depending on t and obtain:

$$\frac{X^{(4)}(x)}{X(x)} = -\frac{T''(t)}{c^2 T(t)} = \lambda, \tag{7.141}$$

where λ is a constant. From the second equality we obtain the second-order equation

$$T''(t) + c^2 \lambda T(t) = 0, \quad t \in [0, \infty) \tag{7.142}$$

This equation must have *bounded solutions* (as $t \to \infty$) and this is possible only if $\lambda > 0$. We take λ of the form: $\lambda = \mu^4$ and solve the equation

$$X^{(4)}(x) - \mu^4 X(x) = 0, \tag{7.143}$$

7.3 Vibrations of a Simply Supported Beam: Fourier Method

with the boundary conditions

$$X(0) = X(l) = 0, \quad X''(0) = X''(l) = 0. \tag{7.144}$$

The general solution of the fourth-order linear equation with constant coefficients (7.143) can be written

$$X(x) = c_1 \exp \mu x + c_2 \exp(-\mu x) + c_3 \cos \mu x + c_4 \sin \mu x, \tag{7.145}$$

or, using the hyperbolic functions

$$\cosh x = \frac{\exp x + \exp(-x)}{2} \quad \text{and} \quad \sinh x = \frac{\exp x - \exp(-x)}{2},$$

as:

$$X(x) = C_1 \cosh \mu x + C_2 \sinh \mu x + C_3 \cos \mu x + C_4 \sin \mu x. \tag{7.146}$$

The boundary conditions (7.144) supply the following linear system in the unknowns C_1, C_2, C_3 and C_3:

$$\begin{cases} C_1 + C_3 = 0 \\ C_1 \cosh \mu l + C_2 \sinh \mu l + C_3 \cos \mu l + C_4 \sin \mu l = 0 \\ C_1 - C_3 = 0 \\ C_1 \cosh \mu l + C_2 \sinh \mu l - C_3 \cos \mu l - C_4 \sin \mu l = 0 \end{cases} \tag{7.147}$$

From the first and the third equation we have $C_1 = C_3 = 0$. So, the remaining system in C_2 and C_4 is:

$$\begin{cases} C_2 \sinh \mu l + C_4 \sin \mu l = 0 \\ C_2 \sinh \mu l - C_4 \sin \mu l = 0 \end{cases}. \tag{7.148}$$

This last system is a homogeneous one. Since we are not interested in the trivial solution, its determinant has to be equal to zero. Thus,

$$\sinh(\mu l) \cdot \sin(\mu l) = 0.$$

Since $\mu l \neq 0$, $\sinh(\mu l) \neq 0$, so $\sin(\mu l) = 0$, which implies that μ must be of the form $\mu = \mu_n = \frac{n\pi}{l}$, $n \in \mathbb{Z}$. From (7.148) we get $C_2 = 0$ and C_4 remains arbitrary. The functions

$$X_n(x) = \sin\left(\frac{n\pi}{l}x\right), \quad n = 1, 2, \ldots \tag{7.149}$$

are said to be the eigenfunctions and $\lambda_n = \mu_n^4 = \left(\frac{n\pi}{l}\right)^4$ are the eigenvalues of our problem. Any solution of the boundary value problem (7.143)–(7.144) with $\mu = \mu_n = \frac{n\pi}{l}$ is of the form $X_n(x)$ multiplied by a constant.

Now, let us solve Eq. (7.142) for each eigenvalue $\lambda_n = \left(\frac{n\pi}{l}\right)^4$, $n = 1, 2, \ldots$:

$$T''(t) + c^2 \left(\frac{n\pi}{l}\right)^4 T(t) = 0.$$

It is easy to see that its general solution is:

$$T_n(t) = A_n \cos\left(\frac{n^2\pi^2 c}{l^2}t\right) + B_n \sin\left(\frac{n^2\pi^2 c}{l^2}t\right). \tag{7.150}$$

Thus, we succeeded to find a sequence $\{u_n(x,t)\}_{n=1,2,\ldots}$ of solutions for the PDE-4 (7.137) with the boundary conditions (7.138), (7.139):

$$u_n(x,t) = \left[A_n \cos\left(\frac{n^2\pi^2 c}{l^2}t\right) + B_n \sin\left(\frac{n^2\pi^2 c}{l^2}t\right)\right] \sin\left(\frac{n\pi}{l}x\right),$$

where A_n and B_n are arbitrary constants. Note that each function, separately, does not satisfy the initial conditions (7.140). This is why we apply again the superposition principle to find a solution $u(x,t)$ that simultaneously satisfies the boundary and the initial conditions:

$$u(x,t) = \sum_{n=1}^{\infty} u_n(x,t) = \sum_{n=1}^{\infty} \left[A_n \cos\left(\frac{n^2\pi^2 c}{l^2}t\right) + B_n \sin\left(\frac{n^2\pi^2 c}{l^2}t\right)\right] \sin\frac{n\pi}{l}x. \tag{7.151}$$

Since $u(x,0) = f(x)$ and $\frac{\partial u}{\partial t}(x,0) = g(x)$, we get:

$$f(x) = \sum_{n=1}^{\infty} A_n \sin\frac{n\pi}{l}x, \quad x \in [0, l],$$

$$g(x) = \sum_{n=1}^{\infty} \frac{n^2\pi^2 c}{l^2} B_n \sin\frac{n\pi}{l}x, \quad x \in [0, l].$$

By the same reasoning as for the finite vibrating string with fixed ends (see (7.97) and (7.98)), we obtain:

$$A_n = \frac{2}{l} \int_0^l f(x) \sin\left(\frac{n\pi}{l}x\right) dx \tag{7.152}$$

and

$$B_n = \frac{2l}{n^2\pi^2 c} \int_0^l g(x) \sin\left(\frac{n\pi}{l} x\right) dx. \quad (7.153)$$

Example 7.13 Let us find the solution $u(x, t)$ for the simply supported beam problem (7.137), (7.138), (7.139) with the length $l = 2$ and the initial conditions $f(x) = 0$, $g(x) = \sin \pi x$.

Since $A_n = 0$, it remains to compute B_n (see formulas (7.152), (7.153)):

$$B_n = \frac{4}{n^2\pi^2 c} \int_0^2 \sin \pi x \cdot \sin\left(\frac{n\pi}{2} x\right) dx = 0,$$

if $n \neq 2$ and

$$B_2 = \frac{4}{4\pi^2 c} \int_0^2 \sin^2(\pi x) dx = \frac{1}{c\pi^2}.$$

Thus,

$$u(x, t) = \frac{1}{c\pi^2} \sin\left(\pi^2 c t\right) \cdot \sin \pi x, \quad x \in [0, 2], t \geq 0.$$

7.4 The Heat Equation

7.4.1 Modeling the Heat Flow from a Body in Space

First of all, we derive the partial differential equation which governs the temperature u in a body in space (see also [21]). Let $u(x, y, z, t)$ denote the temperature at the point (x, y, z) at time t. The *law of heat conduction*, also known as **Fourier's law**, states that heat flows in the direction of decreasing temperature, and the rate of flow, **q**, is proportional to the gradient of the temperature, grad $u = \nabla u = \frac{\partial u}{\partial x}\mathbf{i} + \frac{\partial u}{\partial y}\mathbf{j} + \frac{\partial u}{\partial z}\mathbf{k}$, that is,

$$\mathbf{q} = -K \nabla u, \quad (7.154)$$

where K is the *thermal conductivity*, considered to be constant (we suppose that the material of the body is homogeneous: the thermal conductivity K, the density ρ and the specific heat σ are all constants).

Let $\Omega \subset \mathbb{R}^3$ be a simple region in the body, and let S be its boundary surface, supposed to be piecewise smooth and with positive (outward) orientation. If **n** denotes the outer unit normal vector to the surface S, the total amount of heat that flows across S from Ω per unit time is given by the surface integral

$$\Phi_S(\mathbf{q}) = \iint_S \mathbf{q} \cdot \mathbf{n}\, d\sigma \stackrel{(7.154)}{=} -\iint_S K \nabla u \cdot \mathbf{n}\, d\sigma. \tag{7.155}$$

Recall that the *divergence* of a vector function $\mathbf{F} = F_1 \mathbf{i} + F_2 \mathbf{j} + F_3 \mathbf{k}$ is the scalar function

$$\operatorname{div} \mathbf{F} = \nabla \cdot \mathbf{F} = \frac{\partial F_1}{\partial x} + \frac{\partial F_2}{\partial y} + \frac{\partial F_3}{\partial z}.$$

The Gauss (divergence) theorem (see [38], 16.9) states that, if $\Omega \subset \mathbb{R}^3$ is a simple solid region whose boundary is the piecewise smooth, positively oriented surface S, and $\mathbf{F}$ is a vector field of class C^1 on an open region containing Ω, then

$$\iint_S \mathbf{F} \cdot \mathbf{n}\, d\sigma = \iiint_\Omega \operatorname{div} \mathbf{F}\, dx\, dy\, dz. \tag{7.156}$$

By applying this formula, the heat flux through the surface S (7.155) can be expressed as the triple integral

$$\Phi_S(\mathbf{q}) = -K \iiint_\Omega \operatorname{div}(\nabla u)\, dx\, dy\, dz = -K \iiint_\Omega \nabla^2 u\, dx\, dy\, dz, \tag{7.157}$$

where

$$\nabla^2 u = \frac{\partial^2 u}{\partial x^2} + \frac{\partial^2 u}{\partial y^2} + \frac{\partial^2 u}{\partial z^2}$$

is the Laplacian of the temperature u.

On the other hand, the total amount of heat at time t in Ω is:

$$Q(t) = \iiint_\Omega \sigma \rho\, u(x, y, z, t)\, dx\, dy\, dz,$$

where σ is the specific heat of the material and ρ is the density of the solid Ω (both are supposed to be constant). Thus, the time rate of decrease of Q is:

$$-\frac{\partial Q}{\partial t} = -\iiint_\Omega \sigma \rho\, \frac{\partial u}{\partial t}\, dx\, dy\, dz.$$

This must be equal to the amount of heat leaving Ω because no heat is produced or dissipated in the body. From (7.157) we obtain:

$$-\iiint_\Omega \sigma\rho \frac{\partial u}{\partial t} \, dx\,dy\,dz = -K \iiint_\Omega \nabla^2 u \, dx\,dy\,dz,$$

or, equivalently,

$$\iiint_\Omega \left(\frac{\partial u}{\partial t} - \frac{K}{\sigma\rho} \nabla^2 u \right) dx\,dy\,dz = 0.$$

Since this relation holds for any region Ω in the body and the integrand is a continuous function, we obtain that it must be zero everywhere and we obtain the following partial differential equation:

$$\frac{\partial u}{\partial t} = c^2 \nabla^2 u, \qquad (7.158)$$

where $c^2 = \frac{K}{\sigma\rho}$ is the *thermal diffusivity*. Equation (7.158) is the three-dimensional *heat equation* or *diffusion equation* (it also models chemical diffusion processes of one substance into another).

7.4.2 Heat Flow in a Finite Rod: Fourier Method

To study the one-dimensional case, we consider the temperature in a thin, homogeneous rod of length l and constant cross section, which is oriented along the x-axis from $x = 0$ to $x = l$ (see Fig. 7.12) and is perfectly insulated laterally. This means that heat flows only in the x-direction: the temperature u depends only on x and t (time) and the Laplacian reduces to $\frac{\partial^2 u}{\partial x^2}$. The Eq. (7.158) becomes the one-dimensional heat equation

$$\frac{\partial u}{\partial t} = c^2 \frac{\partial^2 u}{\partial x^2}, \quad x \in (0, l),\, t > 0. \qquad (7.159)$$

This equation seems to differ only very little from the one-dimensional wave Eq. (7.64), but this "small" difference makes it a *parabolic*-type equation (instead of *hyperbolic*-type) and its solutions behave quite differently from the solutions of the wave equation.

As boundary conditions, we assume that the ends of the rod are kept at constant temperature, equal to zero:

$$u(0, t) = u(l, t) = 0, \; t \geq 0. \qquad (7.160)$$

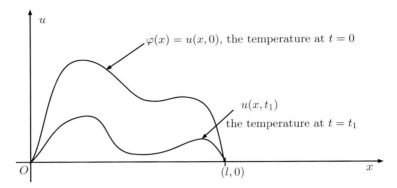

Fig. 7.12 The temperature in the bar at time $t = 0$ and at $t = t_1$

We also suppose that the initial temperature (at time $t = 0$) is

$$u(x, 0) = \varphi(x), \quad x \in [0, l] \tag{7.161}$$

where φ is a continuous function on $[0, l]$ and of class C^2 on $[0, l]$ possibly except for a finite number of points in this interval.

It can be proved (see [22], Theorems 31.3, 31.4) that the solution $u(x, t)$ of this problem is unique and continuously depends on the initial condition (7.161). In order to find the solution of the Eq. (7.159) with the boundary conditions (7.160) and the initial condition (7.161) we use the same Fourier method as in the case of the finite vibrating string with fixed ends.

First of all, we use the **separation of variables** and search for a solution of the form $u(x, t) = X(x) \cdot T(t)$, which satisfies the boundary conditions (7.160). Introducing this expression of $u(x, t)$ in (7.159) and separating the functions of x from the functions of t, we obtain:

$$\frac{X''(x)}{X(x)} = \frac{1}{c^2} \frac{T'(t)}{T(t)} = \lambda = \text{constant.} \tag{7.162}$$

These equalities give rise to the following linear, homogeneous ODEs:

$$X'' - \lambda X = 0 \tag{7.163}$$

$$T' - c^2 \lambda T = 0. \tag{7.164}$$

As in the case of the finite vibrating string, when we try to obtain the solutions of (7.163) that also satisfy the boundary conditions that follows from (7.160),

$$X(0) = X(l) = 0, \tag{7.165}$$

7.4 The Heat Equation

we find that λ must be of the form $\lambda = \lambda_n = -\left(\frac{n\pi}{l}\right)^2$, $n = 1, 2, \ldots$ and any solution of the boundary value problem (7.163)–(7.165) is a constant multiplied by an eigenfunction X_n,

$$X_n(x) = \sin\left(\frac{n\pi}{l}x\right), \quad n = 1, 2, \ldots. \tag{7.166}$$

For $\lambda = -\left(\frac{n\pi}{l}\right)^2$, $n = 1, 2, \ldots$, the general solution of the first-order Eq. (7.164) is

$$T_n(t) = A_n \exp\left(-\frac{n^2\pi^2 c^2}{l^2}t\right) \tag{7.167}$$

Thus, we obtained a sequence of solutions of the Eq. (7.159) with the boundary conditions (7.160):

$$u_n(x, t) = A_n \exp\left(-\frac{n^2\pi^2 c^2}{l^2}t\right) \cdot \sin\left(\frac{n\pi}{l}x\right), \quad n = 1, 2, \ldots. \tag{7.168}$$

These solutions do not satisfy the initial condition (7.161). To obtain a solution that satisfies this initial condition we use the **superposition principle** and search for a solution of the form

$$u(x, t) = \sum_{n=1}^{\infty} u_n(x, t) = \sum_{n=1}^{\infty} A_n \exp\left(-\frac{n^2\pi^2 c^2}{l^2}t\right) \cdot \sin\left(\frac{n\pi}{l}x\right). \tag{7.169}$$

From the initial condition $u(x, 0) = \varphi(x)$ we obtain:

$$\varphi(x) = \sum_{n=1}^{\infty} A_n \sin\left(\frac{n\pi}{l}x\right), \quad x \in [0, l].$$

This is the Fourier expansion of the function $\varphi(x)$ in series of sines, so the coefficients A_n are given by the formula:

$$A_n = \frac{2}{l} \int_0^l \varphi(x) \sin\left(\frac{n\pi}{l}x\right) dx. \tag{7.170}$$

If $\varphi(x)$ is a function of class C^2 on the interval $(0, l)$, then the series (7.169) with the coefficients given by (7.170) is uniformly convergent, as well as its termwise derivatives ([22], 31.4), so it is a solution for the heat flow Eq. (7.159) with the boundary conditions (7.160) and the initial condition (7.161). This solution is unique (see [22], Theorem 31.3). But, even if the function $\varphi(x)$ is only continuous (not of class C^2) on $[0, l]$, we can still construct a generalized solution $u(x, t)$ for the heat flow problem (7.159)–(7.161) as shown in the next example.

Example 7.14 Let us find the temperature function $u(x, t)$ in a rod of length $l = 2$, knowing that $c = 2$, $u(0, t) = u(2, t) = 0$ for $t \geq 0$ and

$$u(x, 0) = \varphi(x) = \begin{cases} x, & \text{if } x \in [0, 1] \\ 2 - x, & \text{if } x \in (1, 2]. \end{cases}$$

We can see here that $\varphi(x)$ is a continuous function, but it is not of class C^1. Let us compute the coefficients A_n using the formula (7.170)

$$A_n = \int_0^2 \varphi(x) \sin\left(\frac{n\pi}{2}x\right) dx = \int_0^1 x \sin\left(\frac{n\pi}{2}x\right) dx + \int_1^2 (2-x) \sin\left(\frac{n\pi}{2}x\right) dx$$

$$= \int_0^1 x \left[-\frac{2}{n\pi} \cos\left(\frac{n\pi}{2}x\right)\right]' dx + \int_1^2 (2-x) \left[-\frac{2}{n\pi} \cos\left(\frac{n\pi}{2}x\right)\right]' dx$$

$$= x \left[-\frac{2}{n\pi} \cos\left(\frac{n\pi}{2}x\right)\right]_0^1 + \frac{2}{n\pi} \int_0^1 \cos\left(\frac{n\pi}{2}x\right) dx$$

$$+ (2-x) \left[-\frac{2}{n\pi} \cos\left(\frac{n\pi}{2}x\right)\right]_1^2 - \frac{2}{n\pi} \int_1^2 \cos\left(\frac{n\pi}{2}x\right) dx$$

$$= -\frac{2}{n\pi} \cos\left(\frac{n\pi}{2}\right) + \frac{4}{n^2\pi^2} \sin\left(\frac{n\pi}{2}\right) + \frac{2}{n\pi} \cos\left(\frac{n\pi}{2}\right) + \frac{4}{n^2\pi^2} \sin\left(\frac{n\pi}{2}\right),$$

so

$$A_n = \frac{8}{n^2\pi^2} \sin\left(\frac{n\pi}{2}\right).$$

It is easy to see that $A_{2k} = 0$ and $A_{2k+1} = (-1)^k \frac{8}{(2k+1)^2\pi^2}$. Thus, from (7.169) we get:

$$u(x, t) = \frac{8}{\pi^2} \sum_{k=0}^{\infty} \frac{(-1)^k}{(2k+1)^2} \exp\left[-(2k+1)^2\pi^2 t\right] \sin\left[\frac{(2k+1)\pi}{2}x\right]. \quad (7.171)$$

This series and its termwise derivatives are uniformly convergent on any subset $[0, l] \times [\varepsilon, M]$ of $[0, l] \times [0, \infty)$ ($\varepsilon > 0$.) So we can differentiate it term by term. This means that $u(x, t)$ is the (unique) solution of the above problem, even if $\varphi(x)$ is not of class C^2 on $[0, l]$.

7.4.3 Heat Flow in an Infinite Rod

We consider now the temperature $u(x, t)$ in a thin, homogeneous rod of infinite length ($x \in (-\infty, \infty)$), which is perfectly laterally insulated In this case, we do not have boundary conditions, so we have to solve the equation

$$\frac{\partial u}{\partial t}(x, t) = c^2 \frac{\partial^2 u}{\partial x^2}(x, t), \quad x \in (-\infty, \infty), \; t > 0, \tag{7.172}$$

with the initial condition

$$u(x, 0) = \varphi(x), \quad x \in (-\infty, \infty), \tag{7.173}$$

where we assume that $u(x, t)$ and $\varphi(x)$ are continuous piecewise smooth functions in the variable x (for all $t \geq 0$) and absolutely integrable on $(-\infty, \infty)$. We denote by $\Phi(\xi)$ the Fourier transform of the function $\varphi(x)$,

$$\Phi(\xi) = \mathcal{F}\{\varphi(x)\}(\xi) = \frac{1}{\sqrt{2\pi}} \int_{-\infty}^{\infty} \varphi(x) \exp(-i\xi x) \, dx,$$

and, for any $t \geq 0$, we denote by $U(\xi, t)$ the Fourier transform of the function $u(x, t)$:

$$U(\xi, t) = \mathcal{F}\{u(x, t)\}(\xi) = \frac{1}{\sqrt{2\pi}} \int_{-\infty}^{\infty} u(x, t) \exp(-i\xi x) \, dx.$$

By Leibniz' formula for integrals with parameters (5.22), we have:

$$\mathcal{F}\left\{\frac{\partial u}{\partial t}\right\} = \frac{\partial U}{\partial t}.$$

On the other hand, from Theorem 5.10 one can write

$$\mathcal{F}\left\{\frac{\partial^2 u}{\partial x^2}\right\} = -\xi^2 U(\xi, t).$$

Hence, the Eq. (7.172) becomes:

$$\frac{dU}{dt}(\xi, t) + \xi^2 c^2 U(\xi, t) = 0, \tag{7.174}$$

and the initial condition (7.173) is written:

$$U(\xi, 0) = \Phi(\xi), \tag{7.175}$$

We solve this linear ODE-1 relative to the variable t, time (ξ is considered to be a parameter). Thus, the solution of the Cauchy problem (7.174)–(7.175) is:

$$U(\xi, t) = \Phi(\xi) \exp(-\xi^2 c^2 t). \tag{7.176}$$

By (5.80), we know that $\mathcal{F}\{\exp(-ax^2)\}(\xi) = \frac{1}{\sqrt{2a}} \exp\left(-\frac{\xi^2}{4a}\right)$ for any $a > 0$. For $\frac{1}{4a} = c^2 t$, that is, $a = \frac{1}{4c^2 t}$, we obtain:

$$\exp(-\xi^2 c^2 t) = \mathcal{F}\left\{\frac{1}{c\sqrt{2t}} \exp\left(-\frac{x^2}{4c^2 t}\right)\right\}.$$

Thus, we can write:

$$\mathcal{F}\{u(x,t)\} = \mathcal{F}\{\varphi(x)\} \mathcal{F}\left\{\frac{1}{c\sqrt{2t}} \exp\left(-\frac{x^2}{4c^2 t}\right)\right\}.$$

By the convolution theorem for the Fourier transform (Theorem 5.12), we have:

$$\mathcal{F}\{u(x,t)\} = \frac{1}{\sqrt{2\pi}} \mathcal{F}\left\{\varphi(x) * \frac{1}{c\sqrt{2t}} \exp\left(-\frac{x^2}{4c^2 t}\right)\right\},$$

hence

$$u(x,t) = \frac{1}{\sqrt{2\pi}} \varphi(x) * \frac{1}{c\sqrt{2t}} \exp\left(-\frac{x^2}{4c^2 t}\right).$$

The convolution of two functions f and g is the function $f * g$ defined by (5.51):

$$(f * g)(x) = \int_{-\infty}^{\infty} f(z) g(x - z) \, dz,$$

so the temperature $u(x, t)$ in the infinite rod is given by the formula:

$$u(x,t) = \frac{1}{2c\sqrt{\pi t}} \int_{-\infty}^{\infty} \varphi(z) \exp\left(-\frac{(x-z)^2}{4c^2 t}\right) dz. \tag{7.177}$$

We can see that when time increases the temperature decreases. It can be proved that this solution is unique in the class of bounded solutions $u(x, t)$ (see [22], Theorem 31.1).

7.4 The Heat Equation

Example 7.15 (Krasnov et al. [22]) Let us use the formula (7.177) to find the solution of the Cauchy problem (7.172)–(7.173) for $c = 1$ and $\varphi(x) = \exp(-\frac{x^2}{2})$. Thus,

$$u(x,t) = \frac{1}{2\sqrt{\pi t}} \int_{-\infty}^{\infty} \exp\left(-\frac{z^2}{2} - \frac{(x-z)^2}{4t}\right) dz =$$

$$= \frac{1}{2\sqrt{\pi t}} \exp\left(-\frac{x^2}{2(1+2t)}\right) \int_{-\infty}^{\infty} \exp\left[-\frac{1+2t}{4t}\left(z - \frac{x}{1+2t}\right)^2\right] dz.$$

In order to use the Gaussian integral (5.11),

$$\int_{-\infty}^{\infty} \exp(-x^2) dx = \sqrt{\pi},$$

we make the change of variable:

$$w = \frac{\sqrt{1+2t}}{2\sqrt{t}} \left(z - \frac{x}{1+2t}\right)$$

and obtain:

$$u(x,t) = \frac{1}{2\sqrt{\pi t}} \exp\left(-\frac{x^2}{2(1+2t)}\right) \frac{2\sqrt{t}}{\sqrt{1+2t}} \int_{-\infty}^{\infty} \exp(-w^2) dw =$$

$$= \frac{1}{\sqrt{\pi(1+2t)}} \exp\left(-\frac{x^2}{2(1+2t)}\right) \sqrt{\pi}.$$

Finally, the solution of the above Cauchy problem is:

$$u(x,t) = \frac{1}{\sqrt{1+2t}} \exp\left(-\frac{x^2}{2(1+2t)}\right).$$

7.4.4 Heat Flow in a Rectangular Plate

Let us consider the temperature $u(x, y, t)$ in a thin, homogeneous, rectangular plate covering the region R in the plane,

$$R: 0 \leq x \leq a, \quad 0 \leq y \leq b.$$

We want to find the temperature $u(x, y, t)$ knowing that the sides of the rectangle are kept at temperature zero and knowing the initial temperature inside the rectangle R. Thus, we have to solve the two-dimensional heat equation

$$\frac{\partial u}{\partial t} = c^2 \nabla^2 u = c^2 \left(\frac{\partial^2 u}{\partial x^2} + \frac{\partial^2 u}{\partial y^2} \right), \quad (x, y) \in R, \ t > 0 \tag{7.178}$$

with the boundary conditions:

$$u(x, 0, t) = u(x, b, t) = 0, \quad x \in [0, a],$$
$$u(0, y, t) = u(a, y, t) = 0, \quad y \in [0, b], \tag{7.179}$$

and the initial condition:

$$u(x, y, 0) = f(x, y), \quad (x, y) \in R, \tag{7.180}$$

where $f(x, y)$ is a given function of class C^2 on R.

We apply the Fourier method. First of all, using the *separation of variables*, we search for functions of the form

$$u(x, y, t) = X(x)Y(y)T(t)$$

that satisfy Eq. (7.178) and the boundary conditions (7.179). Thus, from (7.178) we have:

$$\frac{T'}{T} = c^2 \left(\frac{X''}{X} + \frac{Y''}{Y} \right) \tag{7.181}$$

and we deduce (as in Sect. 7.2.4) that the functions X''/X, Y''/Y, T''/T are constants:

$$\frac{X''}{X} = \lambda, \quad \frac{Y''}{Y} = \mu, \quad \frac{T'}{T} = c^2(\lambda + \mu).$$

Thus, we obtained three linear ODEs with constant coefficients:

$$X'' - \lambda X = 0 \tag{7.182}$$
$$Y'' - \mu Y = 0 \tag{7.183}$$
$$T' - c^2(\lambda + \mu)T = 0 \tag{7.184}$$

On the other hand, from the boundary conditions (7.179) it follows that

$$X(0) = X(a) = 0, \quad Y(0) = Y(b) = 0. \tag{7.185}$$

7.4 The Heat Equation

As we have seen above, the Eqs. (7.182) and (7.183) with the boundary conditions (7.185) have non-trivial solutions if and only if

$$\lambda = \lambda_m = -\left(\frac{m\pi}{a}\right)^2, \quad m = 1, 2, \ldots \text{ and}$$
$$\mu = \mu_n = -\left(\frac{n\pi}{b}\right)^2, \quad n = 1, 2, \ldots,$$
(7.186)

and these solutions are the eigenfunctions (multiplied by constants)

$$X_m = \sin\left(\frac{m\pi}{a}x\right), \quad m = 1, 2, \ldots$$
$$Y_n = \sin\left(\frac{n\pi}{b}y\right), \quad n = 1, 2, \ldots.$$
(7.187)

The general solution of the Eq. (7.184) with λ and μ defined by (7.186) is

$$T_{m,n} = A_{m,n} \exp\left[-c^2\pi^2\left(\frac{m^2}{a^2} + \frac{n^2}{b^2}\right)t\right], \quad m, n = 1, 2, \ldots,$$
(7.188)

where $A_{m,n}$ are arbitrary constants, $m, n = 1, 2, \ldots$. Thus, we have obtained the following solutions of the two-dimensional heat Eq. (7.178) with the boundary conditions (7.179):

$$u_{m,n}(x, y, t) = A_{m,n} \exp\left[-c^2\pi^2\left(\frac{m^2}{a^2} + \frac{n^2}{b^2}\right)t\right] \sin\left(\frac{m\pi}{a}x\right) \sin\left(\frac{n\pi}{b}y\right).$$

The second step is to find the coefficients $A_{m,n}$ such that the double series formed with these functions,

$$u(x, y, t) = \sum_{m=1}^{\infty}\sum_{n=1}^{\infty} A_{m,n} \exp\left[-c^2\pi^2\left(\frac{m^2}{a^2} + \frac{n^2}{b^2}\right)t\right] \sin\left(\frac{m\pi}{a}x\right) \sin\left(\frac{n\pi}{b}y\right)$$
(7.189)

satisfies the initial condition (7.180) as well. Thus, for $t = 0$ we have:

$$u(x, y, 0) = f(x, y) = \sum_{m=1}^{\infty}\sum_{n=1}^{\infty} A_{m,n} \sin\left(\frac{m\pi}{a}x\right) \sin\left(\frac{n\pi}{b}y\right).$$
(7.190)

As shown in Sect. 7.2.4, the coefficients of this *double Fourier series* of sines are given by the formula (7.134):

$$A_{m,n} = \frac{4}{ab} \int_0^b \int_0^a f(x, y) \sin\left(\frac{m\pi}{a}x\right) \sin\left(\frac{n\pi}{b}y\right) dx dy.$$
(7.191)

Example 7.16 Solve the two-dimensional heat equation

$$\frac{\partial u}{\partial t} = 0.01 \left(\frac{\partial^2 u}{\partial x^2} + \frac{\partial^2 u}{\partial y^2} \right), \quad 0 < x < 1, \ 0 < y < 1, \ t > 0,$$

with the boundary conditions

$$u(x, 0, t) = u(x, 1, t) = 0, \ x \in [0, 1],$$
$$u(0, y, t) = u(1, y, t) = 0, \ y \in [0, 1]$$

and the initial condition:

$$u(x, y, 0) = x(1 - x^2) \sin \pi y, \ x \in [0, 1], \ y \in [0, 1].$$

We have $a = b = 1$ and $c = 0.1$. We calculate the coefficients $A_{m,n}$ using the formula (7.191):

$$A_{m,n} = 4 \int_0^1 \int_0^1 x(1-x^2) \sin \pi y \sin m\pi x \sin n\pi y \, dx \, dy =$$

$$= 4 \int_0^1 \sin \pi y \sin n\pi y \, dy \int_0^1 x(1-x^2) \sin m\pi x \, dx.$$

So $A_{m,n} = 0$ for $n \neq 1$ and, for $n = 1$,

$$A_{m,1} = 2 \int_0^1 x(1-x^2) \sin m\pi x \, dx = \frac{12(-1)^{m+1}}{m^3 \pi^3}.$$

By (7.189), the solution is

$$u(x, y, t) = \sum_{m=1}^{\infty} \frac{12(-1)^{m+1}}{m^3 \pi^3} \exp\left(-0.01\pi^2 \left(m^2 + 1\right) t\right) \sin m\pi x \sin \pi y.$$

In Fig. 7.13 is represented the initial temperature in the plate and in Fig. 7.14, the temperature at time $t = 5$.

7.4 The Heat Equation

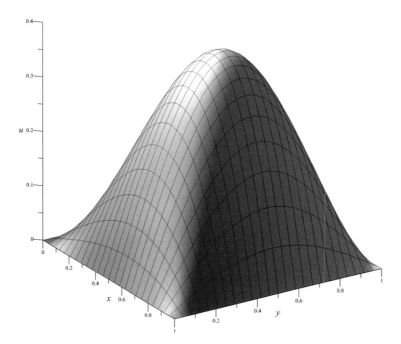

Fig. 7.13 The temperature in the plate at time $t = 0$

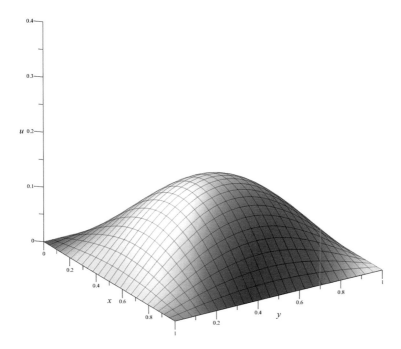

Fig. 7.14 The temperature in the plate at time $t = 5$

7.5 The Laplace's Equation

As we have seen in Sect. 7.1, the elliptic-type equation

$$\frac{\partial^2 u}{\partial x^2} + \frac{\partial^2 u}{\partial y^2} = 0 \tag{7.192}$$

is called *Laplace's equation* in two dimensions. In three dimensions the equation is written

$$\frac{\partial^2 u}{\partial x^2} + \frac{\partial^2 u}{\partial y^2} + \frac{\partial^2 u}{\partial z^2} = 0. \tag{7.193}$$

Using the Laplacian ∇^2 defined by (7.7) or (7.8) respectively, Laplace's equation can be written:

$$\nabla^2 u = 0. \tag{7.194}$$

A function of class C^2 that satisfies Laplace's equation (7.194) in a region R is said to be *harmonic* on that region. For instance, the functions

$$u(x, y) = x^2 - y^2, \quad v(x, y) = e^x \sin y$$

are both harmonic on the whole plane, while the function

$$w(x, y, z) = \frac{1}{\sqrt{x^2 + y^2 + z^2}}$$

is harmonic on the region $R = \mathbb{R}^3 - (0, 0, 0)$.

Laplace's equation appears in problems involving potentials (such as gravitational potential, electrostatic potential, etc). It is also known as the *steady-state heat equation*, because the heat equation

$$\frac{\partial u}{\partial t} = c^2 \nabla^2 u$$

becomes the Laplace's equation (7.194) in the steady-state case (when u is independent of time, $\frac{\partial u}{\partial t} = 0$).

The problem of solving Eq. (7.194) in a region R (either in plane or in space) such that u has prescribed values on the boundary of the region, ∂R, is called the *Dirichlet problem* for Laplace's equation. In the two-dimensional case, the most usual regions encountered in Dirichlet problems are rectangles and disks. It can be proved (see [37], p. 577) that the Dirichlet problem can have only one solution.

7.5 The Laplace's Equation

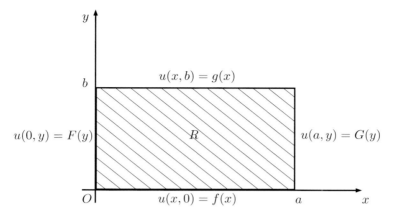

Fig. 7.15 Dirichlet problem for a rectangle

7.5.1 Dirichlet Problem for a Rectangle

We consider the following Dirichlet problem on the rectangular region $R = [0, a] \times [0, b]$ (see Fig. 7.15):

$$\frac{\partial^2 u}{\partial x^2} + \frac{\partial^2 u}{\partial y^2} = 0 \quad \text{for } 0 < x < a,\ 0 < y < b, \tag{7.195}$$

$$u(x, 0) = f(x), \quad u(x, b) = g(x) \quad \text{for } 0 \le x \le a, \tag{7.196}$$

$$u(0, y) = F(y), \quad u(a, y) = G(y) \quad \text{for } 0 \le y \le b. \tag{7.197}$$

To find the solution of this Dirichlet problem, we split it into two problems:

Problem 1

$$\frac{\partial^2 u_1}{\partial x^2} + \frac{\partial^2 u_1}{\partial y^2} = 0 \quad \text{for } 0 < x < a,\ 0 < y < b, \tag{7.198}$$

$$u_1(x, 0) = f(x), \quad u_1(x, b) = g(x) \quad \text{for } 0 \le x \le a, \tag{7.199}$$

$$u_1(0, y) = u_1(a, y) = 0 \quad \text{for } 0 \le y \le b. \tag{7.200}$$

Problem 2

$$\frac{\partial^2 u_2}{\partial x^2} + \frac{\partial^2 u_2}{\partial y^2} = 0 \quad \text{for } 0 < x < a,\ 0 < y < b, \tag{7.201}$$

$$u_2(x, 0) = u_2(x, b) = 0 \quad \text{for } 0 \le x \le a, \tag{7.202}$$

$$u_2(0, y) = F(y), \quad u_2(a, y) = G(y) \quad \text{for } 0 \le y \le b. \tag{7.203}$$

As can be readily seen, if $u_1(x, y)$ is a solution of *Problem 1* and $u_2(x, y)$ is a solution of *Problem 2*, then

$$u(x, y) = u_1(x, y) + u_2(x, y)$$

is a solution of our initial problem (7.195)–(7.197).

Let us solve *Problem 1* using the Fourier method. The first step is to search for solutions of Laplace's equation of the form

$$u_1(x, y) = X(x)Y(y),$$

which satisfy the boundary conditions (7.200). Thus, we have:

$$\frac{X''}{X} = -\frac{Y''}{Y} = \lambda,$$

and we obtain the linear ODEs with constant coefficients:

$$X'' - \lambda X = 0 \tag{7.204}$$

$$Y'' + \lambda Y = 0. \tag{7.205}$$

As shown in Sect. 7.2.2, the Eq. (7.204) with the boundary conditions $X(0) = X(a) = 0$ (which follow from (7.200)) has nontrivial solutions only if $\lambda = \lambda_n = -\left(\frac{n\pi}{a}\right)^2$, $n = 1, 2, \ldots$, and the solutions are the eigenfunctions

$$X_n = \sin\left(\frac{n\pi}{a} x\right)$$

multiplied by constants. For $\lambda = -\left(\frac{n\pi}{a}\right)^2$, the general solution of the Eq. (7.205) (expressed in terms of hyperbolic functions) is

$$Y_n = A_n \cosh\left(\frac{n\pi}{a} y\right) + B_n \sinh\left(\frac{n\pi}{a} y\right).$$

The second step is to use the superposition principle and find a solution of the form

$$u_1(x, y) = \sum_{n=1}^{\infty} \left(A_n \cosh\left(\frac{n\pi}{a} y\right) + B_n \sinh\left(\frac{n\pi}{a} y\right) \right) \sin\left(\frac{n\pi}{a} x\right), \tag{7.206}$$

which satisfies (7.199) as well. For $y = 0$ we get:

$$u_1(x, 0) = f(x) = \sum_{n=1}^{\infty} A_n \sin\left(\frac{n\pi}{a} x\right),$$

7.5 The Laplace's Equation

so the coefficients A_n are given by the formula:

$$A_n = \frac{2}{a} \int_0^a f(x) \sin\left(\frac{n\pi}{a}x\right) dx. \tag{7.207}$$

For $y = b$ we obtain:

$$u_1(x, b) = g(x) = \sum_{n=1}^{\infty} \left(A_n \cosh \frac{n\pi b}{a} + B_n \sinh \frac{n\pi b}{a}\right) \sin\left(\frac{n\pi}{a}x\right),$$

so

$$A_n \cosh \frac{n\pi b}{a} + B_n \sinh \frac{n\pi b}{a} = \frac{2}{a} \int_0^a g(x) \sin\left(\frac{n\pi}{a}x\right) dx.$$

Hence the coefficients B_n are given by the formula:

$$B_n = \frac{1}{\sinh \frac{n\pi b}{a}} \left(\frac{2}{a} \int_0^a g(x) \sin\left(\frac{n\pi}{a}x\right) dx - A_n \cosh \frac{n\pi b}{a}\right). \tag{7.208}$$

Following the same steps, we obtain the solution of *Problem 2* as

$$u_2(x, y) = \sum_{n=1}^{\infty} \left(C_n \cosh\left(\frac{n\pi}{b}x\right) + D_n \sinh\left(\frac{n\pi}{b}x\right)\right) \sin\left(\frac{n\pi}{b}y\right), \tag{7.209}$$

where

$$C_n = \frac{2}{b} \int_0^b F(y) \sin\left(\frac{n\pi}{b}y\right) dy, \tag{7.210}$$

$$D_n = \frac{1}{\sinh \frac{n\pi a}{b}} \left(\frac{2}{b} \int_0^b G(y) \sin\left(\frac{n\pi}{b}y\right) dy - C_n \cosh \frac{n\pi a}{b}\right). \tag{7.211}$$

7.5.2 Dirichlet Problem for a Disk

Consider the open disk of radius r centered at the origin (r is a positive number):

$$D = \{(x, y) \in \mathbb{R}^2 : x^2 + y^2 < r^2\}.$$

Its boundary is the circle of radius r,

$$\partial D = \{(x, y) \in \mathbb{R}^2 : x^2 + y^2 = r^2\}.$$

The *Dirichlet problem for the disk* D is the problem to find a function $u(x, y)$ of class C^2 on D and continuous on $D \cup \partial D$, which satisfies the *Laplace's equation*:

$$\nabla^2 u = \frac{\partial^2 u}{\partial x^2}(x, y) + \frac{\partial^2 u}{\partial y^2}(x, y) = 0, \quad (x, y) \in D \tag{7.212}$$

and the boundary condition

$$u(x, y) = g(x, y), \quad (x, y) \in \partial D, \tag{7.213}$$

where $g(x, y)$ is a given continuous function.

We use the polar coordinates (ρ, θ):

$$\begin{cases} x = \rho \cos \theta \\ y = \rho \sin \theta \end{cases},$$

and denote

$$u(x, y) = u(\rho \cos \theta, \rho \sin \theta) = \bar{u}(\rho, \theta).$$

The open disk D and its boundary ∂D are written (see Fig. 7.16):

$$D : \rho \in [0, r), \theta \in [0, 2\pi),$$

$$\partial D : \rho = r, \theta \in [0, 2\pi).$$

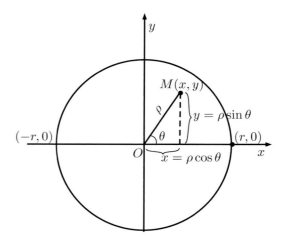

Fig. 7.16 Polar coordinates (ρ, θ)

7.5 The Laplace's Equation

We compute the partial derivatives of the function $\bar{u}(\rho, \theta)$:

$$\frac{\partial \bar{u}}{\partial \rho} = \frac{\partial u}{\partial x} \cdot \frac{\partial x}{\partial \rho} + \frac{\partial u}{\partial y} \cdot \frac{\partial y}{\partial \rho} = \frac{\partial u}{\partial x} \cos\theta + \frac{\partial u}{\partial y} \sin\theta,$$

$$\frac{\partial \bar{u}}{\partial \theta} = \frac{\partial u}{\partial x} \cdot \frac{\partial x}{\partial \theta} + \frac{\partial u}{\partial y} \cdot \frac{\partial y}{\partial \theta} = \frac{\partial \bar{u}}{\partial x}(-\rho \sin\theta) + \frac{\partial u}{\partial y}\rho \cos\theta,$$

$$\frac{\partial^2 \bar{u}}{\partial \rho^2} = \frac{\partial^2 u}{\partial x^2} \cos^2\theta + 2\frac{\partial^2 u}{\partial x \partial y} \sin\theta \cos\theta + \frac{\partial^2 u}{\partial y^2} \sin^2\theta,$$

$$\frac{\partial^2 \bar{u}}{\partial \theta^2} = \frac{\partial u}{\partial x^2}\rho^2 \sin^2\theta - 2\frac{\partial^2 u}{\partial x \partial y}\rho^2 \sin\theta \cos\theta + \frac{\partial^2 u}{\partial y^2}\rho^2 \sin^2\theta$$

$$- \frac{\partial \bar{u}}{\partial x}\rho \cos\theta - \frac{\partial u}{\partial y}\rho \sin\theta,$$

so

$$\frac{\partial^2 \bar{u}}{\partial \rho^2} + \frac{1}{\rho^2} \cdot \frac{\partial^2 \bar{u}}{\partial \theta^2} = \frac{\partial^2 u}{\partial x^2} + \frac{\partial^2 u}{\partial y^2} - \frac{1}{\rho} \cdot \frac{\partial \bar{u}}{\partial \rho}.$$

Thus, the expression of the Laplacian in polar coordinates is

$$\nabla^2 u = \frac{\partial^2 u}{\partial x^2} + \frac{\partial^2 u}{\partial y^2} = \frac{\partial^2 \bar{u}}{\partial \rho^2} + \frac{1}{\rho^2} \cdot \frac{\partial^2 \bar{u}}{\partial \theta^2} + \frac{1}{\rho} \cdot \frac{\partial \bar{u}}{\partial \rho} \quad (7.214)$$

and Laplace's equation $\nabla^2 u = 0$ becomes in polar coordinates:

$$\rho^2 \frac{\partial^2 \bar{u}}{\partial \rho^2} + \rho \frac{\partial \bar{u}}{\partial \rho} + \frac{\partial^2 \bar{u}}{\partial \theta^2} = 0, \quad (7.215)$$

for any $\rho \in (0, r)$ and $\theta \in \mathbb{R}$. We denote $f(\theta) = g(r\cos\theta, r\sin\theta)$, $\theta \in \mathbb{R}$. The function $f(\theta)$ is continuous, periodic (of period 2π) and the boundary condition (7.213) is written:

$$\bar{u}(r, \theta) = f(\theta), \ \theta \in \mathbb{R}. \quad (7.216)$$

Since

$$\bar{u}(\rho, \theta) = \bar{u}(\rho, \theta + 2\pi), \ \forall \rho \in [0, r], \theta \in \mathbb{R}, \quad (7.217)$$

the Dirichlet problem for the disk D written in polar coordinates is to find a function $\bar{u}(\rho, \theta)$ continuous on $[0, r] \times \mathbb{R}$, periodic in the variable θ, which satisfies Laplace's equation in polar coordinates (7.215) and the boundary condition (7.216).

We use the Fourier method: first of all, we search for a solution with separable variables,

$$\overline{u}(\rho, \theta) = R(\rho) \cdot T(\theta),$$

where R and T are functions of class C^2 of one variable and T is a periodic function of period 2π. Now, the Eq. (7.215) becomes:

$$\rho^2 R''(\rho) \cdot T(\theta) + \rho R'(\rho) \cdot T(\theta) = -R(\rho) \cdot T''(\theta),$$

or, after separating functions of ρ from functions of θ,

$$\frac{\rho^2 R''(\rho) + \rho R'(\rho)}{R(\rho)} = -\frac{T''(\theta)}{T(\theta)} = k = \text{constant.} \tag{7.218}$$

The last equality in (7.218) give rise to the ODE-2:

$$T'' + kT = 0. \tag{7.219}$$

We have three possible general solutions of this equation:

$$T(\theta) = C_1 \cosh \lambda\theta + C_2 \sinh \lambda\theta \quad \text{for } k = -\lambda^2 < 0 \tag{7.220}$$

$$T(\theta) = C_1 + C_2 \theta \quad \text{for } k = 0 \tag{7.221}$$

$$T(\theta) = C_1 \cos \lambda\theta + C_2 \sin \lambda\theta \quad \text{for } k = \lambda^2 > 0 \tag{7.222}$$

We can see that (7.220) is always a non-periodic function (except for the trivial case $C_1 = C_2 = 0$). The solution in the second case, (7.221), is also non-periodic if $C_2 \neq 0$, so if $k = 0$ we have to take the constant functions $T(\theta) = C_1$. Finally, in the last case, $T(\theta)$ is periodic of period 2π if and only if $\lambda = n, n = 1, 2, \ldots$. Including the case $\lambda = 0$, we can write the solutions of the Eq. (7.219) as:

$$T_n(\theta) = C_{1,n} \cos n\theta + C_{2,n} \sin n\theta, \ n = 0, 1, 2, \ldots, \tag{7.223}$$

where $C_{1,n}, C_{2,n}$ are arbitrary constants. Now, the equality

$$\frac{\rho^2 R''(\rho) + \rho R'(\rho)}{R(\rho)} = k = n^2, \ n = 0, 1, 2, \ldots$$

from (7.218) give rise to the *Euler equation* (see Sect. 2.6):

$$\rho^2 R''(\rho) + \rho R'(\rho) - n^2 R(\rho) = 0.$$

7.5 The Laplace's Equation

We make the usual Euler substitution $\rho = \exp(s)$ and denote

$$\widetilde{R}(s) = R(\exp(s)).$$

Since

$$R'(\rho) = \exp(-s)\widetilde{R}'(s), \quad R''(\rho) = \exp(-2s)\left(\widetilde{R}''(s) - \widetilde{R}'(s)\right),$$

we finally get the following second-order linear differential equations:

$$\widetilde{R}''(s) - n^2 \widetilde{R}(s) = 0, \quad n = 0, 1, 2, \ldots$$

The general solutions for $n = 1, 2, \ldots$ are:

$$\widetilde{R}_n(s) = D_{1,n} \exp(ns) + D_{2,n} \exp(-ns),$$

and, for $n = 0$,

$$\widetilde{R}_0(s) = D_{1,0} + D_{2,0} s,$$

where $D_{1,n}$, $D_{2,n}$ are arbitrary constants. Coming back to the initial variable ρ we obtain:

$$R_n(\rho) = D_{1,n} \rho^n + D_{2,n} \rho^{-n}, \quad n = 1, 2, \ldots,$$
$$R_0(\rho) = D_{1,0} + D_{2,0} \ln \rho.$$

Since $\lim_{\rho \to 0} \rho^{-n} = \infty$, $\lim_{\rho \to 0} \ln \rho = -\infty$ and the function $\bar{u}(\rho, \theta)$ is bounded, it is compulsory that $D_{2,n} = 0, n = 0, 1, 2, \ldots$, so

$$R_n(\rho) = D_{1,n} \rho^n, \quad n = 0, 1, 2, \ldots. \tag{7.224}$$

Thus, we have found the following sequence of solutions of (7.215):

$$\begin{aligned}\bar{u}_0(\rho, \theta) &= A_0, \\ \bar{u}_n(\rho, \theta) &= (A_n \cos n\theta + B_n \sin n\theta)\rho^n, \quad n = 1, 2, \ldots,\end{aligned} \tag{7.225}$$

where $A_n = C_{1,n} D_{1,n}, n = 0, 1, \ldots$ and $B_n = C_{2,n} D_{1,n}, n = 1, 2, \ldots$.

Using the superposition principle, we search for a solution of the Dirichlet problem (7.215)–(7.216) of the form

$$\bar{u}(\rho, \theta) = A_0 + \sum_{n=1}^{\infty} \bar{u}_n(\rho, \theta) = A_0 + \sum_{n=1}^{\infty} (A_n \cos n\theta + B_n \sin n\theta)\rho^n. \tag{7.226}$$

For $\rho = r$, we get:

$$\bar{u}(r, \theta) = f(\theta) = A_0 + \sum_{n=1}^{\infty} \left(A_n r^n \cos n\theta + B_n r^n \sin n\theta\right). \qquad (7.227)$$

This is the Fourier expansion of the periodic function $f(\theta)$, of period 2π. Thus, the coefficients of the series (7.226) are given by the formulas:

$$A_0 = \frac{1}{2\pi} \int_0^{2\pi} f(\theta)\, d\theta = \frac{1}{2\pi} \int_{-\pi}^{\pi} f(\theta)\, d\theta,$$

$$A_n = \frac{1}{\pi r^n} \int_0^{2\pi} f(\theta) \cos n\theta\, d\theta = \frac{1}{\pi r^n} \int_{-\pi}^{\pi} f(\theta) \cos n\theta\, d\theta, \qquad (7.228)$$

$$B_n = \frac{1}{\pi r^n} \int_0^{2\pi} f(\theta) \sin n\theta\, d\theta = \frac{1}{\pi r^n} \int_{-\pi}^{\pi} f(\theta) \sin n\theta\, d\theta,$$

$n = 1, 2, \ldots$. Finally, we replace these coefficients in (7.226) and find the unique solution of the above Dirichlet problem.

Remark 7.5 The function $\bar{u}(\rho, \theta)$ can be written:

$$\bar{u}(\rho, \theta) = \frac{1}{2\pi} \int_{-\pi}^{\pi} f(\xi)\, d\xi +$$

$$+ \frac{1}{\pi} \sum_{n=1}^{\infty} \left(\frac{\rho}{r}\right)^n \left(\int_{-\pi}^{\pi} f(\xi) \cos n\xi\, d\xi \cdot \cos n\theta + \int_{-\pi}^{\pi} f(\xi) \sin n\xi\, d\xi \cdot \sin n\theta\right)$$

$$= \frac{1}{2\pi} \int_{-\pi}^{\pi} f(\xi)\, d\xi + \frac{1}{\pi} \sum_{n=1}^{\infty} \left(\frac{\rho}{r}\right)^n \int_{-\pi}^{\pi} f(\xi) \cos n(\xi - \theta)\, d\xi$$

$$= \frac{1}{2\pi} \int_{-\pi}^{\pi} \left(1 + 2 \sum_{n=1}^{\infty} \left(\frac{\rho}{r}\right)^n \cos n(\xi - \theta)\right) f(\xi)\, d\xi.$$

Let us denote

$$z = \frac{\rho}{r} \left(\cos(\xi - \theta) + i \sin(\xi - \theta)\right).$$

Since $\sum_{n=1}^{\infty} z^n = \frac{z}{1-z}$ (because $|z| = \frac{\rho}{r} < 1$), we obtain:

$$1 + 2 \sum_{n=1}^{\infty} \left(\frac{\rho}{r}\right)^n \cos n(\xi - \theta) = \mathrm{Re}\left(1 + 2 \sum_{n=1}^{\infty} z^n\right) =$$

$$= \mathrm{Re}\left(1 + 2 \frac{z}{1-z}\right) = \mathrm{Re}\left(\frac{1+z}{1-z}\right)$$

7.5 The Laplace's Equation

For any complex number $z = a(\cos\varphi + i\sin\varphi)$, one can write:

$$\frac{1+z}{1-z} = \frac{1 + a\cos\varphi + ia\sin\varphi}{1 - a\cos\varphi - ia\sin\varphi} =$$

$$= \frac{(1 - a\cos\varphi + ia\sin\varphi)(1 + a\cos\varphi + ia\sin\varphi)}{(1 - a\cos\varphi)^2 + a^2\sin^2\varphi} = \frac{1 - a^2 + 2ia\sin\varphi}{1 + a^2 - 2a\cos\varphi}.$$

Hence

$$\operatorname{Re}\left(\frac{1+z}{1-z}\right) = \frac{1 - a^2}{1 + a^2 - 2a\cos\varphi}$$

and the solution $\bar{u}(\rho, \theta)$ can be written as:

$$\bar{u}(\rho, \theta) = \frac{1}{2\pi}\int_{-\pi}^{\pi} \frac{r^2 - \rho^2}{r^2 + \rho^2 - 2r\rho\cos(\xi - \theta)} f(\xi)\,d\xi. \tag{7.229}$$

This formula is known as *Poisson's integral formula*. Although in most cases the integral cannot be evaluated in closed form, it is often more suitable for numerical approximations than the formula (7.226) which expresses the solution as an infinite series [26].

Example 7.17 Let us find the solution of the Dirichlet problem:

$$\nabla^2 u = 0, \quad x^2 + y^2 < 4,$$

$$u(2, \theta) = \sin 3\theta, \quad \theta \in [0, 2\pi).$$

Usually, the notation $u(\rho, \theta)$ is used instead of $\bar{u}(\rho, \theta)$.

The radius of the disk is $r = 2$. We compute A_n, B_n from the formulas (7.228) and find $A_n = 0$ for $n = 0, 1, \ldots$, $B_n = 0$ for $n \neq 3$, and $B_3 = \frac{1}{8}$. Hence,

$$u(\rho, \theta) = \frac{1}{8}\rho^3 \sin 3\theta.$$

By de Moivre's formula,

$$(\cos\theta + i\sin\theta)^n = \cos n\theta + i\sin n\theta,$$

we obtain that

$$\sin 3\theta = 3\cos^2\theta \sin\theta - \sin^3\theta,$$

so the function $u(\rho, \theta)$ can be written:

$$u(\rho, \theta) = \frac{1}{8}\left(3\rho^2 \cos^2\theta \cdot \rho \sin\theta - \rho^3 \sin^3\theta\right).$$

By replacing $x = \rho\cos\theta$ and $y = \rho\sin\theta$ the expression of $u(x, y)$ is:

$$u(x, y) = \frac{1}{8}(3x^2 y - y^3).$$

7.6 Exercises

1. Find the canonical form of the following PDEs-2. Find also the general solutions of the first three equations:

   (a) $3\dfrac{\partial^2 u}{\partial x^2} - 5\dfrac{\partial^2 u}{\partial x \partial y} - 2\dfrac{\partial^2 u}{\partial y^2} + 3\dfrac{\partial u}{\partial x} + \dfrac{\partial u}{\partial y} = 0.$

   (b) $x\dfrac{\partial^2 u}{\partial x^2} - 4x^3 \dfrac{\partial^2 u}{\partial y^2} - \dfrac{\partial u}{\partial x} = 0, x > 0.$

   (c) $\dfrac{\partial^2 u}{\partial x^2} + 2\dfrac{\partial^2 u}{\partial x \partial y} + \dfrac{\partial^2 u}{\partial y^2} + 4\dfrac{\partial u}{\partial x} + 4\dfrac{\partial u}{\partial y} + 4u = 0;$

   (d) $\dfrac{\partial^2 u}{\partial x^2} - 4\dfrac{\partial^2 u}{\partial x \partial y} + 5\dfrac{\partial^2 u}{\partial y^2} - \dfrac{\partial u}{\partial x} + 2\dfrac{\partial u}{\partial y} = 0;$

2. Solve the following Cauchy problems:

   (a) $x^2 \dfrac{\partial^2 u}{\partial x^2} + 2xy \dfrac{\partial^2 u}{\partial x \partial y} + y^2 \dfrac{\partial^2 u}{\partial y^2} = 0;\ u(1, y) = y^2,\ \dfrac{\partial u}{\partial x}(1, y) = y^2 + y;$

   (b) $x^4 \dfrac{\partial^2 u}{\partial x^2} + 2x^2 y^2 \dfrac{\partial^2 u}{\partial x \partial y} + y^4 \dfrac{\partial^2 u}{\partial y^2} + 2y^2(y - x)\dfrac{\partial u}{\partial y} = 0,$
   $u(1, y) = 0,\ \dfrac{\partial u}{\partial x}(1, y) = y;$

   (c) $\dfrac{\partial^2 u}{\partial x^2} + 6\dfrac{\partial^2 u}{\partial x \partial y} + 5\dfrac{\partial^2 u}{\partial y^2} = 0,\ u(0, y) = y^2,\ \dfrac{\partial u}{\partial x}(0, y) = 2y;$

   (d) $\dfrac{\partial^2 u}{\partial x^2} + 2\cos x \dfrac{\partial^2 u}{\partial x \partial y} - \sin^2 x \dfrac{\partial^2 u}{\partial y^2} - \sin x \dfrac{\partial u}{\partial y} = 0,$
   $u(x, \sin x) = 2x + \cos x,\ \dfrac{\partial u}{\partial y}(x, \sin x) = \sin x;$

   (e) $\dfrac{\partial^2 u}{\partial x \partial y} + \dfrac{\partial u}{\partial y} = 0,\ u(x, x) = 0,\ \dfrac{\partial u}{\partial y}(x, x) = \exp(-x);$

   (f) $\dfrac{\partial^2 u}{\partial y^2} = x + y,\ u(x, 0) = x^2,\ \dfrac{\partial u}{\partial y}(x, 0) = \exp x.$

7.6 Exercises

3. Find $f(t)$ in the following cases:
   (a) $u(x, y) = f\left(\frac{y}{x}\right)$ is a solution of the Laplace equation, $\nabla^2 u = \frac{\partial^2 u}{\partial x^2} + \frac{\partial^2 u}{\partial y^2} = 0$;
   (b) $u(x, y) = f(x^2 + y^2)$ is a solution of the Laplace equation.

4. Find the displacement $u(x, t)$ for the following infinite vibrating strings:
   (a) $\frac{\partial^2 u}{\partial t^2} = 4\frac{\partial^2 u}{\partial x^2}$, $u(x, 0) = x$, $\frac{\partial u}{\partial t}(x, 0) = \exp x$;
   (b) $\frac{\partial^2 u}{\partial t^2} = \frac{\partial^2 u}{\partial x^2}$, $u(x, 0) = x$, $\frac{\partial u}{\partial t}(x, 0) = -x$;
   (c) $\frac{\partial^2 u}{\partial t^2} = \frac{\partial^2 u}{\partial x^2}$, $u(x, 0) = x^2$, $\frac{\partial u}{\partial t}(x, 0) = 0$;

5. Prove that the shape of the infinite string $\frac{\partial^2 u}{\partial t^2} = a^2 \frac{\partial^2 u}{\partial x^2}$, $u(x, 0) = \sin x$, $\frac{\partial u}{\partial t}(x, 0) = 1$, at the moment $t = \frac{\pi}{2a}$ is a straight line.

6. Find the displacement $u(x, t)$ for a finite vibrating string with fixed ends, $u(0, t) = u(l, t) = 0$ for any $t \geq 0$, if:
   (a) $l = 6$, $a = 1$, $u(x, 0) = \begin{cases} x, & 0 \leq x < 5 \\ 5(6-x), & 5 \leq x \leq 6 \end{cases}$, $\frac{\partial u}{\partial t}(x, 0) = 0$.
   (b) $l = \pi$, $a = 1$, $u(x, 0) = c(\pi x - x^2)$, $\frac{\partial u}{\partial t}(x, 0) = 0$.
   (c) $l = \pi$, $a = 1$, $u(x, 0) = 0.1 \sin x$, $\frac{\partial u}{\partial t}(x, 0) = -0.2 \sin 2x$.
   (d) $l = 6$, $a = 2$, $u(x, 0) = 0$, $\frac{\partial u}{\partial t}(x, 0) = \begin{cases} 0, & x \in [0, 2) \cup (4, 6] \\ v, & x \in [2, 4] \end{cases}$.

7. A rod of length l has $0°C$ at its extremities: $u(0, t) = u(l, t) = 0$, $t \geq 0$. Find its temperature $u(x, t)$ at time t, if the initial temperature is:
   (a) $u(x, 0) = \begin{cases} x, & 0 \leq x < \frac{l}{2} \\ l - x, & \frac{l}{2} \leq x < l \end{cases}$;
   (b) $u(x, 0) = \begin{cases} x, & x \in [0, 1] \\ \frac{4-x}{3}, & x \in (1, 4] \end{cases}$, $l = 4$, $c = 4$;
   (c) $u(x, 0) = 2 \sin 3x$, $l = \pi$, $c = 1$;
   (d) $u(x, 0) = 3 \sin \frac{\pi x}{l} - 5 \sin \frac{2\pi x}{l}$, $c = 1$;
   (e) $u(x, 0) = \begin{cases} u_0, & x \in [x_1, x_2] \subset (0, l) \\ 0, & x \in [0, l] \setminus [x_1, x_2] \end{cases}$, $c = 1$;
   (f) $u(x, 0) = \frac{x(l-x)}{l^2}$, $c = 1$.

8. An infinite homogeneous rod has the initial temperature $u(x, 0) = \varphi(x) = A \exp(-Bx^2)$, $x \in (-\infty, \infty)$, $A > 0$, $B > 0$. If $c = 0.5$, find the temperature $u(x, t)$ at the point $x = 0$, at the moment $t = \frac{1}{A^2 B}$. Compute it for $A = 2\sqrt{6}$.

9. Solve the Dirichlet problem
$$\nabla^2 u = \frac{\partial^2 u}{\partial x^2} + \frac{\partial^2 u}{\partial y^2} = 0,$$

for the disk

$$D(0, r) : \{(x, y) : x^2 + y^2 \leq r^2\}$$

in the following cases:

(a) $r = 1$, $u|_{x^2+y^2=1} = 3y$;
(b) $r = 2$, $u|_{x^2+y^2=4} = x^2$;
(c) $r = 4$, $u|_{x^2+y^2=16} = y^2 + 3xy$;
(d) $r = 2$, $u(2, \theta) = \sin 3\theta + 4\cos 2\theta$, $\theta \in [0, 2\pi)$.

Chapter 8
Introduction to the Calculus of Variations

This chapter presents the basic theory of Calculus of Variations applied to fundamental types of variational problems with applications in Physics and Engineering. We begin by stating several classical problems (such as: the brachistochrone problem, the minimal surface of revolution, Dido's problem). Then we introduce the general frame of Calculus of Variation, focusing on necessary conditions of extremum of a functional. We deduce the basic differential equations of Calculus of Variations and apply them to solve some classical variational problems.

8.1 Classical Variational Problems

The calculus of variations is one of the classical field of mathematics, dedicated to the study of *variational problems*: problems related to finding the maximum or the minimum values of some special quantities called *functionals*. A functional is a mapping from a set of *admissible* functions to the set of real numbers. Functionals are usually expressed as integrals involving one or more functions and their derivatives.

Variational problems arise in Geometry, Mechanics and other domains. Some of them may have a very intuitive solution (as "the shortest path problem"—Problem 8.1), some of them may have the origins in Antiquity (as "Dido's problem"—Problem 8.4), but the general mathematical frame for solving the variational problems was founded by the Swiss mathematician Leonhard Euler (1707–1783).

Problem 8.1 (The Shortest Path) Find the shortest curve joining two fixed points A and B.

Let $A(a, a_0)$, $B(b, b_0)$, $a < b$ be two distinct points in the Cartesian plane xOy and $y = y(x)$, $x \in [a, b]$ be a function of class $C^1[a, b]$ whose graph is a smooth plane curve connecting the points A and B, that is, $y(a) = a_0$ and $y(b) = b_0$. The length of this curve is expressed by the following definite integral (see [38], 8.1):

$$L = \int_a^b \sqrt{1 + y'(x)^2}\, dx.$$

Let S denote the set of smooth curves joining the points A and B,

$$S = \{y : [a, b] \to \mathbb{R} : y \in C^1[a, b],\, y(a) = a_0,\, y(b) = b_0\}.$$

Thus, our problem is to find in S the function y that minimizes the *functional* $\mathcal{F}$: $S \to \mathbb{R}$

$$\mathcal{F}[y] = \int_a^b \sqrt{1 + y'(x)^2}\, dx. \tag{8.1}$$

The problem can be also posed for the space curves,

$$r(t) = x(t)\mathbf{i} + y(t)\mathbf{j} + z(t)\mathbf{k}, \quad t \in [a, b],$$

joining the points $A(a_1, a_2, a_3)$ and $B(b_1, b_2, b_3)$. In this case, the set of admissible functions is $S = \{(x, y, z) : [a, b] \to \mathbb{R} : x, y, z \in C^1[a, b],\, x(a) = a_1,\, x(b) = b_1,\, y(a) = a_2,\, y(b) = b_2,\, z(a) = a_3,\, z(b) = b_3\}$ and we have to find $(x, y, z) \in S$ such that the value of the functional $\mathcal{F} : S \to \mathbb{R}$,

$$\mathcal{F}[x, y, z] = \int_a^b \sqrt{x'(t)^2 + y'(t)^2 + z'(t)^2}\, dt \tag{8.2}$$

is minimum.

The intuitive answer (in both cases) is that the curve of minimal length is the straight line, and we shall see later that this is the solution found by using the methods of calculus of variations as well. But, in the second case, if we impose the additional condition that the curve is lying on a given surface, the answer is not so intuitive anymore. This is another problem, regarding *geodesic curves* (see Example 8.7).

Problem 8.2 (The Minimal Surface of Revolution) Let $A(a, a_0)$ and $B(b, b_0)$, $a < b$ be two distinct points in the plane xOy. Let $y = y(x)$, $x \in [a, b]$, be a smooth curve joining the points A and B and lying in the upper half-plane: $y(x) > 0$ for all $x \in (a, b)$. If we revolve this curve around the x-axis we obtain a revolution surface (see Fig. 8.1). We want to find the curve $y(x)$ that yields a surface of smallest area. The area of the revolution surface is given by the integral (see [38], 8.2):

$$A = 2\pi \int_a^b y(x)\sqrt{1 + y'(x)^2}\, dx.$$

8.1 Classical Variational Problems

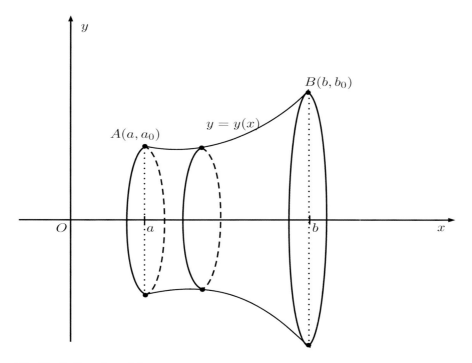

Fig. 8.1 Surface of revolution

Let S denote the set of smooth curves joining the points A and B and lying in the upper half plane,

$$S = \{y : [a, b] \to \mathbb{R} : y \in C^1[a, b], y(a) = a_0, y(b) = b_0, y(x) > 0, \forall x \in (a, b)\}.$$

Since 2π is a constant, our problem is to find in S the function y that minimizes the functional $\mathcal{F} : S \to \mathbb{R}$

$$\mathcal{F}[y] = \int_a^b y(x)\sqrt{1 + y'(x)^2}\, dx. \tag{8.3}$$

As we shall see later, the curve that minimizes this functional is a *catenary* (the curve introduced in Example 2.6) and the revolution surface generated is a *catenoid*.

Problem 8.3 (The Brachistochrone) In 1696 Johann Bernoulli published a letter in which he posed the following problem: Given two points A and B (not on the same vertical line), what is the curve of the fastest descent from A to B ? The word "brachistochrone" means "the shortest time" in Greek (brakhistos = the shortest, khronos = time). We suppose that a particle is moving along the curve $y = y(x)$ without friction, under constant gravity: $G = mg$, where $g = 9.8\,\text{m/s}^2$ is the gravitational acceleration and m is the mass of the particle. Note that the curve $y(x)$ is not the straight line: the shortest path is not equivalent to the shortest time, since the velocity of motion is not a constant. The solution turns out to be a cycloid lying in the vertical plane and passing through the points A and B. The problem was solved by Johann Bernoulli, Jacob Bernoulli, Leibniz, Newton and L'Hospital and it played an important role in the development of the calculus of variations.

We take the coordinate system with the origin at the point A and the y-axis directed downward, so we have $A(0, 0)$ and $B(a, b)$, with $a, b > 0$ (see Fig. 8.2). By the law of conservation of energy, when the particle moves from the point $A(0, 0)$ to the point $M(x, y)$ the potential energy $U = mgy$ is transformed into kinetic energy: $\frac{mv^2}{2} = mgy$, so the velocity at the point $M(x, y)$ is $v = \sqrt{2gy}$. Denote by ds the element of arc on the curve $y = y(x)$ Since the velocity of motion along the curve is

$$v = \frac{ds}{dt} = \sqrt{1 + y'^2}\,\frac{dx}{dt},$$

we can write

$$dt = \frac{\sqrt{1 + y'^2}}{v}\,dx = \frac{\sqrt{1 + y'(x)^2}}{\sqrt{2gy(x)}}\,dx.$$

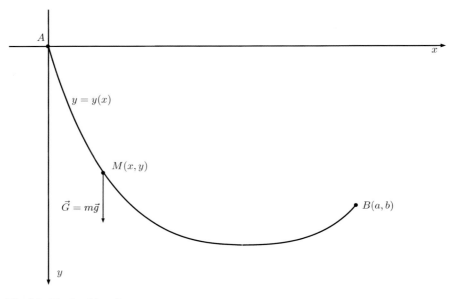

Fig. 8.2 The brachistochrone

8.1 Classical Variational Problems

Thus, the total time spent by the particle to slide down from A to B along the curve $y(x)$ is given by the integral

$$T = \int_0^a \frac{\sqrt{1+[y'(x)]^2}}{\sqrt{2gy}} dx.$$

Since $\frac{1}{\sqrt{2g}}$ is a constant, we have to find, in the set

$$S = \{y : [0, a] \to \mathbb{R} : y \in C^1[0, a], y(0) = 0, y(a) = b\},$$

the function $y(x)$ that minimizes the functional

$$\mathcal{T}[y] = \int_0^a \sqrt{\frac{1+y'(x)^2}{y}} dx. \tag{8.4}$$

The next example is one of the oldest problem in the calculus of variations and belongs to the class of *isoperimetric problems*, corresponding to the problems of conditional extremum in elementary analysis.

Problem 8.4 (Dido's Problem) The legend says that Dido was a Phoenician princess who arrived (arround 850 BC) on the coast of Tunisia, and asked for a piece of land. A local chief agreed to give her as much land as an oxhide could contain. Being very intelligent, she cut the oxhide into narrow strips and joined them together to form a very long thread of oxhide. Then, she bounded the maximum possible area with this thread and the shape of the land turned out to be a circle. The city of Carthage was built on this land and Dido became its queen.

Dido had to solve the following problem: among all the closed curves of fixed length L, which one encloses the maximum possible area? Many mathematicians tried to prove that the circle is the solution of the problem, but the first complete proof was given only in the XIX-th century, by Karl Weierstrass.

In the following, we formulate this problem using functionals.

Let γ be a closed curve defined by the parametric equations:

$$\gamma : \quad x = x(t), \quad y = y(t), \quad t \in [a, b],$$

where $x, y : [a, b] \to \mathbb{R}$ are continuously differentiable functions such that $x(a) = x(b)$ and $y(a) = y(b)$. The length of the curve γ is

$$\mathcal{L}[x, y] = \int_a^b \sqrt{x'(t)^2 + y'(t)^2} \, dt$$

and the area bounded by γ is

$$A[x, y] = \int_\gamma x\,dy = \int_a^b x(t)y'(t)\,dt.$$

Thus, our variational problem is to find the curve γ which satisfies the constraint $\mathcal{L}[x, y] = L$ and maximizes the functional $\mathcal{A}[x, y]$.

Problem 8.5 (Hamilton's Principle (Principle of Least Action)) Pierre-Louis Moreau de Maupertuis (1698–1759) observed that during its motion, a particle chooses the trajectory which minimizes its "action". He could not find a rigorous mathematical expression of this principle, but, later on, William Rowan Hamilton (1805–1865), defined the *action* to be the integral of the *Lagrangian* function, that is, the difference $T - U$ between the kinetic energy T and the potential energy U.

Consider the motion of a particle of mass m in the 3-dimensional space. Let $\mathbf{r}(t) = (x(t), y(t), z(t))$ be the position vector at time t. Since the velocity is $\mathbf{v}(t) = \mathbf{r}'(t)$, the kinetic energy of the particle is

$$T = \frac{1}{2}m \left\| r'(t) \right\|^2 = \frac{1}{2}m \left(x'(t)^2 + y'(t)^2 + z'(t)^2 \right).$$

The potential energy of the particle is a function which depends only on time and position, $U = U(t, x, y, z)$ such that the force acting on the particle, $\mathbf{F} = (f_1, f_2, f_3)$ has the components

$$f_1 = -\frac{\partial U}{\partial x}, \quad f_2 = -\frac{\partial U}{\partial y}, \quad f_3 = -\frac{\partial U}{\partial z}.$$

Let $\mathcal{S}$ denote the *action*, that is, the functional

$$\mathcal{S}[x, y, z] = \int_{t_1}^{t_2} L\left(t, x(t), y(t), z(t), x'(t), y'(t), z'(t)\right) dt$$

$$= \int_{t_1}^{t_2} \left[T\left(x'(t), y'(t), z'(t)\right) - U\left(t, x(t), y(t), z(t)\right) \right] dt. \tag{8.5}$$

Hamilton's principle states that the trajectory $(x(t), y(t), z(t))$ chosen by the particle in the time interval $[t_1, t_2]$ is the one for which the functional S is *stationary*. The functional takes the *minimum* value only for a sufficiently small interval of time, therefore the principle is also known as the *principle of stationary action* (see [15], p. 83–85).

Problem 8.6 (Plateau's Problem) Given a closed curve in space Γ, we have to find the minimal surface bounded by Γ. The problem was raised by Joseph-Louis Lagrange in 1760, but its name is related to the blind physicist Joseph Plateau who demonstrated in 1849 that the minimal surface can be obtained experimentally as a soap film in a wire frame (the *minimal surface* corresponds to the *minimal energy*).

8.1 Classical Variational Problems

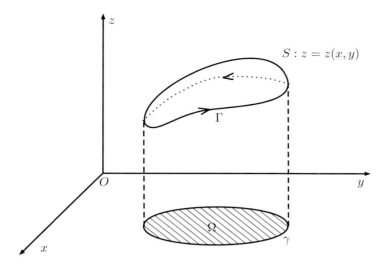

Fig. 8.3 PLateau's problem

Let γ be the projection of Γ on the plane xOy and Ω be the plane closed region bounded by γ (see Fig. 8.3). We suppose that each point (x, y) on the curve γ corresponds to a single point $(x, y, \varphi(x, y))$ on the curve Γ. Consider a surface S defined by $z = z(x, y)$, where $z : \Omega \to \mathbb{R}$ is a function of class C^1 such that

$$z(x, y) = \varphi(x, y), \quad \forall\, (x, y) \in \gamma. \tag{8.6}$$

The area of this surface is given by the formula (see [38], 16.6):

$$\mathcal{F}[z] = \iint_\Omega \sqrt{1 + \left(\frac{\partial z}{\partial x}(x, y)\right)^2 + \left(\frac{\partial z}{\partial y}(x, y)\right)^2}\, dx dy. \tag{8.7}$$

So, the problem is to find the function $z(x, y)$ that minimizes this functional and satisfies (8.6).

If we use the approximation (by the first-order Taylor polynomial)

$$\sqrt{1 + t} \approx 1 + \frac{1}{2}t,$$

the integral (8.7) is approximately equal to

$$\iint_\Omega dx dy + \frac{1}{2} \iint_\Omega \left[\left(\frac{\partial z}{\partial x}(x, y)\right)^2 + \left(\frac{\partial z}{\partial y}(x, y)\right)^2\right] dx dy.$$

Thus, since the first integral is a constant (the area of the domain Ω), we need to find the function $z(x, y)$ that minimizes the second integral,

$$\mathcal{U}[z] = \frac{1}{2}\iint_\Omega \left[\left(\frac{\partial z}{\partial x}(x, y)\right)^2 + \left(\frac{\partial z}{\partial y}(x, y)\right)^2\right] dxdy. \tag{8.8}$$

But this last integral is the expression of the elastic (potential) energy of a homogeneous, thin membrane having the shape S and the modulus of elasticity $\mu = 1$. So, this is why minimizing the area functional (8.7) is equivalent to minimizing the energy functional (8.8). Later we shall see that the functional (8.8) is easier to work with, and a minimizer of it is a solution of the Dirichlet problem for Laplace equation.

Problem 8.7 (Geodesics) Consider the surface $S \in \mathbb{R}^3$, implicitly defined by the equation

$$S: \quad G(x, y, z) = 0,$$

where G is a continuously differentiable function. Let $A(a_1, a_2, a_3)$ and $B(b_1, b_2, b_3)$ be two distinct points on S. The **geodesic** (curve) determined by A and B is the shortest curve lying on S and connecting the points A and B.

Let $\gamma: \ x = x(t), \ y = y(t), \ z = z(t), \ t \in [t_1, t_2]$ be a smooth curve on the surface S, so $G(x(t), y(t), z(t)) = 0$ for any $t \in [t_1, t_2]$. Suppose that γ connects the points A and B: $x(t_i) = a_i$, $y(t_i) = b_i$ and $z(t_i) = c_i$, $i = 1, 2$ (see Fig. 8.4). Since the geodesic curve has the minimum length, we have to find the curve γ that minimizes the functional

$$\mathcal{L}[x, y, z] = \int_{t_1}^{t_2} \sqrt{x'(t)^2 + y'(t)^2 + z'(t)^2}\, dt, \tag{8.9}$$

and satisfies the constraint $G(x(t), y(t), z(t)) = 0, \ t \in [t_1, t_2]$.

Fig. 8.4 Geodesic on the surface S

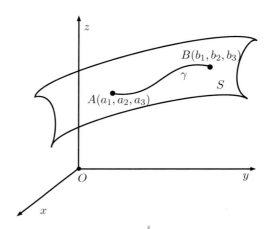

8.2 General Frame of Calculus of Variations

Functionals are defined on some vector spaces of functions. To define the notion of local extremum, we need to define the *distance* between objects in the vector space. The best frame to study such extremum problems is offered by the normed vector spaces. A *norm* on a real vector space V is a real-valued function, $\|\ \| : V \to \mathbb{R}$ with the following properties:

(1) $\|x\| \geq 0$ for all $x \in V$ and $\|x\| = 0$ if and only if $x = \mathbf{0}_V$.
(2) $\|\lambda x\| = |\lambda|\,\|x\|$, for any $x \in V$ and $\lambda \in \mathbb{R}$.
(3) $\|x + y\| \leq \|x\| + \|y\|$, for all $x, y \in V$.
  A vector space equipped with a norm, $(V, \|\cdot\|)$ is called a *normed (vector) space* (see also Sect. 11.1 of Chap. 11).

Example 8.1 In the space $V = \mathbb{R}^n$ we can define, for any $\mathbf{x} = (x_1, \ldots, x_n)$,

$$\|x\| = \sqrt{x_1^2 + \ldots + x_n^2}.$$

It is not difficult to prove that this function is a norm (it has the above properties). It is called the *Euclidean norm*.

Other norms on $V = \mathbb{R}^n$ are defined as follows:

$$\|x\|_1 = |x_1| + \ldots + |x_n|,$$

$$\|x\|_\infty = \max_{i=\overline{1,n}} |x_i|.$$

Example 8.2 Let $V = C[a, b]$ be the vector space of all continuous functions f defined on the interval $[a, b]$ with real values. We can define on V the sup-norm:

$$\|f\|_\infty = \sup_{x \in [a,b]} |f(x)|.$$

It can be easily proved that it is a norm on V.

Example 8.3 Let $V = C^1[a, b]$ be the vector space of the functions of class C^1 on $[a, b]$ (recall that a function f is of class C^k if it has a continuous derivative of order k). We define, for any $f \in V$,

$$\|f\|_{\infty,1} = \sup_{x \in [a,b]} |f(x)| + \sup_{x \in [a,b]} |f'(x)|.$$

It can be proved that this is a norm on the vector space $V = C^1[a, b]$.

Example 8.4 In general, if n is a natural number and $V = C^n[a, b]$ is the vector space of the functions of class C^n defined on $[a, b]$ with real values, then the mapping

$$f \to \|f\|_{\infty,n} = \sup_{x\in[a,b]} |f(x)| + \sup_{x\in[a,b]} |f'(x)| + \ldots + \sup_{x\in[a,b]} |f^{(n)}(x)|$$

is a norm on V.

Example 8.5 Let V be the vector space of the functions of k variables, $f(x_1, \ldots, x_k)$, which are of class C^1 on a closed domain Ω of $\mathbb{R}^k$. Then, the mapping

$$f \to \|f\|_{\infty,1} = \sup_\Omega |f(x_1, \ldots, x_k)| +$$

$$+\sup_\Omega \left|\frac{\partial f}{\partial x_1}(x_1, \ldots, x_k)\right| + \ldots + \sup_\Omega \left|\frac{\partial f}{\partial x_k}(x_1, \ldots, x_k)\right|$$

is a norm on V.

On a normed space $(V, \|\cdot\|)$ one can readily define a *distance function* (the metric induced by the norm $\|\cdot\|$) The distance between two elements $x, y \in V$ is defined to be

$$dist(x, y) = \|x - y\|,$$

so V is also a *metric space* (see Sect. 11.1 of Chap. 11).

In the following, we consider $(V, \|\cdot\|)$ a normed space of functions. Let $\mathcal{F}: V \to \mathbb{R}$ be a functional on V and let $S \subset V$ be a subset of admissible functions (satisfying certain boundary conditions).

We say that the functional $\mathcal{F}$ has a *local extremum* in S at $y_0 \in S$ if there exists $\varepsilon > 0$ such that the expression $\mathcal{F}[y] - \mathcal{F}[y_0]$ keeps the same sign, for all $y \in S$ such that $\|y - y_0\| < \varepsilon$. The functional has a *local minimum* in S at y_0 if $\mathcal{F}[y] - \mathcal{F}[y_0] \geq 0$ for all $y \in S$ such that $\|y - y_0\| < \varepsilon$ and a local maximum, if $\mathcal{F}[y] - \mathcal{F}[y_0] \leq 0$ for all $y \in S$ such that $\|y - y_0\| < \varepsilon$, respectively.

Let $H \subset V$ be the set of functions

$$H = \{h \in V : y_0 + th \in S, t \in \mathbb{R}\}.$$

For instance, if $V = C^2[a, b]$ and $\mathcal{F}$ is the functional defined by

$$\mathcal{F}[y] = \int_a^b F(x, y(x), y'(x))\, dx$$

where $F: \mathbb{R}^3 \to \mathbb{R}$ is a function of class C^2, then the set S of *admissible functions* is

$$S = \{y \in V : y(a) = a_0,\ y(b) = b_0\}$$

($a_0, b_0 \in \mathbb{R}$ are fixed), and the set H of *admissible directions* is

$$H = \{h \in V : h(a) = h(b) = 0\}.$$

Definition 8.1 Consider the functional $\mathcal{F} : S \to \mathbb{R}$, $y_0 \in S$ and $h \in H$, $h \neq 0$. If there exists (as a finite number) the limit

$$\delta_h \mathcal{F}[y_0] = \lim_{t \to 0} \frac{\mathcal{F}[y_0 + th] - \mathcal{F}[y_0]}{t}, \qquad (8.10)$$

then $\delta_h \mathcal{F}[y_0]$ is said to be the **first variation** of $\mathcal{F}$ at y_0 along the direction h.

Remark 8.1 For any $h \in H$ one can define the real function $\varphi_h(t)$,

$$\varphi_h(t) = \mathcal{F}[y_0 + th], \quad t \in \mathbb{R}. \qquad (8.11)$$

Then, the limit (8.10) exists if and only if the function $\varphi_h(t)$ is differentiable at 0 and

$$\delta_h \mathcal{F}[y_0] = \varphi_h'(0) = \frac{d}{dt} (\mathcal{F}[y_0 + th])\bigg|_{t=0}. \qquad (8.12)$$

It can be noticed that the first variation of a functional (also known as the *Gateaux derivative*) has similar properties to the first derivative of a numerical function. Indeed, if $\mathcal{F}, \mathcal{G}$ are two functionals defined on the same admissible subset $S \subset V$ and there exist $\delta_h \mathcal{F}[y_0]$ and $\delta_h \mathcal{G}[y_0]$, then the following properties can be proved:

$$\delta_h (\alpha \mathcal{F} + \beta \mathcal{G})[y_0] = \alpha \delta_h \mathcal{F}[y_0] + \beta \delta_h \mathcal{G}[y_0], \quad \forall \alpha, \beta \in \mathbb{R} \qquad (8.13)$$

$$\delta_h (\mathcal{F}\mathcal{G})[y_0] = \delta_h \mathcal{F}[y_0] \cdot \mathcal{G}[y_0] + \delta_h \mathcal{G}[y_0] \cdot \mathcal{F}[y_0] \qquad (8.14)$$

$$\delta_h \left(\frac{\mathcal{F}}{\mathcal{G}}\right)[y_0] = \frac{\delta_h \mathcal{F}[y_0] \cdot \mathcal{G}[y_0] - \delta_h \mathcal{G}[y_0] \cdot \mathcal{F}[y_0]}{\mathcal{G}[y_0]^2}, \text{ if } \mathcal{G}[y_0] \neq 0. \qquad (8.15)$$

The well-known Fermat's theorem states that, if the function $f : (a, b) \to \mathbb{R}$ has a local extremum at the point $x_0 \in (a, b)$ and it is differentiable at x_0, then $f'(x_0) = 0$. A similar result holds for functionals:

Theorem 8.1 (Fermat's Theorem for Functionals) *If the functional $\mathcal{F} : S \to \mathbb{R}$ has a local extremum in S at $y_0 \in S$ and $h \in H$, $h \neq 0$ is an admissible direction for which there exists the first variation of $\mathcal{F}$ at y_0, $\delta_h \mathcal{F}[y_0]$, then*

$$\delta_h \mathcal{F}[y_0] = 0. \qquad (8.16)$$

Proof Consider the function $\varphi(t)$ defined by (8.11) and assume that $\mathcal{F}$ has a local maximum in S at y_0. Then, there exists $\varepsilon > 0$ such that

$$\mathcal{F}[y] \leq \mathcal{F}[y_0], \text{ for all } y \in S \text{ s.t. } \|y - y_0\| \leq \varepsilon. \tag{8.17}$$

Let $h \in H$, $h \neq \mathbf{0}$ be an admissible direction for which there exists $\delta_h \mathcal{F}[y_0]$. By writing (8.17) for $y = y_0 + th$, it follows that

$$\varphi_h(t) \leq \varphi_h(0), \text{ for all } t \in \left(-\frac{\varepsilon}{\|h\|}, \frac{\varepsilon}{\|h\|}\right),$$

so $t = 0$ is a local maximum point for the function $\varphi_h(t)$. By Fermat's theorem and using Remark 8.1, we obtain that

$$\delta_h \mathcal{F}[y_0] = \varphi_h'(0) = 0.$$

The case when $\mathcal{F}$ has a local minimum at y_0 can be treated in a similar way, so the theorem is proved. □

Remark 8.2 As in the elementary analysis, the condition (8.16) from the Fermat's theorem is only a necessary (but not sufficient) condition for extremum. If $\delta_h \mathcal{F}[y_0] = 0$ for all $h \in H$, we say that $\mathcal{F}$ is **stationary** at y_0 and the function y_0 is called an **extremal** of $\mathcal{F}$ in S, even though **it may not produce a local extremum** in S for the functional $\mathcal{F}$. Anyway, in most practical applications, the functional is known to have a unique local minimum (or maximum) in the set of admissible functions so it is sufficient to find the extremal of $\mathcal{F}$. The methods for setting sufficient conditions for a functional to have minimum (or maximum) at an extremal y_0 are above the level of this book. The reader interested in such investigations can consult [13], Chapter 8, or [15], Chapter 5.

Now, let us examine the "isoperimetric problems"—variational problems that involve searching for local extremum of a functional $\mathcal{F}$ under some constraints. The word "isoperimetric" literally means "having the same perimeter" and it comes from first problem of this kind—Dido's problem (Problem 8.4) which requests to find the closed curve that encloses the maximum area and has a fixed length, L (the constraint). This class of variational problems corresponds to the problems of conditional extremum in elementary analysis, which can be solved by Lagrange's elegant method ([1], Theorem 13.12):

Theorem 8.2 (Lagrange's Multipliers) *Let $\Omega \subset \mathbb{R}^n$ be an open region and consider the functions of class C^1 $f, g_1, \ldots, g_m : \Omega \to \mathbb{R}$, $m < n$. Let X_0 be the subset of Ω on which all the restrictions $g_i(\mathbf{x}) = 0$ are satisfied:*

$$X_0 = \{\mathbf{x} \in \Omega : g_i(\mathbf{x}) = 0, i = 1, \ldots, m\}.$$

If $\mathbf{x}_0 \in X_0$ is a local extremum point of the function f in X_0 and if the Jacobian matrix at $\mathbf{x}_0$, $J(\mathbf{x}_0) = \left(\frac{\partial g_i}{\partial x_j}(\mathbf{x}_0)\right)_{i=1,\ldots,m,\,j=1,\ldots,n}$ has the rank equal to m, then there exist m real numbers $\lambda_1, \ldots, \lambda_m$ such that

$$\nabla f(\mathbf{x}_0) + \lambda_1 \nabla g_1(\mathbf{x}_0) + \ldots + \lambda_m \nabla g_m(\mathbf{x}_0) = \mathbf{0}. \qquad (8.18)$$

The real numbers $\lambda_1, \ldots, \lambda_m$ are called the **Lagrange's multipliers**.

Lagrange's multipliers method can be also formulated in terms of functionals, and used for solving isoperimetric problems ([8]). Since all the isoperimetric problems presented in this chapter deal with a single constraint, we state and prove the theorem for the case $m = 1$.

Theorem 8.3 (Lagrange's Multipliers Method for Functionals) *Let $S \subset V$ be a set of admissible functions in the normed vector space $(V, \|\cdot\|)$, consider the functionals $\mathcal{F}, \mathcal{G} : S \to \mathbb{R}$, and $k \in \mathbb{R}$ a constant. We denote by S_0 the set of all admissible functions that satisfy the constraint $\mathcal{G}[y] = k$. Assume that $\mathcal{F}$ has a local extremum in S_0 at y_0 and y_0 is not an extremal for $\mathcal{G}$. Then, there exists a real number λ_0 such that y_0 is an extremal of the auxiliary functional*

$$\mathcal{F} + \lambda_0 \mathcal{G}. \qquad (8.19)$$

Proof Since y_0 is not an extremal for $\mathcal{G}$, there exists an admissible direction $h_0 \in H$, $h_0 \neq \mathbf{0}$, such that $\delta_{h_0}\mathcal{G}[y_0] \neq 0$.

Let $h \in H$, $h \neq \mathbf{0}$, be an arbitrary admissible direction. We define the functions:

$$f(s,t) = \mathcal{F}[y_0 + sh + th_0], \quad g(s,t) = \mathcal{G}[y_0 + sh + th_0] - k, \; s, t \in \mathbb{R}.$$

As can be readily seen, $(0, 0)$ is a conditional extremum point for the function $f(s, t)$ with the constraint $g(s, t) = 0$. We also notice that $\nabla g(0, 0) \neq \mathbf{0}$, because

$$\frac{\partial g}{\partial t}(0,0) = \lim_{t \to 0} \frac{g(0,t) - g(0,0)}{t} =$$

$$= \lim_{t \to 0} \frac{\mathcal{G}[y_0 + th_0] - \mathcal{G}[y_0]}{t} = \delta_{h_0}\mathcal{G}[y_0] \neq 0.$$

By Theorem 8.2, there exists $\lambda_0 \in \mathbb{R}$ such that

$$\frac{\partial f}{\partial s}(0,0) + \lambda_0 \frac{\partial g}{\partial s}(0,0) = 0, \qquad (8.20)$$

$$\frac{\partial f}{\partial t}(0,0) + \lambda_0 \frac{\partial g}{\partial t}(0,0) = 0. \qquad (8.21)$$

But

$$\frac{\partial f}{\partial s}(0,0) = \lim_{s \to 0} \frac{\mathcal{F}[y_0 + sh] - \mathcal{F}[y_0]}{s} = \delta_h \mathcal{F}[y_0],$$

and

$$\frac{\partial g}{\partial s}(0,0) = \lim_{s \to 0} \frac{\mathcal{G}[y_0 + sh] - \mathcal{G}[y_0]}{s} = \delta_h \mathcal{G}[y_0],$$

so we obtain from (8.20) that

$$\delta_h \mathcal{F}[y_0] + \lambda_0 \delta_h \mathcal{G}[y_0] = 0.$$

Using the property (8.13) of the first variation, it follows that

$$\delta_h (\mathcal{F} + \lambda_0 \mathcal{G})[y_0] = 0,$$

for any admissible direction $h \in H$, $h \neq \mathbf{0}$. Hence y_0 is an extremal of the auxiliary functional (8.19) and the theorem is proved. From (8.21) we can write the value of the Lagrange multiplier $\lambda_0 = -\frac{\delta_{h_0} \mathcal{F}[y_0]}{\delta_{h_0} \mathcal{G}[y_0]}$. $\square$

Remark 8.3 If we have $m > 1$ independent constraints $\mathcal{G}_i[y] = k_i$, $i = 1, \ldots, m$, and the functional $\mathcal{F}[y]$ has a local extremum (with the m constraints) at y_0, then there exist the Lagrange multipliers $\lambda_1, \ldots, \lambda_m \in \mathbb{R}$ such that y_0 is an extremal for the functional $\mathcal{F} + \sum_{i=1}^{m} \lambda_i \mathcal{G}_i$.

In the next sections we discuss the most common types of functionals and show the methods to find their extremals. Recall that a function y_0 is an **extremal** of $\mathcal{F}$ if the functional $\mathcal{F}$ is **stationary** at y_0, that is, if the first variation is 0: $\delta_h \mathcal{F}[y_0] = 0$, for any admissible direction h (see also Remark 8.2).

8.3 The Case $\mathcal{F}[y] = \int_a^b F(x, y, y') dx$

We begin with a basic result in Calculus of variations, attributed to Joseph-Louis Lagrange:

Lemma 8.1 (Fundamental Lemma of Calculus of Variations) *Let $f : [a, b] \to \mathbb{R}$ be a continuous function such that*

$$\int_a^b f(x) h(x) dx = 0$$

8.3 The Case $\mathcal{F}[y] = \int_a^b F(x, y, y')dx$

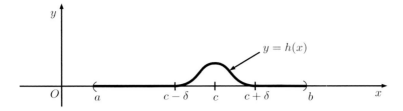

Fig. 8.5 The function $h(x)$

for any function $h : [a, b] \to \mathbb{R}$ of class C^2 with $h(a) = h(b) = 0$. Then

$$f(x) = 0 \quad \text{for all } x \in [a, b].$$

Proof We use the proof by contradiction: suppose that there exists a point $c \in (a, b)$ with $f(c) \neq 0$, say $f(c) > 0$. Since f is continuous, we can take a small number $\delta > 0$ such that $f(x) > 0$ for any $x \in [c-\delta, c+\delta] \subset (a, b)$. The following function (plotted in Fig. 8.5) is obviously of class C^2 on the interval $[a, b]$ and satisfies the condition $h(a) = h(b) = 0$:

$$h(x) = \begin{cases} (x - c + \delta)^3 (c + \delta - x)^3, & \text{if } x \in [c - \delta, c + \delta] \\ 0, & \text{if } x \in [a, c - \delta) \cup (c + \delta, b]. \end{cases}$$

Since $h(x) = 0$ outside the interval $(c - \delta, c + \delta)$ and the functions $f(x)$, $h(x)$ are both positive inside this interval, one can write:

$$\int_a^b f(x)h(x)dx = \int_{c-\delta}^{c+\delta} f(x)h(x)dx > 0,$$

which contradicts the hypothesis. Hence $f(x) = 0$ for any $x \in (a, b)$ and, since the function is continuous, we actually have $f(x) = 0$ on the closed interval $[a, b]$. □

Let F be a function of class C^2 defined on $\mathbb{R}^3$, $V = C^2[a, b]$ be the normed vector space of twice continuously differentiable functions on the interval $[a, b]$, and S, the set of admissible functions

$$S = \{y \in V : y(a) = a_0, \ y(b) = b_0\},$$

where $a_0, b_0 \in \mathbb{R}$ are fixed real numbers.

Theorem 8.4 (Euler-Lagrange) *Consider the functional* $\mathcal{F} : V \to \mathbb{R}$,

$$\mathcal{F}[y] = \int_a^b F(x, y(x), y'(x))dx. \tag{8.22}$$

An admissible function $y_0 \in S$ is an extremal of $\mathcal{F}$ in S if and only if y_0 satisfies the following second-order differential equation (the Euler-Lagrange equation):

$$\frac{\partial F}{\partial y} - \frac{d}{dx}\left(\frac{\partial F}{\partial y'}\right) = 0. \qquad (8.23)$$

Proof Let H denote the set of admissible directions

$$H = \{h \in V : h(a) = h(b) = 0\}$$

and, for any $h \in H$, $h \neq 0$, let $\varphi_h(t)$ denote the function defined by:

$$\varphi_h(t) = \mathcal{F}[y_0 + th] = \int_a^b F\left(x, y_0(x) + th(x), y_0'(x) + th'(x)\right) dx.$$

The function y_0 is an extremal of $\mathcal{F}$ if and only if

$$\delta_h \mathcal{F}[y_0] = \varphi_h'(0) = 0$$

for any $h \in H$, $h \neq 0$. Let us calculate the derivative $\varphi_h'(t)$, using the Leibniz' formula for differentiation under the integral sign (1.72):

$$\varphi_h'(t) = \frac{d}{dt} \int_a^b F\left(x, y_0(x) + th(x), y_0'(x) + th'(x)\right) dx =$$

$$= \int_a^b \frac{d}{dt} F\left(x, y_0(x) + th(x), y_0'(x) + th'(x)\right) dx =$$

$$= \int_a^b \frac{\partial F}{\partial y}\left(x, y_0(x) + th(x), y_0'(x) + th'(x)\right) h(x)\, dx +$$

$$+ \int_a^b \frac{\partial F}{\partial y'}\left(x, y_0(x) + th(x), y_0'(x) + th'(x)\right) h'(x)\, dx.$$

So,

$$0 = \delta_h \mathcal{F}[y_0] = \varphi_h'(0) = \int_a^b \frac{\partial F}{\partial y}\left(x, y_0(x), y_0'(x)\right) h(x)\, dx +$$

$$+ \int_a^b \frac{\partial F}{\partial y'}\left(x, y_0(x), y_0'(x)\right) h'(x)\, dx \stackrel{by\ parts}{=}$$

8.3 The Case $\mathcal{F}[y] = \int_a^b F(x, y, y')dx$

$$= \int_a^b \frac{\partial F}{\partial y}(x, y_0(x), y_0'(x))\, h(x)\, dx +$$

$$+ \frac{\partial F}{\partial y'}(x, y_0(x), y_0'(x))\, h(x) \Big|_a^b - \int_a^b \frac{d}{dx}\left(\frac{\partial F}{\partial y'}(x, y_0(x), y_0'(x))\right) h(x)\, dx.$$

Since $h(a) = h(b) = 0$, we obtain:

$$\int_a^b \left[\frac{\partial F}{\partial y}(x, y_0(x), y_0'(x)) - \frac{d}{dx}\left(\frac{\partial F}{\partial y'}(x, y_0(x), y_0'(x))\right)\right] h(x)\, dx = 0,$$

for any $h \in C^2[a, b]$ with $h(a) = h(b) = 0$. Thus, by applying Lemma 8.1, we get:

$$\frac{\partial F}{\partial y}(x, y_0(x), y_0'(x)) - \frac{d}{dx}\left(\frac{\partial F}{\partial y'}(x, y_0(x), y_0'(x))\right) = 0,$$

for any $x \in [a, b]$, so the function $y_0(x)$ satisfies the Eq. (8.23). □

If we calculate the expression $\frac{d}{dx}\left(\frac{\partial F}{\partial y'}(x, y(x), y'(x))\right)$ in the Euler-Lagrange equation (8.23), we obtain:

$$\frac{\partial F}{\partial y}(x, y(x), y'(x)) - \frac{\partial^2 F}{\partial x \partial y'}(x, y(x), y'(x)) - \quad (8.24)$$

$$- \frac{\partial^2 F}{\partial y \partial y'}(x, y(x), y'(x))\, y'(x) - \frac{\partial^2 F}{\partial y'^2}(x, y(x), y'(x))\, y''(x) = 0$$

and we notice that, if $\frac{\partial^2 F}{\partial y'^2} \neq 0$, the Euler-Lagrange equation is a second-order differential equation.

Remark 8.4 Beltrami identity. If $F = F(y, y')$, i.e. if the variable x does not appear explicitly in the expression of F, then, the Eq. (8.24) becomes:

$$\frac{\partial F}{\partial y}(y(x), y'(x)) - \frac{\partial^2 F}{\partial y \partial y'}(y(x), y'(x))\, y'(x) - \frac{\partial^2 F}{\partial y'^2}(y(x), y'(x))\, y''(x) = 0.$$
(8.25)

It is easy to see that this last equality is equivalent to the following equation:

$$\frac{d}{dx}\left(F(y(x), y'(x)) - y'(x)\frac{\partial F}{\partial y'}(y(x), y'(x))\right) = 0,$$

if $y'(x)$ is not equal to zero on $[a, b]$. Thus,

$$F - y'\frac{\partial F}{\partial y'} = C_1 = \text{constant} \tag{8.26}$$

and so, $F - y'\frac{\partial F}{\partial y'}$ is a first integral for our Euler-Lagrange equation, i.e. the expression

$$F\left(y(x), y'(x)\right) - y'(x)\frac{\partial F}{\partial y'}\left(y(x), y'(x)\right)$$

becomes a constant whenever $y(x)$ is a solution of the Euler-Lagrange equation (8.23). Equation (8.26) is known as the *Beltrami identity*.

Now we can solve some of the problems given at the beginning of this chapter.

Example 8.6 (Solution to the Shortest Path Problem—Problem 8.1) To find the shortest path between the points $A(a, a_0)$, $B(b, b_0)$, $a < b$ means to find the extremal $y(x)$ of the functional

$$\mathcal{F}[y] = \int_a^b \sqrt{1 + y'(x)^2}\, dx,$$

such that $y(a) = a_0$ and $y(b) = b_0$.

Since

$$F(x, y, y') = \sqrt{1 + y'^2},$$

the Euler-Lagrange equation (8.23) becomes:

$$\frac{d}{dx}\left(\frac{\partial F}{\partial y'}\right) = 0 \Rightarrow \frac{\partial F}{\partial y'} = C.$$

But

$$\frac{\partial F}{\partial y'} = \frac{y'}{\sqrt{1 + y'^2}} = C,$$

so

$$y'^2 = \frac{C^2}{1 - C^2}$$

and it follows that $y'(x) = C_1$, a constant. By direct integration we find the family of straight lines

$$y(x) = C_1 x + C_2.$$

8.3 The Case $\mathcal{F}[y] = \int_a^b F(x, y, y')dx$

It remains to find the (uniquely determined) constants C_1 and C_2 such that $y(a) = a_0$ and $y(b) = b_0$ and we obtain the straight line connecting the points A and B.

Example 8.7 (Solution to Problem 8.2—The Minimal Surface of Revolution) If S is the set of smooth curves joining the points A and B and lying in the upper half-plane,

$$S = \{y : [a, b] \to \mathbb{R} : y \in C^1[a, b], y(a) = a_0, y(b) = b_0, y(x) > 0, \forall x \in (a, b)\},$$

we have to find the function $y \in S$ which generates the minimal surface of revolution. Thus, we need to find the extremals of the functional

$$\mathcal{F}[y] = \int_a^b y(x)\sqrt{1 + y'(x)^2}\,dx. \tag{8.27}$$

We have:

$$F(x, y, y') = y\sqrt{1 + y'^2},$$

so we can use the Beltrami identity (8.26) (instead of Euler-Lagrange equation). Since

$$\frac{\partial F}{\partial y'} = \frac{yy'}{\sqrt{1 + y'^2}},$$

the equality (8.26) is written:

$$y\sqrt{1 + y'^2} - \frac{yy'^2}{\sqrt{1 + y'^2}} = C_1,$$

hence

$$y = C_1\sqrt{1 + y'^2}. \tag{8.28}$$

To solve this differential equation, the best idea is to use the hyperbolic functions,

$$\sinh t = \frac{\exp t - \exp(-t)}{2},$$

$$\cosh t = \frac{\exp t + \exp(-t)}{2},$$

which have the following properties:

$$1 + \sinh^2 t = \cosh^2 t \tag{8.29}$$

and

$$(\sinh t)' = \cosh t, \quad (\cosh t)' = \sinh t. \tag{8.30}$$

We put $y'(x) = \sinh t$ in (8.28) and, using (8.29), we find:

$$y = C_1 \cosh t.$$

Since

$$y' = \frac{dy}{dx} = \sinh t, \quad \frac{dy}{dt} = C_1 \sinh t = y' \cdot \frac{dx}{dt},$$

we obtain that $\frac{dx}{dt} = C_1$, so $x = C_1 t + C_2$. Hence, the extremals of the functional (8.27) are of the form

$$y = C_1 \cosh\left(\frac{x - C_2}{C_1}\right), \tag{8.31}$$

i.e. a family of *catenaries*. A *catenary* is the equilibrium shape of a homogeneous chain of given length, suspended at the end points A and B (see also Example 8.17). The resulting revolution surface is called a *catenoid* (see Fig. 8.6). The constants C_1, C_2 are determined from the boundary conditions $y(a) = a_0$ and $y(b) = b_0$.

Example 8.8 (Solution to the Brachistochrone Problem—Problem 8.3) We have seen that, in order to find the curve of fastest descent from the point $A(0, 0)$ to the point $B(a, b)$ we need to search in the set

$$S = \{y : [0, a] \to \mathbb{R} : y \in C^1[0, a], y(0) = 0, y(a) = b\},$$

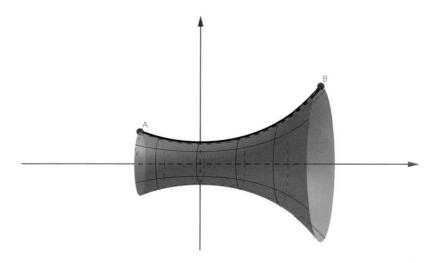

Fig. 8.6 The catenoid

8.3 The Case $\mathcal{F}[y] = \int_a^b F(x, y, y')dx$

an extremal of the functional

$$T[y] = \int_0^a \sqrt{\frac{1+y'(x)^2}{y}}\, dx.$$

In this case also, the function $F(x, y, y')$ does not explicitly depends on x:

$$F(x, y, y') = \frac{\sqrt{1+y'^2}}{\sqrt{y}},$$

so we can use again the prime integral (8.26):

$$\frac{\sqrt{1+y'^2}}{\sqrt{y}} - \frac{y'^2}{\sqrt{y}\sqrt{1+y'^2}} = C \Rightarrow \frac{1}{\sqrt{y}\sqrt{1+y'^2}} = C$$

and we obtain that

$$y = \frac{C_1}{1+y'^2},$$

where $C_1 = \frac{1}{C^2}$. Let $t \in (0, \pi)$ be a parameter s.t. $y' = \cot t$. Then

$$y = C_1 \sin^2 t = \frac{C_1}{2}(1 - \cos 2t). \tag{8.32}$$

It follows that

$$\frac{dy}{dt} = C_1 \sin 2t.$$

But

$$\frac{dy}{dt} = y' \cdot \frac{dx}{dt} = \cot t \, \frac{dx}{dt},$$

so

$$\frac{dx}{dt} = \frac{C_1 \sin 2t}{\cot t} = 2C_1 \sin^2 t = C_1(1 - \cos 2t).$$

By integration we obtain that

$$x = \frac{C_1}{2}(2t - \sin 2t) + C_2. \tag{8.33}$$

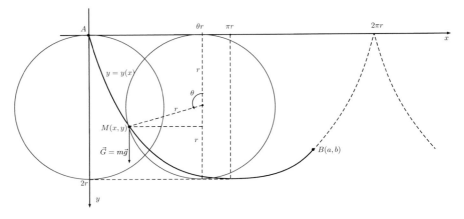

Fig. 8.7 The brachistochrone

The curve $y(x)$ has to pass through the origin, so $y(0) = 0$, which means that $C_2 = 0$. From the condition $y(a) = b$ we can find the constant C_1. We denote $r = \frac{C_1}{2}$ and $\theta = 2t$, $\theta \in (0, 2\pi)$, so the Eqs. (8.32) and (8.33) become:

$$\begin{cases} x = r(\theta - \sin\theta) \\ y = r(1 - \cos\theta). \end{cases} \quad (8.34)$$

These are the equations of a *cycloid*—the curve described by a point of the circle of radius r when the circle is rolling along a straight line (the x-axis). Thus, the brachistochrone is an arc of the cycloid which passes through the given points A and B (see Fig. 8.7). For any $a, b > 0$ there is a unique cycloid defined by (8.34) that passes through the point (a, b).

8.4 The Case $\mathcal{F}[y] = \int_a^b F(x, y, y', \ldots, y^{(n)}) dx$

This is a natural generalization of the previous case (where $n = 1$). First of all, we remark that Lemma 8.1 can be written in a more general form as follows:

Lemma 8.2 *Let n be a positive integer and $f : [a, b] \to \mathbb{R}$ be a continuous function such that*

$$\int_a^b f(x) h(x) dx = 0$$

8.4 The Case $\mathcal{F}[y] = \int_a^b F(x, y, y', \ldots, y^{(n)}) dx$

for any function $h : [a, b] \to \mathbb{R}$ *of class* C^{2n} *with* $h^{(k)}(a) = h^{(k)}(b) = 0$, *for* $k = 0, 1, \ldots, n-1$. *Then*

$$f(x) = 0 \quad \text{for all } x \in [a, b].$$

The proof is similar to the proof of Lemma 8.1, but in this case we need a function $h(x)$ of class C^{2n}, so we take

$$h(x) = \begin{cases} [(x-c+\delta)(c+\delta-x)]^{2n+1}, & \text{if } x \in [c-\delta, c+\delta] \\ 0, & \text{if } x \in [a, c-\delta) \cup (c+\delta, b]. \end{cases}$$

Let F be a function of class C^{2n} defined on $\mathbb{R}^{n+2}$, $V = C^{2n}[a, b]$ and S, the set of admissible functions

$$S = \left\{ y \in V : y^{(k)}(a) = a_k, \ y^{(k)}(b) = b_k, \ k = 0, 1, \ldots, n-1 \right\},$$

where $a_k, b_k \in \mathbb{R}$, $k = 0, 1, \ldots, n-1$ are $2n$ fixed real numbers.

Theorem 8.5 (Euler-Poisson) *Consider the functional* $\mathcal{F} : V \to \mathbb{R}$,

$$\mathcal{F}[y] = \int_a^b F(x, y(x), y'(x), \ldots, y^{(n)}(x)) \, dx. \tag{8.35}$$

An admissible function $y_0 \in S$ *is an extremal of* $\mathcal{F}$ *in* S *if and only if* y_0 *satisfies the following differential equation of order* $2n$ *(the Euler-Poisson equation):*

$$\frac{\partial F}{\partial y} - \frac{d}{dx}\left(\frac{\partial F}{\partial y'}\right) + \frac{d^2}{dx^2}\left(\frac{\partial F}{\partial y''}\right) - \cdots + (-1)^n \frac{d^n}{dx^n}\left(\frac{\partial F}{\partial y^{(n)}}\right) = 0. \tag{8.36}$$

Proof Let H denote the set of admissible directions

$$H = \left\{ h \in V : h^{(k)}(a) = h^{(k)}(b) = 0, k = 0, 1, \ldots, n-1 \right\}$$

and, for any $h \in H$, $h \neq 0$, let $\varphi_h(t)$ denote the function defined by:

$$\varphi_h(t) = \mathcal{F}[y_0 + th] =$$

$$= \int_a^b F\left(x, y_0(x) + th(x), y_0'(x) + th'(x), \ldots, y_0^{(n)}(x) + th^{(n)}(x)\right) dx.$$

We compute the derivative $\varphi'_h(t)$, using the Leibniz' formula (1.72):

$$\varphi'_h(t) = \int_a^b \frac{d}{dt} F\left(x, y_0(x) + th(x), y'_0(x) + th'(x), \ldots, y_0^{(n)}(x) + th^{(n)}(x)\right) dx =$$

$$= \int_a^b \frac{\partial F}{\partial y}\left(x, y_0(x) + th(x), y'_0(x) + th'(x), \ldots, y_0^{(n)}(x) + th^{(n)}(x)\right) h(x) \, dx +$$

$$+ \sum_{k=1}^n \int_a^b \frac{\partial F}{\partial y^{(k)}}\left(x, y_0(x) + th(x), y'_0(x) + th'(x), \ldots, y_0^{(n)}(x) + th^{(n)}(x)\right) h^{(k)}(x) \, dx.$$

The function y_0 is an extremal of $\mathcal{F}$ if and only if, for any $h \in H$, $h \neq \mathbf{0}$, we have:

$$0 = \delta_h \mathcal{F}[y_0] = \varphi'_h(0) = \int_a^b \frac{\partial F}{\partial y}\left(x, y_0(x), y'_0(x), \ldots, y_0^{(n)}(x) + th^{(n)}(x)\right) h(x) \, dx +$$
(8.37)

$$+ \sum_{k=1}^n \int_a^b \frac{\partial F}{\partial y^{(k)}}\left(x, y_0(x), y'_0(x), \ldots, y_0^{(n)}(x)\right) h^{(k)}(x) \, dx.$$

Each integral in the sum above can be computed using integration by parts:

$$\int_a^b \frac{\partial F}{\partial y^{(k)}}\left(x, y_0(x), y'_0(x), \ldots, y_0^{(n)}(x)\right) h^{(k)}(x) \, dx =$$

$$= \frac{\partial F}{\partial y^{(k)}}\left(x, y_0(x), y'_0(x), \ldots, y_0^{(n)}(x)\right) h^{(k-1)}(x) \Big|_a^b -$$

$$- \int_a^b \frac{d}{dx}\left[\frac{\partial F}{\partial y^{(k)}}\left(x, y_0(x), y'_0(x), \ldots, y_0^{(n)}(x)\right)\right] h^{(k-1)}(x) \, dx$$

Since $h^{(k-1)}(a) = h^{(k-1)}(b) = 0$, the first term is 0 and we apply again integration by parts. After k such steps we finally obtain that

$$\int_a^b \frac{\partial F}{\partial y^{(k)}}\left(x, y_0(x), y'_0(x), \ldots, y_0^{(n)}(x)\right) h^{(k)}(x) \, dx =$$

$$= (-1)^k \int_a^b \frac{d^k}{dx^k}\left[\frac{\partial F}{\partial y^{(k)}}\left(x, y_0(x), y'_0(x), \ldots, y_0^{(n)}(x)\right)\right] h(x) \, dx.$$

8.4 The Case $\mathcal{F}[y] = \int_a^b F(x, y, y', \ldots, y^{(n)})dx$

Fig. 8.8 Flexible loaded beam with fixed ends

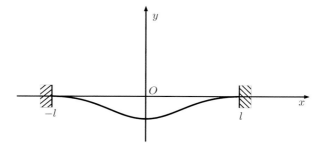

Thus, by replacing in (8.37) we find that $y_0(x)$ satisfies the equality

$$\int_a^b \left[\frac{\partial F}{\partial y} - \frac{d}{dx}\left(\frac{\partial F}{\partial y'}\right) + \frac{d^2}{dx^2}\left(\frac{\partial F}{\partial y''}\right) - \ldots + (-1)^n \frac{d^n}{dx^n}\left(\frac{\partial F}{\partial y^{(n)}}\right) \right] h(x)\, dx = 0$$

for any function $h \in H$. Using Lemma 8.2 the equation (8.36) is obtained. □

Example 8.9 (Flexible Loaded Beam [13], p.323, [5] p.140) Let us consider, in the Cartesian system of coordinates xOy, a flexible homogeneous beam of length $2l$ which is fixed at its ends $A(-l, 0)$, $B(l, 0)$ (see Fig. 8.8). The beam is loaded and the function $w : [-l, l] \to \mathbb{R}$ represents the load per unit length. The beam is bent under this load, and we denote by $y : [-l, l] \to \mathbb{R}$ the shape of its axis. If the beam is clamped at its ends, then we have the following boundary conditions:

$$y(-l) = y(l) = y'(-l) = y'(l) = 0. \tag{8.38}$$

Assuming small deflections, the potential energy due to the elastic forces is given by the formula

$$U_1 = \frac{k}{2} \int_{-l}^{l} y''(x)^2 dx,$$

where $k > 0$ is a constant, while the potential energy due to the gravity force is

$$U_2 = \int_{-l}^{l} w(x) y(x) dx.$$

The shape of the beam is given by the curve $y = y(x)$ that minimizes the total energy (at equilibrium we have only potential energy), which is expressed by the functional

$$\mathcal{F}[y] = \int_{-l}^{l} \left(\frac{k}{2} y''(x)^2 + w(x) y(x) \right) dx. \tag{8.39}$$

This is a variational problem of the type (8.35) with

$$F = F(x, y, y', y'') = \frac{k}{2}y''^2 + wy.$$

So, the associated Euler-Poisson equation (8.36) is written:

$$ky^{(4)} + w(x) = 0.$$

Suppose that the beam is *uniformly* loaded: $w(x) = w =$ constant. Then, by successive integration, we obtain:

$$y(x) = -\frac{w}{24k}x^4 + \frac{C_1}{6}x^3 + \frac{C_2}{2}x^2 + C_3 x + C_4, \qquad (8.40)$$

where C_1, C_2, C_3, C_4 are arbitrary constants. Thus, the extremals of the functional (8.39) are a family of polynomials of degree 4. The particular polynomial that satisfies the boundary conditions (8.38) is the solution of our problem. Thus, we obtain the following linear system in the variables C_1, C_2, C_3, C_4 :

$$\begin{cases} -\frac{w}{24k}l^4 - \frac{C_1}{6}l^3 + \frac{C_2}{2}l^2 - C_3 l + C_4 = 0 \\ -\frac{w}{24k}l^4 + \frac{C_1}{6}l^3 + \frac{C_2}{2}l^2 + C_3 l + C_4 = 0 \\ \frac{w}{6k}l^3 + \frac{C_1}{2}l^2 - C_2 l + C_3 = 0 \\ -\frac{w}{6k}l^3 + \frac{C_1}{2}l^2 + C_2 l + C_3 = 0. \end{cases}$$

By adding the last two equations we find: $C_3 = -\frac{C_1}{2}l^2$. By subtracting them, we get: $C_2 = \frac{l^2 w}{6k}$. Doing the same thing with the first two equations, we obtain: $C_3 = -\frac{C_1}{3}l^2$ and so, $C_1 = C_3 = 0$. Moreover, we also get $C_4 = -\frac{l^4 w}{24k}$. By replacing in (8.40), we finally obtain:

$$y(x) = -\frac{w}{24k}\left(x^2 - l^2\right)^2.$$

8.5 The Case $\mathcal{F}[y_1, \ldots, y_n] = \int_a^b F(x, y_1, \ldots, y_n, y'_1, \ldots, y'_n)dx$

Another direction to generalize the Euler-Lagrange equation is to take a *vector function* $\mathbf{y} = (y_1, \ldots, y_n) : [a, b] \to \mathbb{R}^n$ instead of the scalar function $y(x)$. In this case, we denote by $V = C^2[a, b] \times \ldots \times C^2[a, b]$ the normed space of the vector functions $\mathbf{y}$ having all components twice continuously differentiable functions on the interval $[a, b]$. Let F be a function of class C^2 defined on $\mathbb{R}^{2n+1}$,

8.5 The Case $\mathcal{F}[y_1, \ldots, y_n] = \int_a^b F(x, y_1, \ldots, y_n, y_1', \ldots, y_n') dx$

and let $\mathbf{a} = (a_1, \ldots, a_n)$, $\mathbf{b} = (b_1, \ldots, b_n)$ be two fixed points in $\mathbb{R}^n$. We denote by S be the set of admissible functions:

$$S = \{\mathbf{y} = (y_1, \ldots, y_n) \in V : \mathbf{y}(a) = \mathbf{a}, \ \mathbf{y}(b) = \mathbf{b}\}.$$

Theorem 8.6 (Euler-Lagrange) *Consider the functional* $\mathcal{F} : V \to \mathbb{R}$,

$$\mathcal{F}[y_1, \ldots, y_n] = \int_a^b F(x, y_1(x), \ldots, y_n(x), y_1'(x), \ldots, y_n'(x)) dx, \qquad (8.41)$$

which can be also written in the "compact" form:

$$\mathcal{F}[\mathbf{y}] = \int_a^b F(x, \mathbf{y}(x), \mathbf{y}'(x)) dx,$$

where $\mathbf{y}'(x) = (y_1'(x), \ldots, y_n'(x))$. *An admissible vector function* $\hat{\mathbf{y}} \in S$ *is an extremal of* $\mathcal{F}$ *in* S *if and only if* $\hat{\mathbf{y}}$ *satisfies the following system of differential equations (the Euler-Lagrange system):*

$$\begin{cases} \frac{\partial F}{\partial y_1} - \frac{d}{dx}\left(\frac{\partial F}{\partial y_1'}\right) = 0 \\ \frac{\partial F}{\partial y_2} - \frac{d}{dx}\left(\frac{\partial F}{\partial y_2'}\right) = 0 \\ \vdots \\ \frac{\partial F}{\partial y_n} - \frac{d}{dx}\left(\frac{\partial F}{\partial y_n'}\right) = 0. \end{cases} \qquad (8.42)$$

Proof Let H denote the set of admissible directions

$$H = \{\mathbf{h} = (h_1, \ldots, h_n) \in V : \mathbf{h}(a) = \mathbf{h}(b) = \mathbf{0}\}$$

and, for any $\mathbf{h} \in H$, $\mathbf{h} \neq \mathbf{0}$, let $\varphi_\mathbf{h}(t)$ denote the function defined by:

$$\varphi_\mathbf{h}(t) = \mathcal{F}[\mathbf{y} + t\mathbf{h}] = \int_a^b F\left(x, \mathbf{y}(x) + t\mathbf{h}(x), \mathbf{y}'(x) + t\mathbf{h}'(x)\right) dx.$$

The vector function $\hat{\mathbf{y}} = (y_1, \ldots, y_n)$ is an extremal of $\mathcal{F}$ if and only if

$$\delta_\mathbf{h} \mathcal{F}[\hat{\mathbf{y}}] = \varphi_\mathbf{h}'(0) = 0$$

for any $\mathbf{h} \in H$, $\mathbf{h} \neq \mathbf{0}$. For $k = 1, \ldots, n$, we consider the admissible directions $\mathbf{h}$ with only one nonzero component: $\mathbf{h} = (0, \ldots, h_k, \ldots, 0)$. Let us calculate in this case the derivative $\varphi'_\mathbf{h}(t)$:

$$\varphi'_\mathbf{h}(t) = \frac{d}{dt}\int_a^b F\left(x, y_1, \ldots, y_k + th_k, \ldots, y_n, y'_1, \ldots, y'_1 + th'_k, \ldots, y'_n\right) dx =$$

$$= \int_a^b \frac{d}{dt} F\left(x, y_1, \ldots, y_k + th_k, \ldots, y_n, y'_1, \ldots, y'_1 + th'_k, \ldots, y'_n\right) dx =$$

$$= \int_a^b \frac{\partial F}{\partial y_k}\left(x, y_1, \ldots, y_k + th_k, \ldots, y_n, y'_1, \ldots, y'_1 + th'_k, \ldots, y'_n\right) h_k(x)\, dx +$$

$$+ \int_a^b \frac{\partial F}{\partial y'_k}\left(x, y_1, \ldots, y_k + th_k, \ldots, y_n, y'_1, \ldots, y'_1 + th'_k, \ldots, y'_n\right) h'_k(x)\, dx.$$

So, by replacing $t = 0$, we get:

$$0 = \varphi'_\mathbf{h}(0) = \int_a^b \frac{\partial F}{\partial y_k}\left(x, y_1(x), \ldots, y_n(x), y'_1(x), \ldots, y'_n(x)\right) h_k(x)\, dx +$$

$$+ \int_a^b \frac{\partial F}{\partial y'_k}\left(x, y_1(x), \ldots, y_n(x), y'_1(x), \ldots, y'_n(x)\right) h'_k(x)\, dx \stackrel{by\ parts}{=}$$

$$= \int_a^b \frac{\partial F}{\partial y_k}\left(x, \hat{\mathbf{y}}(x), \hat{\mathbf{y}}'(x)\right) h_k(x)\, dx +$$

$$+ \frac{\partial F}{\partial y'_k}\left(x, \hat{\mathbf{y}}(x), \hat{\mathbf{y}}'(x)\right) h_k(x) \bigg|_a^b - \int_a^b \frac{d}{dx}\left(\frac{\partial F}{\partial y'_k}\left(x, \hat{\mathbf{y}}(x), \hat{\mathbf{y}}'(x)\right)\right) h_k(x)\, dx.$$

Since $h_k(a) = h_k(b) = 0$, we obtain:

$$\int_a^b \left[\frac{\partial F}{\partial y_k}\left(x, \hat{\mathbf{y}}(x), \hat{\mathbf{y}}'(x)\right) - \frac{d}{dx}\left(\frac{\partial F}{\partial y'_k}\left(x, \hat{\mathbf{y}}(x), \hat{\mathbf{y}}'(x)\right)\right)\right] h_k(x)\, dx = 0,$$

for any $h_k \in C^2[a, b]$ with $h_k(a) = h_k(b) = 0$. Thus, by applying Lemma 8.1, we get:

$$\frac{\partial F}{\partial y_k}\left(x, \hat{\mathbf{y}}(x), \hat{\mathbf{y}}'(x)\right) - \frac{d}{dx}\left(\frac{\partial F}{\partial y'_k}\left(x, \hat{\mathbf{y}}(x), \hat{\mathbf{y}}'(x)\right)\right) = 0,$$

for any $x \in [a, b]$, and for any $k = 1, \ldots, n$, so $\hat{\mathbf{y}}(x)$ satisfies the system of Eqs. (8.42). □

8.5 The Case $\mathcal{F}[y_1, \ldots, y_n] = \int_a^b F(x, y_1, \ldots, y_n, y'_1, \ldots, y'_n)dx$

Example 8.10 (The Shortest Path Between Two Points in a 3-D Space—See Problem 8.1) Let $A(a_1, a_2, a_3)$, $B(b_1, b_2, b_3)$ be two distinct points in the 3-D space $Oxyz$ and let $\gamma : x = x(t)$, $y = y(t)$, $z = z(t)$, $t \in [a, b]$ be a smooth curve connecting the points A and B. The length of this curve is given by the functional

$$\mathcal{L}[x, y, z] = \int_a^b \sqrt{x'(t)^2 + y'(t)^2 + z'(t)^2}\, dt \tag{8.43}$$

In this case, $F(t, x, y, z, x', y', z') = \sqrt{x'^2 + y'^2 + z'^2}$, so, the Euler-Lagrange system is written:

$$\begin{cases} \frac{d}{dx}\left(\frac{x'}{\sqrt{x'^2+y'^2+z'^2}}\right) = 0 \\ \frac{d}{dx}\left(\frac{y'}{\sqrt{x'^2+y'^2+z'^2}}\right) = 0 \\ \frac{d}{dx}\left(\frac{z'}{\sqrt{x'^2+y'^2+z'^2}}\right) = 0 \end{cases} \tag{8.44}$$

Hence,

$$\begin{cases} \frac{x'}{\sqrt{x'^2+y'^2+z'^2}} = C_1 \\ \frac{y'}{\sqrt{x'^2+y'^2+z'^2}} = C_2 \\ \frac{z'}{\sqrt{x'^2+y'^2+z'^2}} = C_3, \end{cases} \tag{8.45}$$

where C_1, C_2, C_3 are some real constants. From (8.45) we can write:

$$C_2 x' - C_1 y' = 0, \quad C_3 x' - C_1 z' = 0.$$

and we obtain, by a simple integration, that x, y, z satisfy the equations:

$$C_2 x - C_1 y = C_4, \quad C_3 x - C_1 z = C_5, \tag{8.46}$$

where C_4 and C_5 are arbitrary constants. Equations (8.46) represent the intersection between two families of planes, i.e. a family of straight lines. Thus, the curve we are looking for is exactly the straight line that passes through the given points A and B.

Example 8.11 Find the extremals of the functional

$$\mathcal{F}[y, z] = \int_0^{\pi/2} \left[2y(x)z(x) + y'(x)^2 + z'(x)^2\right] dx$$

with the boundary conditions:

$$y(0) = z(0) = 0, \quad y(\pi/2) = 1, \quad z(\pi/2) = -1. \qquad (8.47)$$

In this case, $F(x, y, z, y', z') = 2yz + y'^2 + z'^2$, so

$$\frac{\partial F}{\partial y} = 2z, \quad \frac{\partial F}{\partial y'} = 2y', \quad \frac{\partial F}{\partial z} = 2y, \quad \frac{\partial F}{\partial z'} = 2z'.$$

Hence, the Euler-Lagrange system of differential equations is written

$$\begin{cases} z - y'' = 0 \\ y - z'' = 0. \end{cases}$$

From the first equation we get:

$$z = y''. \qquad (8.48)$$

If we replace this expression of z in the second equation we obtain:

$$y^{(4)} - y = 0.$$

The general solution of this linear differential equation with constant coefficients is:

$$y(x) = C_1 \exp x + C_2 \exp(-x) + C_3 \cos x + C_4 \sin x.$$

Coming back to (8.48) we finally obtain:

$$z(x) = C_1 \exp x + C_2 \exp(-x) - C_3 \cos x - C_4 \sin x.$$

To find C_1, C_2, C_3, C_4 we use the boundary conditions (8.47) which lead us to the linear system:

$$\begin{cases} C_1 + C_2 + C_3 = 0 \\ C_1 \exp \frac{\pi}{2} + C_2 \exp(-\frac{\pi}{2}) + C_4 = 1 \\ C_1 + C_2 - C_3 = 0 \\ C_1 \exp \frac{\pi}{2} + C_2 \exp(-\frac{\pi}{2}) - C_4 = -1. \end{cases} \qquad (8.49)$$

From the first and the third equations we find that $C_3 = 0$ and $C_1 + C_2 = 0$. By adding the second and the fourth equations, we find that $C_1 \exp \frac{\pi}{2} + C_2 \exp(-\frac{\pi}{2}) = 0$. It follows that $C_1 = C_2 = 0$, and $C_4 = 1$. The solution of the system that satisfies the boundary conditions is

$$y(x) = \sin x, \quad z(x) = -\sin x.$$

8.5 The Case $\mathcal{F}[y_1, \ldots, y_n] = \int_a^b F(x, y_1, \ldots, y_n, y_1', \ldots, y_n') dx$

Remark 8.5 (See Also Remark 8.4) If the variable x does not explicitly appear in the expression of the function F, i.e. $F = F(y_1, \ldots, y_n, y_1', \ldots, y_n')$, then

$$\frac{d}{dx} F(y_1, \ldots, y_n, y_1', \ldots, y_n') = \sum_{k=1}^n \left(\frac{\partial F}{\partial y_k} y_k' + \frac{\partial F}{\partial y_k'} y_k'' \right).$$

If $(y_1, y_2, \ldots, y_n)$ is a solution of the Euler-Lagrange system (8.42), then one can write:

$$\frac{dF}{dx} = \sum_{k=1}^n \left(\frac{d}{dx} \left(\frac{\partial F}{\partial y_k'} \right) y_k' + \frac{\partial F}{\partial y_k'} y_k'' \right) = \sum_{k=1}^n \frac{d}{dx} \left(y_k' \frac{\partial F}{\partial y_k'} \right),$$

so

$$\frac{d}{dx} \left(F - \sum_{k=1}^n y_k' \frac{\partial F}{\partial y_k'} \right) = 0$$

Thus, we have proved that for any extremal of the functional (8.41), the following expression is constant:

$$F - \sum_{k=1}^n y_k' \frac{\partial F}{\partial y_k'} = C, \quad (8.50)$$

which means that (8.50) is a first integral of the Euler-Lagrange system (8.42).

Example 8.12 (Hamilton's Principle (See Problem 8.5)) Hamilton's principle states that the trajectory $(x(t), y(t), z(t))$ chosen by a particle of mass m in the time interval $[t_1, t_2]$ is an extremal of the *action* functional

$$S[x, y, z] = \int_{t_1}^{t_2} L(t, x, y, z, x', y', z') \, dt, \quad (8.51)$$

where $L = T - U$ is the Lagrangian function defined as the difference between the kinetic energy,

$$T(x', y', z') = \frac{m}{2} \left(x'(t)^2 + y'(t)^2 + z'(t)^2 \right),$$

and the potential energy of the particle, $U(t, x, y, z)$. Note that, if the force acting on the particle is $\mathbf{F} = (f_1, f_2, f_3)$, then

$$f_1 = -\frac{\partial U}{\partial x}, \quad f_2 = -\frac{\partial U}{\partial y}, \quad f_3 = -\frac{\partial U}{\partial z}. \quad (8.52)$$

Let us write the Euler-Lagrange system (8.42) for the functional $\mathcal{S}$:

$$\begin{cases} \dfrac{\partial L}{\partial x} - \dfrac{d}{dt}\left(\dfrac{\partial L}{\partial x'}\right) = 0 \\[6pt] \dfrac{\partial L}{\partial y} - \dfrac{d}{dt}\left(\dfrac{\partial L}{\partial y'}\right) = 0 \\[6pt] \dfrac{\partial L}{\partial z} - \dfrac{d}{dt}\left(\dfrac{\partial L}{\partial z'}\right) = 0. \end{cases} \quad (8.53)$$

Since

$$L = \frac{m}{2}\left(x'^2 + y'^2 + z'^2\right) - U(t, x, y, z), \quad (8.54)$$

the system (8.53) is written:

$$\begin{cases} -\dfrac{\partial U}{\partial x} - mx'' = 0 \\[6pt] -\dfrac{\partial U}{\partial y} - my'' = 0 \\[6pt] -\dfrac{\partial U}{\partial z} - mz'' = 0 \end{cases}$$

and from (8.52), the Euler-Lagrange system reduces to

$$mx'' = f_1, \quad my'' = f_2, \quad mz'' = f_3,$$

which is just the Newton's law of motion, $m\mathbf{r}'' = \mathbf{F}$ (see [5], p.64).

Now, if we suppose that the potential energy U does not depend explicitly on the time t, then we can apply Remark 8.5 and write:

$$L - x'\frac{\partial L}{\partial x'} - y'\frac{\partial L}{\partial y'} - z'\frac{\partial L}{\partial z'} = C.$$

Thus, using (8.54), we obtain the *law of conservation of energy*:

$$T + U = \text{constant}.$$

8.6 The Case $\mathcal{F}[z] = \iint_D F\left(x, y, z, \frac{\partial z}{\partial x}, \frac{\partial z}{\partial y}\right) dxdy$

Let D be a simply connected closed region in $\mathbb{R}^2$ with a continuous and piecewise smooth boundary $\gamma = \partial D \subset D$. Let $V = C^2(D)$ be the normed vector space of functions $z : D \to \mathbb{R}$ having continuous partial derivatives of second order.

We consider a functional of the form

$$\mathcal{F}[z] = \iint_D F\left(x, y, z(x, y), \frac{\partial z}{\partial x}(x, y), \frac{\partial z}{\partial y}(x, y)\right) dxdy, \qquad (8.55)$$

where F is a function of class C^2 defined on $\mathbb{R}^5$.

Let S be the set of admissible functions

$$S = \{z \in V : z(x, y) = \varphi(x, y), \text{ for all } (x, y) \in \gamma\},$$

where $\varphi(x, y)$ is a given function of class C^2 defined on an open neighborhood of the curve γ.

The counterpart of Lemma 8.1 in this case is the following result:

Lemma 8.3 *Let $f : D \to \mathbb{R}$ be a continuous function such that*

$$\iint_D f(x, y) h(x, y) \, dxdy = 0$$

for any function $h : D \to \mathbb{R}$ of class C^2 with $h(x, y) = 0$ for $(x, y) \in \gamma$. Then

$$f(x, y) = 0 \quad \text{for all } (x, y) \in D.$$

Proof We use again the proof by contradiction: suppose that there exists an interior point in D, (x_0, y_0) with $f(x_0, y_0) \neq 0$, say $f(x_0, y_0) > 0$. Since f is continuous, we can take a small number $\delta > 0$ such that $f(x, y) > 0$ for any $(x, y) \in \Delta \subset D$, where Δ is the closed disc of radius δ centered at (x_0, y_0):

$$\Delta = \left\{(x, y) : (x - x_0)^2 + (y - y_0)^2 \leq \delta^2\right\}.$$

The following function is of class $C^2(D)$ and satisfies the condition $h(x, y) = 0$ for all $(x, y) \in \gamma$:

$$h(x, y) = \begin{cases} \left[\delta^2 - (x - x_0)^2 - (y - y_0)^2\right]^3, & \text{if } (x, y) \in \Delta \\ 0, & \text{if } (x, y) \in D - \Delta. \end{cases}$$

We also notice that $h(x, y) > 0$ for any (x, y) in the open disc $(x-x_0)^2+(y-y_0)^2 < \delta^2$. Since $h(x, y) = 0$ outside the disc Δ, one can write:

$$\iint_D f(x, y)h(x, y)\, dxdy = \iint_\Delta f(x, y)h(x, y)\, dxdy > 0,$$

where the last inequality follows by the Mean Value Theorem for multiple integrals (see [1], Theorem 14.16): we know that there exists a point (ξ, η) in the interior of Δ such that

$$\iint_\Delta f(x, y)h(x, y)\, dxdy = f(\xi, \eta)h(\xi, \eta) \cdot Area(\Delta) > 0.$$

But this inequality contradicts the hypothesis. Hence $f(x, y) = 0$ for any point (x, y) in the interior of D and, since the function is continuous, we actually have $f(x, y) = 0$ on the closed domain D. □

For simplification, we shall use in the following the Monge's notation:

$$\frac{\partial z}{\partial x}(x, y) = p(x, y) = p, \quad \frac{\partial z}{\partial y}(x, y) = q(x, y) = q. \tag{8.56}$$

Theorem 8.7 (Euler–Ostrogradsky) *Let $\mathcal{F} : V \to \mathbb{R}$ be a functional defined by:*

$$\mathcal{F}[z] = \iint_D F(x, y, z(x, y), p(x, y), q(x, y))\, dxdy, \tag{8.57}$$

where $F(x, y, z, p, q)$ is a function of class C^2. Let S be the set of admissible functions

$$S = \{z \in V : z(x, y) = \varphi(x, y), \text{ for all } (x, y) \in \gamma\},$$

where $\varphi(x, y)$ is a given function of class C^2 defined on an open neighborhood of the curve γ.

Then, a function $\hat{z}(x, y)$ is an extremal of the functional $\mathcal{F}$ if and only if it satisfies the following second-order partial differential equation (called Euler–Ostrogradsky equation):

$$\frac{\partial F}{\partial z} - \frac{\partial}{\partial x}\left(\frac{\partial F}{\partial p}\right) - \frac{\partial}{\partial y}\left(\frac{\partial F}{\partial q}\right) = 0. \tag{8.58}$$

8.6 The Case $\mathcal{F}[z] = \iint_D F\left(x, y, z, \frac{\partial z}{\partial x}, \frac{\partial z}{\partial y}\right) dx dy$

Proof Let H denote the set of admissible directions

$$H = \{h \in V : h(x, y) = 0, \text{ for all } (x, y) \in \gamma\}$$

and, for any $h \in H$, $h \neq 0$, let $\varphi_h(t)$ denote the function defined by:

$$\varphi_h(t) = \mathcal{F}[\hat{z} + th] = \iint_D F\left(x, y, \hat{z} + th, \frac{\partial \hat{z}}{\partial x} + t\frac{\partial h}{\partial x}, \frac{\partial \hat{z}}{\partial y} + t\frac{\partial h}{\partial y}\right) dx dy.$$

The derivative of this function is

$$\varphi_h'(t) = \iint_D \frac{d}{dt}\left[F\left(x, y, \hat{z} + th, \frac{\partial \hat{z}}{\partial x} + t\frac{\partial h}{\partial x}, \frac{\partial \hat{z}}{\partial y} + t\frac{\partial h}{\partial y}\right)\right] dx dy =$$

$$= \iint_D \frac{\partial F}{\partial z}\left(x, y, \hat{z} + th, \frac{\partial \hat{z}}{\partial x} + t\frac{\partial h}{\partial x}, \frac{\partial \hat{z}}{\partial y} + t\frac{\partial h}{\partial y}\right) h(x, y) dx dy +$$

$$+ \iint_D \frac{\partial F}{\partial p}\left(x, y, \hat{z} + th, \frac{\partial \hat{z}}{\partial x} + t\frac{\partial h}{\partial x}, \frac{\partial \hat{z}}{\partial y} + t\frac{\partial h}{\partial y}\right) \frac{\partial h}{\partial x} dx dy +$$

$$+ \iint_D \frac{\partial F}{\partial q}\left(x, y, \hat{z} + th, \frac{\partial \hat{z}}{\partial x} + t\frac{\partial h}{\partial x}, \frac{\partial \hat{z}}{\partial y} + t\frac{\partial h}{\partial y}\right) \frac{\partial h}{\partial y} dx dy.$$

Thus,

$$\varphi'(0) = \iint_D \frac{\partial F}{\partial z}\left(x, y, \hat{z}, \frac{\partial \hat{z}}{\partial x}, \frac{\partial \hat{z}}{\partial y}\right) h(x, y) dx dy + \qquad (8.59)$$

$$+ \iint_D \frac{\partial F}{\partial p}\left(x, y, \hat{z}, \frac{\partial \hat{z}}{\partial x}, \frac{\partial \hat{z}}{\partial y}\right) \frac{\partial h}{\partial x} dx dy + \iint_D \frac{\partial F}{\partial q}\left(x, y, \hat{z}, \frac{\partial \hat{z}}{\partial x}, \frac{\partial \hat{z}}{\partial y}\right) \frac{\partial h}{\partial y} dx dy.$$

But,

$$\frac{\partial F}{\partial p}\left(x, y, \hat{z}, \frac{\partial \hat{z}}{\partial x}, \frac{\partial \hat{z}}{\partial y}\right) \frac{\partial h}{\partial x} + \frac{\partial F}{\partial q}\left(x, y, \hat{z}, \frac{\partial \hat{z}}{\partial x}, \frac{\partial \hat{z}}{\partial y}\right) \frac{\partial h}{\partial y} =$$

$$= \frac{\partial}{\partial x}\left(\frac{\partial F}{\partial p} h\right) + \frac{\partial}{\partial y}\left(\frac{\partial F}{\partial q} h\right) - \frac{\partial}{\partial x}\left(\frac{\partial F}{\partial p}\right) h - \frac{\partial}{\partial y}\left(\frac{\partial F}{\partial q}\right) h.$$

So,

$$\varphi'(0) = \iint_D \left[\frac{\partial F}{\partial z} - \frac{\partial}{\partial x}\left(\frac{\partial F}{\partial p}\right) - \frac{\partial}{\partial y}\left(\frac{\partial F}{\partial q}\right) \right] h\, dx dy + \quad (8.60)$$

$$+ \iint_D \left[\frac{\partial}{\partial x}\left(\frac{\partial F}{\partial p} h\right) + \frac{\partial}{\partial y}\left(\frac{\partial F}{\partial q} h\right) \right] dx dy.$$

In the last double integral we denote: $Q = \frac{\partial F}{\partial p} h$, $P = -\frac{\partial F}{\partial q} h$ and apply Green's Theorem 8.68 (see [38], 16.4):

$$\iint_D \left[\frac{\partial Q}{\partial x} - \frac{\partial P}{\partial y} \right] dx dy = \oint_\gamma P\, dx + Q\, dy =$$

$$= \oint_\gamma \left(-\frac{\partial F}{\partial q} h\right) dx + \left(\frac{\partial F}{\partial p} h\right) dy = 0,$$

because $h(x, y) = 0$ on γ.

So

$$\varphi'(0) = \iint_D \left[\frac{\partial F}{\partial z} - \frac{\partial}{\partial x}\left(\frac{\partial F}{\partial p}\right) - \frac{\partial}{\partial y}\left(\frac{\partial F}{\partial q}\right) \right] h(x, y)\, dx dy. \quad (8.61)$$

The function $\hat{z} \in S$ is an extremal of the functional $\mathcal{F}$ if and only if

$$\delta_h \mathcal{F}[\hat{z}] = \varphi'(0) = 0.$$

Hence, by (8.61) and Lemma 8.3, we obtain that $\hat{z}(x, y)$ satisfies the Eq. (8.58). □

Example 8.13 Solution of the *Plateau's problem* (Problem 8.6). We have to find the surface of minimal area bordered by a given closed curve Γ. The physicist Joseph Plateau proved that the minimal surface can be obtained experimentally as a soap film in a wire frame as the *minimal surface* corresponds to the *minimal energy*.

Let γ be the projection of Γ on the plane xOy and Ω be the plane closed region bounded by γ. We suppose that each point (x, y) on the curve γ corresponds to a single point $(x, y, \varphi(x, y))$ on the curve Γ. Consider a surface S defined by $z = z(x, y)$, where $z : \Omega \to \mathbb{R}$ is a function of class C^1 such that

$$z(x, y) = \varphi(x, y), \quad \forall\, (x, y) \in \gamma. \quad (8.62)$$

8.6 The Case $\mathcal{F}[z] = \iint_D F\left(x, y, z, \frac{\partial z}{\partial x}, \frac{\partial z}{\partial y}\right) dx dy$

The area of this surface is given by the formula

$$\mathcal{F}[z] = \iint_\Omega \sqrt{1 + \left(\frac{\partial z}{\partial x}(x, y)\right)^2 + \left(\frac{\partial z}{\partial y}(x, y)\right)^2} \, dx dy. \tag{8.63}$$

So, the problem is to find an extremal of this functional and satisfies (8.6).

As shown in Problem 8.6, this functional has the same extremals as the following functional, which is the expression of the potential energy of a homogeneous membrane of shape S:

$$\mathcal{U}[z] = \frac{1}{2} \iint_\Omega \left[\left(\frac{\partial z}{\partial x}(x, y)\right)^2 + \left(\frac{\partial z}{\partial y}(x, y)\right)^2\right] dx dy. \tag{8.64}$$

Obviously, the Euler–Ostrogradsky equation (8.58) for the functional $\mathcal{U}[z]$ is written:

$$\frac{\partial^2 z}{\partial x^2} + \frac{\partial^2 z}{\partial y^2} = 0 \tag{8.65}$$

Thus, to find the surface of minimal energy (hence, minimal area) we have to find the solution of the *Laplace equation* (8.65) with the boundary conditions (8.62). This is a *Dirichlet problem* for the Laplace equation.

Example 8.14 (Free Vibrations of a String) Let us use the Hamilton's principle of stationary (least) action to derive the differential equation of free vibrations of a string ([13], p.334).

Consider that the segment $[OA]$ of the x-axis (where $O(0, 0)$ and $A(l, 0)$) represents the equilibrium position of the string. Let $u(x, t)$, $x \in [0, l]$, $t \geq 0$ be the displacement of a point x at time t. We use for the partial derivatives of u the notation: $u_x = \frac{\partial u}{\partial x}$ and $u_t = \frac{\partial u}{\partial t}$.

In the deformed state, a segment dx of the string has the length $ds = \sqrt{1 + u_x^2} \, dx$. The potential energy of the element dx is proportional to the extension

$$ds - dx = \left(\sqrt{1 + u_x^2} - 1\right) dx \approx \frac{1}{2} u_x^2 \, dx,$$

using for the square root the approximation by the first-order Taylor polynomial $\sqrt{1 + t} \approx 1 + \frac{1}{2}t$.

It follows that the total potential energy of the string is

$$U = \frac{1}{2} \int_0^l k u_x^2 \, dx.$$

On the other hand, the kinetic energy of the string is

$$T = \frac{1}{2} \int_0^l \rho u_t^2 \, dx,$$

where ρ is the density (the mass per unit length), so the *action* functional from Hamilton's principle is written in this case as:

$$\mathcal{S}[u] = \int_{t_1}^{t_2} (T - U) dt = \int_{t_1}^{t_2} \int_0^l \left(\frac{1}{2} \rho u_t^2 - \frac{1}{2} k u_x^2 \right) dx \, dt$$

The Euler–Ostrogradsky equation (8.58) for the above functional is written:

$$\frac{\partial}{\partial t} (\rho u_t) - \frac{\partial}{\partial x} (k u_x) = 0.$$

If the string is homogeneous (if ρ and k are constants), then the above equation is written:

$$\rho \frac{\partial^2 u}{\partial t^2} = k \frac{\partial^2 u}{\partial x^2},$$

which is exactly the one-dimensional wave equation (7.62), with $a^2 = \frac{k}{\rho}$.

8.7 Isoperimetric Problems and Geodesic Problems

8.7.1 Isoperimetric Problems

Example 8.15 Dido's problem (I) As presented in Problem 8.4, the simplest form of Dido's problem consists in finding the closed curve γ of a given length, L, such that the area of the domain Ω enclosed by the curve is maximal. If γ is defined by the parametric equations:

$$\gamma: \quad x = x(t), \quad y = y(t), \quad t \in [t_1, t_2], \tag{8.66}$$

where $x, y : [t_1, t_2] \to \mathbb{R}$ are continuously differentiable functions, then the length of the curve is

$$\mathcal{L}[x, y] = \int_{t_1}^{t_2} \sqrt{x'(t)^2 + y'(t)^2} \, dt. \tag{8.67}$$

Using the Green's theorem,

$$\int_\gamma P(x, y) dx + Q(x, y) dy = \iint_\Omega \left(\frac{\partial Q}{\partial x} - \frac{\partial P}{\partial y} \right) dx dy, \tag{8.68}$$

8.7 Isoperimetric Problems and Geodesic Problems

for $P = 0$ and $Q = x$ we obtain the area of the domain Ω expressed by a line integral of second type:

$$Area(\Omega) = \iint_\Omega dx\,dy = \int_\gamma x\,dy, \qquad (8.69)$$

so $Area(\Omega)$ is given by the functional

$$\mathcal{A}[x, y] = \int_{t_1}^{t_2} x(t) y'(t)\,dt. \qquad (8.70)$$

Thus, our variational problem is to find the extremals (x, y) of the functional $\mathcal{A}[x, y]$ under the constraint $\mathcal{L}[x, y] = L$. To solve this isoperimetric problem, we have to search for the extremals of the following auxiliary functional (see Theorem 8.3):

$$\mathcal{F}[x, y] = \mathcal{A} + \lambda\mathcal{L} = \int_{t_1}^{t_2} \left(x(t) y'(t) + \lambda\sqrt{x'(t)^2 + y'(t)^2} \right) dt. \qquad (8.71)$$

Since $F(t, x, y, x', y') = xy' + \lambda\sqrt{x'^2 + y'^2}$, the Euler-Lagrange system of the functional 8.71 is written:

$$\begin{cases} \dfrac{\partial F}{\partial x} - \dfrac{d}{dt}\left(\dfrac{\partial F}{\partial x'}\right) = y' - \dfrac{d}{dt}\left(\dfrac{\lambda x'}{\sqrt{x'^2 + y'^2}}\right) = 0 \\[2ex] \dfrac{\partial F}{\partial y} - \dfrac{d}{dt}\left(\dfrac{\partial F}{\partial y'}\right) = -\dfrac{d}{dt}\left(x + \dfrac{\lambda y'}{\sqrt{x'^2 + y'^2}}\right) = 0 \end{cases} \qquad (8.72)$$

and, by a simple integration with respect to t, we find:

$$\begin{cases} y - \dfrac{\lambda x'}{\sqrt{x'^2 + y'^2}} = c_1 \\[2ex] x + \dfrac{\lambda y'}{\sqrt{x'^2 + y'^2}} = c_2, \end{cases}$$

where c_1, c_2 are arbitrary constants. Thus, we obtain the equation

$$(x - c_2)^2 + (y - c_1)^2 = \lambda^2, \qquad (8.73)$$

which is the equation of a circle of radius λ, centered at an arbitrary point $C(c_2, c_1)$. Thus, the curve of fixed length L that encloses a domain of maximum area is the circle of radius $\lambda = \frac{L}{2\pi}$.

Example 8.16 Dido's problem (II) As a matter of fact, Dido had to solve a more complicated problem, because the domain whose area she wanted to maximize had also a fixed (north) boundary, on the Mediterranean coast. The mathematical formulation of this problem is: *Given a simple curve γ^* with the endpoints A end B and a positive number L (greater than the distance AB), find the curve γ of length L and endpoints B and A such that the area of the domain Ω bounded by the two curves γ and γ^* is maximum.*

We solve this problem when the curve γ^* is the line segment $[AB]$.

Let $A(-a, 0)$ and $B(a, 0)$ be the two fixed points and γ be a curve of length $L > 2a$ with the endpoints B and A. If the curve is defined by the parametric equations (8.66), then we have the boundary conditions:

$$x(t_1) = a, \ y(t_1) = 0, \ x(t_2) = -a, \ y(t_2) = 0.$$

We denote $\Gamma = \gamma \cup [AB]$ and let Ω be the domain enclosed by Γ. Then, the formula (8.69) is written:

$$Area\,(\Omega) = \int_\Gamma x dy = \int_\gamma x dy + \int_{[AB]} x dy,$$

But, since $y =$ constant on the line segment $[AB]$, the second integral is 0, so the area of Ω is expressed by the same functional (8.70). Thus, the problem in this case is very similar to the one above: we have the same Euler-Lagrange system (8.72) with the same general solution:

$$(x - c_2)^2 + (y - c_1)^2 = \lambda^2, \tag{8.74}$$

Still, there are two main differences: first, in this case γ is not the whole circle, but just the arc $\widehat{BA}$, so it is not as easy as above to determine the radius of the circle λ and, second, the coordinates of the center $C(c_2, c_1)$ cannot be arbitrary, like in the problem above, so we have to find them.

From the condition that $A(-a, 0)$ and $B(a, 0)$ satisfy the Eq. (8.74) we find that $c_2 = 0$, so the center of the circle is on the y-axis.

Let D be the midpoint of the arc $\widehat{AB}$, $D(0, c_1 + \lambda)$ and let $\alpha = \widehat{BCD}, \alpha \in (0, \pi)$ (see Fig. 8.9). The length of the arc $\widehat{BDA}$ (the curve γ) is

$$L = 2\alpha\lambda. \tag{8.75}$$

On the other hand, in the triangle OCB one can write

$$\sin \alpha = \frac{a}{\lambda}$$

8.7 Isoperimetric Problems and Geodesic Problems

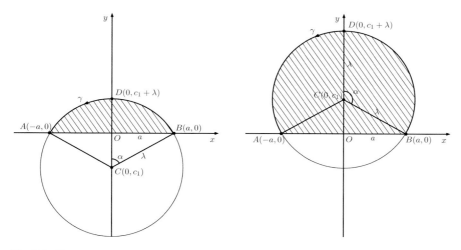

Fig. 8.9 The two cases: $L < a\pi$ and $L > a\pi$

(if $\alpha > \pi/2$ we write $\sin\alpha = \sin(\pi - \alpha) = \frac{a}{\lambda}$). Thus, using (8.75), we obtain:

$$\sin\alpha = \frac{2a}{L}\alpha. \tag{8.76}$$

This is a transcendental equation, which has a unique solution $\alpha \in (0, \pi)$, since $\frac{2a}{L} \in (0, 1)$. We invite the reader to prove it and also, to prove that

$$\alpha < \frac{\pi}{2} \text{ if } \frac{L}{a} < \pi, \quad \alpha = \frac{\pi}{2} \text{ if } \frac{L}{a} = \pi, \quad \text{and } \alpha > \frac{\pi}{2} \text{ if } \frac{L}{a} > \pi.$$

Knowing the solution α of Eq. (8.76), one can write the radius λ of the circle and the ordinate c_1 of the center:

$$\lambda = \frac{L}{2\alpha}, \quad C(0, -a\cot\alpha).$$

The parametric equations of the arc of circle $\gamma = \widehat{BDA}$ are:

$$\begin{cases} x = \frac{L}{2\alpha}\cos t, \\ y = -a\cot\alpha + \frac{L}{2\alpha}\sin t, \end{cases} \quad t \in \left[\frac{\pi}{2} - \alpha, \frac{\pi}{2} + \alpha\right],$$

or, equivalently,

$$\begin{cases} x = \frac{L}{2\alpha}\sin t, \\ y = -a\cot\alpha - \frac{L}{2\alpha}\cos t, \end{cases} \quad t \in [-\alpha, \alpha].$$

Example 8.17 The *catenary (equilibrium shape of a suspended cable)*. Find the position of an absolutely flexible, inextensible, homogeneous cable (chain) of length L suspended at the points $A(x_0, y_0)$ and $B(x_1, y_1)$ (the length L is greater than the distance AB). Suppose that the gravity acts in the negative y-direction. Since the center of gravity must be as low as possible in the equilibrium position and since

$$y_G = \frac{\int_\gamma y \, ds}{\int_\gamma ds} = \frac{1}{L} \int_{x_0}^{x_1} y\sqrt{1 + y'^2} \, dx,$$

we arrive at the following isoperimetric problem: Find the function $y(x)$ such that the functional

$$\mathcal{F}[y] = \int_{x_0}^{x_1} y\sqrt{1 + y'^2} \, dx$$

has the smallest possible value, provided that

$$\mathcal{L}[y] = \int_{x_0}^{x_1} \sqrt{1 + y'^2} \, dx = L,$$

and given the boundary values $y(x_0) = y_0$, $y(x_1) = y_1$.

We introduce the auxiliary functional

$$\mathcal{F}[y] + \lambda \mathcal{L}[y] = \int_{x_0}^{x_1} (y + \lambda)\sqrt{1 + y'^2} \, dx. \tag{8.77}$$

By denoting $y(x) + \lambda = z(x)$, the functional (8.77) is written:

$$\mathcal{G}[z] = \int_{x_0}^{x_1} z\sqrt{1 + z'^2} \, dx. \tag{8.78}$$

This is the functional studied in Example 8.7: its extremals are the family of catenaries (see (8.31)):

$$z(x) = C_1 \cosh \frac{x - C_2}{C_1}.$$

Hence, the function we are searching for is

$$y(x) = C_1 \cosh \frac{x - C_2}{C_1} - \lambda,$$

8.7 Isoperimetric Problems and Geodesic Problems

where the constants C_1, C_2 and λ are determined from the boundary conditions and the constraint:

$$y(x_0) = y_0, \quad y(x_1) = y_1, \quad \int_{x_0}^{x_1} \sqrt{1 + y'^2}\, dx = L.$$

8.7.2 Geodesic Problems

Given two points on a surface S, the shortest curve lying on S and connecting the two points is said to be a *geodesic (curve)*.

Let S be a surface defined by the implicit equation

$$S: \quad G(x, y, z) = 0,$$

where $G(x, y, z)$ is a function of class C^2 and let $A(x_0, y_0, z_0)$, $B(x_1, y_1, z_1)$ be two points on this surface. We have to find the smooth curves γ of parametric equations:

$$\gamma: \quad x = x(t), \quad y = y(t), \quad z = z(t), \quad t \in [t_0, t_1],$$

lying on the surface S (hence $G(x(t), y(t), z(t)) = 0$ for all $t \in [t_0, t_1]$), connecting the points A and B (so $x(t_0) = x_0$, $y(t_0) = y_0$, $z(t_0) = z_0$, and $x(t_1) = x_1$, $y(t_1) = y_1$, $z(t_1) = z_1$)) and having the length as small as possible. Thus, we have to find the extremals (x, y, z) of the functional:

$$\mathcal{L}[x, y, z] = \int_{t_0}^{t_1} \sqrt{x'^2(t) + y'^2(t) + z'^2(t)}\, dt, \tag{8.79}$$

given the constraint:

$$G(x(t), y(t), z(t)) = 0, \quad t \in [t_0, t_1]. \tag{8.80}$$

Let $\Omega \subset \mathbb{R}^3$ be the domain of definition of the function G ($S \subset \Omega$). We assume that $G(x, y, z) \geq 0$ on Ω (otherwise, we can substitute G with G^2). With this condition satisfied, the equality (8.80) is logically equivalent to

$$\int_{t_0}^{t_1} G(x(t), y(t), z(t))\, dt = 0. \tag{8.81}$$

We apply Theorem 8.3 and search for the extremals of the auxiliary functional

$$\mathcal{F}[x, y, z] = \int_{t_0}^{t_1} \left[\sqrt{x'^2(t) + y'^2(t) + z'^2(t)} + \lambda G(x(t), y(t), z(t)) \right] dt.$$

The Euler-Lagrange system in this case is written:

$$\begin{cases} \lambda \dfrac{\partial G}{\partial x} - \dfrac{d}{dt}\left(\dfrac{x'}{\sqrt{x'^2 + y'^2 + z'^2}}\right) = 0 \\ \lambda \dfrac{\partial G}{\partial y} - \dfrac{d}{dt}\left(\dfrac{y'}{\sqrt{x'^2 + y'^2 + z'^2}}\right) = 0 \\ \lambda \dfrac{\partial G}{\partial z} - \dfrac{d}{dt}\left(\dfrac{z'}{\sqrt{x'^2 + y'^2 + z'^2}}\right) = 0. \end{cases} \quad (8.82)$$

Let us consider the *normal parametrization* of the curve γ,

$$x = \tilde{x}(s), \quad y = \tilde{y}(s), \quad z = \tilde{z}(s),$$

where

$$s = s(t) = \int_{t_0}^{t} \sqrt{x'(u)^2 + y'(u)^2 + z'(u)^2}\, du.$$

Since $s'(t) = \frac{ds}{dt} = \sqrt{x'(t)^2 + y'(t)^2 + z'(t)^2}$, the system (8.82) can be written:

$$\frac{\frac{d}{dt}\left(\frac{x'}{s'}\right)}{\frac{\partial G}{\partial x}} = \frac{\frac{d}{dt}\left(\frac{y'}{s'}\right)}{\frac{\partial G}{\partial y}} = \frac{\frac{d}{dt}\left(\frac{x'}{s'}\right)}{\frac{\partial G}{\partial z}}. \quad (8.83)$$

On the other hand, we can see that

$$x'(t) = \frac{dx}{dt} = \frac{d}{dt}\tilde{x}(s(t)) = \frac{d\tilde{x}}{ds} \cdot \frac{ds}{dt},$$

so

$$\frac{d\tilde{x}}{ds} = \frac{\frac{dx}{dt}}{\frac{ds}{dt}} = \frac{x'}{s'},$$

and it follows that

$$\frac{d}{dt}\left(\frac{x'}{s'}\right) = \frac{d}{dt}\left(\frac{d\tilde{x}}{ds}\right) = \frac{d^2\tilde{x}}{ds^2} \cdot \frac{ds}{dt} = \frac{d^2\tilde{x}}{ds^2} \cdot s'.$$

Similarly, we have:

$$\frac{d}{dt}\left(\frac{y'}{s'}\right) = \frac{d^2\tilde{y}}{ds^2} \cdot s' \quad \text{and} \quad \frac{d}{dt}\left(\frac{z'}{s'}\right) = \frac{d^2\tilde{z}}{ds^2} \cdot s',$$

8.7 Isoperimetric Problems and Geodesic Problems

so the system (8.83) is written:

$$\frac{\frac{d^2\tilde{x}}{ds^2}}{\frac{\partial G}{\partial x}} = \frac{\frac{d^2\tilde{y}}{ds^2}}{\frac{\partial G}{\partial y}} = \frac{\frac{d^2\tilde{z}}{ds^2}}{\frac{\partial G}{\partial z}}. \qquad (8.84)$$

Since the vector $\frac{d^2\tilde{x}}{ds^2}\mathbf{i} + \frac{d^2\tilde{y}}{ds^2}\mathbf{j} + \frac{d^2\tilde{z}}{ds^2}\mathbf{k}$ has the same direction as the *principal normal* to the curve γ and since $\frac{\partial G}{\partial x}\mathbf{i} + \frac{\partial G}{\partial y}\mathbf{j} + \frac{\partial G}{\partial z}\mathbf{k}$ is a *normal vector to the surface S*, we obtain that the principal normal to any point of a geodesic curve lies along the normal to the surface at that point.

Example 8.18 Let us study the case when the surface S is a sphere. Consider the sphere of radius a, centered at the origin:

$$S = \left\{(x, y, z) : x^2 + y^2 + z^2 = a^2\right\},$$

Since $G(x, y, z) = x^2 + y^2 + z^2 - a^2$, we have

$$\frac{\partial G}{\partial x} = 2x, \quad \frac{\partial G}{\partial y} = 2y, \quad \frac{\partial G}{\partial z} = 2z,$$

so the system (8.83) is written:

$$\frac{x''s' - x's''}{2xs'^2} = \frac{y''s' - y's''}{2ys'^2} = \frac{z''s' - z's''}{2zs'^2},$$

or, equivalently,

$$\frac{s''}{s'} = \frac{x''y - xy''}{x'y - xy'} = \frac{y''z - yz''}{y'z - yz'},$$

The last equality can be also written:

$$\frac{\frac{d}{dt}(x'y - xy')}{x'y - xy'} = \frac{\frac{d}{dt}(y'z - yz')}{y'z - yz'}$$

and we obtain by integration:

$$\ln|x'y - xy'| = \ln|y'z - yz'| + \ln|C_1|,$$

or

$$x'y - xy' = C_1(yz' - y'z),$$

where C_1 is a constant. If we divide this equality by y^2 we get:

$$\frac{d}{dt}\left(\frac{x}{y}\right) = C_1 \frac{d}{dt}\left(\frac{z}{y}\right),$$

so, integrating once again, we find:

$$\frac{x}{y} = C_1 \frac{z}{y} + C_2,$$

which is the equation of a plane passing through the origin:

$$x - C_2 y - C_1 z = 0.$$

Thus, the geodesic curve on the sphere is given by the intersection of the sphere with the plane (OAB): it is the great circle on the sphere passing through the points A and B (see Fig. 8.10).

Example 8.19 Let us find the geodesic curves on the cylinder

$$x^2 + y^2 = a^2.$$

In this case, we have: $G(x, y, z) = x^2 + y^2 - a^2$, so $\frac{\partial G}{\partial z} = 0$ in (8.82). Hence,

$$\frac{z'}{\sqrt{x'^2 + y'^2 + z'^2}} = C. \tag{8.85}$$

Suppose that we have to find the geodesic curve corresponding to the points $A(x_0, y_0, z_0)$ and $B(x_1, y_1, z_1)$. If $x_0 = x_1$ and $y_0 = y_1$, then the geodesic is the

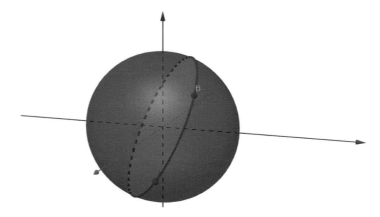

Fig. 8.10 Geodesic on a sphere

8.7 Isoperimetric Problems and Geodesic Problems

straight line $M_0 M_1$. Otherwise, we can write the parametric equation of the geodesic curve as

$$x = a\cos t, \quad y = a\sin t, \quad z = z(t),$$

where $t \in [t_0, t_1]$ is the angle between x-axis and the projection of the position vector on the plane (xOy).

Since $x' = -a\sin t$ and $y' = a\cos t$, the Eq. (8.85) can be written:

$$z'^2 = C^2(a^2 + z'^2),$$

so $z' = C_1$, where $C_1 = \frac{Ca}{\sqrt{1-C^2}}$. It follows that

$$z = C_1 t + C_2.$$

If $C_1 = 0$ (when $z_0 = z_1$) then the corresponding geodesic curve is a circle (the intersection of the cylinder with the horizontal plane $z = z_0$ (Fig. 8.11a):

$$x = a\cos t, \quad y = a\sin t, \quad z = z_0.$$

If $C_1 \neq 0$ then the geodesic curve is a *circular helix* (see Fig. 8.11b):

$$x = a\cos t, \quad y = a\sin t, \quad z = C_1 t + C_2.$$

If we unfold the cylinder up to a rectangular plane surface R, the geodesic curves become straight lines in R.

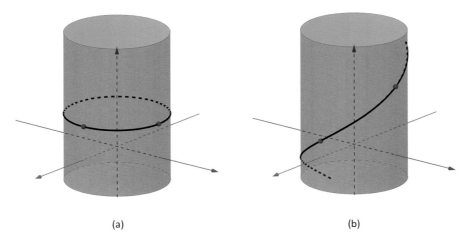

(a)   (b)

Fig. 8.11 Geodesics on a cylinder

8.8 Exercises

1. Consider the space $C^1[a, b]$ with the metric induced by the norm from Example 8.3:

$$d_1(f, g) = \|f - g\|_\infty + \|f' - g'\|_\infty = \sup_{x \in [a,b]} |f(x) - g(x)| + \sup_{x \in [a,b]} |f'(x) - g'(x)|.$$

   Calculate the distance between the functions f and g in the following cases:
   (a) $f(x) = x$, $g(x) = x^2$, $x \in [0, 1]$;
   (b) $f(x) = x \exp(-x)$, $g(x) = 1$, $x \in [0, 2]$.

2. Consider the functional $\mathcal{F} : C^1\left[0, \frac{\pi}{2}\right] \to \mathbb{R}$,

$$\mathcal{F}[y] = \int_0^{\frac{\pi}{2}} \left(y^2 + y'^2\right) dx.$$

   Calculate:
   (a) $\mathcal{F}[x]$; (b) $\mathcal{F}[\sin x]$; (c) $\delta_h \mathcal{F}[\cos x]$ and $\delta_h \mathcal{F}[\sin 2x]$, for $h = \sin 2x$.

3. Find the extremals of the following functionals:
   (a) $\mathcal{F}[y] = \int_0^1 (y'^2 + 2y^2 + 4y) \exp x \, dx$, $y(0) = 0$, $y(1) = e - 1$.
   (b) $\mathcal{F}[y] = \int_0^{\frac{\pi}{2}} \left(y^2 - 4y \cos x - y'^2\right) dx$, $y(0) = 1$, $y(\frac{\pi}{2}) = \frac{\pi}{2}$.
   (c) $\mathcal{F}[y] = \int_1^2 \left(x^2 y'^2 + 6y^2\right) dx$, $y(1) = 1$, $y(2) = 4$.
   (d) $\mathcal{F}[y] = \int_1^e \left(xy'^2 - 2y'\right) dx$, $y(1) = 0$, $y(e) = 1$.
   (e) $\mathcal{F}[y] = \int_1^2 \sqrt{x(1 + y'^2)} \, dx$, $y(1) = 0$, $y(2) = 2$.
   (f) $\mathcal{F}[y] = \int_0^2 \sqrt{y(1 + y'^2)} \, dx$, $y(0) = 1$, $y(2) = 2$.
   (g) $\mathcal{F}[y] = \int_0^{\frac{\pi}{2}} \left(4y^2 - 5y'^2 + y''^2\right) dx$, $y(0) = 3$, $y'(0) = 4$, $y(\frac{\pi}{2}) = 1$, $y'(\frac{\pi}{2}) = 0$.
   (h) $\mathcal{F}[y] = \int_0^1 \left(2yy' - 2y^2 - y'^2 + y''^2\right) dx$, $y(0) = 1$, $y'(0) = 0$, $y(1) = \cosh 1$, $y'(1) = \sinh 1$.
   (i) $\mathcal{F}[y] = \int_0^1 \left(yy' + y'^2 + yy'' + y'y'' + y''^2\right) dx$, $y(0) = 0$, $y'(0) = 0$, $y(1) = 1$, $y'(1) = 2$.

8.8 Exercises

(j) $\mathcal{F}[y, z] = \int_1^2 \left(y'^2 + z^2 + z'^2\right) dx$, $y(1) = 1$, $y(2) = 2$, $z(1) = e$, $z(2) = e^2$.

(k) $\mathcal{F}[y, z] = \int_0^{\frac{\pi}{2}} \left(8y\cos x + 2z^2 + 2y'z' + y'^2 - z'^2\right) dx$, $y(0) = 0$, $y\left(\frac{\pi}{2}\right) = 4$, $z(0) = 0$, $z\left(\frac{\pi}{2}\right) = \frac{\pi}{2}$.

4. Find the smooth curve γ which connects the points $O(0, 0, 0)$ and $A(1, 1, 2/3)$ such that a moving particle in the field $\mathbf{F} = y^2\mathbf{i} + z\mathbf{j} + x^2\mathbf{k}$ yields the least possible work.

5. Write the Euler–Ostrogradsky equation for the functional:

$$\mathcal{F}[z(x, y)] = \iint_D \left[\left(\frac{\partial z}{\partial y}\right)^2 - \left(\frac{\partial z}{\partial x}\right)^2\right] dx\,dy.$$

6. Find the extremals of the functional

$$\mathcal{F}[y] = \int_0^1 \left(y'^2 + x^2\right) dx,$$

given the constraint

$$\int_0^1 y^2 dx = 2$$

and the boundary conditions: $y(0) = 0$, $y(1) = 0$.

7. Find the extremals of the functional

$$\mathcal{F}[y, z] = \int_0^1 \left(y'^2 + z'^2 - 4xz' - 4z\right) dx$$

with the constraint:

$$\int_0^1 \left(y'^2 - xy' - z'^2\right) dx = 2$$

and the boundary conditions: $y(0) = z(0) = 0$, $y(1) = z(1) = 1$.

Chapter 9
Elements of Probability Theory

Probability theory is the branch of Mathematics concerned with random phenomena. The modern theory of probability was developed by great mathematicians like: P. Fermat, B. Pascal, J. Bernoulli, P. Laplace, C. F. Gauss, S. D. Poisson, P.L. Chebyshev, A. Markov, A. M. Lyapunov, E. Borel, A. N. Kolmogorov, P. Lévy. In this chapter we make an elementary introduction to probability theory. We present Laplace's and Kolmogorov's definitions of probability and show how they can be applied to solve practical problems. We introduce the conditional probability and the Bayes formula, also providing several applications. We define the discrete random variables and the continuous random variables, emphasizing the most used distributions of discrete type (Bernoulli, binomial, Poisson), and of continuous type (normal, gamma, chi-squared, student). We also present the most important limit theorems: the weak law and the strong law of large numbers and the central limit theorem, highlighting the role played by the normal distribution.

9.1 Sample Space: Event Space

An *experiment* $\mathcal{E}$ is a process of observation, measurement, etc., that may produce various results (outcomes). The set of all possible outcomes of the experiment is called the *sample space* (usually denoted by Ω). Its elements (the possible outcomes of the experiment) are called *sample points*.

For instance, when a coin is tossed, we have two possible results: H (Head) or T (Tail), so the sample space is $\Omega = \{H, T\}$. If the experiment consists of tossing a coin three times, the sample space is:

$$\Omega = \{HHH, HHT, HTH, HTT, THH, THT, TTH, TTT\}.$$

If we toss a coin 100 times, the sample space will have 2^{100} elements.

When rolling a die, the set of possible outcomes is: $\Omega = \{1, 2, 3, 4, 5, 6\}$. When rolling two dice, the sample space will have 36 elements:

$$\Omega = \{(1, 1), (1, 2), (1, 3), (1, 4), (1, 5), (1, 6),$$

$$(2, 1), (2, 2), (2, 3), (2, 4), (2, 5), (2, 6),$$

$$(3, 1), (3, 2), (3, 3), (3, 4), (3, 5), (3, 6),$$

$$(4, 1), (4, 2), (4, 3), (4, 4), (4, 5), (4, 6),$$

$$(5, 1), (5, 2), (5, 3), (5, 4), (5, 5), (5, 6),$$

$$(6, 1), (6, 2), (6, 3), (6, 4), (6, 5), (6, 6)\}.$$

An *event* is a set of outcomes (a subset of Ω). For instance, in the experiment of tossing a coin three times, the event of obtaining at least two heads is

$$A = \{HHT, HTH, THH, HHH\}.$$

When rolling two dice, the event that the sum of the numbers obtained equals 6 is

$$B = \{(1, 5), (2, 4), (3, 3), (4, 2), (5, 1)\}.$$

We denote by $\emptyset$ the event that contains no sample point (the *impossible event*).

The event consisting of all points not contained in the event A is called the *complementary event* of A (or the *opposite* of A) and is denoted by $\overline{A}$. Note that $\overline{\overline{A}} = A$. In particular, $\overline{\emptyset} = \Omega$ and $\overline{\Omega} = \emptyset$.

Let A and B be two events (two subsets of Ω). We write $A = B$ if and only if the two events consists of exactly the same points. We write $A \subset B$ (or, equivalently, $B \supset A$ if every point of A is contained in B. In this case, we say that A implies B. Note that $A = B$ if and only if $A \subset B$ and $B \subset A$.

The event that both A and B occur is denoted by $A \cap B$ and consists of all the points contained in both sets A and B. If $A \cap B = \emptyset$ then we say that A and B are *disjoint events*.

The event that A or B occur is denoted by $A \cup B$ and consists of all the points contained in at least one of the events A and B (they may be contained in both sets).

The event that A occurs and B does not occur is denoted by $A \setminus B$ and consists of all points that are contained in A and not contained in B. If $A \subset B$, then $A \setminus B = \emptyset$.

For instance, in the experiment of rolling two dice, let us find the event that both numbers are less than 4 and their sum is a multiple of 4. The event that both numbers are less than 4 is

$$A = \{(1, 1), (1, 2), (1, 3), (2, 1), (2, 2), (2, 3), (3, 1), (3, 2), (3, 3)\},$$

9.1 Sample Space: Event Space

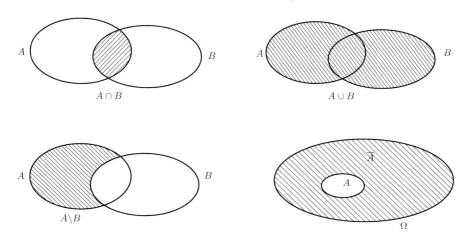

Fig. 9.1 Venn diagrams

while the event that the sum is a multiple of 4 is the set

$$B = \{(1, 3), (2, 2), (2, 6), (3, 1), (3, 5), (4, 4), (5, 3), (6, 2), (6, 6)\}.$$

Hence the required event is

$$A \cap B = \{(1, 3), (2, 2), (3, 1)\}.$$

Thus, when we are working with *events*, we use *set operations*, which can be represented (as in Fig. 9.1) by the well-known *Venn diagrams*, introduced by the English mathematician John Venn (1834–1923).

Proposition 9.1 (De Morgan's Laws)

(i) Any events A and B satisfy the following relations:

$$\begin{aligned} \overline{A \cup B} &= \overline{A} \cap \overline{B} \\ \overline{A \cap B} &= \overline{A} \cup \overline{B}. \end{aligned} \tag{9.1}$$

(ii) Let A_i, $i \in I$ be a collection of events, where I is an indexing set, which may be finite or infinite countable or even infinite uncountable. Then,

$$\overline{\bigcup_{i \in I} A_i} = \bigcap_{i \in I} \overline{A_i}, \tag{9.2}$$

$$\overline{\bigcap_{i \in I} A_i} = \bigcup_{i \in I} \overline{A_i}. \tag{9.3}$$

Proof Obviously, it is sufficient to prove the general case (ii).

Let $x \in \overline{\bigcup_{i \in I} A_i}$. Then $x \notin \bigcup_{i \in I} A_i$, which means that $x \notin A_i$, for any $i \in I$. It follows that $x \in \overline{A_i}$, for any $i \in I$, so $x \in \bigcap_{i \in I} \overline{A_i}$, so we have proved that $\overline{\bigcup_{i \in I} A_i} \subset \bigcap_{i \in I} \overline{A_i}$.

Conversely, let $x \in \bigcap_{i \in I} \overline{A_i}$. Then, $x \notin A_i$, for any $i \in I$. It follows that $x \notin \bigcup_{i \in I} A_i$, so we have $\bigcap_{i \in I} \overline{A_i} \subset \overline{\bigcup_{i \in I} A_i}$ and the equality (9.2) is proved by double inclusion.

To prove (9.3), we choose a point $x \in \overline{\bigcap_{i \in I} A_i}$. Then $x \notin \bigcap_{i \in I} A_i$, which means that there exists some $i_0 \in I$ such that $x \in \overline{A_{i_0}}$. It follows that $x \in \bigcup_{i \in I} \overline{A_i}$, so we have proved that $\overline{\bigcap_{i \in I} A_i} \subset \bigcup_{i \in I} \overline{A_i}$.

Conversely, let $x \in \bigcup_{i \in I} \overline{A_i}$. Then, there exists some $i_0 \in I$ such that $x \notin A_{i_0}$. It follows that $x \notin \bigcap_{i \in I} A_i$, so we have $\bigcup_{i \in I} \overline{A_i} \subset \overline{\bigcap_{i \in I} A_i}$ and so the equality (9.3) is proved. □

The *events* involved in an experiment $\mathcal{E}$ with the sample space Ω are not necessarily *all* the subsets of Ω. If $P(\Omega)$ denotes the set of all subsets of Ω (including the empty subset $\emptyset$ and the entire set Ω), then the set of events $\mathcal{K}_\mathcal{E}$ is a subset of $P(\Omega)$ which contains the empty set $\emptyset$ and the sample space Ω and is closed under complement, under countable unions and under countable intersections (see Definition 9.1 and Remark 9.1).

Definition 9.1 Let Ω be a nonempty set. A nonempty set of subsets of Ω, $\mathcal{K} \subset P(\Omega)$ is called an *event space* on Ω (a σ-algebra on Ω) if it has the following properties:

(1) $A \in \mathcal{K} \Rightarrow \overline{A} \in \mathcal{K}$ ($\mathcal{K}$ is closed under complement);
(2) $A_i \in \mathcal{K}, i = 1, 2, \ldots \Rightarrow \bigcup_{i \geq 1} A_i \in \mathcal{K}$ ($\mathcal{K}$ is closed under countable unions).

Remark 9.1 From this definition, one can easily derive the following properties of an event space:

(3) $\emptyset, \Omega \in \mathcal{K}$;
(4) $A_i \in \mathcal{K}, i = 1, 2, \ldots \Rightarrow \bigcap_{i \geq 1} A_i \in \mathcal{K}$ ($\mathcal{K}$ is closed under countable intersections);
(5) $A, B \in \mathcal{K} \Rightarrow A \setminus B \in \mathcal{K}$.

In order to prove (3), we can take an arbitrary event $A \in \mathcal{K}$. From (1) it follows that $\overline{A} \in \mathcal{K}$ and from (2) we obtain that $\Omega = A \cup \overline{A} \in \mathcal{K}$. Hence, $\emptyset = \overline{\Omega} \in \mathcal{K}$.

The property (4) follows from de Morgan's Laws: Since $A_i \in \mathcal{K}$ for all $i = 1, 2, \ldots$, it follows by (1) that $\overline{A_i} \in \mathcal{K}$ for all $i \geq 1$, so we obtain by (2), that $\bigcup_{i \geq 1} \overline{A_i} \in \mathcal{K}$. But, from (9.3), we have

$$\bigcup_{i \geq 1} \overline{A_i} = \overline{\bigcap_{i \geq 1} A_i} \in \mathcal{K},$$

9.1 Sample Space: Event Space

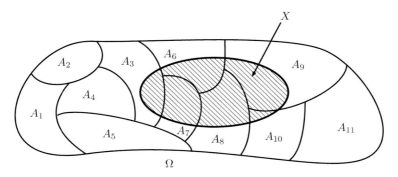

Fig. 9.2 A partition of Ω

so using (1) we get $\bigcap\limits_{i \geq 1} A_i \in \mathcal{K}$.

To prove (5), we write $A \setminus B = A \cap \overline{B}$ and use the properties (1) and (4).

Given a family $A_1, \ldots, A_k$ of subsets of Ω, the least event space on Ω that contains these subsets is the intersection of all the event spaces $\mathcal{K}_\alpha$ containing $A_1, \ldots, A_k$: $\mathcal{K}(A_1, \ldots, A_k) = \bigcap\limits_\alpha \mathcal{K}_\alpha$. For instance, the least event space that contains a given event $A \neq \emptyset$ is: $\mathcal{K}(A) = \{\emptyset, A, \overline{A}, \Omega\}$. (Try to describe $\mathcal{K}(A, B)$ if $A, B \neq \emptyset$ and $A \cap B = \emptyset$).

Definition 9.2 A family $\{A_i\}_{i \in I}$ of nonempty subsets of Ω is said to be a *partition* of Ω if $\bigcup\limits_{i \in I} A_i = \Omega$ (they cover Ω) and, for any $i \neq j$, $A_i \cap A_j = \emptyset$ (they are piecewise disjoint).

We also say that Ω is a *disjoint union* of the subsets $\{A_i\}_{i \in I}$. An equivalent definition of the partition is that every element $x \in \Omega$ is exactly in one of the subsets $\{A_i\}_{i \in I}$ (Fig. 9.2).

Definition 9.3 Let $\mathcal{K}$ be an event space on Ω. An event $A \in \mathcal{K}$, $A \neq \emptyset$ is called a *simple (elementary) event* if for any $B \in \mathcal{K}$ such that $B \subset A$ we have either $B = \emptyset$ or $B = A$.

An event space $\mathcal{K}$ on Ω is said to be *finite* if it has a finite number of events. In particular, if the sample space Ω is a finite set, then any event space on Ω is finite. The event spaces of experiments like: rolling a die, tossing a coin, counting daily traffic accidents in a city, inspecting light bulbs (defective, non defective), etc., are finite since their corresponding sample spaces Ω are finite.

Proposition 9.2 *Let $\mathcal{K}$ be a finite event space over Ω. Then, any event $A \in \mathcal{K}$, $A \neq \emptyset$ contains at least one simple event. The set $\{A_i\}_{i=1,2,\ldots,k}$ of all simple events of $\mathcal{K}$ is a partition of Ω. Moreover, any event $A \in \mathcal{K}$, $A \neq \emptyset$, can be written (in a unique way) as a finite union of some simple events.*

Proof Let $A \in \mathcal{K}$ be a possible event ($A \neq \emptyset$). If A is a simple event, the statement is verified. If not, A contains an event $B_1 \neq \emptyset$ such that $B_1 \subsetneq A$. If B_1 is simple, the statement is verified. If not, B_1 contains an event $B_2 \neq \emptyset$ such that $B_2 \subsetneq B_1 \subsetneq A$, and so on. If doing this we do not find a simple event (and stop the process!), we would get an infinite sequence of events:

$$\emptyset \subsetneq \ldots \subsetneq B_n \subsetneq \ldots \subsetneq B_2 \subsetneq B_1 \subsetneq A$$

which is impossible because $\mathcal{K}$ is finite. Thus, there exists a natural number n such that B_n is a simple event.

Let now $\{A_1, \ldots, A_k\}$ be the set of all simple events in $\mathcal{K}$. If $\bigcup_{i=1}^{k} A_i \subsetneq \Omega$, then the event $C = \Omega \setminus \bigcup_{i=1}^{k} A_i$ would contain a simple event $A_{k+1} \in \mathcal{K}$, which is impossible since we assumed that $A_1, \ldots, A_k$ are all the simple events of $\mathcal{K}$. Thus, $\bigcup_{i=1}^{k} A_i = \Omega$ and it remains to prove that any two different simple events are disjoint, i.e. $A_i \cap A_j = \emptyset$ for $i \neq j$. If $A_i \cap A_j \neq \emptyset$, since $A_i \cap A_j \subset A_i$, $A_i \cap A_j \subset A_j$ and A_i, A_j are simple events, it follows that $A_i \cap A_j = A_i = A_j$, i.e. $i = j$, a contradiction! Hence, $\{A_i\}_{i=1,2,\ldots,k}$ is a partition of Ω.

Consider an arbitrary event $A \in \mathcal{K}$, $A \neq \emptyset$. Suppose that $A_{i_1}, \ldots, A_{i_p}$ are all the simple events contained in A. Then $\bigcup_{j=1}^{p} A_{i_j} \subset A$. If we suppose that $A \setminus \bigcup_{j=1}^{p} A_{i_j} \neq \emptyset$, then it follows that this event contains another simple event $A_{i_{p+1}}$, which is a contradiction, because we supposed that $A_{i_1}, \ldots, A_{i_p}$ are all the simple events contained in A. □

Definition 9.4 Two event spaces: $\mathcal{K}_1$ over Ω_1 and $\mathcal{K}_2$ over Ω_2 are said to be *isomorphic* ($\mathcal{K}_1 \simeq \mathcal{K}_2$) if there exists a one-to-one and onto function $s : \mathcal{K}_1 \to \mathcal{K}_2$ with the following properties:

(1) $s(\overline{A}) = \overline{s(A)}$, for every $A \in \mathcal{K}_1$.
(2) $s(\bigcup_{i \geq 1} A_i) = \bigcup_{i \geq 1} s(A_i)$, for any $A_i \in \mathcal{K}$, $i = 1, 2, \ldots$.

The function s is called an isomorphism of event spaces.

Remark 9.2 The isomorphism s also has the following properties:

(3) $s(\Omega_1) = \Omega_2$, $s(\emptyset) = \emptyset$;
(4) $s(\bigcap_{i \geq 1} A_i) = \bigcap_{i \geq 1} s(A_i)$, for any $A_i \in \mathcal{K}_1$, $i = 1, 2, \ldots$.
(5) $s(A \setminus B) = s(A) \setminus s(B)$, for any $A, B \in \mathcal{K}_1$;
(6) For any $A, B \in \mathcal{K}_1$, $s(A) \subset s(B)$ if and only if $A \subset B$.
(7) $A \in \mathcal{K}_1$ is a simple event if and only if $s(A)$ is a simple event in $\mathcal{K}_2$.

We prove only (6), but we encourage the reader to prove the other statements too.

Let $A, B \in \mathcal{K}_1$ such that $A \subset B$. Since $B \smallsetminus A \in \mathcal{K}_1$ (see Remark 9.1), by Definition 9.4, (2) we can write: $s(B) = s(A \cup (B \smallsetminus A)) = s(A) \cup s(B \smallsetminus A)$, so $s(A) \subset s(B)$.

Conversely, suppose that $A, B \in \mathcal{K}_1$ such that $s(A) \subset s(B)$. Since $s(B) \smallsetminus s(A) \in \mathcal{K}_2$ and s is a surjective function, it follows that there exists $C \in \mathcal{K}_1$ such that $s(C) = s(B) \smallsetminus s(A)$. We have: $s(B) = s(A) \cup s(C) = s(A \cup C)$, so $B = A \cup C$ (since s is injective), hence $A \subset B$.

Example 9.1 Let $\Omega_1 = \{1, 2, 3, 4, 5, 6\}$ be the sample space of the experiment which consists in rolling a die. Suppose that we are testing the parity of the number obtained. Thus, our event space is:

$$\mathcal{K}_1 = \{\emptyset, O = \{1, 3, 5\}, E = \{2, 4, 6\}, \Omega_1\}.$$

Let $\Omega_2 = \{H, T\}$ be the sample space of the experiment which consist in tossing a coin. In this case, the event space is $\mathcal{K}_2 = \{\emptyset, \{H\}, \{T\}, \{H, T\}\}$.

The two event spaces are isomorphic because we can define the bijection

$$s : \mathcal{K}_1 \to \mathcal{K}_2, s(\emptyset) = \emptyset, s(O) = \{H\}, s(E) = \{T\}, s(\Omega_1) = \Omega_2,$$

which has the properties (1) and (2) from Definition 9.4. (Note that we could also take $s(O) = \{T\}$ and $s(E) = \{H\}$).

Remark 9.3 The example above give rise to the following question: Can we tell that any two **finite** event spaces with the same number of events are isomorphic? We show that the answer is "Yes" (we also prove that the cardinality of a finite event space is not an arbitrary natural number).

Let $\mathcal{K}$ be a finite event space over Ω. By Proposition 9.2 we know that there exists a unique partition of Ω with simple events: $A_1, A_2, \ldots, A_n$.

Consider the sample space $\Omega_0 = \{1, 2, \ldots, n\}$ and the event space $\mathcal{K}_0 = P(\Omega_0)$ containing all the subsets of Ω_0. We define the function $s : \mathcal{K}_0 \to \mathcal{K}$,

$$s(\emptyset) = \emptyset, \ s(\{k\}) = A_k, \ k = 1, 2, \ldots, n,$$

and, for any other subset of Ω_0,

$$s(\{k_1, \ldots, k_p\}) = \bigcup_{i=1}^{p} A_{k_i}.$$

It is easy to see that the function s has the properties (1) and (2) from Definition 9.4. By Proposition 9.2, s is a bijection since any event $A \in \mathcal{K}$ can be written in a unique form as a union of simple events, $A = \bigcup_{i=1}^{p} A_{k_i}$. It follows that s is an isomorphism of event spaces and $\mathcal{K} \simeq P(\Omega_0)$.

A direct consequence is that the number of events of any finite event space is a power of 2.

9.2 Probability Space

Probability functions provide mathematical models of practical situations governed by "chance effects", such as: quality of technical products, traffic problems, life insurance, reliability of working systems, weather forecasting, games of chance with cards and dice, economical decisions, etc. All these probability functions come from suitable statistical observations and are also tested by statistical experiments. The probability of an event A measures how frequently this event is about to occur in several trials.

The mathematical definition of the probability function was given by the great Russian mathematician A. N. Kolmogorov (1903–1987):

Definition 9.5 Let $\mathcal{K}$ be an event space over the sample space Ω. A *probability function* on $\mathcal{K}$ is a function $\mathcal{P} : \mathcal{K} \to [0, 1]$, which satisfies the following conditions:

(i) $\mathcal{P}(\Omega) = 1$
(ii) If $\{A_i\}_{i=1,2,\ldots} \subset \mathcal{K}$ is a countable collection of pairwise disjoint events, that is, $A_i \cap A_j = \emptyset, \forall i \neq j$, then

$$\mathcal{P}\left(\bigcup_{i \geq 1} A_i\right) = \sum_{i \geq 1} \mathcal{P}(A_i). \tag{9.4}$$

The mathematical triplet $(\Omega, \mathcal{K}, \mathcal{P})$ is called a *probability space*.

On a nontrivial event space, $\mathcal{K} \neq \{\emptyset, \Omega\}$, different probability functions can be defined. For instance, if $\mathcal{K} = \{\emptyset, A, \overline{A}, \Omega\}$ and $\lambda \in (0, 1)$ is an arbitrary real number, then the function $\mathcal{P}_\lambda : \mathcal{K} \to [0, 1]$, $\mathcal{P}_\lambda(A) = \lambda$ and $\mathcal{P}_\lambda(\overline{A}) = 1 - \lambda$ is a probability function on $\mathcal{K}$.

Proposition 9.3 *Let $(\Omega, \mathcal{K}, \mathcal{P})$ be a probability space and let A, B, C be three arbitrary events of $\mathcal{K}$. Then the following relations hold:*

(1) $\mathcal{P}(\emptyset) = 0$;
(2) $\mathcal{P}(\overline{A}) = 1 - \mathcal{P}(A)$;
(3) $A \subset B \Longrightarrow \mathcal{P}(A) \leq \mathcal{P}(B)$;
(4) $\mathcal{P}(A) = \mathcal{P}(A \setminus B) + \mathcal{P}(A \cap B)$;
(5) $\mathcal{P}(A \cup B) = \mathcal{P}(A) + \mathcal{P}(B) - \mathcal{P}(A \cap B)$;
(6) $\mathcal{P}(A \cup B \cup C) = \mathcal{P}(A) + \mathcal{P}(B) + \mathcal{P}(C) - \mathcal{P}(A \cap B) - \mathcal{P}(A \cap C) - \mathcal{P}(B \cap C) + \mathcal{P}(A \cap B \cap C)$.

Proof

(1) $1 = \mathcal{P}(\Omega) = \mathcal{P}(\Omega \cup \emptyset) = \mathcal{P}(\Omega) + \mathcal{P}(\emptyset) = 1 + \mathcal{P}(\emptyset)$, so $\mathcal{P}(\emptyset) = 0$;
(2) Since $A \cup \overline{A} = \Omega$ and $A \cap \overline{A} = \emptyset$, from Definition 9.5 we obtain: $1 = \mathcal{P}(A \cup \overline{A}) = \mathcal{P}(A) + \mathcal{P}(\overline{A})$, so $\mathcal{P}(\overline{A}) = 1 - \mathcal{P}(A)$;

9.2 Probability Space

(3) Since $B = A \cup (B \smallsetminus A)$ and since $A \cap (B \smallsetminus A) = \emptyset$, we obtain that $\mathcal{P}(B) = \mathcal{P}(A) + \mathcal{P}(B \smallsetminus A) \geq \mathcal{P}(A)$;

(4) Since $A = (A \smallsetminus B) \cup (A \cap B)$ and $(A \smallsetminus B) \cap (A \cap B) = \emptyset$, from Definition 9.5, (ii), we find: $\mathcal{P}(A) = \mathcal{P}(A \smallsetminus B) + \mathcal{P}(A \cap B)$.

(5) Since $A \cup B = (A \smallsetminus B) \cup (A \cap B) \cup (B \smallsetminus A)$ and it is a disjoint union, one can write:

$$\mathcal{P}(A \cup B) = \mathcal{P}(A \smallsetminus B) + \mathcal{P}(A \cap B) + \mathcal{P}(B \smallsetminus A) \stackrel{4)}{=}$$

$$= \mathcal{P}(A) - \mathcal{P}(A \cap B) + \mathcal{P}(A \cap B) + \mathcal{P}(B) - \mathcal{P}(A \cap B) =$$

$$= \mathcal{P}(A) + \mathcal{P}(B) - \mathcal{P}(A \cap B);$$

(6) We can prove the equality by using (5) with $B \cup C$ in place of B:

$$\mathcal{P}(A \cup (B \cup C)) = \mathcal{P}(A) + \mathcal{P}(B \cup C) - \mathcal{P}(A \cap (B \cup C)).$$

But $A \cap (B \cup C) = (A \cap B) \cup (A \cap C)$ and, by using again (5) we obtain:

$$\mathcal{P}(A \cup B \cup C) = \mathcal{P}(A) + \mathcal{P}(B) + \mathcal{P}(C) - \mathcal{P}(B \cap C) -$$

$$- \mathcal{P}(A \cap B) - \mathcal{P}(A \cap C) + \mathcal{P}((A \cap B) \cap (A \cap C)),$$

which is equivalent to the equality (6).

□

Remark 9.4 (Inclusion-Exclusion Formula) The properties (5) and (6) are particular cases (for $n = 2$ and $n = 3$) of the following formula, which can be easily proved by mathematical induction on n:

$$\mathcal{P}\left(\bigcup_{i=1}^{n} A_i\right) = \sum_{i=1}^{n} P(A_i) - \sum_{1 \leq i < j \leq n} P(A_i \cap A_j) + \quad (9.5)$$

$$+ \sum_{1 \leq i < j < k \leq n} P(A_i \cap A_j \cap A_k) + \ldots + (-1)^{n-1} P(A_1 \cap A_2 \cap \ldots \cap A_n),$$

for any events $A_1, \ldots, A_n \in \mathcal{K}$.

Proposition 9.4 (Laplace) *Let $\mathcal{K}$ be a **finite** event space over Ω, and let $A_1, A_2, \ldots, A_n$ be its simple events (see Proposition 9.2). Consider $p_1, p_2, \ldots, p_n$ $\in [0, 1]$ n real numbers such that $\sum_{i=1}^{n} p_i = 1$. Then, there exists a unique probability function $\mathcal{P}$ on $\mathcal{K}$ such that $\mathcal{P}(A_i) = p_i$, for all $i = 1, 2, \ldots, n$.*

*In particular, suppose that $\mathcal{P}(A_1) = \ldots = \mathcal{P}(A_n) = \frac{1}{n}$ (the simple events are **equally probable or equally likely**). By Proposition 9.2, any event $B \in \mathcal{K}$ can be* uniquely written as a finite union of simple events: $B = A_{i_1} \cup A_{i_2} \cup \ldots \cup A_{i_k}$. Then,

$$\mathcal{P}(B) = \frac{k}{n}, \tag{9.6}$$

i.e. the probability of B is the number of "favorable" cases over the number of all "possible" cases (Laplace's definition of probability).

Proof If $B = A_{i_1} \cup \ldots \cup A_{i_k}$, then we define the probability of B as

$$\mathcal{P}(B) = p_{i_1} + \ldots + p_{i_k} \tag{9.7}$$

Let us verify that this function is indeed a probability function. It is clear that $\mathcal{P}(\Omega) = \sum_{i=1}^{n} p_i = 1$, because $\Omega = A_1 \cup A_2 \cup \ldots \cup A_n$. Let now $A, B \in \mathcal{K}$ such that $A \cap B = \emptyset$ and $A = A_{j_1} \cup A_{j_2} \cup \ldots \cup A_{j_k}$, $B = A_{i_1} \cup A_{i_2} \cup \ldots \cup A_{i_m}$. The condition $A \cap B = \emptyset$ implies that $A_{i_t} \neq A_{j_h}$ for any $t \in \{1, 2, \ldots, m\}$ and $h \in \{1, 2, \ldots, k\}$. Thus,

$$\mathcal{P}(A \cup B) = p_{j_1} + \ldots + p_{j_k} + p_{i_1} + \ldots + p_{i_m} = \mathcal{P}(A) + \mathcal{P}(B).$$

Since $\mathcal{K}$ is finite, the unions are finite, so we can prove the general formula (9.4) by mathematical induction. Formula (9.6) can be easily derived from (9.7). $\square$

Laplace's formula (9.6) is extremely useful in different problems involving a finite sample space Ω. It says that

$$\mathcal{P}(A) = \frac{\text{number of elements of } A}{\text{number of elements of } \Omega} = \frac{|A|}{|\Omega|} \tag{9.8}$$

For instance, when rolling a die, the simple events are: $\{1\}, \ldots, \{6\}$. If the die is fair, all these events are equally probable, so $\mathcal{P}(\{i\}) = \frac{1}{6}$ and, for any subset $A \subset \{1, 2, \ldots, 6\}$, $\mathcal{P}(A) = \frac{|A|}{6}$. This function $\mathcal{P}$ is the unique probability function that can be associated to this experiment.

If the event space is not finite, in general, the probabilities of simple events does not determine the entire probability function as shown in the following example.

Example 9.2 (Geometric Probability) Two students X and Y agreed to meet in the University Square, between 12 and 13 o'clock. They also established that the first who comes waits for 20 min and then leaves the place. Find the probability that the two students meet each other.

If we take a Cartesian coordinate system xOy and consider the square $\Omega = [0, 60] \times [0, 60]$, each point $M(x, y)$ corresponds to a simple event, namely, the student X arrives at the meeting place at the time 12 and x minutes and the student Y arrives at 12 and y minutes. All these points of the sample space Ω are equally probable. Since their number is infinity, and since for any probability function $\mathcal{P}$,

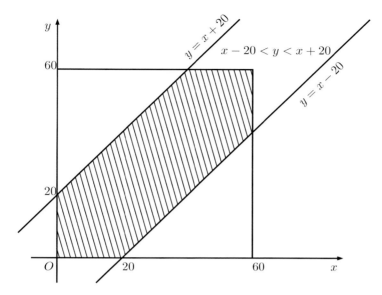

Fig. 9.3 Defining probability as the ratio of two areas

$\mathcal{P}(\Omega) = 1$, we see that $\mathcal{P}(\{M(x, y)\}) = 0$, i.e. the probability of the simple events cannot determine the probability function $\mathcal{P}$. Nevertheless, we can construct a probability function by using the area of plane figures. If A is a subset of Ω which has an area, then we define its probability as:

$$\mathcal{P}(A) = \frac{area(A)}{area(\Omega)}.$$

It is easy to see that the two students meet each other if and only if $|x - y| \leq 20$. Thus, the figure $B \subset \Omega$ that corresponds to the event "X and Y meet each other" is the region of the square between the lines $y - x = -20$ and $y - x = 20$ (see Fig. 9.3).

The area of B is equal to 2000. Thus, the probability of B is:

$$\mathcal{P}(B) = \frac{area(B)}{area(\Omega)} = \frac{2000}{3600} = \frac{5}{9}.$$

Note that geometric probability models can be also constructed (in a three-dimensional space) using volumes instead of areas.

Remark 9.5 A *Bernoulli trial* is a random experiment with exactly two possible outcomes: "success" (A) and "failure" ($\overline{A}$) which is performed repeatedly such that the probability of success is the same at every performance. In the seventeenth century (before Laplace), the Swiss mathematician Jacob Bernoulli analyzed this kind of experiments and proposed the following definition for probability: if $f_n(A)$ denotes the number of times that A occurs in n independent trials (the *absolute frequency* of A), then the probability of A is equal to the limit of the *relative frequency* $\frac{f_n(A)}{n}$:

$$\mathcal{P}(A) = \lim_{n \to \infty} \frac{f_n(A)}{n}.$$

As we shall see in Sect. 9.6, this definition of the probability is justified by the *law of large numbers*.

Definition 9.6 Let $(\Omega_1, \mathcal{K}_1, \mathcal{P}_1)$ and $(\Omega_2, \mathcal{K}_2, \mathcal{P}_2)$ be two probability spaces such that $\mathcal{K}_1 \simeq \mathcal{K}_2$. If $s : \mathcal{K}_1 \to \mathcal{K}_2$ is an isomorphism of event spaces such that

$$\mathcal{P}_1(A) = \mathcal{P}_2(s(A)), \text{ for all } A \in \mathcal{K}_1,$$

then s is an *isomorphism of the probability spaces* and the spaces $(\Omega_1, \mathcal{K}_1, \mathcal{P}_1)$ and $(\Omega_2, \mathcal{K}_2, \mathcal{P}_2)$ are said to be isomorphic. The class $\{(\Omega, \mathcal{K}, \mathcal{P})\}$ of all the probability spaces that are isomorphic with a given probability space $(\Omega, \mathcal{K}, \mathcal{P})$ (usually called a *standard probability space*) is called a *probabilistic mathematical model*.

Remark 9.6 We have shown in Remark 9.3 that two finite event spaces with the same number of simple events are isomorphic. On the other hand, by Proposition 9.4 we know that the probability function can be defined by the probabilities of the simple events. Thus, if $(\Omega_1, \mathcal{K}_1, \mathcal{P}_1)$ and $(\Omega_2, \mathcal{K}_2, \mathcal{P}_2)$ are two finite probability spaces such that $A_1, \ldots, A_n$ are the simple events of $\mathcal{K}_1$ and $B_1, \ldots, B_n$ are the simple events of $\mathcal{K}_2$ and if there exists a permutation σ of $\{1, \ldots, n\}$ such that

$$\mathcal{P}_1(A_i) = \mathcal{P}_2(B_{\sigma(i)}), \ i = 1, \ldots, n,$$

then the two probability spaces are isomorphic.

If we suppose that the simple events are equally probable (with the probability $p(A_i) = p(B_i) = \frac{1}{n}$), $i = 1, \ldots, n$, then the resulting probability spaces are isomorphic.

Example 9.3 Three dice are rolled. What is the probability of getting the sum greater than 15? What is the probability that the sum equals 15?

In this experiment, the set of possible outcomes is

$$\Omega = \{1, 2, \ldots, 6\}^3.$$

Let A denotes the event "the sum is greater than 15". Then, A is the disjoint union of the following events:

$A_{18} = \{(6, 6, 6)\}$ ("the sum equals 18");
$A_{17} = \{(6, 6, 5), (6, 5, 6), (5, 6, 6)\}$ ("the sum equals 17");

$A_{16} = \{(6, 5, 5), (5, 6, 5), (5, 5, 6), (6, 6, 4), (6, 4, 6), (4, 6, 6)\}$ ("the sum equals 16").

It follows that $|A| = |A_{18}| + |A_{17}| + |A_{16}| = 10$, and by the formula (9.8) we have:

$$P(A) = \frac{|A|}{|\Omega|} = \frac{10}{216}.$$

Now, if B denotes the event "the sum equals 15", then B is the disjoint union of the following events:

$B_{\{6,6,3\}} = \{(6, 6, 3), (6, 3, 6), (3, 6, 6)\}$;
$B_{\{6,5,4\}} = \{(6, 5, 4), (6, 4, 5), (5, 6, 4), (5, 4, 6), (4, 6, 5), (4, 5, 6)\}$;
$B_{\{5,5,5\}} = \{(5, 5, 5)\}$;

It follows that $|B| = |B_{\{6,6,3\}}| + |B_{\{6,5,4\}}| + |B_{\{5,5,5\}}| = 10$, so the probability of B is

$$P(B) = \frac{|B|}{|\Omega|} = \frac{10}{216} = P(A).$$

Example 9.4 (Sampling With/Without Replacement) Two numbers are chosen at random from $\{0, 1, \ldots, 9\}$ such that:

(a) the numbers may be equal (sampling *with* replacement: the first number a is chosen from $\{0, 1, \ldots, 9\}$, the second number b is chosen from the same set);
(b) the numbers may not be equal (sampling *without* replacement: the difference is that the second number b is chosen from $\{0, 1, \ldots, 9\} \setminus \{a\}$, or, equivalently, the two numbers are simultaneously chosen).

Find the probability of the following events: A: "both numbers are odd", and B: "at least one number is odd" in the case with replacement and in the case without replacement, respectively.

In the case *with replacement* the set of possible outcomes is the set of all the pairs of numbers (a, b), with $a, b \in \{0, 1, \ldots, 9\}$:

$$\Omega_1 = \{0, 1, \ldots, 9\}^2.$$

The event $A = A_1$ is formed by all the pairs of odd numbers:

$$A_1 = \{1, 3, 5, 7, 9\}^2,$$

so

$$P(A_1) = \frac{|A_1|}{|\Omega_1|} = \frac{25}{100} = \frac{1}{4}.$$

The event $B = B_1$ is the complement of the event that "all the numbers are even". Thus, since

$$\overline{B_1} = \{0, 2, 4, 6, 8\}^2,$$

it follows that

$$P(B_1) = 1 - P\left(\overline{B_1}\right) = 1 - \frac{1}{4} = \frac{3}{4}.$$

In the case *without replacement* the set Ω_2 of possible outcomes is the set of all the subsets of 2 distinct numbers $\{a, b\}$, with $a, b \in \{0, 1, \ldots, 9\}$. The number of these subsets is $|\Omega_2| = \binom{10}{2} = 45$. The event $A = A_2$ is formed by all the subsets of 2 distinct odd numbers, $\{a, b\}$, with $a, b \in \{1, 3, 5, 7, 9\}$, hence $|A_2| = \binom{5}{2} = 10$, so

$$P(A_2) = \frac{|A_2|}{|\Omega_2|} = \frac{10}{45} = \frac{2}{9}.$$

As above, the event $B = B_2$ is the complement of the event that "all the numbers are even". Since $|\overline{B_2}| = |A_2| = \frac{2}{9}$, it follows that

$$P(B_2) = 1 - P\left(\overline{B_2}\right) = 1 - \frac{2}{9} = \frac{7}{9}.$$

Remark 9.7 (How to Count?) We summarize here some basic combinatorics results:

(i) The number of permutations of a set with n elements (the number of ways one can arrange n distinct objects in a line) is $n!$.
(ii) The number of k-tuples $(a_1, \ldots, a_k)$ with $a_j \in S$, $j = 1, \ldots, k$ and $|S| = n$ (the number of ways one can choose k objects from n distinct objects if the same object can be chosen repeatedly) is n^k.
(iii) The number of subsets with k distinct elements chosen from a set with n elements (the number of ways one can choose k distinct objects from n distinct objects, when the order of selection does not matter) is the binomial coefficient

$$\binom{n}{k} = \frac{n!}{k!(n-k)!}.$$

(iv) The number of k-tuples $(a_1, \ldots, a_k)$ with $a_j \in S$, $j = 1, \ldots, k$, $a_i \neq a_j$ for $i \neq j$ and $|S| = n$ (the number of ways one can choose k distinct objects from n distinct objects when the order of selection matters) is

$$\frac{n!}{(n-k)!} = n(n-1)\ldots(n-k+1).$$

9.2 Probability Space

(v) The number of partitions of a set with n elements into k subsets, each subset having a fixed number of elements, $S = S_1 \cup \ldots \cup S_k$, such that $|S_i| = n_i$, $i = 1, \ldots, k$ (where $n_1 + \ldots + n_k = n$), is

$$\binom{n}{n_1}\binom{n-n_1}{n_2}\cdots\binom{n-n_1-\ldots-n_{k-1}}{n_k} = \frac{n!}{n_1!n_2!\cdots n_k!}.$$

Note that, for $k = 2$, we get (iii).

Example 9.5 Four coins are tossed.

(a) Find the probability of getting exactly two heads.
(b) Find the probability of getting at least two heads.
(c) Find the probability of getting at least two heads when 10 coins are tossed.

When four coins are tossed, the set of possible outcomes is $\Omega = \{H, T\}^4$.

(a) Let A_k denote the event of getting exactly k heads ($k = 0, 1, 2, 3, 4$). Thus,

$$A_0 = \{(T, T, T, T)\},$$

$$A_1 = \{(H, T, T, T), (T, H, T, T), (T, T, H, T), (T, T, T, H)\}, \ldots$$

$$\ldots, A_4 = \{(H, H, H, H)\}.$$

Hence,

$$P(A_2) = \frac{|A_2|}{|\Omega|} = \frac{\binom{4}{2}}{2^4} = \frac{6}{16} = \frac{3}{8}.$$

(b) The event of getting at least two heads is the disjoint union of the events A_2, A_3 and A_4. So

$$P(B) = P(A_2 \cup A_3 \cup A_4) = \frac{\binom{4}{2} + \binom{4}{3} + \binom{4}{4}}{2^4} = \frac{6 + 4 + 1}{16} = \frac{11}{16}.$$

(c) In this case, the set of possible outcomes is $\Omega = \{H, T\}^{10}$. If B is the event of getting at least 2 heads, we notice that it is easier to calculate first the probability of $\overline{B}$ (of getting at most one head). Thus,

$$P(B) = 1 - P(\overline{B}) = 1 - \frac{\binom{10}{0} + \binom{10}{1}}{2^{10}} = 1 - \frac{11}{1024} = \frac{1013}{1024}.$$

Example 9.6 (Urn Model. Sampling with Replacement) We consider a urn (a box) with a white balls and b black balls. One ball is drawn randomly from the urn and its color observed. It is then placed back in the urn and the selection process is repeated.

Let p denote the probability to draw a white ball and $q = 1 - p$ the probability to draw a black ball. Obviously, we have $p = \frac{a}{a+b}$ and $q = \frac{b}{a+b}$.

Now, we consider the experiment of drawing (with replacement) n balls from the urn. What is the probability of the event A_k: "among the n balls there are exactly k white balls"?

Consider $S = \{x_1, \ldots, x_a, y_1, \ldots, y_b\}$ the set of the $a+b$ balls (where $x_1, \ldots, x_a$ are the white balls and $y_1, \ldots, y_b$ are the black balls). Then, the set of possible outcomes is the set of n-tuples formed with elements of S, $\Omega = S^n$. For each subset with k elements $\{i_1, \ldots, i_k\} \subset \{1, 2, \ldots, n\}$, we denote by $A_{\{i_1, \ldots, i_k\}}$ the event that the k balls chosen in the selections $i_1, i_2, \ldots, i_k$ are white and all the other $n - k$ balls are black. For instance, if $n = 4$ and $k = 2$, we draw (with replacement) 4 balls from the box and want to know the probability that exactly 2 of them are white. In this case $A_{\{2,3\}}$ denotes the event that the second and the third balls are white and the first and the last balls chosen are black. By Remark 9.7(ii), the number of ways we can choose (with replacement) k white balls out of a balls is a^k. For each selection of k white balls, we can choose $n - k$ black balls from b balls in b^{n-k} ways, so $|A_{\{i_1, \ldots, i_k\}}| = a^k b^{n-k}$ and

$$\mathcal{P}\left(A_{\{i_1, \ldots, i_k\}}\right) = \frac{|A_{\{i_1, \ldots, i_k\}}|}{\Omega} = \frac{a^k b^{n-k}}{(a+b)^n} = p^k q^{n-k}.$$

Since the event A_k is the disjoint union of all the events $A_{\{i_1, \ldots, i_k\}}$ and, by Remark 9.7(iii) we have $\binom{n}{k}$ such distinct events, it follows that

$$\mathcal{P}(A_k) = \binom{n}{k} p^k q^{n-k}. \tag{9.9}$$

Notice that, from Newton's binomial formula, one can write:

$$\sum_{k=0}^{n} \mathcal{P}(A_k) = \sum_{k=0}^{n} \binom{n}{k} p^k q^{n-k} = (p+q)^n = 1,$$

which also derives from the fact that $A_0, A_1, \ldots, A_n$ form a partition of Ω.

Finally, we remark that a more general case can be also considered, when the urn contains $t > 2$ kinds of balls ($t = 2$ in our example with black and white balls). Suppose that we have a_j balls of kind j, $j = 1, \ldots, t$. The probability that a ball of kind j is drawn (at a single trial) is $p_j = \frac{a_j}{a_1 + \ldots + a_t}$, $j = 1, \ldots, t$. Let n be a positive integer and $k_1, \ldots, k_t \geq 0$ such that $k_1 + \ldots + k_t = n$. We consider the event $A_{n;k_1,\ldots,k_t}$: "from n balls drawn from the urn at random (with replacement) exactly k_j are of kind j, $j = 1, \ldots, t$". Its probability is:

$$\mathcal{P}(A_{n;k_1,\ldots,k_t}) = \frac{n!}{k_1! \ldots k_t!} p_1^{k_1} p_2^{k_2} \ldots p_t^{k_t}. \tag{9.10}$$

Example 9.7 (Urn Model. Sampling Without Replacement) We consider the same urn with a white balls and b black balls as above, but now we draw n balls simultaneously ($n < a + b$). Let us find the probability of the event A_k: "among the n balls there are exactly k white balls".

Let $S = \{x_1, \ldots, x_a, y_1, \ldots, y_b\}$ be the set of the $a + b$ balls. Now, the set Ω of the possible outcomes is the set of all the subsets of S with k elements. Thus, $|\Omega| = \binom{a+b}{n}$. A favorable outcome (an element of A_k) is a union $X \cup Y$ between a subset X of $\{x_1, \ldots, x_a\}$ with k elements and a subset Y of $\{y_1, \ldots, y_b\}$ with $n - k$ elements. Thus, using Remark 9.7(iii), the number of favorable outcomes is $|A_k| = \binom{a}{k}\binom{b}{n-k}$ and the probability of A_k is

$$P(A_k) = \frac{|A_k|}{|\Omega|} = \frac{\binom{a}{k}\binom{b}{n-k}}{\binom{a+b}{n}}. \tag{9.11}$$

We remark that, since $A_0, A_1, \ldots, A_n$ form a partition of Ω, we have:

$$1 = P(A_0 \cup \ldots \cup A_n) = \sum_{k=0}^{n} P(A_k) = \sum_{k=0}^{n} \frac{\binom{a}{k}\binom{b}{n-k}}{\binom{a+b}{n}},$$

so we obtain the identity

$$\sum_{k=0}^{n} \binom{a}{k}\binom{b}{n-k} = \binom{a+b}{n},$$

which can also be derived by equating the coefficients of x^n on the left and on the right side of the equality: $(1 + x)^a (1 + x)^b = (1 + x)^{a+b}$.

The model can be also generalized as in Example 9.6 for a urn containing t kinds of balls. We draw n balls (without replacement), $n < a_1 + \ldots + a_t$. Then, for any $0 \leq k_j \leq a_j$, $j = 1, \ldots, t$ such that $k_1 + \ldots + k_t = n$, the probability of the event $A_{n;k_1,\ldots,k_t}$ (from the n balls drawn, exactly k_j are of kind j, $j = 1, \ldots, t$) is

$$P(A_{n;k_1,\ldots,k_t}) = \frac{\binom{a_1}{k_1}\binom{a_2}{k_2}\cdots\binom{a_t}{k_t}}{\binom{a_1+\ldots+a_t}{n}}. \tag{9.12}$$

In the following we give some examples of probability spaces which are isomorphic to one of the standard probability spaces presented in Examples 9.6 and 9.7.

Example 9.8 A metal bar does not deform during a resistance trial (with replacement) with the probability $p = 0.96$ (this means that from 100 bars, approximately

4 bars does not resist). What is the probability of the event A: "from 5 trials, at least 4 bars resist"?

We consider a box with 96 white balls and 4 black balls and randomly draw (with replacement) 5 balls. Obviously, the probability to get a white ball (at a single selection) is $p = 0.96$. Thus, the probability space of this experiment is isomorphic with the probability space of our problem: a bar that does not deform during the trial corresponds to the selection of a white ball, while a bar which does not resist corresponds to a black ball drawn from the box.

We denote by A_4 the event that exactly 4 bars resist during 5 trials and by A_5 the event that all the 5 bars resist. So, $A = A_4 \cup A_5$ and $A_4 \cap A_5 = \emptyset$. We can apply the formula (9.9) and find:

$$P(A) = P(A_4) + P(A_5) = \binom{5}{4}(0.96)^4(0.04) + (0.96)^5 = 0.985.$$

Example 9.9 In a batch of 200 iron rods, 50 rods are oversized, 50 are undersized, and 100 are the right length. If two rods are drawn at random without replacement, what is the probability of obtaining:

(a) two rods of the right length,
(b) exactly one rod with the right length or oversized,
(c) no undersized rod,
(d) at least one rod of the right length.

This experiment is equivalent to the sampling without replacement from an urn containing three kinds of balls (see the general case of Example 9.7 for $t = 3$). We have $a_1 = 50$ balls of kind 1 (corresponding to oversized rods), $a_2 = 50$ balls of kind 2 (undersized rods) and $a_3 = 100$ balls of kind 3 (right-dimensioned rods). We draw $n = 2$ balls from the urn. Denote by $A_{n;k_1,k_2,k_3}$ the event that from n balls drawn (without replacement), exactly k_1 are of kind 1, k_2 are of kind 2 and $k_3 = n - k_1 - k_2$ are of kind 3.

(a) $P(A_{2;0,0,2}) = \dfrac{\binom{50}{0}\binom{50}{0}\binom{100}{2}}{\binom{200}{2}} = \dfrac{99}{398} = 0.249.$

(b) We have to find the probability of the event $A_{2;1,1,0} \cup A_{2;0,1,1}$:

$$P(A_{2;1,1,0} \cup A_{2;0,1,1}) = P(A_{2;1,1,0}) + P(A_{2;0,1,1}) =$$

$$= \dfrac{\binom{50}{1}\binom{50}{1}\binom{100}{0}}{\binom{200}{2}} + \dfrac{\binom{50}{0}\binom{50}{1}\binom{100}{1}}{\binom{200}{2}} = \dfrac{75}{199} = 0.377.$$

9.2 Probability Space 503

(c) In this case, our event is the (disjoint) union $A_{2;0,0,2} \cup A_{2;1,0,1} \cup A_{2;2,0,0}$:

$$\mathcal{P}(A_{2;0,0,2} \cup A_{2;1,0,1} \cup A_{2;2,0,0}) = \mathcal{P}(A_{2;0,0,2}) + \mathcal{P}(A_{2;1,0,1}) + \mathcal{P}(A_{2;2,0,0}) =$$

$$= \frac{\binom{50}{0}\binom{50}{0}\binom{100}{2}}{\binom{200}{2}} + \frac{\binom{50}{1}\binom{50}{0}\binom{100}{1}}{\binom{200}{2}} + \frac{\binom{50}{2}\binom{50}{0}\binom{100}{0}}{\binom{200}{2}} = \frac{447}{796} = 0.562.$$

(d) Now, we have to find the probability of the (disjoint) union $A_{2;1,0,1} \cup A_{2;0,1,1} \cup A_{2;0,0,2}$:

$$\mathcal{P}(A_{2;1,0,1} \cup A_{2;0,1,1} \cup A_{2;0,0,2}) = \mathcal{P}(A_{2;1,0,1}) + \mathcal{P}(A_{2;0,1,1}) + \mathcal{P}(A_{2;0,0,2}) =$$

$$= \frac{\binom{50}{1}\binom{50}{0}\binom{100}{1}}{\binom{200}{2}} + \frac{\binom{50}{0}\binom{50}{1}\binom{100}{1}}{\binom{200}{2}} + \frac{\binom{50}{0}\binom{50}{0}\binom{100}{2}}{\binom{200}{2}} = \frac{299}{398} = 0.751.$$

Example 9.10 (The Birthday Problem [11]) What is the probability that in a gathering of n people we can find two or more persons with the same birthday? Assume that a calendar year has 365 days and $n \leq 365$.

There are 365^n possible choices of birthdays for the set of n people.

Let A_n denote the event that two or more people have the same birthday. We calculate the probability of the complementary event $\overline{A_n}$: all the n birthdays are distinct. In this case, for the birthday of the first person we have 365 possible choices, for the second person we can choose anyone of the remaining 364 days, and so on, up to the last (n) person whose birthday can be anyone of the $366 - n$ days left after choosing the previous $n - 1$ distinct birthdays. This means that $|\overline{A_n}| = 365 \cdot 364 \cdot \ldots \cdot (366 - n)$ so we have:

$$\mathcal{P}(A_n) = 1 - \mathcal{P}(\overline{A_n}) = 1 - \frac{\frac{365!}{(365-n)!}}{365^n}.$$

For large numbers, the factorials can be approximated by the *Stirling's formula*:

$$k! \approx \exp(-k) k^{k+\frac{1}{2}} \sqrt{2\pi}, \qquad (9.13)$$

where the sign $\approx$ means that the ratio converges to 1 as $k \to \infty$. We apply this formula to approximate $q_n = \mathcal{P}(\overline{A_n})$:

$$q_n = \frac{365!}{(365-n)! \, 365^n} \approx \frac{\exp(-365) 365^{365+\frac{1}{2}}}{\exp(-365+n)(365-n)^{365-n+\frac{1}{2}} 365^n} =$$

$$= \exp(-n) \left(\frac{365}{365-n}\right)^{365-n+\frac{1}{2}}.$$

We can write:

$$\ln q_n \approx -n + \left(365 - n + \frac{1}{2}\right) \ln\left(1 + \frac{n}{365 - n}\right)$$

and use the approximation $\ln(1+x) \approx x - \frac{1}{2}x^2$ (for small values of x, the higher powers of x in the Taylor series can be neglected):

$$\ln q_n \approx -n + \left(365 - n + \frac{1}{2}\right)\left(\frac{n}{365-n} - \frac{n^2}{2(365-n)^2}\right) \approx -\frac{\binom{n}{2}}{365-n},$$

where we neglected also the term $\frac{n^2}{4(365-n)^2}$ because we assume that n is small compared to 365. This is why we can also write 365 instead of $365 - n$ at the denominator of the fraction, so we obtain that

$$p_n = 1 - q_n \approx 1 - \exp\left(-\frac{\binom{n}{2}}{365}\right).$$

We denote by $\tilde{p}_n = 1 - \exp\left(-\frac{\binom{n}{2}}{365}\right)$ this approximation of p_n. In Table 9.1 we present the real values p_n as well as the approximated values $\tilde{p}_n$. Notice that (see the bold values in Table 9.1), for $n = 23$ the probability p_n becomes greater than 0.5 (in a set of only 23 people it is more likely that two persons have the same birthday than all the people have different birthdays).

Example 9.11 (The Matching Problem [11]) At a party of n couples, the men are paired up at random with the women for a dance. What is the probability that at least one man dances with his wife?

Let us find a mathematical formulation of the problem: Suppose that the numbers $1, 2, \ldots, n$ are arranged at random on a line and $\pi(i)$ denotes the number placed on the ith position, for every $i = 1, 2, \ldots, n$. We have to find the probability that for at least one i, $\pi(i) = i$.

The set of all possible outcomes is $\Omega = S_n$, the set of all permutations of $\{1, 2, \ldots, n\}$. For every $i = 1, \ldots, n$, we define the event A_i as the set of all permutations $\pi \in S_n$ such that $\pi(i) = i$. We need to find the probability that at least one event A_i occurs, that is, the probability of the union $A_1 \cup \ldots \cup A_n$. To compute it, we use the inclusion-exclusion formula (9.5). Since $|A_i| = (n-1)!$, we

Table 9.1 The real values p_n and the approximated values $\tilde{p}_n$

n	5	10	15	20	23	30	35	40	50	60	70
p_n	0.027	0.117	0.253	0.411	**0.507**	0.706	0.814	0.891	0.97	0.994	0.999
$\tilde{p}_n$	0.027	0.116	0.25	0.406	**0.5**	0.696	0.804	0.882	0.965	0.992	0.999

9.2 Probability Space

have, for all $i = 1, \ldots, n$:

$$P(A_i) = \frac{|A_i|}{|S_n|} = \frac{(n-1)!}{n!} = \frac{1}{n}.$$

Now, since $|A_i \cap A_j| = (n-2)!$ for $1 \le i < j \le n$, the probability of the intersection is:

$$P(A_i \cap A_j) = \frac{(n-2)!}{n!}.$$

In general, for every $k = 1, 2, \ldots, n$ and $1 \le i_1 < i_2 < \ldots < i_k \le n$,

$$P(A_{i_1} \cap \ldots \cap A_{i_k}) = \frac{(n-k)!}{n!}$$

and we have $\binom{n}{k}$ such intersection of events. Thus, from (9.5) we get:

$$P(A_1 \cup \ldots \cup A_n) = \binom{n}{1}\frac{1}{n} - \binom{n}{2}\frac{(n-2)!}{n!} + \ldots + (-1)^{n-1}\frac{1}{n!}$$

$$= 1 - \frac{1}{2!} + \frac{1}{3!} - \ldots + \frac{(-1)^{n-1}}{n!}.$$

Thus, if we denote by p_n the probability that at least one number from the set $\{1, 2, \ldots, n\}$ is on its place ($\pi(i) = i$), we have:

$$p_n = P(A_1 \cup \ldots \cup A_n) = 1 - \left(\frac{1}{2!} - \frac{1}{3!} + \ldots + \frac{(-1)^n}{n!}\right). \tag{9.14}$$

Recall that the exponential function has the Taylor series expansion:

$$\exp(x) = 1 + \frac{x}{1!} + \frac{x^2}{2!} + \ldots + \frac{x^n}{n!} + \ldots.$$

Consequently, for $x = -1$ one can write:

$$\lim_{n \to \infty} \left(\frac{1}{2!} - \frac{1}{3!} + \ldots + \frac{(-1)^n}{n!}\right) = \exp(-1),$$

which means that, for large values of n, we can approximate:

$$\frac{1}{2!} - \frac{1}{3!} + \ldots + \frac{(-1)^n}{n!} \approx \frac{1}{e}.$$

Using this approximation in (9.14) we obtain:

$$p_n \approx 1 - \frac{1}{e} = 0.6321.$$

This is a good approximation, even for small values of n: for $n = 5$, the exact probability given by the formula (9.14) is $p_5 = 0.6333$, while for $n = 7$, the exact formula already gives $p_7 = 0.6321$.

9.3 Conditional Probability: Bayes Formula

The conditional probability is used in situations when additional information is available. Thus, when we know that a certain event A occurred, to find the probability of another event B, we should take the sample space A instead of Ω and $A \cap B$ instead of B.

Example 9.12 Two dice are rolled. What is the probability to obtain a sum greater than 10?

The sample space is the set of pairs $\Omega = \{1, 2, \ldots, 6\}^2$ and the event B of getting a sum greater than 10 is $B = \{(5, 6), (6, 5), (6, 6)\}$, so

$$\mathcal{P}(B) = \frac{|B|}{\Omega} = \frac{3}{36} = \frac{1}{12}.$$

But suppose that one die is *unfair*: it always comes up with 6. Knowing this, what is the probability to get a sum greater than 10?

In this case, we know that we can obtain only one of the following pairs:

$$A = \{(6, 1), (6, 2), (6, 3), (6, 4), (6, 5), (6, 6)\},$$

so the probability to obtain the sum at least 11 is $\frac{2}{6} = \frac{1}{3}$.

We call this updated probability the *conditional probability* of B given A and denote it $\mathcal{P}(B|A)$ or $\mathcal{P}_A(B)$.

Definition 9.7 Let $(\Omega, \mathcal{K}, \mathcal{P})$ be a probability space and let $A \in \mathcal{K}$ be an event with $\mathcal{P}(A) > 0$. Then, for any $B \in \mathcal{K}$, the *conditional probability* of B given A is defined as

$$\mathcal{P}_A(B) = \mathcal{P}(B|A) = \frac{\mathcal{P}(A \cap B)}{\mathcal{P}(A)}. \tag{9.15}$$

Proposition 9.5 *If $A \in \mathcal{K}$ is an event with $\mathcal{P}(A) > 0$, the function $\mathcal{P}_A : \mathcal{K} \to [0, 1]$, $\mathcal{P}_A(B) = \frac{\mathcal{P}(A \cap B)}{\mathcal{P}(A)}$ for all $B \subset \mathcal{K}$, is a probability function on $\mathcal{K}$ (the conditional probability function relative to A).*

9.3 Conditional Probability: Bayes Formula

Proof We have to prove that the function $\mathcal{P}_A$ has the two properties from Definition 9.5:

$$\mathcal{P}_A(\Omega) = \frac{\mathcal{P}(A \cap \Omega)}{\mathcal{P}(A)} = \frac{\mathcal{P}(A)}{\mathcal{P}(A)} = 1,$$

so the first condition is fulfilled.

Now, let Let $\{A_i\}_{i=1,2,\ldots}$ be a countable collection of disjoint events ($A_i \cap A_j = \emptyset$ $\forall i \neq j$). Then,

$$\mathcal{P}_A \left(\bigcup_{i \geq 1} A_i \right) = \frac{\mathcal{P}\left(A \cap (\bigcup_{i \geq 1} A_i)\right)}{\mathcal{P}(A)} = \frac{\mathcal{P}\left(\bigcup_{i \geq 1} A \cap A_i\right)}{\mathcal{P}(A)} = \sum_{i \geq 1} \frac{\mathcal{P}(A \cap A_i)}{\mathcal{P}(A)}. \tag{9.16}$$

Thus,

$$\mathcal{P}_A \left(\bigcup_{i \geq 1} A_i \right) = \sum_{i \geq 1} \mathcal{P}_A(A_i),$$

so the condition (ii) from Definition 9.5 is also satisfied. □

The next proposition follows at once from Definition 9.7:

Proposition 9.6 (The Multiplicative Rule) *Let $A, B \in \mathcal{K}$ be two events with $\mathcal{P}(A) > 0, \mathcal{P}(B) > 0$. Then,*

$$\mathcal{P}(A \cap B) = \mathcal{P}_A(B) \cdot \mathcal{P}(A) = \mathcal{P}_B(A) \cdot \mathcal{P}(B). \tag{9.17}$$

If $A, B \in \mathcal{K}$ are two events such that $\mathcal{P}(A) \neq 0$ and

$$\mathcal{P}_A(B) = \mathcal{P}(B), \tag{9.18}$$

then we remark that the information whether A has occurred has no influence upon the probability of B, so we can say that the events A and B are *independent*. Equation (9.18) and Proposition 9.6 lead us to the following definition of independent events:

Definition 9.8 Two events $A, B \in \mathcal{K}$ are called *independent* if

$$\mathcal{P}(A \cap B) = \mathcal{P}(A) \cdot \mathcal{P}(B) \tag{9.19}$$

In general, a collection of events $A_1, A_2, \ldots, A_n$ are said to be *independent* if, for any $k = 2, 3, \ldots, n$ and any k events $A_{i_1}, \ldots, A_{i_k}$, we have

$$\mathcal{P}(A_{i_1} \cap \ldots \cap A_{i_k}) = \mathcal{P}(A_{i_1}) \cdot \ldots \cdot \mathcal{P}(A_{i_k}). \tag{9.20}$$

If this property holds just for $k = 2$, they are called *pairwise independent*.

Proposition 9.7 *If $A, B \in \mathcal{K}$ are independent events, then the following pair of events are also independent:*

a) A and $\overline{B}$, b) $\overline{A}$ and B, c) $\overline{A}$ and $\overline{B}$.

Proof Note that it is sufficient to prove (a) (the other statements easily follows from this one).

Since $A = (A \cap B) \cup (A \cap \overline{B})$, using the independence of A and B, one can write:

$$\mathcal{P}\left(A \cap \overline{B}\right) = \mathcal{P}(A) - \mathcal{P}(A \cap B) = \mathcal{P}(A) - \mathcal{P}(A) \cdot \mathcal{P}(B)$$

$$= \mathcal{P}(A) \cdot (1 - \mathcal{P}(B)) = \mathcal{P}(A) \cdot \mathcal{P}\left(\overline{B}\right)$$

so A and $\overline{B}$ are independent. □

Example 9.13 This is the generalization of the "urn model" from Example 9.6. Thus, we consider n urns that contain black and white balls, in different proportions. For each $i = 1, 2, \ldots, n$, let p_i denote the probability to draw a white ball from the ith urn, and $q_i = 1 - p_i$, the probability to draw a black ball. (As discussed in Example 9.6, p_i is equal to the number of white balls divided by the total number of balls in the ith urn).

We randomly select a ball from each box. What is the probability of getting exactly k white balls ($k = 0, 1, \ldots, n$)?

Let A_i denote the event of sampling a white ball from the box i.

Our event is a disjoint union of all events of the form $E_1 \cap E_2 \cap \ldots \cap E_n$, where each event E_i is either A_i or $\overline{A}_i$ and we have exactly k events E_i such that $E_i = A_i$. Since the events $E_1, \ldots, E_n$ are independent, the probability $\mathcal{P}(E_1 \cap E_2 \cap \ldots \cap E_n)$ is equal to the product of probabilities $\mathcal{P}(E_1) \cdot \mathcal{P}(E_2) \cdot \ldots \cdot \mathcal{P}(E_n)$. It is not difficult to see that the sum of all these possible products is exactly the coefficient of the monomial x^k in the polynomial

$$(p_1 x + q_1)(p_2 x + q_2) \cdots (p_n x + q_n).$$

For $p_1 = p_2 = \ldots = p_n$ we obtain the sampling with replacement model from Example 9.6.

A practical application of this abstract model is given in the next example.

9.3 Conditional Probability: Bayes Formula

Example 9.14 Three factories F_1, F_2 and F_3 produce a type of "good" tires with the probabilities $p_1 = 0.99$, $p_2 = 0.98$ and $p_3 = 0.96$. Let us take 3 tires, one from each factory. What is the probability to have at least 2 "good" tires?

Following Example 9.13, this probability is the sum of the coefficients of x^2 and x^3 in the polynomial $(0.99x + 0.01)(0.98x + 0.02)(0.96x + 0.04)$, i.e.

$$0.99 \cdot 0.98 \cdot 0.04 + 0.99 \cdot 0.96 \cdot 0.02 + 0.98 \cdot 0.96 \cdot 0.01+$$

$$+0.99 \cdot 0.98 \cdot 0.96 = 0.999.$$

Let $(\Omega, \mathcal{K}, \mathcal{P})$ be a probability space and suppose that $C_1, \ldots, C_n \in \mathcal{K}$ with $\mathcal{P}(C_i) > 0$, $i = 1, \ldots, n$, form a partition of Ω (i.e. $\Omega = C_1 \cup \ldots \cup C_n$ and $C_i \cap C_j = \emptyset$ if $i \neq j$).

Given an event $X \in \mathcal{K}$ with known conditional probabilities $\mathcal{P}_{C_i}(X)$ and knowing all the probabilities $\mathcal{P}(C_i)$, how can we calculate the probability of X? The answer to this question is given by the next proposition.

Proposition 9.8 (Law of Total Probability) *Let $C_1, \ldots, C_n \in \mathcal{K}$ be a partition of Ω such that $\mathcal{P}(C_i) > 0$, for all $i = 1, \ldots, n$ and let $X \in \mathcal{K}$ be an arbitrary event. Then:*

$$\mathcal{P}(X) = \sum_{i=1}^{n} \mathcal{P}_{C_i}(X)\mathcal{P}(C_i). \tag{9.21}$$

Proof Since $X = \bigcup_{i=1}^{n}(X \cap C_i)$ and $(X \cap C_i) \cap (X \cap C_j) = \emptyset$ for $i \neq j$, one can write:

$$\mathcal{P}(X) = \sum_{i=1}^{n} \mathcal{P}(X \cap C_i). \tag{9.22}$$

But, $\mathcal{P}(X \cap C_i) = \mathcal{P}_{C_i}(X)\mathcal{P}(C_i)$ (see Proposition 9.6) for $i = 1, 2, \ldots, n$. Thus, we obtain the formula (9.21). □

Proposition 9.9 (Bayes' Rule) *Let $C_1, \ldots, C_n \in \mathcal{K}$ be a partition of Ω such that $\mathcal{P}(C_i) > 0$, for all $i = 1, \ldots, n$ and let $X \in \mathcal{K}$ be an arbitrary event. Then, for any $i = 1, \ldots, n$, we have:*

$$\mathcal{P}_X(C_i) = \frac{\mathcal{P}_{C_i}(X)\mathcal{P}(C_i)}{\sum_{j=1}^{n} \mathcal{P}_{C_j}(X)\mathcal{P}(C_j)}. \tag{9.23}$$

Proof Since $\mathcal{P}_X(C_i) = \frac{\mathcal{P}(X \cap C_i)}{\mathcal{P}(X)}$, and $\mathcal{P}(X \cap C_i) = \mathcal{P}_{C_i}(X)\mathcal{P}(C_i)$ (see (9.17)), by applying the formula (9.21) for $\mathcal{P}(X)$, we obtain (9.23). □

We present in the following a frequent application of the Bayes' rule: Suppose that a certain device may be defective because of the disjoint events $C_1, C_2, \ldots, C_n$ with the probabilities $p_1, p_2, \ldots, p_n$ respectively (if D is the event that our device is defective, than $p_i = \mathcal{P}_{C_i}(D)$). We know that the frequencies of the occurrence of $C_1, C_2, \ldots, C_n$ are $f_1, f_2, \ldots, f_n$. Knowing that the device failed, the probability that its failure occurred due to the event C_i ($i = 1, 2, \ldots, n$) is:

$$\mathcal{P}_D(C_i) = \frac{p_i f_i}{p_1 f_1 + p_2 f_2 + \ldots + p_n f_n}. \tag{9.24}$$

Example 9.15 Three factories F_1, F_2, F_3 produce some items and we know that 1%, 2% and 4%, respectively, of their products are defective. They ensure the supplying of a company F in proportion of 20%, 30% and 50% respectively. What is the probability that an item delivered by the company F is defective?

Knowing that an item is defective, what is the probability that it was produced by the factory F_3? Prove that it is ten times greater than the probability that the defective item was produced by F_1.

Let X denote the event that the item is defective and C_i denote the event that it was produced by the factory F_i, $i = 1, 2, 3$. We also denote by p_i the probability that an item produced by the factory F_1 is defective: $p_i = \mathcal{P}_{C_i}(X), i = 1, 2, 3$. From the hypothesis we know that $p_1 = 0.01$, $p_2 = 0.02$ and $p_3 = 0.04$. On the other hand, if $f_i = \mathcal{P}(C_i)$ denotes the probability that an item delivered by the company F was produced by the factory F_i, then $f_1 = 0.2$, $f_2 = 0.3$ and $f_3 = 0.5$.

By applying the total probability formula (9.21) we find

$$\mathcal{P}(X) = \sum_{i=1}^{3} \mathcal{P}_{C_i}(X)\mathcal{P}(C_i) = p_1 f_1 + p_2 f_2 + p_3 f_3 =$$

$$= 0.01 \cdot 0.2 + 0.02 \cdot 0.3 + 0.04 \cdot 0.5 = 0.028,$$

so we conclude that 2.8% of the items delivered by the company F are defective.

For the second question, we need to calculate the conditional probability $\mathcal{P}_X(C_3)$. We apply the Bayes' formula (9.23), but we already computed the denominator, $\mathcal{P}(X)$, so we can write:

$$\mathcal{P}_X(C_3) = \frac{\mathcal{P}_{C_3}(X)\mathcal{P}(C_3)}{\mathcal{P}(X)} = \frac{0.04 \cdot 0.5}{0.028} = \frac{10}{14} = 0.714.$$

Now, the probability that the defective item was produced by F_1 is

$$\mathcal{P}_X(C_1) = \frac{\mathcal{P}_{C_1}(X)\mathcal{P}(C_1)}{\mathcal{P}(X)} = \frac{0.01 \cdot 0.2}{0.028} = \frac{1}{14} = 0.0714.$$

Example 9.16 (Inaccurate Blood Tests [11]) A certain blood test for a disease gives a positive result for 90% of the patients having the disease (we say that the *sensitivity* of the test is 90%). But it also gives a positive result for 25% of the patients who do not have the disease (the test has a *specificity* of 75%). It is believed that 30% of the population has this disease. What is the probability that a person with a positive test result indeed has the disease?

Let A denote the event that the person has the disease and B denote the event that the result of their blood test is positive. We know that $\mathcal{P}(A) = 0.3$, $\mathcal{P}_A(B) = 0.9$ and $\mathcal{P}_{\overline{A}}(B) = 0.25$. To compute the conditional probability $\mathcal{P}_B(A)$, we apply the Bayes' formula for the partition formed by A and $\overline{A}$:

$$\mathcal{P}_B(A) = \frac{\mathcal{P}_A(B)\mathcal{P}(A)}{\mathcal{P}_A(B)\mathcal{P}(A) + \mathcal{P}_{\overline{A}}(B)\mathcal{P}(\overline{A})} =$$

$$= \frac{0.9 \cdot 0.3}{0.9 \cdot 0.3 + 0.25 \cdot 0.7} = 0.607.$$

9.4 Discrete Random Variables

9.4.1 Random Variables

Let $(\Omega, \mathcal{K}, \mathcal{P})$ be the probability space associated to a random experiment. Most of the time, we are not interested in each outcome of the experiment, but only in the value of some numerical quantity determined by the outcome. This quantity of interest determined by the results of an experiment is said to be a *random variable*. Thus, a random variable is a real-valued function defined on the sample space Ω, $X : \Omega \to \mathbb{R}$. We shall see in Definition 9.9 the specific properties of such a function. Now let us present some examples.

Example 9.17 Consider the experiment of tossing three fair coins. We denote the possible outcomes of this experiment by 3-letter strings:

$$\Omega = \{HHH, HHT, HTH, HTT, THH, THT, TTH, TTT\}.$$

If we let X denote the number of heads appearing, then X is a random variable that takes on one of the values 0, 1, 2, and 3. Thus, we have:

$$X(TTT) = 0, \ X(HTT) = X(THT) = X(TTH) = 1,$$

$$X(HHT) = X(HTH) = X(THH) = 2, \ X(HHH) = 3.$$

For $k = 0, 1, 2, 3$ we denote by $\{X = k\}$ the event that exactly k heads appear:

$$\{X = k\} = \{\omega \in \Omega : X(\omega) = k\}. \tag{9.25}$$

Let us calculate the probabilities of these events:

$$\{X = 0\} = \{TTT\} \Rightarrow P\{X = 0\} = \tfrac{1}{8},$$
$$\{X = 1\} = \{HTT, THT, TTH\} \Rightarrow P\{X = 1\} = \tfrac{3}{8},$$
$$\{X = 2\} = \{HHT, HTH, THH\} \Rightarrow P\{X = 2\} = \tfrac{3}{8},$$
$$\{X = 3\} = \{HHH\} \Rightarrow P\{X = 3\} = \tfrac{1}{8}.$$

Since the random variable X takes on one of the values 0, 1, 2, 3, we must have

$$1 = P\left(\bigcup_{k=0}^{3} \{X = k\}\right) = \sum_{k=0}^{3} P\{X = k\},$$

which can be easily checked.

Example 9.18 Consider the experiment that consists in rolling two fair dice. We denote the possible outcomes of this experiment by two-digit numbers (we have 36 possible outcomes):

$$\Omega = \{11, 12, 13, \ldots, \ldots, 65, 66\}.$$

We define two random variables on the sample space Ω: Y, which denotes the sum the two dice, and Z, which denotes the absolute value of their difference. Thus, Y may take on any value from 2 to 12, and Z takes on the values $0, 1, \ldots, 5$. Let us calculate the probability of each event $\{Y = k\}$, $k = 2, \ldots, 12$, and $\{Z = j\}$, $j = 0, \ldots, 5$ respectively. We have:

$$\{Y = 2\} = \{11\} \Rightarrow P\{Y = 2\} = \tfrac{1}{36},$$
$$\{Y = 3\} = \{12, 21\} \Rightarrow P\{Y = 3\} = \tfrac{2}{36},$$
$$\{Y = 4\} = \{13, 22, 31\} \Rightarrow P\{Y = 4\} = \tfrac{3}{36},$$
$$\{Y = 5\} = \{14, 23, 32, 41\} \Rightarrow P\{Y = 5\} = \tfrac{4}{36},$$
$$\{Y = 6\} = \{15, 24, 33, 42, 51\} \Rightarrow P\{Y = 6\} = \tfrac{5}{36},$$
$$\{Y = 7\} = \{16, 25, 34, 43, 52, 61\} \Rightarrow P\{Y = 7\} = \tfrac{6}{36},$$
$$\{Y = 8\} = \{26, 35, 44, 53, 62\} \Rightarrow P\{Y = 8\} = \tfrac{5}{36},$$
$$\{Y = 9\} = \{36, 45, 54, 63\} \Rightarrow P\{Y = 9\} = \tfrac{4}{36},$$

9.4 Discrete Random Variables

$$\{Y = 10\} = \{46, 55, 64\} \Rightarrow \mathcal{P}\{Y = 10\} = \tfrac{3}{36},$$
$$\{Y = 11\} = \{56, 65\} \Rightarrow \mathcal{P}\{Y = 11\} = \tfrac{2}{36},$$
$$\{Y = 12\} = \{66\} \Rightarrow \mathcal{P}\{Y = 12\} = \tfrac{1}{36}.$$

and

$$\{Z = 0\} = \{11, 22, 33, 44, 55, 66\} \Rightarrow \mathcal{P}\{Z = 0\} = \tfrac{6}{36},$$
$$\{Z = 1\} = \{12, 21, 23, 32, 34, 43, 45, 54, 56, 65\} \Rightarrow \mathcal{P}\{Z = 1\} = \tfrac{10}{36},$$
$$\{Z = 2\} = \{13, 31, 24, 42, 35, 53, 46, 64\} \Rightarrow \mathcal{P}\{Z = 2\} = \tfrac{8}{36},$$
$$\{Z = 3\} = \{14, 41, 25, 52, 36, 63\} \Rightarrow \mathcal{P}\{Z = 3\} = \tfrac{6}{36},$$
$$\{Z = 4\} = \{15, 51, 26, 62\} \Rightarrow \mathcal{P}\{Z = 4\} = \tfrac{4}{36},$$
$$\{Z = 5\} = \{16, 61\} \Rightarrow \mathcal{P}\{Z = 5\} = \tfrac{2}{36}.$$

It can be easily verified that

$$1 = \mathcal{P}\left(\bigcup_{k=2}^{12}\{Y = k\}\right) = \sum_{k=2}^{12} \mathcal{P}\{Y = k\},$$

and

$$1 = \mathcal{P}\left(\bigcup_{j=0}^{5}\{Z = j\}\right) = \sum_{j=0}^{5} \mathcal{P}\{Z = j\}.$$

Example 9.19 Consider the experiment of tossing a (biased) coin until the first head appears. Let p denote the probability of coming heads and let X denote the number of times the coin is tossed. Then X is a random variable which takes on the values $1, 2, \ldots$ with the following probabilities:

$$\mathcal{P}\{X = 1\} = \mathcal{P}\{H\} = p,$$
$$\mathcal{P}\{X = 2\} = \mathcal{P}\{TH\} = (1-p)p,$$
$$\mathcal{P}\{X = 3\} = \mathcal{P}\{TTH\} = (1-p)^2 p,$$
$$\vdots$$

$$P\{X = n\} = P\{\underbrace{T \ldots T}_{n-1} H\} = (1-p)^{n-1} p,$$

$\vdots$

Note that the following series is convergent and its sum equals 1:

$$P\left(\bigcup_{n=1}^{\infty} \{X = n\}\right) = \sum_{n=1}^{\infty} (1-p)^{n-1} p = \frac{p}{1-(1-p)} = 1.$$

Definition 9.9 Let $(\Omega, \mathcal{K}, \mathcal{P})$ be a probability space. A function $X : \Omega \to \mathbb{R}$ is said to be a *random variable* if and only if $\{X \leq x\} \in \mathcal{K}$, for any $x \in \mathbb{R}$, where $\{X \leq x\}$ denotes the subset of Ω defined by:

$$\{X \leq x\} = \{\omega \in \Omega : X(\omega) \leq x\} = X^{-1}(-\infty, x].$$

If the set of possible values of the random variable is either finite (as in Examples 9.17 and 9.18) or infinite but countable (as in Example 9.19), then it is said to be a *discrete random variable*. Otherwise, if the set of values is an interval, or a union of intervals, or the whole set of real numbers $\mathbb{R}$, the random variable is said to be *continuous*. For instance, the lifetime of an electronic device is a continuous random variable assumed to take on values in some interval (a, b). The mass, the energy or the temperature of a physical system can also be considered as continuous random variables.

For both types of random variables we can define the *cumulative distribution function* (or more simply the *distribution function*) F.

Definition 9.10 The *(cumulative) distribution function* of the random variable X is the function $F : \mathbb{R} \to [0, 1]$ defined by:

$$F(x) = \mathcal{P}\{X \leq x\}, \quad \forall x \in \mathbb{R}. \tag{9.26}$$

In order to prove the properties of the distribution function we need the following lemma (see [31]):

Lemma 9.1 *If* $\{A_n\}_{n \geq 1}$, *is an increasing sequence of events in the probability space* $(\Omega, \mathcal{K}, \mathcal{P})$, *that is,*

$$A_1 \subset A_2 \subset \cdots \subset A_n \subset \ldots,$$

9.4 Discrete Random Variables

then,

$$\lim_{n \to \infty} \mathcal{P}(A_n) = \mathcal{P}\left(\bigcup_{n \geq 1} A_n\right). \tag{9.27}$$

If $\{B_n\}_{n \geq 1}$ is a decreasing sequence of events, that is,

$$B_1 \supset B_2 \supset \cdots \supset B_n \supset \ldots,$$

then

$$\lim_{n \to \infty} \mathcal{P}(B_n) = \mathcal{P}\left(\bigcap_{n \geq 1} B_n\right). \tag{9.28}$$

Proof If $\{A_n\}_{n \geq 1}$ is an increasing sequence of events, then we can define the following sequence of disjoint events:

$$C_1 = A_1, \ C_2 = A_2 \setminus A_1, \ldots, C_n = A_n \setminus A_{n-1}, \ldots.$$

Since $\bigcup_{n \geq 1} A_n = \bigcup_{n \geq 1} C_n$, one can write:

$$\mathcal{P}\left(\bigcup_{n \geq 1} A_n\right) = \mathcal{P}\left(\bigcup_{n \geq 1} C_n\right) = \sum_{n=1}^{\infty} \mathcal{P}(C_n).$$

The partial sums of the series $\sum_{n=1}^{\infty} \mathcal{P}(C_n)$ can be written

$$\sum_{k=1}^{n} \mathcal{P}(C_k) = \mathcal{P}(A_1) + \mathcal{P}(A_2) - \mathcal{P}(A_1) + \ldots + \mathcal{P}(A_n) - \mathcal{P}(A_{n-1}) = \mathcal{P}(A_n),$$

so we find that

$$\mathcal{P}\left(\bigcup_{n \geq 1} A_n\right) = \sum_{n=1}^{\infty} \mathcal{P}(C_n) = \lim_{n \to \infty} \mathcal{P}(A_n).$$

For the second statement, we denote $A_n = \overline{B_n}$, $n = 1, 2, \ldots$. Since $\{B_n\}_{n \geq 1}$ is a decreasing sequence of events, it follows that $\{A_n\}_{n \geq 1}$ is an increasing sequence of events, so we can apply (9.27):

$$\lim_{n \to \infty} \mathcal{P}\left(\overline{B_n}\right) = \mathcal{P}\left(\bigcup_{n \geq 1} \overline{B_n}\right).$$

Using De Morgan's laws (Proposition 9.1), the equality above can be written:

$$1 - \lim_{n \to \infty} \mathcal{P}(B_n) = \mathcal{P}\left(\overline{\bigcap_{n \geq 1} B_n}\right) = 1 - \mathcal{P}\left(\bigcap_{n \geq 1} B_n\right)$$

and (9.28) follows. □

Theorem 9.1 *If $F : \mathbb{R} \to [0, 1]$ is the distribution function of a random variable X, then:*

(i) *For any $a < b$, we have:*

$$\mathcal{P}\{a < X \leq b\} = F(b) - F(a). \tag{9.29}$$

(ii) *F is an increasing function, that is,*

$$F(x) \leq F(y), \quad \forall x < y.$$

(iii) *F is right-continuous, that is, for any $a \in \mathbb{R}$,*

$$\lim_{x \searrow a} F(x) = F(a).$$

(iv) $\lim_{x \to \infty} F(x) = 1, \quad \lim_{x \to -\infty} F(x) = 0.$

Proof

(i) Since $\{X \leq b\} = \{X \leq a\} \cup \{a < X \leq b\}$ and since the events $\{X \leq a\}$ and $\{a < X \leq b\}$ are disjoint, we have:

$$\mathcal{P}\{X \leq b\} = \mathcal{P}\{X \leq a\} + \mathcal{P}\{a < X \leq b\},$$

and the relation (9.29) follows by the definition (9.26) of $F(x)$.

(ii) Let $x < y$. Since the probability of an event is always nonnegative, it follows by (9.29) that

$$F(y) - F(x) = \mathcal{P}\{x < X \leq y\} \geq 0.$$

(iii) Let $\{x_n\}_{n \geq 1}$ be a sequence decreasing to a. Then $\{X \leq x_n\}$, $n = 1, 2, \ldots$ is a decreasing sequence of events and $\bigcap_{n \geq 1} \{X \leq x_n\} = \{X \leq a\}$. By Lemma 9.1 we have:

$$\lim_{n \to \infty} F(x_n) = \lim_{n \to \infty} \mathcal{P}\{X \leq x_n\} = \mathcal{P}\{X \leq a\} = F(a).$$

(iv) Let $\{x_n\}_{n \geq 1}$ be a sequence increasing to ∞. Then $\{X \leq x_n\}$, $n = 1, 2, \ldots$ is an increasing sequence of events and $\bigcup_{n \geq 1} \{X \leq x_n\} = \Omega$. By applying Lemma 9.1

9.4 Discrete Random Variables

we obtain:

$$\lim_{n\to\infty} F(x_n) = \lim_{n\to\infty} \mathcal{P}\{X \leq x_n\} = \mathcal{P}(\Omega) = 1.$$

In a similar way, we consider $\{x_n\}_{n\geq 1}$ a sequence decreasing to $-\infty$. Then $\{X \leq x_n\}$, $n = 1, 2, \ldots$ is a decreasing sequence of events and we have: $\bigcap_{n\geq 1} \{X \leq x_n\} = \emptyset$. By Lemma 9.1 we can write:

$$\lim_{n\to\infty} F(x_n) = \lim_{n\to\infty} \mathcal{P}\{X \leq x_n\} = \mathcal{P}(\emptyset) = 0.$$

$\square$

The rest of this section deals mainly with discrete random variables. Continuous random variables will be studied in Sect. 9.5. *Simple* random variables are a special type of discrete random variables having a *finite* set of possible values.

Definition 9.11 Let $X : \Omega \to \mathbb{R}$ be a discrete random variable in the probability space $(\Omega, \mathcal{K}, \mathcal{P})$ and let $X(\Omega) = \{x_1, x_2, x_3, \ldots\}$ be the set of values of X (a finite or at most countable set). The function $f(x)$ defined by

$$f(x_i) = \mathcal{P}\{X = x_i\} = p_i, \quad i = 1, 2, \ldots$$
$$f(x) = 0, \text{ all other values of } x$$

is called the *(probability) distribution* or the *probability mass function* (pmf) of the random variable X.

A discrete random variable X taking on the values $x_1 < x_2 < \ldots$ with the probabilities $p_i = \mathcal{P}\{X = x_i\} = f(x_i)$, $i = 1, 2, \ldots$ can be represented in the following form:

$$X \sim \begin{pmatrix} x_1 & x_2 & \ldots & x_n & \ldots \\ p_1 & p_2 & \ldots & p_n & \ldots \end{pmatrix}. \tag{9.30}$$

Note that $p_i \in [0, 1]$, for all $i = 1, 2, \ldots$ and, since the events $\{X = x_i\}$, $i = 1, 2, \ldots$ form a partition of Ω, we must have

$$\sum_{i\geq 1} p_i = \sum_{i\geq 1} f(x_i) = 1. \tag{9.31}$$

We also remark that the distribution function $F(x)$ of a discrete random variable is a step function: the value of $F(x)$ is constant in every interval $[x_{i-1}, x_i)$ and takes a step of size $p_i = f(x_i)$ at the point x_i, $i = 1, 2, \ldots$:

$$F(x) = \mathcal{P}\{X \le x\} = \begin{cases} 0, & \text{if } x \in (-\infty, x_1) \\ p_1, & \text{if } x \in [x_1, x_2) \\ p_1 + p_2, & \text{if } x \in [x_2, x_3) \\ p_1 + p_2 + p_3, & \text{if } x \in [x_3, x_4) \\ \vdots \\ p_1 + p_2 + \ldots + p_{n-1}, & \text{if } x \in [x_{n-1}, x_n) \\ \vdots \end{cases} \quad (9.32)$$

Example 9.20 Consider again the discrete random variable from Example 9.17. Its probability distribution (probability mass function) is the function $f(x)$ defined by:

$$f(0) = \frac{1}{8}, \quad f(1) = \frac{3}{8}, \quad f(2) = \frac{3}{8}, \quad f(3) = \frac{1}{8}$$

(and $f(x) = 0$ for any other values of x). Thus, we can write:

$$X \sim \begin{pmatrix} 0 & 1 & 2 & 3 \\ \frac{1}{8} & \frac{3}{8} & \frac{3}{8} & \frac{1}{8} \end{pmatrix}.$$

Now, using (9.32), let us calculate the distribution function, $F(x)$:

$$F(x) = \mathcal{P}\{X \le x\} = \begin{cases} 0, & \text{if } x \in (-\infty, 0) \\ 1/8, & \text{if } x \in [0, 1) \\ 1/2, & \text{if } x \in [1, 2) \\ 7/8, & \text{if } x \in [2, 3) \\ 1, & \text{if } x \in [3, \infty). \end{cases}$$

The graph of the probability distribution $f(x)$ and the graph of the (cumulative) distribution function $F(x)$ are presented in Fig. 9.4.

9.4.2 Expected Value; Moments

Definition 9.12 Let X be a discrete random variable taking on the possible values $x_1, x_2, \ldots$ with the probabilities $f(x_1), f(x_2), \ldots$. The *expected value* of X (also known as the *mean* or the *expectation* of X) is the number $\mu = E(x)$ defined by

$$E(X) = \sum_{k \ge 1} x_k f(x_k) \quad (9.33)$$

9.4 Discrete Random Variables

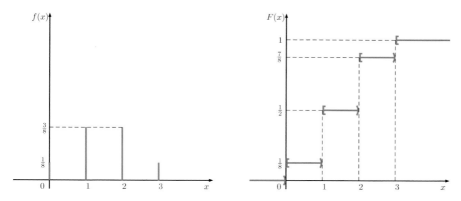

Fig. 9.4 The probability mass function $f(x)$ and the cumulative distribution function $F(x)$

provided that the series converges absolutely. In this case we say that X has a finite expectation. If the series $\sum_{k \geq 1} |x_k| f(x_k)$ is divergent, then we say that X has no finite expectation.

For instance, the expected value of the random variable X from Example 9.20 is

$$\mu = E(X) = 0 \cdot \frac{1}{8} + 1 \cdot \frac{3}{8} + 2 \cdot \frac{3}{8} + 3 \cdot \frac{1}{8} = \frac{12}{8} = 1.5.$$

Example 9.21 A simple random variable is said to be a *uniform discrete random variable* if all the values are equally probable, that is, $p_1 = p_2 = \ldots = p_n = \frac{1}{n}$. For instance, when rolling a fair die, the associated random variable is uniform:

$$X \sim \begin{pmatrix} 1 & 2 & 3 & 4 & 5 & 6 \\ \frac{1}{6} & \frac{1}{6} & \frac{1}{6} & \frac{1}{6} & \frac{1}{6} & \frac{1}{6} \end{pmatrix}$$

The expected value of X is

$$E(X) = \sum_{k=1}^{6} k \cdot \frac{1}{6} = 3.5.$$

If $X : \Omega \to \mathbb{R}$ is a discrete random variable with the set of values $X(\Omega) = \{x_1, x_2, \ldots\}$ and the probability distribution $f(x_i) = p_i, i = 1, 2, \ldots$, and $g : \mathbb{R} \to \mathbb{R}$ is a real-valued function, then $g(X) = g \circ X : \Omega \to \mathbb{R}$ is also a discrete random variable. If $g(x_i) \neq g(x_j)$ for any different numbers $x_i, x_j \in X(\Omega)$, then the probability distribution of $g(X)$ is written:

$$g(X) \sim \begin{pmatrix} g(x_1) & g(x_2) & \ldots & g(x_n) & \ldots \\ p_1 & p_2 & \ldots & p_n & \ldots \end{pmatrix}. \tag{9.34}$$

Otherwise, we denote by $y_1, y_2, \ldots$ the values of $g(X)$ and the probability distribution of $g(X)$ is written

$$g(X) \sim \begin{pmatrix} y_1 & y_2 & \ldots & y_n & \ldots \\ q_1 & q_2 & \ldots & q_n & \ldots \end{pmatrix}, \tag{9.35}$$

where

$$q_j = \mathcal{P}\{g(x) = y_j\} = \sum_{i: g(x_i) = y_j} p_i, \quad j = 1, 2, \ldots.$$

Suppose, for example, that $g(x) = ax + b$, for all $x \in \mathbb{R}$, where $a \neq 0$ and b are arbitrary constants. Since g is an injective function, the probability distribution of $g(X) = aX + b$ can be written:

$$aX + b \sim \begin{pmatrix} ax_1 + b & ax_2 + b & \ldots & ax_n + b & \ldots \\ p_1 & p_2 & \ldots & p_n & \ldots \end{pmatrix}. \tag{9.36}$$

It can be easily proved that, if X has a finite expected value, $E(X)$, then

$$E(aX + b) = aE(X) + b. \tag{9.37}$$

By the definition formula (9.33), we can write:

$$E(aX + b) = \sum_{k \geq 1} (ax_k + b) p_k = a \sum_{k \geq 1} x_k p_k + b \sum_{k \geq 1} p_k = aE(X) + b.$$

Example 9.22 Let X be a random variable taking on the integer values from -2 to 2 with equal probabilities:

$$X \sim \begin{pmatrix} -2 & -1 & 0 & 1 & 2 \\ \frac{1}{5} & \frac{1}{5} & \frac{1}{5} & \frac{1}{5} & \frac{1}{5} \end{pmatrix}.$$

Obviously, $E(X) = 0$.

Consider the function $g(x) = x^2$, $x \in \mathbb{R}$. The random variable $g(X) = X^2$ takes on the values $0, 1, 4$ with the following probabilities:

$$X^2 \sim \begin{pmatrix} 0 & 1 & 4 \\ \frac{1}{5} & \frac{2}{5} & \frac{2}{5} \end{pmatrix}.$$

9.4 Discrete Random Variables

The expected value of X^2 is

$$E(X^2) = 0 \cdot \frac{1}{5} + 1 \cdot \frac{2}{5} + 4 \cdot \frac{2}{5} = 2$$

and it can be also computed in the following way:

$$E(X^2) = (-2)^2 \cdot \frac{1}{5} + (-1)^2 \cdot \frac{1}{5} + 0^2 \cdot \frac{1}{5} + 1^2 \cdot \frac{1}{5} + 2^2 \cdot \frac{1}{5} = 2$$

Proposition 9.10 *If X is a discrete random variable taking on the values x_i with the probabilities p_i, $i = 1, 2, \ldots$, and g is a real-valued function such that the series $\sum_{k \geq 1} |g(x_k)| p_k$ converges, then the random variable $g(X)$ has a finite expected value, given by the formula*

$$E(g(X)) = \sum_{k \geq 1} g(x_k) p_k. \tag{9.38}$$

Proof If $g(x_i) \neq g(x_j)$ for any $i \neq j$, the formula follows instantly by (9.34).

If g is not one-to-one on the set $\{x_1, x_2, \ldots\}$ then the distribution of $g(X)$ is given by (9.35). We group together the terms in $\sum_{k \geq 1} g(x_k) p_k$ with the same value $g(x_k)$:

$$\sum_{k \geq 1} g(x_k) p_k = \sum_{j \geq 1} \sum_{k: g(x_k) = y_j} g(x_k) p_k = \sum_{j \geq 1} y_j \sum_{k: g(x_k) = y_j} p_k$$

$$= \sum_{j \geq 1} y_j q_j = E(g(X)).$$

□

Definition 9.13 Let X be a discrete random variable and $n \geq 1$ a positive integer. The expected value of X^n (if exists) is called the *n-th moment* of X. (Note that the first moment is the expected value of X). By Proposition 9.10, if X takes on the values x_i with the probabilities p_i, $i = 1, 2, \ldots$, then the n-th moment of X is

$$E(X^n) = \sum_{i \geq 1} x_i^n p_i, \tag{9.39}$$

provided that the series converges absolutely. If the series is not absolutely convergent, the n-th moment of X does not exist.

Proposition 9.11 *If the n-th moment of X exists, then the k-th moment exists, for all $k = 1, 2, \ldots, n - 1$.*

Proof For any $x \in \mathbb{R}$ and $k = 1, 2, \ldots, n - 1$ the following inequality holds:

$$|x|^k \leq |x|^n + 1$$

(if $|x| > 1$, then $|x|^k < |x|^n$; otherwise, $|x|^k \leq 1$).
If the n-th moment of X exists, since $\sum_{i \geq 1} p_i = 1$, the series

$$\sum_{i \geq 1} \left(|x_i|^n + 1\right) p_i$$

converges. Therefore, by the inequality above, the series $\sum_{i \geq 1} |x_i|^k p_i$ is also convergent, hence the k-th moment of X exists, for all $k = 1, 2, \ldots, n - 1$. □

Example 9.23 Let $X \sim \begin{pmatrix} -1 & 0 & 1 \\ p_1 & p_2 & p_3 \end{pmatrix}$ be a random variable.

(a) Find p_1, p_2, p_3 knowing that $M(X) = -1/4$ and $M_2(X) = 3/4$.
(b) For what values of $\mu_1 = M(X)$ and $\mu_2 = M_2(X)$ there exists such a random variable X?

(a) We have to solve the following linear system:

$$\begin{cases} p_1 + p_2 + p_3 = 1 \\ -p_1 + p_3 = -1/4 \\ p_1 + p_3 = 3/4 \end{cases}.$$

From the last two equations we easily find $p_1 = 1/2$ and $p_3 = 1/4$. Now, from the first equation we find $p_2 = 1/4$.

(b) The linear system

$$\begin{cases} p_1 + p_2 + p_3 = 1 \\ -p_1 + p_3 = \mu_1 \\ p_1 + p_3 = \mu_2 \end{cases}.$$

has always a unique solution, for any values of μ_1 and μ_2. However, we should find the values of μ_1 and μ_2 for which $p_1, p_2, p_3 \in (0, 1)$.

From the last two equations we get: $p_1 = \frac{\mu_2 - \mu_1}{2}$, $p_3 = \frac{\mu_2 + \mu_1}{2}$ and from the first equation $p_2 = 1 - \mu_2$. The conditions $p_1, p_2, p_3 \in (0, 1)$ give rise to the constraints:

$$\begin{cases} 0 < \mu_2 - \mu_1 < 2 \\ 0 < \mu_2 + \mu_1 < 2 \\ 0 < \mu_2 < 1. \end{cases} \tag{9.40}$$

9.4 Discrete Random Variables

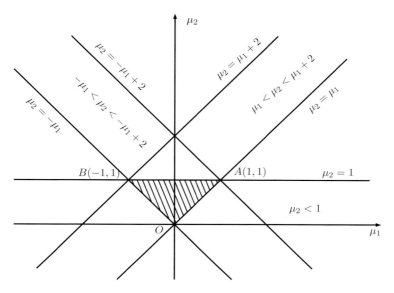

Fig. 9.5 The set of the points (μ_1, μ_2) defined by the constraints (9.40)

It is not difficult to see (in Fig. 9.5) that the available region for a point (μ_1, μ_2), in the Cartesian plane $\mu_1 O \mu_2$, is the interior of the triangle OAB, where $O(0, 0)$, $A(1, 1)$ and $B(-1, 1)$.

Definition 9.14 Let $X, Y : \Omega \to \mathbb{R}$ be two discrete random variables in the probability space $(\Omega, \mathcal{K}, \mathcal{P})$ taking on the values $x_1, x_2, \ldots$ and $y_1, y_2, \ldots$ respectively. The *joint probability distribution* of X and Y is defined by:

$$r_{ij} = \mathcal{P}\{X = x_i, Y = y_j\} = \mathcal{P}(A_i \cap B_j), \ i, j = 1, 2, \ldots,$$

where A_i and B_j are the events defined by:

$$A_i = \{X = x_i\} = \{\omega \in \Omega : X(\omega) = x_i\}, \ i = 1, 2, \ldots$$
$$B_j = \{Y = y_j\} = \{\omega \in \Omega : Y(\omega) = y_j\}, \ j = 1, 2, \ldots.$$

Suppose that $\mathcal{P}(A_i) = p_i$ and $\mathcal{P}(B_j) = q_j$, for every $i, j = 1, 2, \ldots$, in other words, X and Y have the distributions:

$$X \sim \begin{pmatrix} x_1 & x_2 & \ldots & x_n & \ldots \\ p_1 & p_2 & \ldots & p_n & \ldots \end{pmatrix}, \ Y \sim \begin{pmatrix} y_1 & y_2 & \ldots & y_n & \ldots \\ q_1 & q_2 & \ldots & q_n & \ldots \end{pmatrix}.$$

Since $\Omega = \bigcup_{i\geq 1} A_i = \bigcup_{j\geq 1} B_j$, we can see that

$$\sum_{j\geq 1} r_{ij} = \mathcal{P}\left(\bigcup_{j\geq 1}(A_i \cap B_j)\right) = \mathcal{P}\left(A_i \cap \left(\bigcup_{j\geq 1} B_j\right)\right) = \mathcal{P}(A_i) = p_i, \quad (9.41)$$

and

$$\sum_{i\geq 1} r_{ij} = \mathcal{P}\left(\bigcup_{i\geq 1}(A_i \cap B_j)\right) = \mathcal{P}\left(B_j \cap \left(\bigcup_{i\geq 1} A_i\right)\right) = \mathcal{P}(B_j) = q_j, \quad (9.42)$$

for every $i, j = 1, 2, \ldots$. Hence, if we know the joint distribution $r_{i,j}$, $i, j = 1, 2, \ldots$, we can find the distributions of X and Y. The reciprocal is not true, in general. We can calculate the joint distribution starting from the distributions of X and Y if and only if they are independent random variables.

Now, consider the random variables $X + Y$ and XY, which take on the values $x_i + y_j$ and $x_i y_j$ respectively with the probabilities r_{ij}, $i, j = 1, 2, \ldots$. Thus,

$$X + Y \sim \begin{pmatrix} x_i + y_j \\ r_{ij} \end{pmatrix} \text{ and } XY \sim \begin{pmatrix} x_i y_j \\ r_{ij} \end{pmatrix}, \quad i, j = 1, 2, \ldots. \quad (9.43)$$

We calculate the expected value of $X + Y$:

$$E(X+Y) = \sum_{i,j\geq 1}(x_i + y_j)r_{ij} = \sum_{i\geq 1} x_i \sum_{j\geq 1} r_{ij} + \sum_{j\geq 1} y_j \sum_{i\geq 1} r_{ij}$$

$$= \sum_{i\geq 1} x_i p_i + \sum_{j\geq 1} y_j q_j = E(X) + E(Y).$$

Note that a similar statement, $E(XY) = E(X)E(Y)$ does not hold for *any* random variables, but only for *independent random variables* (see the definition below).

Definition 9.15 With the notations above, two discrete random variables $X, Y : \Omega \to \mathbb{R}$ are said to be *independent* if the events A_i and B_j are independent for any $i, j = 1, 2, \ldots$, that is, if

$$\mathcal{P}\{X = x_i, Y = y_j\} = \mathcal{P}\{X = x_i\}\mathcal{P}\{Y = y_j\} = p_i q_j,$$

or, equivalently,

$$r_{ij} = \mathcal{P}(A_i \cap B_j) = \mathcal{P}(A_i)\mathcal{P}(B_j) = p_i q_j,$$

for any $i, j = 1, 2, \ldots$.

9.4 Discrete Random Variables

Let us calculate the expected value of XY when X and Y are independent random variables.

$$E(XY) = \sum_{i,j \geq 1} (x_i y_j) r_{ij} = \sum_{i \geq 1} \sum_{j \geq 1} x_i y_j p_i q_j$$

$$= \sum_{i \geq 1} x_i p_i \underbrace{\sum_{j \geq 1} y_j q_j}_{E(Y)} = E(Y) \sum_{i \geq 1} x_i p_i = E(Y) E(X).$$

The following theorem summarizes the main properties of the expected value we have proved above.

Theorem 9.2 *In the probability space $(\Omega, \mathcal{K}, \mathcal{P})$, let X and Y be two discrete random variables with finite expected values and let $a, b \in \mathbb{R}$. Then the following equalities hold:*

(i) $E(aX + b) = aE(X) + b$.
(ii) $E(X + Y) = E(X) + E(Y)$.
(iii) *If X, Y are independent, then $E(XY) = E(X)E(Y)$.*

Corollary 9.1 *If $X_1, X_2, \ldots, X_n$ are discrete random variables in the probability space $(\Omega, \mathcal{K}, \mathcal{P})$, and $c_1, c_2, \ldots, c_n \in \mathbb{R}$, then*

$$E(c_1 X_1 + c_2 X_2 + \ldots + c_n X_n) = c_1 E(X_1) + c_2 E(X_2) + \ldots + c_n E(X_n).$$

Example 9.24 Consider the following *independent* random variables:

$$X \sim \begin{pmatrix} 0 & 1 & 2 \\ 1/2 & 1/4 & 1/4 \end{pmatrix} \text{ and } Y \sim \begin{pmatrix} -2 & -1 & 1 \\ 1/5 & 2/5 & 2/5 \end{pmatrix}$$

For instance,

$$\mathcal{P}\{X = 2, Y = -1\} = \mathcal{P}\{X = 2\}\mathcal{P}\{Y = -1\} = \frac{1}{4} \cdot \frac{2}{5} = \frac{2}{20}.$$

Let us find the distribution of $X + Y$ and XY. Thus, from (9.43) we obtain:

$$X + Y \sim \begin{pmatrix} -2 & -1 & 1 & -1 & 0 & 2 & 0 & 1 & 3 \\ \frac{1}{10} & \frac{1}{5} & \frac{1}{5} & \frac{1}{20} & \frac{1}{10} & \frac{1}{10} & \frac{1}{20} & \frac{1}{10} & \frac{1}{10} \end{pmatrix},$$

$$= \begin{pmatrix} -2 & -1 & 0 & 1 & 2 & 3 \\ \frac{1}{10} & \frac{1}{4} & \frac{3}{20} & \frac{3}{10} & \frac{1}{10} & \frac{1}{10} \end{pmatrix},$$

$$XY \sim \begin{pmatrix} 0 & 0 & 0 & -2 & -1 & 1 & -4 & -2 & 2 \\ \frac{1}{10} & \frac{1}{5} & \frac{1}{5} & \frac{1}{20} & \frac{1}{10} & \frac{1}{10} & \frac{1}{20} & \frac{1}{10} & \frac{1}{10} \end{pmatrix}$$

$$= \begin{pmatrix} -4 & -2 & -1 & 0 & 1 & 2 \\ \frac{1}{20} & \frac{3}{20} & \frac{1}{10} & \frac{1}{2} & \frac{1}{10} & \frac{1}{10} \end{pmatrix}.$$

Suppose, for instance, that we have to calculate $\mathcal{P}(-1.5 < XY \leq 1.5)$. The values of XY in the interval $(-1.5, 1.5]$ are $\{-1, 0, 1\}$. So,

$$\mathcal{P}(-1.5 < XY \leq 1.5) = \frac{1}{10} + \frac{1}{2} + \frac{1}{10} = \frac{7}{10}.$$

Example 9.25 Random placements of 3 balls, $\{a, b, c\}$ into 3 cells, $\{1, 2, 3\}$ [14]. The sample space Ω has $3^3 = 27$ outcomes, equally probable (each one of probability $\frac{1}{27}$):

$\omega_1 = [abc| - |-]$, $\omega_2 = [ab | c |-]$, $\omega_3 = [ab | - | c]$,
$\omega_4 = [ac | b | -]$, $\omega_5 = [a | bc |-]$, $\omega_6 = [a | b | c]$,
$\omega_7 = [ac | - | b]$, $\omega_8 = [a | c | b]$, $\omega_9 = [a | - | bc]$,
$\omega_{10} = [bc | a |-]$, $\omega_{11} = [b |ac |-]$, $\omega_{12} = [b | a | c]$,
$\omega_{13} = [c |ab |-]$, $\omega_{14} = [-|abc|-]$, $\omega_{15} = [-|ab | c]$,
$\omega_{16} = [c | a | b]$, $\omega_{17} = [-|ac | b]$, $\omega_{18} = [-| a |bc]$,
$\omega_{19} = [bc | - | a]$, $\omega_{20} = [b | c | a]$, $\omega_{21} = [b | - |ac]$,
$\omega_{22} = [c | b | a]$, $\omega_{23} = [-| bc| a]$, $\omega_{24} = [-| b |ac]$,
$\omega_{25} = [c | - |ab]$, $\omega_{26} = [-| c |ab]$, $\omega_{27} = [-| - |abc]$.

Let X_i denote the number of balls in the i-th cell, $i = 1, 2, 3$. We want to find the probability distribution of X_1 (obviously, X_1, X_2 and X_3 have the same distribution):

$\{X_1 = 0\} = \{\omega_{14}, \omega_{15}, \omega_{17}, \omega_{18}, \omega_{23}, \omega_{24}, \omega_{26}, \omega_{27}\} \Rightarrow \mathcal{P}\{X_1 = 0\} = \dfrac{8}{27}$

$\{X_1 = 1\} = \{\omega_5, \omega_6, \omega_8, \omega_9, \omega_{11}, \omega_{12}, \omega_{13}, \omega_{16}, \omega_{20}, \omega_{21}, \omega_{22}, \omega_{25}\}$

$\Rightarrow \mathcal{P}\{X_1 = 1\} = \dfrac{12}{27}$

$\{X_1 = 2\} = \{\omega_2, \omega_3, \omega_4, \omega_7, \omega_{10}, \omega_{19}\} \Rightarrow \mathcal{P}\{X_1 = 2\} = \dfrac{6}{27}$

$\{X_1 = 3\} = \{\omega_1\} \Rightarrow \mathcal{P}\{X_1 = 3\} = \dfrac{1}{27}.$

9.4 Discrete Random Variables

Thus, the distribution of X_i, $i = 1, 2, 3$ is:

$$X_i \sim \begin{pmatrix} 0 & 1 & 2 & 3 \\ \frac{8}{27} & \frac{12}{27} & \frac{6}{27} & \frac{1}{27} \end{pmatrix}$$

and the expected values are:

$$E(X_1) = E(X_2) = E(X_3) = 0 \cdot \frac{8}{27} + 1 \cdot \frac{2}{27} + 2 \cdot \frac{6}{27} + 3 \cdot \frac{1}{27} = 1.$$

The random variables X_1 and X_2 are not independent. For instance, since $\{X_1 = 0, X_2 = 0\} = \{\omega_{27}\}$, we have:

$$P\{X_1 = 0, X_2 = 0\} = \frac{1}{27} \neq P\{X_1 = 0\}P\{X_2 = 0\} = \frac{64}{729}.$$

The table bellow contains all the probabilities $p_{ij} = P\{X_1 = i, X_2 = j\}$, $i, j = 0, 1, 2, 3$ (the joint distribution of X_1 and X_2:

	$X_2 = 0$	$X_2 = 1$	$X_2 = 2$	$X_2 = 3$
$X_1 = 0$	$\frac{1}{27}$	$\frac{3}{27}$	$\frac{3}{27}$	$\frac{1}{27}$
$X_1 = 1$	$\frac{3}{27}$	$\frac{6}{27}$	$\frac{3}{27}$	0
$X_1 = 2$	$\frac{3}{27}$	$\frac{3}{27}$	0	0
$X_1 = 3$	$\frac{1}{27}$	0	0	0

The probability distributions of $X_1 + X_2$ and $X_1 X_2$ are presented below. Note that $X_1 + X_2 = 3 - X_3$:

$$X_1 + X_2 \sim \begin{pmatrix} 0 & 1 & 2 & 3 \\ \frac{1}{27} & \frac{6}{27} & \frac{12}{27} & \frac{8}{27} \end{pmatrix},$$

$$X_1 X_2 \sim \begin{pmatrix} 0 & 1 & 2 \\ \frac{15}{27} & \frac{6}{27} & \frac{6}{27} \end{pmatrix}.$$

Note that

$$E(X_1 + X_2) = 2 = E(X_1) + E(X_2),$$

but

$$E(X_1 X_2) = \frac{2}{3} \neq E(X_1)E(X_2).$$

So, if X and Y are not independent, the equality *(iii)* from Theorem 9.2 does not hold. It can be proved (see Proposition 9.12) that the expectation of XY does exist if X and Y have finite second moments.

Proposition 9.12 *If X and Y are two random variables in the probability space $(\Omega, \mathcal{K}, \mathcal{P})$ such that $E(X^2), E(Y^2) < \infty$, then XY has a finite expectation: $E(|XY|) < \infty$.*

Proof Suppose that X and Y have the distributions:

$$X \sim \begin{pmatrix} x_1 & x_2 & \cdots & x_n & \cdots \\ p_1 & p_2 & \cdots & p_n & \cdots \end{pmatrix} \text{ and } Y \sim \begin{pmatrix} y_1 & y_2 & \cdots & y_m & \cdots \\ q_1 & q_2 & \cdots & q_m & \cdots \end{pmatrix}.$$

We denote by $r_{i,j} = \mathcal{P}\{X = x_i, Y = y_j\}$, for every $i = 1, 2, \ldots$ and $j = 1, 2, \ldots$. Since the following inequality holds for every $i = 1, 2, \ldots$ $j = 1, 2, \ldots$

$$|x_i y_j| r_{i,j} \leq \tfrac{1}{2}\left(x_i^2 + y_j^2\right) r_{i,j},$$

and

$$\sum_{i \geq 1} \sum_{j \geq 1} \tfrac{1}{2}\left(x_i^2 + y_j^2\right) r_{i,j} = \tfrac{1}{2}\left(\sum_{i \geq 1} x_i^2 \sum_{j \geq 1} r_{i,j} + \sum_{j \geq 1} y_j^2 \sum_{i \geq 1} r_{i,j}\right) =$$

$$= \tfrac{1}{2}\left(\sum_{i \geq 1} x_i^2 p_i + \sum_{j \geq 1} y_j^2 q_j\right) = \tfrac{1}{2}\left(E(X^2) + E(Y^2)\right) < \infty,$$

we obtain that the series

$$\sum_{i \geq 1} \sum_{j \geq 1} x_i y_j r_{i,j} < \infty$$

is absolutely convergent, so there exists the expected value $E(XY)$. □

The following theorem states an inequality similar to the Cauchy–Schwarz inequality in an inner-product space.

Theorem 9.3 (Cauchy–Schwarz) *If X and Y are two random variables in the probability space $(\Omega, \mathcal{K}, \mathcal{P})$ such that $E(X^2), E(Y^2) < \infty$, then*

$$[E(XY)]^2 \leq E(X^2) \cdot E(Y^2). \tag{9.44}$$

If $X, Y \neq 0$, the equality takes place if and only if there exists $a \in \mathbb{R}$ such that $Y = aX$.

9.4 Discrete Random Variables

Proof It is very easy to see that $E(X^2) = 0$ if and only if X is constantly equal to 0. In this case we have also $XY \equiv 0$, so the relation (9.44) holds (as equality).

Now, let us suppose that $E(X^2) \neq 0$. For any $t \in \mathbb{R}$, the random variable

$$(tX - Y)^2 = t^2 X^2 - 2t XY + Y^2$$

has a finite expectation (by Proposition 9.12) and we can write:

$$E\left((tX - Y)^2\right) = t^2 E\left(X^2\right) - 2t E(XY) + E\left(Y^2\right) \geq 0,$$

for any $t \in \mathbb{R}$. It follows that

$$\Delta = 4\left([E(XY)]^2 - E(X^2) \cdot E(Y^2)\right) \leq 0$$

and we obtain the inequality (9.44).

If $\Delta = 0$, this means that there exists a real number t such that

$$t^2 E\left(X^2\right) - 2t E(XY) + E\left(Y^2\right) = 0.$$

Thus, for $t = \dfrac{E(XY)}{E(X^2)}$ we have $E\left((tX - Y)^2\right) = 0$, which means that $tX - Y = 0$, or, equivalently, $Y = tX$. □

9.4.3 Variance

Imagine a gambler who bets on a coin tossing: if *Head* comes up, he gains 1, otherwise, he loses 1. Assuming the coin is fair, the mathematical expression of this game is given by the random variable:

$$X \sim \begin{pmatrix} -1 & 1 \\ \frac{1}{2} & \frac{1}{2} \end{pmatrix}. \tag{9.45}$$

If he gains 100 for *Head* and loses 100 for *Tail*, the random variable will have the distribution:

$$Y \sim \begin{pmatrix} -100 & 100 \\ \frac{1}{2} & \frac{1}{2} \end{pmatrix}. \tag{9.46}$$

Both random variables have the same expected value: $E(X) = E(Y) = 0$, but, obviously, the second game is much riskier. A measure of the risk is the *variance* of the random variable, defined to be the second moment of the deviation from the expected value, $X - \mu$, where $\mu = E(X)$.

Definition 9.16 Let X be a random variable with a finite mean $\mu = E(X)$. The *variance* (or the *dispersion*) of X, denoted by $\sigma^2 = Var(X)$, is defined by

$$Var(X) = E\left((X - \mu)^2\right), \qquad (9.47)$$

if this expected value is finite. In this case, $\sigma = \sqrt{Var(X)}$ is said to be the *standard deviation* of X.

If the discrete random variable X has the distribution

$$X \sim \begin{pmatrix} x_1 & x_2 & \ldots & x_n & \ldots \\ p_1 & p_2 & \ldots & p_n & \ldots \end{pmatrix},$$

and $\mu = E(X)$ is the mean of X, then its variance is:

$$Var(X) = \sum_{n \geq 1}(x_n - \mu)^2 p_n. \qquad (9.48)$$

Let us calculate the variance and the standard deviation of the random variables (9.45) and (9.46). Both have the mean equal to 0: $\mu_X = \mu_Y = 0$, so

$$Var(X) = E(X^2) = (-1)^2 \cdot \tfrac{1}{2} + 1^2 \cdot \tfrac{1}{2} = 1 \Rightarrow \sigma_X = 1$$
$$Var(Y) = E(Y^2) = (-100)^2 \cdot \tfrac{1}{2} + 100^2 \cdot \tfrac{1}{2} = 10000 \Rightarrow \sigma_Y = 100.$$

Note that the standard deviation has the same unit of measurement as the random variable.

Proposition 9.13 *The variance of X is finite if and only if the second moment $E(X^2)$ is finite and in this case, the following formula holds:*

$$Var(X) = E\left(X^2\right) - E(X)^2. \qquad (9.49)$$

Proof Let $\mu = E(X)$ be the expected value of X. By Definition 9.16 we have:

$$Var(X) = E\left((X - \mu)^2\right) = E\left(X^2 - 2\mu X + \mu^2\right) =$$
$$= E\left(X^2\right) - 2\mu E(X) + \mu^2 = E\left(X^2\right) - \mu^2.$$

□

9.4 Discrete Random Variables

The variance is computed in most cases using the formula (9.49) (known in the French literature as the König-Huygens formula). Let us calculate the variance of the uniform random variable from Example 9.21 (rolling a fair die). We have seen above that the mean is $E(x) = 3.5$, so

$$Var(X) = E(X^2) - \left(\frac{7}{2}\right)^2 = \sum_{k=1}^{6} k^2 \cdot \frac{1}{6} - \frac{49}{4} = \frac{35}{12} = 2.92.$$

Theorem 9.4 *In the probability space* $(\Omega, \mathcal{K}, \mathcal{P})$, *let* X *and* Y *be two discrete random variables with finite second moments,* $E(X^2), E(Y^2) < \infty$, *and let* $a \in \mathbb{R}$. *Then, the following equalities hold:*

(i) $Var(aX) = a^2 Var(X)$.
(ii) $Var(X + a) = Var(X)$.
(iii) *If* X, Y *are independent, then* $Var(X + Y) = Var(X) + Var(Y)$.
(iv) $Var(X) \geq 0$ *and* $Var(X) = 0$ *if and only if* X *is a constant random variable (it takes a constant value,* $X = a$ *with the probability 1).*
(v) $E\left((X-a)^2\right) \geq Var(X)$, *for any* $a \in \mathbb{R}$ *(equality is obtained for* $a = E(X)$).

Proof

(i) We use the formula (9.13) and the properties of the expected value (Theorem 9.2):

$$Var(aX) = E(a^2 X^2) - (E(aX))^2 = a^2 E(X^2) - a^2 (E(X))^2 = a^2 Var(X).$$

(ii) Since $E(X + a) = E(X) + a$, we have:

$$Var(X + a) = E\left((X + a - E(X + a))^2\right) = E\left((X - E(X))^2\right) = Var(X).$$

(iii) Since X and Y are independent, $E(XY) = E(X)E(Y)$ (see Theorem 9.2) and we have:

$$Var(X + Y) = E\left((X + Y)^2\right) - (E(X + Y))^2 =$$

$$= E(X^2) + 2E(XY) + E(Y^2) - (E(X) + E(Y))^2 =$$

$$= E(X^2) - E(X)^2 + E(Y^2) - E(Y)^2 = Var(X) + Var(Y).$$

(iv) By the formula (9.48), the inequality $Var(X) \geq 0$ is evident and also, we have that $Var(X) = 0$ if and only if $x_n = \mu$, for any $n = 1, 2, \ldots$, so X is constant.

(v) Since $(E(X) - a)^2 = E(X)^2 - 2aE(X) + a^2 \geq 0$, we have:

$$E\left((X-a)^2\right) = E\left(X^2 - 2aX + a^2\right) = E\left(X^2\right) - 2aE(X) + a^2 \geq$$

$$\geq E\left(X^2\right) - E(X)^2 = Var(X).$$

$\square$

Remark 9.8 (Interpretation in Mechanics [14]) Suppose that a unit mass is distributed on the x-axis such that the mass $p_i = f(x_i)$ is concentrated at the point x_i, $i = 1, 2, \ldots$. Then the mean $\mu = E(x)$ is the abscissa of the *center of gravity*, and the variance is the *moment of inertia*.

We know that

$$E(X + Y) = E(X) + E(Y)$$

for *any* random variables X and Y (see Theorem 9.2). On the other hand, Theorem 9.4 states a similar relation for variance,

$$Var(X + Y) = Var(X) + Var(Y),$$

only for *independent* random variables X and Y. For instance, if $X = Y$ the equality cannot be true, because

$$Var(X + X) = Var(2X) \stackrel{i)}{=} 4Var(X)$$
$$Var(X) + Var(X) = 2Var(X).$$

We introduce the notion of *covariance* of two random variables, $Cov(X, Y)$, which exists for any X and Y with finite variances (recall that, by Proposition 9.13, $Var(X) < \infty$ if and only if $E(X^2) < \infty$).

Definition 9.17 Let X and Y be two random variables in the probability space $(\Omega, \mathcal{K}, \mathcal{P})$ such that $E(X^2), E(Y^2) < \infty$. We denote by $\mu_X = E(X)$ and $\mu_Y = E(Y)$ the expected values of X and Y. The *covariance* of X and Y, $Cov(X, Y)$ is defined as

$$Cov(X, Y) = E\left((X - \mu_X)(Y - \mu_Y)\right). \tag{9.50}$$

Note that $(X - \mu_X)(Y - \mu_Y)$ has a finite expected value by Proposition 9.12.

From the definition formula (9.50) one can write

$$Cov(X, Y) = E(XY) - \mu_Y E(X) - \mu_X E(Y) + \mu_X \mu_Y,$$

9.4 Discrete Random Variables

so the covariance can be also written in the form:

$$Cov(X, Y) = E(XY) - E(X)E(Y). \tag{9.51}$$

Now, let us calculate the variance of the sum $X + Y$, using the formula (9.49):

$$Var(X + Y) = E\left(X^2 + 2XY + Y^2\right) - (E(X) + E(Y))^2 =$$

$$= E(X^2) - E(X)^2 + E(Y^2) - E(Y)^2 + 2\left(E(XY) - E(X)E(Y)\right) =$$

$$= Var(X) + Var(Y) + 2Cov(X, Y)$$

and the following theorem is proved.

Theorem 9.5 *If X and Y are two random variables in the probability space $(\Omega, \mathcal{K}, \mathcal{P})$ such that $E(X^2), E(Y^2) < \infty$, then*

$$Var(X + Y) = Var(X) + Var(Y) + 2Cov(X, Y). \tag{9.52}$$

By Theorem 9.4 (iii) we know that $Var(X + Y) = Var(X) + Var(Y)$ if X and Y are independent, so the next corollary follows.

Corollary 9.2 *Let X and Y be two random variables in the probability space $(\Omega, \mathcal{K}, \mathcal{P})$ such that $E(X^2), E(Y^2) < \infty$. If X and Y are **independent**, then then $Cov(X, Y) = 0$.*

Remark 9.9 The converse is not true: there exist random variables which are not independent but have the covariance equal to 0. For instance, we can take

$$X \sim \begin{pmatrix} -1 & 0 & 1 \\ \frac{1}{3} & \frac{1}{3} & \frac{1}{3} \end{pmatrix} \quad \text{and} \quad Y = X^2 \sim \begin{pmatrix} 0 & 1 \\ \frac{1}{3} & \frac{2}{3} \end{pmatrix}$$

Obviously, they are not independent, but their covariance is 0:

$$Cov(X, Y) = E(XY) - E(X)E(Y) = E(X^3) - E(X)E(X^2) = 0.$$

Notice that $Cov(X, X) = Var(X)$ and $Cov(X, -X) = -Var(X)$.
By applying Theorem 9.3 for $X - \mu_X$ and $Y - \mu_Y$, one can write:

$$|Cov(X, Y)| \leq \sqrt{Var(X)Var(Y)} \tag{9.53}$$

Definition 9.18 Let X and Y be two non-constant random variables in the probability space $(\Omega, \mathcal{K}, \mathcal{P})$ such that $E(X^2), E(Y^2) < \infty$. The *correlation coefficient* of

X and Y is defined as

$$Corr(X,Y) = \frac{Cov(X,Y)}{\sqrt{Var(X)Var(Y)}} \qquad (9.54)$$

By the inequality (9.53), the correlation coefficient is always in the interval $[-1, 1]$. The next theorem states that the extreme values -1 and 1 are attained if and only if the random variables are *linearly correlated*, i.e. there exist $\alpha, \beta \in \mathbb{R}$, $\alpha \neq 0$, such that $Y = \alpha X + \beta$.

Theorem 9.6 *Let X and Y be two non-constant random variables in the probability space $(\Omega, \mathcal{K}, \mathcal{P})$ such that $E(X^2), E(Y^2) < \infty$. Then $Corr(X,Y) = \pm 1$ if and only if there exist $\alpha, \beta \in \mathbb{R}$, $\alpha \neq 0$, such that $Y = \alpha X + \beta$. We have $Corr(X,Y) = 1$ if $\alpha > 0$, and $Corr(X,Y) = -1$ if $\alpha < 0$.*

Proof Suppose that $Y = \alpha X + \beta$, with $\alpha \neq 0$. Then $E(Y) = \alpha E(X) + \beta$ and $Var(Y) = \alpha^2 Var(X)$. By denoting $\mu_X = E(X)$ and using the formula (9.50), we can write the covariance as

$$Cov(X,Y) = E((X - \mu_X) \cdot \alpha(X - \mu_X)) = \alpha Var(X).$$

So the correlation coefficient is

$$Corr(X,Y) = \frac{\alpha Var(X)}{\sqrt{Var(X)\alpha^2 Var(X)}} = \frac{\alpha}{|\alpha|} = \pm 1$$

and we have: $Corr(X,Y) = 1$ if $\alpha > 0$ and $Corr(X,Y) = -1$ if $\alpha < 0$.

Now, let us prove that the extreme values of the correlation coefficient, ± 1, are attained only when X and Y are linearly correlated.

$$Corr(X,Y) = \pm 1 \Leftrightarrow (Cov(X,Y))^2 = Var(X)Var(Y)$$

$$\Leftrightarrow [E((X - \mu_X)(Y - \mu_Y))]^2 = E\left((X - \mu_X)^2\right) E\left((Y - \mu_Y)^2\right).$$

By Theorem 9.3 we know that inequality (9.53) becomes equality if and only if there exists $\alpha \neq 0$ such that $Y - \mu_Y = \alpha(X - \mu_X)$. We denote by $\beta = \mu_Y - \alpha \mu_X$ and obtain that $Y = \alpha X + \beta$. □

Theorem 9.5 can be generalized for a sum of n random variables.

Theorem 9.7 *If $X_1, X_2, \ldots, X_n$ are n random variables in the probability space $(\Omega, \mathcal{K}, \mathcal{P})$ such that $E(X_i^2) < \infty$, $i = 1, \ldots, n$, then*

$$Var\left(\sum_{i=1}^{n} X_i\right) = \sum_{i=1}^{n} Var(X_i) + 2 \sum_{1 \leq i < j \leq n} Cov(X_i, X_j). \qquad (9.55)$$

9.4 Discrete Random Variables

Corollary 9.3 *If $X_1, X_2, \ldots, X_n$ are independent random variables in the probability space $(\Omega, \mathcal{K}, \mathcal{P})$ such that $E(X_i^2) < \infty$, $i = 1, \ldots, n$, then*

$$Var(X_1 + \ldots + X_n) = Var(X_1) + \ldots + Var(X_n). \tag{9.56}$$

Remark 9.10 An important particular case is when $X_1, X_2, \ldots, X_n$ are *independent and identically distributed* (**iid**) random variables. Let $\mu = E(X_i)$ and $\sigma^2 = Var(X_i)$ for all $i = 1, \ldots, n$. We denote by S_n and $\overline{X}_n$ the random variables

$$S_n = X_1 + \ldots + X_n, \quad \overline{X}_n = \frac{X_1 + \ldots + X_n}{n}.$$

Then,

$$E(S_n) = n\mu, \quad Var(S_n) = n\sigma^2, \tag{9.57}$$

$$E(\overline{X}_n) = \mu, \quad Var(\overline{X}_n) = \frac{\sigma^2}{n}. \tag{9.58}$$

We present in the following some discrete distributions that arise very frequently in applications.

9.4.4 Discrete Uniform Distribution

Let X be a random variable that takes on the values $1, 2, \ldots, n$ with equal probabilities, $\mathcal{P}\{X = k\} = \frac{1}{n}$:

$$X \sim \begin{pmatrix} 1 & 2 & \ldots & n \\ \frac{1}{n} & \frac{1}{n} & \ldots & \frac{1}{n} \end{pmatrix}.$$

We also denote $X \sim Unif\{1, \ldots, n\}$. Let us compute the mean and the variance of X:

$$E(X) = \sum_{k=1}^{n} k \cdot \frac{1}{n} = \frac{n(n+1)}{2} \cdot \frac{1}{n} = \frac{n+1}{2}.$$

$$Var(X) = E(X^2) - E(X)^2 = \sum_{k=1}^{n} k^2 \cdot \frac{1}{n} - \frac{(n+1)^2}{4} =$$

$$= \frac{n(n+1)(2n+1)}{6} \cdot \frac{1}{n} - \frac{(n+1)^2}{4} = \frac{n^2 - 1}{12}.$$

9.4.5 Bernoulli Distribution

Let X be a random variable which takes only 2 values: 0 and 1. If $p = \mathcal{P}\{X = 1\}$ and $q = 1 - p = \mathcal{P}\{X = 0\}$, then we can represent the distribution of X as

$$X \sim \begin{pmatrix} 0 & 1 \\ q & p \end{pmatrix}$$

This is the random variable associated to an experiment with only two possible outcomes: "success" ($X = 1$), with probability p, or "failure" ($X = 0$), with probability $q = 1 - p$. For instance, when tossing a coin, we can take $X = 1$ for head and $X = 0$ for tail.

We compute the expected value and the variance of X:

$$E(X) = 0 \cdot (1 - p) + 1 \cdot p = p,$$

$$Var(X) = E(X^2) - E(X)^2 = p - p^2 = p(1 - p) = pq.$$

9.4.6 Binomial Distribution

Suppose that an experiment as above is independently performed n times and let X be the random variable representing the number of successes that occur in these *Bernoulli trials*. Then X is said to be a binomial random variable with parameters (n, p) and we denote $X \sim Bin(n, p)$. Note that the Bernoulli distribution is a particular case (for $n = 1$) of the binomial distribution.

To find the distribution of X, we need to find, for every $k = 0, 1, \ldots, n$ the probability of having exactly k successes in n trials. This is similar to the *urn model—sampling with replacement* discussed in Example 9.6. Since the n trials are independent, the probability of any particular sequence of the n outcomes containing k successes and $n - k$ failures is $p^k q^{n-k}$. There are $\binom{n}{k}$ different sequences containing exactly k successes (and $n - k$ failures), so

$$p_k = \mathcal{P}\{X = k\} = \binom{n}{k} p^k q^{n-k} = \binom{n}{k} p^k (1 - p)^{n-k} \qquad (9.59)$$

and the binomial distribution of parameters (n, p) is written:

$$X \sim \begin{pmatrix} 0 & 1 & \cdots & k & \cdots & n \\ \binom{n}{0} p^0 q^n & \binom{n}{1} p q^{n-1} & \cdots & \binom{n}{k} p^k q^{n-k} & \cdots & \binom{n}{n} p^n q^0 \end{pmatrix} \qquad (9.60)$$

9.4 Discrete Random Variables

Note that the sum of the probabilities equals 1:

$$\sum_{k=0}^{n} p_k = \sum_{k=0}^{n} \binom{n}{k} p^k (1-p)^{n-k} = (p+1-p)^n = 1,$$

The binomial random variable X with parameters (n, p) represents the number of successes in n independent trials, each one having success probability p, so we can write:

$$X = \sum_{k=1}^{n} X_k, \tag{9.61}$$

where

$$X_k = \begin{cases} 1, & \text{if the } k\text{-th trial is a success} \\ 0, & \text{if the } k\text{-th trial is a failure.} \end{cases}$$

Thus, $X_1, X_2, \ldots, X_n$ are Bernoulli random variables with $\mathcal{P}\{X_k = 1\} = p$ for every $k = 1, 2, \ldots, n$. As proved above,

$$E(X_k) = p \quad \text{and} \quad Var(X_k) = p(1-p),$$

so

$$E(X) = \sum_{k=1}^{n} E(X_k) = np, \tag{9.62}$$

and, because $X_1, X_2, \ldots, X_n$ are *independent* random variables,

$$Var(X) = \sum_{k=1}^{n} Var(X_k) = np(1-p). \tag{9.63}$$

We remark that the expected value $E(X)$ and the variance $Var(X)$ can be also obtained by direct calculation. Thus, we use the binomial formula

$$(px + q)^n = \sum_{k=0}^{n} \binom{n}{k} p^k x^k q^{n-k}, \tag{9.64}$$

which holds for any $x \in \mathbb{R}$. By differentiation we get

$$np(px + q)^{n-1} = \sum_{k=1}^{n} k \binom{n}{k} p^k x^{k-1} q^{n-k}, \tag{9.65}$$

which becomes for $x = 1$:

$$np = \sum_{k=1}^{n} k \binom{n}{k} p^k q^{n-k} = E(X).$$

By differentiating (9.65) we obtain:

$$n(n-1)p^2 (px+q)^{n-2} = \sum_{k=1}^{n} k(k-1) \binom{n}{k} p^k x^{k-2} q^{n-k},$$

and for $x = 1$ we find:

$$n(n-1)p^2 = \sum_{k=1}^{n} k^2 \binom{n}{k} p^k q^{n-k} - \sum_{k=1}^{n} k \binom{n}{k} p^k q^{n-k} = E(X^2) - E(X),$$

so the second moment of X can be written $E(X^2) = n^2 p^2 - np^2 + np$ and the variance is

$$Var(X) = E(X^2) - E(X)^2 = n^2 p^2 - np^2 + np - n^2 p^2 = np(1-p).$$

We present in Figs. 9.6, 9.7, and 9.8 the binomial distributions for $n = 10$ and $p = 0.3$, $p = 0.5$ and $p = 0.6$, respectively.

Example 9.26 Find the probability of obtaining at least 3 "six" when rolling a fair die 4 times. Here, the probability of "success" (since the die is fair) is $p = 1/6$. So,

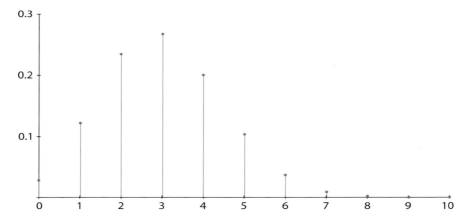

Fig. 9.6 Binomial distribution for $n = 10$, $p = 0.3$

9.4 Discrete Random Variables

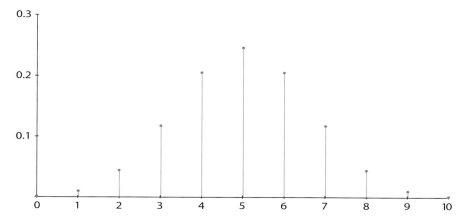

Fig. 9.7 Binomial distribution for $n = 10$, $p = 0.5$

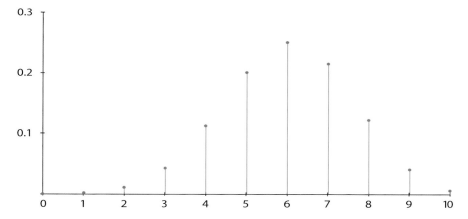

Fig. 9.8 Binomial distribution for $n = 10$, $p = 0.6$

the random variable that counts the number of "six" occurring in $n = 4$ trials is

$$X \sim Bin\left(4, \tfrac{1}{6}\right) = \begin{pmatrix} 0 & 1 & 2 & 3 & 4 \\ \left(\tfrac{5}{6}\right)^4 & 4\left(\tfrac{1}{6}\right)\left(\tfrac{5}{6}\right)^3 & 6\left(\tfrac{1}{6}\right)^2\left(\tfrac{5}{6}\right)^2 & 4\left(\tfrac{1}{6}\right)^3\left(\tfrac{5}{6}\right) & \left(\tfrac{1}{6}\right)^4 \end{pmatrix}.$$

If C denotes the event of occurring at least 3 "six", then

$$\mathcal{P}(C) = \mathcal{P}\{X = 3\} + \mathcal{P}\{X = 4\} = 4 \cdot \frac{5}{6^4} + \frac{1}{6^4} = \frac{21}{6^4} = 0.016.$$

9.4.7 Poisson Distribution

Theorem 9.8 (Poisson Limit Theorem) *Let $X_n \sim Bin(n, p_n)$ be a sequence of binomial random variables having all the same expected value, $np_n = \lambda$, where $\lambda > 0$ is a positive number. Then, for any fixed integer $k \geq 0$ we have:*

$$\lim_{n \to \infty} \mathcal{P}\{X_n = k\} = \exp(-\lambda) \frac{\lambda^k}{k!}. \tag{9.66}$$

Proof Using the binomial distribution formula (9.59) with $p = p_n = \frac{\lambda}{n}$ one can write:

$$\mathcal{P}\{X_n = k\} = \binom{n}{k} p_n^k (1 - p_n)^{n-k} =$$

$$= \frac{n!}{k!(n-k)!} \frac{\lambda^k}{n^k} \left(1 - \frac{\lambda}{n}\right)^{n-k} =$$

$$= \frac{\lambda^k}{k!} \cdot \frac{n(n-1)\ldots(n-k+1)}{n^k} \left(1 - \frac{\lambda}{n}\right)^{n-k} \xrightarrow[n \to \infty]{} \frac{\lambda^k}{k!} \exp(-\lambda).$$

$\square$

The limit distribution of the sequence $X_n \sim Bin(n, p_n)$,

$$X \sim \begin{pmatrix} 0 & 1 & 2 & \cdots & k & \cdots \\ \frac{\lambda^0}{0!} \exp(-\lambda) & \frac{\lambda}{1!} \exp(-\lambda) & \frac{\lambda^2}{2!} \exp(-\lambda) & \cdots & \frac{\lambda^k}{k!} \exp(-\lambda) & \cdots \end{pmatrix}, \tag{9.67}$$

is said to be the *Poisson distribution* of parameter λ. Note that

$$\sum_{n=0}^{\infty} \frac{\lambda^n}{n!} \exp(-\lambda) = \exp(-\lambda) \sum_{n=0}^{\infty} \frac{\lambda^n}{n!} = \exp(-\lambda) \exp(\lambda) = 1,$$

so (9.67) is correctly defined. The random variable $X \sim Poi(\lambda)$ expresses the number of times a very rare event occurs in a specified period of time. It can be used, for instance, to model the number of traffic accidents occurring weekly on a highway, the number of misprints on a page of a book, the number of people in a community living to 100 years of age, and so on.

9.4 Discrete Random Variables

We prove that the mean and the variance of the random variable $X \sim Poi(\lambda)$ are both equal to λ.

$$E(X) = \sum_{k=0}^{\infty} k \frac{\lambda^k}{k!} \exp(-\lambda) = \lambda \exp(-\lambda) \sum_{k=1}^{\infty} \frac{\lambda^{k-1}}{(k-1)!} =$$

$$= \lambda \exp(-\lambda) \sum_{k=0}^{\infty} \frac{\lambda^k}{k!} = \lambda \exp(-\lambda) \exp \lambda = \lambda,$$

because $\sum_{k=0}^{\infty} \frac{\lambda^k}{k!} = \exp \lambda$. Hence, the expected value is

$$E(X) = \lambda. \tag{9.68}$$

To find the variance of X, we calculate the second moment:

$$E(X^2) = \exp(-\lambda) \sum_{k=1}^{\infty} k^2 \frac{\lambda^k}{k!} = \lambda \exp(-\lambda) \sum_{k=1}^{\infty} k \frac{\lambda^{k-1}}{(k-1)!} =$$

$$= \lambda \exp(-\lambda) \sum_{k=1}^{\infty} (k-1+1) \frac{\lambda^{k-1}}{(k-1)!} =$$

$$= \lambda^2 \exp(-\lambda) \sum_{k=2}^{\infty} \frac{\lambda^{k-2}}{(k-2)!} + \lambda \exp(-\lambda) \sum_{k=1}^{\infty} \frac{\lambda^{k-1}}{(k-1)!} =$$

$$= \lambda^2 \exp(-\lambda) \sum_{k=0}^{\infty} \frac{\lambda^k}{k!} + \lambda \exp(-\lambda) \sum_{k=0}^{\infty} \frac{\lambda^k}{k!} =$$

$$= \lambda^2 \exp(-\lambda) \exp \lambda + \lambda \exp(-\lambda) \exp \lambda = \lambda^2 + \lambda.$$

Thus, since $Var(X) = E(X^2) - E(X)^2$, we get:

$$Var(X) = \lambda. \tag{9.69}$$

Example 9.27 A box contains 10 white balls and 40 black balls. We draw out 20 balls with replacement. What is the probability to obtain at most 4 white balls? Compute it using: (i) the binomial distribution; (ii) the Poisson distribution.

The probability to draw a white ball is $p = 0.2$.

Let X be the random variable representing the number of white balls that occurs when 20 balls are drawn with replacement from the box.

(i) Suppose that $X \sim Bin(20, 0.2)$. Then, the probability to obtain at most 4 white balls is:

$$\mathcal{P}\{X \leq 4\} = \mathcal{P}\{X = 0\} + \mathcal{P}\{X = 1\} + \mathcal{P}\{X = 2\} + \mathcal{P}\{X = 3\} + \mathcal{P}\{X = 4\} =$$

$$= 0.8^{20} + \binom{20}{1} 0.2 \cdot 0.8^{19} + \binom{20}{2} 0.2^2 \cdot 0.8^{18} +$$

$$+ \binom{20}{3} 0.2^3 \cdot 0.8^{17} + \binom{20}{4} 0.2^4 \cdot 0.8^{16} = 0.629648.$$

(ii) Now, let us suppose that $X \sim Poi(\lambda)$, where $\lambda = np = 20 \cdot 0.2 = 4$:

$$\mathcal{P}\{X = k\} = \exp(-4) \cdot \frac{4^k}{k!} \implies$$

$$\mathcal{P}\{X \leq 4\} = \exp(-4) \left(1 + \frac{4}{1} + \frac{4^2}{2!} + \frac{4^3}{3!} + \frac{4^4}{4!} \right) = 0.628837.$$

We present in Fig. 9.9 the binomial distributions for $n = 20$ and $p = 0.2$ and the Poisson distribution with the same expected value, $\lambda = np = 4$.

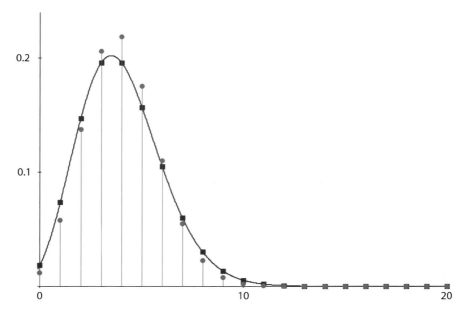

Fig. 9.9 Binomial distribution for $n = 20$, $p = 0.2$ (solid red circles) and Poisson distribution for $\lambda = 4$ (solid blue squares)

9.4 Discrete Random Variables

Example 9.28 The average number of earthquakes (of magnitude ≥ 5) per year in the island of Java is 4. Find the probability to have:

(i) at most 2 earthquakes in a year;
(ii) more than 3 earthquakes in a year.

The number of earthquakes per year is given by the random variable $X \sim Poi(4)$.

(i) The probability of having "at most 2 earthquakes in a year" is:

$$P\{X=0\} + P\{X=1\} + P\{X=2\} = e^{-4}\left(1 + \frac{4}{1} + \frac{4^2}{2!}\right) = \frac{13}{e^4} = 0.238.$$

(ii) The event of having "more than 3 earthquakes in a year" is the complementary of the event to have "at most 3 earthquakes in a year":

$$P\{X > 3\} = 1 - P\{X \leq 3\} =$$

$$= 1 - (P\{X=0\} + P\{X=1\} + P\{X=2\} + P\{X=3\}) =$$

$$= 1 - e^{-4}\left(1 + \frac{4}{1} + \frac{4^2}{2!} + \frac{4^3}{3!}\right) = 1 - \frac{71}{3e^4} = 0.567.$$

9.4.8 Geometric Distribution

Let us consider again the experiment from Example 9.19: suppose that a (biased) coin is tossed repeatedly until a head is obtained for the first time. Let p be the probability of getting a head in a single trial, $p \in (0, 1)$ and X be the number of the toss when the first head is obtained. Then, since the tosses are independent, the probability of getting successively $k - 1$ tails and a head at the k-th trial is

$$P\{X = k\} = (1-p)^{k-1} p, \quad k = 1, 2, \ldots.$$

This is a geometric progression of ratio $1 - p$ and we say that X has a *geometric distribution* with parameter p, $X \sim Geo(p)$. If we denote by $q = 1 - p$, the geometric distribution can be written:

$$X \sim \begin{pmatrix} 1 & 2 & 3 & \cdots & n & \cdots \\ p & pq & pq^2 & \cdots & pq^{n-1} & \cdots \end{pmatrix}. \tag{9.70}$$

We notice that $\sum_{n=1}^{\infty} pq^{n-1} = p(1 + q + q^2 + \ldots) = \frac{p}{1-q} = 1$.

To find the expected value of X,

$$E(X) = \sum_{n=1}^{\infty} npq^{n-1},$$

we have to calculate the sum of the power series $s(x) = \sum_{n=1}^{\infty} nx^{n-1}$, for $x \in (0, 1)$. We integrate on both sides of the equality and obtain:

$$\int s(x)dx = \sum_{n=1}^{\infty} x^n = \frac{x}{1-x}.$$

It follows that

$$s(x) = \left(\frac{x}{1-x}\right)' = \frac{1}{(1-x)^2},$$

so

$$E(X) = p \cdot \frac{1}{(1-q)^2} = \frac{1}{p}.$$

Now, we calculate the second moment of X as

$$E(X^2) = E(X^2 + X) - E(X) = E(X(X+1)) - E(X) =$$

$$= \sum_{n=1}^{\infty} n(n+1)pq^{n-1} - \frac{1}{p}.$$

We apply the same method as above to compute $\sum_{n=1}^{\infty} n(n+1)x^{n-1}$. Thus, by integrating twice (and differentiating twice the result) we obtain

$$\sum_{n=1}^{\infty} n(n+1)x^{n-1} = \left(\frac{x^2}{1-x}\right)'' = \frac{2}{(1-x)^3}.$$

It follows that

$$E(X^2) = p \cdot \frac{2}{(1-q)^3} - \frac{1}{p} = \frac{2}{p^2} - \frac{1}{p},$$

so the variance of the geometric random variable is

$$Var(X) = E(X^2) - E(X)^2 = \frac{2}{p^2} - \frac{1}{p} - \frac{1}{p^2} = \frac{q}{p^2}.$$

9.5 Continuous Random Variables

9.5.1 The Probability Density Function; The Distribution Function

Recall that a random variable in the probability space $(\Omega, \mathcal{K}, \mathcal{P})$ is a function $X : \Omega \to \mathbb{R}$ such that $\{X \leq x\} \in \mathcal{K}$, for any $x \in \mathbb{R}$ (see Definition 9.9). In Sect. 9.4 we discussed about *discrete random variables*. This section deals with *continuous random variables*, which take on all the values in some nonempty interval, or union of intervals, or the entire real line.

Definition 9.19 Let $X : \Omega \to \mathbb{R}$ be a continuous random variable in the probability space $(\Omega, \mathcal{K}, \mathcal{P})$. The piecewise continuous function $f : \mathbb{R} \to [0, \infty)$ with the property that

$$\mathcal{P}\{a \leq X \leq b\} = \int_a^b f(x)\,dx, \tag{9.71}$$

for any $-\infty \leq a \leq b \leq \infty$, is called the *probability density function* of X (or simply, the *density function* of X).

Since $\mathcal{P}\{-\infty < X < \infty\} = \mathcal{P}(\Omega) = 1$, a density function must satisfy the relation:

$$\int_{-\infty}^{\infty} f(x)\,dx = 1. \tag{9.72}$$

From Definition 9.19 it follows that

$$\mathcal{P}\{X = a\} = \int_a^a f(x)\,dx = 0,$$

so the density function $f(x)$ is *not* the probability $\mathcal{P}\{X = x\}$. However, since

$$\mathcal{P}\{x - \varepsilon \leq X \leq x + \varepsilon\} = \int_{x-\varepsilon}^{x+\varepsilon} f(t)\,dt \approx 2\varepsilon f(x),$$

the density function $f(x)$ provides an estimation of the probability that X is near x.

Remark 9.11 If $f(x)$ is the density function of the continuous random variable X, then the (cumulative) distribution function $F(x)$ is given by the formula:

$$F(x) = \mathcal{P}\{X \leq x\} = \int_{-\infty}^{x} f(t)\,dt. \tag{9.73}$$

Since $\mathcal{P}\{X = x\} = 0$, we have:

$$F(x) = \mathcal{P}\{X \leq x\} = \mathcal{P}\{X < x\},$$

so the distribution function $F(x)$ is *continuous*, not just *right-continuous* as in the case of discrete random variables. There are no "gaps", which would correspond to numbers with a nonzero probability of occurring (Fig. 9.10).

If the density function f is continuous at the point x, then F is differentiable at x and

$$F'(x) = f(x). \tag{9.74}$$

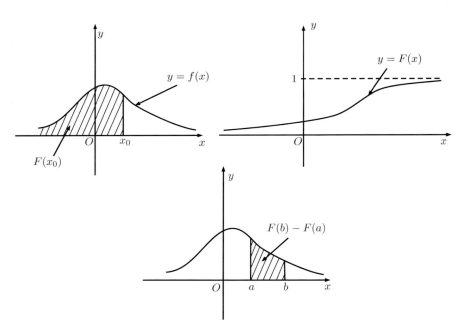

Fig. 9.10 Probability density function $f(x)$ and cumulative distribution function $F(x)$

9.5 Continuous Random Variables

Example 9.29 Let X be a continuous random variable with the probability density function

$$f(x) = \begin{cases} C(2x - x^2), & \text{if } 0 < x < 2, \\ 0 & \text{otherwise.} \end{cases}$$

(a) What is the value of C?
(b) Find $P\{X > 1\}$.
(c) Find the cumulative distribution function of X.

(a) Since $f(x)$ is a probability density function, it has to satisfy (9.72), so we can write:

$$1 = \int_{-\infty}^{\infty} f(x)dx = \int_{0}^{2} C(2x - x^2)dx \Rightarrow$$

$$1 = C\left(x^2 - \frac{x^3}{3}\right)\Big|_{0}^{2} = C \cdot \frac{4}{3} \Rightarrow C = \frac{3}{4}.$$

(b)

$$P\{X > 1\} = \int_{1}^{\infty} f(x)dx = \int_{1}^{2} \frac{3}{4}(2x - x^2) = \frac{1}{2}.$$

(c) Since, for all $x \in [0, 2]$, $\int_{0}^{x} \frac{3}{4}(2t - t^2)dt = \frac{3t^2 - t^3}{4}\Big|_{0}^{x}$, one can write:

$$F(x) = P\{X \leq x\} = \int_{-\infty}^{x} f(t)dt = \begin{cases} 0, & \text{if } x \leq 0, \\ \frac{1}{4}(3x^2 - x^3), & \text{if } 0 < x < 2, \\ 1, & \text{if } x \geq 2. \end{cases}$$

The graphs of the density function $f(x)$ and the distribution function $F(x)$ are presented in Fig. 9.11.

Example 9.30 Find the constant C such that the function $f : \mathbb{R} \to \mathbb{R}$,

$$f(x) = \frac{C}{x^2 + 1}, \quad x \in \mathbb{R}$$

is a density function. If X is a random variable with the density function $f(x)$, find the probability that $|X| \leq 1$.

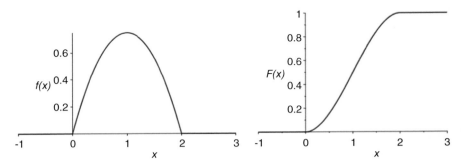

Fig. 9.11 The density function $f(x)$ and the distribution function $F(x)$ in Example 9.29

By (9.72) we have:

$$1 = \int_{-\infty}^{\infty} f(x)dx = C \int_{-\infty}^{\infty} \frac{dx}{x^2+1} =$$

$$C \arctan x \Big|_{-\infty}^{\infty} = C \left[\frac{\pi}{2} - \left(-\frac{\pi}{2}\right)\right] \Rightarrow C = \frac{1}{\pi}.$$

The function $f(x) = \dfrac{1}{\pi(x^2+1)}$ is called the *Cauchy density*.
The probability that $-1 \leq X \leq 1$ is

$$\mathcal{P}\{-1 \leq X \leq 1\} = \int_{-1}^{1} \frac{dx}{\pi(x^2+1)} = \frac{1}{\pi} \arctan x \Big|_{-1}^{1} = \frac{1}{\pi}\left[\frac{\pi}{4} - \left(-\frac{\pi}{4}\right)\right] = \frac{1}{2}.$$

9.5.2 Expected Value, Moments and Variance for Continuous Random Variables

Recall that the mean (or expected value) of a *discrete* random variable X taking on the values x_i with the probabilities $f(x_i)$, $i = 1, 2, \ldots$ is defined as the sum

$$E(X) = \sum_{i=1}^{\infty} x_i f(x_i),$$

provided that the series absolutely converges. The expected value of a *continuous* random variable is defined using the (improper) integral instead of the sum (of a series):

9.5 Continuous Random Variables

Definition 9.20 If X is a continuous random variable with the probability density function $f(x)$, then the mean (or expected value) of X is defined by

$$E(X) = \int_{-\infty}^{\infty} x f(x)\, dx, \qquad (9.75)$$

provided that the integral absolutely converges. If the integral is not absolutely convergent, then the mean value of X is undefined.

For instance, the mean of the random variable X from Example 9.30 is not defined because the integral

$$\int_{-\infty}^{\infty} |x| f(x)\, dx = \int_{-\infty}^{\infty} \frac{|x|}{\pi(x^2+1)}\, dx$$

is divergent.

Example 9.31 The probability density function of the random variable X is given by

$$f(x) = \begin{cases} 1, & \text{if } x \in [0, 1], \\ 0, & \text{otherwise.} \end{cases}$$

(a) Find the expected value $E(X)$ and the distribution function $F(x)$.
(b) Consider the random variable $Y = \exp(X)$. Find its cumulative distribution function $F_Y(x)$, its probability density function $f_Y(x)$ and its mean $E(Y)$.

(a) By Definition 9.20, the expected value of X is

$$E(X) = \int_{-\infty}^{\infty} x f(x)\, dx = \int_0^1 x\, dx = \frac{1}{2}.$$

The cumulative distribution function is given by (9.73):

$$F(x) = \int_{-\infty}^{x} f(t)\, dt = \begin{cases} 0, & \text{if } x < 0, \\ x, & \text{if } x \in [0, 1], \\ 1, & \text{if } x > 1. \end{cases}$$

(b) The cumulative distribution function of $Y = \exp(X)$ is defined as:

$$F_Y(x) = \mathcal{P}\{\exp(X) \leq x\} = \mathcal{P}\{X \leq \ln x\} = F(\ln x) =$$

$$= \begin{cases} 0, & \text{if } \ln x < 0, \\ \ln x, & \text{if } \ln x \in [0, 1], \\ 1, & \text{if } \ln x > 1. \end{cases}$$

for any $x > 0$. Since the exponential function takes only positive values, $F_Y(x) = 0$ for $x \leq 0$, so we find:

$$F_Y(x) = \begin{cases} 0, & \text{if } x < 1, \\ \ln x, & \text{if } x \in [1, e], \\ 1, & \text{if } x > e. \end{cases}$$

Consequently, the probability density function of Y is

$$f_Y(x) = F_Y'(x) = \begin{cases} \frac{1}{x}, & \text{if } x \in [1, e], \\ 0, & \text{otherwise} \end{cases}$$

and the expected value of Y is

$$E(Y) = \int_{-\infty}^{\infty} f_Y(x)\, dx = \int_{1}^{e} x \cdot \frac{1}{x}\, dx = e - 1.$$

We remark that the expected value of the random variable $Y = \exp(X)$ could be also calculated by using the formula

$$E(Y) = \int_{-\infty}^{\infty} \exp(x) f(x)\, dx.$$

As stated in Proposition 9.10, if X is a discrete random variable and $g : \mathbb{R} \to \mathbb{R}$, then the expected value of $g(X)$ (if exists) is given by

$$E(Y) = \sum_{k \geq 1} g(x_k) \mathcal{P}\{X = x_k\}.$$

A similar result holds for continuous random variables. To prove it, we need the following lemma (see [31]).

Lemma 9.2 *If X is a continuous **nonnegative** random variable with a finite expectation $E(X)$, then*

$$E(X) = \int_{0}^{\infty} \mathcal{P}\{X > y\}\, dy. \tag{9.76}$$

If X is a continuous random variable with a finite expectation $E(X)$, then

$$E(X) = \int_{0}^{\infty} \mathcal{P}\{X > y\}\, dy - \int_{0}^{\infty} \mathcal{P}\{X < -y\}\, dy. \tag{9.77}$$

9.5 Continuous Random Variables

Proof Let $f(x)$ be the density function of X. For any $y > 0$, the probability

$$\mathcal{P}\{X > y\} = \int_y^\infty f(x)dx$$

is known as the *survival probability* and we can write:

$$\int_0^\infty \mathcal{P}\{X > y\}dy = \int_0^\infty \int_y^\infty f(x)dx\,dy.$$

By changing the order of integration in the last double integral, we obtain:

$$\int_0^\infty \int_y^\infty f(x)dx\,dy = \int_0^\infty \int_0^x f(x)dy\,dx = \int_0^\infty f(x) \int_0^x dy\,dx,$$

so

$$\int_0^\infty \mathcal{P}\{X > y\}dy = \int_0^\infty xf(x)\,dx.$$

If $X \geq 0$, then $f(x) = 0$ for all $x < 0$, so the formula (9.76) follows. Otherwise, one can prove, in a similar manner as above, that

$$\int_{-\infty}^0 xf(x)\,dx = -\int_0^\infty \mathcal{P}\{X < -y\}\,dy$$

and so, since

$$E(X) = \int_{-\infty}^0 xf(x)\,dx + \int_0^\infty xf(x)\,dx,$$

the Eq. (9.77) readily follows. □

Theorem 9.9 *If X is a continuous random variable with the probability density function $f(x)$ and $g : \mathbb{R} \to \mathbb{R}$ is an arbitrary function, then the random variable $Y = g(X)$ has a finite expected value if and only if the integral $\int_{-\infty}^\infty |g(x)|f(x)\,dx$ converges and in this case we have:*

$$E(g(X)) = \int_{-\infty}^\infty g(x)f(x)\,dx. \qquad (9.78)$$

Proof By Lemma 9.2 we can write:

$$E(g(X)) = \int_0^\infty \mathcal{P}\{g(X) > y\}\,dy - \int_0^\infty \mathcal{P}\{g(X) < -y\}\,dy =$$

$$= \int_0^\infty \int_{\{x:g(x)>y\}} f(x)\,dx\,dy - \int_0^\infty \int_{\{x:g(x)<-y\}} f(x)\,dx\,dy$$

and so, by changing the order of integration, we obtain:

$$E(g(X)) = \int_{\{x:g(x)\geq 0\}} \int_0^{g(x)} f(x)\,dy\,dx - \int_{\{x:g(x)<0\}} \int_0^{-g(x)} f(x)\,dy\,dx =$$

$$= \int_{\{x:g(x)\geq 0\}} f(x) \int_0^{g(x)} dy\,dx - \int_{\{x:g(x)<0\}} f(x) \int_0^{-g(x)} dy\,dx =$$

$$= \int_{\{x:g(x)\geq 0\}} f(x)g(x)\,dx + \int_{\{x:g(x)<0\}} f(x)g(x)\,dx = \int_{-\infty}^\infty g(x)f(x)\,dx.$$

$\square$

If $g(x) = ax + b$, then

$$E(aX + b) = \int_{-\infty}^\infty (ax+b)f(x)\,dx =$$

$$= a\int_{-\infty}^\infty xf(x)\,dx + b\int_{-\infty}^\infty f(x)\,dx = aE(X) + b,$$

and the next corollary follows.

Corollary 9.4 *If X is a continuous random variable with finite expected value $E(X)$, then, for any real constants a, b, the random variable $Y = aX + b$ has the expected value*

$$E(aX + b) = aE(X) + b. \tag{9.79}$$

The *moments* of a continuous random variable are defined in the same way as the moments of a discrete random variable, but in the continuous case the expected value is calculated by an improper integral (instead of the sum of a series, like in the discrete case)

9.5 Continuous Random Variables

Definition 9.21 If X is a continuous random variable with the probability density function $f(x)$, then the k-th moment of X is defined by

$$E\left(X^k\right) = \int_{-\infty}^{\infty} x^k f(x)\, dx, \tag{9.80}$$

provided that the integral absolutely converges. If the integral is not absolutely convergent, then the k-th moment of X is undefined.

Proposition 9.14 *If the n-th moment of X exists, then the k-th moment exists, for all $k = 1, \ldots, n$.*

Proof Let $f(x)$ be the probability density function of X. Since there exists the n-th moment, the integral

$$E\left(X^n\right) = \int_{-\infty}^{\infty} x^n f(x)\, dx$$

is absolutely convergent. Now, by the same inequality we used in proof of Proposition 9.11,

$$|x|^k \le |x|^n + 1,$$

for all $x \in \mathbb{R}$ and $k < n$, one can write:

$$\int_{-\infty}^{\infty} |x|^k f(x)\, dx \le \underbrace{\int_{-\infty}^{\infty} |x|^n f(x)\, dx}_{<\infty} + \underbrace{\int_{-\infty}^{\infty} f(x)\, dx}_{=1} < \infty.$$

$\square$

The variance of a continuous random variable is defined in the same way as the variance of a discrete one.

Definition 9.22 If X is a continuous random variable having the expected value $\mu = E(X)$, then the variance of X is defined by

$$Var(X) = E\left((X - \mu)^2\right). \tag{9.81}$$

If $E\left((X - \mu)^2\right) = \infty$, we say that the variance of X does not exist.

Proposition 9.15 *If X is a continuous random variable, then the variance $Var(X)$ exists if and only if the second moment $E(X^2)$ exists. In this case, the following formula holds:*

$$Var(X) = E\left(X^2\right) - E(X)^2. \tag{9.82}$$

Proof By Proposition 9.14, if the second moment $E(X^2)$ exists, then the expected value $E(X)$ exists, too. Let $f(x)$ be the probability density function of X and $\mu = E(X)$. Then, by Theorem 9.9, we have:

$$Var(X) = \int_{-\infty}^{\infty} (x - \mu)^2 f(x)\,dx =$$

$$= \int_{-\infty}^{\infty} x^2 f(x)\,dx - 2\mu \int_{-\infty}^{\infty} x f(x)\,dx + \mu^2 \int_{-\infty}^{\infty} f(x)\,dx =$$

$$= E\left(X^2\right) - 2\mu^2 + \mu^2 = E\left(X^2\right) - \mu^2.$$

□

Proposition 9.16 *If X is a continuous random variable with the variance $Var(X)$, then, for any real constants a, b, the random variable $Y = aX + b$ has the variance*

$$Var(aX + b) = a^2 Var(X). \tag{9.83}$$

Proof By Corollary 9.4 we know that $E(aX + b) = a\mu + b$, so

$$Var(aX + b) = E\left((aX - a\mu)^2\right) = a^2 E\left((X - \mu)^2\right) = a^2 Var(X).$$

□

Theorem 9.10 *If X is a continuous random variable with the probability density function $f(x)$ and a, b are two real numbers, $a \neq 0$, then the probability density function of the random variable $Y = aX + b$ is*

$$f_Y(x) = \frac{1}{|a|} f\left(\frac{x-b}{a}\right). \tag{9.84}$$

Proof First, we suppose that $a > 0$. Let us calculate the cumulative distribution function of Y,

$$F_Y(x) = \mathcal{P}\{Y \leq x\} = \mathcal{P}\{aX + b \leq x\} =$$

$$= \mathcal{P}\left\{X \leq \frac{x-b}{a}\right\} = F\left(\frac{x-b}{a}\right).$$

It follows that

$$f_Y(x) = F_Y'(x) = \left[F\left(\frac{x-b}{a}\right)\right]' = \frac{1}{a} f\left(\frac{x-b}{a}\right).$$

9.5 Continuous Random Variables

Now, if $a < 0$, we have:

$$F_Y(x) = \mathcal{P}\{aX + b \le x\} =$$

$$= \mathcal{P}\left\{X \ge \frac{x-b}{a}\right\} = 1 - F\left(\frac{x-b}{a}\right).$$

It follows that

$$f_Y(x) = F'_Y(x) = \left[1 - F\left(\frac{x-b}{a}\right)\right]' = -\frac{1}{a} f\left(\frac{x-b}{a}\right),$$

and the theorem is proved. □

We discussed in Sect. 9.4 about jointly distributed discrete random variables and we defined the independence of two (or more) discrete random variables. Now we extend these notions to continuous random variables.

Definition 9.23 Let X and Y be two continuous random variables in the probability space $(\Omega, \mathcal{K}, \mathcal{P})$. The *joint cumulative distribution function* of X and Y is the function $F : \mathbb{R}^2 \to \mathbb{R}$,

$$F(x, y) = \mathcal{P}\{X \le x, Y \le y\}. \tag{9.85}$$

The *joint probability density function* of X and Y is the function $f : \mathbb{R}^2 \to \mathbb{R}$ with the property that, for every $D \subset \mathbb{R}^2$,

$$\mathcal{P}\{(X, Y) \in D\} = \iint_D f(x, y)\, dx\, dy. \tag{9.86}$$

If we take $D = (-\infty, x] \times (-\infty, y]$, then we can write:

$$F(x, y) = \mathcal{P}\{(X, Y) \in (-\infty, x] \times (-\infty, y]\} = \int_{-\infty}^{y} \int_{-\infty}^{x} f(u, v)\, du\, dv. \tag{9.87}$$

It follows, by differentiation, that

$$f(x, y) = \frac{\partial^2 F}{\partial x \partial y}, \tag{9.88}$$

for any $(x, y) \in \mathbb{R}^2$ where the partial derivative (9.88) is defined.

Remark 9.12 Given the joint cumulative distribution function $F(x, y)$, we can find the distribution function of X, $F_X(x)$ and the distribution function of Y, $F_Y(y)$.

Thus, by (9.85), we can write:

$$F(x, \infty) = \mathcal{P}\{X \leq x, Y \leq \infty\} = \mathcal{P}\{X \leq x\} = F_X(x),$$
$$F(\infty, y) = \mathcal{P}\{X \leq \infty, Y \leq y\} = \mathcal{P}\{Y \leq y\} = F_Y(y).$$

Similarly, given the joint probability density function $f(x, y)$, we can calculate the probability density function of X, $f_X(x)$ and the probability density function of Y, $f_Y(x)$, using the formula (9.86) for domains of the form $D = [a, b] \times (-\infty, \infty)$ and $D = (-\infty, \infty) \times [a, b]$, respectively:

$$\mathcal{P}\{X \in [a, b]\} = \mathcal{P}\{X \in [a, b], Y \in (-\infty, \infty)\}$$
$$= \int_a^b \int_{-\infty}^{\infty} f(x, y) dy\, dx$$
$$= \int_a^b f_X(x) dx.$$

so the probability density function of X is

$$f_X(x) = \int_{-\infty}^{\infty} f(x, y) dy. \tag{9.89}$$

Similarly, the probability density function of Y is given by

$$f_Y(y) = \int_{-\infty}^{\infty} f(x, y) dx. \tag{9.90}$$

Definition 9.24 The random variables X and Y are said to be *independent* if

$$\mathcal{P}\{X \leq x, Y \leq y\} = \mathcal{P}\{X \leq x\}\mathcal{P}\{Y \leq y\},$$

for any $x, y \in \mathbb{R}$.

The random variables $X_1, X_2, \ldots, X_n$ are said to be independent if for any $1 \leq i_1 < i_2 < \ldots < i_k \leq n$ and any real numbers $x_1, x_2, \ldots, x_k$, we have

$$\mathcal{P}\{X_{i_1} \leq x_1, \ldots, X_{i_k} \leq x_k\} = \prod_{j=1}^{k} \mathcal{P}\{X_{i_j} \leq x_j\}.$$

Using the joint cumulative distribution function, we say that X and Y are independent if and only if

$$F(x, y) = F_X(x) F_Y(y), \quad \text{for all } x, y \in \mathbb{R}.$$

9.5 Continuous Random Variables

Using the joint probability density function, the condition of independence is written:

$$f(x, y) = f_X(x) f_Y(y), \quad \text{for all } x, y \in \mathbb{R}.$$

As shown in the case of discrete random variables, the joint distribution function and the joint density function of some continuous random variables X and Y can be used to define new random variables like $X + Y$, XY, X/Y and so on.

Example 9.32 Suppose that X and Y are independent random variables with the same density function,

$$f(x) = \begin{cases} \exp(-x) & \text{if } x \geq 0 \\ 0 & \text{otherwise.} \end{cases}$$

Find the density function of the random variable X/Y.

First, we determine the distribution function of X/Y. We remark that, since X and Y are independent, their joint density function is $\varphi(x, y) = f(x)f(y) = \exp(-x)\exp(-y)$, for $x > 0$, $y > 0$, and $\varphi(x, y) = 0$ if $x \leq 0$ or $y \leq 0$. Thus, $F_{X/Y}(z) = 0$ if $z \leq 0$, and, for any $z > 0$, we have:

$$F_{X/Y}(z) = \mathcal{P}\{X/Y \leq z\} = \iint\limits_{x/y \leq z} \varphi(x, y) \, dx dy$$

$$= \int_0^\infty \int_0^{yz} \exp(-x) \exp(-y) dx \, dy = \int_0^\infty \exp(-y) \left(-\exp(-x) \Big|_0^{yz} \right) dy$$

$$= \int_0^\infty [\exp(-y) - \exp(-y(z+1))] \, dy = \left[-\exp(-y) + \frac{\exp(-y(z+1))}{z+1} \right]_0^\infty$$

$$\Rightarrow F_{X/Y}(z) = 1 - \frac{1}{z+1}, \quad \text{if } z > 0.$$

We find by differentiation that the density function of X/Y is

$$f_{X/Y}(z) = \begin{cases} \frac{1}{(z+1)^2} & \text{if } z \geq 0 \\ 0 & \text{otherwise.} \end{cases}$$

Using the joint density function, the results stated by Theorems 9.2 and 9.4 regarding the expected value and the variance of the sum of two discrete random variables can be extended for continuous random variables:

Theorem 9.11 *Let X and Y be two random variables in the probability space $(\Omega, \mathcal{K}, \mathcal{P})$, with finite expected values. Then*

(i) $E(X + Y) = E(X) + E(Y)$.

Moreover, if X and Y are independent and $E\left(X^2\right), E\left(Y^2\right) < \infty$, then

(ii) $E(XY) = E(X)E(Y)$.

(iii) $Var(X + Y) = Var(X) + Var(Y)$.

Proof

(i) Let $f(x, y)$ denote the joint density function of X and Y. Then,

$$E(X + Y) = \int_{-\infty}^{\infty}\int_{-\infty}^{\infty}(x + y)f(x, y)\,dx\,dy$$

$$= \int_{-\infty}^{\infty} x \int_{-\infty}^{\infty} f(x, y)\,dy\,dx + \int_{-\infty}^{\infty} y \int_{-\infty}^{\infty} f(x, y)\,dx\,dy$$

$$= \int_{-\infty}^{\infty} x f_X(x)\,dx + \int_{-\infty}^{\infty} y f_Y(y)\,dy = E(X) + E(Y).$$

(ii) Since X and Y are independent, we have $f(x, y) = f_X(x) f_Y(y)$ for any $x, y \in \mathbb{R}$. Thus, we can write:

$$E(XY) = \int_{-\infty}^{\infty}\int_{-\infty}^{\infty} xy f(x, y)\,dx\,dy$$

$$= \int_{-\infty}^{\infty} x f_X(x)\,dx \int_{-\infty}^{\infty} y f_Y(y)\,dy = E(X)E(Y).$$

(iii) By Proposition 9.15 we have:

$$Var(X + Y) = E\left((X + Y)^2\right) - E(X + Y)^2$$

$$\stackrel{i)}{=} E(X^2) + 2E(XY) + E(Y^2) - E(X)^2 - 2E(X)E(Y) - E(Y)^2$$

$$\stackrel{ii)}{=} E(X^2) - E(X)^2 + E(Y^2) - E(Y)^2$$

$$= Var(X) + Var(Y).$$

□

Remark 9.13 As shown in the discrete case, Definition 9.23 and Theorem 9.11 can be extended for n random variables ($n > 2$). Thus, for any continuous random variables $X_1, X_2, \ldots, X_n$, we have:

$$E(X_1 + X_2 + \ldots + X_n) = E(X_1) + E(X_2) + \ldots + E(X_n).$$

9.5 Continuous Random Variables

If, in addition, $X_1, X_2, \ldots, X_n$ are *independent*, then

$$Var(X_1 + X_2 + \ldots + X_n) = Var(X_1) + Var(X_2) + \ldots + Var(X_n).$$

Moreover, if $X_1, X_2, \ldots, X_n$ are *independent* and *identically distributed* (**iid**) random variables with $\mu = E(X_i)$, $\sigma^2 = Var(X_i)$ for all $i = 1, \ldots, n$, and

$$S_n = X_1 + \ldots + X_n, \quad \overline{X}_n = \frac{X_1 + \ldots + X_n}{n},$$

then

$$E(S_n) = n\mu, \quad Var(S_n) = n\sigma^2, \tag{9.91}$$

$$E(\overline{X}_n) = \mu, \quad Var(\overline{X}_n) = \frac{\sigma^2}{n}. \tag{9.92}$$

9.5.3 Characteristic Function

In addition to the probability distribution function and the cumulative distribution function, there is another function that completely characterizes a random variable, namely the *characteristic function*.

We have seen that if X is a random variable (either discrete or continuous) and $g : \mathbb{R} \to \mathbb{R}$ is a real function, then $g(X)$ is also a random variable. If X is continuous, with the probability distribution $f(x)$, then the expected value of $g(X)$ is given by the formula

$$E[g(X)] = \int_{-\infty}^{\infty} g(x) f(x) dx,$$

whenever the integral is absolutely convergent (see Theorem 9.9). If X is discrete, with the probability distribution $f(x)$, where $f(x) \neq 0$ only for an at most countable set of numbers, $x_1, x_2, \ldots$, then the expected value of $g(X)$ is given by the formula

$$E[g(X)] = \sum_{k \geq 1} g(x_k) f(x_k),$$

whenever the series is absolutely convergent (see Proposition 9.10).

To define the characteristic function, we shall apply these results for *complex functions* of the form $g : \mathbb{R} \to \mathbb{C}$, $g(x) = \exp(itx) = \cos tx + i \sin tx$ (where $t \in \mathbb{R}$).

Definition 9.25 Let X be a random variable in the probability space $(\Omega, \mathcal{K}, \mathcal{P})$. The *characteristic function* of X is the complex-valued function $\varphi_X : \mathbb{R} \to \mathbb{C}$ defined by:

$$\varphi_X(t) = E\left[\exp(itX)\right] \qquad (9.93)$$

Thus, if X is a continuous random variable with the density function $f(x)$, then

$$\varphi_X(t) = \int_{-\infty}^{\infty} f(x) \exp(itx)\, dx. \qquad (9.94)$$

If X is a discrete random variable taking on the values x_k with the probabilities $f(x_k), k = 1, 2, \ldots$ ($f(x) = 0$ for $x \neq x_k, k \geq 1$), then

$$\varphi_X(t) = \sum_{k \geq 1} f(x_k) \exp(itx_k). \qquad (9.95)$$

We remark that $\varphi_X(t)$ is correctly defined for any $t \in \mathbb{R}$: since

$$|\exp(itx)| = |\cos tx + i \sin tx| = 1,$$

the improper integral in (9.94) as well as the series in (9.95) are *absolutely convergent*:

$$\int_{-\infty}^{\infty} |f(x) \exp(itx)|\, dx = \int_{-\infty}^{\infty} f(x)\, dx = 1,$$

$$\sum_{k \geq 1} |p_k \exp(itx_k)| = \sum_{k \geq 1} p_k = 1.$$

Remark 9.14 By the formulas (9.94) and (9.95) it follows that, for any random variable X we have:

$$\varphi_X(0) = 1$$

and

$$|\varphi_X(t)| \leq 1, \text{ for all } t \in \mathbb{R}.$$

(the function φ_X takes values inside the disc $\{z \in \mathbb{C} : |z| \leq 1\}$).

Theorem 9.12 *If X is a random variable in the probability space $(\Omega, \mathcal{K}, \mathcal{P})$ with the characteristic function φ_X, and $a, b \in \mathbb{R}$, $a \neq 0$, then the characteristic function*

9.5 Continuous Random Variables

of the random variable $aX + b$ is given by the formula

$$\varphi_{aX+b}(t) = \exp(ibt)\varphi_X(at). \tag{9.96}$$

Proof By the definition of the characteristic function (9.93), since $E(aX) = aE(x)$ for any constant a, we can write:

$$E\left[\exp(it(aX + b))\right] = E\left[\exp(itb)\exp(itaX)\right] = \exp(itb)\varphi_X(at).$$

$\square$

Corollary 9.5 *If X is a random variable with the characteristic function φ_X, then the characteristic function of the random variable $-X$ is*

$$\varphi_{-X}(t) = \overline{\varphi_X(t)}, \text{ for all } t \in \mathbb{R}. \tag{9.97}$$

Proof By (9.96) we can write:

$$\varphi_{-X}(t) = \varphi_X(-t) = E\left[\exp(-itX)\right]$$
$$= \int_{-\infty}^{\infty} \exp(-itx) f(x)\, dx = \int_{-\infty}^{\infty} \overline{\exp(itx) f(x)}\, dx =$$
$$= \overline{\varphi_X(t)}.$$

$\square$

Theorem 9.13 *If X and Y are two independent random variables in the probability space $(\Omega, \mathcal{K}, \mathcal{P})$ having the characteristic functions $\varphi_X(t)$ and $\varphi_Y(t)$ respectively, then the characteristic function of the random variable $X + Y$ is*

$$\varphi_{X+Y}(t) = \varphi_X(t)\varphi_Y(t). \tag{9.98}$$

Proof Since X and Y are independent, it follows that the random variables $\exp(itX)$ and $\exp(itY)$ are also independent, so

$$E\left[\exp(itX)\exp(itY)\right] = E\left[\exp(itX)\right] E\left[\exp(itY)\right]$$

and the Eq. (9.98) follows immediately by the definition of the characteristic function (9.93). $\square$

Corollary 9.6 *If $X_1, X_2, \ldots, X_n$ are n independent random variables in the probability space $(\Omega, \mathcal{K}, \mathcal{P})$ having the characteristic functions $\varphi_{X_k}(t)$, $k = 1, 2, \ldots, n$,* then the characteristic function of the random variable $X_1 + X_2 + \ldots + X_n$ is

$$\varphi_{X_1+\ldots+X_n}(t) = \varphi_{X_1}(t)\varphi_{X_2}(t)\ldots\varphi_{X_n}(t). \tag{9.99}$$

Theorem 9.14 *If X is a random variable which has a moment of order n, then the characteristic function of X is differentiable n times and*

$$\varphi_X^{(k)}(0) = i^k E\left(X^k\right) \tag{9.100}$$

for all $k = 1, \ldots, n$.

Proof Suppose that X is continuous, so φ_X is given by (9.94). By (formal) differentiating this equation k times, we obtain:

$$\varphi_X^{(k)}(t) = i^k \int_{-\infty}^{\infty} x^k f(x) \exp(itx)\, dx. \tag{9.101}$$

Since the n-th moment of X exists, it follows that all the moments of order k, $k \leq n$ exist, so we have:

$$\int_{-\infty}^{\infty} \left|x^k f(x) \exp(itx)\right| dx = \int_{-\infty}^{\infty} |x|^k f(x)\, dx < \infty.$$

Therefore, the improper integral in (9.101) absolutely converges, so the relation holds indeed for every $t \in \mathbb{R}$ and $k \leq n$. For $t = 0$ we obtain the Eq. (9.100).

Now, if X is a discrete random variable, its characteristic function is given by (9.95):

$$\varphi_X(t) = \sum_{j \geq 1} f(x_j) \exp(itx_j).$$

By (formal) differentiating this equation k times, we obtain:

$$\varphi_X^{(k)}(t) = i^k \sum_{j \geq 1} x_j^k f(x_j) \exp(itx_j) \tag{9.102}$$

Again, since all the moments of order k, $k \leq n$ exist, we can write:

$$\sum_{j \geq 1} \left|x_j^k f(x_j) \exp(itx_j)\right| = \sum_{j \geq 1} |x_j|^k f(x_j) < \infty,$$

so the series in (9.102) absolutely converges and the relation holds indeed for every $t \in \mathbb{R}$ and $k \leq n$. For $t = 0$ we obtain the Eq. (9.100). □

9.5 Continuous Random Variables

We have seen that, given the distribution function of a random variable X, it is always possible to find its characteristic function. The converse result holds as well: a distribution function is uniquely determined by the characteristic function $\varphi_X(t)$ (see Theorem 9.15 for the continuous random variables and Theorem 9.16 for the discrete case).

Theorem 9.15 *If X is a continuous random variable with the characteristic function $\varphi_X : \mathbb{R} \to \mathbb{C}$, then its probability density function $f(x)$ is given by the formula*

$$f(x) = \frac{1}{2\pi} \int_{-\infty}^{\infty} \varphi_X(t) \exp(-itx) \, dt, \qquad (9.103)$$

for any $x \in \mathbb{R}$ where the function f is continuous.

Proof Since the density function of a random variable is always absolutely integrable (we know that $\int_{-\infty}^{\infty} f(x) dx = 1$), we can apply the Fourier Integral Theorem (Theorem 5.5). Thus, at any point x where f is continuous, by the formula (5.33) one can write:

$$f(x) = \frac{1}{2\pi} \int_{-\infty}^{\infty} \int_{-\infty}^{\infty} f(u) \exp(iv(x-u)) \, du \, dv$$

$$\stackrel{t=-v}{=} \frac{1}{2\pi} \int_{-\infty}^{\infty} \int_{-\infty}^{\infty} f(u) \exp(it(u-x)) \, du \, dt$$

$$= \frac{1}{2\pi} \int_{-\infty}^{\infty} \exp(-itx) \underbrace{\int_{-\infty}^{\infty} f(u) \exp(itu) \, du}_{\varphi_X(t)} \, dt$$

and the formula (9.103) follows. □

Theorem 9.16 *Let X be a discrete random variable which takes on the integer values $0, 1, \ldots$ with the probabilities $f(k) = \mathcal{P}\{X = k\}$ ($f(x)=0$ if $x \notin \mathbb{N}$). If $\varphi_X(t)$ is the characteristic function of X, then its probability distribution function $f(x)$ is given by the formula*

$$f(n) = \frac{1}{2\pi} \int_{-\pi}^{\pi} \varphi_X(t) \exp(-itn) \, dt \qquad (9.104)$$

for all $n = 1, 2, \ldots$.

Proof By the definition formula (9.95), the characteristic function $\varphi_X(t)$ is:

$$\varphi_X(t) = \sum_{k=0}^{\infty} f(k) \exp(itk). \qquad (9.105)$$

For every $n = 0, 1, \ldots$, we multiply by $\exp(-itn)$ the relation (9.105) and, since the resulting series absolutely converges, we can integrate term by term on the interval $[-\pi, \pi]$ and find:

$$\int_{-\pi}^{\pi} \varphi_X(t) \exp(-itn) \, dt = \sum_{k=0}^{\infty} f(k) \int_{-\pi}^{\pi} \exp(it(k-n)) \, dt.$$

We notice that the series above has only one nonzero term, for $k = n$:

$$\int_{-\pi}^{\pi} \exp(it(k-n)) \, dt = \int_{-\pi}^{\pi} \cos(k-n)t \, dt + i \int_{-\pi}^{\pi} \sin(k-n)t \, dt$$

$$= \begin{cases} 2\pi, & \text{if } k = n \\ 0, & \text{if } k \neq n. \end{cases}$$

Hence

$$\int_{-\pi}^{\pi} \varphi_X(t) \exp(-itn) \, dt = 2\pi f(n)$$

and (9.104) follows. □

Example 9.33 Find the characteristic function of a binomial random variable, $X \sim Bin(n, p)$. Use it to find the formulas for the mean and the variance of X. Prove that, if $X \sim Bin(n, p)$ and $Y \sim Bin(m, p)$, then $X + Y \sim Bin(m+n, p)$.

The probability distribution function of X is:

$$X \sim \begin{pmatrix} 0 & 1 & \cdots & k & \cdots & n \\ \binom{n}{0} p^0 q^n & \binom{n}{1} p q^{n-1} & \cdots & \binom{n}{k} p^k q^{n-k} & \cdots & \binom{n}{n} p^n q^0 \end{pmatrix},$$

so the characteristic function is given by:

$$\varphi_X(t) = \sum_{k=0}^{n} \binom{n}{k} p^k q^{n-k} \exp(itk)$$

$$= (p \exp(it) + q)^n$$

Note that the same result could be obtained if we saw the binomial random variable X as the sum of n independent Bernoulli random variables,

$$X_k \sim \begin{pmatrix} 0 & 1 \\ q & p \end{pmatrix}$$

9.5 Continuous Random Variables

having the characteristic function $\varphi_{X_k}(t) = p\exp(it) + q$. Thus, since

$$X = X_1 + X_2 + \ldots + X_n,$$

by Corollary 9.6 we can write:

$$\varphi_X(t) = \varphi_{X_1}(t)\varphi_{X_2}(t)\ldots\varphi_{X_n}(t) = (p\exp(it) + q)^n.$$

Let us apply Theorem 9.14 to find $E(X)$ and $E(X^2)$. The first and the second derivatives of $\varphi_X(t)$ are given by:

$$\varphi'_X(t) = npi\,(p\exp(it) + q)^{n-1}\exp(it),$$

$$\varphi''_X(t) = n(n-1)(pi)^2(p\exp(it) + q)^{n-2}\exp(2it) + npi^2(p\exp(it) + q)^{n-1}\exp(it).$$

Therefore, the expected value of X is (the same as in (9.62)):

$$E(X) = \frac{1}{i}\varphi'_X(0) = np,$$

while the second moment is given by

$$E(X^2) = \frac{1}{i^2}\varphi''_X(0) = n(n-1)p^2 + np,$$

so we obtain for the variance the same expression as in (9.63):

$$Var(X) = E(X^2) - E(X)^2 = np(1-p).$$

Now, if $X \sim Bin(n, p)$ and $Y \sim Bin(m, p)$, then the characteristic function of $X + Y$ is

$$\varphi_{X+Y}(t) = \varphi_X(t)\varphi_Y(t) = (p\exp(it) + q)^n (p\exp(it) + q)^m$$
$$= (p\exp(it) + q)^{n+m},$$

which means that $X + Y \sim Bin(m+n, p)$.

Example 9.34 Find the characteristic function of a Poisson random variable, $X \sim Poi(\lambda)$. Use it to find the formulas for the mean and the variance of X. Prove that, if $X \sim Poi(\lambda)$ and $Y \sim Poi(\mu)$, then $X + Y \sim Poi(\lambda + \mu)$.

The probability distribution function of X is:

$$X \sim \begin{pmatrix} 0 & 1 & \cdots & k & \cdots \\ \exp(-\lambda) & \frac{\lambda}{1!}\exp(-\lambda) & \cdots & \frac{\lambda^k}{k!}\exp(-\lambda) & \cdots \end{pmatrix},$$

so the characteristic function of X is given by:

$$\varphi_X(t) = \sum_{k=0}^{\infty} \frac{\lambda^k}{k!} \exp(-\lambda) \exp(itk)$$

$$= \exp(-\lambda) \sum_{k=0}^{\infty} \frac{[\lambda \exp(it)]^k}{k!}$$

$$= \exp[\lambda(\exp(it) - 1)].$$

Let us apply Theorem 9.14 to find $E(X)$ and $E(X^2)$. The first and the second derivatives of $\varphi_X(t)$ are given by:

$$\varphi'_X(t) = \exp[\lambda(\exp(it) - 1)] \lambda i \exp(it),$$

$$\varphi''_X(t) = i^2 \exp[\lambda(\exp(it) - 1)] \exp(it) \left(\lambda^2 \exp(it) + \lambda\right).$$

Therefore, the expected value of X is:

$$E(X) = \frac{1}{i} \varphi'_X(0) = \lambda,$$

while the second moment is given by

$$E\left(X^2\right) = \frac{1}{i^2} \varphi''_X(0) = \lambda^2 + \lambda,$$

so the variance is:

$$Var(X) = E\left(X^2\right) - E(X)^2 = \lambda$$

(see (9.68) and (9.69)).

Let $X \sim Poi(\lambda)$ and $Y \sim Poi(\mu)$ be two Poisson random variables. The characteristic function of $X + Y$ is

$$\varphi_{X+Y}(t) = \varphi_X(t) \varphi_Y(t) = \exp[\lambda(\exp(it) - 1)] \exp[\mu(\exp(it) - 1)]$$

$$= \exp[(\lambda + \mu)(\exp(it) - 1)],$$

so $X + Y$ is also a Poisson random variable with parameter $\lambda + \mu$.

9.5 Continuous Random Variables

Example 9.35 Now, let us find the characteristic function of a continuous random variable X having the following (uniform) probability density function:

$$f(x) = \begin{cases} \frac{1}{2}, & \text{if } x \in [-1, 1], \\ 0, & \text{if } x \notin [-1, 1]. \end{cases}$$

By Eq. (9.94) we can write:

$$\varphi_X(t) = \int_{-\infty}^{\infty} f(x) \exp(itx)\, dx = \frac{1}{2} \int_{-1}^{1} \exp(itx)\, dx =$$

$$= \frac{\exp(itx)}{2it} \Big|_{-1}^{1} = \frac{\exp(it) - \exp(-it)}{2it},$$

hence

$$\varphi_X(t) = \frac{\sin t}{t}.$$

We notice that, in this case, the characteristic function takes only real values. The next proposition shows that this result holds for random variables with symmetric distribution functions (see [16], p.228).

Proposition 9.17 *The characteristic function of a continuous random variable is a real-valued function if and only if its cumulative distribution function is symmetric, that is,*

$$F(-x) = 1 - F(x), \text{ for all } x \in \mathbb{R}. \tag{9.106}$$

*If, in addition, the probability density function $f(x)$ is continuous, then the characteristic function takes only real values if and only if $f(x)$ is an **even** function.*

Proof The characteristic function φ_X is a *real* function if and only if

$$\varphi_X(t) = \overline{\varphi_X(t)}, \text{ for all } t \in \mathbb{R}.$$

Since, by Corollary 9.5,

$$\overline{\varphi_X(t)} = \varphi_{-X}(t) \text{ for all } t \in \mathbb{R},$$

we find that X and $-X$ have the same distribution function, that is,

$$F(x) = \mathcal{P}\{X \leq x\} = \mathcal{P}\{-X \leq x\}$$
$$= \mathcal{P}\{X \geq -x\} = 1 - \mathcal{P}\{X < -x\}$$
$$= 1 - F(-x)$$

and the relation (9.106) follows.

Now, let us suppose that the density function $f(x)$ is continuous. If the characteristic function is a real function, since $f(x) = F'(x)$ for any $x \in \mathbb{R}$, we obtain by differentiating (9.106) that

$$f(x) = f(-x), \text{ for all } x \in \mathbb{R}.$$

Conversely, if we suppose that $f(x)$ is even, then

$$\varphi_X(t) = \int_{-\infty}^{\infty} f(x)\exp(itx)\,dx = \int_{-\infty}^{\infty} f(x)\cos tx\,dx + i\underbrace{\int_{-\infty}^{\infty} f(x)\sin tx\,dx}_{0}$$

$$= \int_{-\infty}^{\infty} f(x)\cos tx\,dx,$$

hence $\varphi_X(t)$ takes only real values. □

The next subsections are devoted to several important types of continuous random variables, which frequently appear in applications.

9.5.4 The Uniform Distribution

The random variable X is said to be uniformly distributed over the interval $[a, b]$ if its probability density function is a positive constant $c > 0$ on the interval $[a, b]$ and 0 outside the interval. We denote $X \sim U[a, b]$.

By the condition that

$$\int_{-\infty}^{\infty} f(x)\,dx = \int_{a}^{b} c\,dx = 1$$

we find that the constant must be $c = \frac{1}{b-a}$, so the density function is:

$$f(x) = \begin{cases} \frac{1}{b-a}, & \text{if } x \in [a,b], \\ 0, & \text{if } x \notin [a,b]. \end{cases}$$

The cumulative distribution function of X is

$$F(x) = \int_{-\infty}^{x} f(t)\,dt = \begin{cases} 0, & \text{if } x < a, \\ \frac{x-a}{b-a}, & \text{if } x \in [a,b], \\ 1, & \text{if } x > b. \end{cases}$$

9.5 Continuous Random Variables

The expected value of X is the midpoint of the interval $[a, b]$:

$$E(X) = \frac{1}{b-a} \int_a^b x\,dx = \frac{1}{b-a} \cdot \frac{b^2 - a^2}{2} \Rightarrow$$

$$E(X) = \frac{a+b}{2}.$$

To find the variance of X, we first compute the second moment:

$$E\left(X^2\right) = \frac{1}{b-a} \int_a^b x^2\,dx = \frac{b^3 - a^3}{3(b-a)} = \frac{a^2 + ab + b^2}{3}.$$

It follows that

$$Var(X) = E\left(X^2\right) - E(X)^2 = \frac{a^2 + ab + b^2}{3} - \frac{(a+b)^2}{4} \Rightarrow$$

$$Var(X) = \frac{(b-a)^2}{12},$$

so the variance of the uniform random variable is the square of the length of the interval divided by 12.

Example 9.36 Consider the uniform random variable $X \sim U[-2, 3]$. Compute the probability $P\left(\sqrt[3]{X^2 + 2} > 2\right)$. Find the cumulative distribution function $F_Y(x)$ and the probability density function $f_Y(x)$ of the random variable $Y = X^2$.

The probability density function of X is

$$f(x) = \begin{cases} \frac{1}{5}, & \text{if } x \in [-2, 3], \\ 0, & \text{if } x \notin [-2, 3], \end{cases}$$

so we have:

$$P\left(\sqrt[3]{X^2 + 2} > 2\right) = P\left(X^2 + 2 > 8\right) = P\left(X^2 > 6\right) =$$

$$= 1 - P\left(X \in [-\sqrt{6}, \sqrt{6}]\right) = 1 - \int_{-\sqrt{6}}^{\sqrt{6}} f(x)\,dx =$$

$$= 1 - \left[\int_{-\sqrt{6}}^{-2} 0\,dx + \int_{-2}^{\sqrt{6}} \frac{1}{5}\,dx\right] = 1 - \frac{\sqrt{6} + 2}{5} = \frac{3 - \sqrt{6}}{5}.$$

The cumulative distribution function of $Y = X^2$ is

$$F_Y(x) = \mathcal{P}\{X^2 \leq x\} =$$

$$= \begin{cases} 0, & \text{if } x \leq 0, \\ \int_{-\sqrt{x}}^{\sqrt{x}} \frac{1}{5} dx, & \text{if } x \in (0, 4], \\ \int_{-2}^{\sqrt{x}} \frac{1}{5} dx, & \text{if } x \in (4, 9], \\ 1, & \text{if } x > 9 \end{cases} = \begin{cases} 0, & \text{if } x \leq 0, \\ \dfrac{2\sqrt{x}}{5}, & \text{if } x \in (0, 4], \\ \dfrac{\sqrt{x}+2}{5}, & \text{if } x \in (4, 9], \\ 1, & \text{if } x > 9. \end{cases}$$

The probability density function, $f_Y(x) = F'_Y(x)$ is given by

$$f_Y(x) = \begin{cases} \dfrac{1}{5\sqrt{x}}, & \text{if } x \in (0, 4), \\ \dfrac{1}{10\sqrt{x}}, & \text{if } x \in (4, 9), \\ 0, & \text{otherwise.} \end{cases}$$

Example 9.37 Consider n independent random variables, uniformly distributed, $X_i \sim U[0, 1]$, $i = 1, \ldots, n$. Find the expected value of the random variables $Y = \max_{i=\overline{1,n}} X_i$ and $Z = \min_{i=\overline{1,n}} X_i$.

The cumulative distribution function of X_i, $i = 1, 2, \ldots, n$ is

$$F(x) = \begin{cases} 0, & \text{if } x < 0, \\ x, & \text{if } x \in [0, 1], \\ 1, & \text{if } x > 1. \end{cases} \tag{9.107}$$

Let us calculate the cumulative distribution function of Y:

$$F_Y(x) = \mathcal{P}\{Y \leq x\} = \mathcal{P}\{X_1 \leq x, X_2 \leq x, \ldots, X_n \leq x\}$$
$$= \mathcal{P}\{X_1 \leq x\}\mathcal{P}\{X_2 \leq x\}\ldots\mathcal{P}\{X_n \leq x\} = F(x)^n,$$

since $X_1, X_2, \ldots, X_n$ are independent. Thus, by Eq. (9.107), one can write:

$$F_Y(x) = \begin{cases} 0, & \text{if } x < 0, \\ x^n, & \text{if } x \in [0, 1], \\ 1, & \text{if } x > 1, \end{cases}$$

9.5 Continuous Random Variables

so the probability density function of Y is

$$f_Y(x) = \begin{cases} nx^{n-1}, & \text{if } x \in [0,1], \\ 0, & \text{otherwise.} \end{cases}$$

The expected value of Y is:

$$E(Y) = \int_0^1 x f_Y(x)\,dx = \int_0^1 nx^n\,dx = \frac{n}{n+1}.$$

To find the cumulative distribution function of Z, we notice that

$$\begin{aligned} F_Z(x) &= \mathcal{P}\{Z \le x\} = 1 - \mathcal{P}\{Z > x\} \\ &= 1 - \mathcal{P}\{X_1 > x, X_2 > x, \ldots, X_n > x\} \\ &= 1 - \mathcal{P}\{X_1 > x\}\mathcal{P}\{X_2 > x\}\ldots\mathcal{P}\{X_n > x\} \\ &= 1 - (1 - F(x))^n. \end{aligned}$$

Thus, we have

$$F_Z(x) = \begin{cases} 0, & \text{if } x < 0, \\ 1 - (1-x)^n, & \text{if } x \in [0,1], \\ 1, & \text{if } x > 1, \end{cases}$$

so the probability density function of Z is

$$f_Z(x) = \begin{cases} n(1-x)^{n-1}, & \text{if } x \in [0,1], \\ 0, & \text{otherwise.} \end{cases}$$

The expected value of Z is:

$$\begin{aligned} E(Z) &= \int_0^1 x f_Z(x)\,dx = \int_0^1 nx(1-x)^{n-1}\,dx \\ &= -x(1-x)^n \Big|_0^1 + \int_0^1 (1-x)^n\,dx \\ &= 0 - \frac{(1-x)^{n+1}}{n+1}\Big|_0^1 = \frac{1}{n+1}. \end{aligned}$$

9.5.5 The Exponential Distribution

A continuous random variable X with the probability density function

$$f(x) = \begin{cases} \lambda \exp(-\lambda x), & \text{if } x \geq 0, \\ 0, & \text{if } x < 0, \end{cases}$$

where $\lambda > 0$, is said to be an exponential random variable with parameter λ. We denote $X \sim Exp(\lambda)$. Note that the condition $\int_{-\infty}^{\infty} f(x)\,dx = 1$ is fulfilled.

The cumulative distribution function of X is 0, for all $x < 0$. For $x \geq 0$ we have:

$$F(x) = \int_0^x \lambda \exp(-\lambda t)dt = -\exp(-\lambda t)\Big|_0^x = 1 - \exp(-\lambda x).$$

We prove that the expected value is $E(X) = \dfrac{1}{\lambda}$:

$$E(X) = \int_0^\infty x\lambda \exp(-\lambda x)dx = -x\exp(-\lambda x)\Big|_0^\infty + \int_0^\infty \exp(-\lambda x)dx =$$

$$= 0 - \dfrac{\exp(-\lambda x)}{\lambda}\Big|_0^\infty = \dfrac{1}{\lambda}.$$

By mathematical induction it can be proved that the n-th order moment is

$$E\left(X^n\right) = \dfrac{n!}{\lambda^n}. \tag{9.108}$$

Thus, supposing that (9.108) holds for $n = k$, we get:

$$E\left(X^{k+1}\right) = \int_0^\infty x^{k+1}\lambda \exp(-\lambda x)dx =$$

$$= -x^{k+1}\exp(-\lambda x)\Big|_0^\infty + \int_0^\infty (k+1)x^k \exp(-\lambda x)dx =$$

$$= \dfrac{k+1}{\lambda}E\left(X^k\right) = \dfrac{(k+1)}{\lambda} \cdot \dfrac{k!}{\lambda^k} = \dfrac{(k+1)!}{\lambda^{k+1}},$$

and it follows that (9.108) holds for any $n = 1, 2, \ldots$.

Consequently, the variance is $Var(X) = \dfrac{1}{\lambda^2}$:

$$Var(X) = E\left(X^2\right) - E(X)^2 = \dfrac{2}{\lambda^2} - \dfrac{1}{\lambda^2} = \dfrac{1}{\lambda^2}.$$

9.5 Continuous Random Variables

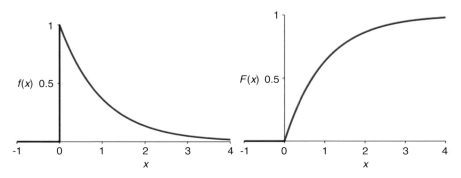

Fig. 9.12 Standard exponential distribution $Exp(1)$

The exponential distribution is used to model the amount of time until a specific event occurs (for instance, the time until an earthquake occurs, or until a new war breaks out, or the time until the first failure of some device). We present in Fig. 9.12 the graph of the probability density function and cumulative distribution function of the standard exponential random variable (with $\lambda = 1$).

The key property of an exponential random variable is the "lack of memory". We say that a nonnegative random variable is *memoryless* if

$$\mathcal{P}\{X > s+t | X > t\} = \mathcal{P}\{X > s\}, \quad \text{for all } s, t \geq 0 \tag{9.109}$$

To understand Eq. (9.109), imagine that X represents the length of time that a certain item functions before failing. The conditional probability $\mathcal{P}\{X > s+t | X > t\}$ is the probability that an item that is functioning at time t will continue to function for (at least) s units of time (until the time $s+t$). By Eq. (9.109), this probability is the same as that of a new item, so the age of an item has no importance, as long as it is still functional it is "as good as new".

If we denote, for all $t \geq 0$,

$$\overline{F}(t) = \mathcal{P}\{X > t\},$$

the Eq. (9.109) can be written in the equivalent form

$$\overline{F}(s+t) = \overline{F}(s)\overline{F}(t), \quad \text{for all } s, t \geq 0. \tag{9.110}$$

Since $\overline{F}(t) = 1 - F(t)$, for an exponential random variable we have

$$\overline{F}(t) = \exp(\lambda t) \text{ for all } t \geq 0, \tag{9.111}$$

and it is very easy to see that this function satisfies (9.110), since

$$\exp(\lambda(s+t)) = \exp(\lambda s)\exp(\lambda t).$$

Moreover, it can be proved that *any* right-continuous function $\overline{F}$ that satisfies (9.110) must be of the form (9.111). Thus, exponential random variables are the only memoryless random variables (see [31], p. 218).

Example 9.38 ([31]) Let X be the number of (thousand of) miles that a car can run before its battery wears out. We suppose that X is exponentially distributed with an average value of 10,000 miles. If somebody starts a 5000-mile trip, what is the probability that the battery will not have to be changed during the trip? What can be said if X was not exponentially distributed?

The random variable X is exponentially distributed with the parameter $\lambda = 0.1$. The desired probability is

$$P\{X > 5\} = 1 - F(5) = \exp(-5\lambda) = \exp(-0.5) \approx 0.6.$$

If X was not exponentially distributed then we have to know t, the number of (thousand of) miles the battery have been used before starting the trip. Thus, the probability that the battery will still function after 5000 miles is

$$P\{X > t+5 \mid X > t\} = \frac{1 - F(t+5)}{1 - F(t)}.$$

Example 9.39 ([11]) Let X be an exponential random variable of parameter λ. Calculate the expected value of the integer part of X, $E([X])$ and of the fractional part of X, $E(\{X\})$. Prove that

$$\lim_{\lambda \to 0} E(\{X\}) = \frac{1}{2}.$$

By Theorem 9.9, we can write:

$$E([X]) = \int_0^\infty [x]\lambda \exp(-\lambda x)\,dx = \sum_{k=0}^\infty \int_k^{k+1} k\lambda \exp(-\lambda x)\,dx =$$

$$= \sum_{k=1}^\infty k\left[\exp(-\lambda k) - \exp(-\lambda(k+1))\right] =$$

$$= \sum_{k=1}^\infty \exp(-\lambda k) = \frac{\exp(-\lambda)}{1 - \exp(-\lambda)} \Rightarrow$$

$$E([X]) = \frac{1}{\exp(\lambda) - 1}.$$

9.5 Continuous Random Variables

Since the fractional part of a real number is $\{x\} = x - [x]$, we have:

$$E(\{X\}) = \int_0^\infty (x - [x])\lambda \exp(-\lambda x)\,dx =$$

$$= \int_0^\infty x\lambda \exp(-\lambda x)\,dx - \int_0^\infty [x]\lambda \exp(-\lambda x)\,dx =$$

$$= E(X) - E([X]) = \frac{1}{\lambda} - \frac{1}{\exp(\lambda) - 1}.$$

Now, let us compute the limit:

$$\lim_{\lambda \to 0} E(\{X\}) = \lim_{\lambda \to 0} \left(\frac{1}{\lambda} - \frac{1}{\exp(\lambda) - 1}\right) =$$

$$= \lim_{\lambda \to 0} \frac{\exp(\lambda) - 1 - \lambda}{\lambda \exp(\lambda) - \lambda} \stackrel{l'H}{=} \lim_{\lambda \to 0} \frac{\exp(\lambda) - 1}{\exp(\lambda)(\lambda + 1) - 1} \stackrel{l'H}{=}$$

$$= \lim_{\lambda \to 0} \frac{\exp(\lambda)}{\exp(\lambda)(\lambda + 2)} = \frac{1}{2}.$$

9.5.6 The Normal Distribution

The normal distribution was introduced by the French mathematician Abraham de Moivre in 1733 in order to approximate the binomial distribution for large values of n. This result was later extended by Pierre-Simon Laplace, Carl Friedrich Gauss and other mathematicians and now it is known as the *central limit theorem*, which gives a theoretical validation to the empirical observation that many random phenomena approximately fit a normal distribution.

Gauss, while using the method of least squares for analyzing astronomical data, noticed that there exists an error μ of greatest frequency and most of the other errors are close to it, in an interval of the type $[\mu-\sigma, \mu+\sigma]$, $\sigma > 0$. Moreover, he observed that, if $f(x)$ denotes the frequency of the error x, then the points $\mu - \sigma$, $\mu + \sigma$ are the inflexion points of the graph of f. The function having the "bell-shaped" graph which gives the most accurate approximation of the errors' distribution is of the form

$$f(x) = k \exp\left(-\frac{(x - \mu)^2}{2\sigma^2}\right), \qquad (9.112)$$

where the constant k can be found by the condition $\int_{-\infty}^{\infty} f(x)dx = 1$. Indeed, using the Gaussian integral $\int_{-\infty}^{\infty} \exp(-t^2)dt = \sqrt{\pi}$, we find:

$$1 = k \int_{-\infty}^{\infty} \exp\left(-\frac{(x-\mu)^2}{2\sigma^2}\right) dx \quad \overset{\frac{x-\mu}{\sigma\sqrt{2}}=t}{=}$$
$$= k\sigma\sqrt{2} \int_{-\infty}^{\infty} \exp(-t^2)dt = k\sigma\sqrt{2\pi},$$

so the constant is $k = \frac{1}{\sigma\sqrt{2\pi}}$.

Definition 9.26 A random variable X is said to have a normal distribution with parameters μ and σ^2 if its density is given by the formula:

$$f(x) = \frac{1}{\sigma\sqrt{2\pi}} \exp\left(-\frac{(x-\mu)^2}{2\sigma^2}\right), \quad x \in \mathbb{R}, \tag{9.113}$$

where $\mu \in \mathbb{R}$ and $\sigma > 0$. We denote $X \sim \mathcal{N}(\mu, \sigma^2)$. If $X \sim \mathcal{N}(0, 1)$, then X is called a *standard normal variable*. The standard normal distribution is

$$f(x) = \frac{1}{\sqrt{2\pi}} \exp\left(-\frac{x^2}{2}\right), \quad x \in \mathbb{R}. \tag{9.114}$$

The normal distribution (9.113) is often called the *Gaussian distribution* (although de Moivre and Laplace worked with it before Gauss). It is a bell-shaped curve which is symmetric about μ, that is,

$$f(\mu - x) = f(\mu + x), \quad \forall x \in \mathbb{R},$$

has two inflection points at $x_{1,2} = \mu \pm \sigma$ and attains its maximum value at $x = \mu$; this maximum value is $f(\mu) = \frac{1}{\sigma\sqrt{2\pi}} \approx \frac{0.399}{\sigma}$ (see Fig. 9.13).

Theorem 9.17 *The mean of a normal random variable $X \sim \mathcal{N}(\mu, \sigma^2)$ is $E(X) = \mu$ and the variance is $Var(X) = \sigma^2$.*

Proof We can write:

$$E(X) = \frac{1}{\sigma\sqrt{2\pi}} \int_{-\infty}^{\infty} x \exp\left(-\frac{(x-\mu)^2}{2\sigma^2}\right) dx$$
$$= \frac{1}{\sigma\sqrt{2\pi}} \int_{-\infty}^{\infty} (\mu + x - \mu) \exp\left(-\frac{(x-\mu)^2}{2\sigma^2}\right) dx$$

9.5 Continuous Random Variables

Fig. 9.13 Normal distribution $\mathcal{N}(\mu, \sigma^2)$

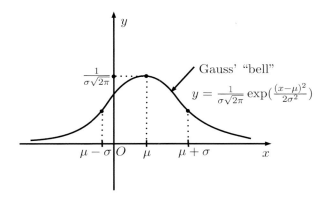

$$= \mu \cdot 1 + \frac{1}{\sigma\sqrt{2\pi}} \int_{-\infty}^{\infty} (x - \mu) \exp\left(-\frac{(x-\mu)^2}{2\sigma^2}\right) dx \quad \stackrel{\frac{x-\mu}{\sigma\sqrt{2}}=t}{=}$$

$$= \mu + \frac{\sigma\sqrt{2}}{\sqrt{\pi}} \int_{-\infty}^{\infty} t \exp(-t^2) dt = \mu + 0 = \mu,$$

where the last integral is 0 because the function $t \exp(-t^2)$ is odd.

Now, let us calculate the second moment of X:

$$E\left(X^2\right) = \frac{1}{\sigma\sqrt{2\pi}} \int_{-\infty}^{\infty} (x - \mu + \mu)^2 \exp\left(-\frac{(x-\mu)^2}{2\sigma^2}\right) dx$$

$$= \frac{1}{\sigma\sqrt{2\pi}} \int_{-\infty}^{\infty} (x - \mu)^2 \exp\left(-\frac{(x-\mu)^2}{2\sigma^2}\right) dx +$$

$$+ \frac{2\mu}{\sigma\sqrt{2\pi}} \int_{-\infty}^{\infty} (x - \mu) \exp\left(-\frac{(x-\mu)^2}{2\sigma^2}\right) dx +$$

$$+ \frac{\mu^2}{\sigma\sqrt{2\pi}} \int_{-\infty}^{\infty} \exp\left(-\frac{(x-\mu)^2}{2\sigma^2}\right) dx.$$

Since $\int_{-\infty}^{\infty} (x - \mu) \exp\left(-\frac{(x-\mu)^2}{2\sigma^2}\right) dx = 0$ (we proved above) and since
$\frac{1}{\sigma\sqrt{2\pi}} \int_{-\infty}^{\infty} \exp\left(-\frac{(x-\mu)^2}{2\sigma^2}\right) dx = 1$ ($f(x)$ is a density function), we get:

$$E\left(X^2\right) = \mu^2 + \frac{1}{\sigma\sqrt{2\pi}} \int_{-\infty}^{\infty} (x - \mu)^2 \exp\left(-\frac{(x-\mu)^2}{2\sigma^2}\right) dx \quad \stackrel{\frac{x-\mu}{\sigma\sqrt{2}}=t}{=}$$

$$= \mu^2 + \frac{2\sigma^2}{\sqrt{\pi}} \int_{-\infty}^{\infty} t^2 \exp\left(-t^2\right) dt = \mu^2 + \frac{4\sigma^2}{\sqrt{\pi}} \int_{0}^{\infty} t^2 \exp\left(-t^2\right) dt$$

$$= \mu^2 - \frac{2\sigma^2}{\sqrt{\pi}} \int_0^\infty t \left(\exp\left(-t^2\right)\right)' dt$$

$$= \mu^2 - \frac{2\sigma^2}{\sqrt{\pi}} \left[t \exp\left(-t^2\right) \Big|_0^\infty + \int_0^\infty \exp\left(-t^2\right) dt \right]$$

$$= \mu^2 + \frac{2\sigma^2}{\sqrt{\pi}} \cdot \frac{\sqrt{\pi}}{2} = \mu^2 + \sigma^2,$$

because $\int_0^\infty \exp\left(-u^2\right) du = \frac{\sqrt{\pi}}{2}$. Finally,

$$Var(X) = E\left(X^2\right) - E(X)^2 = \mu^2 + \sigma^2 - \mu^2 = \sigma^2.$$

□

Remark 9.15 If $X \sim \mathcal{N}(\mu, \sigma^2)$ is a normal random variable of parameters μ and σ^2, then, by Theorem 9.10, $Z = \dfrac{X - \mu}{\sigma}$ is a *standard* normal variable. Conversely, if $Z \sim \mathcal{N}(0, 1)$, then $X = \sigma Z + \mu$ is a normal random variable of parameters μ and σ^2.

The cumulative distribution function of the standard normal variable $Z \sim \mathcal{N}(0, 1)$ is denoted by $\Phi(x)$,

$$\Phi(x) = \mathcal{P}\{Z \leq x\} = \mathcal{P}\{Z < x\} = \frac{1}{\sqrt{2\pi}} \int_{-\infty}^x \exp\left(-\frac{t^2}{2}\right) dt, \quad (9.115)$$

Knowing the values of the standard normal distribution function $\Phi(x)$, we can find the values of the distribution function $F(x)$ of any normal random variable $X \sim \mathcal{N}(\mu, \sigma^2)$. Thus, one can write:

$$F(x) = \mathcal{P}(X \leq x) = \mathcal{P}\left\{\frac{X - \mu}{\sigma} \leq \frac{x - \mu}{\sigma}\right\} = \mathcal{P}\left\{Z \leq \frac{x - \mu}{\sigma}\right\},$$

hence

$$F(x) = \Phi\left(\frac{x - \mu}{\sigma}\right). \quad (9.116)$$

and

$$\mathcal{P}(a \leq X \leq b) = \mathcal{P}(a < X < b) = \Phi\left(\frac{b - \mu}{\sigma}\right) - \Phi\left(\frac{a - \mu}{\sigma}\right). \quad (9.117)$$

9.5 Continuous Random Variables

The standard normal distribution function $\Phi(x)$ cannot be written in terms of elementary functions but can be calculated at a given value x. The values of $\Phi(x)$ for x between 0 and 3.6 can be found in Table 9.2. For $x > 3.6$ the values are close to 1 ($\Phi(3.6) = 0.99984$); for negative values of x the values of $\Phi(x)$ can be found by the symmetry of the function, as the next proposition shows.

Proposition 9.18 *The graph of the standard normal distribution function $\Phi(x)$ is symmetric w.r.t. the point $\left(0, \frac{1}{2}\right)$, that is,*

$$\Phi(-x) = 1 - \Phi(x), \quad \forall x \in \mathbb{R}. \tag{9.118}$$

Proof

$$\Phi(-x) = \frac{1}{\sqrt{2\pi}} \int_{-\infty}^{-x} \exp\left(-\frac{t^2}{2}\right) dt \stackrel{t=-u}{=} -\frac{1}{\sqrt{2\pi}} \int_{\infty}^{x} \exp\left(-\frac{u^2}{2}\right) du$$

$$= \frac{1}{\sqrt{2\pi}} \int_{x}^{\infty} \exp\left(-\frac{u^2}{2}\right) du$$

$$= \frac{1}{\sqrt{2\pi}} \int_{-\infty}^{\infty} \exp\left(-\frac{u^2}{2}\right) du - \frac{1}{\sqrt{2\pi}} \int_{-\infty}^{x} \exp\left(-\frac{u^2}{2}\right) du$$

$$= 1 - \Phi(x).$$

$\square$

Let $X \sim \mathcal{N}(\mu, \sigma^2)$ be a normal random variable of mean μ and variance σ^2. By (9.117) and (9.118) we have that, for any $\alpha > 0$,

$$\mathcal{P}(\mu - \alpha < X < \mu + \alpha) = 2\Phi\left(\frac{\alpha}{\sigma}\right) - 1. \tag{9.119}$$

In particular, for $\alpha = 3\sigma$, we have:

$$\mathcal{P}(\mu - 3\sigma < X < \mu + 3\sigma) = 2\Phi(3) - 1 = .9974\ldots \approx 1,$$

i.e. it is extremely probable that a measured value of X is in the interval $(\mu - 3\sigma, \mu + 3\sigma)$ (*the three σ rule*—see Fig. 9.14).

Example 9.40 If X is a normal random variable with mean $\mu = 3$ and variance $\sigma^2 = 16$, compute the following the probabilities using Table 9.2:
(a) $\mathcal{P}\{X < 11\}$;
(b) $\mathcal{P}\{X > -1\}$;
(c) $\mathcal{P}\{0 < X < 6\}$.

Table 9.2 Standard normal distribution function $\Phi(x)$

x	0.00	0.01	0.02	0.03	0.04	0.05	0.06	0.07	0.08	0.09
0.0	0.5000	0.5040	0.5080	0.5120	0.5160	0.5199	0.5239	0.5279	0.5319	0.5359
0.1	0.5398	0.5438	0.5478	0.5517	0.5557	0.5596	0.5636	0.5675	0.5714	0.5753
0.2	0.5793	0.5832	0.5871	0.5910	0.5948	0.5987	0.6026	0.6064	0.6103	0.6141
0.3	0.6179	0.6217	0.6255	0.6293	0.6331	0.6368	0.6406	0.6443	0.6480	0.6517
0.4	0.6554	0.6591	0.6628	0.6664	0.6700	0.6736	0.6772	0.6808	0.6844	0.6879
0.5	0.6915	0.6950	0.6985	0.7019	0.7054	0.7088	0.7123	0.7157	0.7190	0.7224
0.6	0.7257	0.7291	0.7324	0.7357	0.7389	0.7422	0.7454	0.7486	0.7517	0.7549
0.7	0.7580	0.7611	0.7642	0.7673	0.7704	0.7734	0.7764	0.7794	0.7823	0.7852
0.8	0.7881	0.7910	0.7939	0.7967	0.7995	0.8023	0.8051	0.8078	0.8106	0.8133
0.9	0.8159	0.8186	0.8212	0.8238	0.8264	0.8289	0.8315	0.8340	0.8365	0.8389
1.0	0.8413	0.8438	0.8461	0.8485	0.8508	0.8531	0.8554	0.8577	0.8599	0.8621
1.1	0.8643	0.8665	0.8686	0.8708	0.8729	0.8749	0.8770	0.8790	0.8810	0.8830
1.2	0.8849	0.8869	0.8888	0.8907	0.8925	0.8944	0.8962	0.8980	0.8997	0.9015
1.3	0.9032	0.9049	0.9066	0.9082	0.9099	0.9115	0.9131	0.9147	0.9162	0.9177
1.4	0.9192	0.9207	0.9222	0.9236	0.9251	0.9265	0.9279	0.9292	0.9306	0.9319
1.5	0.9332	0.9345	0.9357	0.9370	0.9382	0.9394	0.9406	0.9418	0.9429	0.9441
1.6	0.9452	0.9463	0.9474	0.9484	0.9495	0.9505	0.9515	0.9525	0.9535	0.9545
1.7	0.9554	0.9564	0.9573	0.9582	0.9591	0.9599	0.9608	0.9616	0.9625	0.9633
1.8	0.9641	0.9649	0.9656	0.9664	0.9671	0.9678	0.9686	0.9693	0.9699	0.9706
1.9	0.9713	0.9719	0.9726	0.9732	0.9738	0.9744	0.9750	0.9756	0.9761	0.9767
2.0	0.9772	0.9778	0.9783	0.9788	0.9793	0.9798	0.9803	0.9808	0.9812	0.9817
2.1	0.9821	0.9826	0.9830	0.9834	0.9838	0.9842	0.9846	0.9850	0.9854	0.9857
2.2	0.9861	0.9864	0.9868	0.9871	0.9875	0.9878	0.9881	0.9884	0.9887	0.9890
2.3	0.9893	0.9896	0.9898	0.9901	0.9904	0.9906	0.9909	0.9911	0.9913	0.9916
2.4	0.9918	0.9920	0.9922	0.9925	0.9927	0.9929	0.9931	0.9932	0.9934	0.9936
2.5	0.9938	0.9940	0.9941	0.9943	0.9945	0.9946	0.9948	0.9949	0.9951	0.9952
2.6	0.9953	0.9955	0.9956	0.9957	0.9959	0.9960	0.9961	0.9962	0.9963	0.9964
2.7	0.9965	0.9966	0.9967	0.9968	0.9969	0.9970	0.9971	0.9972	0.9973	0.9974
2.8	0.9974	0.9975	0.9976	0.9977	0.9977	0.9978	0.9979	0.9979	0.9980	0.9981
2.9	0.9981	0.9982	0.9982	0.9983	0.9984	0.9984	0.9985	0.9985	0.9986	0.9986
3.0	0.9987	0.9987	0.9987	0.9988	0.9988	0.9989	0.9989	0.9989	0.9990	0.9990
3.1	0.9990	0.9991	0.9991	0.9991	0.9992	0.9992	0.9992	0.9992	0.9993	0.9993
3.2	0.9993	0.9993	0.9994	0.9994	0.9994	0.9994	0.9994	0.9995	0.9995	0.9995
3.3	0.9995	0.9995	0.9995	0.9996	0.9996	0.9996	0.9996	0.9996	0.9996	0.9997
3.4	0.9997	0.9997	0.9997	0.9997	0.9997	0.9997	0.9997	0.9997	0.9997	0.9998
3.5	0.9998	0.9998	0.9998	0.9998	0.9998	0.9998	0.9998	0.9998	0.9998	0.9998

9.5 Continuous Random Variables

Fig. 9.14 The three σ rule

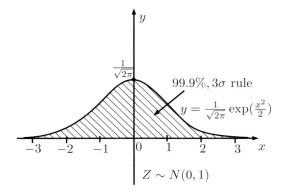

Using the formula (9.117) we can write:

(a) $P\{X < 11\} = P\left\{\frac{X-3}{4} < \frac{11-3}{4}\right\} = \Phi(2) = .9772.$

(b) $P\{X > -1\} = P\left\{\frac{X-3}{4} > \frac{-1-3}{4}\right\} = 1 - \Phi(-1) = \Phi(1) = .8413.$

(c) $P\{0 < X < 6\} = P\left\{-.75 < \frac{X-3}{4} < .75\right\} = 2\Phi(.75) - 1 = .5468.$

Proposition 9.19 *The moments of a standard normal random variable* $Z \sim \mathcal{N}(0, 1)$ *are given by:*

$$E\left(Z^{2n-1}\right) = 0,$$

$$E\left(Z^{2n}\right) = \frac{(2n)!}{2^n n!},$$

for all $n = 1, 2, \ldots$.

Proof For any $n = 1, 2, \ldots$, since the function $g(x) = x^{2n-1} \exp\left(-\frac{x^2}{2}\right)$ is odd, we have:

$$E\left(Z^{2n-1}\right) = \frac{1}{\sqrt{2\pi}} \int_{-\infty}^{\infty} x^{2n-1} \exp\left(-\frac{x^2}{2}\right) dx = 0.$$

To compute the even moments of Z, we use the gamma function (see (5.96)–(5.99)). The function $h(x) = x^{2n} \exp\left(-\frac{x^2}{2}\right)$ is even, hence

$$E\left(Z^{2n}\right) = \frac{2}{\sqrt{2\pi}} \int_{0}^{\infty} x^{2n} \exp\left(-\frac{x^2}{2}\right) dx \stackrel{t=x^2/2}{=\!=\!=}$$

$$= \frac{2^n}{\sqrt{\pi}} \int_{0}^{\infty} t^{n-\frac{1}{2}} \exp(-t)\, dx =$$

$$= \frac{2^n}{\sqrt{\pi}} \Gamma\left(n + \frac{1}{2}\right) \quad \Gamma(x+1) = x\Gamma(x)$$

$$= \frac{2^n}{\sqrt{\pi}} \left(n - \frac{1}{2}\right)\left(n - \frac{3}{2}\right) \cdots \frac{1}{2}\Gamma\left(\frac{1}{2}\right) \quad \Gamma\left(\frac{1}{2}\right) = \sqrt{\pi}$$

$$= (2n-1)(2n-3) \cdots 3 \cdot 1 = \frac{(2n)!}{2^n n!}.$$

□

Theorem 9.18 *The characteristic function of a standard normal random variable $Z \sim \mathcal{N}(0, 1)$ is*

$$\varphi_Z(t) = \exp\left(-\frac{t^2}{2}\right). \tag{9.120}$$

The characteristic function of a normal random variable $X \sim \mathcal{N}(\mu, \sigma^2)$ is

$$\varphi_X(t) = \exp\left(i\mu t - \frac{\sigma^2 t^2}{2}\right). \tag{9.121}$$

Proof The density function of Z is $f(x) = \frac{1}{\sqrt{2\pi}} \exp\left(-\frac{x^2}{2}\right)$, so the characteristic function is:

$$\varphi_Z(t) = \frac{1}{\sqrt{2\pi}} \int_{-\infty}^{\infty} \exp(itx) \exp\left(-\frac{t^2}{2}\right)$$

$$= \frac{1}{\sqrt{2\pi}} \int_{-\infty}^{\infty} \exp(-itx) \exp\left(-\frac{t^2}{2}\right)$$

$$= \mathcal{F}\left\{\exp\left(-\frac{x^2}{2}\right)\right\}(t),$$

the Fourier transform of the function $\exp\left(-\frac{x^2}{2}\right)$ (see Definition 5.7). By the formula (5.81) we know that

$$\mathcal{F}\left\{\exp\left(-\frac{x^2}{2}\right)\right\}(t) = \exp\left(-\frac{t^2}{2}\right)$$

so the Eq. (9.120) follows.

As noted in Remark 9.15, a normal random variable $X \sim \mathcal{N}(\mu, \sigma^2)$ can be written as $X = \sigma Z + \mu$, where Z is a standard normal variable. By Theorem 9.12,

9.5 Continuous Random Variables

the characteristic function of X is:

$$\varphi_X(t) = \exp(i\mu t)\varphi_Z(\sigma t) = \exp\left(i\mu t - \frac{\sigma^2 t^2}{2}\right).$$

□

Corollary 9.7 *Let $X_1, X_2, \ldots, X_n$ be n independent normally distributed random variables, $X_k \sim \mathcal{N}(\mu_k, \sigma_k^2)$, for $k = 1, 2, \ldots, n$, and let $a_1, a_2, \ldots, a_n$ be n arbitrary real numbers. Then, the random variable $Y = \sum_{k=1}^{n} a_k X_k$ is also normally distributed,*

$$Y \sim \mathcal{N}\left(\sum_{k=1}^{n} a_k \mu_k, \sum_{k=1}^{n} a_k^2 \sigma_k^2\right). \tag{9.122}$$

If, in addition, $X_1, X_2, \ldots, X_n$ are identically distributed, $X_k \sim \mathcal{N}(\mu, \sigma^2)$, for $k = 1, 2, \ldots, n$, then

$$\overline{X} = \frac{X_1 + X_2 + \ldots + X_n}{n} \sim \mathcal{N}\left(\mu, \frac{\sigma^2}{n}\right). \tag{9.123}$$

Proof Since $X_1, X_2, \ldots, X_n$ are independent, the characteristic function $\varphi_Y(t)$ is the product of the characteristic functions $\varphi_{a_k X_k}(t)$, $k = 1, 2, \ldots, n$ (see Corollary 9.6). On the other hand, by Theorem 9.12 and using the formula (9.121) one can write:

$$\varphi_Y(t) = \prod_{k=1}^{n} \varphi_{X_k}(a_k t)$$

$$= \prod_{k=1}^{n} \exp\left(i a_k \mu_k t - \frac{a_k^2 \sigma_k^2 t^2}{2}\right)$$

$$= \exp\left(it \sum_{k=1}^{n} a_k \mu_k - \frac{t^2}{2} \sum_{k=1}^{n} a_k^2 \sigma_k^2\right),$$

which is the characteristic function of a normal random variable of mean $\sum_{k=1}^{n} a_k \mu_k$ and variance $\sum_{k=1}^{n} a_k^2 \sigma_k^2$.

In the particular case $\mu_1 = \ldots = \mu_n = \mu$, $\sigma_1 = \ldots = \sigma_n = \sigma$ and $a_1 = \ldots = a_n = \frac{1}{n}$ we obtain the relation (9.123). □

Example 9.41 A machine cuts cylindrical iron rods with an average length of 2 cm. We know that 60% of the rods have the length <2.1 cm.

(a) Assuming that the length of rods is a random variable $X \sim \mathcal{N}(2, \sigma^2)$, find σ, the standard deviation of X.
(b) Find the probability that the length of a rod is greater than 2 and less than 2.02 cm.
(c) We make "square" frames with these rods. What is the probability that the perimeter of such a frame is greater than or equal to 8.2 cm?

(a) We can write:

$$0.6 = \mathcal{P}(X < 2.1) = \mathcal{P}\left(Z = \frac{X-2}{\sigma} < \frac{2.1-2}{\sigma}\right) = \Phi\left(\frac{0.1}{\sigma}\right).$$

Thus, $\frac{0.1}{\sigma} = \Phi^{-1}(0.6)$. From Table 9.2 we deduce that $\frac{0.1}{\sigma} = 0.25$, hence $\sigma = \frac{0.1}{0.25} = 0.4$ and the variance is $\sigma^2 = 0.16$.

(b) Using the formula (9.117) and Table 9.2 we get:

$$\mathcal{P}(2 < X < 2.02) = \mathcal{P}\left(\frac{2-2}{0.4} < Z = \frac{X-2}{0.4} < \frac{2.02-2}{0.4}\right)$$
$$= \mathcal{P}(0 < Z < 0.05) = \Phi(0.05) - \Phi(0) = 0.52 - 0.5 = 0.02,$$

i.e. 2% of the rods have the length between 2 and 2.02 centimeters.

(c) By Corollary 9.7, since X_1, X_2, X_3, X_4 are independent identically distributed normal random variables of mean $\mu = 2$ and variance $\sigma^2 = 0.16$, the perimeter $P = X_1 + X_2 + X_3 + X_4$ has also a normal distribution of mean $E(P) = 4\mu = 8$ and variance $Var(P) = 4\sigma^2 = 0.64$. It follows that

$$\mathcal{P}(P \geq 8.2) = \mathcal{P}\left(Z = \frac{P-8}{0.8} \geq \frac{8.2-8}{0.8}\right)$$
$$= \mathcal{P}(Z \geq 0.25) = 1 - \mathcal{P}(Z < 0.25)$$
$$= 1 - \Phi(0.25) = 1 - 0.599 = 0.401.$$

9.5.7 Gamma Distribution

A random variable is said to have a *gamma distribution* with parameters α and β (where $\alpha, \beta > 0$), if its density function is given by

$$f(x) = \begin{cases} \frac{1}{\Gamma(\alpha)} \beta^\alpha x^{\alpha-1} \exp(-\beta x), & x > 0, \\ 0, & x \leq 0, \end{cases} \quad (9.124)$$

9.5 Continuous Random Variables

where $\Gamma(\alpha) = \int_0^\infty u^{\alpha-1} \exp(-u)\, du$ is the gamma function (see (5.96)–(5.99)). We denote $X \sim \Gamma(\alpha, \beta)$.

First of all, let us show that the function defined by (9.124) is a density function indeed:

$$\int_{-\infty}^{\infty} f(x)\, dx = \frac{1}{\Gamma(\alpha)} \int_0^\infty (\beta x)^{\alpha-1} \exp(-\beta x) \beta\, dx \stackrel{u=\beta x}{=}$$

$$= \frac{1}{\Gamma(\alpha)} \int_0^\infty u^{\alpha-1} \exp(-u)\, du = 1.$$

We remark that that for $\alpha = 1$ the gamma distribution reduces to the exponential distribution of parameter β: $f(x) = \beta \exp(-\beta x)$, for $x > 0$.

We present in Fig. 9.15 the graph of the gamma distribution for $\beta = 2$ and different values of α, and in Fig. 9.16, the graph of the gamma distribution for $\alpha = 2$ and different values of β.

Since the characteristic function of a random variable is a powerful tool in the study of that variable, we shall compute it for $X \sim \Gamma(\alpha, \beta)$:

$$\varphi_X(t) = \frac{\beta^\alpha}{\Gamma(\alpha)} \int_0^\infty \exp(itx)\, x^{\alpha-1} \exp(-\beta x)\, dx$$

$$= \frac{\beta^\alpha}{\Gamma(\alpha)} \int_0^\infty x^{\alpha-1} \exp(-(\beta - it)x)\, dx.$$

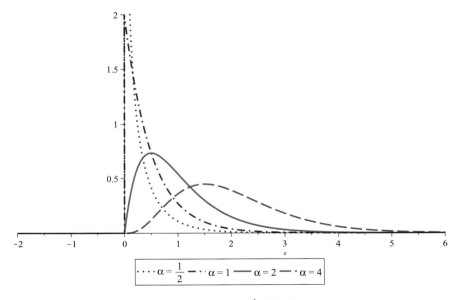

Fig. 9.15 Gamma distributions with $\beta = 2$ and $\alpha = \frac{1}{2}, 1, 2, 4$

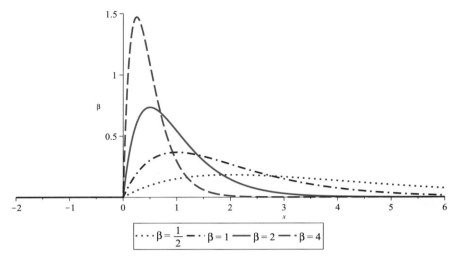

Fig. 9.16 Gamma distributions with $\alpha = 2$ and $\beta = \frac{1}{2}, 1, 2, 4$

If we denote $p = \beta - it$, we can see that the last integral is the Laplace transform of the function

$$g(x) = \begin{cases} x^{\alpha-1}, & x > 0, \\ 0, & x \leq 0, \end{cases}$$

(see Definition 6.2). By the formula (6.9), the Laplace transform of the function $g(x)$ is given by

$$\mathcal{L}\{g(x)\}(p) = \frac{\Gamma(\alpha)}{p^\alpha}, \text{ for all } p \text{ with } \operatorname{Re} p > 0.$$

Since $\operatorname{Re} p = \beta > 0$ in our case, it follows that

$$\varphi_X(t) = \frac{\beta^\alpha}{\Gamma(\alpha)} \cdot \frac{\Gamma(\alpha)}{(\beta - it)^\alpha},$$

hence

$$\varphi_X(t) = \left(\frac{\beta}{\beta - it}\right)^\alpha. \tag{9.125}$$

Proposition 9.20 *If $X \sim \Gamma(\alpha, \beta)$ is a gamma distributed random variable, then the expected value and the variance of X are given by:*

$$E(X) = \frac{\alpha}{\beta}, \quad Var(X) = \frac{\alpha}{\beta^2}.$$

9.5 Continuous Random Variables

Proof From Theorem 9.14 we know that $E(X) = \frac{1}{i}\varphi'_X(0)$. By Eq. (9.125), the derivative of the characteristic function is

$$\varphi'_X(t) = \frac{i\alpha\beta^\alpha}{(\beta - it)^{\alpha+1}},$$

so we obtain that $E(X) = \frac{\alpha}{\beta}$.

The second derivative of the characteristic function is

$$\varphi''_X(t) = \frac{i^2\beta^\alpha \alpha(\alpha+1)}{(\beta - it)^{\alpha+2}},$$

so

$$E\left(X^2\right) = \frac{1}{i^2}\varphi''_X(0) = \frac{\alpha(\alpha+1)}{\beta^2}$$

and the variance is

$$Var(X) = E\left(X^2\right) - E(X)^2 = \frac{\alpha}{\beta^2}.$$

□

Proposition 9.21 *If $X_1, X_2, \ldots, X_n$ are independent gamma random variables with the same parameter β, $X_k \sim \Gamma(\alpha_k, \beta)$, $k = 1, \ldots, n$, then $Y = X_1 + X_2 + \ldots + X_n$ is also a gamma distributed random variable, $Y \sim \Gamma\left(\sum_{k=1}^{n} \alpha_k, \beta\right)$.*

In particular, if $X_1, X_2, \ldots, X_n$ are independent exponential random variables with the same parameter β, $X_k \sim Exp(\beta)$, $k = 1, \ldots, n$, then $Y = X_1 + X_2 + \ldots + X_n$ is a gamma distributed random variable, $Y \sim \Gamma(n, \beta)$.

Proof Since $X_1, X_2, \ldots, X_n$ are independent, by Corollary 9.6 one can write:

$$\varphi_Y(t) = \varphi_{X_1}(t) \ldots \varphi_{X_n}(t)$$

$$= \left(\frac{\beta}{\beta - it}\right)^{\alpha_1} \ldots \left(\frac{\beta}{\beta - it}\right)^{\alpha_n}$$

$$= \left(\frac{\beta}{\beta - it}\right)^{\alpha_1 + \ldots + \alpha_n},$$

which is the characteristic function of a gamma distributed random variable of parameters $\sum_{k=1}^{n} \alpha_k$ and β.

The second statement follows immediately by the remark that an exponential random variable with parameter β is a gamma random variable of parameters 1 and β. □

9.5.8 Chi-Squared Distribution

The χ^2 distribution was discovered by the German statistician F. R. Helmert (1843–1917) while analyzing the behavior of the sum of squares of measurement errors. Later on, the English mathematicians K. Pearson (1857–1936) and R. A. Fisher (1890–1962) deepened the study of this distribution, named it "chi-squared distribution" (after the Greek letter χ) and applied it extensively in Statistics.

Definition 9.27 If $Z_1, Z_2, \ldots, Z_n$ are independent standard normal random variables, then the random variable X defined by

$$X = Z_1^2 + Z_2^2 + \ldots + Z_n^2$$

is said to have a *chi-squared distribution with n degrees of freedom*. We denote

$$X \sim \chi_n^2.$$

Remark 9.16 It readily follows by the definition above the additive property of the chi-squared distribution: if $X_1 \sim \chi_{n_1}^2$ and $X_2 \sim \chi_{n_2}^2$ are independent chi-squared random variables with n_1 and n_2 degrees of freedom, respectively, then $X_1 + X_2$ is a chi-squared random variable with $n_1 + n_2$ degrees of freedom: $X_1 + X_2 \sim \chi_{n_1+n_2}^2$.

Theorem 9.19 *If $Z \sim \mathcal{N}(0, 1)$ is a standard normal random variable, then Z^2 has a gamma distribution with parameters $\alpha = \frac{1}{2}$ and $\beta = \frac{1}{2}$.*

A chi-squared random variable with n degrees of freedom has a Gamma distribution with parameters $\alpha = \frac{n}{2}$ and $\beta = \frac{1}{2}$, so its density function is

$$f(x) = \begin{cases} \frac{1}{\Gamma(\frac{n}{2}) \cdot 2^{\frac{n}{2}}} x^{\frac{n}{2}-1} \exp\left(-\frac{x}{2}\right), & x > 0, \\ 0, & x \leq 0, \end{cases} \quad (9.126)$$

Proof We calculate the cumulative distribution function of Z^2, $F(x) = \mathcal{P}\left(Z^2 \leq x\right)$. Obviously, $F(x) = 0$ for $x \leq 0$. For $x > 0$ we have:

$$F(x) = \mathcal{P}(-\sqrt{x} \leq Z \leq \sqrt{x}) = \int_{-\sqrt{x}}^{\sqrt{x}} \frac{1}{\sqrt{2\pi}} \exp\left(-\frac{t^2}{2}\right) dt.$$

9.5 Continuous Random Variables

Using the fact that the function $\exp\left(-\frac{t^2}{2}\right)$ is even, and making the change of variable $t^2 = u$, we can write:

$$F(x) = 2\int_0^{\sqrt{x}} \frac{1}{\sqrt{2\pi}} \exp\left(-\frac{t^2}{2}\right) dt$$

$$= \int_0^x \frac{1}{\sqrt{\pi}} \cdot \frac{1}{\sqrt{2}} \cdot \frac{1}{\sqrt{u}} \exp\left(-\frac{u}{2}\right) du$$

$$= \int_0^x \frac{1}{\Gamma(\frac{1}{2})} \left(\frac{1}{2}\right)^{\frac{1}{2}} u^{\frac{1}{2}-1} \exp\left(-\frac{1}{2}u\right) du,$$

so the density function of Z^2 is

$$f(x) = F'(x) = \begin{cases} \frac{1}{\Gamma(\frac{1}{2})}\left(\frac{1}{2}\right)^{\frac{1}{2}} u^{\frac{1}{2}-1} \exp\left(-\frac{1}{2}u\right), & x > 0, \\ 0, & x \leq 0, \end{cases}$$

which means that $Z^2 \sim \Gamma\left(\frac{1}{2}, \frac{1}{2}\right)$.

Now, let $Z_1, Z_2, \ldots, Z_n$ be n independent standard normal random variables. Since $Z_k^2 \sim \Gamma\left(\frac{1}{2}, \frac{1}{2}\right)$ for $k = 1, 2, \ldots, n$, it follows, by Proposition 9.21, that the χ^2 random variable with n degrees of freedom,

$$X = Z_1^2 + Z_2^2 + \ldots + Z_n^2,$$

is also a gamma distributed random variable with parameters $\alpha = \frac{n}{2}$ and $\beta = \frac{1}{2}$, and the formula (9.126) follows from (9.124). □

By Proposition 9.20, the mean and the variance of the χ^2 random variable with n degrees of freedom are:

$$E(X) = n,$$
$$Var(X) = 2n.$$

We represent in Fig. 9.17 the graph of the χ^2 distributions with $n = 1, 2, 3$ and 4 degrees of freedom.

9.5.9 Student t-Distribution

This distribution was introduced by the English statistician W. S. Gosset (1876–1937) in a paper published in 1908 under the pseudonym "Student".

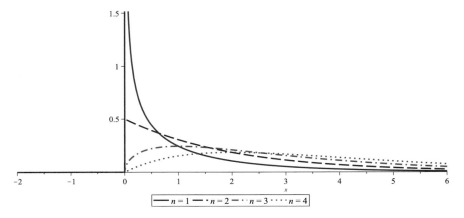

Fig. 9.17 χ^2 distributions with $n = 1, 2, 3, 4$

Definition 9.28 Let $Y \sim \chi_n^2$ be a chi-squared random variable with n degrees of freedom and $Z \sim \mathcal{N}(0, 1)$ be a standard normal variable. The random variable T_n defined by

$$T_n = \frac{Z}{\sqrt{Y/n}}$$

is said to have a Student **t**-distribution with n degrees of freedom.

In order to deduce the density function of T_n, we first prove the following lemma.

Lemma 9.3 *Let X and Y be two continuous random variables with the density functions f_X, f_Y respectively. Then, the density function of the random variable $\frac{Y}{X}$ is given by:*

$$f_{Y/X}(t) = \int_{-\infty}^{\infty} |x| \, f_X(x) f_Y(tx) \, dx, \tag{9.127}$$

provided that the integral is uniformly convergent with respect to the parameter t.

Proof Let F_X, F_Y, $F_{Y/X}$ be the distribution functions of X, Y and Y/X respectively. Then

$$F_{Y/X}(t) = \mathcal{P}\left\{\frac{Y}{X} \leq t\right\} = \mathcal{P}\{Y \leq tX, X > 0\} + \mathcal{P}\{Y \geq tX, X < 0\} =$$

$$= \int_0^\infty f_X(x) \left(\int_{-\infty}^{tx} f_Y(y) \, dy\right) dx + \int_{-\infty}^0 f_X(x) \left(\int_{tx}^\infty f_Y(y) \, dy\right) dx$$

$$= \int_0^\infty f_X(x) F_Y(tx) \, dx + \int_{-\infty}^0 f_X(x) \left(1 - F_Y(tx)\right) dx.$$

9.5 Continuous Random Variables

These are convergent improper integral and we can apply the Leibniz' formula to differentiate them with respect to the parameter t:

$$F_{Y/X}(t) = F'_{Y/X}(t) = \int_0^\infty f_X(x) f_Y(tx) x \, dx + \int_{-\infty}^0 f_X(x) f_Y(tx)(-x) \, dx$$

$$= \int_{-\infty}^\infty f_X(x) f_Y(tx) |x| \, dx.$$

□

The following theorem gives the expression of the density function of a Student t random variable (see also [10], 18.2).

Theorem 9.20 *If $Y \sim \chi_n^2$ and $Z \sim \mathcal{N}(0, 1)$, then the density function of the Student t random variable with n degrees of freedom,*

$$T_n = \frac{Z}{\sqrt{Y/n}}$$

is given by the formula:

$$f_{T_n}(t) = \frac{\Gamma\left(\frac{n+1}{2}\right)}{\sqrt{\pi n}\, \Gamma\left(\frac{n}{2}\right)} \left(\frac{n}{t^2+n}\right)^{\frac{n+1}{2}}. \tag{9.128}$$

The mean and the variance of T_n are $E(T_n) = 0$ for every $n > 1$, and $Var(T_n) = \frac{n}{n-2}$ for all $n > 2$.

Proof First of all, let us find the density function of the random variable $X = \sqrt{\frac{Y}{n}}$, the denominator of T_n. Since Y is a χ^2 random variable with n degrees of freedom, we can write:

$$F_X(x) = P\left\{\sqrt{\frac{Y}{n}} < x\right\} = \begin{cases} P(Y < nx^2) = F_Y(nx^2), & \text{if } x > 0, \\ 0, & \text{if } x \leq 0. \end{cases}$$

Consequently, for $x > 0$, we have:

$$f_X(x) = F'_X(x) = 2nx F'_Y(nx^2) = 2nx f_Y(nx^2) \overset{(9.126)}{=}$$

$$= \frac{2nx}{\Gamma\left(\frac{n}{2}\right) \cdot 2^{\frac{n}{2}}} \left(nx^2\right)^{\frac{n}{2}-1} \exp\left(-\frac{nx^2}{2}\right) =$$

$$= \frac{n^{\frac{n}{2}}}{\Gamma\left(\frac{n}{2}\right) \cdot 2^{\frac{n}{2}-1}} x^{n-1} \exp\left(-\frac{nx^2}{2}\right),$$

and $f_X(x) = 0$ for $x \leq 0$.

Let us use now Lemma 9.3 to find the density function f_{T_n} of the random variable $T_n = \frac{Z}{X}$:

$$f_{T_n}(t) = \int_{-\infty}^{\infty} |x| f_X(x) f_Z(tx) \, dx =$$

$$= \int_0^{\infty} \frac{n^{\frac{n}{2}}}{\Gamma\left(\frac{n}{2}\right) \cdot 2^{\frac{n}{2}-1}} x^n \exp\left(-\frac{nx^2}{2}\right) \frac{1}{\sqrt{2\pi}} \exp\left(-\frac{t^2 x^2}{2}\right) dx =$$

$$= \frac{n^{\frac{n}{2}}}{\Gamma\left(\frac{n}{2}\right) \cdot 2^{\frac{n-1}{2}} \sqrt{\pi}} \int_0^{\infty} x^n \exp\left(-\frac{(t^2+n)x^2}{2}\right) dx.$$

To compute the last integral, we make the change of variable

$$\frac{(t^2+n)x^2}{2} = u \Leftrightarrow x = \left(\frac{2u}{t^2+n}\right)^{\frac{1}{2}}.$$

Thus, we can write:

$$f_{T_n}(t) = \frac{n^{\frac{n}{2}}}{\Gamma\left(\frac{n}{2}\right) \cdot 2^{\frac{n-1}{2}} \sqrt{\pi}} \int_0^{\infty} \left(\frac{2u}{t^2+n}\right)^{\frac{n}{2}} \exp(-u) \left(\frac{2}{t^2+n}\right)^{\frac{1}{2}} \frac{1}{2} u^{-\frac{1}{2}} du =$$

$$= \frac{n^{\frac{n}{2}}}{\Gamma\left(\frac{n}{2}\right) \sqrt{\pi}(t^2+n)^{\frac{n+1}{2}}} \int_0^{\infty} u^{\frac{n-1}{2}} \exp(-u) \, du.$$

By the definition of gamma function, $\Gamma(x) = \int_0^{\infty} u^{x-1} \exp(-u) du$, we find that the last integral is equal to $\Gamma\left(\frac{n+1}{2}\right)$ and the formula (9.128) follows.

We remark that Eq. (9.128) can be also written using the beta function (see the formulas (6.40), (6.42)),

$$B(\alpha, \beta) = \int_0^1 x^{\alpha-1}(1-x)^{\beta-1} dx$$

$$= \frac{\Gamma(\alpha)\Gamma(\beta)}{\Gamma(\alpha+\beta)},$$

as

$$f_{T_n}(t) = \frac{1}{B\left(\frac{n}{2}, \frac{1}{2}\right) \sqrt{n}} \left(\frac{n}{t^2+n}\right)^{\frac{n+1}{2}}. \tag{9.129}$$

9.5 Continuous Random Variables

Now, since the integral

$$\int_{-\infty}^{\infty} t \left(\frac{n}{t^2 + n} \right)^{\frac{n+1}{2}} dt$$

is absolutely convergent for $n > 1$ and f_{T_n} is an even function, $f_{T_n}(t) = f_{T_n}(-t)$ for every $t \in \mathbb{R}$, it follows that the mean of T_n is 0:

$$E(T_n) = \int_{-\infty}^{\infty} t f_T(t) \, dt = 0.$$

We compute the variance:

$$Var(T_n) = E\left(T_n^2\right) - E(T_n)^2 = \int_{-\infty}^{\infty} t^2 f_{T_n}(t) \, dt$$

$$= 2 \int_0^{\infty} t^2 f_{T_n}(t) \, dt$$

$$= \frac{2}{B\left(\frac{n}{2}, \frac{1}{2}\right) \sqrt{n}} \int_{-\infty}^{\infty} t^2 \left(\frac{n}{t^2 + n} \right)^{\frac{n+1}{2}} dt.$$

The integral above is convergent for $n > 2$ and we make the change of variable

$$\frac{n}{t^2 + n} = x \iff t = \sqrt{n} \left(\frac{1-x}{x} \right)^{\frac{1}{2}}$$

(note that $t = 0 \Rightarrow x = 1$ and $t = \infty \Rightarrow x = 0$) and obtain:

$$Var(T_n) = \frac{2}{B\left(\frac{n}{2}, \frac{1}{2}\right) \sqrt{n}} \int_1^0 n \frac{1-x}{x} x^{\frac{n+1}{2}} \left(-\frac{\sqrt{n}}{2} \right) (1-x)^{-\frac{1}{2}} x^{-\frac{3}{2}} dx =$$

$$= \frac{n}{B\left(\frac{n}{2}, \frac{1}{2}\right)} \int_0^1 x^{\frac{n-2}{2} - 1} (1-x)^{\frac{3}{2} - 1} dx =$$

$$= \frac{n}{B\left(\frac{n}{2}, \frac{1}{2}\right)} \cdot B\left(\frac{n-2}{2}, \frac{3}{2}\right) =$$

$$= n \cdot \frac{\Gamma\left(\frac{n+1}{2}\right)}{\Gamma\left(\frac{n}{2}\right)\Gamma\left(\frac{1}{2}\right)} \cdot \frac{\Gamma\left(\frac{n-2}{2}\right)\Gamma\left(\frac{3}{2}\right)}{\Gamma\left(\frac{n+1}{2}\right)} \quad \underline{\underline{\Gamma(x+1)=x\Gamma(x)}}$$

$$= \frac{n}{n-2}.$$

□

Proposition 9.22 *If $f_{T_n}(t)$ is the student **t**-distribution with n degrees of freedom, $n = 1, 2, \ldots$, and $f(t)$ is the standard normal distribution, then, for any $t \in \mathbb{R}$, we have:*

$$\lim_{n \to \infty} f_{T_n}(t) = f(t).$$

Proof The gamma function can be approximated by the *Stirling's formula*:

$$\Gamma(x) \approx \sqrt{2\pi} x^{x-\frac{1}{2}} e^{-x}, \qquad (9.130)$$

where the approximation sign means that

$$\lim_{x \to \infty} \frac{\Gamma(x)}{\sqrt{2\pi} x^{x-\frac{1}{2}} e^{-x}} = 1.$$

Consequently, we can write:

$$\lim_{n \to \infty} f_{T_n}(t) = \lim_{n \to \infty} \frac{\Gamma\left(\frac{n+1}{2}\right)}{\sqrt{\pi n}\, \Gamma\left(\frac{n}{2}\right)} \left(\frac{n}{t^2+n}\right)^{\frac{n+1}{2}} =$$

$$= \lim_{n \to \infty} \frac{\sqrt{2\pi} \left(\frac{n+1}{2}\right)^{\frac{n}{2}} \exp\left(-\frac{n+1}{2}\right)}{\sqrt{\pi n}\, \sqrt{2\pi} \left(\frac{n}{2}\right)^{\frac{n-1}{2}} \exp\left(-\frac{n}{2}\right)} \lim_{n \to \infty} \left(1 - \frac{t^2}{t^2+n}\right)^{\frac{n+1}{2}} =$$

$$= \lim_{n \to \infty} \frac{1}{\sqrt{2\pi}} \left(\frac{n+1}{n}\right)^{\frac{n}{2}} \exp\left(-\frac{1}{2}\right) \exp\left(-\lim_{n \to \infty} \frac{nt^2+t^2}{2n+2t^2}\right) =$$

$$= \frac{1}{\sqrt{2\pi}} \exp\left(-\frac{t^2}{2}\right).$$

□

The student **t**-distribution has (approximately) the same "bell" shape as the normal distribution, but it has thicker "tails" than the normal density, indicating a greater variability. The graph of the density function of T_n is given in Fig. 9.18 for $n = 1, 2,$ and 8. It can be noticed that the student **t**-distribution with n degrees of

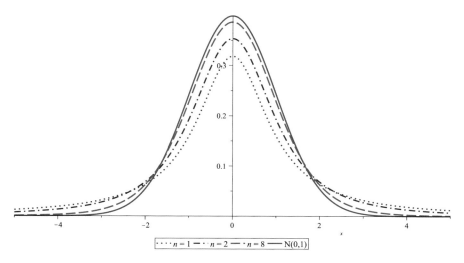

Fig. 9.18 Student t-distribution T_n with $n = 1, 2, 8$ and standard normal distribution (solid line)

freedom tends to the standard normal distribution (also represented with solid line) as n increases.

9.6 Limit Theorems

We introduce in this section two key theorems in probability, the *law of large numbers* and the *central limit theorem* which both describe the results obtained when the same experiment is performed a large number of times.

The law of large numbers basically states that the average of the results obtained from a large number of trials tends to be closer and closer to the expected value as more trials are performed. From a mathematical point of view, it says that the average of n independent identically distributed random variables (with mean μ) converges to the expected value μ as n goes to infinity.

The central limit theorem states that the sum of a large number of independent identically distributed random variables has an approximately normal distribution. It provides a simple method to approximate probabilities for sums of identically distributed random variables and it also gives a mathematical explanation to the observed "omnipresence" of the bell-shaped normal distribution in various phenomena.

Before stating these theorems, we present two important inequalities, Markov's inequality (Lemma 9.4) and Chebyshev's inequality (Theorem 9.21).

Lemma 9.4 (Markov's Inequality) *If X is a random variable in the probability space $(\Omega, \mathcal{K}, \mathcal{P})$ which takes only nonnegative values, then, for any $a > 0$, we* have:

$$P\{X \geq a\} \leq \frac{E(X)}{a}. \tag{9.131}$$

Proof We prove that the inequality holds for both continuous random variables and discrete random variables.

Suppose that X is a continuous random variable having the density function $f(x)$, where $f(x) = 0$ for $x < 0$. Then,

$$E(X) = \int_0^\infty x f(x)\, dx \geq \int_a^\infty x f(x)\, dx \geq$$
$$\geq \int_a^\infty a f(x)\, dx = a P\{X \geq a\}$$

for any $a > 0$, and (9.131) readily follows.

Now, if X is a discrete random variable, taking on the values $0 \leq x_1 < x_2 < \ldots$ with the probabilities $f(x_1), f(x_2), \ldots$, and if we denote by $k_a = \min\{k \in \mathbb{N} : x_k \geq a\}$, for any $k > 0$, then we can write:

$$E(X) = \sum_{k \geq 1} x_k f(x_k) \geq \sum_{k \geq k_a} x_k f(x_k) \geq$$
$$\geq \sum_{k \geq k_a} a f(x_k) = a P\{X \geq a\}$$

and so (9.131) holds for discrete random variables as well. □

A random variable X can be roughly approximated by its expected value, $\mu = E(X)$. An important problem is to estimate the error $|X - \mu|$. Chebyshev's inequality gives an estimation for the probability that $|X - \mu| < \varepsilon$, for every $\varepsilon > 0$.

Theorem 9.21 (Chebyshev's Inequality) *If X is a random variable in the probability space $(\Omega, \mathcal{K}, \mathcal{P})$, having the expected value $\mu = E(X)$ and the variance $\sigma^2 = Var(X)$, then, for any $\varepsilon > 0$, the following inequality holds:*

$$\mathcal{P}\{|X - \mu| \geq \varepsilon\} \leq \frac{\sigma^2}{\varepsilon^2}. \tag{9.132}$$

Another form of Chebyshev's inequality is:

$$\mathcal{P}\{|X - \mu| \geq k\sigma\} \leq \frac{1}{k^2}, \quad \forall k > 0. \tag{9.133}$$

9.6 Limit Theorems

Proof Since $(X - \mu)^2$ is a nonnegative random variable, we can apply Markov's inequality (9.131) (for $a = \varepsilon^2$) to obtain

$$\mathcal{P}\left\{(X - \mu)^2 \geq \varepsilon^2\right\} \leq \frac{E\left[(X - \mu)^2\right]}{\varepsilon^2}.$$

But $(X - \mu)^2 \geq \varepsilon^2$ if and only if $|X - \mu| \geq \varepsilon$, hence

$$\mathcal{P}\{|X - \mu| \geq \varepsilon\} = \mathcal{P}\{(X - \mu)^2 \geq \varepsilon^2\} \leq \frac{Var(X)}{\varepsilon^2}$$

and (9.132) is proved.

For $\varepsilon = k\sigma$ the second inequality (9.133) follows. □

The law of large numbers states that, if $\{X_n\}_{n \geq 1}$ is a sequence of identically distributed random variables (with mean μ), then the sequence $\{\overline{X}_n\}_{n \geq 1}$ defined by $\overline{X}_n = \frac{X_1 + X_2 + \ldots + X_2}{n}$ converges to the expected value μ as n goes to infinity. But what means that a sequence of random variables converges? There are different types of convergence for sequences of random variables. In the following we define three types of convergence: convergence in distribution, convergence in probability, and almost sure convergence.

Definition 9.29 Let X and $X_1, X_2, \ldots, X_n, \ldots$ be random variables in the probability space $(\Omega, \mathcal{K}, \mathcal{P})$. We denote by $F(x)$ the distribution function of X, and by $F_n(x)$ the distribution function of X_n, $n = 1, 2, \ldots$.

(a) The sequence $\{X_n\}_{n \geq 1}$ converges *in distribution* to X, $X_n \xrightarrow{d} X$, if

$$\lim_{n \to \infty} F_n(x) = F(x),$$

for all the points x where the function $F(x)$ is continuous.

(b) The sequence $\{X_n\}_{n \geq 1}$ converges *in probability* to X, $X_n \xrightarrow{p} X$, if

$$\lim_{n \to \infty} \mathcal{P}\{|X_n - X| \geq \varepsilon\} = 0,$$

for any $\varepsilon > 0$.

(c) The sequence $\{X_n\}_{n \geq 1}$ converges *almost sure* to X, $X_n \xrightarrow{as} X$, if

$$\mathcal{P}\left\{\omega \in \Omega : \lim_{n \to \infty} X_n(\omega) = X(\omega)\right\} = 1.$$

The strongest type of convergence is the almost sure convergence, and the weaker type is the convergence in distribution. This means that almost sure convergence implies the convergence in probability, which implies the convergence in distribution:

$$X_n \xrightarrow{as} X \implies X_n \xrightarrow{p} X \implies X_n \xrightarrow{d} X. \qquad (9.134)$$

Remark 9.17 (Scheffé's Theorem) Let $f_n(x)$ denote the density function of X_n for every $n = 1, 2, \ldots$, and $f(x)$, the density functions of X. If

$$\lim_{n \to \infty} f_n(x) = f(x)$$

almost everywhere (that is, for all $x \in \mathbb{R}$, possibly except an at most countable set of points), then

$$X_n \xrightarrow{d} X.$$

Thus, by Proposition 9.22, the sequence of student's **t** random variables with n degrees of freedom, T_n, $n = 1, 2, \ldots$, converges in distribution to a standard normal random variable $Z \sim \mathcal{N}(0, 1)$.

In the following theorem we state and prove the relations (9.134). We also show that convergence in probability and convergence in distribution are equivalent when when the limit of the sequence $\{X_n\}$ is a constant.

Theorem 9.22 *Let X and $X_1, X_2, \ldots, X_n, \ldots$ be random variables in the probability space $(\Omega, \mathcal{K}, \mathcal{P})$.*

(i) *If $X_n \xrightarrow{as} X$ then $X_n \xrightarrow{p} X$.*
(ii) *If $X_n \xrightarrow{p} X$ then $X_n \xrightarrow{d} X$.*
(iii) *If $X_n \xrightarrow{d} c$, where $c \in \mathbb{R}$ is a constant, then $X_n \xrightarrow{p} c$.*

Proof

(i) Let O be the subset of Ω defined by:

$$O = \left\{ \omega \in \Omega : \lim_{n \to \infty} X_n(\omega) \neq X(\omega) \right\}.$$

Since $X_n \xrightarrow{as} X$, we know that $\mathcal{P}(O) = 0$.

Let $\varepsilon > 0$. We consider the subsets of Ω defined by:

$$A_n = \bigcup_{m \geq n} \{\omega \in \Omega : |X_m - X| \geq \varepsilon\},$$

for every $n = 1, 2, \ldots$. We denote $A_\infty = \bigcap_{n \geq 1} A_n$. It is easy to see that $A_n \supset A_{n+1}$ for all n, hence $\{A_n\}_{n \geq 1}$ is a decreasing sequence of events. By

9.6 Limit Theorems

Lemma 9.1, it follows that

$$\lim_{n\to\infty} \mathcal{P}(A_n) = \mathcal{P}(A_\infty).$$

Let $\omega \in \overline{O}$. We have $\lim_{n\to\infty} X_n(\omega) = X(\omega)$, so there exists $n \geq 1$ such that $|X_m - X| < \varepsilon$, for all $m \geq n$. Hence $\omega \notin A_\infty$ and we have just proved that $A_\infty \subset O$. It follows that $\mathcal{P}(A_\infty) = 0$ and, since $\{|X_n - X| \geq \varepsilon\} \subset A_n$, we obtain that

$$\lim_{n\to\infty} \mathcal{P}\{|X_n - X| \geq \varepsilon\} = 0,$$

which means that $X_n \xrightarrow{P} X$.

(ii) First, we prove that, for any random variables X, Y, any $a \in \mathbb{R}$ and $\varepsilon > 0$, the following inequality holds:

$$\mathcal{P}\{Y \leq a\} \leq \mathcal{P}\{X \leq a + \varepsilon\} + \mathcal{P}\{|Y - X| \geq \varepsilon\}. \quad (9.135)$$

To prove this inequality, we write:

$$\mathcal{P}\{Y \leq a\} = \mathcal{P}\{Y \leq a, X \leq a + \varepsilon\} + \mathcal{P}\{Y \leq a, X > a + \varepsilon\} \leq$$
$$\leq \mathcal{P}\{X \leq a + \varepsilon\} + \mathcal{P}\{|Y - X| \geq \varepsilon\},$$

because $\{Y \leq a, X \leq a + \varepsilon\} \subset \{X \leq a + \varepsilon\}$ and $\{Y \leq a, X > a + \varepsilon\} \subset \{|Y - X| \geq \varepsilon\}$.

Now, let $F_n(x)$ be the distribution function of X_n, $n = 1, 2, \ldots$, and $F(x)$ be the distribution function of X. Let $a \in \mathbb{R}$ be a point of continuity for $F(x)$ and consider $\varepsilon > 0$. We apply the inequality (9.135) for X, X_n and obtain:

$$F_n(a) = \mathcal{P}\{X_n \leq a\} \leq \mathcal{P}\{X \leq a + \varepsilon\} + \mathcal{P}\{|X_n - X| \geq \varepsilon\}.$$

If we apply the same inequality for X_n, X and the real number $a - \varepsilon$ (instead of a), we find:

$$\mathcal{P}\{X \leq a - \varepsilon\} \leq \mathcal{P}\{X_n \leq a\} + \mathcal{P}\{|X - X_n| \geq \varepsilon\},$$

which can be also written as:

$$F_n(a) = \mathcal{P}\{X_n \leq a\} \geq \mathcal{P}\{X \leq a - \varepsilon\} - \mathcal{P}\{|X - X_n| \geq \varepsilon\}.$$

Thus, we have obtained that

$$F(a - \varepsilon) - \mathcal{P}\{|X_n - X| \geq \varepsilon\} \leq F_n(a) \leq F(a + \varepsilon) + \mathcal{P}\{|X_n - X| \geq \varepsilon\},$$

for all $n = 1, 2, \ldots$. Since $X_n \xrightarrow{P} X$, by taking the limit as $n \to \infty$, we get:

$$F(a - \varepsilon) \le \lim_{n \to \infty} F_n(a) \le F(a + \varepsilon). \tag{9.136}$$

The inequalities (9.136) hold for every $\varepsilon > 0$. Therefore, by taking the limit as $\varepsilon \to 0$ and using the continuity of F at a, we obtain that

$$\lim_{n \to \infty} F_n(a) = F(a),$$

so $X_n \xrightarrow{d} X$.

(iii) The distribution function of the constant random variable $X = c$ is

$$F(x) = \begin{cases} 0, & x < c \\ 1, & x \ge c. \end{cases}$$

and we notice that it has a single point of discontinuity, $x = c$. Since $X_n \xrightarrow{d} X$, it follows that $\lim_{n \to \infty} F_n(x) = F(x)$, for any $x \ne c$ ($F_n(x)$ is the distribution of X_n, $n = 1, 2, \ldots$). Therefore, for any $\varepsilon > 0$, one can write:

$$\lim_{n \to \infty} F_n(c - \varepsilon) = 0, \tag{9.137}$$

$$\lim_{n \to \infty} F_n\left(c + \frac{\varepsilon}{2}\right) = 1. \tag{9.138}$$

Since

$$\mathcal{P}\{|X_n - c| \ge \varepsilon\} = \mathcal{P}\{X_n \le c - \varepsilon\} + \mathcal{P}\{X_n \ge c + \varepsilon\} \le$$

$$\le \mathcal{P}\{X_n \le c - \varepsilon\} + \mathcal{P}\{X_n > c + \frac{\varepsilon}{2}\} =$$

$$= F_n(c - \varepsilon) + 1 - F_n\left(c + \frac{\varepsilon}{2}\right),$$

by (9.137) and (9.138) we obtain that

$$\lim_{n \to \infty} \mathcal{P}\{|X_n - c| \ge \varepsilon\} = 0,$$

which means that $X_n \xrightarrow{P} X$.

□

Due to the existence of different types of convergence, the law of large numbers has two forms: the *weak* law of large numbers, which is related to the *convergence in probability*, and the *strong* law of large numbers, related to the *almost sure convergence*. Both refer to sequences of independent identically distributed random

9.6 Limit Theorems

variables that have expected value. By Theorem 9.22, the strong law implies the weak law, but we will prove only the weak law (Theorem 9.23) under the supplementary assumption that the variance is finite (see [14], Ch. X, for the proof of the strong law).

Theorem 9.23 (The Weak Law of Large Numbers) *Let $X_1, X_2, \ldots$ be a sequence of independent identically distributed random variables with the mean $E(X_n) = \mu$, $n = 1, 2, \ldots$. Then, the sequence of averages,*

$$\overline{X}_n = \frac{X_1 + \ldots + X_n}{n}, \quad n = 1, 2, \ldots,$$

converges in probability to the constant random variable μ: $\overline{X}_n \xrightarrow{P} \mu$. This means that for any positive real number $\varepsilon > 0$, we have:

$$\lim_{n \to \infty} \mathcal{P}\left\{\left|\frac{X_1 + X_2 + \ldots + X_n}{n} - \mu\right| \geq \varepsilon\right\} = 0, \quad (9.139)$$

or, equivalently,

$$\lim_{n \to \infty} \mathcal{P}\left\{\left|\frac{X_1 + X_2 + \ldots + X_n}{n} - \mu\right| < \varepsilon\right\} = 1. \quad (9.140)$$

Proof If we make the supplementary assumption that the variance of X_n is finite, $Var(X_n) = \sigma^2$, $n = 1, 2, \ldots$, then the theorem easily follows from the Chebyshev's inequality. The result holds without this assumption as well, but the proof is more difficult in this case (see [14], Ch. X).

The expected value of $\overline{X}_n$ is:

$$E\left(\overline{X}_n\right) = \frac{1}{n}[E(X_1) + E(X_2) + \ldots + E(X_n)] = \mu,$$

and, since $X_1, X_2, \ldots, X_n$ are independent, the variance is:

$$Var\left(\overline{X}_n\right) = \frac{1}{n^2}[Var(X_1) + Var(X_2) + \ldots + Var(X_n)] = \frac{\sigma^2}{n}.$$

By Chebyshev's inequality (9.132) we can write:

$$\mathcal{P}\{|\overline{X}_n - E(\overline{X}_n)| \geq \varepsilon\} \leq \frac{Var\left(\overline{X}_n\right)}{\varepsilon^2},$$

hence

$$P\{|\overline{X}_n - \mu| \geq \varepsilon\} \leq \frac{\sigma^2}{n\varepsilon^2} \to 0 \text{ as } n \to \infty.$$

The second relation easily follows by considering the probability of the complementary event: $P\{|\overline{X}_n - \mu| < \varepsilon\} = 1 - P\{|\overline{X}_n - \mu| \geq \varepsilon\}$. □

Theorem 9.24 (The Strong Law of Large Numbers) *Let $X_1, X_2, \ldots$ be a sequence of independent identically distributed random variables with the mean $E(X_n) = \mu$, $n = 1, 2, \ldots$. Then, the sequence of averages,*

$$\overline{X}_n = \frac{X_1 + \ldots + X_n}{n}, \quad n = 1, 2, \ldots,$$

converges almost sure to the constant random variable μ: $\overline{X}_n \overset{as}{\to} \mu$. This means that we have:

$$P\left\{\omega \in \Omega : \lim_{n \to \infty} \frac{X_1(\omega) + X_2(\omega) + \ldots + X_n(\omega)}{n} = \mu\right\} = 1. \quad (9.141)$$

The following theorem is a particular case of the weak law of large numbers when $X_1, X_2, \ldots$ are independent identically distributed Bernoulli random variables. It was originally proved by Jacob Bernoulli in his book "Ars Conjectandi" (published in 1713, 8 years after his death), and says that the relative frequency of success in a sequence of Bernoulli trials tends to the probability of success as the number of trials increases to infinity (see Remark 9.5). Let

$$X_n \sim \begin{pmatrix} 0 & 1 \\ q & p \end{pmatrix}, \quad n = 1, 2, \ldots \quad (9.142)$$

be a sequence of independent, identically distributed Bernoulli random variables. Then, for every $n = 1, 2, \ldots$, $S_n = X_1 + X_2 + \ldots + X_n$ is the random variable representing the number of successes (1) in n independent trials (we know that $S_n \sim Bin(n, p)$). Since $E(X_n) = p$, the Bernoulli's theorem directly follows from Theorem 9.23.

Theorem 9.25 (Bernoulli's Theorem) *Let $B_n \sim Bin(n, p)$ be the random variable representing the number of successes in n Bernoulli trials, $n = 1, 2, \ldots$. Then, for any positive real number $\varepsilon > 0$, we have:*

$$\lim_{n \to \infty} P\left\{\left|\frac{B_n}{n} - p\right| \geq \varepsilon\right\} = 0, \quad (9.143)$$

or, equivalently,

$$\lim_{n\to\infty} \mathcal{P}\left\{\left|\frac{B_n}{n} - p\right| < \varepsilon\right\} = 1, \tag{9.144}$$

Now, let us introduce the *central limit theorem*, one of the most remarkable results in probability. It establishes that, if $X_1, X_2, \ldots$ are independent identically distributed random variables with mean μ and variance σ^2, then the sum $S_n = X_1 + X_2 + \ldots + X_n$ tends to have a normal distribution (of mean $n\mu$ and variance $n\sigma^2$), even if the original variables themselves are not normally distributed. While the weak law of large numbers deals with convergence in probability and the strong law of large numbers, with almost sure convergence, the central limit theorem involve the *convergence in distribution*. We present only the statement of the theorem (see [10] 17.4 for the proof).

Theorem 9.26 (The Central Limit Theorem) *Let $X_1, X_2, \ldots$ be a sequence of independent identically distributed random variables with mean μ and variance σ^2. Let $S_n = X_1 + \ldots + X_n$ and $\overline{X}_n = \dfrac{X_1 + \ldots + X_n}{n}$, $n = 1, 2, \ldots$. Then, for any $x \in \mathbb{R}$,*

$$\lim_{n\to\infty} \mathcal{P}\left\{\frac{S_n - n\mu}{\sigma\sqrt{n}} \leq x\right\} = \Phi(x), \tag{9.145}$$

$$\lim_{n\to\infty} \mathcal{P}\left\{\frac{\sqrt{n}(\overline{X}_n - \mu)}{\sigma} \leq x\right\} = \Phi(x), \tag{9.146}$$

where $\Phi(x)$ denotes the standard normal distribution function.
This means that $S_n \approx \mathcal{N}(n\mu, n\sigma^2)$ and $\overline{X}_n \approx \mathcal{N}\left(\mu, \dfrac{\sigma^2}{n}\right)$.

The first version of the central limit theorem was proved by the French mathematician Abraham de Moivre around 1733 for the special case when X_n are Bernoulli random variables with $p = 1/2$. Later, the result was extended by Laplace to the case of an arbitrary $p \in (0, 1)$. Since the sum of n Bernoulli random variables (9.142) is a binomial random variable, $B_n = X_1 + \ldots + X_n \sim Bin(n, p)$ the central limit theorem has in this case the following form (known as *de Moivre-Laplace Theorem*):

Theorem 9.27 (de Moivre-Laplace Theorem) *Let $B_n \sim Bin(n, p)$, $n = 1, 2, \ldots$ be a sequence of binomial distributed random variables representing the number of successes in n Bernoulli trials. Then, for any $x \in \mathbb{R}$,*

$$\lim_{n\to\infty} \mathcal{P}\left\{\frac{B_n - np}{\sqrt{np(1-p)}} \leq x\right\} = \Phi(x), \tag{9.147}$$

Remark 9.18 If $B_n \sim Bin(n, p)$, the probability that B_n takes value in a given interval $(a, b]$ can be approximated (for large values of n) by the formula

$$\mathcal{P}\{a < B_n \leq b\} \approx \Phi\left(\frac{b - np}{\sqrt{np(1-p)}}\right) - \Phi\left(\frac{a - np}{\sqrt{np(1-p)}}\right). \qquad (9.148)$$

Consider now a sequence of independent Poisson distributed random variables $X_n \sim Poi(1)$, $n = 1, 2, \ldots$. We know that $E(X_n) = 1$ and $Var(X_n) = 1$ and, because they are independent, the sum $S_n = X_1 + \ldots + X_n$ is also Poisson distributed (see Example 9.34), $S_n \sim Poi(n)$. By the central limit theorem, we have:

$$\lim_{n \to \infty} \mathcal{P}\left\{\frac{S_n - n}{\sqrt{n}} \leq x\right\} = \Phi(x).$$

It can be proved that this equality holds for any Poisson distributions (of an arbitrary parameter $\lambda > 0$, not necessarily natural, like above).

Theorem 9.28 *Let $P_\lambda \sim Poi(\lambda)$ be a Poisson distributed random variable. Then, for any $x \in \mathbb{R}$,*

$$\lim_{\lambda \to \infty} \mathcal{P}\left\{\frac{P_\lambda - \lambda}{\sqrt{\lambda}} \leq x\right\} = \Phi(x). \qquad (9.149)$$

Remark 9.19 If $P_\lambda \sim Poi(\lambda)$, the probability that P_λ takes value in a given interval $(a, b]$ can be approximated (for large values of λ) by the formula

$$\mathcal{P}\{a < P_\lambda \leq b\} \approx \Phi\left(\frac{b - \lambda}{\sqrt{\lambda}}\right) - \Phi\left(\frac{a - \lambda}{\sqrt{\lambda}}\right), \qquad (9.150)$$

Example 9.42 If the probability of hitting a target is 90%, find the probability that the target is hit at least 30 times out of 35 trials.

The random variable counting the number of "successful hits" is a binomial random variable, $X \sim Bin(n, p)$, with $p = 0.9$, and $n = 35$. The required probability can be approximated as:

$$\mathcal{P}\{30 \leq X \leq 35\} = \mathcal{P}\{29 < X \leq 35\} \approx$$

$$\approx \Phi\left(\frac{35 - 35 \cdot 0.9}{\sqrt{35 \cdot 0.9 \cdot 0.1}}\right) - \Phi\left(\frac{29 - 35 \cdot 0.9}{\sqrt{35 \cdot 0.9 \cdot 0.1}}\right) =$$

$$= \Phi(1.97) - \Phi(-1.4) = \Phi(1.97) - [1 - \Phi(1.4)]$$

We use now Table 9.2 for the values of the standard normal distribution $\Phi(x)$ and find:

$$\mathcal{P}\{30 \leq X \leq 35\} \approx 0.972 - 1 + 0.919 = 0.891.$$

We remark that the "exact probability" is

$$\mathcal{P}\{30 \leq X \leq 35\} = \sum_{k=0}^{5} \binom{35}{k} 0.1^k 0.9^{35-k} = 0.868.$$

Example 9.43 The average number of cars entering a certain parking lot per hour is 30. What is the probability that during a given hour at least 35 cars enter the parking lot?

We use the Poisson distribution with $\lambda = 30$, so our random variable is $X \sim Poi(30)$. By using the approximation formula (9.150), the required probability is:

$$\mathcal{P}\{35 \leq X < \infty\} = \mathcal{P}\{34 < X < \infty\} \approx$$
$$= 1 - \Phi\left(\frac{34 - 30}{\sqrt{30}}\right) =$$
$$= 1 - \Phi(0.73) = 1 - 0.767 = 0.233.$$

On the other hand, if we compute the probability using the Poisson distribution formula, we get:

$$\mathcal{P}\{35 \leq X < \infty\} = 1 - \exp(-30) \sum_{k=0}^{34} \frac{30^k}{k!} = 0.203$$

Example 9.44 ([11]) We assume that the probability of having any nuclear accidents in any nuclear plant during a given year is 0.0005. A country has 100 such nuclear plants. What is the probability that there will be at least six nuclear accidents in that country during the next 250 years?

We denote by $X_{i,j}$ the random variable representing the number of accidents in the ith year at the jth plant, $i = 1, \ldots, 250$, $j = 1, \ldots, 100$. These are independent identically Poisson distributed random variables, $X_{i,j} \sim Poi(\theta)$, where θ is determined from the condition that $\mathcal{P}\{X_{i,j} \geq 1\} = 0.0005$. This is equivalent to $1 - \exp(-\theta) = 0.0005$, so $\theta = -\ln 0.9995 = 0.0005$.

Since $X_{i,j}$ are independent, the number of accidents in the country during the next 250 years, $T = \sum_{i=1}^{250} \sum_{j=1}^{100} X_{i,j}$, is a Poisson random variable, $T \sim Poi(\lambda)$, where

$$\lambda = 100 \cdot 250 \cdot \theta = 12.5$$

The probability of having at least 6 accidents can be approximated by

$$\mathcal{P}\{6 \leq X < \infty\} = \mathcal{P}\{5 < X < \infty\} \approx 1 - \Phi\left(\frac{5 - 12.5}{\sqrt{12.5}}\right) =$$
$$= 1 - \Phi(-2.12) =$$
$$= \Phi(2.12) = 0.983.$$

Note that this approximate value is very close to the probability computed using the Poisson distribution formula:

$$\mathcal{P}\{6 \leq X < \infty\} = 1 - \exp(-12.5) \sum_{k=0}^{5} \frac{12.5^k}{k!} = 0.985$$

9.7 Exercises

1. Prove that $\mathcal{P}(A \cap \overline{B}) - \mathcal{P}(\overline{A} \cap B) = \mathcal{P}(A) - \mathcal{P}(B)$.
2. Prove that (A, B) are independent if and only if $(\overline{A}, \overline{B})$ are independent.
3. We know that $\mathcal{P}(A) = 1/5$, $\mathcal{P}(\overline{B}) = 5/6$ and $\mathcal{P}_A(B) = 1/6$. Compute $\mathcal{P}(A \cup B)$. Are A and B independent?
4. It is known that 95% of the products sold by a manufacturer are good. If we buy 5 such products, what is the probability to get:

    (a) exactly 3 good products;
    (b) at least one good product;
    (c) 5 good products.

5. A fair coin is tossed 10 times. Find the probability that:

    (a) at most 8 *heads* are obtained;
    (b) exactly 3 *tails* are obtained;
    (c) the first *head* appears at the 4-th trial.

6. In a classroom there are 7 boys and 9 girls. Three students leave the classroom. Find the probability that:

    (a) exactly 2 of them are boys;
    (b) all of them are girls;
    (c) there is at least one boy between them.

7. Consider the set of digits $\{0, 1, \ldots, 9\}$. We randomly take two elements: (a) with replacement; (b) without replacement.
    Find, in each case, the probability to get the numbers 3 and 4 (in this order).

9.7 Exercises

8. We have two boxes with black balls and white balls. The first box contains one black ball and two white balls. The second box contains two black balls and two white balls. We transfer one ball from the first box into the second box and then take one ball from the second box. What is the probability to take a white ball?

9. Three sugar refineries A, B, C supply sugar to a supermarket D in the following monthly quantities: 500, 300 and 200 bags respectively. It is also known that 2% of the sugar bags produced by A and 1% of the sugar bags produced by B and C are defective.

    (a) We randomly take a sugar bag from the supermarket D. What is the probability that it is defective?
    (b) Now, suppose that we took a defective bag from the supermarket. What is the probability that it was supplied by the refinery B?

10. Let X be the random variable representing the number of girls in a family with 3 children.

    (a) Find the probability distribution of X.
    (b) Find the distribution function $F_X(x)$.
    (c) Compute the expected value and the variance of X.

11. Let $X \sim \begin{pmatrix} 1 & 2 & 3 \\ 0.4 & 0.1 & 0.5 \end{pmatrix}$ and $Y \sim \begin{pmatrix} 1 & 2 & 3 \\ 0.1 & 0.6 & 0.3 \end{pmatrix}$ be two independent random variables.

    (a) Find the probability distributions of XY, $X+Y$, and $Z = 3X + Y$.
    (b) Compute the mean and the variance of X, Y and Z.
    (c) What is the probability that the random variable Z takes values in the interval $(\mu_Z - 2\sigma_Z, \mu_Z + 2\sigma_Z)$?

12. Suppose that a pair of fair dice are rolled. Let X and Y be the smaller and the larger of the two values. Find the joint distribution of X and Y. Are they independent? Compute their correlation coefficient.

13. For each of the following functions, find C such that $f(x)$ is a density function. Compute, in each case, the mean, the variance and the cumulative distribution function of a random variable with the density function $f(x)$:

    (a) $f(x) = \begin{cases} Cx^3, & x \in (0, 2) \\ 0, & \text{otherwise.} \end{cases}$

    (b) $f(x) = \begin{cases} Cx^n, & x \in (0, 1) \\ 0, & \text{otherwise} \end{cases}$ where $n > 0$ is given.

    (c) $f(x) = \begin{cases} C(1 - x^2), & x \in (-1, 1) \\ 0, & \text{otherwise.} \end{cases}$

    (d) $f(x) = \begin{cases} Cx \exp(-\lambda x), & x \in (0, \infty) \\ 0, & x \in (-\infty, 0] \end{cases}$ where $\lambda > 0$ is given.

14. Let X be a random variable with the distribution function

$$F(x) = \begin{cases} 1 - \exp(-x^2), & x \in (0, \infty) \\ 0, & \text{otherwise.} \end{cases}$$

Compute: (a) $\mathcal{P}\{X > 2\}$; (b) $\mathcal{P}\{1 < X < 3\}$; (c) the mean and the variance of X.

15. A random variable X with the expected value $E(X) = 0.6$ has the density function

$$f(x) = \begin{cases} ax^2 + bx, & x \in (0, 1) \\ 0, & \text{otherwise.} \end{cases}$$

(a) Find a and b and the variance $Var(X)$;
(b) Compute $\mathcal{P}\left\{X < \frac{1}{2}\right\}$;
(c) Compute the cumulative distribution function $F(x)$.

16. A Cauchy random variable has the density function

$$f(x) = \frac{k}{\pi(x^2 + k^2)}, \quad x \in \mathbb{R}, \; k > 0.$$

(a) Prove that $f(x)$ is indeed a probability density function.
(b) Compute the distribution function of a Cauchy random variable.
(c) Let X be a *standard* Cauchy random variable ($k = 1$). Find the probability that $X^2 - 2X - 3 \leq 0$.
(d) If X is a standard Cauchy random variable, prove that $Y = 1/X$ is also a standard Cauchy random variable.

17. Let X be the lifetime of a certain type of electronic device (measured in hours). We suppose that the probability density function of X is:

$$f(x) = \begin{cases} \dfrac{C}{x^2}, & x > 10 \\ 0, & \text{otherwise.} \end{cases}$$

(a) Find C and the probability that $X > 20$.
(b) Find the cumulative distribution function of X.
(c) What can you say about the expected value of X?
(d) What is the probability that of 6 such devices, at least 3 will be still working after 15 hours?

9.7 Exercises

18. Let X be a normal random variable with the mean $\mu = 10$ and the variance $\sigma^2 = 36$. Using Table 9.2 for the values of the standard normal distribution function $\Phi(x)$, compute the following probabilities:
 (a) $P\{X > 4\}$; (b) $P\{X < 8\}$; (c) $P\{4 < X < 16\}$; (d) $P\{X < 20\}$.
19. Let X be a normal random variable with the mean $\mu = 70$ and the standard deviation $\sigma = 5$. Find $a \in \mathbb{R}$ such that

$$P\{|X - 70| < a\} = 0.8.$$

20. Let X be a normal random variable with mean 5. Estimate the variance of X, knowing that $P\{X > 9\} = 0.2$.
21. The lengths of some metal bars vary according to a normal law with mean μ and standard deviation 6 mm. We know that 5% of these bars have their length ≥ 82 mm. Find the average length μ of the bars.
22. On each bet, a gambler loses 1 with probability 0.7, loses 2 with probability 0.2, or wins 10 with probability 0.1. What is the probability that the gambler will be losing after 3 bets? Approximate the probability that the gambler will be losing after 100 bets.

Chapter 10
Answers and Solutions to Exercises

This chapter contains the detailed solutions to the exercises and problems proposed in the final section of each chapter.

10.1 Chapter 1

1. Since $y'(x) = C - \cos x + x \sin x$, it follows that $xy' - y = x^2 \sin x$, $\forall x \in \mathbb{R}$. By the initial condition $y(\pi) = 4$ we obtain that $C = 3$, so the required particular solution is $y(x) = 3x - x \cos x$.

2. Since $y'(x) = \dfrac{C \exp(-x)}{(C \exp(-x) + 1)^2}$, we obtain that

$$y - y^2 = \dfrac{1}{C\exp(-x)+1} - \dfrac{1}{(C\exp(-x)+1)^2} = y',$$

for all $x \in \mathbb{R}$. By the initial condition $y(0) = 0.2$ we get $C = 4$, so the required particular solution is $y(x) = \dfrac{1}{4\exp(-x)+1}$.

3. (a) This is a homogeneous equation: if we denote by $f(x, y) = \dfrac{3x-y}{x+y}$ then we can see this function is homogeneous, because $f(tx, ty) = \dfrac{3tx-ty}{tx+ty} = \dfrac{t(3x-y)}{t(x+y)} = f(x, y)$. We use the substitution $u(x) = \dfrac{y(x)}{x}$. We have $y(x) = xu(x)$, so its derivative is $y'(x) = u(x) + xu'(x)$. By replacing in the initial equation we obtain $u + u'x = \dfrac{3-u}{1+u}$, or $u' = \dfrac{du}{dx} = -\dfrac{u^2+2u-3}{x(u+1)}$, which is a separable equation, we can write $\dfrac{u+1}{u^2+2u-3}du = -\dfrac{1}{x}dx$. We have: $\int \dfrac{u+1}{u^2+2u-3}du = \int\left(-\dfrac{1}{x}\right)dx$ and it follows that $\tfrac{1}{2}\ln(u^2 + 2u - 3) = -\ln|x| + \ln c$, or, equivalently, $u^2 + 2u - 3 = \dfrac{c^2}{x^2}$. If we denote by $C = c^2$ and

replace u with $\frac{y}{x}$, we obtain the general solution of the differential equation in an implicit form: $y^2 + 2xy - 3x^2 = C$.

(b) The equation can be written in the normal form: $y' = \frac{x^2+y^2}{2xy}$, so it is also a homogeneous equation, therefore we use the substitution $u(x) = \frac{y(x)}{x}$. The equation becomes $u + u'x = \frac{1+u^2}{2u}$, or, equivalently, $u' = \frac{du}{dx} = -\frac{u^2-1}{2ux}$, a separable equation. We have $\int \frac{2u}{u^2-1} du = \int \left(-\frac{1}{x}\right) dx \Leftrightarrow \ln(u^2 - 1) = \ln \frac{C}{x} \Leftrightarrow \frac{y^2}{x^2} - 1 = \frac{C}{x}$. Hence, the general solution of the differential equation in an implicit form is $y^2 = x^2 + Cx$.

(c) The normal form is $\frac{dy}{dx} = y' = \frac{x^2+xy+y^2}{x^2}$, so it is also homogeneous, we need the same substitution $u(x) = \frac{y(x)}{x}$. The equation becomes $u + u'x = 1 + u + u^2$, or, equivalently, $u' = \frac{du}{dx} = \frac{1+u^2}{x}$, a separable equation. We have $\int \frac{du}{u^2+1} = \int \frac{dx}{x} \Leftrightarrow \tan^{-1} u = \ln |Cx| \Leftrightarrow u = \tan(\ln |Cx|) \Rightarrow y = x \tan(\ln |Cx|)$.

(d) This is an equation of the type $y' = f\left(\frac{a_1 x + b_1 y + c_1}{a_2 x + b_2 y + c_2}\right)$, the case when the lines $a_1 x + b_1 y + c_1 = 0$ and $a_2 x + b_2 y + c_2 = 0$ have a common point, $M(x_0, y_0)$. Let us find this common point and then use the substitutions $x - x_0 = u$, $y - y_0 = v(u)$. The solution of the system of equations $4x - y = 0$; $2x + y - 3 = 0$ is $(x_0, y_0) = \left(\frac{1}{2}, 2\right)$. We denote $x - \frac{1}{2} = u$, $y - 2 = v(u) = v\left(x - \frac{1}{2}\right)$; from this last equality we obtain $y' = v'$, so the equation becomes $v' = \frac{4u-v}{2u+v}$, which is a homogeneous equation. By using the substitution $z(u) = \frac{v(u)}{u}$, a separable equation is obtained: $z' = \frac{dz}{du} = \frac{4-3z-z^2}{u(2+z)}$. Let us integrate this equation: $\int \frac{z+2}{z^2+3z-4} dz = -\int \frac{du}{u}$. For the first integral we use the partial fraction decomposition: $\int \frac{(z+2)dz}{(z-1)(z+4)} = \frac{3}{5} \int \frac{dz}{z-1} + \frac{2}{5} \int \frac{dz}{z+4} = \frac{3}{5} \ln(z-1) + \frac{2}{5} \ln(z+4)$. Hence, $\frac{1}{5} \ln(z-1)^3 (z+4)^2 = \ln c - \ln u \Rightarrow (z-1)^3 (z+4)^2 = \left(\frac{c}{u}\right)^5 = \frac{C}{u^5}$. But we used the substitutions $z = \frac{v}{u}$ and $u = x - \frac{1}{2}$, $v = y - 2$. If we change back, we get the general solution of the differential equation (in an implicit form): $(v - u)^3 (v + 4u)^2 = C$, which can be written $\left(y - x - \frac{3}{2}\right)^3 (y + 4x - 4)^2 = C$.

(e) The normal form is $y' = \frac{2x-y}{x-2y+3}$, so it is the same type of equation. The common point of the lines $2x - y = 0$ and $x - 2y + 3 = 0$ is $(x_0, y_0) = (1, 2)$. We denote $x - 1 = u$, $y - 2 = v(u)$ and obtain the homogeneous equation $v' = \frac{2u-v}{u-2v}$. By using the substitution $z(u) = \frac{v(u)}{u}$, a separable equation is obtained: $\frac{dz}{du} = \frac{2-2z-2z^2}{u(1-2z)}$. We integrate this equation: $\int \frac{2z-1}{z^2-z+1} dz = -2 \int \frac{du}{u} \Rightarrow \ln(z^2 - z + 1) = \ln C - 2 \ln u \Rightarrow z^2 - z + 1 = \frac{C}{u^2} \Rightarrow v^2 - uv + u^2 = C$. The general solution (in an implicit form) is: $(y-2)^2 - (y-2)(x-1) + (x-1)^2 = C$.

(f) The normal form of the equation is $y' = \frac{x+y}{x+y-1}$ and we notice that the straight lines $x + y = 0$ and $x + y - 1 = 0$ are parallel, so we use the substitution $x + y(x) = v(x)$. It follows that $y' = v' - 1 = \frac{v}{v-1}$ and a separable equation is obtained: $v' = \frac{dv}{dx} = \frac{2v-1}{v-1}$. We integrate this equation: $\int \frac{v-1}{2v-1} dv = \int dx \Rightarrow \int \frac{2v-1-1}{2v-1} dv = 2\int dx \Rightarrow v - \frac{1}{2}\ln(2v - 1) = 2x + \frac{1}{2}\ln C$, which is equivalent to $2(v-2x) = \ln C(2v-1)$. The general solution of the initial equation is: $2(y - x) = \ln C(2x + 2y - 1)$, or, equivalently, $\exp 2(y - x) = C(2x + 2y - 1)$.

(g) The lines $2x + y - 4 = 0$ and $4x + 2y + 2 = 0$ are parallel, so we use the substitution $2x + y(x) = v(x)$. It follows that $y' = v' - 2 = \frac{v-4}{2v+2}$ and a separable equation is obtained: $v' = \frac{dv}{dx} = \frac{5v}{2(v+1)}$. We integrate this equation: $2 \int \frac{v+1}{v} dv = 5 \int dx \Rightarrow 2v + 2\ln v = 5x + \ln C$, which is equivalent to $2v - 5x = \ln \frac{C}{v^2} \Leftrightarrow v^2 \exp(2v - 5x) = C$. The general solution of the initial equation is: $(2x + y)^2 \exp(2y - x) = C$.

(h) The equation can be written $y dy = -x dx$, so it is a separable equation. By integrating it, we obtain $\int y dy = - \int x dx \Leftrightarrow \frac{y^2}{2} = -\frac{x^2}{2} + \frac{C}{2}$. The general solution is $x^2 + y^2 = C$ (circles with the centre in the origin).

(i) We can write: $x dy = (y^2 + y) dx$, or $\frac{dy}{y^2+y} = \frac{dx}{x}$. Let us integrate this separable equation, using for the first integral partial fraction decomposition: $\frac{1}{y(y+1)} = \frac{1}{y} - \frac{1}{y+1} \Rightarrow \int \frac{dy}{y(y+1)} = \int \frac{dy}{y} - \int \frac{dy}{y+1} = \int \frac{dx}{x} \Rightarrow \ln y - \ln(y+1) = \ln x + \ln C$. It follows that $\ln \frac{y}{y+1} = \ln Cx$, so the solution is $\frac{y}{y+1} = Cx$, or (in an explicit form) $y = \frac{Cx}{1-Cx}$. Note that the equation could be solved like a Bernoulli equation as well (with $\alpha = 2$): $\frac{dy}{dx} = \frac{1}{x}y + \frac{1}{x}y^2$.

(j) It is a separable equation: $y' = \frac{dy}{dx} = \frac{y^3+y}{x} \Rightarrow \int \frac{dy}{y(y^2+1)} = \int \frac{dx}{x}$. The first integral can be computed by using partial fraction decomposition: $\int \frac{dy}{y(y^2+1)} = \int \frac{dy}{y} - \int \frac{y dy}{y^2+1} = \ln|y| - \frac{1}{2}\ln(y^2+1) = \frac{1}{2}\ln \frac{y^2}{y^2+1}$. So we have $\ln \frac{y^2}{y^2+1} = 2\int \frac{dx}{x} = 2\ln|x| + \ln C = \ln Cx^2$. The general solution (in an implicit form) is $\frac{y^2}{y^2+1} = Cx^2$. We could also consider it as a Bernoulli equation (with $\alpha = 3$).

(k) It is a second-order ODE where the independent variable is t. In order to find the solution $y = y(t)$, we need to integrate twice, so the general solution will depend on two constants: $y' = \frac{t^{101}}{101} + C_1 \Rightarrow y = \frac{t^{102}}{101 \cdot 102} + C_1 t + C_2$.

(l) Now we have a 5th-order ODE; consequently, the general solution will depend on 5 constants: $y^{(4)} = x + c_1 \Rightarrow y''' = \frac{x^2}{2} + c_1 x + c_2 \Rightarrow \ldots \Rightarrow y = \frac{x^5}{5!} + C_1 x^4 + C_2 x^3 + C_3 x^2 + C_4 x + C_5$ (we denoted $C_1 = \frac{c_1}{4!}$, $C_2 = \frac{c_2}{3!}, \ldots$ etc). In fact, the solution of the linear equation $y^{(5)} = 1$ is the sum of the general solution of the homogeneous equation $y^{(5)} = 0$, $y_{GH} = C_1 x^4 + C_2 x^3 + C_3 x^2 + C_4 x + C_5$, and a particular solution of the initial equation, $y_P = \frac{x^5}{5!}$.

(m) This is a 4th-order differential equation where the independent variable is t and the dependent variable, x. To find the general solution $x = x(t)$, we have to integrate the equation four times: $x^{(4)} = \cos 3t \Rightarrow x''' = \frac{\sin 3t}{3} + c_1 \Rightarrow x'' = -\frac{\cos 3t}{3^2} + c_1 t + c_2 \Rightarrow x' = -\frac{\sin 3t}{3^3} + c_1 \frac{t^2}{2} + c_2 t + c_3 \Rightarrow x = \frac{\cos 3t}{3^4} + c_1 \frac{t^3}{6} + c_2 \frac{t^2}{2} + c_3 t + c_4$. By denoting $C_1 = \frac{c_1}{6}$, $C_2 = \frac{c_2}{2}$, $C_3 = c_3$ and $C_4 = c_4$ one obtains: $x = \frac{\cos 3t}{3^4} + C_1 t^3 + C_2 t^2 + C_3 t + C_4$. In fact, this is the sum of the general solution of the homogeneous equation $x^{(4)} = 0$, $x_{GH} = C_1 t^3 + C_2 t^2 + C_3 t + C_4$, and a particular solution of the initial equation, $x_P = \frac{\cos 3t}{3^4}$.

(n) It is a first-order linear equation, i.e. an equation of the type $y' + a(x)y = b(x)$, in our case $a(x) = \frac{2}{x}$ and $b(x) = x^3$. Firstly we solve the homogeneous associated equation: $\frac{dy}{dx} + \frac{2y}{x} = 0$, which is a separable equation: $\frac{dy}{y} = -\frac{2dx}{x}$. We integrate and obtain $\int \frac{dy}{y} = -\int \frac{2dx}{x} \Rightarrow \ln y = \ln C - 2\ln x = \ln \frac{C}{x^2}$, so the general solution of the homogeneous equation is $y_{GH} = \frac{C}{x^2}$. Now, let us find a particular solution of the non-homogeneous equation by using Lagrange method: we search for a solution of the form $y_P = \frac{C(x)}{x^2}$, where $C(x)$ is an unknown function. In order to find this function we require that y_P satisfies the initial equation. Since $y'_P = \frac{C'(x)x - 2C(x)}{x^3}$, we obtain $\frac{C'(x)x - 2C(x)}{x^3} + \frac{2C(x)}{x^3} = x^3$, therefore $C'(x) = x^5 \Rightarrow C(x) = \frac{x^6}{6}$ and so $y_P = \frac{x^4}{6}$. The general solution is $y = y_P + y_{GH} = \frac{x^4}{6} + \frac{C}{x^2}$.

(o) It is also a first-order linear equation. Let us integrate the homogeneous equation $y' + xy = 0$. We have: $\int \frac{dy}{y} = -\int x\,dx \Rightarrow \ln y = -\frac{x^2}{2} + \ln C = \ln\left(C \exp\left(-\frac{x^2}{2}\right)\right)$, so the general solution of the homogeneous equation is $y_{GH} = C \exp\left(-\frac{x^2}{2}\right)$. By using Lagrange method we search for a particular solution of the non-homogeneous equation, $y_P = C(x) \exp\left(-\frac{x^2}{2}\right)$. Since $y'_P = C'(x) \exp\left(-\frac{x^2}{2}\right) - xC(x) \exp\left(-\frac{x^2}{2}\right)$ and $y'_P + xy_P = 1 + x^2$, we obtain $C'(x) = (1 + x^2) \exp\left(\frac{x^2}{2}\right)$. It follows that $C(x) = x \exp\left(\frac{x^2}{2}\right)$ and so, $y_P = x$. The general solution is $y = y_P + y_{GH} = x + C \exp\left(-\frac{x^2}{2}\right)$.

(p) A linear differential equation, too. The homogeneous associated equation $\frac{dy}{dx} - \frac{y}{x} = 0$ has the general solution $y_{GH} = Cx$. We search for $y_P = C(x)x$, a particular solution of the non-homogeneous (initial) equation. Since $y'_P = C'(x)x + C(x)$, we obtain $C'(x) = -\sin x - \cos x \Rightarrow C(x) = \cos x - \sin x$ and $y_P = x(\cos x - \sin x)$. The general solution is $y = y_{GH} + y_P = Cx + x(\cos x - \sin x)$.

(q) Note that if we try to put this equation in the "classic" form $y' = \frac{dy}{dx} = f(x, y)$ we obtain $y' = \frac{dy}{dx} = \frac{y}{2x + y^3}$, which cannot be classified in any of the types of equations studied in this chapter. But if we consider x as a function of y, $x = x(y)$ then we can write $x' = \frac{dx}{dy} = \frac{2x+y^3}{y} = \frac{2}{y}x + y^2$, which is a linear equation. The homogeneous associated equation is $\frac{dx}{dy} = \frac{2}{y}x \Leftrightarrow \int \frac{dx}{x} = 2\int \frac{dy}{y}$, whose general solution is $x_{GH} = Cy^2$. We search for a particular solution $x_p = C(y)y^2$ for the initial equation. Since $x'_p = C'(y)y^2 + 2yC(y)$, we obtain $C'(y) = 1 \Rightarrow C(y) = y$, so $x_p = y^3$ and the general solution is $x = x_{GH} + x_p = Cy^2 + y^3$.

(r) It is a Bernoulli equation: an equation of the form $y' + a(x)y = b(x)y^\alpha$, in our case $\alpha = 2$, $a(x) = \frac{1}{x}$, $b(x) = -x$. We divide by $y^\alpha = y^2$: $\frac{y'}{y^2} + \frac{1}{x} \cdot \frac{1}{y} = -x$ and use the substitution $z = y^{1-\alpha} = \frac{1}{y}$. Since $z' = -\frac{y'}{y^2}$, we obtain the linear equation $z' - \frac{1}{x} \cdot z = x$. The general solution of the homogeneous equation $z' - \frac{1}{x} \cdot z = 0$ is $z_{GH} = Cx$ and a particular solution of the non-homogeneous equation is $z_p = x^2$, so the linear equation in z has the general solution $z = z_{GH} + z_p = Cx + x^2$. Since $y = \frac{1}{z}$, the general solution of the initial equation is $y = \frac{1}{Cx+x^2}$.

(s) This is a Bernoulli equation with $\alpha = \frac{1}{2}$. We use the substitution $z = y^{1-\alpha} = \sqrt{y}$. Since $z' = \frac{y'}{2\sqrt{y}}$, we divide the equation by $2\sqrt{y}$ and obtain a linear equation in z: $\frac{y'}{2\sqrt{y}} = \frac{2}{x}\sqrt{y} + \frac{x}{2} \Rightarrow z' = \frac{2}{x}z + \frac{x}{2}$. The general solution of this linear equation is $z = z_p + z_{GH} = x^2\left(\ln\sqrt{x} + C\right)$. Since $y = z^2$, it follows that $y = x^4\left(\ln\sqrt{x} + C\right)^2$.

(t) If we divide the equation by $3xy^2$, it becomes: $y' + \frac{1}{3x}y = \frac{2}{3}y^{-2}$, and one can see this is a Bernoulli equation with $\alpha = -2$. We make the substitution $z = y^{1-\alpha} = y^3$. Since $z' = 3y^2 y'$, we divide the initial equation by x to obtain a linear equation in z: $3y^2 y' + \frac{1}{x}y^3 = 2 \Rightarrow z' + \frac{1}{x}z = 2$. The homogeneous equation, $\frac{dz}{dx} = -\frac{z}{x}$, has the general solution $z_{GH} = \frac{C}{x}$. We search for a particular solution of the non-homogeneous equation, $z_P = \frac{C(x)}{x}$ and we obtain $C(x) = x^2$, so $z_P = x$, and the general solution of the linear equation in z is $z = \frac{C}{x} + x$. Since $z = y^3$, the general solution of the initial equation is $y = \sqrt[3]{\frac{C+x^2}{x}}$.

(u) A differential equation of the form $P(x, y)dx + Q(x, y)dy = 0$ where the functions P and Q satisfy the condition $\frac{\partial P}{\partial y} = \frac{\partial Q}{\partial x}$ is an exact differential equation. In our case, $P(x, y) = 2x + 3x^2 y$, $Q(x, y) = x^3 - 3y^2$ and we notice that $\frac{\partial P}{\partial y} = 3y^2 = \frac{\partial Q}{\partial x}$, so the equation is exact. This means there exists a "potential function" U, a function such that $\frac{\partial U}{\partial x} = P(x, y) = 2x + 3x^2 y$ and $\frac{\partial U}{\partial y} = Q(x, y) = x^3 - 3y^2$. Let us integrate the first equation; we obtain $U(x, y) = \int \left(2x + 3x^2 y\right) dx = x^2 + x^3 y + C(y)$, where $C(y)$ is a function depending on y (it is a "constant" with respect to x). Therefore, $\frac{\partial U}{\partial y} = x^3 + C'(y)$ and we find the function $C(y)$ from the second equality: $\frac{\partial U}{\partial y} = x^3 - 3y^2 = x^3 + C'(y) \Rightarrow C'(y) = -3y^2 \Rightarrow C(y) = -y^3$. Hence, the potential function is $U(x, y) = x^2 + x^3 y - y^3$. Since our equation can be written $dU = 0$, its general solution is $U(x, y) = C$, or $x^2 + x^3 y - y^3 = C$, where C is a real constant. Note that in fact, $C(y)$ could be $C(y) = -y^3 + C_1$, with C_1 a real constant, so the potential function can be any function of the form $U(x, y) = x^2 + x^3 y - y^3 + C_1$. Since $dU = 0$, we have $U(x, y) = C_2$, where C_2 is a real constant, so the general solution is the same: $x^2 + x^3 y - y^3 = C$ (where $C = C_2 - C_1$ is a real constant.)

(v) Since $\frac{\partial P}{\partial y} = \exp(-y) = \frac{\partial Q}{\partial x}$, we have an exact differential equation. Let us find the function $U(x, y)$ such that $\frac{\partial U}{\partial x} = P(x, y) = 2x + \exp(-y)$ and $\frac{\partial U}{\partial y} = Q(x, y) = -2y - x \exp(-y)$. By integrating (with respect to x) the first equation we obtain $U(x, y) = x^2 + x \exp(-y) + C(y)$ and from the second equation we find $C(y) = -y^2$. The solution of the differential equation is $x^2 + x \exp(-y) - y^2 = C$.

(w) Since $\frac{\partial P}{\partial y} = 2xy - 3y^2 \neq -y^2 = \frac{\partial Q}{\partial x}$, it is not an exact equation. Let us find an integrating factor $\mu = \mu(y)$, i.e. a function of y such that, when the initial equation is multiplied by $\mu = \mu(y)$, an exact equation is obtained. The equation $(xy^2 - y^3)\mu(y)dx + (1 - xy^2)\mu(y)dy = 0$ is exact if and only if $\frac{\partial P_1}{\partial y} = \frac{\partial Q_1}{\partial x}$, where $P_1(x, y) = (xy^2 - y^3)\mu(y)$ and $Q_1(x, y) = (1 - xy^2)\mu(y)$. Since $\frac{\partial P_1}{\partial y} = (2xy - 3y^2)\mu(y) + (xy^2 - y^3)\mu'(y)$ and $\frac{\partial Q_1}{\partial x}(x, y) = -y^2\mu(y)$, we obtain $2\mu(y) + y\mu'(y) = 0$, which is a separable equation: $\mu' = \frac{d\mu}{dy} = -\frac{2\mu}{y} \Rightarrow \int \frac{d\mu}{\mu} = -2\int \frac{dy}{y} \Rightarrow \mu = \frac{1}{y^2}$. Now we can replace μ in the exact equation above and solve it: $(x - y)dx + \left(\frac{1}{y^2} - x\right)dy = 0$. Let us find the function $U(x, y)$ such that $\frac{\partial U}{\partial x} = x - y$ and $\frac{\partial U}{\partial y} = \frac{1}{y^2} - x$. It follows that $U(x, y) = \frac{x^2}{2} - xy - \frac{1}{y}$, so the solution is $\frac{x^2}{2} - xy - \frac{1}{y} = C$.

(x) It is a separable equation: $ydy = 2dx$. We integrate and obtain the general solution $y^2 = 4x + C$.

(y) It is a second-order differential equation. But, if we denote by $z = y'$, we obtain a first-order differential equation: $x^2 z' = z^2$. It is a separable equation, by integration we find: $\int \frac{dz}{z^2} = \int \frac{dx}{x^2} \Rightarrow -\frac{1}{z} = -\frac{1}{x} - \frac{1}{C_1} \Rightarrow z = \frac{C_1 x}{x + C_1}$. Since $z = y'$, we obtain $y' = \frac{C_1 x}{x + C_1}$ and integrate once again. The solution is $y = C_1 x + C_1^2 \ln(x + C_1) + C_2$.

(z) Since $(y^2)' = 2yy'$ we have: $y'' = (y')' = (y^2)'$, and, consequently, $y' = y^2 + C$. Let us solve now the separable equation $\frac{dy}{y^2 + C} = dx$. If $C = 0$ then we obtain by integration $-\frac{1}{y} = x + C_1$, so $y = -\frac{1}{x + C_1}$. If $C > 0$, or, equivalent, $C = C_1^2$, then $\frac{1}{C_1} \arctan \frac{y}{C_1} = x + C_2 \Rightarrow y = C_1 \tan C_1 (x + C_2)$. If $C < 0$, $C = -C_1^2$, then the solution is: $\frac{1}{2C_1} \ln \frac{y - C_1}{y + C_1} = x + C_2$

4. Since $y_1' = -\frac{1}{x^2}$, we have: $-\frac{2}{x^2} + \frac{1}{x^2} + \frac{1}{x^\alpha} = 0$, which implies $\alpha = 2$. Now we have to solve a Riccati equation, knowing a particular solution, y_1. By using the substitution $y(x) = y_1(x) + \frac{1}{u(x)}$, we'll obtain a linear equation in $u(x)$.

10.1 Chapter 1

Indeed, since $y' = -\frac{1}{x^2} - \frac{u'}{u^2}$, by replacing in the initial equation we obtain $-\frac{2}{x^2} - \frac{2u'}{u^2} + \frac{1}{x^2} + \frac{2}{xu} + \frac{1}{u^2} + \frac{1}{x^2} = 0$, which is equivalent to $u' - \frac{1}{x}u = \frac{1}{2}$. The general solution of this equation is $u = u_{GH} + u_P = Cx + \frac{1}{2}x \ln x$. So $y = \frac{1}{x} + \frac{1}{Cx + x \ln \sqrt{x}}$ is the general solution of the initial equation.

5. (a) This is a separable equation because it can be written $y' = \frac{dy}{dx} = -\frac{y}{x}$. Let us separate the variables and integrate: $\int \frac{dy}{y} = -\int \frac{dx}{x} \Rightarrow \ln y = -\ln x + \ln C$, so the general solution is $y = \frac{C}{x}$. Now use the initial condition $y(1) = 2$ and find $C = 2$. Hence, the solution of the Cauchy problem is $y = \frac{2}{x}$.

(b) It is also a separable equation: $\frac{dy}{dx} = \frac{y \exp x}{1 + \exp x} \Rightarrow \int \frac{dy}{y} = \int \frac{\exp x}{1 + \exp x} dx \Rightarrow \ln y = \ln(1 + \exp x) + \ln C$, so the general solution is $y = C(1 + \exp x)$. Since $y(0) = 2$, the constant is $C = 1$, so the solution of the Cauchy problem is $y = 1 + \exp x$.

(c) This is a linear equation. Let us solve the homogeneous equation $\frac{dy}{dx} = -y \cos x$. We have $\int \frac{dy}{y} = -\int \cos x \, dx \Rightarrow \ln y = -\sin x + \ln C$, so the general solution of the homogeneous equation is $y_{GH} = C \exp(-\sin x)$. Using Lagrange method, we search for a particular solution $y_P = C(x) \exp(-\sin x)$ of the non-homogeneous equation and obtain $C'(x) = \cos x \exp(\sin x)$, which means $C(x) = \exp(\sin x)$. So $y_P = 1$ and the general solution is $y = y_{GH} + y_P = C \exp(-\sin x) + 1$. If we compel it to satisfy the initial condition $y(0) = 1$, we obtain $C = 0$, so the solution of the Cauchy problem is the constant function $y(x) = 1$.

(d) The equation can be written $y' - \frac{1}{x}y = -\frac{\ln x}{x}$, so it is also a linear differential equation. Firstly we solve the homogeneous equation: $\frac{dy}{dx} = \frac{y}{x} \Leftrightarrow \int \frac{dy}{y} = \int \frac{dx}{x} \Rightarrow \ln y = \ln x + \ln C$, so the general solution of the homogeneous ecuation is $y_{GH} = Cx$. The second step is to search for a particular solution, $y_P = C(x)x$, of the non-homogeneous equation. Since $y'_P = C'(x)x + C(x)$, by replacing in the initial equation we obtain $C'(x) = -\frac{\ln x}{x^2}$. In order to compute $C(x)$ we need to use integration by parts: $C(x) = -\int \frac{\ln x}{x^2} dx = \frac{\ln x}{x} - \int \frac{1}{x^2} = \frac{\ln x}{x} + \frac{1}{x}$. Hence $y_P = \ln x + 1$ and the general solution of our differential equation is $y = y_{GH} + y_P = Cx + \ln x + 1$. From the initial condition $y(1) = 1$ we obtain $C = 0$, so the solution of the Cauchy problem is $y = \ln x + 1$.

(e) It is a linear equation. Let us solve the homogeneous equation $\frac{dy}{dx} = -xy \Rightarrow \int \frac{dy}{y} = -\int x \, dx \Rightarrow \ln y = -\frac{x^2}{2} + \ln C$, so the general solution of the homogeneous equation is $y_{GH} = C \exp\left(-\frac{x^2}{2}\right)$. We search for a particular solution of the non-homogeneous equation $y_P = C(x) \exp\left(-\frac{x^2}{2}\right)$. We obtain $C'(x) = (1 + x^2) \exp\left(\frac{x^2}{2}\right) \Rightarrow C(x) = x \exp\left(\frac{x^2}{2}\right)$, so $y_P = x$ and the general solution is $y = y_{GH} + y_P = C \exp\left(-\frac{x^2}{2}\right) + x$. Since $y(0) = 1$, the constant is $C = 1$, so the requested solution is $y = \exp\left(-\frac{x^2}{2}\right) + x$.

(f) This is a Bernoulli equation with $\alpha = 2$: $y' + \frac{1}{x}y = 2xy^2$. We use the substitution $z = y^{1-\alpha} = \frac{1}{y}$, but first, let us divide the equation by $-y^2$: $-\frac{y'}{y^2} - \frac{1}{x} \cdot \frac{1}{y} = -2x^2 \Rightarrow z' - \frac{1}{x}z = -2x^2$, which is a linear equation in $z(x)$. The general solution of this linear equation is $z = z_{GH} + z_P = Cx - x^3$. Since $y = \frac{1}{z}$, the general solution of the equation is $y = \frac{1}{Cx-x^3}$. From the initial condition $y(1) = 1$ we obtain that $C = 2$, so the solution of the Cauchy problem is $y = \frac{1}{2x-x^3}$.

(g) Note that the equation can be written as $y' + \frac{1}{x}y = 3x^3 y^{-1}$, so it is a Bernoulli equation with $\alpha = -1$. We make the substitution $z = y^{1-\alpha} = y^2$ and obtain the linear equation in $z(x)$: $z' + \frac{2}{x}z = 6x^3$, which has the general solution $z = z_{GH} + z_P = \frac{C}{x^2} + x^4$. Since $y = \pm\sqrt{z}$, the general solution of the Bernoulli equation is $y = \pm\sqrt{\frac{C+x^6}{x^2}}$. Because of the initial condition $y(1) = 1 > 0$ we have to choose $y = \sqrt{\frac{C+x^6}{x^2}}$ and we get $C = 0$, so the required solution is $y = x^2$.

(h) It is a Bernoulli equation with $\alpha = 2$. We divide it by y^2 and make the substitution $z = \frac{1}{y}$. Since $z' = -\frac{y'}{y^2}$, we obtain the linear equation $z' - 2z = -\exp x$ whose general solution is $z = z_P + z_{GH} = \exp x + C \exp 2x$. It follows that $y = \frac{1}{\exp x(1+C \exp x)}$ is the general solution. The particular solution which satisfies the condition $y(0) = \frac{1}{2}$ is obtained for $C = 1$: $y = \frac{1}{\exp x + \exp 2x}$.

(i) This is a homogeneous equation (because the function $f(x,y) = \frac{y-x}{x}$ is homogeneous: $f(tx, ty) = f(x, y)$), so we can use the substitution $u(x) = \frac{y(x)}{x}$. Since $y'(x) = u + xu'(x)$, the equation becomes: $u + xu' = u - 1 \Rightarrow u' = -\frac{1}{x}$, so $u = \ln \frac{C}{x}$. Hence, the general solution of the initial equation is $y = x \ln \frac{C}{x}$ and the particular solution which satisfies the condition $y(1) = 0$ is obtained for $C = 1$: $y = -x \ln x$. Note that we could also treat it like a linear equation.

(j) It is also a homogeneous equation. By using the substitution $u(x) = \frac{y(x)}{x}$ the equation becomes: $xu' + u = u - u^2 \Rightarrow \frac{du}{dx} = -\frac{u^2}{x} \Rightarrow -\int \frac{du}{u^2} = \int \frac{dx}{x}$, so $\frac{1}{u} = \ln x + C$. The general solution of the initial equation is $y = \frac{x}{\ln x + C}$. The condition $y(1) = 1$ fulfilled for $C = 1$, so the required particular solution is $y = \frac{x}{1+\ln x}$. Note that we could also solve it like a Bernoulli equation.

6. (a) This is a Clairaut equation, i.e. an equation of the type $y = xy' + b(y')$. By using the notation $p = y'$, the equation can be written as $y = xp - (2 + p^2)$. Let us differentiate this equation: $y' = p = p + xp' - 2pp' \Rightarrow p'(x - 2p) = 0$. We have either $p' = 0$, so $p = y' = C$ (C is a constant) and we obtain

the general solution, the family of straight lines $y = Cx - (2 + C^2)$, or $x - 2p = 0$, so $x = 2p$, $y = 2p^2 - 2 - p^2 = p^2 - 2$ and, by replacing p with $\frac{x}{2}$, we obtain a singular solution, the parabola $y = \frac{x^2}{4} - 2$ which is the envelope of the family of straight lines.

(b) It is also a Clairaut equation, we use the same notation and, by differentiating the equation $y = xp + 4\sqrt{p}$, we obtain $p'\left(x + \frac{2}{\sqrt{p}}\right) = 0$. So, the general solution (obtained when $p' = 0$) is the family of straight lines $y = Cx + 4\sqrt{C}$ (C is a positive constant), and the singular solution (obtained when $x = -\frac{2}{\sqrt{p}}$) is the hyperbola $y = -\frac{4}{x}$, $x < 0$.

(c) The same type of equation. By differentiating the equation $y = xp + \sqrt{1 + p^2}$ we obtain $p'\left(x + \frac{p}{\sqrt{1+p^2}}\right) = 0$. The general solution is the family of straight lines $y = Cx + \sqrt{1 + C^2}$, while the singular solution is the envelope of this family, the curve defined by the parametric equations

$$\begin{cases} x = \frac{-p}{\sqrt{1+p^2}} \\ y = \frac{1}{\sqrt{1+p^2}} \end{cases} \quad p \in \mathbb{R}.$$

The cartesian equation of this curve is: $x^2 + y^2 = 1$, so it is the circle of radius 1 with the center at the origin.

(d) It is also a Clairaut equation, we use the same notation and, by differentiating the equation $y = xp - \ln p$, we obtain $p'\left(x - \frac{1}{p}\right) = 0$. So, the general solution (obtained when $p' = 0$) is the family of straight lines $y = Cx - \ln C$ (where C is a positive constant), and the singular solution (obtained for $x = \frac{1}{p}$) is the curve $y = 1 + \ln x$, $x > 0$.

(e) This is a Lagrange equation, an equation of the type $y = xa(y') + b(y')$. Let us introduce the same notation, $y' = p$. We have $y = 2xp - 4p^3$ and, by differentiating we get $p = 2p + 2xp' - 12p^2p'$, or, equivalently, $-p = \frac{dp}{dx}(2x - 12p^2)$. If $\frac{dp}{dx} = 0$ then p is a constant, but, from the above equation, it follows that $p = 0$, and so we found the singular solution, the constant function $y = 0$. If $\frac{dp}{dx} \neq 0$ then we can write $\frac{dx}{dp} = -\frac{2}{p}x + 12p$, which is a linear equation in $x = x(p)$. The homogeneous linear equation $\frac{dx}{dp} = -\frac{2}{p}x$ has the general solution $x_{GH} = \frac{C}{p^2}$, so we search for a particular solution of the form $x_P = \frac{C(p)}{p^2}$. Since $x'_P = \frac{C'(p)p^2 - C(p) \cdot 2p}{p^4}$, we obtain $C'(p) = 12p^3$, hence $C(p) = 3p^4$ and $x_P = 3p^2$. The general solution of the linear equation above is $x = x_P + x_{GH} = 3p^2 + \frac{C}{p^2}$. Since $y = 2xp - 4p^3$, we obtain the parametric form of the general solution of the initial Lagrange

equation:
$$\begin{cases} x(p) = 3p^2 + \frac{C}{p^2} \\ y(p) = 2p^3 - \frac{2C}{p} \end{cases} \quad p \in \mathbb{R}.$$

where C is a constant.

(f) It is also a Lagrange equation, with $a(y') = -y' + 2$ and $b(y') = y'^2 - 2y'$. We denote $y' = p$ and differentiate the equation $y = -xp + p^2 - 2p + 2x$. Thus, we obtain $p = -p - xp' + 2pp' - 2p' + 2$, or, equivalently, $2p - 2 = p'(-x + 2p - 2)$. If $\frac{dp}{dx} = 0$ then p is a constant; since $2p - 2 = 0$, it follows that $p = 1$ and, by replacing in the above equation, we find the singular solution, the linear function $y = x - 1$. If $\frac{dp}{dx} \neq 0$ then we can write $\frac{dx}{dp} = -\frac{1}{2p-2}x + 1$, which is a linear equation in $x = x(p)$. The homogeneous linear equation $\frac{dx}{dp} = -\frac{1}{2(p-1)}x$ has the general solution $x_{GH} = \frac{C}{\sqrt{p-1}}$, so we search for a particular solution of the form $x_P = \frac{C(p)}{\sqrt{p-1}}$. We obtain $C'(p) = \sqrt{p-1}$, hence $C(p) = \frac{2}{3}(p-1)^{\frac{3}{2}}$ and $x_P = \frac{2}{3}(p-1)$. The general solution of the linear equation above is $x = x_P + x_{GH} = \frac{2}{3}(p-1) + \frac{C}{\sqrt{p-1}}$. Since $y = -xp + p^2 - 2p + 2x$, we obtain the parametric form of the general solution of the Lagrange equation:

$$\begin{cases} x(p) = \frac{2}{3}(p-1) + \frac{C}{\sqrt{p-1}} \\ y(p) = \frac{p^2-4}{3} + \frac{C(2-p)}{\sqrt{p-1}} \end{cases} \quad p \in \mathbb{R}.$$

where C is a constant.

7.(a) Picard's method of successive approximations constructs a sequence of functions, $\{y_n(x)\}_{n \geq 0}$, uniformly convergent to the unique solution of the Cauchy problem $y' = f(x, y), y(x_0) = y_0$. This sequence is built by the following recurrence formula: $y_n(x) = \int_{x_0}^x f(t, y_{n-1}(t))dt$, for any $n \geq 1$ and the initial approximation is the constant function $y_0(x) = y_0$. In our case, $f(x, y) = x^2 + y, x_0 = 0, y_0 = 0$, so we have:

$$y_1(x) = y_0 + \int_0^x (t^2 + y_0)dt = \frac{t^3}{3}\Big|_0^x = \frac{x^3}{3}$$

$$y_2(x) = y_0 + \int_0^x (t^2 + y_1(t))dt = \int_0^x \left(t^2 + \frac{t^3}{3}\right)dt = \left(\frac{t^3}{3} + \frac{t^4}{12}\right)\Big|_0^x = \frac{x^3}{3} + \frac{x^4}{12}$$

$$y_3(x) = y_0 + \int_0^x (t^2 + y_2(t))dt = \int_0^x \left(t^2 + \frac{t^3}{3} + \frac{t^4}{12}\right)dt = \frac{x^3}{3} + \frac{x^4}{12} + \frac{x^5}{60}.$$

Remark In fact, by mathematical induction, one can prove that

$$y_n(x) = 2 \sum_{k=3}^{n+2} \frac{x^k}{k!}$$

Let us solve the linear differential equation above. The general solution of the homogeneous equation $y' = y$ is $y_{GH} = \exp x$. Because the free term is a polynomial of degree 2 (and, obviously, we don't have resonance), we search for a particular solution of the same type, $y_P = Ax^2 + Bx + C$. We'll find $y_P = -x^2 - 2x - 2$, so the general solution is $y = y_{GH} + y_P = C \exp x - 2x^2 - 2x - 2$ and from the initial condition we obtain $C = 2$, i.e. the solution of the Cauchy problem is $y = 2 \exp x - x^2 - 2x - 2$. Using the well-known expansion of the exponential function, $\exp x = \sum_{n=0}^{\infty} \frac{x^n}{n!}$ one can write $y(x) = 2 \sum_{n=3}^{\infty} \frac{x^n}{n!}$ and it is obvious that the sequence of partial sums of this series is just the sequence $\{y_n(x)\}_{n \geq 0}$ constructed above.

(b) The equation can be written $y' = \frac{x-y}{x}$, so in this case the function $f(x, y)$ is $f(x, y) = \frac{x-y}{x}$. Since $x_0 = 1$ and $y_0 = 1$, we construct the sequence $\{y_n(x)\}_{n \geq 0}$ in the following way:

$$y_1(x) = 1 + \int_1^x \frac{t-1}{t} dt = 1 + (t - \ln t)\Big|_1^x = x - \ln x$$

$$y_2(x) = 1 + \int_1^x \frac{t - t + \ln t}{t} dt = 1 + \int_1^x \frac{\ln t}{t} dt = 1 + \frac{\ln^2 t}{2}\Big|_1^x = 1 + \frac{\ln^2 x}{2}$$

$$y_3(x) = 1 + \int_1^x \frac{t - 1 - \frac{1}{2}\ln^2 t}{t} dt = 1 + \left(t - \ln t - \frac{\ln^3 t}{3!}\right)\Big|_1^x = x - \ln x - \frac{\ln^3 x}{3!}$$

Remark By mathematical induction it can be proved that

$$y_{2n}(x) = \sum_{k=0}^{n} \frac{\ln^{2k} x}{(2k)!} \; ; \; y_{2n+1}(x) = x - \sum_{k=0}^{n} \frac{\ln^{2k+1} x}{(2k+1)!}$$

Try to find the solution of the Cauchy problem $y' = \frac{x-y}{y}$, $y(1) = 1$ and prove that the sequences of functions above are both convergent to this function (Hint: the solution is $y(x) = \frac{1}{2}\left(x + \frac{1}{x}\right)$; to prove the convergence, make the notation $\ln x = t \Rightarrow x = \exp t$ and use the expansion in power series of the exponential function).

10.2 Chapter 2

1. Since $y'(x) = -2C_1 \sin 2x + 2C_2 \cos 2x$ and $y''(x) = -4C_1 \cos 2x - 4C_2 \sin 2x$, it follows that $y''(x) + 4y(x) = 0$, $\forall x \in \mathbb{R}$.
2. Since $y'(x) = 2C_1 \exp 2x - 2C_2 \exp(-2x)$ and $y''(x) = 4C_1 \exp 2x + 4C_2 \exp(-2x)$, it follows that $y''(x) - 4y(x) = 0$, $\forall x \in \mathbb{R}$.
3. (a) It is the general solution of a linear homogeneous 2nd order equation. To find this equation, we write the polynomial which has the roots $r_1 = 3$ and $r_2 = -5$: $(r-3)(r+5) = r^2 + 2r - 15$. Hence the differential equation having the basis of solutions $\{y_1 = \exp(3x), y_2 = \exp(-5x)\}$ is $y'' + 2y' - 15y = 0$.
    (b) We have to find the linear homogeneous equation having the basis of solutions $\{y_1 = \cos 4x, y_2 = \sin 4x\}$. The algebraic equation associated to the differential equation has the complex roots $r_{1,2} = \pm 4i$, so this equation is: $r^2 + 16 = 0$. Hence, the differential equation is $y'' + 16y = 0$.
    (c) We search for the linear (third-order) differential equation with the basis of solutions $\{y_1 = 1, y_2 = x, y_3 = e^x\}$. The associated algebraic equation has the roots $r_1 = r_2 = 0$, $r_3 = 1$, so this equation is: $r^3 - r^2 = 0$. The linear homogeneous differential equation corresponding to this characteristic equation is $y''' - y'' = 0$.
4. (a) We write the characteristic (algebraic) equation related to this homogeneous linear differential equation: $r^2 + r - 2 = 0$, which has the real distinct roots $r_1 = 1$ and $r_2 = -2$. It follows that $\{y_1 = \exp(r_1 x) = \exp x, y_2 = \exp(r_2 x) = \exp(-2x)\}$ is a basis of solutions for the differential equation, so the general solution is $y = C_1 \exp x + C_2 \exp(-2x)$.
    (b) The characteristic equation associated with the linear differential equation is $r^2 - 6r + 9 = 0$, which has one (real) root, $r = 3$, with multiplicity 2. So $\{y_1 = \exp(3x), y_2 = x \exp(3x)\}$ is a basis of solutions for the differential equation and the general solution is $y = C_1 \exp(3x) + C_2 x \exp(3x)$.
    (c) The characteristic equation $r^2 + 2r + 10 = 0$ has the complex (conjugate) roots $r_{1,2} = -1 \pm 3i$. Hence, $\{y_1 = \exp(-x) \cos(3x), y_2 = \exp(-x) \sin(3x)\}$ is a basis of solutions for the differential equation and the general solution is $y = C_1 \exp(-x) \cos(3x) + C_2 \exp(-x) \sin(3x)$.
    (d) The characteristic equation is an algebraic equation of degree 3: $r^3 - r = 0$, which can be written $r(r-1)(r+1) = 0$, so it has 3 real distinct roots: $r_1 = 0, r_2 = 1$ and $r_3 = -1$. Consequently, a basis of solutions for the homogeneous linear differential equation is $\{y_1 = \exp(0 \cdot x) = 1, y_2 = \exp x, y_3 = \exp(-x)\}$ and the general solution is written: $y = C_1 + C_2 \exp x + C_3 \exp(-x)$.
    (e) The characteristic equation $r^3 + r = 0$, can be written $r(r^2 + 1) = 0$, so it has one real root, $r_1 = 0$, and two complex conjugate roots: $r_{2,3} = \pm i$. Consequently, a basis of solutions for the homogeneous linear differential equation is $\{y_1 = 1, y_2 = \exp(0 \cdot x) \cos(1 \cdot x) = \cos x, y_3 = \exp(0 \cdot x) \sin(1 \cdot x) = \sin x\}$ and the general solution is: $y = C_1 + C_2 \cos x + C_3 \sin x$.

(f) In this case, the algebraic equation associated is $r^3 + 1 = 0$, which can be written $(r+1)(r^2 - r + 1) = 0$ and one can see that it has one real root, $r_1 = -1$, and two complex conjugate roots: $r_{2,3} = \frac{1}{2} \pm \frac{\sqrt{3}}{2}i$. Consequently, $\{y_1 = \exp(-x),\ y_2 = \exp\left(\frac{1}{2}x\right)\cos\left(\frac{\sqrt{3}}{2}x\right),\ y_3 = \exp\left(\frac{1}{2}x\right)\cos\left(\frac{\sqrt{3}}{2}x\right)\}$ is a basis of solutions for the differential equation and the general solution is: $y = C_1 \exp(-x) + C_2 \exp\left(\frac{1}{2}x\right)\cos\left(\frac{\sqrt{3}}{2}x\right) + C_3 \exp\left(\frac{1}{2}x\right)\cos\left(\frac{\sqrt{3}}{2}x\right)$.

(g) The algebraic equation $r^4 - 1 = 0$ can be written $(r-1)(r+1)(r^2+1) = 0$, so it has two real distinct roots, $r_1 = 1$ and $r_2 = -1$ and two complex (conjugate) roots: $r_{3,4} = \pm i$. So $\{y_1 = \exp x,\ y_2 = \exp(-x),\ y_3 = \cos x,\ y_4 = \sin x\}$ is a basis of solutions for our homogeneous linear differential equation and the general solution is $y = C_1 \exp x + C_2 \exp(-x) + C_3 \cos x + C_4 \sin x$.

(h) The algebraic equation related to the differential equation is $r^4 - 5r^2 + 4 = 0$. Since $t^2 - 5t + 4 = (t-1)(t-4)$, we can write the characteristic equation in the form $(r^2 - 1)(r^2 - 4) = 0$, or, equivalent, $(r-1)(r+1)(r-2)(r+2) = 0$. Now it is obvious that the equation has 4 real distinct roots: $r_1 = 1$, $r_2 = -1$, $r_3 = 2$, $r_4 = -2$, therefore a basis of solutions is $\{y_1 = \exp x,\ y_2 = \exp(-x),\ y_3 = \exp(2x),\ y_4 = \exp(-2x)\}$ and the general solution of the linear differential equation is $y = C_1 \exp x + C_2 \exp(-x) + C_3 \exp(2x) + C_4 \exp(-2x)$.

(i) The algebraic equation $r^4 + 5r^2 + 4 = 0$ can be written $(r^2 + 1)(r^2 + 4) = 0$, and one can see that it has 4 complex roots, in complex conjugate pairs: $r_{1,2} = \pm i$ and $r_{3,4} = \pm 2i$. A basis of solutions for the linear homogeneous differential equation is $\{y_1 = \cos x,\ y_2 = \sin x,\ y_3 = \cos(2x),\ y_4 = \sin(2x)\}$ and the general solution is $y = C_1 \cos x + C_2 \sin x + C_3 \cos(2x) + C_4 \sin(2x)$.

(j) The characteristic equation is $r^4 - 2r^2 + 1 = 0$ which can be written $(r^2 - 1)^2 = 0$, or, equivalent, $(r-1)^2(r+1)^2 = 0$. All the roots are real: $r_1 = 1$, with multiplicity 2 and $r_2 = -1$, also with multiplicity 2. A basis of solutions is $\{y_1 = \exp x,\ y_2 = x \exp x,\ y_3 = \exp(-x),\ y_4 = x\exp(-x)\}$ and the general solution is $y = C_1 \exp x + C_2 x \exp x + C_3 \exp(-x) + C_4 x \exp(-x)$.

(k) The characteristic equation $r^4 + 2r^2 + 1 = 0$ can be written $(r^2 + 1)^2 = 0$, and one can see that it has two complex conjugate roots of multiplicity 2: $r_{1,2} = \pm i$. A basis of solutions for the linear homogeneous differential equation is $\{y_1 = \cos x,\ y_2 = \sin x,\ y_3 = x\cos x,\ y_4 = x \sin x\}$ and the general solution is $y = C_1 \cos x + C_2 \sin x + C_3 x \cos x + C_4 x \sin x$.

(l) The characteristic equation is $r^5 + 3r^4 + 3r^3 + r^2 = 0$ which can be written $r^2(r+1)^3 = 0$. All the roots are real: $r_1 = 0$, with multiplicity 2 and $r_2 = -1$, with multiplicity 3. A basis of solutions is $\{y_1 = 1,\ y_2 = x,\ y_3 = \exp(-x),\ y_4 = x \exp(-x),\ y_5 = x^2 \exp(-x)\}$ and so the general solution is $y = C_1 + C_2 x + C_3 \exp(-x) + C_4 x \exp(-x) + C_5 x^2 \exp(-x)$.

5. (a) A set of functions is linearly independent if and only if their Wronskian is non zero. The Wronskian of the functions $\{y_1 = \exp x, \; y_2 = \exp(-x)\}$ is

$$W[y_1(x), y_2(x)] = \begin{vmatrix} \exp x & \exp(-x) \\ \exp x & -\exp(-x) \end{vmatrix} = -2 \neq 0,$$

so the functions are linearly independent. Since $r_1 = 1$ and $r_2 = -1$ are the roots of the algebraic equation $r^2 - 1 = 0$, the linear homogeneous differential equation having this basis of solutions is $y'' - y = 0$.

(b) The Wronskian of the functions $\{y_1 = \exp x, \; y_2 = x \exp x, \; y_3 = \exp(-x)\}$ is

$$W[y_1(x), y_2(x), y_3(x)] = \begin{vmatrix} \exp x & x \exp x & \exp(-x) \\ \exp x & (x+1) \exp x & -\exp(-x) \\ \exp x & (x+2) \exp x & \exp(-x) \end{vmatrix} = 4 \exp x \neq 0,$$

so the functions are linearly independent. The algebraic equation having the double root $r_1 = 1$ and the simple root $r_2 = -1$ is $(r-1)^2(r+1) = 0 \Leftrightarrow r^3 - r^2 - r + 1 = 0$, so the linear homogeneous differential equation which have this basis of solutions is $y''' - y'' - y' + y = 0$.

(c) Since the Wronskian of the functions $\{y_1 = 1, \; y_2 = x, \; y_3 = x^2, \; y_4 = \exp 2x, \; y_5 = \cos x, \; y_6 = \sin x\}$ is a determinant of order 6, let us find another (easier) way to prove that the functions are linearly independent. Let $\alpha_1, \alpha_2, \cdots, \alpha_6$ be six real numbers such that

$$\alpha_1 + \alpha_2 x + \alpha_3 x^2 + \alpha_4 \exp(2x) + \alpha_5 \cos x + \alpha_6 \sin x = 0, \; \forall x \in \mathbb{R}.$$

If we divide this equation by $\exp(2x)$ and make $x \to \infty$, we find $\alpha_4 = 0$. Hence, we can write:

$$\alpha_2 x + \alpha_3 x^2 = -\alpha_1 - \alpha_5 \cos x - \alpha_6 \sin x, \; \forall x \in \mathbb{R}.$$

Since the left side of this equation is a bounded expression, it follows that $\alpha_2 = \alpha_3 = 0$ and we have:

$$\alpha_1 + \alpha_5 \cos x + \alpha_6 \sin x = 0, \; \forall x \in \mathbb{R}.$$

If we write this equation for $x = 0$ and $x = \pi$, we obtain $\alpha_1 + \alpha_5 = 0$ and $\alpha_1 - \alpha_5 = 0$. Hence, $\alpha_1 = \alpha_5 = 0$ and also, α_6 must be 0. Since all the coefficients of the linear combination have to be 0, the six functions are linearly independent. The algebraic equation having the triple root $r_1 = 0$, the simple root $r_2 = 2$ and the complex conjugate roots $r_{3,4} = \pm i$ is $r^3(r-2)(r^2+1) = 0 \Leftrightarrow r^6 - 2r^5 + r^4 - 2r^3 = 0$, so the linear homogeneous equation which have this basis of solutions is $y^{(6)} - 2y^{(5)} + y^{(4)} - 2y''' = 0$.

6. (a) First, we need to solve the homogeneous equation, $y'' - 4y = 0$. The characteristic equation $r^2 - 4 = 0$ has the roots $r_1 = 2$ and $r_2 = -2$, so the general solution of the linear homogeneous equation is $y_{GH} = C_1 \exp(2x) + C_2 \exp(-2x)$. The second step is to find a particular solution of the non-homogeneous equation. Let us note that the free term is a function of the form $f(x) = P(x) \exp(\lambda x)$, where $P(x) = 3x$ is a polynomial of degree 1 and $\lambda = 1$ is not a root of the characteristic equation. Therefore, we search for a particular solution of the non-homogeneous equation of the form $y_P = Q(x) \exp x$ where $Q(x)$ is a polynomial of the same degree as $P(x)$, so $Q(x) = Ax + B$. We have: $y_P = (Ax + B) \exp x$, $y'_P = (Ax + B + A) \exp x$ and $y''_P = (Ax + B + 2A) \exp x$. By replacing in the initial equation we obtain: $(-3Ax + 2A - 3B) \exp x = 3x \exp x \Rightarrow A = -1$ and $B = -\frac{2}{3}$, so $y_P = \left(-x - \frac{2}{3}\right) \exp x$ and the general solution is $y = y_{GH} + y_P = C_1 \exp(2x) + C_2 \exp(-2x) - \left(x + \frac{2}{3}\right) \exp x$.

(b) The homogeneous equation is the same as above. The free term has the same form $f(x) = P(x) \exp(\lambda x)$ where $P(x) = 4$ is a polynomial of degree 0 and $\lambda = 2$ is a root of the characteristic equation (with multiplicity 1). So we search for a particular solution of the non-homogeneous equation of the form $y_P = x^m Q(x) \exp(\lambda x)$, where $m = 1$ is the multiplicity of the root $\lambda = 2$ and $Q(x) = A$ is a polynomial of the same degree (0) as $P(x)$. So $y_P = Ax \exp(2x)$, $y'_P = A(1 + 2x) \exp(2x)$, $y''_P = A(4 + 4x) \exp(2x)$ and, by replacing in the initial equation, we obtain $A = 1$. So $y_P = x \exp(2x)$ and the general solution is $y = y_{GH} + y_P = C_1 \exp(2x) + C_2 \exp(-2x) + x \exp(2x)$.

(c) The homogeneous equation is $y'' + 9y = 0$. Since the algebraic equation $r^2 + 9 = 0$ has the complex conjugate roots $r_{1,2} = \pm 3i$, the general solution of the homogeneous equation is $y_{GH} = C_1 \cos(3x) + C_2 \sin(3x)$. The free term is a function of the form $f(x) = \exp(\alpha x)(P_1(x) \cos \beta x + P_2(x) \sin \beta x)$, where $\alpha = 0$, $\beta = 1$ and $P_1(x) = 0$, $P_2(x) = 1$. Since $\alpha + \beta i = i$ is not a root of the characteristic equation, we search for a particular solution of the form $y_P = Q_1(x) \cos x + Q_2(x) \sin x$ where $\deg Q_1 = \deg Q_2 = \max\{\deg P_1, \deg P_2\} = 0$. So Q_1 and Q_2 are constant polynomials: $Q_1(x) = A$ and $Q_2(x) = B$. Let us find the constant A and B. We have: $y_P = A \cos x + B \sin x$, $y'_P = -A \sin x + B \cos x$, $y''_P = -A \cos x - B \sin x$ and, by replacing in the initial equation, we obtain $8A \cos x + 8B \sin x = \sin x \Rightarrow A = 0$, $B = \frac{1}{8}$. So $y_P = \frac{1}{8} \sin x$ and the general solution is $y = y_{GH} + y_P = C_1 \cos(3x) + C_2 \sin(3x) + \frac{1}{8} \sin x$.

(d) We have the same homogeneous equation and the free term has the same form as above, but, in this case, $\alpha = 0$, $\beta = 3$ and $\alpha + \beta i = 3i$ is a root of the characteristic equation (we have resonance of order 1, the multiplicity of the root $3i$). So this time we search for a particular solution of the form $y_P = x(A \cos 3x + B \sin 3x)$. Since $y'_P = (A + 3Bx) \cos 3x + (B - 3Ax) \sin 3x$ and $y''_P = (6B - 9Ax) \cos 3x + (-6A - 9Bx) \sin 3x$, by replacing in the initial equation, we find $A = 0$ and $B = \frac{1}{6}$, so $y_P = \frac{1}{6}x \sin 3x$ and the general solution is $y = y_{GH} + y_P = C_1 \cos 3x + C_2 \sin 3x + \frac{1}{6}x \sin 3x$.

(e) We use the superposition principle: since the free term is $f = f_1 + f_2$, a particular solution can be written $y_P = y_{P_1} + y_{P_2}$ where y_{P_1} is a particular solution of the differential equation $y'' + 9y = f_1(x) = \sin x$ and y_{P_2} is a particular solution of the differential equation $y'' + 9y = f_2(x) = \cos 3x$. The general solution is $y = y_{GH} + y_{P_1} + y_{P_2} = C_1 \cos 3x + C_2 \sin 3x + \frac{1}{8} \sin x + \frac{1}{6}x \sin 3x$.

(f) First, we solve the homogeneous equation $y'' + 2y' + 5y = 0$. The characteristic equation $r^2 + 2r + 5 = 0$ has the complex (conjugate) roots $r_{1,2} = -1 \pm 2i$, so the general solution of the homogeneous equation is $y_{GH} = C_1 \exp(-x) \cos 2x + C_2 \exp(-x) \sin 2x$. The free term is a function of the form $f(x) = P(x) \exp(\lambda x)$, where $P(x) = 5x^2 - x$ is a polynomial of degree 2 and $\lambda = 0$ is not a root of the characteristic equation (we have no resonance). We search for a particular solution of the non-homogeneous equation of the form $y_P = Q(x) \exp(0 \cdot x) = Ax^2 + Bx + C$ (because $Q(x)$ must be a polynomial of the same degree as $P(x)$). We have: $y'_P = 2Ax + B$, $y''_P = 2A$, and, by replacing in the initial equation we find $A = 1$, $B = -1$ and $C = 0$, so $y_P = x^2 - x$ and the general solution is $y = y_{GH} + y_P = C_1 \exp(-x) \cos 2x + C_2 \exp(-x) \sin 2x + x^2 - x$.

(g) Let us solve the homogeneous equation $y''+2y'+2y = 0$. The characteristic equation $r^2 + 2r + 2 = 0$ has the complex (conjugate) roots $r_{1,2} = -1 \pm i$, so the general solution of the homogeneous equation is $y_{GH} = C_1 \exp(-x) \cos x + C_2 \exp(-x) \sin x$. The free term is a function of the form $f(x) = \exp(\alpha x)(P_1(x) \cos \beta x + P_2(x) \sin \beta x$, where $P_1(x) = 0$, $P_2(x) = 8$ and $\alpha + \beta i = 1 + i$ is not a root of the characteristic equation (we have no resonance). So we need to search for a particular solution of the non-homogeneous equation of the form $y_P = \exp x (A \cos x + B \sin x)$. We have: $y'_P = \exp x [(A+B) \cos x + (B-A) \sin x]$ and $y''_P = \exp x (2B \cos x - 2A \sin x)$ and, by replacing in the initial equation we find $A = -1$ and $B = 1$, so $y_P = \exp x (- \cos x + \sin x)$ and the general solution is $y = y_{GH} + y_P = C_1 \exp(-x) \cos x + C_2 \exp(-x) \sin x + \exp x (-\cos x + \sin x)$.

(h) The characteristic equation of the linear homogeneous equation $y'' - 5y' = 0$ is $r^2 - 5r = 0$. Since this algebraic equation has real distinct roots, $r_1 = 0$ and $r_2 = 5$, the general solution of the homogeneous equation is $y_{GH} = C_1 + C_2 \exp(5x)$. In order to find a particular solution of the non-homogeneous equation, we use the superposition principle: since the free term is $f = f_1 + f_2$, a particular solution can be written $y_P = y_{P_1} + y_{P_2}$ where y_{P_1} is a particular solution of the differential equation $y'' - 5y' = f_1(x) = 5x^2$ and y_{P_2} is a particular solution of the differential equation $y'' - 5y' = f_2(x) = \exp(5x)$. The first function, f_1, can be written $f_1(x) = 5x^2 = \exp(\lambda_1 x) P_1(x)$, where $P_1(x) = 5x^2$ and $\lambda_1 = 0$ is a root of the characteristic equation (with multiplicity 1). We have a resonance of order 1, so we search for a particular solution $y_{P_1} = x(Ax^2 + Bx + C)$. We calculate

y'_{P_1} and y''_{P_1} and replace in the equation $y'' - 5y' = 5x^2$. We get $A = -\frac{1}{3}$, $B = -\frac{1}{5}$, $C = -\frac{2}{25}$, and so $y_{P_1} = -\frac{1}{3}x^3 - \frac{1}{5}x^2 - \frac{2}{25}x$. The second function, f_2, can be written $f_2(x) = \exp(5x) = \exp(\lambda_2 x)P_2(x)$, where $P_2(x) = 1$ and $\lambda_2 = 5$ is also a root of the characteristic equation (with multiplicity 1). So we search for a particular solution $y_{P_2} = Cx \exp(5x)$. By computing y'_{P_2} and y''_{P_2} and replacing in the equation $y'' - 5y' = \exp(5x)$, we find $C = \frac{1}{5}$, so $y_{P_2} = \frac{1}{5}x \exp(5x)$. The general solution is $y = y_{GH} + y_{P_1} + y_{P_2} = C_1 + C_2 \exp(5x) - \frac{1}{3}x^3 - \frac{1}{5}x^2 - \frac{2}{25}x + \frac{1}{5}x \exp(5x)$. i) The linear homogeneous equation $y'' - 2y' + y = 0$ has the characteristic equation $r^2 - 2r + 1 = 0$ which has a real root, $r_1 = 1$ with multiplicity 2. The general solution of the homogeneous equation is $y_{GH} = C_1 \exp x + C_2 x \exp x$. For the particular solution, let us use again the superposition principle: $y_P = y_{P_1} + y_{P_2} + y_{P_3}$ where y_{P_1} is a particular solution of the differential equation $y'' - 2y' + y = f_1(x) = \sin x$, y_{P_2} is a particular solution of $y'' - 2y' + y = f_2(x) = \exp x$, and y_{P_3} is a particular solution of $y'' - 2y' + y = f_3(x) = \exp(-x)$. In the first case, $f_1(x) = \sin x = \exp \alpha x (P_1(x) \cos \beta x + P_2(x) \sin \beta x)$ where $P_1(x) = 0$, $P_2(x) = 1$, $\alpha = 0$ and $\beta = 1$. Since $\alpha + \beta i = i$ is not a root of the characteristic equation, we have no resonance and we search for $y_{P_1} = A \cos x + B \sin x$. The second function has the form $f_2(x) = P_3(x) \exp \lambda x$, where $P_3(x) = 1$ and $\lambda = 1$ is a root of the characteristic equation (with multiplicity 2), so in this case we have a resonance of order 2. We need to search for a particular solution of the form $y_{P_2} = Cx^2 \exp x$. And, finally, $f_3(x) = P_4(x) \exp \mu x$, where $P_4(x) = 1$ and $\mu = -1$ is not a root of the characteristic equation. Since we have no resonance in this case, we search for $y_{P_3} = D \exp(-x)$. We calculate the derivatives, replace in each equation and so we obtain $A = \frac{1}{2}$, $B = 0$, $C = \frac{1}{2}$ and $D = \frac{1}{4}$. The general solution of the initial equation is $y = y_{GH} + y_{P_1} + y_{P_2} + y_{P_3} = C_1 \exp x + C_2 x \exp x + \frac{1}{2} \cos x + \frac{1}{2}x^2 \exp x + \frac{1}{4} \exp(-x)$.

(j) The roots of the algebraic equation $r^2 - 3r + 2 = 0$ are $r_1 = 1$ and $r_2 = 2$, therefore the general solution of the homogeneous equation $y'' - 3y' + 2y = 0$ is $y_{GH} = C_1 \exp x + C_2 \exp 2x$. The free term is a function of the form $f(x) = \exp \alpha x (P_1(x) \cos \beta x + P_2(x) \sin \beta x)$, where $P_1(x) = 10x$, $P_2(x) = 0$ and $\alpha + i\beta = i$. Since i is not a root of the characteristic equation, we have no resonance, so we search for a particular solution of the form $y_P = Q_1(x) \cos x + Q_2(x) \sin x$, where the degree of the polynomials $Q_1(x)$ and $Q_2(x)$ is equal to $\max\{\deg P_1, \deg P_2\} = 1$, so we have: $y_P = (Ax + B) \cos x + (Cx + D) \sin x$. We calculate the derivatives, y'_P and y''_P

and replace them in the initial equation. Thus we obtain the linear system:

$$\begin{cases} A - 3C = 10 \\ 3A + C = 0 \\ -3A + 2C + B - 3D = 0 \\ -2A - 3C + 3B + D = 0 \end{cases}$$

The solution of this system is: $A = 1$, $C = -3$, $B = -1.2$ and $D = -3.4$. So, $y = y_{GH} + y_P = C_1 \exp x + C_2 \exp 2x + (x-1.2) \cos x - (3x+3.4) \sin x$.

(k) The characteristic equation is $r^4 - 2r^3 + r^2 = 0$, which can be written $r^2(r-1)^2 = 0$ and has the roots $r_1 = 0$ and $r_2 = 1$ (both with multiplicity 2). So, the general solution of the homogeneous equation is $y_{GH} = C_1 + C_2 x + C_3 \exp x + C_4 x \exp x$. The free term can be written $f = f_1 + f_2$, where $f_1(x) = x = \exp(\lambda_1 x) P_1(x)$ (with $P_1(x) = x$ and $\lambda_1 = 0$) and $f_2(x) = \exp 2x = \exp(\lambda_2 x) P_2(x)$ (with $P_2(x) = 1$ and $\lambda_2 = 2$). We have a resonance (of order 2) only in the first case, so we search for a particular solution of the form $y_P = y_{P_1} + y_{P_2} = x^2(Ax + B) + C \exp 2x$. After computing the derivatives (up to order 4) and replacing in the initial equation, we obtain $A = \frac{1}{6}$, $B = 1$ and $C = \frac{1}{4}$, so the general solution is $y = y_P + y_{GH} = C_1 + C_2 x + C_3 \exp x + C_4 x \exp x + \frac{x^3}{6} + x^2 + \frac{\exp 2x}{4}$.

(l) We have to solve a Cauchy problem: after finding the general solution (like above), we need to get the particular solution which fulfills the initial conditions. The general solution is $y = y_{GH} + y_P = C_1 + C_2 \exp 2x + C_3 \exp(-2x) + \exp(-x)$. From the initial conditions we have:

$$\begin{cases} y(0) = 1 \Rightarrow C_1 + C_2 + C_3 + 1 = 1 \\ y'(0) = 1 \Rightarrow 2C_2 - 2C_3 - 1 = 1 \\ y''(0) = 1 \Rightarrow 4C_2 + 4C_3 + 1 = 1 \end{cases}$$

By solving the system of equations we obtain $C_1 = 0$, $C_2 = \frac{1}{2}$ and $C_3 = -\frac{1}{2}$, so the solution of the Cauchy problem is $y = \frac{\exp 2x - \exp(-2x)}{2} + \exp(-x)$.

(m) The general solution of the linear differential equation is $y = y_{GH} + y_P = C_1 \exp 2x + C_2 x \exp 2x + \exp x$. From the initial conditions we have:

$$\begin{cases} y(0) = 0 \Rightarrow C_1 + 1 = 0 \\ y'(0) = 1 \Rightarrow 2C_1 + C_2 + 1 = 0 \end{cases} \Rightarrow \begin{cases} C_1 = -1 \\ C_2 = 1 \end{cases}$$

So the solution is $y = (x-1) \exp 2x + \exp x$.

(n) The general solution of the linear differential equation is $y = y_{GH} + y_P = C_1 + C_2 x + C_3 \cos x + C_4 \sin x - x \sin x$. From the initial conditions we have:

$$\begin{cases} y(0) = -2 \Rightarrow C_1 + C_3 = -2 \\ y'(0) = 1 \Rightarrow C_2 + C_4 = 1 \\ y''(0) = 0 \Rightarrow -C_3 - 2 = 0 \\ y'''(0) = 0 \Rightarrow -C_4 = 0 \end{cases} \Longrightarrow \begin{cases} C_1 = 0 \\ C_2 = 1 \\ C_3 = -2 \\ C_4 = 0 \end{cases}$$

So the solution is $y = x - 2\cos x - x \sin x$.

7. (a) By solving the homogeneous equation, $y'' + 9y = 0$, we obtain that $\{y_1 = \cos 3x, \ y_2 = \sin 3x\}$ is a basis of solutions for this equation. Now let us apply Lagrange method of variation of constants in order to find a particular solution of the non-homogeneous equation. (Anyway, one can see that all the free terms of these equations are not functions of the type $\exp \alpha x (P_1(x) \cos \beta x + P_2(x) \sin \beta x)$, so the method used in the previous exercise cannot be applied for these equations). We search for a particular solution of the form $y_P = C_1(x) y_1 + C_2(x) y_2$ where $C_1(x)$ and $C_2(x)$ are functions whose derivatives satisfy the following equations:

$$\begin{cases} C_1'(x) y_1 + C_2'(x) y_2 = 0 \\ C_1'(x) y_1' + C_2'(x) y_2' = f(x) \end{cases} \Longleftrightarrow \begin{cases} C_1'(x) \cos 3x + C_2'(x) \sin 3x = 0 \\ -3 C_1'(x) \sin 3x + 3 C_2'(x) \cos 3x = \frac{1}{\sin 3x} \end{cases}$$

Let us solve this system of equations: we can multiply the first equality by $3 \cos 3x$, the second, by $(-3 \sin 3x)$ and, if we add them, we get $C_1'(x) = -\frac{1}{3}$ and, after replacing $C_1'(x)$ in the first equation, $C_2'(x) = \frac{\cos 3x}{3 \sin 3x}$. The primitives of these functions are $C_1(x) = -\frac{x}{3}$ and $C_2(x) = \frac{1}{9} \ln |\sin 3x|$, so we obtained the particular solution $y_P = -\frac{x}{3} \cos 3x + \frac{\sin 3x}{9} \ln |\sin 3x|$. Hence, the general solution of the equation is $y = y_{GH} + y_P = C_1 \cos 3x + C_2 \sin 3x - \frac{x}{3} \cos 3x + \frac{\sin 3x}{9} \ln |\sin 3x|$.

(b) A basis of solutions for the homogeneous equation $y'' - 2y' + y = 0$ is $\{y_1 = \exp x, \ y_2 = x \exp x\}$, so the general solution of the homogeneous equation is $y_{GH} = C_1 \exp x + C_2 x \exp x$. We use the method of variation of constants and search for a particular solution of the form $y_P = C_1(x) \exp x + C_2(x) x \exp x$, where the derivatives of the functions $C_1(x)$ and $C_2(x)$ satisfy the equations

$$\begin{cases} C_1'(x) \exp x + C_2'(x) x \exp x = 0 \\ C_1'(x) \exp x + C_2'(x)(x+1) \exp x = \frac{\exp x}{\sqrt{1-x^2}} \end{cases}$$

Both equations can be divided by $\exp x$ and, by subtracting them, we obtain $C_2'(x) = \frac{1}{\sqrt{1-x^2}}$ and then $C_1'(x) = \frac{-x}{\sqrt{1-x^2}}$. The primitives of these functions are $C_1(x) = \sqrt{1-x^2}$ and $C_2(x) = \sin^{-1} x$, so $y_P = \exp x \cdot \sqrt{1-x^2} + x \exp x \cdot \sin^{-1} x$ and the general solution of the equation is $y = y_{GH} + y_P = C_1 \exp x + C_2 x \exp x + \exp x \cdot \sqrt{1-x^2} + x \exp x \cdot \sin^{-1} x$.

(c) A basis of solutions for the homogeneous equation $y'' + 4y = 0$ is $\{y_1 = \cos 2x, \ y_2 = \sin 2x\}$, so the general solution of the homogeneous equation is $y_{GH} = C_1 \cos 2x + C_2 \sin 2x$. We search for a particular solution of the form $y_P = C_1(x) \cos 2x + C_2(x) \sin 2x$, where $C_1'(x)$ and $C_2'(x)$ satisfy the equations

$$\begin{cases} C_1'(x) \cos 2x + C_2'(x) \sin 2x = 0 \\ -2C_1'(x) \sin 2x + 2C_2'(x) \cos 2x = 2 \tan 2x \end{cases}$$

We multiply the first equality by $\sin 2x$, the second by $\frac{1}{2} \cos 2x$ and add them. We obtain $C_2'(x) = \sin 2x$ and then $C_1'(x) = -\frac{\sin^2 2x}{\cos 2x}$. The primitives of these functions are $C_2(x) = -\frac{1}{2} \cos 2x$ and $C_1(x) = -\int \frac{\sin^2 2x}{\cos 2x} dx = -\frac{1}{2} \int \frac{\sin^2 2x}{\cos^2 2x} 2 \cos 2x\, dx = -\frac{1}{2} \int \frac{t^2}{1-t^2} dt = \frac{1}{2} \int \frac{t^2-1+1}{t^2-1} dt = \frac{t}{2} + \frac{1}{4} \ln \left| \frac{t-1}{t+1} \right| = \frac{1}{2} \sin 2x + \frac{1}{4} \ln \left(\frac{1-\sin 2x}{1+\sin 2x} \right)$, so $y_P = \frac{\cos 2x}{4} \ln \left(\frac{1-\sin 2x}{1+\sin 2x} \right)$.

(d) A basis of solutions for the homogeneous equation $y'' + 3y' + 2y = 0$ is $\{y_1 = \exp(-x), \ y_2 = \exp(-2x)\}$, so the general solution of the homogeneous equation is $y_{GH} = C_1 \exp(-x) + C_2 \exp(-2x)$. We search for a particular solution of the form $y_P = C_1(x) \exp(-x) + C_2(x) \exp(-2x)$, where the derivatives of the functions $C_1(x)$ and $C_2(x)$ satisfy the equations

$$\begin{cases} C_1'(x) \exp(-x) + C_2'(x) \exp(-2x) = 0 \\ -C_1'(x) \exp(-x) - 2C_2'(x) \exp(-2x) = \frac{1}{\exp x + 1} \end{cases}$$

If we add these equations, we obtain $C_2'(x) = -\frac{\exp 2x}{\exp x + 1}$ and then, $C_1'(x) = \frac{\exp x}{\exp x + 1}$. The primitives of these functions are $C_1(x) = \int \frac{\exp x\, dx}{\exp x + 1} = \ln(\exp x + 1)$ and $C_2(x) = -\int \frac{\exp 2x}{\exp x + 1} = -\int \frac{\exp x}{\exp x + 1} \exp x\, dx = -\int \frac{t}{t+1} dt = -t + \ln(t+1) = -\exp x + \ln(\exp x + 1)$, so $y_P = \exp(-x) \cdot \ln(\exp x + 1) + \exp(-x) + \exp(-2x) \cdot \ln(\exp x + 1)$.

(e) The general solution of the homogeneous equation is $y_{GH} = C_1 \exp x + C_2 x \exp x$. We search for a particular solution of the form $y_P = C_1(x) \exp x + C_2(x) x \exp x$ such that

$$\begin{cases} C_1'(x) \exp x + C_2'(x) x \exp x = 0 \\ C_1'(x) \exp x + C_2'(x)(x+1) \exp x = \frac{\exp x}{x} \end{cases}$$

We obtain $C_1'(x) = -1 \Rightarrow C_1(x) = -x$ and $C_2'(x) = \frac{1}{x} \Rightarrow C_2(x) = \ln|x|$. The general solution is $y = y_{GH} + y_P = C_1 \exp x + C_2 x \exp x - x \exp x + x \exp x \cdot \ln|x|$. Now we have to find the constants C_1 and C_2 such that the initial conditions are fulfilled. We obtain: $C_1 e + C_2 e - e = 0$ and $C_1 e + 2C_2 e - e = e$, so $C_1 = 0$, $C_2 = 1$ and the solution of the Cauchy problem is $y = x \exp x \cdot \ln|x|$.

8. (a) An equation of the type $a_0 x^n y^{(n)} + a_1 x^n y^{(n-1)} + \ldots + a_{n-1} xy' + a_n y = f(x)$ with $a_0, \ldots, a_n$ real numbers ($a_0 \neq 0$) is called an Euler equation. To solve such an equation we use the substitution $x = \exp t$ if $x > 0$, or $x = -\exp t$ if $x < 0$, and we obtain a linear differential equation of nth order in the new function $\tilde{y}(t) = y(\exp t)$ (or, respectively, $\tilde{y}(t) = y(-\exp t)$).

In our case, we have a second-order Euler equation in the domain $x > 0$, so we put $x = \exp t$ (equivalent to $t = \ln x$). The derivatives of y (with respect to x) can be expressed in terms of derivatives of $\tilde{y}$ (with respect to t) in the following way: $y' = \exp(-t)\tilde{y}'$, $y'' = \exp(-2t)(\tilde{y}'' - \tilde{y}')$. The general expression (for the k-th derivative) is $y^{(k)} = \exp(-kt) L_k[\tilde{y}]$, where L_k is the polynomial differential operator $L_k = \frac{d}{dt}\left(\frac{d}{dt} - I\right)\left(\frac{d}{dt} - 2I\right)\ldots\left(\frac{d}{dt} - (k-1)I\right)$, corresponding to $P_k(r) = r(r-1)(r-2)\ldots(r-k+1)$.

The equation becomes: $\exp 2t \exp(-2t)(\tilde{y}'' - \tilde{y}') - 2\exp t \exp(-t)\tilde{y}' + 2\tilde{y} = \frac{t}{\exp t}$, which is equivalent to $\tilde{y}'' - 3\tilde{y}' + 2\tilde{y} = t\exp(-t)$. Since $\lambda = -1$ is not a root of the characteristic equation $r^2 - 3r + 2 = 0$, we don't have resonance, so we search for a particular solution $\tilde{y}_P = (At + B)\exp(-t)$. The general solution of the linear equation is $\tilde{y} = \tilde{y}_{GH} + \tilde{y}_P = C_1 \exp t + C_2 \exp 2t + \frac{6t+5}{36}\exp(-t)$, so the general solution of the Euler equation is $y = C_1 x + C_2 x^2 + \frac{6\ln x + 5}{36x}$.

(b) The domain of this equation is $x < 0$, so we use the substitution $x = -\exp t$ (equivalent to $t = \ln(-x)$). In this case, the general expression of the derivatives is $y^{(k)} = (-1)^k \exp(-kt) L_k[\tilde{y}]$, so only the first derivative is changed: $y' = -\exp(-t)\tilde{y}'$, but the linear equation will have the same form (excepting the free term): $\tilde{y}'' - 3\tilde{y}' + 2\tilde{y} = -\exp t$. Now we have resonance ($\lambda = 1$ is a simple root of the characteristic equation), so we search for $\tilde{y}_P = At \exp t$ and obtain the general solution $\tilde{y} = \tilde{y}_{GH} + \tilde{y}_P = C_1 \exp t + C_2 \exp 2t + t \exp t$. The general solution of the Euler equation is $y = -C_1 x + C_2 x^2 - x \ln(-x)$. By using the initial conditions $y(-1) = -1$, $y'(-1) = 0$ we obtain $C_1 = -1$, $C_2 = 0$, so the solution of the Cauchy problem is $y = x - x \ln(-x)$.

(c) We use the substitution $x = \exp t$. Note that $P_3(r) = r(r-1)(r-2) = r^3 - 3r^2 + 2r$, so the expression of the third derivative is $y''' = \exp(-3t)(\tilde{y}''' - 3\tilde{y}'' + 2\tilde{y}')$ and we obtain the linear equation: $\tilde{y}''' - 2\tilde{y}'' - \tilde{y}' + \tilde{y} = 12\exp(-2t)$. The characteristic equation $r^3 - 2r^2 - r + 1 = 0$ can be written $(r-2)(r-1)(r+1) = 0$, so the roots are $r_1 = 2$, $r_2 = 1$ and $r_3 = -1$ (we don't have resonance: $\lambda = -2$ is not a root). The general solution of the linear equation is $\tilde{y} = C_1 \exp 2t + C_2 \exp t + C_3 \exp(-t) - \exp(-2t)$, so the general solution of the Euler equation is $y = C_1x^2 + C_2x + \frac{C_3}{x} - \frac{1}{x^2}$. From the initial conditions we obtain $C_1 = 1$, $C_2 = -2$, $C_3 = 2$, so the solution of the Cauchy problem is $y = x^2 - 2x + \frac{2}{x} - \frac{1}{x^2}$.

(d) We use the substitution $x = -\exp t$ and denote $\tilde{y}(t) = y(-\exp t)$, so the equation becomes $\tilde{y}''' - 3\tilde{y}'' + 3\tilde{y}' - \tilde{y} = \exp 2t$. The characteristic equation $r^3 - 3r^2 + 3r - 1 = 0 \Leftrightarrow (r-1)^3 = 0$ has one root, $r_1 = 1$, with multiplicity 3. We have no resonance and the general solution of the linear equation is $\tilde{y} = \tilde{y}_{GH} + \tilde{y}_P = C_1 \exp t + C_2 t \exp t + C_3 t^2 \exp t + \exp 2t$. The general solution of the Euler equation is $y = -C_1 x - C_2 x \ln(-x) - C_3 x \ln^2(-x) + x^2$.

(e) By using the substitution $x = \exp t$ and denoting $\tilde{y}(t) = y(\exp t)$, the equation becomes $\tilde{y}'' - 2\tilde{y}' + \tilde{y} = t\exp(-t) + \frac{\exp t}{t}$. The characteristic equation $r^2 - 2r + 1 = 0$ has the root $r_1 = 1$ with multiplicity 2, so the solution of the homogeneous equation is $\tilde{y}_{GH} = C_1 \exp t + C_2 t \exp t$. We search for $\tilde{y}_P = \tilde{y}_{P_1} + \tilde{y}_{P_2}$, where $\tilde{y}_{P_1}$ is a solution of the equation $\tilde{y}'' - 2\tilde{y}' + \tilde{y} = t \exp(-t)$ and $\tilde{y}_{P_2}$ is a solution of $\tilde{y}'' - 2\tilde{y}' + \tilde{y} = \frac{\exp t}{t}$. In the first case, we take $\tilde{y}_{P_1} = (At + B)\exp(-t)$ (we don't have resonance) and find $A = B = \frac{1}{4}$. For the second equation we have to use Lagrange method, so we search for $\tilde{y}_{P_2} = C_1(t)\exp t + C_2(t) t \exp t$ where the functions $C_1(t)$ and $C_2(t)$ satisfy the equations:

$$\begin{cases} C_1'(t)\exp t + C_2'(t) t \exp t = 0 \\ C_1'(t)\exp t + C_2'(t)(t+1)\exp t = \frac{\exp t}{t} \end{cases}$$

We obtain $C_1(t) = -t$ and $C_2(t) = \ln t$, so the solution of the linear equation is $\tilde{y} = \tilde{y}_{GH} + \tilde{y}_{P_1} + \tilde{y}_{P_2} = C_1 \exp t + C_2 t \exp t + \frac{t+1}{4}\exp(-t) - t \exp t + \ln t \cdot t \exp t$. If we put $C_2 - 1 = K_2$ and $C_1 = K_1$, the solution of the Euler equation is written $y = K_1 x + K_2 x \ln x + \frac{1+\ln x}{4x} + x \ln x \ln(\ln x)$.

9. Let $P(x, y(x))$ be a point on the curve $y = y(x)$. Then $y'(x)$ is the slope of the curve at this point (the slope of the tangent line to the curve at the point P). So we have to solve the following Cauchy problem: $y'(x) = xy$, $y(0) = 1$. It is a separable differentiable equation: $\frac{dy}{dx} = xy \Leftrightarrow \int \frac{dy}{y} = \int x dx \Leftrightarrow \ln y = \frac{x^2}{2} + \ln C$, where C is a positive constant. So the general solution is $y = C \exp(\frac{x^2}{2})$. From the initial condition $y(0) = 1$ we obtain $C = 1$, so the solution of the Cauchy problem is $y = \exp(\frac{x^2}{2})$.

10. Let us denote the function by $y(x)$. We have to solve the following Cauchy problem: $y' = y - y^2$, $y(0) = 1$. The differential equation is a Bernoulli equation with $\alpha = 2$, so we use the substitution $z = y^{1-\alpha} = \frac{1}{y}$ and the equation becomes $z' = -z + 1$. The general solution of this linear equation is $z = C\exp(-x) + 1$, so the general solution of the initial equation is $y = \frac{1}{z} = \frac{\exp x}{C + \exp x}$. From the initial condition we obtain $C = 0$, so the solution of the Cauchy problem is the constant function $y(x) = 1$, $\forall x \in \mathbb{R}$.

11. By using the notation $u'(r) = v(r)$, the second-order differential equation is transformed into a first-order differential equation with separable variables: $v' + \frac{2}{r}v = 0 \Leftrightarrow \frac{dv}{dr} = -\frac{2}{r}v \Leftrightarrow \int \frac{dv}{v} = -2\int \frac{dr}{r} \Leftrightarrow \ln v = -2\ln r + \ln C_1 \Leftrightarrow v = \frac{C_1}{r^2}$, so $u' = \frac{C_1}{r^2}$ and the general solution of the initial equation is $u = -\frac{C_1}{r} + C_2$. Now we have to find the constants C_1 and C_2 such that the *boundary conditions* are fulfilled: $u(1) = -C_1 + C_2 = 15$, $u(2) = -\frac{C_1}{2} + C_2 = 25 \Rightarrow C_1 = 20$ and $C_2 = 35$ and the solution of the *boundary value problem* is $u(r) = -\frac{10}{r} + 35$.

12. Using the formula $\frac{d^2y}{dt^2} = \frac{dv}{dt} = v\frac{dv}{dy}$, the second-order differential equation in $y = y(t)$ can be written as a first-order differential equation in $v = v(y)$: $v\frac{dv}{dy} + \frac{gR^2}{(y+R)^2} = 0$. It is a separable equation: $\int v dv = \int \frac{-gR^2}{(y+R)^2}dy$, so the general solution is $\frac{v^2}{2} = \frac{gR^2}{y+R} + C$. We know that the maximum height reached by the rocket is h, so $v(h) = 0$ and we obtain the constant $C = -\frac{gR^2}{h+R}$. So, the velocity at the height y is $v(y) = \sqrt{2gR^2\left(\frac{1}{y+R} - \frac{1}{h+R}\right)}$. Therefore, the initial velocity is $v_0 = v(0) = \sqrt{2gR^2\left(\frac{1}{R} - \frac{1}{h+R}\right)} = \sqrt{\frac{2gRh}{R+h}}$.

The *escape velocity* is $v_e = \lim_{h\to\infty} v_0 = \lim_{h\to\infty} \sqrt{\frac{2gRh}{R+h}} = \sqrt{2gR} \simeq 11.2$ km/s.

13. If we put $z = y'$, the ODE-2 above is transformed into a first-order equation: $z' = k\sqrt{1+z^2}$. By integrating this (separable) equation one obtains: $\int \frac{dz}{\sqrt{1+z^2}} = k\int dx \Leftrightarrow \ln(z + \sqrt{1+z^2}) = kx + \ln C_1 \Leftrightarrow z + \sqrt{1+z^2} = C_1 \exp(kx)$, where C_1 is a positive constant. The last equation can be written: $1+z^2 = (C_1\exp(kx) - z)^2$ and it follows the expression of z: $z = \frac{C_1}{2}\exp(kx) - \frac{1}{2C_1}\exp(-kx)$. Since $z = y'$, we have: $y = \frac{C_1}{2k}\exp(kx) + \frac{1}{2kC_1}\exp(-kx) + C_2$. We can find the constants C_1 and C_2 from the conditions $y(-a) = y(a) = h$. We obtain $C_1 = 1$ and $C_2 = h - \frac{1}{2k}(\exp(ka) + \exp(-ka)) = h - \frac{1}{k}\cosh(ka)$. So the sought function is $y(x) = \frac{1}{k}(\cosh kx - \cosh ka) + h$. In order to compute the length of the curve, we have to compute the following integral: $L = \int_{-a}^{a} \sqrt{1 + y'(x)^2}dx$. Since $(\cosh kx)' = k\sinh x$ and $\cosh^2 x - \sinh^2 x = 1, \forall x \in \mathbb{R}$, we have $L = \int_{-a}^{a} \sqrt{1 + \sinh^2 kx}dx = \int_{-a}^{a} \cosh kxdx = 2\int_0^a \cosh kxdx = \frac{2}{k}\sinh ka$.

14. By differentiating both sides with respect to x one obtains $2y(x)y'(x) = y^2(x) + y'^2(x)$, which can be written $(y(x) - y'(x))^2 = 0$. It follows $y' = y$, so $y(x) = C\exp x$, where C is a constant which can be found by using the initial equation: $C^2\exp 2x = 2 + C^2 \int_0^x 2\exp 2t dt = 2 + C^2\exp 2x - C^2 \Rightarrow C = \pm\sqrt{2}$.

15. (a) Let $C(x(t), y(t))$ be the position of the cat at the moment t. The tangent line at C to the curve $y = y(x)$ has the equation:

$$y - y(x(t)) = y'(x(t))(x - x(t)).$$

This straight line passes through the point $M(0, v_m t)$, so we have: $v_m t = y(x(t)) - y'(x(t))x(t)$, at any moment t, or:

$$v_m t = y(x) - x y'(x). \tag{10.1}$$

The length of the curve $\widehat{C_0 C}$ can be written in two ways:

$$L(\widehat{C_0 C}) = v_c t = \int_x^{x_0} \sqrt{1 + y'^2(\xi)} d\xi. \tag{10.2}$$

But, from the equation (10.1) we find $t = \frac{y(x) - x y'(x)}{v_m}$. Hence, by replacing t in (10.2), we get:

$$\frac{v_c}{v_m}(y(x) - x y'(x)) = \int_x^{x_0} \sqrt{1 + y'^2(\xi)} d\xi. \tag{10.3}$$

We differentiate the equation (10.3) with respect to x (by using Leibniz' formula) and obtain: $\frac{v_c}{v_m}(y'(x) - y'(x) - x y''(x)) = 0 - \sqrt{1 + y'^2(x)}$, or:

$$k x y''(x) = \sqrt{1 + y'^2(x)}, \tag{10.4}$$

where $k = \frac{v_c}{v_m}$. In our case, $k = 1$, so we have $x y''(x) = \sqrt{1 + y'^2(x)}$.

(b) Let us solve the differential equation (10.4): we denote $y'(x) = z(x)$ and obtain a first-order (separable) equation: $z' = \frac{1}{kx}\sqrt{1 + z^2}$. Let us integrate this equation: $\int \frac{dz}{\sqrt{1+z^2}} = \int \frac{dx}{kx} \Leftrightarrow \ln(z + \sqrt{1 + z^2}) = \frac{1}{k}\ln x + \ln C_1$. So, $z + \sqrt{1 + z^2} = C_1 x^{\frac{1}{k}} \Leftrightarrow 1 + z^2 = (C_1 x^{\frac{1}{k}} - z)^2 = C_1^2 x^{\frac{2}{k}} - 2 C_1 x^{\frac{1}{k}} z + z^2$ and we obtain

$$y' = z = \frac{C_1}{2} x^{\frac{1}{k}} - \frac{1}{2 C_1} x^{-\frac{1}{k}}. \tag{10.5}$$

For $k = 1$ the formula (10.5) becomes $y' = \frac{C_1}{2} x - \frac{1}{2 C_1} \cdot \frac{1}{x}$ and we obtain the general solution:

$$y = \frac{C_1}{4} x^2 - \frac{1}{2 C_1} \ln x + C_2. \tag{10.6}$$

We can find the constants C_1 and C_2 by using the initial conditions: $y(x_0) = 0$ and $y'(x_0) = 0$ (because the initial position of the cat is $C_0(x_0, 0)$ and the initial position of the mouse is $O(0, 0)$, so the tangent line to the curve at this point is the Ox-axis). We obtain $C_1 = \frac{1}{x_0}$ and $C_2 = -\frac{x_0}{4} + \frac{x_0}{2} \ln x_0$, so the sought function is $y(x) = \frac{x^2 - x_0^2}{4x_0} - \frac{x_0}{2} \ln \frac{x}{x_0}$. Now it is obvious that $y \to \infty$ when $x \to 0$, so the cat never catches the mouse.

(c) Let us integrate the equation (10.5) for $k \neq 1$:

$$y = \frac{C_1}{2} \cdot \frac{k}{k+1} x^{\frac{k+1}{k}} - \frac{1}{2C_1} \cdot \frac{k}{k-1} x^{\frac{k-1}{k}} + C_2. \tag{10.7}$$

From this expression of $y(x)$ we can see that, if $k > 1$ (i.e. $v_c > v_m$), then $y(0) = C_2$, so the cat catches the mouse at the point $M_1(0, C_2)$. If $k < 1$ (i.e. $v_c < v_m$), then $y(0) = \infty$, so the cat doesn't catch the mouse.

The constants C_1 and C_2 can be found from the initial conditions $y(x_0) = 0$ and $y'(x_0) = 0$:

$$y'(x_0) = 0 \Rightarrow \frac{C_1}{2} x_0^{\frac{1}{k}} = \frac{1}{2C_1} x_0^{-\frac{1}{k}} \Rightarrow C_1 = x_0^{-\frac{1}{k}}.$$

By replacing C_1 in (10.7) and using $y(x_0) = 0$, we obtain

$$y(x_0) = \frac{k}{k+1} \cdot \frac{x_0}{2} - \frac{k}{k-1} \cdot \frac{x_0}{2} + C_2 = 0 \Rightarrow C_2 = \frac{kx_0}{k^2 - 1}.$$

In our case, $k = 2$, so $C_2 = \frac{2x_0}{3}$: the cat catches the mouse at the point $M_1(0, \frac{2x_0}{3})$.

(d) If we write the formula (10.7) for $k = \frac{1}{2}$ we obtain $y = y = \frac{C_1}{6} x^3 - \frac{1}{2C_1 x} + C_2$. Since $C_1 = x_0^{-\frac{1}{k}} = \frac{1}{x_0^2}$ and $C_2 = \frac{kx_0}{k^2 - 1} = -\frac{2x_0}{3}$, we obtain $y = \frac{x^3}{6x_0^2} + \frac{x_0^2}{2x} - \frac{2x_0}{3}$. Obviously, $y \to \infty$ when $x \to 0$, so the cat doesn't catch the mouse (as we could see above, this is true for any $k < 1$), but we can find the minimum distance between them. The distance between the cat and the mouse at the moment t is equal to $\sqrt{x(t)^2 + (v_m t - y(t))^2}$. Since $v_m t = y(t) - y'(x(t))x(t)$, we have to minimize the function $f(x) = \sqrt{x^2 + x^2 y'^2(x)}$. But $y'(x) = \frac{1}{2}\left(\frac{x^2}{x_0^2} - \frac{x_0^2}{x^2}\right)$, so the function above can be written $f(x) = \frac{x}{2}\left(\frac{x^2}{x_0^2} + \frac{x_0^2}{x^2}\right)$. The derivative of this function is $f(x) = \frac{3x^2}{2x_0^2} - \frac{x_0^2}{2x^2}$ and has the (positive) root $x_1 = \frac{x_0}{\sqrt[4]{3}}$. Since $f''(x_1) > 0$, this point is a minimum point of $f(x)$.

10.3 Chapter 3

1. (a) *Elimination Method* (E.M.). By differentiating the first equation we obtain $x'' = x' + y'$. We replace y' with $-2x - 2y$ (from the second equation) and obtain $x'' = x' - 2x - 2y$. But, from the first equation, we can write $y = x' - x$, so we have $x'' = x' - 2x - 2x' + 2x$. Thus, we obtained a second-order linear differential equation in $x = x(t) : x'' + x' = 0$.

   Another method to obtain this equation in $x = x(t)$ is to write y in terms of x and x' from the first equation: $y = x' - x$, and then, replace this y in the second equation: $x'' - x' = -2x - 2(x' - x)$, or, equivalently, $x'' + x' = 0$.

   It is a homogeneous equation and its characteristic equation is $\lambda^2 + \lambda = 0$, with the (real, distinct) roots $\lambda_1 = -1$ and $\lambda_2 = 0$. Hence, the general solution is $x = C_1 \exp(-t) + C_2 \exp(0t) = C_1 \exp(-t) + C_2$. Since $y = x' - x$, we have $y = -2C_1 \exp(-t) - C_2$ and the system is solved.

   Algebraic Method (A.M.). The matrix of the system is

   $$A = \begin{pmatrix} 1 & 1 \\ -2 & -2 \end{pmatrix}.$$

   The eigenvalues of A are the roots of the characteristic polynomial $P(\lambda) = \det(A - \lambda I_2)$. It is easy to prove (you can try!) that for a square matrix of order 2 this polynomial is $P(\lambda) = \lambda^2 - Tr(A)\lambda + \det(A)$, where $Tr(A)$ is the trace of the matrix (the sum of the elements on the main diagonal). In our case, $Tr(A) = -1$ and $\det(A) = 0$, so the eigenvalues of A are the solutions of the equation $\lambda^2 + \lambda = 0 : \lambda_1 = -1$ and $\lambda_2 = 0$. (Note that we obtained the same equation as above.) Now, let us find two eigenvectors $\mathbf{v}_1$ and $\mathbf{v}_2$ (one eigenvector for each eigenvalue). Since $\mathbf{v}_1 = (\alpha, \beta)^t$ is an eigenvector corresponding to λ_1 if and only if $A\mathbf{v}_1 = \lambda_1 \mathbf{v}_1$, or, equivalently, $(A - \lambda_1 I_2)\mathbf{v}_1 = 0$, we need to solve a homogeneous system (the matrix of the system is $A - \lambda_1 I_2 = A + I_2$) :

   $$\begin{pmatrix} 2 & 1 \\ -2 & -1 \end{pmatrix} \begin{pmatrix} \alpha \\ \beta \end{pmatrix} = \mathbf{0} \Leftrightarrow \begin{cases} 2\alpha + \beta = 0 \\ -2\alpha - \beta = 0 \end{cases}$$

   The set of solutions of this system is the eigenspace of $\lambda_1 = -1$, $V_{-1} = \{(\alpha, -2\alpha)^t : \alpha \in \mathbb{R}\}$, so we can take $\mathbf{v}_1 = (1, -2)^t$. In the same way, by solving the system $(A - \lambda_2 I_2)\mathbf{v}_2 = 0$, i.e.

   $$\begin{pmatrix} 1 & 1 \\ -2 & -2 \end{pmatrix} \begin{pmatrix} \alpha \\ \beta \end{pmatrix} = \mathbf{0} \Leftrightarrow \begin{cases} \alpha + \beta = 0 \\ -2\alpha - 2\beta = 0 \end{cases}$$

   we find the eigenspace of $\lambda_2 = 0 : V_0 = \{(\alpha, -\alpha)^t : \alpha \in \mathbb{R}\}$, and we take $\mathbf{v}_2 = (1, -1)^t$. From Proposition 3.4, a fundamental system of solutions is $\{\mathbf{Y}_1 = \mathbf{v}_1 \exp(\lambda_1 t), \mathbf{Y}_2 = \mathbf{v}_2 \exp(\lambda_2 t)\}$, so the general solution is $\mathbf{Y} =$

10.3 Chapter 3

$C_1 \mathbf{Y}_1 + C_2 \mathbf{Y}_2$

$$\Rightarrow \mathbf{Y} = \begin{pmatrix} x(t) \\ y(t) \end{pmatrix} = C_1 \begin{pmatrix} 1 \\ -2 \end{pmatrix} \exp(-t) + C_2 \begin{pmatrix} 1 \\ -1 \end{pmatrix}.$$

We obtained the same solution as above:

$$\begin{cases} x(t) = C_1 \exp(-t) + C_2 \\ y(t) = -2C_1 \exp(-t) - C_2 \end{cases}.$$

(b) Let us write the system in the normal form

$$\begin{cases} x' = x + y \\ y' = -2x + 4y \end{cases}.$$

(E.M.) By differentiating the first equation we obtain (using the second equation) $x'' = x' + y' = x' - 2x + 4y$. But, from the first equation, we can write $y = x' - x$, so we obtain a second-order linear differential equation: $x'' - 5x' + 6x = 0$. It is a homogeneous equation and its characteristic equation is $\lambda^2 - 5\lambda + 6 = 0$, with the (real, distinct) roots $\lambda_1 = 2$ and $\lambda_2 = 3$. Hence, the general solution is $x = C_1 \exp(2t) + C_2 \exp(3t)$ and, from the relation $y = x' - x$, we can write $y = C_1 \exp(2t) + 2C_2 \exp(3t)$.

(A.M.) The matrix of the system is

$$A = \begin{pmatrix} 1 & 1 \\ -2 & 4 \end{pmatrix}.$$

Its eigenvalues are $\lambda_1 = 2$ and $\lambda_2 = 3$ (the roots of the characteristic polynomial $P(\lambda) = \lambda^2 - 5\lambda + 6$). Let $\mathbf{v}_1 = (\alpha, \beta)^t$ be an eigenvector corresponding to $\lambda_1 = 2$ so $(A - \lambda_1 I_2)\mathbf{v}_1 = 0$, and we have the homogeneous system:

$$\begin{pmatrix} -1 & 1 \\ -2 & 2 \end{pmatrix} \begin{pmatrix} \alpha \\ \beta \end{pmatrix} = \mathbf{0} \Leftrightarrow \begin{cases} -\alpha + \beta = 0 \\ -2\alpha + 2\beta = 0 \end{cases}$$

The set of solutions of this system is the eigenspace of $\lambda_1 = 2$, $V_2 = \{(\alpha, \alpha)^t : \alpha \in \mathbb{R}\}$, so we can take $\mathbf{v}_1 = (1, 1)^t$. In the same way we find the eigenspace of $\lambda_2 = 3$, $V_3 = \{(\alpha, 2\alpha)^t : \alpha \in \mathbb{R}\}$, so we can take $\mathbf{v}_2 = (1, 2)^t$. From Proposition 3.4, a fundamental system of solutions is $\{\mathbf{Y}_1 = \mathbf{v}_1 \exp(\lambda_1 t), \mathbf{Y}_2 = \mathbf{v}_2 \exp(\lambda_2 t)\}$, so the general solution is

$$\mathbf{Y} = \begin{pmatrix} x(t) \\ y(t) \end{pmatrix} = C_1 \begin{pmatrix} 1 \\ 1 \end{pmatrix} \exp(2t) + C_2 \begin{pmatrix} 1 \\ 2 \end{pmatrix} \exp(3t).$$

and we obtained the same general solution as above.

Now, let us solve the Cauchy problem: we have to find the particular solution which satisfies the initial conditions.

$$\begin{cases} x(0) = 0 \Rightarrow C_1 + C_2 = 0 \\ y(0) = -1 \Rightarrow C_1 + 2C_2 = -1 \end{cases} \Rightarrow C_1 = 1, \ C_2 = -1,$$

so the solution of the Cauchy problem is

$$\begin{cases} x = \exp(2t) - \exp(3t) \\ y = \exp(2t) - 2\exp(3t) \end{cases}.$$

(c) We write the system in the normal form:

$$\begin{cases} x' = 3x - y \\ y' = 10x - 4y + 2 \end{cases}.$$

(E.M.) From the first equation we get $y = 3x - x'$ and we replace this y in the second equation: $3x' - x'' - 10x + 4(3x - x') = 2$, so we obtain the non-homogeneous equation $x'' + x' - 2x = -2$. First, we solve the homogeneous equation $x'' + x' - 2x = 0$. The characteristic equation is $\lambda^2 + \lambda - 2 = 0$, with the roots $\lambda_1 = 1$ and $\lambda_2 = -2$. Hence, the general solution of the homogeneous equation is $x_{GH} = C_1 \exp t + C_2 \exp(-2t)$. Since the free term, $f(t) = -2\exp(0 \cdot t)$ is a polynomial of degree 0 and we have no resonance (0 is not a solution of the characteristic equation), we search for a particular solution of the form $x_P = A$. We obtain $A = 1$, so the general solution of the non-homogeneous equation is $x = x_{GH} + x_P = C_1 \exp t + C_2 \exp(-2t) + 1$ and the other function is $y = -x' + 3x = 2C_1 \exp t + 5C_2 \exp(3t) + 3$.

(A.M.) We have to solve the non-homogeneous system: $\mathbf{Y}' = A\mathbf{Y} + \mathbf{G}$, where

$$\mathbf{Y} = \begin{pmatrix} x(t) \\ y(t) \end{pmatrix}, \quad A = \begin{pmatrix} 3 & -1 \\ 10 & -4 \end{pmatrix} \quad \text{and} \quad \mathbf{G} = \begin{pmatrix} 0 \\ 2 \end{pmatrix}.$$

First, let us solve the homogeneous system $\mathbf{Y}' = A\mathbf{Y}$. Since the eigenvalues of A are $\lambda_1 = 1$ and $\lambda_2 = -2$ and two corresponding eigenvectors are $\mathbf{v}_1 = (1, 2)^t$ and $\mathbf{v}_2 = (1, 5)^t$, the general solution of the homogeneous system is $\mathbf{Y}_{GH} = C_1 \begin{pmatrix} 1 \\ 2 \end{pmatrix} \exp t + C_2 \begin{pmatrix} 1 \\ 5 \end{pmatrix} \exp(-2t)$. Now, let us find $\mathbf{Y}_P$, a particular solution of the non-homogeneous system. Since the free term is of the form $\mathbf{G}(t) = \mathbf{P}(t) \exp(\alpha t)$, where $\mathbf{P}(t) = (0, 2)^t$ is a polynomial vector of degree 0 and $\alpha = 0$ is not an eigenvalue of A, we can apply the formula (3.87), so we search for a particular solution of the form $\mathbf{Y}_P(t) =$

$\mathbf{Q}(t)\exp(0x) = \begin{pmatrix} a \\ b \end{pmatrix}$, because $\deg \mathbf{Q}(t) = \deg \mathbf{P}(t) = 0$. By replacing $x(t) = a$ and $y(t) = b$ in the initial system, we obtain

$$\begin{cases} 0 = 3a - b \\ 0 = 10a - 4b + 2 \end{cases} \Rightarrow a = 1, \ b = 3$$

and the general solution of the non-homogeneous system is $\mathbf{Y} = \mathbf{Y}_{GH} + \mathbf{Y}_P$,

$$\mathbf{Y} = \begin{pmatrix} x(t) \\ y(t) \end{pmatrix} = \begin{pmatrix} C_1 \exp t + C_2 \exp(-2t) + 1 \\ 2C_1 \exp t + 5C_2 \exp(-2t) + 3 \end{pmatrix}.$$

Now, let us solve the Cauchy problem: we have to find the particular solution that satisfies the initial conditions $x(0) = C_1 + C_2 + 1 = 1$ and $y(0) = 2C_1 + 5C_2 + 3 = 6$. We get $C_1 = -1$ and $C_2 = 1$, so the solution is

$$\mathbf{Y} = \begin{pmatrix} x(t) \\ y(t) \end{pmatrix} = \begin{pmatrix} \exp(-2t) - \exp t + 1 \\ 5\exp(-2t) - 2\exp t + 3 \end{pmatrix}.$$

(d) Let us write the system in the normal form

$$\begin{cases} \dot{x} = -2x - 4y + 1 + 4t \\ \dot{y} = -x + y + \frac{3}{2}t^2 \end{cases}.$$

(E.M.) By differentiating the first equation we obtain $\ddot{x} = -2\dot{x} - 4\dot{y} + 4 = -2\dot{x} - 4(-x + y + \frac{3}{2}t^2) + 4$. From the first equation we can write $y = \frac{1}{4}(-\dot{x} - 2x + 1 + 4t)$ and, after replacing y in the equation above, we obtain a linear differential equation in $x(t)$: $\ddot{x} + \dot{x} - 6x = -6t^2 - 4t + 3$. Since the characteristic equation $\lambda^2 + \lambda - 6 = 0$ has the roots $\lambda_1 = 2$ and $\lambda_2 = -3$, the general solution of the homogeneous equation is $x_{GH} = C_1 \exp(2t) + C_2 \exp(-3t)$. The free term, $f(t) = (-6t^2 - 4t + 3)\exp(0 \cdot t)$ is a polynomial of degree 2 and we have no resonance (0 is not a root of the characteristic equation), so we search for a particular solution of the non-homogeneous equation of the form $x_P = At^2 + Bt + C$ and we get $x_P = t^2 + t$. Hence, the general solution is $x = x_{GH} + x_P = C_1 \exp(2t) + C_2 \exp(-3t) + t^2 + t$. and $y = \frac{1}{4}(-\dot{x} - 2x + 1 + 4t) = -C_1 \exp(2t) + \frac{1}{4}C_2 \exp(-3t) - \frac{1}{2}t^2$.

(A.M.) We have to solve the non-homogeneous system: $\mathbf{Y}' = A\mathbf{Y} + \mathbf{G}$, where

$$A = \begin{pmatrix} -2 & -4 \\ -1 & 1 \end{pmatrix} \quad \text{and} \quad \mathbf{G} = \begin{pmatrix} 1 + 4t \\ \frac{3}{2}t^2 \end{pmatrix}.$$

First, we solve the homogeneous system $\mathbf{Y}' = A\mathbf{Y}$. The eigenvalues of A are $\lambda_1 = 2$ and $\lambda_2 = -3$ and two corresponding eigenvectors are $\mathbf{v}_1 = (1, -1)^t$

and $\mathbf{v}_2 = (4, 1)^t$, so the general solution of the homogeneous system is $\mathbf{Y}_{GH} = C_1 \begin{pmatrix} 1 \\ -1 \end{pmatrix} \exp(2t) + C_2 \begin{pmatrix} 4 \\ 1 \end{pmatrix} \exp(-3t)$. Now, let us find $\mathbf{Y}_P$, a particular solution of the non-homogeneous system. Since the free term is of the form $\mathbf{G}(t) = \mathbf{P}(t) \exp(\alpha t)$, where $\mathbf{P}(t) = \begin{pmatrix} 1 + 4t \\ \frac{3}{2}t^2 \end{pmatrix}$ is a polynomial vector of degree 2 and $\alpha = 0$ is not an eigenvalue of A, we can apply the formula (3.87), so we search for a particular solution of the form

$$\mathbf{Y}_P(t) = \mathbf{Q}(t) \exp(0x) = \begin{pmatrix} At^2 + Bt + C \\ Dt^2 + Et + F \end{pmatrix} = \begin{pmatrix} x(t) \\ y(t) \end{pmatrix}.$$

By replacing $x(t)$ and $y(t)$ in the initial system we obtain $\mathbf{Y}_P(t) = \begin{pmatrix} t^2 + t \\ -\frac{1}{2}t^2 \end{pmatrix}$, so the general solution is

$$\mathbf{Y} = \mathbf{Y}_{GH} + \mathbf{Y}_P = \begin{pmatrix} C_1 \exp(2t) + 4C_2 \exp(-3t) + t^2 + t \\ -C_1 \exp(2t) + C_2 \exp(-3t) - \frac{1}{2}t^2 \end{pmatrix}.$$

The solution is slightly different from that obtained by the elimination method, but in fact, they are equivalent. Indeed, C_2 from the above solution is an arbitrary real number, so we can write it as $C_2 = \frac{1}{4}C'_2$, where C'_2 is also an arbitrary real number and so we get the solution obtained by the first method.

(e) (E.M.) By differentiating the first equation and using also the other two equations, we obtain $x'' = y' + z' = 2x + y + z$. From the first equation, $y + z = x'$, so we get $x'' - x' - 2x = 0$, a linear homogeneous equation in $x(t)$. Its solution is $x = C_1 \exp(2t) + C_2 \exp(-t)$. We replace it in the last two equations and obtain the system

$$\begin{cases} y' = z + C_1 \exp(2t) + C_2 \exp(-t) \\ z' = y + C_1 \exp(2t) + C_2 \exp(-t) \end{cases}.$$

We differentiate the first equation and obtain the following non-homogeneous equation in $y(t)$: $y'' - y = 3C_1 \exp(2t)$. The general solution of the homogeneous equation is $y_{GH} = C_3 \exp(-t) + C_4 \exp t$ and a particular solution of the non-homogeneous equation is $y_P = C_1 \exp(2t)$. Hence, $y = C_3 \exp(-t) + C_4 \exp t + C_1 \exp(2t)$ and $z = y' - C_1 \exp(2t) - C_2 \exp(-t) = -(C_2 + C_3) \exp(-t) + C_4 \exp t + C_1 \exp(2t)$. But y and z must also satisfy the first equation; $y + z = x' = 2C_1 \exp(2t) - C_2 \exp(-t)$, so we obtain that $C_4 = 0$. The solution of the system is:

$$\begin{cases} x = C_1 \exp(2t) + C_2 \exp(-t) \\ y = C_1 \exp(2t) + C_3 \exp(-t) \\ z = C_1 \exp(2t) - (C_2 + C_3) \exp(-t) \end{cases}.$$

10.3 Chapter 3

(A.M.) The system can be written $\mathbf{Y}' = A\mathbf{Y}$, where

$$\mathbf{Y} = \begin{pmatrix} x(t) \\ y(t) \\ z(t) \end{pmatrix} \quad \text{and} \quad A = \begin{pmatrix} 0 & 1 & 1 \\ 1 & 0 & 1 \\ 1 & 1 & 0 \end{pmatrix}.$$

The characteristic polynomial of the matrix is

$$P(\lambda) = \det(A - \lambda I_3) = \begin{vmatrix} -\lambda & 1 & 1 \\ 1 & -\lambda & 1 \\ 1 & 1 & -\lambda \end{vmatrix} = -\lambda^3 + 3\lambda + 2.$$

Since $P(\lambda) = -(\lambda - 2)(\lambda + 1)^2$, the eigenvalues of the matrix are $\lambda_1 = 2$ and $\lambda_2 = \lambda_3 = -1$. Now, let us find 3 corresponding eigenvectors. The eigenspace corresponding to $\lambda_1 = 2$ i.e. the set of solutions of the system $(A - \lambda_1 I_3)X = \mathbf{0}$ is $V_2 = \{(\alpha, \alpha, \alpha)^t : \alpha \in \mathbb{R}\}$, so we can take the eigenvector $\mathbf{v}_1 = (1, 1, 1)^t$. The eigenspace corresponding to the other eigenvalue, $\lambda_2 = -1$ (the set of solutions of the system $(A + I_3)X = \mathbf{0}$) is $V_{-1} = \{(\alpha, \beta, -\alpha - \beta)^t : \alpha, \beta \in \mathbb{R}\} = \{\alpha(1, 0, -1)^t + \beta(0, 1, -1)^t : \alpha, \beta \in \mathbb{R}\}$, so we can take the linearly independent eigenvectors $\mathbf{v}_2 = (1, 0, -1)^t$ and $\mathbf{v}_3 = (0, 1, -1)^t$. From Proposition 3.4, a fundamental system of solutions is $\{\mathbf{Y}_1 = \mathbf{v}_1 \exp(2t), \mathbf{Y}_2 = \mathbf{v}_2 \exp(-t), \mathbf{Y}_3 = \mathbf{v}_3 \exp(-t)\}$, so the general solution is

$$\mathbf{Y} = C_1 \begin{pmatrix} 1 \\ 1 \\ 1 \end{pmatrix} \exp(2t) + C_2 \begin{pmatrix} 1 \\ 0 \\ -1 \end{pmatrix} \exp(-t) + C_3 \begin{pmatrix} 0 \\ 1 \\ -1 \end{pmatrix} \exp(-t).$$

2. (a) The system can be written $\mathbf{Y}' = A\mathbf{Y}$, where

$$\mathbf{Y} = \begin{pmatrix} x(t) \\ y(t) \\ z(t) \end{pmatrix} \quad \text{and} \quad A = \begin{pmatrix} -1 & 1 & 1 \\ 1 & -1 & 1 \\ 1 & 1 & -1 \end{pmatrix}.$$

The characteristic polynomial of the matrix is

$$P(\lambda) = \begin{vmatrix} -1-\lambda & 1 & 1 \\ 1 & -1-\lambda & 1 \\ 1 & 1 & -1-\lambda \end{vmatrix} = -\lambda^3 - 3\lambda^2 + 4.$$

Since $P(\lambda) = -(\lambda - 1)(\lambda + 2)^2$, the eigenvalues of the matrix are $\lambda_1 = 1$ and $\lambda_2 = \lambda_3 = -2$. Now, let us find 3 corresponding eigenvectors. The eigenspace corresponding to $\lambda_1 = 1$ i.e. the set of solutions of the system $(A - I_3)X = \mathbf{0}$ is $V_1 = \{(\alpha, \alpha, \alpha)^t : \alpha \in \mathbb{R}\}$, so we can take the eigenvector $\mathbf{v}_1 = (1, 1, 1)^t$. The eigenspace corresponding to $\lambda_2 = -2$ (the set of solutions of the system $(A + 2I_3)X = \mathbf{0}$) is $V_{-2} = \{(\alpha, \beta, -\alpha - \beta)^t : \alpha, \beta \in \mathbb{R}\} = \{\alpha(1, 0, -1)^t + \beta(0, 1, -1)^t : \alpha, \beta \in \mathbb{R}\}$, so we can take the linearly independent eigenvectors $\mathbf{v}_2 = (1, 0, -1)$ and $\mathbf{v}_3 = (0, 1, -1)$. From Proposition 3.4, a fundamental system of solutions is $\{\mathbf{Y}_1 = \mathbf{v}_1 \exp t, \mathbf{Y}_2 = \mathbf{v}_2 \exp(-2t), \mathbf{Y}_3 = \mathbf{v}_3 \exp(-2t)\}$, so the general solution is

$$\mathbf{Y} = C_1 \begin{pmatrix} 1 \\ 1 \\ 1 \end{pmatrix} \exp t + C_2 \begin{pmatrix} 1 \\ 0 \\ -1 \end{pmatrix} \exp(-2t) + C_3 \begin{pmatrix} 0 \\ 1 \\ -1 \end{pmatrix} \exp(-2t),$$

where C_1, C_2, and C_3 are arbitrary constants. We can also write the solution in the form:

$$\begin{cases} x(t) = C_1 \exp t + C_2 \exp(-2t) \\ y(t) = C_1 \exp t \qquad\qquad\ \ + C_3 \exp(-2t) \\ z(t) = C_1 \exp t - C_2 \exp(-2t) - C_3 \exp(-2t) \end{cases}$$

(b) The matrix of the system is

$$A = \begin{pmatrix} 2 & 0 & -1 \\ -1 & 2 & 0 \\ -1 & 0 & 2 \end{pmatrix}.$$

The characteristic polynomial of the matrix is

$$P(\lambda) = \begin{vmatrix} 2-\lambda & 0 & -1 \\ -1 & 2-\lambda & 0 \\ -1 & 0 & 2-\lambda \end{vmatrix} = -\lambda^3 + 6\lambda^2 - 11\lambda + 6.$$

Since $P(\lambda) = -(\lambda - 1)(\lambda - 2)(\lambda - 3)$, the eigenvalues of the matrix are $\lambda_1 = 1$, $\lambda_2 = 2$ and $\lambda_3 = 3$. We search for 3 corresponding eigenvectors. The eigenspace corresponding to $\lambda_1 = 1$ i.e. the set of solutions of the system $(A - I_3)X = \mathbf{0}$ is $V_1 = \{(\alpha, \alpha, \alpha)^t : \alpha \in \mathbb{R}\}$, so we can take the eigenvector $\mathbf{v}_1 = (1, 1, 1)^t$. The eigenspace corresponding to $\lambda_2 = 2$ (the set of solutions of the system $(A - 2I_3)X = \mathbf{0}$) is $V_2 = \{0, \alpha, 0)^t : \alpha \in \mathbb{R}\}$, so we can take $\mathbf{v}_2 = (0, 1, 0)$. Finally, the eigenspace corresponding to $\lambda_3 = 3$ is $V_3 = \{\alpha, -\alpha, -\alpha)^t : \alpha \in \mathbb{R}\}$, so we take $\mathbf{v}_3 = (1, -1, -1)^t$. From Proposition 3.4, a fundamental system of solutions is $\{\mathbf{Y}_1 = \mathbf{v}_1 \exp t, \mathbf{Y}_2 =$

$\mathbf{v}_2 \exp(2t)$, $\mathbf{Y}_3 = \mathbf{v}_3 \exp(3t)\}$, so the general solution is

$$\mathbf{Y} = C_1 \begin{pmatrix} 1 \\ 1 \\ 1 \end{pmatrix} \exp t + C_2 \begin{pmatrix} 0 \\ 1 \\ 0 \end{pmatrix} \exp(2t) + C_3 \begin{pmatrix} 1 \\ -1 \\ -1 \end{pmatrix} \exp(3t),$$

where C_1, C_2, and C_3 are arbitrary constants. We can also write the solution in the form:

$$\begin{cases} x(t) = C_1 \exp t & + C_3 \exp(3t) \\ y(t) = C_1 \exp t + C_2 \exp(2t) - C_3 \exp(3t) \\ z(t) = C_1 \exp t & - C_3 \exp(3t) \end{cases}.$$

(c) This non-homogeneous system can be written $\mathbf{Y}' = A\mathbf{Y} + \mathbf{G}$, where

$$A = \begin{pmatrix} 1 & 1 & -2 \\ 1 & 0 & 1 \\ -2 & 1 & 1 \end{pmatrix}, \quad \mathbf{G} = \begin{pmatrix} \exp(3t) \\ 0 \\ -\exp(3t) \end{pmatrix} = \begin{pmatrix} 1 \\ 0 \\ -1 \end{pmatrix} \exp(3t).$$

The characteristic polynomial of the matrix is $P(\lambda) = \det(A - \lambda I_3) = -\lambda^3 + 2\lambda^2 + 5\lambda^2 - 6 = -(\lambda - 3)(\lambda - 1)(\lambda + 2)$, so the eigenvalues of the matrix are $\lambda_1 = 3$, $\lambda_2 = 1$, $\lambda_3 = -2$ and $\mathbf{v}_1 = (1, 0, -1)^t$, $\mathbf{v}_2 = (1, 2, 1)^t$ $\mathbf{v}_3 = (1, -1, 1)^t$ are 3 linearly independent eigenvectors. From Proposition 3.4, a fundamental system of solutions for the homogeneous system $\mathbf{Y}' = A\mathbf{Y}$ is $\{\mathbf{Y}_1 = \mathbf{v}_1 \exp(3t), \mathbf{Y}_2 = \mathbf{v}_2 \exp t, \mathbf{Y}_3 = \mathbf{v}_3 \exp(-2t)\}$, so the general solution is

$$\mathbf{Y}_{GH} = C_1 \begin{pmatrix} 1 \\ 0 \\ -1 \end{pmatrix} \exp(3t) + C_2 \begin{pmatrix} 1 \\ 2 \\ 1 \end{pmatrix} \exp t + C_3 \begin{pmatrix} 1 \\ -1 \\ 1 \end{pmatrix} \exp(-2t).$$

Now, let us find a particular solution for the non-homogeneous system. Since $\mathbf{G}(t) = \mathbf{P}(t) \exp(\alpha t)$ where $\mathbf{P}(t)$ is a vector polynomial of degree 0 and $\alpha = 3$ is an eigenvalue of A with multiplicity 1, by using the formula (3.88), we search for a particular solution of the form $\mathbf{Y}_P = (a\mathbf{v}_1 t + b\mathbf{v}_2 + c\mathbf{v}_3) \exp(3t)$, where a, b, c are real constants:

$$\begin{cases} x(t) = (at + b + c) \exp(3t) \\ y(t) = \quad\quad (2b - c) \exp(3t) \\ z(t) = (-at + b + c) \exp(3t) \end{cases}.$$

We replace in the initial system and find the constants $a = 1, b = c = 0$, so the general solution of the system is

$$\begin{cases} x(t) = C_1 \exp(3t) + C_2 \exp t + C_3 \exp(-2t) + t \exp(3t) \\ y(t) = 2C_2 \exp t - C_3 \exp(-2t) \\ z(t) = -C_1 \exp(3t) + C_2 \exp t + C_3 \exp(-2t) - t \exp(3t) \end{cases},$$

where C_1, C_2, and C_3 are arbitrary constants.

(d) We have to solve an non-homogeneous system: $\mathbf{Y}' = A\mathbf{Y} + \mathbf{G}$, where

$$A = \begin{pmatrix} 1 & 1 & 1 \\ 1 & 1 & 1 \\ 1 & 1 & 1 \end{pmatrix}, \ \mathbf{G} = \begin{pmatrix} 3 \\ -\exp t \\ -\exp t \end{pmatrix} = \begin{pmatrix} 3 \\ 0 \\ 0 \end{pmatrix} + \begin{pmatrix} 0 \\ -1 \\ -1 \end{pmatrix} \exp t.$$

First, we solve the homogeneous system $\mathbf{Y}' = A\mathbf{Y}$. The characteristic polynomial of the matrix A is $P(\lambda) = \det(A - \lambda I_3) = -\lambda^3 + 3\lambda^2 = -\lambda^2(\lambda - 3)$, so the eigenvalues of the matrix are $\lambda_1 = \lambda_2 = 0$, $\lambda_3 = 3$ and $\mathbf{v}_1 = (1, 0, -1)^t$, $\mathbf{v}_2 = (0, 1, -1)^t$ $\mathbf{v}_3 = (1, 1, 1)^t$ are 3 linearly independent eigenvectors. Hence, the general solution of the homogeneous system is

$$\mathbf{Y}_{GH} = C_1 \begin{pmatrix} 1 \\ 0 \\ -1 \end{pmatrix} + C_2 \begin{pmatrix} 0 \\ 1 \\ -1 \end{pmatrix} + C_3 \begin{pmatrix} 1 \\ 1 \\ 1 \end{pmatrix} \exp(3t).$$

Now, let us find a particular solution for the non-homogeneous system. Since $\mathbf{G}(t) = \mathbf{P}_1(t) \exp(\alpha_1 t) + \mathbf{P}_2(t) \exp(\alpha_2 t)$ where $\mathbf{P}_1(t) = (3, 0, 0)^t$ and $\mathbf{P}_2(t) = (0, -1, -1)^t$ are vector polynomials of degree 0 and $\alpha_1 = 0$, $\alpha_2 = 1$, from the superposition principle (see Proposition 3.5), a particular solution of the system can be written $\mathbf{Y}_P = \mathbf{Y}_{P,1} + \mathbf{Y}_{P,2}$, where $\mathbf{Y}_{P,1}$ is a particular solution of the system $\mathbf{Y}' = A\mathbf{Y} + \mathbf{P}_1(t) \exp(\alpha_1 t)$ and $\mathbf{Y}_{P,2}$ is a particular solution of the system $\mathbf{Y}' = A\mathbf{Y} + \mathbf{P}_2(t) \exp(\alpha_2 t)$. Since $\alpha_1 = 0$ is an eigenvalue of A with multiplicity 2, by using the formula (3.88), one can write $\mathbf{Y}_{P,1} = at\mathbf{v}_1 + bt\mathbf{v}_2 + c\mathbf{v}_3$, where a, b, c are real constants:

$$\mathbf{Y}_{P,1} = \begin{pmatrix} x(t) \\ y(t) \\ z(t) \end{pmatrix} = \begin{pmatrix} at + c \\ bt + c \\ -(a+b)t + c \end{pmatrix}.$$

By replacing in the non-homogeneous system

$$\begin{cases} \dot{x} = x + y + z + 3 \\ \dot{y} = x + y + z \\ \dot{z} = x + y + z \end{cases},$$

we obtain $a = 2, b = -1$ and $c = -\frac{1}{3}$.

10.3 Chapter 3

Now, we notice that $\alpha_2 = 1$ is not an eigenvalue of A, so we use the formula (3.87) and search for $\mathbf{Y}_{P,2} = \mathbf{Q}(t) \exp t$, where $\mathbf{Q}(t) = (d, e, f)^t$ is a vector polynomial of degree 0. By replacing in the non-homogeneous system

$$\begin{cases} \dot{x} = x + y + z \\ \dot{y} = x + y + z - \exp t \\ \dot{z} = x + y + z - \exp t \end{cases},$$

we obtain $d = 1$ and $e = f = 0$, so a particular solution is

$$\mathbf{Y}_P = \mathbf{Y}_{P,1} + \mathbf{Y}_{P,2} = \begin{pmatrix} 2t - \frac{1}{3} \\ -t - \frac{1}{3} \\ -t - \frac{1}{3} \end{pmatrix} + \begin{pmatrix} \exp t \\ 0 \\ 0 \end{pmatrix}$$

and the general solution of the system is

$$\mathbf{Y} = \mathbf{Y}_{GH} + \mathbf{Y}_P = \begin{pmatrix} C_1 + C_3 \exp(3t) + 2t - \frac{1}{3} + \exp t \\ C_2 + C_3 \exp(3t) - t - \frac{1}{3} \\ -C_1 - C_2 + C_3 \exp(3t) - t - \frac{1}{3} \end{pmatrix},$$

where C_1, C_2, C_3 are arbitrary constants.

Now, let us find the particular solution that verifies the initial conditions $x(0) = 1$, $y(0) = 0$, $z(0) = -1$. We find the constants: $C_1 = C_2 = \frac{1}{3}$, $C_3 = 0$, so the solution of the Cauchy problem is $\begin{cases} x(t) = 2t + \exp t \\ y(t) = -t \\ z(t) = -t - 1 \end{cases}$.

3. (a) First, we solve the homogeneous system $\begin{cases} \dot{x} = -2x + y \\ \dot{y} = -3x + 2y \end{cases}$. The matrix $A = \begin{pmatrix} -2 & 1 \\ -3 & 2 \end{pmatrix}$ has the characteristic polynomial $P(\lambda) = \lambda^2 - 1$, so its eigenvalues are $\lambda_1 = 1$ and $\lambda_2 = -1$. The corresponding eigenspaces are $V_1 = \{(\alpha, 3\alpha)^t : \alpha \in \mathbb{R}\}$ and $V_{-1} = \{(\alpha, \alpha)^t : \alpha \in \mathbb{R}\}$, so we can take the eigenvectors $\mathbf{v}_1 = (1, 3)^t$ and $\mathbf{v}_2 = (1, 1)^t$. From Proposition 3.4, a fundamental system of solutions is $\{\mathbf{Y}_1 = \mathbf{v}_1 \exp t, \mathbf{Y}_2 = \mathbf{v}_2 \exp(-t)\}$, so the general solution of the homogeneous system is

$$\mathbf{Y}_{GH} = \begin{pmatrix} x(t) \\ y(t) \end{pmatrix} = C_1 \mathbf{Y}_1 + C_2 \mathbf{Y}_2 = C_1 \begin{pmatrix} 1 \\ 3 \end{pmatrix} \exp t + C_2 \begin{pmatrix} 1 \\ 1 \end{pmatrix} \exp(-t).$$

Now, we use the Lagrange method of variation of constants to find $\mathbf{Y}_P$, a particular solution of the non-homogeneous system, which is of the form

$$\mathbf{Y}_P = C_1(t)\mathbf{Y}_1 + C_2(t)\mathbf{Y}_2 = C_1(t) \begin{pmatrix} 1 \\ 3 \end{pmatrix} \exp t + C_2(t) \begin{pmatrix} 1 \\ 1 \end{pmatrix} \exp(-t),$$

where the functions $C_1(t)$ and $C_2(t)$ verify the equations:

$$C_1'\mathbf{Y}_1 + C_2'\mathbf{Y}_2 = \begin{pmatrix} -\exp(2t) \\ 3\exp(2t) \end{pmatrix}.$$

So, we obtained the following system of equations in the unknowns C_1', C_2':

$$\begin{cases} C_1' \exp t + C_2' \exp(-t) = -\exp(2t) \\ 3C_1' \exp t + C_2' \exp(-t) = 3\exp(2t) \end{cases}.$$

The solution of this system is $C_1' = 2\exp t$, $C_2' = -3\exp(3t)$, so $C_1(t) = 2\exp t$, $C_2(t) = -\exp(3t)$ and the particular solution is $\mathbf{Y}_P = \begin{pmatrix} 1 \\ 5 \end{pmatrix} \exp(2t)$. The general solution is $\mathbf{Y} = \mathbf{Y}_{GH} + \mathbf{Y}_P$:

$$\begin{cases} x(t) = C_1 \exp t + C_2 \exp(-t) + \exp(2t) \\ y(t) = 3C_1 \exp t + C_2 \exp(-t) + 5\exp(2t) \end{cases}.$$

(b) First, we solve the homogeneous system $\begin{cases} \dot{x} = 2y \\ \dot{y} = x - y \end{cases}$. The matrix $A = \begin{pmatrix} 0 & 2 \\ 1 & -1 \end{pmatrix}$ has the characteristic polynomial $P(\lambda) = \lambda^2 + \lambda - 2$, so its eigenvalues are $\lambda_1 = 1$ and $\lambda_2 = -2$. The corresponding eigenspaces are $V_1 = \{(2\alpha, \alpha)^t : \alpha \in \mathbb{R}\}$ and $V_{-2} = \{(\alpha, -\alpha)^t : \alpha \in \mathbb{R}\}$, so we can take the eigenvectors $\mathbf{v}_1 = (2, 1)^t$ and $\mathbf{v}_2 = (1, -1)^t$. From Proposition 3.4, a fundamental system of solutions is $\{\mathbf{Y}_1 = \mathbf{v}_1 \exp t, \mathbf{Y}_2 = \mathbf{v}_2 \exp(-2t)\}$, so the general solution of the homogeneous system is

$$\mathbf{Y}_{GH} = C_1\mathbf{Y}_1 + C_2\mathbf{Y}_2 = C_1 \begin{pmatrix} 2 \\ 1 \end{pmatrix} \exp t + C_2 \begin{pmatrix} 1 \\ -1 \end{pmatrix} \exp(-2t).$$

Now, we use the Lagrange method of variation of constants to find $\mathbf{Y}_P$, a particular solution of the non-homogeneous system, of the form

$$\mathbf{Y}_P = C_1(t)\mathbf{Y}_1 + C_2(t)\mathbf{Y}_2 = C_1(t) \begin{pmatrix} 2 \\ 1 \end{pmatrix} \exp t + C_2(t) \begin{pmatrix} 1 \\ -1 \end{pmatrix} \exp(-2t),$$

where the functions $C_1(t)$ and $C_2(t)$ verify the equations:

$$C'_1 \mathbf{Y}_1 + C'_2 \mathbf{Y}_2 = \begin{pmatrix} \cos t + 2\sin t \\ -\cos t + \sin t \end{pmatrix}.$$

So, we obtained the following system of equations:

$$\begin{cases} 2C'_1 \exp t + C'_2 \exp(-2t) = \cos t + 2\sin t \\ C'_1 \exp t - C'_2 \exp(-2t) = -\cos t + \sin t \end{cases}.$$

The solution of this system is $C'_1 = \exp(-t)\sin t$, $C'_2 = \exp(2t)\cos t$. We integrate (using integration by parts) and obtain: $C_1(t) = \int \exp(-t)\sin t\, dt = -\frac{1}{2}\exp(-t)(\sin t + \cos t)$ and $C_2(t) = \int \exp(2t)\cos t\, dt = \frac{1}{5}\exp(2t)(\sin t + 2\cos t)$ and the particular solution is $\mathbf{Y}_P = \begin{pmatrix} -\frac{1}{5}(4\sin t + 3\cos t) \\ -\frac{1}{10}(7\sin t + 9\cos t) \end{pmatrix}$. The general solution is $\mathbf{Y} = \mathbf{Y}_{GH} + \mathbf{Y}_P$:

$$\begin{cases} x(t) = 2C_1 \exp t + C_2 \exp(-2t) - \frac{1}{5}(4\sin t + 3\cos t) \\ y(t) = 3C_1 \exp t - C_2 \exp(-2t) - \frac{1}{10}(7\sin t + 9\cos t) \end{cases}.$$

(c) First, we solve the homogeneous system $\begin{cases} \dot{x} = -4x - 2y \\ \dot{y} = 6x + 3y \end{cases}$. The matrix $A = \begin{pmatrix} -4 & -2 \\ 6 & 3 \end{pmatrix}$ has the characteristic polynomial $P(\lambda) = \lambda^2 + \lambda$, so its eigenvalues are $\lambda_1 = 0$ and $\lambda_2 = -1$. The corresponding eigenspaces are $V_0 = \{(\alpha, -2\alpha)^t : \alpha \in \mathbb{R}\}$ and $V_{-1} = \{(2\alpha, -3\alpha)^t : \alpha \in \mathbb{R}\}$, so we can take the eigenvectors $\mathbf{v}_1 = (1, -2)^t$ and $\mathbf{v}_2 = (2, -3)^t$. From Proposition 3.4, a fundamental system of solutions is $\{\mathbf{Y}_1 = \mathbf{v}_1, \mathbf{Y}_2 = \mathbf{v}_2 \exp(-t)\}$, so the general solution of the homogeneous system is

$$\mathbf{Y}_{GH} = C_1 \mathbf{Y}_1 + C_2 \mathbf{Y}_2 = C_1 \begin{pmatrix} 1 \\ -2 \end{pmatrix} + C_2 \begin{pmatrix} 2 \\ -3 \end{pmatrix} \exp(-t).$$

Now, we use the Lagrange method of variation of constants to find $\mathbf{Y}_P$, a particular solution of the non-homogeneous system, which is of the form

$$\mathbf{Y}_P = C_1(t)\mathbf{Y}_1 + C_2(t)\mathbf{Y}_2 = C_1(t) \begin{pmatrix} 1 \\ -2 \end{pmatrix} + C_2(t) \begin{pmatrix} 2 \\ -3 \end{pmatrix} \exp(-t),$$

where the functions $C_1(t)$ and $C_2(t)$ verify the equations:

$$C'_1 \mathbf{Y}_1 + C'_2 \mathbf{Y}_2 = \begin{pmatrix} \frac{2}{\exp t - 1} \\ -\frac{3}{\exp t - 1} \end{pmatrix}.$$

So, we obtained the following system of equations:

$$\begin{cases} C_1' + 2C_2' \exp(-t) = \frac{2}{\exp t - 1} \\ -2C_1' - 3C_2' \exp(-t) = -\frac{3}{\exp t - 1} \end{cases}.$$

The solution of this system is $C_1' = 0$, $C_2' = \frac{\exp t}{\exp t - 1}$, so $C_1(t)$ is a constant, we can take $C_1(t) = 0$, $C_2(t) = \ln(\exp t - 1)$, and we have $\mathbf{Y}_P = \begin{pmatrix} 2\exp(-t)\ln(\exp t - 1) \\ -3\exp(-t)\ln(\exp t - 1) \end{pmatrix}$. The general solution is $\mathbf{Y} = \mathbf{Y}_{GH} + \mathbf{Y}_P$:

$$\begin{cases} x(t) = C_1 + 2C_2 \exp(-t) + 2\exp(-t)\ln(\exp t - 1) \\ y(t) = -2C_1 - 3C_2 \exp(-t) - 3\exp(-t)\ln(\exp t - 1) \end{cases}.$$

4. (a) From the first equation one can write $y = -x' - 2x + \sin t$. By replacing this y in the second equation we obtain $-x'' - 2x' + \cos t = 4x - 2x' - 4x + 2\sin t + \cos t$, or, equivalently, $x'' = -2\sin t$. We integrate (twice) and get $x = C_1 t + C_2 + 2\sin t$. Since $y = -x' - 2x + \sin t$, the general solution of the system is

$$\begin{cases} x = C_1 t + C_2 + 2\sin t \\ y = -2C_1 t - (C_1 + 2C_2) - 3\sin t - 2\cos t \end{cases},$$

where C_1, C_2 are arbitrary constants.

(b) Let us differentiate the second equation: we obtain $\ddot{y} = 2\dot{x} - \dot{y} = 2(x - y + \frac{1}{\cos t}) - \dot{y}$. Since $2x = \dot{y} + y$, we obtain the following equation in $y(t)$: $\ddot{y} + y = \frac{2}{\cos t}$. The characteristic equation, $\lambda^2 + 1 = 0$ has the complex (conjugate) roots $\lambda_{1,2} = \pm i$. Hence, $\{y_1 = \cos t, \ y_2 = \sin t\}$ is a fundamental system of solutions for the homogeneous equation $\ddot{y} + y = 0$ and the general solution of this equation is $y_{GH} = C_1 \cos t + C_2 \sin t$. We use the Lagrange method of variation of constants to find a particular solution of the non-homogeneous equation: we search for $y_P = C_1(t)\cos t + C_2(t)\sin t$, where the functions $C_1(t)$ and $C_2(t)$ verify the equations

$$\begin{cases} C_1' y_1 + C_2' y_2 = 0 \\ C_1' y_1' + C_2' y_2' = \frac{2}{\cos t} \end{cases} \Leftrightarrow \begin{cases} C_1' \cos t + C_2' \sin t = 0 \\ -C_1' \sin t + C_2' \cos t = \frac{2}{\cos t} \end{cases}$$

We obtain $C_1' = -\frac{2\sin t}{\cos t}$ and $C_2' = 2$, so $C_1 = 2\ln|\cos t|$ and $C_2 = 2t$. The general solution of the non-homogeneous equation is $y = y_{GH} + y_P = C_1 \cos t + C_2 \sin t + 2\cos t \ln|\cos t| + 2t \sin t$. Since $x = \frac{1}{2}(\dot{y} + y)$, the

10.3 Chapter 3

general solution of the system is

$$\begin{cases} x = \frac{C_1+C_2}{2}\cos t + \frac{C_2-C_1}{2}\sin t + (\cos t - \sin t)\ln|\cos t| + t(\cos t + \sin t) \\ y = C_1\cos t + C_2\sin t + 2\cos t \ln|\cos t| + 2t\sin t. \end{cases},$$

where C_1, C_2 are arbitrary constants.

(c) We differentiate the second equation and obtain $\ddot{y} = -\dot{x} + \frac{1}{\cos^2 t} = -y - \frac{\sin^2 t}{\cos^2 t} + 1 + \frac{1}{\cos^2 t} = -y$. So we obtained the homogeneous equation $\ddot{y} + y = 0$ which has the general solution $y = C_1\cos t + C_2\sin t$. Since $x = -\dot{y} + \tan t$, the general solution of the system is

$$\begin{cases} x = -C_2\cos t + C_1\sin t + \tan t \\ y = C_1\cos t + C_2\sin t. \end{cases},$$

where C_1, C_2 are arbitrary constants.

(d) We differentiate the first equation and obtain $x'' + y' - z' = 0$. From the last two equations we have $y' = z$ and $z' = z - x$, so we obtain the homogeneous equation $x'' + x = 0$. Its general solution is $x = C_1\cos t + C_2\sin t$. Since $z' - z = -x$, the function z is the solution of the non-homogeneous linear equation $z' - z = -C_1\cos t - C_2\sin t$. The homogeneous associated equation $z' - z = 0$ has the general solution $z_{GH} = C_3\exp t$ and, since we have no resonance, we search for a particular solution of the non-homogeneous equation $z_P = A\cos t + B\sin t$. We have $-A\sin t + B\cos t - A\cos t - B\sin t = -C_1\cos t - C_2\sin t$, so $A + B = C_2$ and $B - A = -C_1$. The solution of this system is $A = \frac{C_1+C_2}{2}$, $B = \frac{C_2-C_1}{2}$, so we get $z = C_3\exp t + \frac{C_1+C_2}{2}\cos t + \frac{C_2-C_1}{2}\sin t$. Since $y = z - x'$, we have $y = C_3\exp t + \frac{C_1-C_2}{2}\cos t + \frac{C_1+C_2}{2}\sin t$.

We obtained the general solution, now we have to find the constants C_1, C_2 and C_3 such that $x(0) = y(0) = 1$ and $z(0) = 0$. Since $x(0) = C_1$, $y(0) = C_3 + \frac{C_1-C_2}{2}$, $z(0) = C_3 + \frac{C_1+C_2}{2}$, we get $C_1 = 1$, $C_2 = -1$, $C_3 = 0$, so the solution of the Cauchy problem is: $x = \cos t - \sin t$, $y = \cos t$, $z = -\sin t$.

(e) We differentiate the first equation and obtain $x'' - 2x' - y' + 2z' = -1$. From the last two equations we have $y' = -x + 1$ and $z' = x + y - z + 1 - t$, so we get $x'' - 2x' + 3x + 2y - 2z = 2t - 2$. By differentiating once again (and using the expressions of y' and z'), one gets $x''' - 2x'' + 3x' - 4x - 2y + 2z = 2 - 2t$. Now, from the system

$$\begin{cases} x' - 2x - y + 2z = 2 - t \\ x'' - 2x' + 3x + 2y - 2z = 2t - 2 \end{cases}$$

we can write y and z as functions of x, x' and x'':

$$\begin{cases} y = -x'' + x' - x + t \\ z = \frac{1}{2}(-x'' + x) + 1 \end{cases}.$$

By replacing y and z in the last (third-order) equation, we obtain:

$$x''' - x'' + x' - x = 0,$$

a linear equation in $x(t)$. Its characteristic equation is $\lambda^3 - \lambda^2 + \lambda - 1 = 0$ and can be written $(\lambda - 1)(\lambda^2 + 1) = 0$, so the roots are $\lambda_1 = 1$, $\lambda_{2,3} = \pm i$. Hence, the general solution of the differential equation is $x = C_1 \exp t + C_2 \cos t + C_3 \sin t$, where C_1, C_2, C_3 are arbitrary (real) constants. By using the above expression of y an z, we can write the general solution of the system:

$$\begin{cases} x = C_1 \exp t + C_2 \cos t + C_3 \sin t \\ y = -C_1 \exp t + C_3 \cos t - C_2 \sin t + t \\ z = \qquad C_2 \cos t + C_3 \sin t + 1 \end{cases}.$$

(f) The system can be written:

$$\begin{cases} \dot{x} = 3x + y \\ \dot{y} = 3y + z \\ \dot{z} = 3z + \exp(3t) \end{cases}$$

and we can solve the last equation (a first-order linear equation in $z(t)$). The general solution of the homogeneous equation $\dot{z} = 3z$ is $z_{GH} = C_1 \exp(3t)$. Since the free term is $\exp(3t)$ and $\lambda = 3$ is a root of the characteristic equation, we search for a particular solution of the form $z_P(t) = at \exp(3t)$ and find $a = 1$, so the general solution is $z = (t + C_1) \exp(3t)$. We replace it in the second equation and obtain the following first-order linear equation in $y(t)$: $\dot{y} = 3y + (t + C_1) \exp(3t)$. The general solution of the homogeneous equation is $y_{GH} = C_2 \exp(3t)$. Since the free term is $P(t) \exp(3t)$ where $P(t)$ is a polynomial of degree 1 and $\lambda = 3$ is a root of the characteristic equation, we search for a particular solution of the form $y_P = t(bt + c) \exp(3t)$ and find $b = \frac{1}{2}$ and $c = C_1$, so the general solution is $y = (\frac{1}{2}t^2 + C_1 t + C_2) \exp(3t)$. Finally, we replace this y in the first equation and obtain the following first-order linear equation in $x(t)$: $\dot{x} = 3x + (\frac{1}{2}t^2 + C_1 t + C_2) \exp(3t)$. Obviously, the general solution of the homogeneous equation has the same form, $x_{GH} = C_3 \exp(3t)$. The free term is $P(t) \exp(3t)$ where $P(t)$ is a polynomial of degree 2, so we search for a particular solution of the form $x_P = t(dt^2 + et + f) \exp(3t)$ and find $d = \frac{1}{6}$, $e = \frac{C_1}{2}$ and $f = C_2$, so the general solution is

$x = (\frac{1}{6}t^3 + \frac{C_1}{2}t^2 + C_2 t + C_3)\exp(3t)$. The general solution of the system is

$$\begin{cases} x = (\frac{1}{6}t^3 + \frac{C_1}{2}t^2 + C_2 t + C_3)\exp(3t) \\ y = (\frac{1}{2}t^2 + C_1 t + C_2)\exp(3t) \\ z = (t + C_1)\exp(3t) \end{cases}$$

The particular solution that satisfies the initial conditions $x(0) = 0$, $y(0) = 0$ and $z(0) = 0$ is obtained for $C_1 = C_2 = C_3 = 0$, so the solution of the Cauchy problem is: $x = \frac{1}{6}t^3 \exp(3t)$, $y = \frac{1}{2}t^2 \exp(3t)$, $z = t\exp(3t)$.

(g) The system can be written:

$$\begin{cases} \dot{x} = -x - 2y + 2z + 2\exp(-t) \\ \dot{y} = -y - z + 1 \\ \dot{z} = -z + 1 \end{cases}$$

and we can solve the last equation (a first-order linear equation in $z(t)$). The general solution of the homogeneous equation $\dot{z} = -z$ is $z_{GH} = C_1 \exp(-t)$ and we search for a particular solution of the form $z_P(t) = a$ (we have no resonance). We get $a = 1$ and the general solution is $z = C_1 \exp(-t) + 1$. We replace this z in the second equation and obtain the following first-order linear equation in $y(t)$: $\dot{y} = -y - C_1 \exp(-t)$. The general solution of the homogeneous equation $\dot{y} = -y$ is $y_{GH} = C_2 \exp(-t)$. Since $\lambda = -1$ is a root of the characteristic equation, we search for a particular solution of the form $y_P = bt\exp(-t)$ and find $b = -C_1$, so the general solution is $y = (-C_1 t + C_2)\exp(-t)$. Now, we replace this y in the first equation and obtain the following first-order linear equation in $x(t)$: $\dot{x} = -x + (2C_1 t + 2C_1 - 2C_2 + 2)\exp(-t) + 2$. The general solution of the homogeneous equation has the same form, $x_{GH} = C_3 \exp(-t)$. The free term is $P(t)\exp(-t) + k$ where $P(t)$ is a polynomial of degree 1 and $k = 2$ is a constant, so we search for a particular solution of the form $x_P = t(ct + d)\exp(-t) + e$ and find $c = C_1$ and $d = 2C_1 - 2C_2 + 2$, $e = 2$ so the general solution is $x = [C_1 t^2 + (2C_1 - 2C_2 + 2)t + C_3]\exp(-t) + 2$. Hence, the general solution of the system is

$$\begin{cases} x = (C_1 t^2 + (2C_1 - 2C_2 + 2)t + C_3)\exp(-t) + 2 \\ y = (-C_1 t + C_2)\exp(-t) \\ z = C_1 \exp(-t) + 1 \end{cases}.$$

The particular solution that satisfies the initial conditions $x(0) = 1$, $y(0) = 1$ and $z(0) = 1$ is obtained for $C_1 = 0$, $C_2 = 1$, $C_3 = -1$, so the solution of the Cauchy problem is:

$$\begin{cases} x = -\exp(-t) + 2 \\ y = \exp(-t) \\ z = 1 \end{cases}.$$

(h) By differentiating the first equation one obtains: $\dddot{x} + \ddot{x} + \ddot{y} - 2\dot{y} = 0$. From the second equation we have $\dot{y} = \dot{x} + x$ and $\ddot{y} = \ddot{x} + \dot{x}$. By replacing in the equation above we obtain a third-order homogeneous equation in $x(t)$: $\dddot{x} + 2\ddot{x} - \dot{x} - 2x = 0$. The characteristic equation $\lambda^3 + \lambda^2 - 2\lambda - 2 = 0 \Leftrightarrow (\lambda + 2)(\lambda - 1)(\lambda + 1) = 0$ has the roots $\lambda_1 = 1, \lambda_2 = -1, \lambda_3 = -2$, so the general solution of the equation is $x = C_1 \exp t + C_2 \exp(-t) + C_3 \exp(-2t)$, where C_1, C_2, C_3 are arbitrary constants. By adding the equations of the system we obtain $y = \frac{1}{2}(\ddot{x} + 2\dot{x} + x)$, so $y = 2C_1 \exp t + \frac{1}{2}C_3 \exp(-2t)$.

Remark If we denoted $\dot{x} = z$ then our system could have been transformed into a first-order system of three equations. Thus, from the second equation one can write $\dot{y} = x + z$ and, By replacing $\dot{x}$ and $\dot{y}$ in the first equation we find $\dot{z} + z + x + z - 2y = 0$ so we obtain the following system:

$$\begin{cases} \dot{x} = z \\ \dot{y} = x + z \\ \dot{z} = -x + 2y - 2z \end{cases}$$

(i) If we differentiate the second equation and subtract it from the first, we obtain $x = 2\dot{y}$. Hence, $\dot{x} = 2\ddot{y}$ and we obtain the following second-order equation in $y(t)$: $2\ddot{y} + 3\dot{y} - 2y = 0$. The characteristic equation is $2\lambda^2 + 3\lambda - 2 = 0$ with the roots $\lambda_1 = -2$ and $\lambda_2 = \frac{1}{2}$, so the general solution of the equation is $y = C_1 \exp(-2t) + C_2 \exp(\frac{1}{2}t)$, where C_1, C_2 are arbitrary constants. Since $x = 2\dot{y}$, we get: $x = -4C_1 \exp(-2t) + C_2 \exp(\frac{1}{2}t)$.

(j) From the first equation we get $y = \frac{1}{2}(\ddot{x} - \exp t)$, so $\ddot{y} = \frac{1}{2}(x^{IV} - \exp t)$. By replacing this $\ddot{y}$ in the second equation we obtain a linear fourth-order equation in $x(t)$: $x^{IV} + 4x = \exp t$. The characteristic equation can be written: $\lambda^4 + 4 = 0 \Leftrightarrow (\lambda^2 + 2)^2 - 4\lambda^2 = 0 \Leftrightarrow (\lambda^2 - 2\lambda + 2)(\lambda^2 + 2\lambda + 2) = 0$ and one can see that all of its roots are complex numbers: $\lambda_{1,2} = 1 \pm i, \lambda_{3,4} = -1 \pm i$. Hence, the general solution of the homogeneous equation $x^{IV} + 4x = 0$ is $x_{GH} = C_1 \exp t \cos t + C_2 \exp t \sin t + C_3 \exp(-t) \cos t + C_4 \exp(-t) \sin t$. We search for a particular solution of the non-homogeneous equation of the form $x_P = a \exp t$ and we get $a = \frac{1}{5}$, so the general solution is $x = C_1 \exp t \cos t + C_2 \exp t \sin t + C_3 \exp(-t) \cos t + C_4 \exp(-t) \sin t + \frac{1}{5} \exp t$. Since $y = \frac{1}{2}(x''')$, we have $y = C_2 \exp t \cos t - C_1 \exp t \sin t - C_4 \exp(-t) \cos t + C_3 \exp(-t) \sin t - \frac{2}{5} \exp t$.

(k) From the first equation we have $y = \frac{1}{4}(x'' - 3x - \sin t)$, so $y'' = \frac{1}{4}(x^{IV} - 3x'' + \sin t)$, By replacing this y'' in the second equation we obtain a fourth-order (homogeneous) equation in $x(t)$: $x^{IV} - 2x'' + x = 0$. The characteristic equation, $\lambda^4 - 2\lambda^2 + 1 = 0 \Leftrightarrow (\lambda - 1)^2(\lambda + 1)^2 = 0$ has the (double) roots $\lambda_1 = \lambda_2 = 1, \lambda_3 = \lambda_4 = -1$. Hence, the general solution of the equation is $x = C_1 \exp t + C_2 t \exp t + C_3 \exp(-t) + C_4 t \exp(-t)$.

10.3 Chapter 3   653

Since $y = \frac{1}{4}(x'' - 3x - \sin t)$, we have $y = \frac{C_2-C_1}{2}\exp t - \frac{C_2}{2}t\exp t - \frac{C_3+C_4}{2}\exp(-t) + \frac{C_4}{2}t\exp(-t) - \frac{1}{4}\sin t$.

(l) From the first equation we can write: $y = t\dot{x} - x$, so $\dot{y} = t\ddot{x} + \dot{x} - \dot{x} = t\ddot{x}$. By replacing this $\dot{y}$ in the second equation we get: $t^2\ddot{x} = 3x - t\dot{x} + x$, so we obtain the following equation in $x(t)$: $t^2\ddot{x} + t\dot{x} - 4x = 0$ (an Euler equation). We use the change of variable $t = \exp z \Leftrightarrow z = \ln t$. We denote by $\tilde{x}$ the function $\tilde{x}(z) = x(\exp z) = x(t)$. Since $\dot{x} = \exp(-z)\tilde{x}'$ and $\ddot{x} = \exp(-2z)(\tilde{x}'' - \tilde{x}')$, the Euler equation in $x(t)$ becomes a linear equation with constant coefficients in $\tilde{x}(z)$: $\tilde{x}'' - 4\tilde{x} = 0$. The general solution of this equation is $\tilde{x} = C_1\exp(2z) + C_2\exp(-2z)$, so the general solution of the Euler equation is $x = C_1 t^2 + C_2 t^{-2}$. Since $y = t\dot{x} - x$, the general solution of the system is:

$$\begin{cases} x(t) = C_1 t^2 + C_2 t^{-2} \\ y(t) = C_1 t^2 - 3C_2 t^{-2} \end{cases},$$

where C_1, C_2 are arbitrary constants. Now we have to find the particular solution that satisfies the initial conditions $x(1) = 0$, $y(1) = 1$. We obtain $C_1 = \frac{1}{4}$ and $C_2 = -\frac{1}{4}$, so the solution of the Cauchy problem is: $x = \frac{t^2}{4} - \frac{1}{4t^2}$, $y = \frac{t^2}{4} + \frac{3}{4t^2}$.

(m) From the first equation we have: $y = \frac{1}{2}(-t\dot{x} + 3x)$, so $\dot{y} = \frac{1}{2}(-t\ddot{x} - \dot{x} + 3\dot{x}) = -\frac{1}{2}t\ddot{x} + \dot{x}$. By replacing $\dot{y}$ in the second equation we obtain the following Euler equation: $t^2\ddot{x} - t\dot{x} + x = -2\ln t$. We use the change of variable $t = \exp z \Leftrightarrow z = \ln t$. We denote by $\tilde{x}$ the function $\tilde{x}(z) = x(\exp z) = x(t)$. Since $\dot{x} = \exp(-z)\tilde{x}'$ and $\ddot{x} = \exp(-2z)(\tilde{x}'' - \tilde{x}')$, the Euler equation in $x(t)$ becomes a linear equation with constant coefficients in $\tilde{x}(z)$: $\tilde{x}'' - 2\tilde{x}' + \tilde{x} = -2z$. The general solution of this equation is $\tilde{x} = C_1\exp z + C_2 z \exp z - 2z - 4$, so the general solution of the Euler equation is $x = C_1 t + C_2 t \ln t - 2\ln t - 4$. Since $y = \frac{1}{2}(-t\dot{x} + 3x)$, the general solution of the system is:

$$\begin{cases} x(t) = C_1 t + C_2 t \ln t - 2\ln t - 4 \\ y(t) = (C_1 - \frac{C_2}{2})t + C_2 t \ln t - 3\ln t - 5 \end{cases},$$

where C_1, C_2 are arbitrary constants. Now we have to find the particular solution that satisfies the initial conditions $x(1) = 1$, $y(1) = 1$. We obtain $C_1 = 5$ and $C_2 = -2$.

5. (a) We write the system in the symmetric form $\frac{dx}{x+2y} = \frac{dy}{\frac{x^2+2xy}{y}}$. If we multiply by x the numerator and the denominator of the first fraction, and by y, the

numerator and the denominator of the second fraction, we obtain:

$$\frac{xdx}{x^2+2xy} = \frac{ydy}{x^2+2xy} = \frac{xdx-ydy}{0}.$$

Hence, $d(x^2 - y^2) = 0$, or, equivalently, $x^2 - y^2 = C$, i.e. the general solution of the system is a family of hyperbolas.

(b) The system can be written: $\frac{dx}{-\frac{1}{y}} = \frac{dy}{-\frac{1}{x}}$, or, equivalently, $\frac{dx}{x} = \frac{dy}{y}$. By integration we obtain $\ln y = \ln x + \ln C = \ln Cx$, so $y = Cx$, i.e. the general solution of the system is a family of straight lines.

(c) We write the system in the symmetric form $\frac{dx}{x^2+y^2} = \frac{dy}{2xy}$. This is equivalent to the first order differential equation $\frac{dy}{dx} = \frac{2xy}{x^2+y^2}$ (if we consider $y = y(x)$). We have to solve a homogeneous equation in x and y, so we denote by $u(x) = \frac{y(x)}{x}$. Since $y(x) = xu(x)$, we can write $\frac{dy}{dx} = u + xu' = \frac{2u}{1+u^2}$. Thus, we obtain a separable equation: $u' = \frac{du}{dx} = -\frac{1}{x} \cdot \frac{u(u^2-1)}{u^2+1}$. Let us integrate this equation: $\int \frac{(u^2+1)du}{u(u-1)(u+1)} = -\int \frac{dx}{x}$. By using partial fraction decomposition, we get: $\int \frac{du}{u-1} + \int \frac{du}{u+1} - \int \frac{du}{u} = -\int \frac{dx}{x}$ and it follows that $\ln \frac{u^2-1}{u} = -\ln x + \ln C = \ln \frac{C}{x}$. Hence, $\frac{u^2-1}{u} = \frac{C}{x}$ and, by replacing u with $\frac{y}{x}$, we obtain: $y^2 - x^2 = Cx$, a family of hyperbolas.

(d) The system is equivalent to the homogeneous equation $\frac{dy}{dx} = \frac{-2x+4y}{x+y}$. We denote by $u(x) = \frac{y(x)}{x}$, so $\frac{dy}{dx} = u + xu' = \frac{-2+4u}{1+u}$. Thus, we get a separable equation: $u' = \frac{du}{dx} = -\frac{1}{x} \cdot \frac{u^2-3u+2}{u+1}$. We integrate this equation: $\int \frac{(u+1)du}{(u-1)(u-2)} = -\int \frac{dx}{x}$ and obtain: $\int \frac{3du}{u-2} - \int \frac{2du}{u-1} = -\int \frac{dx}{x}$, so $\ln \frac{(u-2)^3}{(u-1)^2} = \ln \frac{C}{x}$. Hence, $\frac{(u-2)^3}{(u-1)^2} = \frac{C}{x}$ and, by replacing u with $\frac{y}{x}$, we obtain: $(y - 2x)^3 = C(y - x)^2$.

(e) From the first equality we get: $xdx = dy$, or, equivalently, $\frac{x^2}{2} - y = C_1$ and we have found a first integral of the system, $\varphi_1(x, y, z) = \frac{x^2}{2} - y$. Now, let us use this first integral in order to find another one: we have $y = \frac{x^2}{2} - C_1$ and from $\frac{dx}{z} = \frac{dz}{y}$ we obtain: $(\frac{x^2}{2} - C_1)dx = zdz$, or, equivalently, $\frac{x^3}{6} - C_1 x = \frac{z^2}{2} + C_2$. Since $C_1 = \frac{x^2}{2} - y$, we have $xy - \frac{x^3}{3} - \frac{z^2}{2} = C_2$, so we obtained another first integral, $\varphi_2(x, y, z) = xy - \frac{x^3}{3} - \frac{z^2}{2}$. Obviously, φ_1 and φ_2 are functionally independent (because φ_2 depends on z while φ_1 does not), so the general solution of the system is the family of trajectories:

$$\begin{cases} \frac{x^2}{2} - y = C_1 \\ xy - \frac{x^3}{3} - \frac{z^2}{2} = C_2 \end{cases}.$$

10.3 Chapter 3

(f) The system can be written:

$$\frac{dx}{y^2+z^2} = \frac{-\frac{dy}{y}}{-x^2-z^2} = \frac{\frac{dz}{z}}{x^2-y^2} = \frac{\frac{dx}{x} - \frac{dy}{y} + \frac{dz}{z}}{0}$$

and we obtain $d(\ln x - \ln y + \ln z) = 0$, or, equivalently, $\frac{xz}{y} = C_1$.

Now, we can write the initial system by using equivalent fractions: we multiply by $-x$ the numerator and the denominator of the first fraction, by y, the numerator and the denominator of the second and by z, the numerator and the denominator of the last fraction, so we obtain:

$$\frac{-xdx}{-x^2y^2-x^2z^2} = \frac{ydy}{y^2x^2+y^2z^2} = \frac{zdz}{z^2x^2-z^2y^2} = \frac{-xdx+ydy+zdz}{0}.$$

It follows that $d(-x^2+y^2+z^2) = 0$, so $y^2+z^2-x^2 = C_2$ and the general solution is the family of curves:

$$\begin{cases} xz = C_1 y \\ y^2 + z^2 - x^2 = C_2 \end{cases}.$$

Note that, if we denote the first integrals by $\varphi_1(x,y,z) = \frac{xz}{y}$ and $\varphi_2(x,y,z) = y^2+z^2-x^2$, the Jacobi matrix

$$\begin{pmatrix} \frac{\partial \varphi_1}{\partial x} & \frac{\partial \varphi_1}{\partial y} & \frac{\partial \varphi_1}{\partial z} \\ \frac{\partial \varphi_2}{\partial x} & \frac{\partial \varphi_2}{\partial y} & \frac{\partial \varphi_2}{\partial z} \end{pmatrix} = \begin{pmatrix} \frac{z}{y} & -\frac{xz}{y^2} & \frac{x}{y} \\ -2x & 2y & 2z \end{pmatrix}$$

has the rank 2 for $x,y,z \neq 0$, so φ_1 and φ_2 are functionally independent.

(g) From the last equality, $\frac{dy}{y} = \frac{dz}{z}$, we obtain $\ln y = \ln z + \ln C_1$, so $\frac{y}{z} = C_1$ and we found a first integral of the autonomous system. Now, we multiply by $2y$ the numerator and the denominator of the second fraction and by $2z$, the numerator and the denominator of the last one:

$$\frac{dx}{x+y^2+z^2} = \frac{2ydy}{2y^2} = \frac{2zdz}{2z^2} = \frac{x-2ydy-2zdz}{x-y^2-z^2} = \frac{d(x-y^2-z^2)}{x-y^2-z^2}.$$

If we denote by $u = x-y^2-z^2$, then we can write $\frac{dz}{z} = \frac{du}{u}$, which is equivalent to $\frac{u}{z} = C_2$, or $\frac{x-y^2-z^2}{z} = C_2$. Obviously, the first integrals we found, $\varphi_1(x,y,z) = \frac{y}{z}$ and $\varphi_2(x,y,z) = \frac{x-y^2-z^2}{z}$ are functionally independent, so the general solution of the system is:

$$\begin{cases} y = C_1 z \\ x - y^2 - z^2 = C_2 z \end{cases}.$$

(h) If we multiply by -1 the numerator and the denominator of the second fraction, the system is written:

$$\frac{dx}{xz - xy} = \frac{-dy}{-y^2 + xy} = \frac{dz}{y^2 - xz} = \frac{d(x - y + z)}{0}$$

and we obtain: $x - y + z = C_1$.

Now, let us multiply by z the numerator and the denominator of the second fraction and by $-y$, the numerator and the denominator of the last one:

$$\frac{dx}{x(z - y)} = \frac{zdy}{y^2z - xyz} = \frac{-ydz}{-y^3 + xyz} = \frac{zdy - ydz}{y^2(z - y)}.$$

We obtain: $\frac{dx}{x} = \frac{zdy - ydz}{y^2}$, or, equivalently, $d(\ln |x|) = -d\left(\frac{z}{y}\right)$. Hence, it follows that $\ln |x| + \frac{z}{y} = C_2$ and we found two independent first integrals because the Jacobi matrix of $\varphi_1(x, y, z) = x - y + z$ and $\varphi_2(x, y, z) = \ln |x| + \frac{z}{y}$,

$$\begin{pmatrix} \frac{\partial \varphi_1}{\partial x} & \frac{\partial \varphi_1}{\partial y} & \frac{\partial \varphi_1}{\partial z} \\ \frac{\partial \varphi_2}{\partial x} & \frac{\partial \varphi_2}{\partial y} & \frac{\partial \varphi_2}{\partial z} \end{pmatrix} = \begin{pmatrix} 1 & -1 & 1 \\ \frac{1}{x} & -\frac{z}{y^2} & \frac{1}{y} \end{pmatrix}$$

has the rank 2 for $x, y \neq 0$, $x \neq y$.

(i) From the first equality we have: $\frac{dx}{x} = \frac{dy}{y}$, so $\frac{y}{x} = C_1$ and we found a first integral, $\varphi_1(x, y, z) = \frac{y}{x}$. Let us use this first integral in order to find another one. Since $y = C_1 x$, one can write: $\frac{dx}{xz} = \frac{dz}{C_1 x^2 \sqrt{z^2+1}}$ and we obtain a separable equation: $C_1 x dx = \frac{z dz}{\sqrt{z^2+1}}$. We integrate it: $C_1 x^2 = 2\sqrt{z^2 + 1} + C_2$ and replace C_1 with $\frac{y}{x}$, so we get: $xy - 2\sqrt{z^2 + 1} = C_2$.

(j) We can write:

$$\frac{dx}{y + z} = \frac{dy}{x + z} = \frac{dz}{x + y} = \frac{dx - dy}{y - x} = \frac{dz - dy}{y - z}$$

and from the last equality we have: $\frac{d(x-y)}{x-y} = \frac{d(z-y)}{z-y}$, therefore $\frac{x-y}{z-y} = C_1$.

In order to find another first integral, we write:

$$\frac{dx}{y + z} = \frac{dy}{x + z} = \frac{dz}{x + y} = \frac{dx + dy + dz}{2(x + y + z)} = \frac{dx - dy}{y - x}.$$

Hence, $\frac{d(x+y+z)}{x+y+z} = -2\frac{d(x-y)}{x-y}$ and we obtain $\ln(x + y + z) = -2\ln(x - y) + \ln C_2$, so $(x + y + z)(x - y)^2 = C_2$.

(k) For this system we have to find three independent first integrals. Since $\frac{dx}{z} = \frac{dz}{x}$, we can write $xdx = zdz$ and we obtain $x^2 - z^2 = C_1$. Similarly, we find $y^2 - u^2 = C_2$ and, from the equalities:

$$\frac{dx}{z} = \frac{dy}{u} = \frac{dz}{x} = \frac{du}{y} = \frac{d(x-z)}{z-x} = \frac{d(y-u)}{u-y}$$

we obtain the third first integral: $\frac{x-z}{y-u} = C_3$. We leave the reader to study the functional independence of these first integrals.

6. (a) The system associated to this linear partial differential equation is: $\frac{dx}{x+2y} = \frac{dy}{-y}$ and we can write:

$$\frac{dx}{x+2y} = \frac{dy}{-y} = \frac{dx+dy}{x+y}.$$

We have: $\frac{d(x+y)}{x+y} = -\frac{dy}{y} \Leftrightarrow \ln(x+y) = -\ln y + \ln C = \ln \frac{C}{y}$, so $y(x+y) = C$ and we obtained a first integral of the system, $\varphi(x,y) = yx + y^2$. Hence, the general solution of the partial differential equation is $z(x,y) = F(yx + y^2)$, where F is a (one variable) arbitrary function of class C^1.

(b) The system associated is: $\frac{dx}{x} = \frac{dy}{y} = \frac{dz}{x+y}$ and we can write:

$$\frac{dx}{x} = \frac{dy}{y} = \frac{dz}{x+y} = \frac{dx+dy-dz}{0}.$$

It follows that $d(x+y-z) = 0$, so $x+y-z = C_1$ and we found a first integral: $\varphi_1(x,y,z) = x+y-z$. In order to find another first integral, we integrate the equation $\frac{dx}{x} = \frac{dy}{y}$ and obtain $\frac{y}{x} = C_2$, so another first integral is $\varphi_2(x,y,z) = \frac{y}{x}$. Obviously, φ_1 and φ_2 are functionally independent, so the general solution of the partial differential equation is $u(x,y,z) = F(x+y-z, \frac{y}{x})$, where F is a (two variables) arbitrary function of class C^1.

(c) The system associated is: $\frac{dx}{yz} = \frac{dy}{xz} = \frac{dz}{xy}$. From the first equality we obtain: $xdx = ydy$, so $x^2 - y^2 = C_1$. Similarly, from the last equality we have: $ydy = zdz$, so $z^2 - y^2 = C_2$. We have found two independent first integrals, $\varphi_1(x,y,z) = x^2 - y^2$ and $\varphi_2(x,y,z) = z^2 - y^2$, so the general solution of the partial differential equation is $u(x,y,z) = F(x^2 - y^2, z^2 - y^2)$, where F is a (two variables) arbitrary function of class C^1.

(d) The system associated is: $\frac{dx}{y^2+z^2-x^2} = \frac{dy}{-2xy} = \frac{dz}{-2xz}$. From the last equality we obtain: $\frac{dy}{y} = \frac{dz}{z}$, so $\frac{y}{z} = C_1$. In order to find another first integral, we can write the system in the equivalent form:

$$\frac{xdx}{xy^2+xz^2-x^3} = \frac{ydy}{-2xy^2} = \frac{zdz}{-2xz^2} = \frac{xdx+ydy+zdz}{-x(x^2+y^2+z^2)}.$$

From the last equality we have: $\frac{dz}{z} = \frac{d(x^2+y^2+z^2)}{x^2+y^2+z^2}$, and we obtain: $\frac{x^2+y^2+z^2}{z} = C_2$. The general solution of the partial differential equation is $u(x, y, z) = F(\frac{y}{z}, \frac{x^2+y^2+z^2}{z})$, where F is a (two variables) arbitrary function of class C^1.

7. (a) Let us find a first integral of the associated system $\frac{dx}{y} = \frac{dy}{-x}$. This is equivalent to $xdx = -ydy$, so we have: $x^2 + y^2 = C$ and the general solution of the partial differential equation is $z(x, y) = F(x^2 + y^2)$, where F is an arbitrary function of class C^1. Now we have to find the function F such that $z(x, 1) = F(x^2 + 1) = x^2$. Hence, $F(u) = u - 1$ and the solution of the Cauchy problem is $z(x, y) = x^2 + y^2 - 1$.

(b) The associated system is $\frac{dx}{1+x^2} = \frac{dy}{xy}$, which can be written $\frac{2xdx}{1+x^2} = \frac{2dy}{y}$. By integrating this separable equation we obtain $\ln(1 + x^2) = 2\ln y + \ln C$, or $1 + x^2 = Cy^2$, so the general solution of the partial differential equation is $z(x, y) = F(\frac{1+x^2}{y^2})$, where F is an arbitrary function of class C^1. Now we have to find the function F such that $z(0, y) = F(\frac{1}{y^2}) = y^2$. Hence, $F(u) = \frac{1}{u}$ and the solution of the Cauchy problem is $z(x, y) = \frac{y^2}{1+x^2}$.

(c) The associated system is $\frac{dx}{1} = \frac{dy}{2\exp x - y}$, which can be written $\frac{dy}{dx} = -y + 2\exp x$. But this is a first-order linear differential equation, and its general solution is $y = C\exp(-x) + \exp x$, so we obtain $(y - \exp x)\exp x = C$ and the general solution of the partial differential equation is $z(x, y) = F(y\exp x - \exp 2x)$ where F is an arbitrary function of class C^1. Now we have to find F such that $z(0, y) = F(y - 1) = y$. Hence, $F(u) = u + 1$ and the solution of the Cauchy problem is $z(x, y) = y\exp x - \exp 2x + 1$.

(d) First, let us find two independent first integrals of the associated c system $\frac{dx}{xz} = \frac{dy}{yz} = \frac{dz}{xy}$. From the first equality we have $\frac{dx}{x} = \frac{dy}{y}$, so $\frac{y}{x} = C_1$. Now, we write the system in the following equivalent form:

$$\frac{ydx}{xyz} = \frac{xdy}{xyz} = \frac{-2zdz}{-2xyz} = \frac{ydx + xdy - 2zdz}{0}.$$

It follows that $d(xy - z^2) = 0$, so $xy - z^2 = C_2$ and the general solution of the partial differential equation is $u(x, y, z) = F(\frac{y}{x}, xy - z^2)$ where F is an arbitrary function of class C^1. We have to find F such that $u(x, y, 2) = F(\frac{y}{x}, xy - 4) = xy$. Hence, $F(v, w) = w + 4$ and the solution of the Cauchy problem is $u(x, y, z) = xy - z^2 + 4$.

(e) The partial differential equation $\frac{\partial u}{\partial x} + \frac{\partial u}{\partial y} + 2\frac{\partial u}{\partial z} = 0$ has the associated system $dx = dy = \frac{dz}{2}$. We solve this system and obtain $y - x = C_1$ and $z - 2x = C_2$, so the general solution of the partial differential equation is $u(x, y, z) = F(y - x, z - 2x)$, where F is an arbitrary function of class C^1. From the condition $u(1, y, z) = F(y - 1, z - 2) = yz$ we obtain the expression of F, namely, $F(v, w) = (v + 1)(w + 2)$, so the solution of the Cauchy problem is $u(x, y, z) = (y - x + 1)(z - 2x + 2)$.

(f) The partial differential equation $x\frac{\partial u}{\partial x} + y\frac{\partial u}{\partial y} + xy\frac{\partial u}{\partial z} = 0$ has the associated system $\frac{dx}{x} = \frac{dy}{y} = \frac{dz}{xy}$. From the first equality we obtain $\frac{y}{x} = C_1$. In order to find another first integral, we write the system in the following equivalent form:

$$\frac{ydx}{xy} = \frac{xdy}{xy} = \frac{-2dz}{-2xy} = \frac{ydx + xdy - 2dz}{0}.$$

Hence, $d(xy - 2z) = 0$, so $xy - 2z = C_2$ and the general solution of the partial differential equation is $u(x, y, z) = F(\frac{y}{x}, xy - 2z)$, where F is an arbitrary function of class C^1. From the condition $u(x, y, 0) = F(\frac{y}{x}, xy) = x^2 + y^2$ we obtain the expression of F, $F(v, w) = \frac{w}{v} + vw$, so the solution of the Cauchy problem is $u(x, y, z) = \frac{x(xy-2z)}{y} + \frac{y(xy-2z)}{x}$.

8. We have to find the function $u(x, y, z)$ which is the solution of the Cauchy problem:

$$x\frac{\partial u}{\partial x} + y\frac{\partial u}{\partial y} + z\frac{\partial u}{\partial z} = 0; \quad u(x, 1, z) = x + z.$$

We solve the associated system, $\frac{dx}{x} = \frac{dy}{y} = \frac{dz}{z}$, and find $\frac{x}{y} = C_1$ and $\frac{z}{y} = C_2$, so the general solution of the partial differential equation is $u(x, y, z) = F(\frac{x}{y}, \frac{z}{y})$, where F is an arbitrary function of class C^1. We determine the function F such that $u(x, 1, z) = F(x, z) = x + z$. Hence, the solution of the Cauchy problem is $u(x, y, z) = \frac{x+z}{y}$ and the required surface is the plane $x + z = 0$.

9. We have to find a surface of equation $u(x, y, z) = 0$, orthogonal to any surface of equation $f(x, y, z) = \frac{x^2+y^2+z^2}{x} - C = 0$. This means that $\nabla u \cdot \nabla f = 0$ and we have to solve the following partial differential equation:

$$\left(1 - \frac{y^2}{x^2} - \frac{z^2}{x^2}\right)\frac{\partial u}{\partial x} + \frac{2y}{x} \cdot \frac{\partial u}{\partial y} + \frac{2z}{x} \cdot \frac{\partial u}{\partial z} = 0.$$

The associated system is

$$\frac{dx}{\frac{x^2-y^2-z^2}{x^2}} = \frac{dy}{\frac{2y}{x}} = \frac{dz}{\frac{2z}{x}}.$$

From the last equality we obtain $\frac{dy}{y} = \frac{dz}{z}$, so $\frac{y}{z} = C_1$. The system above can be written:

$$\frac{xdx}{x^2 - y^2 - z^2} = \frac{ydy}{2y^2} = \frac{zdz}{2z^2} = \frac{xdx + ydy + zdz}{x^2 + y^2 + z^2}.$$

and from the last equality we obtain $\frac{dz}{z} = \frac{d(x^2+y^2+z^2)}{x^2+y^2+z^2}$, so $\frac{x^2+y^2+z^2}{z} = C_2$ and the general solution is $u(x, y, z) = F(\frac{y}{z}, \frac{x^2+y^2+z^2}{z})$.

Now, we have to solve a Cauchy problem: to find the function F such that the surface of equation $F(\frac{y}{z}, \frac{x^2+y^2+z^2}{z}) = 0$ passes through the line $y = x$, $z = 1$. We denote $v = \frac{y}{z}$, $w = \frac{x^2+y^2+z^2}{z}$ and try to eliminate x, y, z between the following relations: $y = x$, $z = 1$, $v = \frac{y}{z}$, $w = \frac{x^2+y^2+z^2}{z}$. We obtain $v = y$ and $w = 2y^2 + 1$, so $w - 2v^2 - 1 = 0$ and the equation of the surface is $\frac{x^2+y^2+z^2}{z} - \frac{2y^2}{z^2} - 1 = 0$, or $z(x^2 + y^2 + z^2) - 2y^2 - z^2 = 0$.

10. (a) We search for a solution $z = z(x, y)$ in an implicit form, $\Phi(x, y, z) = 0$. The quasilinear PDE becomes a linear PDE in the function Φ (see (3.185) and (3.186)): $y\frac{\partial\Phi}{\partial x} + x\frac{\partial\Phi}{\partial y} + (x - 2y)\frac{\partial\Phi}{\partial z} = 0$. The associated system is $\frac{dx}{y} = \frac{dy}{x} = \frac{dz}{x-2y}$. From the first equality we can find a first integral $xdx = ydy \Rightarrow x^2 - y^2 = C_1$. In order to find another first integral, we write the system in the form: $\frac{2dx}{2y} = \frac{-dy}{-x} = \frac{dz}{x-2y} = \frac{2dx-dy+dz}{0}$. Hence, $2x - y + z = C_2$ and we found two independent first integrals: $\varphi_1(x, y, z) = x^2 - y^2$ and $\varphi_2(x, y, z) = 2x - y + z$. The general solution of the linear PDE is $\Phi(x, y, z) = F(x^2 - y^2, 2x - y + z)$, where F is an arbitrary function of class C^1 and so the general solution of the initial quasilinear PDE is $F(x^2 - y^2, 2x - y + z) = 0$.

(b) We use the same method: we search for two independent first integrals for the autonomous system $\frac{dx}{xy} = \frac{dy}{-y^2} = \frac{dz}{x^2+1}$. The general solution of the quasilinear equation will be expressed in the implicit form: $F(\varphi_1(x, y, z), \varphi_2(x, y, z)) = 0$. The first equation of the system can be written $\frac{dx}{x} = \frac{dy}{-y}$, so we obtain $xy = C_1$ and a first integral of the system is $\varphi_1(x, y, z) = xy$. We can use it in order two find another first integral. Thus, from the system above, we can write $\frac{dx}{C_1} = \frac{dz}{x^2+1}$, or $(x^2+1)dx = C_1 dz$. We obtain $\frac{x^3}{3} + x = C_1 z + C_2$ and, since $C_1 = xy$, we have $\frac{x^3}{3} + x - xyz = C_2$, so $\varphi_2(x, y, z) = x^3 + 3x - 3xyz$ is another first integral of the autonomous system. The general solution of the initial PDE is $F(xy, x^3+3x-3xyz) = 0$.

(c) The associated system is $\frac{dx}{\sin^2 x} = \frac{dy}{\tan z} = \frac{dz}{\cos^2 z}$. From the last equality we have $\int dy = \int \frac{\tan z}{\cos^2 z} dz = \int \frac{\sin z}{\cos^3 z} dz$, so we obtain $y - \frac{1}{2\cos^2 z} = C_1$. By integrating the equation $\frac{dx}{\sin^2 x} = \frac{dz}{\cos^2 z}$ we get $\tan z + \frac{1}{\tan x} = C_2$. The general solution of the quasilinear equation is $F(y - \frac{1}{2\cos^2 z}, \tan z + \frac{1}{\tan x}) = 0$.

(d) The associated system is $\frac{dx}{x} = \frac{dy}{z+u} = \frac{dz}{y+u} = \frac{du}{y+z}$. We have to find three independent first integrals for this system. We can write $\frac{dx}{x} = \frac{dy}{z+u} = \frac{dz}{y+u} = \frac{dy-dz}{z+u-y-u}$ and we obtain $\frac{dx}{x} = -\frac{d(y-z)}{y-z}$, so $x(y - z) = C_1$. Similarly, from

10.3 Chapter 3

$\frac{dx}{x} = \frac{dz}{y+u} = \frac{du}{y+z} = \frac{dz-du}{u-z}$ we find $x(u-z) = C_2$. Finally, if we write

$$\frac{dx}{x} = \frac{dy}{z+u} = \frac{dz}{y+u} = \frac{du}{y+z} = \frac{dy+dz+du}{2(y+z+u)},$$

then we can see that $\frac{y+z+u}{x^2} = C_3$. Let us prove the functional independence of the first integrals $\varphi_1(x, y, z, u) = xy - xz$, $\varphi_2(x, y, z, u) = xu - xz$ and $\varphi_3(x, y, z, u) = \frac{y+z+u}{x^2}$. The Jacobi matrix is

$$\begin{pmatrix} \frac{\partial \varphi_1}{\partial x} & \frac{\partial \varphi_1}{\partial y} & \frac{\partial \varphi_1}{\partial z} & \frac{\partial \varphi_1}{\partial u} \\ \frac{\partial \varphi_2}{\partial x} & \frac{\partial \varphi_2}{\partial y} & \frac{\partial \varphi_2}{\partial z} & \frac{\partial \varphi_2}{\partial u} \\ \frac{\partial \varphi_3}{\partial x} & \frac{\partial \varphi_3}{\partial y} & \frac{\partial \varphi_3}{\partial z} & \frac{\partial \varphi_3}{\partial u} \end{pmatrix} = \begin{pmatrix} y-x & x & -x & 0 \\ u & 0 & -x & x \\ -\frac{2(y+z+u)}{x^3} & \frac{1}{x^2} & \frac{1}{x^2} & \frac{1}{x^2} \end{pmatrix}$$

and, since the determinant

$$\begin{vmatrix} x & -x & 0 \\ 0 & -x & x \\ \frac{1}{x^2} & \frac{1}{x^2} & \frac{1}{x^2} \end{vmatrix} = \begin{vmatrix} 1 & -1 & 0 \\ 0 & -1 & 1 \\ 1 & 1 & 1 \end{vmatrix} = -3 \neq 0$$

if $x \neq 0$, the rank of the Jacobi matrix is 3, so the first integrals are functionally independent. The general solution is $F(xy - xz, xu - xz, \frac{y+z+u}{x^2}) = 0$.

11. (a) We have to solve the quasilinear equation $z\frac{\partial z}{\partial x} - xy\frac{\partial z}{\partial y} = 2xz$. The associated system is $\frac{dx}{z} = \frac{dy}{-xy} = \frac{dz}{2xz}$. From $\frac{dx}{z} = \frac{dz}{2xz}$ we obtain $2xdx = dz$, so $x^2 - z = C_1$ and from $\frac{dy}{-xy} = \frac{dz}{2xz}$ we get $\frac{2dy}{y} = -\frac{dz}{z}$, hence $y^2z = C_2$. The general solution of the equation is $F(x^2 - z, y^2z) = 0$, where $F(w_1, w_2)$ is an arbitrary function of class C^1. Now, we have to solve the Cauchy problem, i.e. to find F such that the conditions $x + y = 2$ and $yz = 1$ are satisfied. We denote $x^2 - z = w_1$, $y^2z = w_2$ and try to eliminate x, y, z between the relations: $x + y = 2$, $yz = 1$, $x^2 - z = w_1$, $y^2z = w_2$. From the second and the last we can write $y = w_2$, so $x = 2 - w_2$, $z = \frac{1}{w_2}$ and, by replacing in the first relation, we get: $(2 - w_2)^2 - \frac{1}{w_2} - w_1 = 0$. Thus, the solution of the Cauchy problem is $y^2z[(y^2z - 2)^2 - x^2 + z] - 1 = 0$.

(b) We have to solve the quasilinear equation $x\frac{\partial z}{\partial x} - y\frac{\partial z}{\partial y} = z^2(x - 3y)$. The associated system is $\frac{dx}{x} = \frac{dy}{-y} = \frac{dz}{z^2(x-3y)}$. We can easily obtain a first integral by integrating the first equation: $xy = C_1$. The system can be written:

$$\frac{dx}{x} = \frac{3dy}{-3y} = \frac{dx + 3dy}{x - 3y} = \frac{dz}{z^2(x - 3y)},$$

so we get $d(x + 3y) = \frac{dz}{z^2}$ and we have $x + 3y + \frac{1}{z} = C_2$. The general solution of the equation is $F(xy, x + 3y + \frac{1}{z}) = 0$, where $F(w_1, w_2)$ is an arbitrary function of class C^1. Let us find F such that the conditions $x = 1$ and $yz = -1$ are satisfied. We denote $xy = w_1$, $x + 3y + \frac{1}{z} = w_2$ and try to eliminate x, y, z between the relations: $x = 1$, $yz = -1$, $xy = w_1$, $x + 3y + \frac{1}{z} = w_2$. We have: $y = w_1$, $z = -\frac{1}{w_1}$ and, by replacing in the last one, we get: $1 + 2w_1 = w_2$. The solution of the Cauchy problem is $1 + xy = x + 3y + \frac{1}{z}$, or $z(x, y) = \frac{1}{1+xy-x-3y}$ (in this case it is possible to write the solution in an explicit form).

(c) We have to solve the quasilinear equation $x\frac{\partial z}{\partial x} + y\frac{\partial z}{\partial y} = z - xy$. The associated system is $\frac{dx}{x} = \frac{dy}{y} = \frac{dz}{z-xy}$. We can easily obtain a first integral by integrating the first equation: $\frac{y}{x} = C_1$. The system can be written:

$$\frac{ydx}{xy} = \frac{xdy}{xy} = \frac{dz}{z-xy} = \frac{ydx + xdy + dz}{xy + xy + z - xy} = \frac{d(xy+z)}{xy+z}$$

so we get $\frac{dx}{x} = \frac{d(xy+z)}{xy+z}$ and we have $\frac{xy+z}{x} = C_2$. The general solution of the equation is $F(\frac{y}{x}, y + \frac{z}{x}) = 0$, where $F(w_1, w_2)$ is an arbitrary function of class C^1. Let us find F such that $x = 2$ and $z = y^2 + 1$. We denote $\frac{y}{x} = w_1$, $y + \frac{z}{x} = w_2$ and try to eliminate x, y, z between the relations: $x = 2$, $z = y^2 + 1$, $\frac{y}{x} = w_1$, $y + \frac{z}{x} = w_2$. We have: $y = 2w_1$, $z = 4w_1^2 + 1$ and, by replacing in the last one, we get: $2w_1 + \frac{4w_1^2+1}{2} = w_2$, or, equivalently, $(2w_1 + 1)^2 = 2w_2$. The solution of the Cauchy problem is $(\frac{2y}{x} + 1)^2 = 2(y + \frac{z}{x})$ and we can also write the solution in an explicit form: $z(x, y) = \frac{2y^2}{x} + 2y + \frac{x}{2} - xy$.

12. First, let us find the general solution of the quasilinear equation $2yz\frac{\partial z}{\partial x} + xz\frac{\partial z}{\partial y} = -xy$. The associated system is $\frac{dx}{2yz} = \frac{dy}{xz} = \frac{dz}{-xy}$. From the first equation we have $xdx = 2ydy \Rightarrow x^2 - 2y^2 = C_1$, and from the last equation we get $ydy = -zdz \Rightarrow y^2 + z^2 = C_2$. The general solution of the equation is $F(x^2 - 2y^2, y^2 + z^2) = 0$, where $F(w_1, w_2)$ is an arbitrary function of class C^1. Let us find F such that $z = 0$ and $x^2 + y^2 = y$. We denote $x^2 - 2y^2 = w_1$, $y^2 + z^2 = w_2$ and try to eliminate x, y, z between the relations: $z = 0$, $x^2 + y^2 = y$, $x^2 - 2y^2 = w_1$, $y^2 + z^2 = w_2$. From the last one we have: $y^2 = w_2$, then $x^2 = w_1 + 2w_2$ and, by replacing in the second relation, we get: $w_1 + 3w_2 = \pm\sqrt{w_2}$, or, equivalently, $(w_1 + 3w_2)^2 = w_2$. The solution of the Cauchy problem is $(x^2 + y^2 + 3z^2)^2 = y^2 + z^2$.

10.4 Chapter 4

1.(a) We use formulas (4.109)-(4.111) for $T = 2\pi$, $\omega = 1$ and find:

$$a_k = \frac{4}{\pi} \int_0^\pi x \cos kx \, dx = \frac{4}{\pi k^2}\left[(-1)^k - 1\right], \quad \text{for } k \neq 0$$

$$a_0 = \frac{4}{\pi} \int_0^\pi x \, dx = 2\pi,$$

$$b_k = \frac{4}{\pi} \int_0^\pi x \sin kx \, dx = \frac{4}{k}(-1)^{k+1}.$$

Hence

$$S_{\widetilde{f}}(x) = \pi - \frac{8}{\pi}\sum_{n=1}^\infty \frac{1}{(2n-1)^2} \cos(2n-1)x + 4\sum_{k=1}^\infty \frac{(-1)^{k+1}}{k}\sin kx.$$

(b) We set $x = 0$ in the equality above:

$$0 = f(0) = S_{\widetilde{f}}(0) = \pi - \frac{8}{\pi}\sum_{n=1}^\infty \frac{1}{(2n-1)^2}.$$

So $\sum_{n=1}^\infty \frac{1}{(2n-1)^2} = \frac{\pi^2}{8}$. In order to find $\sum_{n=1}^\infty \frac{1}{(2n-1)^4}$ we use the Parseval's identity (4.58):

$$2\pi^2 + \frac{64}{\pi^2}\sum_{n=1}^\infty \frac{1}{(2n-1)^4} + 16\sum_{k=1}^\infty \frac{1}{k^2} = \frac{16\pi^2}{3}$$

Since $\sum_{k=1}^\infty \frac{1}{k^2} = \frac{\pi^2}{6}$ (see (4.107)) we finally obtain:

$$\sum_{n=1}^\infty \frac{1}{(2n-1)^4} = \frac{\pi^4}{96}. \tag{10.8}$$

(c) The set of points at which $\widetilde{f}$ is not continuous is $\{x_m = (2m-1)\pi, \ m \in \mathbb{Z}\}$. At these points, the function $\widetilde{f}$ has side limits:

$$\widetilde{f}((2m-1)\pi - 0) = 4\pi, \quad \widetilde{f}((2m-1)\pi + 0) = 0.$$

From Dirichlet Theorem, we have:

$$S_{\tilde{f}}(x) = \begin{cases} \tilde{f}(x), & \text{if } x \neq (2m-1)\pi,\ m \in \mathbb{Z} \\ 2\pi, & \text{if } x = (2m-1)\pi,\ m \in \mathbb{Z} \end{cases},$$

so the function defined by $S_{\tilde{f}}(x)$ is not continuous on $\mathbb{R}$. Hence, the series of functions $S_{\tilde{f}}(x)$ cannot be uniformly convergent on $\mathbb{R}$.

2. For finding the sin Fourier expansion we have to extend the function f by oddness to the interval $(-\pi, \pi)$, $\bar{f}(x) = 4x$, $x \in (-\pi, \pi)$. Let $\tilde{f}$ be the extension by periodicity of this odd function to the whole $\mathbb{R}$. By applying the formula (4.87) we obtain:

$$b_k = \frac{2}{\pi} \int_0^\pi 4x \sin kx\, dx = \frac{8}{k}(-1)^{k+1},\ k = 1, 2, \ldots,$$

$$a_k = 0,\ k = 0, 1, \ldots.$$

So

$$f(x) = S_f(x) = 8\sum_{k=1}^{\infty} \frac{(-1)^{k+1}}{k} \sin kx,\ \forall x \in (0, \pi). \tag{10.9}$$

For $x = \frac{\pi}{2}$ we get

$$\sum_{n=1}^{\infty} \frac{(-1)^{n+1}}{2n-1} = \frac{\pi}{4}.$$

If we write the Parseval's identity for the Fourier series (10.9), we obtain again the formula $\sum_{k=1}^{\infty} \frac{1}{k^2} = \frac{\pi^2}{6}$.

To find the cosine Fourier expansion of f, we consider its extension by evenness, to $(-\pi, \pi)$, $\bar{f}(x) = |4x|$, $x \in (-\pi, \pi)$ and then extend this even function to the unique periodic continuous function $\tilde{f}$ ($T = 2\pi$). By using the formula (4.85) we obtain:

$$a_k = \frac{2}{\pi} \int_0^\pi 4x \cos kx\, dx = \frac{8}{k^2\pi}\left[(-1)^k - 1\right],\ k = 1, 2, \ldots,$$

$$a_0 = 4\pi,\ b_k = 0,\ k = 1, 2, \ldots.$$

Hence,

$$f(x) = S_f(x) = 2\pi - \frac{16}{\pi} \sum_{n=1}^{\infty} \frac{1}{(2n-1)^2} \cos(2n-1)x. \tag{10.10}$$

By setting $x = 0$, we get again that $\sum_{n=1}^{\infty} \frac{1}{(2n-1)^2} = \frac{\pi^2}{8}$. If we apply the Parseval's formula for the Fourier series (10.10), we obtain again (10.8).

3. (a) Since $f(-x) = f(x)$ for any $x \in [-\pi, \pi]$, the function is even. So $b_k = 0$ for every $k = 1, 2, \ldots$ and, by applying the formula (4.85), we obtain:

$$a_k = \frac{2}{\pi} \cdot \frac{1}{12} \int_0^{\pi} \left(\pi^2 - 3x^2\right) \cos kx \, dx = \frac{(-1)^{k+1}}{k^2}, \quad k = 1, 2, \ldots$$

$$a_0 = \frac{2}{\pi} \cdot \frac{1}{12} \int_0^{\pi} \left(\pi^2 - 3x^2\right) dx = 0.$$

Hence

$$f(x) = \sum_{k=1}^{\infty} \frac{(-1)^{k+1}}{k^2} \cos kx. \tag{10.11}$$

(b) For $x = 0$ we get:

$$\sum_{k=1}^{\infty} \frac{(-1)^{k+1}}{k^2} = \frac{\pi^2}{12}.$$

By applying the Parseval's identity we obtain:

$$\sum_{k=1}^{\infty} \frac{1}{k^4} = \frac{\pi^4}{90}.$$

(c) Since $S_n(x) = \sum_{k=1}^{n} \frac{(-1)^{k+1}}{k^2} \cos kx$, for any $x \in [-\pi, \pi]$ we can write:

$$|f(x) - S_n(x)| \le \sum_{k=n+1}^{\infty} \frac{1}{k^2} < \sum_{k=n+1}^{\infty} \frac{1}{k(k-1)} = \sum_{k=n+1}^{\infty} \left(\frac{1}{k-1} - \frac{1}{k}\right) = \frac{1}{n},$$

so it is sufficient to take $n = 100$. Therefore, if $f(x)$ is approximated by

$$S_{100}(x) = \sum_{k=1}^{100} \frac{(-1)^{k+1}}{k^2} \cos kx$$

the error of approximation is less than 0.01, for any x in $[-\pi, \pi]$.

4. (a) Since $f(-2) = 5$ and $\lim_{x \nearrow 2} f(x) = 1$, the function $\widetilde{f}$ is not continuous (but it has side limits) at the points of the form $x_m = 4m + 2$, $m \in \mathbb{Z}$. Dirichlet

theorem states that $\tilde{f}(x) = S_{\tilde{f}}(x)$ for any $x \in \mathbb{R}$, except the points $x_m = 4m + 2$, $m \in \mathbb{Z}$. At these points we have: $S_{\tilde{f}}(x_m) = \frac{5+1}{2} = 3$ and $\tilde{f}(x_m) = 5$.

(b) The coefficients of the Fourier series are:

$$a_0 = \frac{2}{4} \int_{-2}^{2} (3 - x) \, dx = 6,$$

$$a_k = \frac{2}{4} \int_{-2}^{2} (3 - x) \cos\left(\frac{k\pi}{2}x\right) dx = 0, \quad k = 1, 2, \ldots.$$

$$b_k = \frac{2}{4} \int_{-2}^{2} (3 - x) \sin\left(\frac{k\pi}{2}x\right) dx = -\frac{1}{2} \int_{-2}^{2} (3 - x) \left[\frac{\cos\left(\frac{k\pi}{2}x\right)}{\frac{k\pi}{2}}\right]' dx =$$

$$= -\frac{1}{2} \left[(3 - x) \frac{2}{k\pi} \cos\left(\frac{k\pi}{2}x\right) \bigg|_{-2}^{2} + \frac{2}{k\pi} \int_{-2}^{2} \cos\left(\frac{k\pi}{2}x\right) dx \right] = \frac{4}{k\pi}(-1)^k.$$

Hence, for any $x \in (-2, 2)$ we get:

$$f(x) = 3 + \frac{4}{\pi} \sum_{k=1}^{\infty} \frac{(-1)^k}{k} \sin\left(\frac{k\pi}{2}x\right). \qquad (10.12)$$

(c) By applying Parseval's identity we have:

$$18 + \frac{16}{\pi^2} \sum_{k=1}^{\infty} \frac{1}{k^2} = \frac{1}{2} \int_{-2}^{2} (3 - x)^2 \, dx = \frac{62}{3} \Rightarrow \sum_{k=1}^{\infty} \frac{1}{k^2} = \frac{\pi^2}{6}.$$

(d) Let us find the complex form of the Fourier series (10.12). For this we use the formulas (4.114) and (4.119):

$$c_k = \frac{1}{4} \int_{-2}^{2} (3 - x) \exp\left(-i\frac{k\pi}{2}x\right) dx = -2i(-1)^k, \quad \text{for } k \neq 0,$$

$$c_0 = 3.$$

$$f(x) = 3 - 2i \sum_{k \in \mathbb{Z}^*} (-1)^k \exp\left(i\frac{k\pi}{2}x\right), \quad x \in (-2, 2).$$

5. (a) It is easy to see that $\widetilde{f}(x) = |x| + 2$ for any $x \in [-1, 1]$. In general,
$$\widetilde{f}(x) = |x - 2m| + 2, \text{ for } x \in [2m - 1, 2m + 1], \; m \in \mathbb{Z}.$$

(b) The set of points x at which $\widetilde{f}(x) \neq S\widetilde{f}(x)$ is the empty set because $\widetilde{f}(2m - 1) = \widetilde{f}(2m + 1) = 3$, i.e. $\widetilde{f}$ is continuous on $\mathbb{R}$. Since $\widetilde{f}$ is even, we have $b_k = 0, k = 1, 2, \ldots$ and

$$a_0 = \frac{4}{2} \int_0^1 (x + 2) \, dx = 5,$$

$$a_k = 2 \int_0^1 (x + 2) \cos(k\pi x) \, dx = \frac{2}{k^2 \pi^2} \left[(-1)^k - 1 \right], \; k = 1, 2, \ldots.$$

So

$$\widetilde{f}(x) = \frac{5}{2} - \frac{4}{\pi^2} \sum_{n=0}^{\infty} \frac{1}{(2n+1)^2} \cos(2n+1)\pi x.$$

(c) For $x = 0$ we get:

$$2 = \frac{5}{2} - \frac{4}{\pi^2} \sum_{n=0}^{\infty} \frac{1}{(2n+1)^2} \Rightarrow \sum_{n=0}^{\infty} \frac{1}{(2n+1)^2} = \frac{\pi^2}{8}.$$

Applying now the Parseval's identity, we get:

$$\frac{25}{2} + \frac{16}{\pi^4} \sum_{n=0}^{\infty} \frac{1}{(2n+1)^4} = \frac{4}{2} \int_0^1 (x+2)^2 \, dx = \frac{38}{3} \Rightarrow \sum_{n=0}^{\infty} \frac{1}{(2n+1)^4} = \frac{\pi^4}{96}.$$

6. Let us apply the formulas (4.82), (4.83) for $\omega = \frac{2\pi}{3}$. Since f is a periodic function of period $T = 3$, we have $f(x) = x(x - 3), \forall x \in [0, 3]$, so

$$a_0 = \frac{2}{3} \int_3^6 (x - 3)(x - 6) \, dx = \frac{2}{3} \int_0^3 x(x - 3) \, dx = -3,$$

$$a_k = \frac{2}{3} \int_0^3 x(x - 3) \cos\left(\frac{2k\pi}{3} x\right) dx = \frac{9}{k^2 \pi^2}, \; k = 1, 2, \ldots,$$

$$b_k = \frac{2}{3} \int_0^3 x(x - 3) \sin\left(\frac{2k\pi}{3} x\right) dx = 0, \; k = 1, 2, \ldots.$$

(we could observe from the beginning that f is an even function, consequently $b_k = 0$). Hence

$$f(x) = -\frac{3}{2} + \frac{9}{\pi^2} \sum_{k=1}^{\infty} \frac{1}{k^2} \cos\left(\frac{2k\pi}{3}x\right).$$

For $x = 0$, we get

$$0 = -\frac{3}{2} + \frac{9}{\pi^2} \sum_{k=1}^{\infty} \frac{1}{k^2} \Rightarrow \sum_{k=1}^{\infty} \frac{1}{k^2} = \frac{\pi^2}{6}.$$

Let us write now the Parseval's identity:

$$\frac{9}{2} + \frac{81}{\pi^4} \sum_{k=1}^{\infty} \frac{1}{k^4} = \frac{2}{3} \int_0^3 x^2(x-3)^2 \, dx = \frac{27}{5} \Rightarrow \sum_{k=1}^{\infty} \frac{1}{k^4} = \frac{\pi^4}{90}.$$

7.(a) We extend f by evenness to the entire interval $[-\pi, \pi]$. So the period will be $T = 2\pi$ and $\omega = 1$. Since $b_k = 0$, $k = 1, 2, \ldots$, it remains to compute a_k.

$$a_k = \frac{2}{\pi} \int_0^{\pi} \sin\frac{x}{2} \cos kx \, dx = \frac{1}{\pi} \int_0^{\pi} \left[\sin\left(\frac{x}{2} + kx\right) + \sin\left(\frac{x}{2} - kx\right)\right] dx =$$

$$= -\frac{1}{\pi}\left[\frac{2\cos\left(\frac{x}{2} + kx\right)}{2k+1} + \frac{2\cos\left(\frac{x}{2} - kx\right)}{-2k+1}\right]\Big|_0^{\pi} = -\frac{4}{\pi}\frac{1}{4k^2 - 1}, \ k \in \mathbb{N},$$

so

$$f(x) = \frac{2}{\pi} - \frac{4}{\pi} \sum_{k=1}^{\infty} \frac{1}{4k^2 - 1} \cos kx.$$

For $x = 0$ we get

$$0 = \frac{2}{\pi} - \frac{4}{\pi} \sum_{k=1}^{\infty} \frac{1}{4k^2 - 1} \Rightarrow \sum_{k=1}^{\infty} \frac{1}{4k^2 - 1} = \frac{1}{2}.$$

By applying the Parseval's identity we obtain

$$\frac{8}{\pi^2} + \frac{16}{\pi^2} \sum_{k=1}^{\infty} \frac{1}{(4k^2 - 1)^2} = \frac{2}{\pi} \int_0^{\pi} \sin^2 \frac{x}{2} \, dx = 1. \Rightarrow \sum_{k=1}^{\infty} \frac{1}{(4k^2 - 1)^2} = \frac{\pi^2 - 8}{16}.$$

10.4 Chapter 4

(b) Now, we extend f by oddness to the whole interval $[-\pi, \pi]$ and compute only b_k because $a_k = 0, k = 0, 1, \ldots$.

$$b_k = \frac{2}{\pi} \int_0^\pi \sin \frac{x}{2} \sin kx \, dx = \frac{1}{\pi} \int_0^\pi \left[\cos \left(\frac{x}{2} - kx \right) - \cos \left(\frac{x}{2} + kx \right) \right] dx =$$

$$= \frac{1}{\pi} \left[\frac{2 \sin \left(\frac{x}{2} - kx \right)}{-2k + 1} - \frac{2 \sin \left(\frac{x}{2} + kx \right)}{2k + 1} \right] \Bigg|_0^\pi = \frac{8}{\pi} (-1)^{k+1} \frac{k}{4k^2 - 1}.$$

Hence

$$f(x) = \frac{8}{\pi} \sum_{k=1}^\infty (-1)^{k+1} \frac{k}{4k^2 - 1} \sin kx.$$

For $x = \frac{\pi}{2}$ we get

$$\frac{1}{\sqrt{2}} = \frac{8}{\pi} \sum_{k=1}^\infty (-1)^{k+1} \frac{k}{4k^2 - 1} \sin k\frac{\pi}{2} = \frac{8}{\pi} \sum_{m=0}^\infty (-1)^m \frac{2m+1}{16m^2 + 16m + 3}.$$

$$\Rightarrow \sum_{m=0}^\infty (-1)^m \frac{2m+1}{16m^2 + 16m + 3} = \frac{\pi}{8\sqrt{2}}.$$

8. Since f is even, we have to compute only a_k ($b_k = 0, k = 1, 2, \ldots$).

$$a_k = \frac{2}{3\pi} \int_0^{3\pi} \cos \frac{x}{3} \cos \left(\frac{k}{3} x \right) dx =$$

$$= \frac{1}{3\pi} \int_0^{3\pi} \left[\cos \left(\frac{x + kx}{3} \right) + \cos \left(\frac{kx - x}{3} \right) \right] dx = 0, \forall k \neq 1$$

and $a_1 = 1$. So ... $f(x) = \cos \frac{x}{3}$ is the Fourier expansion of f. Indeed, we could say this without any computation because the Fourier expansion is unique and $f(x) = \cos \frac{x}{3}$ is a periodic function of period $T = 3 \cdot 2\pi = 6\pi$, exactly the length of the interval of definition of f.

9. We simply substitute the function $\ln(x+1)$ with $f(x) = x - \frac{x^2}{2}$, $x \in [0, 1]$. The next step is to extend $f(x)$ by periodicity to a periodic function $\widetilde{f} : \mathbb{R} \to \mathbb{R}$ of period $T = 1, \omega = 2\pi$. So

$$a_0 = 2 \int_0^1 \left(x - \frac{x^2}{2} \right) dx = \frac{2}{3}.$$

$$a_k = 2\int_0^1 \left(x - \frac{x^2}{2}\right) \cos(2\pi kx)\, dx = -\frac{1}{2\pi^2 k^2}, k = 1, 2, \ldots.$$

$$b_k = 2\int_0^1 \left(x - \frac{x^2}{2}\right) \sin(2\pi kx)\, dx = -\frac{1}{2\pi k}, k = 1, 2, \ldots.$$

Hence,

$$g(x) \approx \frac{1}{3} - \sum_{k=1}^{\infty}\left[\frac{1}{2\pi^2 k^2} \cos 2\pi kx + \frac{1}{2\pi k} \sin 2\pi kx\right]. \tag{10.13}$$

To find its complex representation we use the formulas (4.114) and (4.119).

$$c_0 = \int_0^1 \left(x - \frac{x^2}{2}\right) dx = \frac{1}{3}.$$

$$c_k = \int_0^1 \left(x - \frac{x^2}{2}\right) \exp(-2\pi ikx)\, dx = -\frac{1}{4\pi^2 k^2} + \frac{1}{4\pi k}i, \text{ for } k \neq 0$$

So

$$g(x) \approx \frac{1}{3} + \sum_{k \in \mathbb{Z}^*}\left(-\frac{1}{4\pi^2 k^2} + \frac{1}{4\pi k}i\right)\exp(2\pi ikx).$$

If we put together the coefficients c_k and c_{-k} for $k = 1, 2, \ldots$, we obtain again formula (10.13). The conclusion is that it is sufficient to find the complex representation and then to obtain the usual "real form" by putting together its symmetric terms.

10.5 Chapter 5

1.(a) We use Definition 5.7 (formula (5.35)):

$$F(\xi) = \frac{1}{\sqrt{2\pi}}\int_{-\infty}^{\infty} f(x)\exp(-i\xi x)\, dx = \frac{1}{\sqrt{2\pi}}\int_{-2}^{2} 3\exp(-i\xi x)\, dx =$$

$$= \frac{3}{\sqrt{2\pi}} \cdot \frac{1}{-i\xi}\exp(-i\xi x)\bigg|_{-2}^{2} = \frac{3i}{\sqrt{2\pi}\xi}(\exp(-2i\xi) - \exp(2i\xi)) =$$

$$= \frac{3i}{\sqrt{2\pi}\xi}(\cos 2\xi - i\sin 2\xi - \cos 2\xi - i\sin 2\xi) = \sqrt{\frac{2}{\pi}} \cdot \frac{3\sin 2\xi}{\xi}.$$

10.5 Chapter 5

Note that we could also apply Proposition 5.2.

(b) From Proposition 5.2 we know that the Fourier transform of the function
$g(x) = \begin{cases} 1, & |x| \leq l \\ 0, & |x| > l \end{cases}$ is the function $G(\xi) = \sqrt{\frac{2}{\pi}} \cdot \frac{\sin l\xi}{\xi}$. Let $c = \frac{a+b}{2}$
be the midpoint of the interval $[a, b]$ and $2l = b - a$ the length of the interval.
Then $f(x)$ can be written: $f(x) = 4g(x - c)$, and, by Theorem 5.7, we have:
$\mathcal{F}\{f(x)\} = 4\mathcal{F}\{g(x - c)\} = 4\exp(-i\xi c)\mathcal{F}\{g(x)\} = 4\sqrt{\frac{2}{\pi}}\exp(-i\xi c)\frac{\sin l\xi}{\xi}$,
where $c = \frac{a+b}{2}$ and $l = \frac{b-a}{2}$.

(c) We can use the definition of the Fourier transform (5.35):

$$F(\xi) = \frac{1}{\sqrt{2\pi}} \int_{-\infty}^{\infty} \exp(-2|x|)\exp(-i\xi x)\,dx =$$

$$= \frac{1}{\sqrt{2\pi}} \left(\int_{-\infty}^{0} \exp((2 - i\xi)x)\,dx + \int_{0}^{\infty} \exp((-2 - i\xi)x)\,dx \right) =$$

$$= \frac{1}{\sqrt{2\pi}} \left(\frac{1}{2 - i\xi}\exp((2 - i\xi)x)\Big|_{-\infty}^{0} + \frac{1}{-2 - i\xi}\exp((-2 - i\xi)x)\Big|_{0}^{\infty} \right)$$

Since $\lim_{x \to \infty} \exp(-2x \pm 2i\xi x) = \lim_{x \to \infty} \exp(-2x)(\cos(\xi x) \pm i\sin(\xi x)) = 0$,
we can write $F(\xi) = \frac{1}{\sqrt{2\pi}}\left(\frac{1}{2-i\xi} + \frac{1}{2+i\xi}\right) = \sqrt{\frac{2}{\pi}} \cdot \frac{2}{4+\xi^2}$. (Note that we could also apply Proposition 5.1 for $a = 2$.)

(d) The function can be written $f(x) = -\frac{1}{2}\left(\frac{1}{x^2+a^2}\right)'$ and we apply Theorem 5.10
for the function $g(x) = \frac{1}{x^2+a^2}$, so we have: $\mathcal{F}\{g'(x)\} = i\xi\mathcal{F}\{g(x)\} = i\xi G(\xi)$, where $G(\xi) = \mathcal{F}\{g(x)\}$ is the Fourier transform of the function
$g(x)$. By Proposition 5.5, $G(\xi) = \frac{1}{a}\sqrt{\frac{\pi}{2}}\exp(-a|\xi|)$, so we can write

$$\mathcal{F}\left\{\frac{x}{(x^2+a^2)^2}\right\} = -\frac{1}{2}i\xi G(\xi) = -\frac{i\xi}{2a}\sqrt{\frac{\pi}{2}}\exp(-a|\xi|).$$

(e) Since $f(x)$ is an even function, we have: $\mathcal{F}\{f(x)\} = \mathcal{F}_c\{f(x)\} =$

$$= \sqrt{\frac{2}{\pi}} \int_{0}^{\pi} \cos x \cos \xi x\,dx = \frac{1}{\sqrt{2\pi}} \int_{0}^{\pi} (\cos(\xi + 1)x + \cos(\xi - 1)x)\,dx =$$

$$= \frac{1}{\sqrt{2\pi}}\left(\frac{1}{\xi + 1}\sin(\xi + 1)x\Big|_{0}^{\pi} + \frac{1}{\xi - 1}\sin(\xi - 1)x\Big|_{0}^{\pi}\right) =$$

$$= \frac{1}{\sqrt{2\pi}} \left(\frac{\sin(\xi\pi + \pi)}{\xi + 1} + \frac{\sin(\xi\pi - \pi)}{\xi - 1} \right) = -\frac{\sin \xi\pi}{\sqrt{2\pi}} \cdot \frac{2\xi}{\xi^2 - 1}.$$

2. Since $f(x)$ is an even function, we have:

$$\mathcal{F}\{f(x)\} = \mathcal{F}_c\{f(x)\} = \sqrt{\frac{2}{\pi}} \int_0^\infty f(t) \cos(\xi t)\, dt =$$

$$= \sqrt{\frac{2}{\pi}} \cdot \frac{1}{1 + \xi^2} = \mathcal{F}\{\exp(-|x|)\},$$

where, for the last equality, we applied Proposition 5.1.
  Hence, $f(x) = \exp(-|x|)$.

3. Let $f(x) = \begin{cases} 1, & |x| < 1 \\ 0, & |x| > 1 \end{cases}$ be the rectangular pulse function. By Proposition 5.2 we know that its Fourier transform is the function

$$F(\xi) = \sqrt{\frac{2}{\pi}} \cdot \frac{\sin \xi}{\xi}.$$

From Parseval's formula we can write:

$$\|F\|^2 = \frac{2}{\pi} \int_{-\infty}^{\infty} \frac{\sin^2 \xi}{\xi^2}\, d\xi = \|f\|^2 = \int_{-1}^{1} dx = 2,$$

so $\int_{-\infty}^{\infty} \frac{\sin^2 \xi}{\xi^2}\, d\xi = \pi$.

4. If $F(\xi) = \mathcal{F}\{f(x)\}$ then

$$F(\xi) = \frac{1}{\sqrt{2\pi}} \int_{-\infty}^{\infty} f(x) \exp(-i\xi x)\, dx = \frac{1}{\sqrt{2\pi}} \int_{-a}^{a} x \exp(-i\xi x)\, dx =$$

$$= -i\sqrt{\frac{2}{\pi}} \int_0^a x \sin \xi x\, dx = -i\sqrt{\frac{2}{\pi}} \left(-\frac{x \cos \xi x}{\xi} \Big|_0^a + \frac{1}{\xi} \int_0^a \cos \xi x\, d\xi \right) =$$

$$= -i\sqrt{\frac{2}{\pi}} \left(-\frac{a \cos \xi a}{\xi} + \frac{\sin \xi x}{\xi^2} \Big|_0^a \right) = i\sqrt{\frac{2}{\pi}} \cdot \frac{a\xi \cos \xi a - \sin \xi a}{\xi^2}.$$

Now, let $g(x) = \begin{cases} 1, & |x| < a \\ 0, & |x| > a \end{cases}$ be the rectangular pulse function, whose Fourier transform is the function

$$G(\xi) = \sqrt{\frac{2}{\pi}} \cdot \frac{\sin a\xi}{\xi}.$$

By Theorem 5.11, since $f(x) = xg(x)$, we obtain that

$$\mathcal{F}\{f\}(\xi) = iG'(\xi) = i\sqrt{\frac{2}{\pi}} \cdot \frac{a\xi \cos a\xi - \sin a\xi}{\xi^2},$$

the same result as above.

5. If $F(\xi) = \mathcal{F}\{f(x)\}$ then

$$F(\xi) = \frac{1}{\sqrt{2\pi}} \int_{-\infty}^{\infty} f(x) \exp(-i\xi x)\, dx = \frac{1}{\sqrt{2\pi}} \int_{-b}^{b} \exp(-i(\xi - a)x)\, dx =$$

$$= \sqrt{\frac{2}{\pi}} \int_0^b \cos(\xi - a)x\, dx = \sqrt{\frac{2}{\pi}} \cdot \frac{\sin(\xi - a)x}{\xi - a}\bigg|_0^b = \sqrt{\frac{2}{\pi}} \cdot \frac{\sin b(\xi - a)}{\xi - a}.$$

Now, let $g(x) = \begin{cases} 1, & |x| < b \\ 0, & |x| > b \end{cases}$ be the rectangular pulse function, whose Fourier transform is the function

$$G(\xi) = \sqrt{\frac{2}{\pi}} \cdot \frac{\sin b\xi}{\xi}.$$

By Theorem 5.8, since $f(x) = \exp(iax)g(x)$, we have $F(\xi) = G(\xi - a)$, so the same result is obtained.

6. By Definition 5.9 we have:

$$f_a * f_a(x) = \int_{-\infty}^{\infty} f_a(t) f_a(x - t)\, dt.$$

Since f_a is the rectangular pulse function, we have:

$$f_a(t) f_a(x - t) = \begin{cases} 1, & |t| < a \text{ and } |x - t| < a \\ 0, & \text{otherwise,} \end{cases}$$

so $f_a(t) f_a(x - t) \neq 0$ if and only if the following inequalities are satisfied:

$$\begin{cases} -a < t < a \\ x - a < t < a \end{cases}.$$

If $|x| > 2a$ the system above has no solution, so $f_a(t) f_a(x - t) = 0$ for all $t \in \mathbb{R}$ and $f_a * f_a(x) = 0$ in this case.

If $x \in (-2a, 0)$ the solution of the system is $t \in (-a, x + a)$, so

$$f_a * f_a(x) = \int_{-a}^{x+a} dt = x + 2a.$$

If $x \in (0, 2a)$ the solution of the system is $t \in (x - a, a)$, so

$$f_a * f_a(x) = \int_{x-a}^{a} dt = 2a - x.$$

and we obtain that

$$f_a * f_a(x) = \begin{cases} 2a - |x|, & |x| < 2a \\ 0, & |x| > 2a \end{cases} = (2a - |x|) f_{2a}.$$

We know that the Fourier transform of the function $f_a(x)$ is $F_a(\xi) = \sqrt{\frac{2}{\pi}} \cdot \frac{\sin a\xi}{\xi}$. By Theorem 5.12 one can write:

$$\mathcal{F}\{f * f\}(\xi) = \sqrt{2\pi} F_a^2(\xi) = \sqrt{\frac{2}{\pi}} \cdot \frac{2 \sin^2 a\xi}{\xi^2} = \sqrt{\frac{2}{\pi}} \cdot \frac{1 - \cos 2a\xi}{\xi^2}.$$

But, using the linearity of the Fourier transform, we get:

$$\mathcal{F}\{f * f\}(\xi) = \mathcal{F}\{2a f_{2a} - g_{2a}\}(\xi) = 2a \cdot \sqrt{\frac{2}{\pi}} \cdot \frac{\sin 2a\xi}{\xi} - G_{2a}(\xi),$$

where $G_a(\xi) = \mathcal{F}\{g_a(x)\}$. Thus, we obtain that

$$G_{2a}(\xi) = \sqrt{\frac{2}{\pi}} \left(\frac{2a \sin 2a\xi}{\xi} - \frac{1 - \cos 2a\xi}{\xi^2} \right),$$

hence

$$G_a(\xi) = \sqrt{2\pi} \cdot \frac{a\xi \sin a\xi + \cos a\xi - 1}{\xi^2}.$$

Now, let us use the definition formula (5.35) to compute $G_a(\xi)$:

$$G_a(\xi) = \frac{1}{\sqrt{2\pi}} \int_{-\infty}^{\infty} g_a(x) \exp(-i\xi x) \, dx = \frac{1}{\sqrt{2\pi}} \int_{-a}^{a} |x| \exp(-i\xi x) \, dx =$$

$$= \sqrt{\frac{2}{\pi}} \int_0^a x \cos\xi x\, dx = \sqrt{\frac{2}{\pi}} \left(\left.\frac{x \sin\xi x}{\xi}\right|_0^a - \frac{1}{\xi}\int_0^a \sin\xi x\, d\xi \right) =$$

$$= \sqrt{\frac{2}{\pi}} \left(\frac{a \sin\xi a}{\xi} + \left.\frac{\cos\xi x}{\xi^2}\right|_0^a \right) = \sqrt{\frac{2}{\pi}} \cdot \frac{a\xi \sin\xi a + \cos\xi a - 1}{\xi^2}.$$

7. In this case $v = \exp\left(-i\frac{\pi}{2}\right) = -i$. By (5.110), we have:

$$W = \frac{1}{2}\begin{pmatrix} 1 & 1 & 1 & 1 \\ 1 & v & v^2 & v^3 \\ 1 & v^2 & v^4 & v^6 \\ 1 & v^3 & v^6 & v^9 \end{pmatrix} = \frac{1}{2}\begin{pmatrix} 1 & 1 & 1 & 1 \\ 1 & -i & -1 & i \\ 1 & -1 & 1 & -1 \\ 1 & i & -1 & -i \end{pmatrix}.$$

(a)

$$F = Wf = \frac{1}{2}\begin{pmatrix} 1 & 1 & 1 & 1 \\ 1 & -i & -1 & i \\ 1 & -1 & 1 & -1 \\ 1 & i & -1 & -i \end{pmatrix}\begin{pmatrix} f_0 \\ f_1 \\ f_2 \\ f_3 \end{pmatrix} = \begin{pmatrix} f_0 + f_1 + f_2 + f_3 \\ f_0 - if_1 - f_2 + if_3 \\ f_0 - f_1 + f_2 - f_3 \\ f_0 + if_1 - f_2 - if_3 \end{pmatrix}.$$

(b)

$$W^{-1} = \overline{W} = \frac{1}{2}\begin{pmatrix} 1 & 1 & 1 & 1 \\ 1 & i & -1 & -i \\ 1 & -1 & 1 & -1 \\ 1 & -i & -1 & i \end{pmatrix}.$$

$$f = \overline{W}F = \frac{1}{2}\begin{pmatrix} 1 & 1 & 1 & 1 \\ 1 & i & -1 & -i \\ 1 & -1 & 1 & -1 \\ 1 & -i & -1 & i \end{pmatrix}\begin{pmatrix} F_0 \\ F_1 \\ F_2 \\ F_3 \end{pmatrix} = \begin{pmatrix} F_0 + F_1 + F_2 + F_3 \\ F_0 + iF_1 - F_2 - iF_3 \\ F_0 - F_1 + F_2 - F_3 \\ F_0 - iF_1 - F_2 + iF_3 \end{pmatrix}.$$

10.6 Chapter 6

1. (a) Since $\sin^2 3t = \frac{1}{2}(1 - \cos 6t)$, using the linearity of $\mathcal{L}$, one can write: $\mathcal{L}\{\sin^2 3t\} = \frac{1}{2}\mathcal{L}\{1\} - \frac{1}{2}\mathcal{L}\{\cos 6t\}$. Now we can apply formula (6.7) for

$\mathcal{L}\{1\} = \mathcal{L}\{\eta(t)\} = \frac{1}{p}$ and formula (6.13) with $\omega = 6$ for $\mathcal{L}\{\cos 6t\} = \frac{p}{p^2+36}$, so we obtain $\mathcal{L}\{\sin^2 3t\} = \frac{1}{2p} - \frac{p}{2(p^2+36)} = \frac{18}{p^2+36}$.

(b) Since $\sin 2t \cos 3t = \frac{1}{2}(\sin 5t - \sin t)$, using the linearity of $\mathcal{L}$, one can write: $\mathcal{L}\{\sin 2t \cos 3t\} = \frac{1}{2}\mathcal{L}\{\sin 5t\} - \frac{1}{2}\mathcal{L}\{\sin t\}$. By applying formula (6.14) for $\omega = 5$ and $\omega = 1$ we obtain $\mathcal{L}\{\sin 2t \cos 3t\} = \frac{5}{2(p^2+25)} - \frac{1}{2(p^2+1)}$.

(c) For $k = 1$ the formula (6.5) can be written: $\mathcal{L}\{tf(t)\} = -F'(p)$. In our case, $f(t) = \cos 2t$, so $F(p) = \frac{p}{p^2+4}$ (see formula (6.13) for $\omega = 2$). It follows that $\mathcal{L}\{t \cos 2t\} = -\left(\frac{p}{p^2+4}\right)' = \frac{p^2-4}{(p^2+4)^2}$.

(d) $\mathcal{L}\{(2t+3)\sin 4t\} = 2\mathcal{L}\{t \sin 4t\} + 3\mathcal{L}\{\sin 4t\} = -2\left(\frac{4}{p^2+16}\right)' + 3\frac{4}{p^2+16} = \frac{16p}{(p^2+16)^2} + \frac{12}{p^2+16}$.

(e) We apply formula (6.24) for $\lambda = 2$ and $\omega = 1$ and obtain $\mathcal{L}\{\exp(2t)\sin t\} = \frac{1}{(p-2)^2+1}$.

(f) We apply formula (6.23) for $\lambda = -2$ and $\omega = 3$ and obtain $\mathcal{L}\{\exp(-2t)\cos 3t\} = \frac{p+2}{(p+2)^2+9}$.

(g) By using formula (6.31) one can write $\mathcal{L}\left\{\frac{\exp(-at)-\exp(-bt)}{t}\right\} = \int_p^\infty F(q)dq$, where $F(p)$ is the Laplace transform of $f(t) = \exp(-at) - \exp(-bt)$. By applying formula (6.11) we get: $F(p) = \mathcal{L}\{\exp(-at)\} - \mathcal{L}\{\exp(-at)\} = \frac{1}{p+a} - \frac{1}{p+b}$, so $\mathcal{L}\left\{\frac{\exp(-at)-\exp(-bt)}{t}\right\} = \int_p^\infty (\frac{1}{q+a} - \frac{1}{q+b})dq = \log\frac{q+a}{q+b}\Big|_p^\infty = \log\frac{p+b}{p+a}$.

2.(a) We use the partial fraction decomposition: $F(p) = \frac{1}{(p+1)(p+2)} = \frac{1}{p+1} - \frac{1}{p+2} = \mathcal{L}\{\exp(-t)\}(p) - \mathcal{L}\{\exp(-2t)\}(p)$ (see formula (6.11)). It follows that $\mathcal{L}^{-1}\{F(p)\}(t) = \exp(-t) - \exp(-2t)$.

(b) $F(p) = \frac{p+3-3}{(p+3)^2} = \frac{1}{p+3} - 3 \cdot \frac{1}{(p+3)^2}$. We know from formula (6.11) that $\mathcal{L}^{-1}\{\frac{1}{p+3}\}(t) = \exp(-3t)$. By writing (6.8) for $k = 1$ we can see that $\mathcal{L}^{-1}\{\frac{1}{p^2}\}(t) = t$ and if we apply Theorem 6.5 for $f(t) = t$ and $p_0 = -3$, we obtain $\mathcal{L}\{t\exp(-3t)\}(p) = \frac{1}{(p+3)^2}$. Hence $\mathcal{L}^{-1}\{F(p)\}(t) = \exp(-3t)(1 - 3t)$.

(c) First of all, we notice that $\frac{p}{(p^2+4)^2} = -\frac{1}{4}\left(\frac{2}{p^2+4}\right)'$. If we apply formula (6.5) for $k = 1$, we have $\mathcal{L}\{tf(t)\}(p) = -F'^{-1}\{F'(p)\} = -tf(t)$. In our case, $f(t) = \mathcal{L}^{-1}\left\{\frac{2}{p^2+4}\right\}(t) = \sin 2t$, so $\mathcal{L}^{-1}\left\{\frac{p}{(p^2+4)^2}\right\}(t) = \frac{t \sin 2t}{4}$.

Note that we could also use the convolution formula (see Theorem 6.11). Thus, we can write $\frac{p}{(p^2+4)^2} = \frac{1}{2} \cdot \frac{p}{p^2+4} \cdot \frac{2}{p^2+4} = \frac{1}{2}\mathcal{L}\{\cos 2t\}(p)\mathcal{L}\{\sin 2t\}(p) = \frac{1}{2}\mathcal{L}\{(\cos 2t) * (\sin 2t)\}(p)$. Now we use formula (6.38): $(\cos 2t) * (\sin 2t) =$

$\int_0^t \cos 2u \cdot \sin 2(t-u)\,du = \frac{1}{2}\int_0^t (\sin 2t + \sin 2(t-2u))\,du = \frac{1}{2}t\sin 2t \Rightarrow\Rightarrow$
$\mathcal{L}^{-1}\left\{\frac{p}{(p^2+4)^2}\right\}(t) = \frac{1}{4}t\sin 2t.$

(d) We use the convolution formula (Theorem 6.11): $\frac{1}{(p^2+1)^2} = \frac{1}{p^2+1} \cdot \frac{1}{p^2+1} = \mathcal{L}\{\sin t\}(p)\mathcal{L}\{\sin t\}(p) = \mathcal{L}\{(\sin t) * (\sin t)\}(p)$. From the formula (6.38) we can write: $(\sin t) * (\sin t) = \int_0^t \sin u \cdot \sin(t-u)\,du =$

$= \frac{1}{2}\int_0^t (\cos(2u-t) - \cos t\,du) = \frac{1}{4}\sin(2u-t)\Big|_0^t - \frac{1}{2}\cos t \cdot u\Big|_0^t =$
$= \frac{1}{2}(\sin t - t\cos t) \Rightarrow \mathcal{L}^{-1}\left\{\frac{1}{(p^2+1)^2}\right\} = \frac{1}{2}(\sin t - t\cos t).$

(e) By applying (6.23) and (6.24) for $\lambda = 2$ and $\omega = 1$ we get: $F(p) = \frac{p}{(p-2)^2+1} = \frac{p-2}{(p-2)^2+1} + \frac{2}{(p-2)^2+1} = \mathcal{L}\{\exp 2t\cos t\}(p) + 2\mathcal{L}\{\exp 2t\sin t\}(p).$
It follows that $\mathcal{L}^{-1}\{F(p)\}(t) = \exp 2t(\cos t + 2\sin t).$

(f) $F(p) = \frac{p}{(p+2)(p^2-2p+4)} = \frac{1}{6}\left(-\frac{1}{p+2} + \frac{p+2}{p^2-2p+4}\right)$. From the formula (6.11) we have $\mathcal{L}^{-1}\{\frac{1}{p+2}\}(t) = \exp(-2t)$. In order to calculate the inverse transform for the second function we use the formulas (6.24) and (6.23) for $\lambda = 1$ and $\omega = \sqrt{3}$. We have: $\mathcal{L}^{-1}\left\{\frac{p+2}{p^2-2p+4}\right\}(t) =$
$= \mathcal{L}^{-1}\left\{\frac{p-1}{(p-1)^2+(\sqrt{3})^2} + \sqrt{3}\frac{\sqrt{3}}{(p-1)^2+(\sqrt{3})^2}\right\}(t) = \exp t(\cos\sqrt{3}t + \sqrt{3}\sin\sqrt{3}t) \Rightarrow \mathcal{L}^{-1}\{F(p)\} = -\frac{1}{6}\exp(-2t) + \frac{1}{6}\exp t(\cos\sqrt{3}t + \sqrt{3}\sin\sqrt{3}t).$
Note that we could also use the formula (6.52) because the denominator $Q(p) = p^3 + 8 = (p+2)(p^2-2p+4)$ has only simple roots: $p_1 = -2$, $p_{2,3} = 1 \pm i\sqrt{3}$. So: $\mathcal{L}^{-1}\{F(p)\}(t) = \sum_{k=1}^{3}\frac{P(p_k)}{Q'(p_k)}\exp(p_k t) = \sum_{k=1}^{3}\frac{1}{3p_k}\exp(p_k t) =$
$\frac{1}{3}\left(\frac{\exp(-2t)}{-2} + \frac{\exp t(1+i\sqrt{3})}{1+i\sqrt{3}} + \frac{\exp t(1-i\sqrt{3})}{1-i\sqrt{3}}\right) = -\frac{1}{6}\exp(-2t) + \frac{1}{6}\exp t(\cos\sqrt{3}t + \sqrt{3}\sin\sqrt{3}t).$

(g) Theorem 6.4 is written: $\exp(-p)\mathcal{L}\{f(t)\}(p) = \mathcal{L}\{f(t-1)\}(p)$ for $\tau = 1$. In our case, $\mathcal{L}\{f(t)\}(p) = \frac{1}{p(p-1)} = \frac{1}{p-1} - \frac{1}{p} = \mathcal{L}\{\exp t\}(p) - \mathcal{L}\{1\}(p)$, so $f(t) = \exp t - 1$. It follows that $\mathcal{L}^{-1}\left\{\exp(-p)\frac{1}{p(p-1)}\right\}(t) = f(t-1) = (\exp(t-1)-1)\eta(t-1).$

3.(a) By applying the Laplace transform the equation becomes: $\mathcal{L}\{y'\} + \mathcal{L}\{y\} = \mathcal{L}\{\exp(-t)\}$. We denote by $Y(p) = \mathcal{L}\{y\}$ the Laplace transform of the function $y(t)$. From Theorem 6.6 we have $\mathcal{L}\{y'\} = pY(p) - y(0)$ and we obtain: $pY(p) - 2 + Y(p) = \frac{1}{p+1}$. It follows that $Y(p) = \frac{2p+3}{(p+1)^2} = \frac{2}{p+1} + \frac{1}{(p+1)^2}$.
Hence, $y(t) = \mathcal{L}^{-1}\{Y(p)\}(t) = 2\mathcal{L}^{-1}\left\{\frac{2}{p+1}\right\}(t) + \mathcal{L}^{-1}\left\{\frac{1}{(p+1)^2}\right\}(t) = 2\exp(-t) + t\exp(-t)$ (using the formula (6.65) for $\lambda = -1$ and $k = 1, 2$).

(b) $\mathcal{L}\{y''\} + 4\mathcal{L}\{y\} = \mathcal{L}\{t\}$. If we denote by $Y(p) = \mathcal{L}\{y\}$ then, by applying Theorem 6.6, we can write: $\mathcal{L}\{y''\} = p^2 Y(p) - py(0) - y'(0)$. Hence,

$p^2 Y(p) - 2 + 4Y(p) = \frac{1}{p^2}$ and we obtain $Y(p) = \frac{2p^2+1}{p^2(p^2+4)}$. In order to find the inverse Laplace transform of this function we use the partial fraction decomposition: $Y(p) = \frac{1}{4} \cdot \frac{1}{p^2} + \frac{7}{8} \cdot \frac{2}{p^2+4} = \frac{1}{4}\mathcal{L}\{t\}(p) + \frac{7}{8}\mathcal{L}\{\sin 2t\}(p) \Rightarrow$
$y(t) = \mathcal{L}^{-1}\{Y(p)\}(t) = \frac{1}{4}t + \frac{7}{8}\sin 2t$.

(c) $\mathcal{L}\{y''\} + 9\mathcal{L}\{y\} = \mathcal{L}\{1\}$. If $Y(p) = \mathcal{L}\{y\}$ then $\mathcal{L}\{y''\}(p) = p^2 Y(p) - py(0) - y'(0) = p^2 Y(p) + p$. Hence, $p^2 Y(p) + p + 9Y(p) = \frac{1}{p}$ and we obtain $Y(p) = \frac{1-p^2}{p(p^2+9)}$. We have: $Y(p) = \frac{1}{9} \cdot \frac{1}{p} - \frac{10}{9} \cdot \frac{p}{p^2+9} = \frac{1}{9}\mathcal{L}\{1\}(p) - \frac{10}{9}\mathcal{L}\{\cos 3t\} \Rightarrow$
$y(t) = \mathcal{L}^{-1}\{Y(p)\}(t) = \frac{1}{9} - \frac{10}{9}\cos 3t$.

(d) $\mathcal{L}\{y'''\} + \mathcal{L}\{y'\} = 4\mathcal{L}\{1\}$. If $Y(p) = \mathcal{L}\{y\}$ then $\mathcal{L}\{y'''\} = p^3 Y(p) - p^2 y(0) - py'(0) - y''(0) = p^3 Y(p)$ and $\mathcal{L}\{y'\} = pY(p) - y(0) = pY(p)$. Hence, $p^3 Y(p) + pY(p) = \frac{4}{p}$ and we obtain $Y(p) = \frac{4}{p^2(p^2+1)}$. We have: $Y(p) = 4 \cdot \frac{1}{p^2} - 4 \cdot \frac{1}{p^2+1} = 4\mathcal{L}\{t\}(p) - 4\mathcal{L}\{\sin t\}(p) \Rightarrow y(t) = 4t - 4\sin t$.

(e) $\mathcal{L}\{y''\} + \mathcal{L}\{y\} = \mathcal{L}\{\sin t\}$. If $Y(p) = \mathcal{L}\{y\}$ then $\mathcal{L}\{y''\} = p^2 Y(p) - py(0) - y'(0) = p^2 Y(p) - p - 1$. Hence, $p^2 Y(p) - p - 1 + Y(p) = \frac{1}{p^2+1}$ and we obtain
$Y(p) = \frac{p}{p^2+1} + \frac{1}{p^2+1} + \frac{1}{(p^2+1)^2}$. So $y(t) = \mathcal{L}^{-1}\left\{\frac{p}{p^2+1}\right\} + \mathcal{L}^{-1}\left\{\frac{1}{p^2+1}\right\} + \mathcal{L}^{-1}\left\{\frac{1}{(p^2+1)^2}\right\} = \cos t + \sin t + \frac{1}{2}(\sin t - t\cos t) = \frac{3}{2}\sin t + (1 - \frac{t}{2})\cos t$.

(f) $\begin{cases} \mathcal{L}\{x'\} + 2\mathcal{L}\{y\} = 0 \\ \mathcal{L}\{y'\} + 2\mathcal{L}\{x\} = 0 \end{cases}$. Let $X(p)$ and $Y(p)$ be the Laplace transforms of the functions $x(t)$ and $y(t)$. Since $\mathcal{L}\{x'\} = pX(p) - 1$ and $\mathcal{L}\{y'\} = pY(p) + 2$, we have: $\begin{cases} pX(p) + 2Y(p) - 1 = 0 \\ pY(p) + 2X(p) + 2 = 0 \end{cases}$. We solve this system of equations and obtain $X(p) = \frac{p+4}{p^2-4}$ and $Y(p) = \frac{-2p-2}{p^2-4}$. The last step is to find the inverse Laplace transforms of these functions. We can write:
$\begin{cases} X(p) = \frac{3}{2}\frac{1}{p-2} - \frac{1}{2}\frac{1}{p+2} = \frac{3}{2}\mathcal{L}\{\exp(2t)\}(p) - \frac{1}{2}\mathcal{L}\{\exp(-2t)\}(p) \\ Y(p) = -\frac{3}{2}\frac{1}{p-2} - \frac{1}{2}\frac{1}{p+2} = -\frac{3}{2}\mathcal{L}\{\exp(2t)\}(p) - \frac{1}{2}\mathcal{L}\{\exp(-2t)\}(p) \end{cases} \Rightarrow$
$\Rightarrow \begin{cases} x(t) = \mathcal{L}^{-1}\{X(p)\}(t) = \frac{3}{2}\exp(2t) - \frac{1}{2}\exp(-2t) \\ y(t) = \mathcal{L}^{-1}\{Y(p)\}(t) = -\frac{3}{2}\exp(2t) - \frac{1}{2}\exp(-2t) \end{cases}$.

(g) $\begin{cases} \mathcal{L}\{x'\} + \mathcal{L}\{x\} - 2\mathcal{L}\{y\} = \mathcal{L}\{t\} \\ \mathcal{L}\{y'\} + 2\mathcal{L}\{x\} - \mathcal{L}\{y\} = \mathcal{L}\{1\} \end{cases}$.

We denote $\begin{cases} X(p) = L\{x\} \\ Y(p) = L\{y\} \end{cases} \Rightarrow \begin{cases} \mathcal{L}\{x'\} = pX(p) - 2 \\ \mathcal{L}\{y'\} = pY(p) - 3 \end{cases} \Rightarrow$

$\begin{cases} pX(p) - 2 + X(p) - 2Y(p) = \frac{1}{p^2} \\ pY(p) - 3 + 2X(p) - Y(p) = \frac{1}{p} \end{cases} \Rightarrow$

$\begin{cases} (p+1)X(p) - 2Y(p) = \frac{2p^2+1}{p^2} \\ (p-1)Y(p) + 2X(p) = \frac{3p+1}{p} \end{cases} \Rightarrow \begin{cases} X(p) = \frac{2p^3+4p^2+3p-1}{p^2(p^2+3)} \\ Y(p) = \frac{3p^3+p-2}{p^2(p^2+3)} \end{cases}$.

The last step is to find the inverse Laplace transforms of these functions. We can write:

$$\begin{cases} X(p) = \frac{1}{p} - \frac{1}{3} \cdot \frac{1}{p^2} + \frac{p}{p^2+3} + \frac{13}{3\sqrt{3}} \cdot \frac{\sqrt{3}}{p^2+3} \\ Y(p) = \frac{1}{3} \cdot \frac{1}{p} - \frac{2}{3} \cdot \frac{1}{p^2} + \frac{8}{3} \cdot \frac{p}{p^2+3} + \frac{2}{3\sqrt{3}} \cdot \frac{\sqrt{3}}{p^2+3} \end{cases} \Rightarrow$$

$$\Rightarrow \begin{cases} x(t) = \mathcal{L}^{-1}\{X(p)\}(t) = 1 - \frac{1}{3}t + \cos(\sqrt{3}t) + \frac{13}{3\sqrt{3}} \sin(\sqrt{3}t) \\ y(t) = \mathcal{L}^{-1}\{Y(p)\}(t) = \frac{1}{3} - \frac{2}{3}t + \frac{8}{3} \cos(\sqrt{3}t) + \frac{2}{3\sqrt{3}} \sin(\sqrt{3}t) \end{cases}.$$

4. We apply the Laplace transform and obtain:

$$\mathcal{L}\{y''\}(p) + \mathcal{L}\{y\}(p) = \mathcal{L}\{\delta_{2\pi}\}(p) = 4\exp(-2\pi p).$$

Denote $\mathcal{L}\{y\}(p) = Y(p)$. Since

$$\mathcal{L}\{y''\}(p) = p^2 Y(p) - py(0) - y'(0) = p^2 Y(p) - 2p,$$

we obtain:

$$(p^2 + 1)Y(p) - 2p = 4\exp(-2\pi p),$$

hence

$$Y(p) = 2\frac{p}{p^2+1} + 4\exp(-2\pi p) \cdot \frac{1}{p^2+1}.$$

It follows that

$$y(t) = \mathcal{L}^{-1}\{Y(p)\} = 2\cos t \cdot \eta(t) + 4\sin(t - 2\pi) \cdot \eta(t - 2\pi) =$$

$$= \begin{cases} \cos t, & 0 \le t \le 2\pi \\ 2\cos t + 4\sin t, & t \ge 2\pi \end{cases}$$

10.7 Chapter 7

1. (a) Since $\Delta > 0$, the equation is of the hyperbolic type. The associated characteristic equation, $3y'^2 + 5y' - 2 = 0$, has the roots $y' = -2$ and $y' = \frac{1}{3}$. The general solutions are $y + 2x = C_1$ and $x - 3y = C_2$, so we make the change of variables: $\xi = 2x + y$ and $\eta = x - 3y$. We denote $\bar{u}(\xi, \eta) = u(x(\xi, \eta), y(\xi, \eta))$ and, by using formulas (7.15), (7.16),

(7.19), (7.21) and (7.20) we get the following expressions for the partial derivatives of u:

$$\frac{\partial u}{\partial x} = \frac{\partial \bar{u}}{\partial \xi} \cdot 2 + \left.\frac{\partial \bar{u}}{\partial \eta}\right| \cdot 3$$

$$\frac{\partial u}{\partial y} = \frac{\partial \bar{u}}{\partial \xi} + \frac{\partial \bar{u}}{\partial \eta} \cdot (-3)$$

$$\frac{\partial^2 u}{\partial x^2} = \frac{\partial^2 \bar{u}}{\partial \xi^2} \cdot 4 + 2\frac{\partial^2 \bar{u}}{\partial \xi \partial \eta} \cdot 2 + \left.\frac{\partial^2 \bar{u}}{\partial \eta^2}\right| \cdot 3,$$

$$\frac{\partial^2 u}{\partial x \partial y} = \frac{\partial^2 \bar{u}}{\partial \xi^2} \cdot 2 + \frac{\partial^2 \bar{u}}{\partial \xi \partial \eta} \cdot (-5) + \left.\frac{\partial^2 \bar{u}}{\partial \eta^2} \cdot (-3)\right| \cdot (-5),$$

$$\frac{\partial^2 u}{\partial y^2} = \frac{\partial^2 \bar{u}}{\partial \xi^2} + 2\frac{\partial^2 \bar{u}}{\partial \xi \partial \eta} \cdot (-3) + \left.\frac{\partial^2 \bar{u}}{\partial \eta^2} \cdot 9\right| \cdot (-2).$$

If we replace these partial derivatives in the initial equation (in fact, we multiply each expression by the corresponding coefficient (3, 1, 3, −5, −2) and add everything by columns), we obtain the following equation in the new function $\bar{u}(\xi, \eta)$:

$$7\frac{\partial^2 \bar{u}}{\partial \xi \partial \eta} + \frac{\partial \bar{u}}{\partial \xi} = 0,$$

which is the canonical form of our hyperbolic equation.

Now, let us solve this equation. We denote by $v = \frac{\partial \bar{u}}{\partial \xi}$, so the equation becomes: $7\frac{\partial v}{\partial \eta} + v = 0$. The general solution of this equation is

$$v = \frac{\partial \bar{u}}{\partial \xi} = C(\xi) \exp\left(-\frac{1}{7}\eta\right),$$

where $C(\xi)$ is an arbitrary function. Let $F(\xi)$ be a primitive of this function. By integrating with respect to ξ, one obtains:

$$\bar{u}(\xi, \eta) = F(\xi) \exp\left(-\frac{1}{7}\eta\right) + G(\eta),$$

where F and G are functions of one variable, of class C^2 on $\mathbb{R}$. Since $\xi = 2x + y$ and $\eta = x - 3y$, we finally get the general solution of the initial equation:

$$u(x, y) = F(2x + y) \exp\left(-\frac{1}{7}(x - 3y)\right) + G(x - 3y).$$

(b) We have $\Delta = 4x^4 > 0$, so the equation is also of the hyperbolic type. The associated characteristic equation, $xy'^2 - 4x^3 = 0$, has the roots $y' = -2x$ and $y' = 2x$. The general solutions are $x^2 + y = C_1$ and $x^2 - y = C_2$, so we make the change of variables: $\xi = x^2 + y$ and $\eta = x^2 - y$. We calculate only the partial derivatives of u that we need in our equation:

$$\frac{\partial u}{\partial x} = 2x\frac{\partial \bar{u}}{\partial \xi} + 2x\frac{\partial \bar{u}}{\partial \eta} \bigg| \cdot (-1),$$

$$\frac{\partial^2 u}{\partial x^2} = 4x^2\frac{\partial^2 \bar{u}}{\partial \xi^2} + 8x^2\frac{\partial^2 \bar{u}}{\partial \xi \partial \eta} + 4x^2\frac{\partial^2 \bar{u}}{\partial \eta^2} + 2\frac{\partial \bar{u}}{\partial \xi} + 2\frac{\partial \bar{u}}{\partial \eta} \bigg| \cdot x,$$

$$\frac{\partial^2 u}{\partial y^2} = \frac{\partial^2 \bar{u}}{\partial \xi^2} - 2\frac{\partial^2 \bar{u}}{\partial \xi \partial \eta} + \frac{\partial^2 \bar{u}}{\partial \eta^2} \bigg| \cdot (-4x^3).$$

If we multiply the first equality by -1, the second, by x and the last one by $-4x^3$ and add everything by columns, we obtain the canonical form:

$$\frac{\partial^2 \bar{u}}{\partial \xi \partial \eta} = 0.$$

The general solution of this equation is (see Example 7.1) $\bar{u}(\xi, \eta) = F(\xi) + G(\eta)$, where F and G are functions of one variable, of class C^2 on $\mathbb{R}$. Since $\xi = x^2 + y$ and $\eta = x^2 - y$, the general solution of the initial equation is $u(x, y) = F(x^2 + y) + G(x^2 - y)$.

(c) Since $\Delta = 0$, the equation is of the parabolic type. The associated characteristic equation, $y'^2 - 2y' + 1 = 0$, has only one root $y' = 1$. The general solution is $x - y = C_1$, so we make the change of variables: $\xi = x - y$ and $\eta = x$.

Note that $\begin{vmatrix} \frac{\partial \xi}{\partial x} & \frac{\partial \xi}{\partial y} \\ \frac{\partial \eta}{\partial x} & \frac{\partial \eta}{\partial y} \end{vmatrix} = \begin{vmatrix} 1 & -1 \\ 1 & 0 \end{vmatrix} = 1 \neq 0.$ We calculate the partial derivatives of u:

$$\frac{\partial u}{\partial x} = \frac{\partial \bar{u}}{\partial \xi} + \frac{\partial \bar{u}}{\partial \eta}$$

$$\frac{\partial u}{\partial y} = -\frac{\partial \bar{u}}{\partial \xi}$$

$$\frac{\partial^2 u}{\partial x^2} = \frac{\partial^2 \bar{u}}{\partial \xi^2} + 2\frac{\partial^2 \bar{u}}{\partial \xi \partial \eta} + \frac{\partial^2 \bar{u}}{\partial \eta^2},$$

$$\frac{\partial^2 u}{\partial x \partial y} = -\frac{\partial^2 \bar{u}}{\partial \xi^2} - \frac{\partial^2 \bar{u}}{\partial \xi \partial \eta},$$

$$\frac{\partial^2 u}{\partial y^2} = \frac{\partial^2 \bar{u}}{\partial \xi^2}.$$

Hence, by replacing in the initial equation, we obtain the following equation in the new function $\bar{u}(\xi, \eta)$:

$$\frac{\partial^2 \bar{u}}{\partial \eta^2} + 4\frac{\partial \bar{u}}{\partial \eta} + 4\bar{u} = 0,$$

which is the canonical form of our parabolic equation.

If we consider ξ as a parameter and η as variable, the above equation becomes a linear homogeneous equation with constant coefficients. Its general solution is

$$\bar{u}(\xi, \eta) = C_1(\xi) \exp(-2\eta) + C_2(\xi)\eta \exp(-2\eta),$$

where $C_1(\xi)$, $C_2(\xi)$ are functions of class C^1 on $\mathbb{R}$. Hence, the general solution of the initial equation is

$$\bar{u}(\xi, \eta) = C_1(x - y) \exp(-2x) + C_2(x - y)x \exp(-2x).$$

(d) We have $\Delta = -1 < 0$, so the equation is of the elliptic type. The associated characteristic equation, $y'^2 + 4y' + 5 = 0$, has the complex conjugate roots $y' = -2 \pm i$. The general solutions are $y + 2x + ix = C_1$, $y + 2x - ix = C_2$, so we make the change of variables: $\xi = 2x + y$ and $\eta = x$. The partial derivatives of u can be written:

$$\frac{\partial u}{\partial x} = 2\frac{\partial \bar{u}}{\partial \xi} + \frac{\partial \bar{u}}{\partial \eta} \bigg| \cdot (-1)$$

$$\frac{\partial u}{\partial y} = \frac{\partial \bar{u}}{\partial \xi} \bigg| \cdot 2$$

$$\frac{\partial^2 u}{\partial x^2} = 4\frac{\partial^2 \bar{u}}{\partial \xi^2} + 4\frac{\partial^2 \bar{u}}{\partial \xi \partial \eta} + \frac{\partial^2 \bar{u}}{\partial \eta^2},$$

$$\frac{\partial^2 u}{\partial x \partial y} = 2\frac{\partial^2 \bar{u}}{\partial \xi^2} + \frac{\partial^2 \bar{u}}{\partial \xi \partial \eta}\bigg| \cdot (-4),$$

$$\frac{\partial^2 u}{\partial y^2} = \frac{\partial^2 \bar{u}}{\partial \xi^2}\bigg| \cdot 5.$$

By replacing in the initial equation, we obtain the chanonical form of our elliptic equation:

$$\frac{\partial^2 \bar{u}}{\partial \xi^2} + \frac{\partial^2 \bar{u}}{\partial \eta^2} - \frac{\partial \bar{u}}{\partial \eta} = 0, \text{ or } \nabla^2 u - \frac{\partial \bar{u}}{\partial \eta} = 0.$$

2.(a) Since $\Delta = 0$, the equation is of the parabolic type. The associated characteristic equation, $x^2 y'^2 - 2xyy' + y^2 = 0$, has one root, $y' = \frac{y}{x}$, so the general solution is $\frac{y}{x} = C_1$. We use the change of variables: $\xi = \frac{y}{x}$ and $\eta = x$. The partial derivatives of u can be written:

$$\frac{\partial^2 u}{\partial x^2} = \frac{\partial^2 \bar{u}}{\partial \xi^2} \cdot \frac{y^2}{x^4} + 2\frac{\partial^2 \bar{u}}{\partial \xi \partial \eta} \cdot \frac{-y}{x^2} + \frac{\partial^2 \bar{u}}{\partial \eta^2} + \frac{\partial \bar{u}}{\partial \xi} \cdot \frac{2y}{x^3},$$

$$\frac{\partial^2 u}{\partial x \partial y} = \frac{\partial^2 \bar{u}}{\partial \xi^2} \cdot \frac{-y}{x^3} + \frac{\partial^2 \bar{u}}{\partial \xi \partial \eta} \cdot \frac{1}{x} + \frac{\partial \bar{u}}{\partial \xi} \cdot \frac{-1}{x^2},$$

$$\frac{\partial^2 u}{\partial y^2} = \frac{\partial^2 \bar{u}}{\partial \xi^2} \cdot \frac{1}{x^2}.$$

If we multiply the first expression by x^2, the second by $2xy$, the last one by y^2 and add everything by columns, we obtain:

$$\frac{\partial^2 \bar{u}}{\partial \eta^2} = 0.$$

Let us find the general solution of this PDE-2:

$$\frac{\partial}{\partial \eta}\left(\frac{\partial \bar{u}}{\partial \eta}\right) = 0 \Rightarrow \frac{\partial \bar{u}}{\partial \eta} = F(\xi) \Rightarrow \bar{u}(\xi, \eta) = \eta F(\xi) + G(\xi)$$

$$\Rightarrow u(x, y) = xF\left(\frac{y}{x}\right) + G\left(\frac{y}{x}\right),$$

where F and G are two arbitrary functions of class C^2. Now we have to find these functions such that the initial conditions are satisfied:

$$u(1, y) = y^2 \Rightarrow F(y) + G(y) = y^2$$

$$\frac{\partial u}{\partial x}(1, y) = y^2 + y \Rightarrow F(y) - yF'(y) - yG'(y) = y^2 + y$$

If we differentiate the first equation we obtain $F'(y) + G'(y) = 2y$. We replace in the second equation and obtain the expression of the function F: $F(y) = 3y^2 + y$. Since $F(y) + G(y) = y^2$, we find $G(y) = -2y^2 - y$. So, the solution of the Cauchy problem is:

$$u(x, y) = x\left(\frac{3y^2}{x^2} + \frac{y}{x}\right) - \frac{2y^2}{x^2} - \frac{y}{x}.$$

(b) We have again an equation of the parabolic type, because $\Delta = 0$. The associated characteristic equation, $x^4 y'^2 - 2x^2 y^2 y' + y^4 = 0$, has one root, $y' = \frac{y^2}{x^2}$, and the general solution is $\frac{1}{y} - \frac{1}{x} = C_1$. We use the change of variables: $\xi = \frac{1}{y} - \frac{1}{x}$ and $\eta = x$. The partial derivatives of u can be written:

$$\frac{\partial^2 u}{\partial x^2} = \frac{\partial^2 \bar{u}}{\partial \xi^2} \cdot \frac{1}{x^4} + 2\frac{\partial^2 \bar{u}}{\partial \xi \partial \eta} \cdot \frac{1}{x^2} + \frac{\partial^2 \bar{u}}{\partial \eta^2} + \frac{\partial \bar{u}}{\partial \xi} \cdot \frac{-2}{x^3},$$

$$\frac{\partial^2 u}{\partial x \partial y} = \frac{\partial^2 \bar{u}}{\partial \xi^2} \cdot \frac{-1}{x^2 y^2} + \frac{\partial^2 \bar{u}}{\partial \xi \partial \eta} \cdot \frac{-1}{y^2},$$

$$\frac{\partial^2 u}{\partial y^2} = \frac{\partial^2 \bar{u}}{\partial \xi^2} \cdot \frac{1}{y^4} + \frac{\partial \bar{u}}{\partial \xi} \cdot \frac{2}{y^3}.$$

$$\frac{\partial u}{\partial y} = \frac{\partial \bar{u}}{\partial \xi} \cdot \frac{-1}{y^2}.$$

If we multiply each expression by the corresponding coefficient, we obtain the same canonical form as above:

$$\frac{\partial^2 \bar{u}}{\partial \eta^2} = 0.$$

So, the general solution is:

$$\bar{u}(\xi, \eta) = \eta F(\xi) + G(\xi) \Rightarrow u(x, y) = xF\left(\frac{1}{y} - \frac{1}{x}\right) + G\left(\frac{1}{y} - \frac{1}{x}\right),$$

where F and G are two arbitrary functions of class C^2. Now, let us find these functions such that the initial conditions are fulfilled:

$$u(1, y) = 0 \Rightarrow F\left(\frac{1}{y} - 1\right) + G\left(\frac{1}{y} - 1\right) = 0.$$

Since

$$\frac{\partial u}{\partial x} = F\left(\frac{1}{y} - \frac{1}{x}\right) + x F'\left(\frac{1}{y} - \frac{1}{x}\right)\frac{1}{x^2} + G'\left(\frac{1}{y} - \frac{1}{x}\right)\frac{1}{x^2},$$

from the second initial condition, we obtain:

$$\frac{\partial u}{\partial x}(1, y) = y \Rightarrow F\left(\frac{1}{y} - 1\right) + F'\left(\frac{1}{y} - 1\right) + G'\left(\frac{1}{y} - 1\right) = 0$$

If we differentiate the first equation we get $F'\left(\frac{1}{y} - 1\right) + G'\left(\frac{1}{y} - 1\right) = 0$, so $F\left(\frac{1}{y} - 1\right) = y$. If we denote $\frac{1}{y} - 1 = z$, we obtain the expression of the function F: $F(z) = \frac{1}{z+1}$. Since $F(z) + G(z) = 0$, we find $G(z) = \frac{-1}{z+1}$. Hence, the solution of the Cauchy problem is:

$$u(x, y) = \frac{x - 1}{\frac{1}{y} - \frac{1}{x} + 1}.$$

(c) Since $\Delta = 4 > 0$, we have an equation of the hyperbolic type. The associated characteristic equation, $y'^2 - 6y' + 5 = 0$, has the roots $y' = 1$ and $y' = 5$. The general solutions are $x - y = C_1$, $5x - y = C_2$, so we use the change of variables: $\xi = x - y$ and $\eta = 5x - y$:

$$\frac{\partial^2 u}{\partial x^2} = \frac{\partial^2 \bar{u}}{\partial \xi^2} + 10\frac{\partial^2 \bar{u}}{\partial \xi \partial \eta} + 25\frac{\partial^2 \bar{u}}{\partial \eta^2},$$

$$\frac{\partial^2 u}{\partial x \partial y} = -\frac{\partial^2 \bar{u}}{\partial \xi^2} - 6\frac{\partial^2 \bar{u}}{\partial \xi \partial \eta} - 5\frac{\partial^2 \bar{u}}{\partial \eta^2},$$

$$\frac{\partial^2 u}{\partial y^2} = \frac{\partial^2 \bar{u}}{\partial \xi^2} - 2\frac{\partial \bar{u}}{\partial \xi} + \frac{\partial^2 \bar{u}}{\partial \eta^2}.$$

We multiply the second expression by 6, the third, by 5 and add everything by columns. We obtain the canonical form:

$$\frac{\partial^2 \bar{u}}{\partial \xi \partial \eta} = 0.$$

The general solution of this equation is (see Example 7.1):

$$\bar{u}(\xi, \eta) = F(\xi) + G(\xi) \Rightarrow u(x, y) = F(x - y) + G(5x - y),$$

where F and G are two arbitrary functions of class C^2. Now, let us find these functions such that the initial conditions are fulfilled:

$$u(0, y) = y^2 \Rightarrow F(-y) + G(-y) = y^2 \Rightarrow -F'(-y) - G'(-y) = 2y,$$

$$\frac{\partial u}{\partial x}(0, y) = 2y \Rightarrow F'(-y) + 5G'(-y) = 2y$$

We obtain $G'(-y) = y$ and $F'(-y) = -3y$, so $F'(y) = 3y$, $G'(y) = -y$. Hence, $F(y) = \frac{3y^2}{2} + C_1$, $G(y) = \frac{-y^2}{2} + C_2$ and from the first initial condition we see that $C_1 + C_2 = 0$. The solution of the Cauchy problem is:

$$u(x, y) = \frac{3(x - y)^2 - (5x - y)^2}{2}.$$

(d) We have an equation of the same hyperbolic type, because $\Delta = \cos^2 x + \sin^2 x = 1 > 0$. The associated characteristic equation, $y'^2 - 2\cos x\, y' - \sin^2 x = 0$, has the roots $y' = \cos x + 1$ and $y' = \cos x - 1$. The general solutions are $\sin x + x - y = C_1$, $\sin x - x - y = C_2$, so we use the change of variables: $\xi = \sin x + x - y$ and $\eta = \sin x - x - y$:

$$\frac{\partial^2 u}{\partial x^2} = \frac{\partial^2 \bar{u}}{\partial \xi^2}(\cos x + 1)^2 - 2\frac{\partial^2 \bar{u}}{\partial \xi \partial \eta}\sin^2 x + \frac{\partial^2 \bar{u}}{\partial \eta^2}(\cos x - 1)^2 +$$

$$+ \frac{\partial \bar{u}}{\partial \xi}(-\sin x) + \frac{\partial \bar{u}}{\partial \eta}(-\sin x),$$

$$\frac{\partial^2 u}{\partial x \partial y} = -\frac{\partial^2 \bar{u}}{\partial \xi^2}(\cos x - 1) - 2\frac{\partial^2 \bar{u}}{\partial \xi \partial \eta}\cos x - \frac{\partial^2 \bar{u}}{\partial \eta^2}(\cos x - 1),$$

$$\frac{\partial^2 u}{\partial y^2} = \frac{\partial^2 \bar{u}}{\partial \xi^2} + 2\frac{\partial \bar{u}}{\partial \xi} + \frac{\partial^2 \bar{u}}{\partial \eta^2}.$$

$$\frac{\partial \bar{u}}{\partial y} = -\frac{\partial \bar{u}}{\partial \xi} - \frac{\partial \bar{u}}{\partial \eta}$$

If we replace these expressions in the initial equation, we obtain:

$$\frac{\partial^2 \bar{u}}{\partial \xi \partial \eta} = 0.$$

We saw that the general solution of this equation is:

$$\bar{u}(\xi, \eta) = F(\xi) + G(\xi)$$

$$\Rightarrow u(x, y) = F(\sin x + x - y) + G(\sin x - x - y)$$

where F and G are two arbitrary functions of class C^2. Now, let us find these functions such that the initial conditions are fulfilled:

$$u(x, \sin x) = 2x + \cos x \Rightarrow F(x) + G(-x) = 2x + \cos x,$$

$$\frac{\partial u}{\partial y}(x, \sin x) = \sin x \Rightarrow -F'(x) - G'(-x) = \sin x,$$

If we differentiate the first initial condition, we obtain

$$F'(x) - G'(-x) = 2 - \sin x$$

Hence, $G'(-x) = -1 \Rightarrow G(-x) = x + C_1 \Rightarrow F(x) = x + \cos x - C_1$ and the solution of the Cauchy problem is:

$$u(x, y) = \sin x + x - y + \cos(\sin x + x - y) - \sin x + x + y =$$

$$= 2x + \cos(\sin x + x - y).$$

(e) The equation can be written: $\frac{\partial}{\partial x}\left(\frac{\partial u}{\partial y}\right) + \frac{\partial u}{\partial y} = 0$. If we denote by $v = \frac{\partial u}{\partial y}$ and consider v as a function of x (and y as a parameter), the above equation becomes a linear homogeneous equation and its solution is $v = C(y)\exp(-x)$, where C is an arbitrary function of one variable. Thus, from $\frac{\partial u}{\partial y} = C(y)\exp(-x)$, by integrating with respect to y, we obtain the general solution: $u(x, y) = F(y)\exp(-x) + G(x)$, where F and G are arbitrary functions of class C^2 (F is a primitive of the function C).

Now, let as find these functions F and G such that u satisfies the initial conditions. From $u(x, x) = 0$, we have: $F(x)\exp(-x) + G(x) = 0$, so $G(x) = -F(x)\exp(-x)$ and $u(x, y) = \exp(-x)(F(y) - F(x))$.

Since $\frac{\partial u}{\partial y} = \exp(-x)F'(y)$, from the second condition we obtain: $\frac{\partial u}{\partial y}(x, x) = \exp(-x)F'(x) = \exp(-x)$, so $F'(x) = 1$, which means that

$F(x) = x + c$ (c is a real constant). Hence, the solution of the Cauchy problem is $u(x, y) = \exp(-x)(y - x)$.

(f) Let us integrate the equation with respect to y (two times):

$$\frac{\partial^2 u}{\partial y^2} = x + y \Rightarrow \frac{\partial u}{\partial y} = xy + \frac{y^2}{2} + F(x) \Rightarrow u(x, y) = \frac{xy^2}{2} + \frac{y^3}{6} + F(x)y + G(x),$$

where F and G are arbitrary functions of class C^2. From the initial conditions we have:

$$u(x, 0) = x^2 \Rightarrow G(x) = x^2,$$

$$\frac{\partial u}{\partial y}(x, 0) = \exp x \Rightarrow F(x) = \exp x,$$

so $u(x, y) = \frac{xy^2}{2} + \frac{y^3}{6} + \exp x \cdot y + x^2$.

3. (a) Let us calculate the partial derivatives of $u(x, y) = f\left(\frac{y}{x}\right)$:

$$\frac{\partial u}{\partial x} = f'\left(\frac{y}{x}\right) \cdot \frac{-y}{x^2}, \quad \frac{\partial u}{\partial y} = f'\left(\frac{y}{x}\right) \cdot \frac{1}{x} \Rightarrow$$

$$\frac{\partial^2 u}{\partial x^2} = f''\left(\frac{y}{x}\right) \cdot \frac{y^2}{x^4} + f'\left(\frac{y}{x}\right) \cdot \frac{2y}{x^3}, \quad \frac{\partial^2 u}{\partial y^2} = f''\left(\frac{y}{x}\right) \cdot \frac{1}{x^2}.$$

By replacing in the Laplace equation, we get:

$$\frac{1}{x^2}\left(f''\left(\frac{y}{x}\right) \cdot \frac{y^2}{x^2} + f'\left(\frac{y}{x}\right) \cdot \frac{2y}{x} + f''\left(\frac{y}{x}\right)\right) = 0.$$

If we use the notation $\frac{y}{x} = t$, we obtain the following ODE:

$$(t^2 + 1) f''(t) + 2t f'(t) = 0.$$

In order to solve this equation, we denote the derivative of f by $g(t) = f'(t)$ and we have to solve a first order differential equation in $g(t)$:

$$(t^2 + 1) g'(t) + 2t \cdot g(t) = 0.$$

The general solution of this separable equation is $g(t) = \frac{C_1}{t^2+1}$, so the searched function $f(t)$ is any primitive of this function: $f(t) = C_1 \arctan t + C_2$.

(b) The partial derivatives of the function $u(x, y) = f(x^2 + y^2)$ are:

$$\frac{\partial u}{\partial x} = f'(x^2 + y^2) \cdot 2x, \quad \frac{\partial u}{\partial y} = f'(x^2 + y^2) \cdot 2y \Rightarrow$$

$$\frac{\partial^2 u}{\partial x^2} = f''(x^2 + y^2) \cdot 4x^2 + 2f'(x^2 + y^2),$$

$$\frac{\partial^2 u}{\partial y^2} = f''(x^2 + y^2) \cdot 4y^2 + 2f'(x^2 + y^2).$$

By replacing in the Laplace equation, we get:

$$f''(x^2 + y^2) \cdot (x^2 + y^2) + f'(x^2 + y^2) = 0.$$

If we use the notation $x^2 + y^2 = t$, we obtain the following ODE:

$$tf''(t) + f'(t) = 0.$$

We denote by $g(t) = f'(t)$ and so we have to solve a first order differential equation in $g(t)$:

$$tg'(t) + g(t) = 0.$$

The general solution of this separable equation is $g(t) = \frac{C_1}{t}$, so the searched function $f(t)$ is any primitive of this function: $f(t) = C_1 \ln t + C_2$.

4. (a) We use the d'Alembert formula (7.77):

$$u(x, t) = \frac{1}{2}[f(x + at) + f(x - at)] + \frac{1}{2a} \int_{x-at}^{x+at} g(z)dz.$$

In our case, $a = 2$, $f(x) = x$, $g(x) = \exp x$, so we have:

$$u(x, t) = \frac{1}{2}(x + 2t + x - 2t) + \frac{1}{4} \int_{x-2t}^{x+2t} \exp(z)dz = x + \frac{\exp x}{2} \sinh(2t)$$

(b) Now, we have $a = 1$, $f(x) = x$, $g(x) = -x$. Therefore,

$$u(x, t) = \frac{1}{2}(x + 2t + x - 2t) + \frac{1}{2} \int_{x-2t}^{x+2t} (-z)dz = x - 2tx$$

(c) We have $a = 1$, $f(x) = x^2$, $g(x) = 0$, so:

$$u(x,t) = \frac{1}{2}[(x+2t)^2 + (x-2t)^2] + \frac{1}{2}\int_{x-2t}^{x+2t} 0 dz = x^2 + 4t^2$$

5. We use the same d'Alembert formula:

$$u(x,t) = \frac{1}{2}[\sin(x+at) + \sin(x-at)] + \frac{1}{2a}\int_{x-at}^{x+at} dz =$$

$$= \sin x \cos at + t.$$

Hence, $u(x, \frac{\pi}{2a}) = \sin x \cos \frac{\pi}{2} + \frac{\pi}{2a} = \frac{\pi}{2a}$. At the moment $\frac{\pi}{2a}$, the string lies along the horizontal straight line $u = \frac{\pi}{2a}$.

6. (a) We use the formula (7.94):

$$u(x,t) = \sum_{n=1}^{\infty}(A_n \cos \omega_n at + B_n \sin \omega_n at)\sin \omega_n x,$$

where $\omega_n = \frac{n\pi}{l} = \frac{n\pi}{6}$ and the coefficients A_n, B_n are given by the formulas (7.97), (7.98):

$$A_n = \frac{2}{l}\int_0^l f(x)\sin\left(\frac{n\pi}{l}x\right)dx, \quad B_n = \frac{2}{n\pi a}\int_0^l g(x)\sin\left(\frac{n\pi}{l}x\right)dx.$$

In our case, $g(x) = 0$, so $B_n = 0$, and $f(x) = \begin{cases} x, & 0 \leq x < 5 \\ 5(6-x), & 5 \leq x \leq 6 \end{cases}$.

$$A_n = \frac{2}{6}\left[\int_0^5 x\sin\left(\frac{n\pi}{6}x\right)dx + 5\int_5^6(6-x)\sin\left(\frac{n\pi}{6}x\right)dx\right] =$$

$$= \frac{1}{3}\cdot\frac{6}{n\pi}\left[-x\cos\left(\frac{n\pi}{6}x\right)\Big|_0^5 + \int_0^5 \cos\left(\frac{n\pi}{6}x\right)dx + $$

$$+ 5\left(-(6-x)\cos\left(\frac{n\pi}{6}x\right)\Big|_5^6 - \int_5^6 \cos\left(\frac{n\pi}{6}x\right)dx\right)\right] =$$

$$= \frac{2}{n\pi}\left[-5\cos\left(\frac{5n\pi}{6}\right) + \frac{6}{n\pi}\sin\left(\frac{5n\pi}{6}\right) + 5\cos\left(\frac{5n\pi}{6}\right) + \frac{30}{n\pi}\sin\left(\frac{5n\pi}{6}\right)\right] =$$

$$= \frac{72}{n^2\pi^2}\sin\left(\frac{5n\pi}{6}\right),$$

so
$$u(x,t) = \frac{72}{\pi^2} \sum_{n=1}^{\infty} \frac{1}{n^2} \sin\left(\frac{5n\pi}{6}\right) \cos\left(\frac{n\pi}{6}t\right) \sin\left(\frac{n\pi}{6}x\right).$$

(b) In this case, $\omega_n = n$, $g(x) = 0$ so we have again $B_n = 0$. Let us calculate the coefficients A_n:

$$A_n = \frac{2c}{\pi} \int_0^{\pi} (\pi x - x^2) \sin nx\, dx =$$

$$= -\frac{2c}{\pi n} \left((\pi x - x^2) \cos nx \Big|_0^{\pi} - \int_0^{\pi} (\pi - 2x) \cos nx\, dx \right) =$$

$$= \frac{2c}{\pi n^2} \left((\pi - 2x) \sin nx \Big|_0^{\pi} + 2 \int_0^{\pi} \sin nx\, dx \right) = -\frac{4c}{\pi n^3} \cos nx \Big|_0^{\pi} =$$

$$= \frac{4c(1 - (-1)^n)}{\pi n^3} = \begin{cases} 0, & \text{if } n = 2k \\ \frac{8c}{\pi(2k+1)^3}, & \text{if } n = 2k+1 \end{cases},$$

so

$$u(x,t) = \frac{8c}{\pi} \sum_{k=0}^{\infty} \frac{1}{(2k+1)^3} \cos(2k+1)t \sin(2k+1)x.$$

(c) Since $f(x) = 0.1 \sin x$, we have:

$$A_n = \frac{0.2}{\pi} \int_0^{\pi} \sin x \sin nx\, dx = 0, \; \forall n > 1, \quad A_1 = \frac{0.2}{\pi} \cdot \frac{\pi}{2} = 0.1.$$

Since $g(x) = -0.2 \sin 2x$, we have:

$$B_n = -\frac{0.4}{n\pi} \int_0^{\pi} \sin 2x \sin nx\, dx = 0, \; \forall n \neq 2, \quad B_2 = -\frac{0.4}{2\pi} \cdot \frac{\pi}{2} = -0.1.$$

$$u(x,t) = 0.1(\cos t \sin x - \sin 2t \sin 2x).$$

(d) In this case, $A_n = 0$ because $f(x) = 0$. We calculate B_n:

$$B_n = \frac{2}{2n\pi} \int_0^6 g(x) \sin\left(\frac{n\pi}{6}x\right) dx = \frac{v}{n\pi} \int_2^4 \sin\left(\frac{n\pi}{6}x\right) dx =$$

$$= -\frac{6v}{n^2\pi^2} \cos\left(\frac{n\pi}{6}x\right)\Big|_2^4 = -\frac{6v}{n^2\pi^2}\left(\cos\frac{2n\pi}{3} - \cos\frac{n\pi}{3}\right) =$$

$$= \frac{12v}{n^2\pi^2} \sin\frac{n\pi}{6} \sin\frac{n\pi}{2} = \begin{cases} 0, & n = 2k \\ \frac{12v(-1)^k}{(2k+1)^2\pi^2} \sin\frac{(2k+1)\pi}{6}, & n = 2k+1. \end{cases}$$

$$u(x,t) = \frac{12v}{\pi^2} \sum_{k=0}^{\infty} \frac{(-1)^k}{(2k+1)^2} \sin\frac{(2k+1)\pi}{6} \sin\frac{(2k+1)\pi t}{3} \sin\frac{(2k+1)\pi x}{6}.$$

7. (a) We apply the formula (7.169):

$$u(x,t) = \sum_{n=1}^{\infty} u_n(x,t) = \sum_{n=1}^{\infty} A_n \exp\left(-\frac{n^2\pi^2 c^2}{l^2}t\right) \cdot \sin\left(\frac{n\pi}{l}x\right),$$

where the coefficients A_n are given by the formula (7.170):

$$A_n = \frac{2}{l} \int_0^l \varphi(x) \sin\left(\frac{n\pi}{l}x\right) dx =$$

$$= \frac{2}{l} \int_0^{\frac{l}{2}} x \sin\left(\frac{n\pi}{l}x\right) dx + \frac{2}{l} \int_{\frac{l}{2}}^l (l-x) \sin\left(\frac{n\pi}{l}x\right) dx =$$

$$= \frac{2}{l} \cdot \frac{-l}{n\pi}\left[x \cos\left(\frac{n\pi}{l}x\right)\Big|_0^{\frac{l}{2}} - \int_0^{\frac{l}{2}} \cos\left(\frac{n\pi}{l}x\right) dx +$$

$$+ (l-x)\cos\left(\frac{n\pi}{l}x\right)\Big|_{\frac{l}{2}}^l + \int_{\frac{l}{2}}^l \cos\left(\frac{n\pi}{l}x\right) dx\right] =$$

$$= \frac{-2}{n\pi}\left(\frac{l}{2}\cos\frac{n\pi}{2} - \frac{l}{n\pi}\sin\frac{n\pi x}{l}\Big|_0^{\frac{l}{2}} - \frac{l}{2}\cos\frac{n\pi}{2} + \frac{l}{n\pi}\sin\frac{n\pi x}{l}\Big|_{\frac{l}{2}}^l\right) =$$

$$= \frac{4l}{n^2\pi^2}\sin\frac{n\pi}{2} = \begin{cases} 0, & n=2k \\ \frac{4l(-1)^k}{(2k+1)^2\pi^2}, & n=2k+1. \end{cases}$$

$$u(x,t) = \frac{4l}{\pi^2}\sum_{k=0}^{\infty}\frac{(-1)^k}{(2k+1)^2}\exp\left(-\frac{(2k+1)^2\pi^2c^2}{l^2}t\right)\cdot\sin\left(\frac{(2k+1)\pi}{l}x\right).$$

(b) Let us compute the coefficients A_n:

$$A_n = \frac{2}{4}\int_0^4 \varphi(x)\sin\left(\frac{n\pi}{4}x\right)dx =$$

$$= \frac{1}{2}\int_0^1 x\sin\left(\frac{n\pi}{4}x\right)dx + \frac{1}{6}\int_1^4 (4-x)\sin\left(\frac{n\pi}{4}x\right)dx =$$

$$= \frac{1}{2}\cdot\frac{-4}{n\pi}\left[x\cos\left(\frac{n\pi}{4}x\right)\Big|_0^1 - \int_0^1 \cos\left(\frac{n\pi}{4}x\right)dx\right] +$$

$$+ \frac{1}{6}\cdot\frac{-4}{n\pi}\left[(4-x)\cos\left(\frac{n\pi}{4}x\right)\Big|_1^4 + \int_1^4 \cos\left(\frac{n\pi}{4}x\right)dx\right] =$$

$$= \frac{-2}{n\pi}\left(\cos\frac{n\pi}{4} - \frac{4}{n\pi}\sin\frac{n\pi x}{4}\Big|_0^1 - \cos\frac{n\pi}{4} + \frac{4}{3n\pi}\sin\frac{n\pi x}{4}\Big|_1^4\right) =$$

$$= \frac{32}{3n^2\pi^2}\sin\frac{n\pi}{4}.$$

$$u(x,t) = \frac{32}{3\pi^2}\sum_{n=1}^{\infty}\frac{1}{n^2}\exp\left(-n^2\pi^2 t\right)\cdot\sin\left(\frac{n\pi}{4}x\right).$$

(c) We calculate the coefficients A_n by the formula (7.170):

$$A_n = \frac{2}{\pi}\int_0^\pi 2\sin 3x \sin nx\,dx = 0, \quad \forall n \neq 3, \quad A_3 = 2$$

$$\Rightarrow u(x,t) = 2\exp(-9t)\sin(3x).$$

(d) Let us calculate the coefficients A_n:

$$A_n = \frac{2}{l} \int_0^l \left(3 \sin \frac{\pi x}{l} - 5 \sin \frac{2\pi x}{l}\right) \sin \left(\frac{n\pi}{l} x\right) dx =$$

$$= \frac{6}{l} \int_0^l \sin \frac{\pi x}{l} \sin \frac{n\pi x}{l} dx - \frac{10}{l} \int_0^l \sin \frac{2\pi x}{l} \sin \frac{n\pi x}{l} dx = 0, \ \forall n > 3,$$

$$A_1 = 3, \ A_2 = -5$$

$$\Rightarrow u(x,t) = 3 \exp\left(-\frac{\pi^2}{l^2} t\right) \sin \frac{\pi x}{l} - 5 \exp\left(-\frac{4\pi^2}{l^2} t\right) \sin \frac{2\pi x}{l}.$$

(e) The coefficients A_n are given by:

$$A_n = \frac{2}{l} \int_{x_1}^{x_2} u_0 \sin\left(\frac{n\pi}{l} x\right) dx = \frac{2 u_0}{l} \cdot \frac{-l}{n\pi} \cos \frac{n\pi x}{l} \Big|_{x_1}^{x_2} =$$

$$= \frac{-2 u_0}{n\pi} \left(\cos \frac{n\pi x_2}{l} - \cos \frac{n\pi x_1}{l}\right) = \frac{4 u_0}{n\pi} \sin \frac{n\pi(x_1+x_2)}{2l} \sin \frac{n\pi(x_2-x_1)}{2l}$$

and

$$u(x,t) = \sum_{n=1}^\infty A_n \exp\left(-\frac{n^2 \pi^2}{l^2} t\right) \cdot \sin\left(\frac{n\pi}{l} x\right).$$

(f) We have:

$$A_n = \frac{2}{l^3} \int_0^l x(l-x) \sin\left(\frac{n\pi}{l} x\right) dx =$$

$$= -\frac{2}{l^2 n\pi} \left[x(l-x) \cos\left(\frac{n\pi}{l} x\right) \Big|_0^l - \int_0^l (l-2x) \cos\left(\frac{n\pi}{l} x\right) dx \right] =$$

$$= \frac{2}{l n^2 \pi^2} \left[(l-2x) \sin\left(\frac{n\pi}{l} x\right) \Big|_0^l + 2 \int_0^l \sin\left(\frac{n\pi}{l} x\right) dx \right] =$$

$$= -\frac{4}{n^3 \pi^3} \cos\left(\frac{n\pi}{l} x\right) \Big|_0^l = -\frac{4((-1)^n - 1)}{n^3 \pi^3} = \begin{cases} 0, & n = 2k \\ \frac{8}{(2k+1)^3 \pi^3}, & n = 2k+1 \end{cases}.$$

$$\Rightarrow u(x,t) = \frac{8}{\pi^3} \sum_{k=0}^\infty \frac{1}{(2k+1)^3} \exp\left(-\frac{(2k+1)^2 \pi^2}{l^2} t\right) \cdot \sin\left(\frac{(2k+1)\pi}{l} x\right).$$

10.7 Chapter 7

8. We use the formula (7.177):

$$u(x,t) = \frac{1}{2c\sqrt{\pi t}} \int_{-\infty}^{\infty} \varphi(z) \exp\left(-\frac{(x-z)^2}{4c^2 t}\right) dz =$$

$$= \frac{A}{\sqrt{\pi t}} \int_{-\infty}^{\infty} \exp\left(-Bz^2 - \frac{(x-z)^2}{t}\right) dz =$$

$$= \frac{A}{\sqrt{\pi t}} \int_{-\infty}^{\infty} \exp\left[-\frac{Bt+1}{t}\left(z^2 - \frac{2xz}{Bt+1} + \left(\frac{x}{Bt+1}\right)^2 + \frac{Btx^2}{(Bt+1)^2}\right)\right] dz =$$

$$= \frac{A}{\sqrt{\pi t}} \exp\left(-\frac{Bx^2}{Bt+1}\right) \int_{-\infty}^{\infty} \exp\left[-\frac{Bt+1}{t}\left(z - \frac{x}{Bt+1}\right)^2\right] dz.$$

We use the change of variable: $w = \sqrt{\frac{Bt+1}{t}}\left(z - \frac{x}{Bt+1}\right)$ and we get:

$$= \frac{A}{\sqrt{\pi t}} \exp\left(-\frac{Bx^2}{Bt+1}\right) \sqrt{\frac{t}{Bt+1}} \int_{-\infty}^{\infty} \exp(-w^2) dw,$$

Since $\int_{-\infty}^{\infty} \exp(-w^2) dw = \sqrt{\pi}$, we obtain:

$$u(x,t) = \frac{A}{\sqrt{Bt+1}} \exp\left(-\frac{Bx^2}{Bt+1}\right).$$

Hence, $u\left(0, \frac{1}{A^2 B}\right) = \frac{A^2}{\sqrt{A^2+1}}$. For $A = 2\sqrt{6}$, we have $u\left(0, \frac{1}{A^2 B}\right) = \frac{24}{5} = 4.8$.

9. (a) We apply the formula (7.226):

$$u(\rho, \theta) = A_0 + \sum_{n=1}^{\infty} u_n(\rho, \theta) = A_0 + \sum_{n=1}^{\infty} (A_n \cos n\theta + B_n \sin n\theta) \rho^n,$$

where the coefficients A_n, B_n are given by the formula (7.228):

$$A_0 = \frac{1}{2\pi} \int_0^{2\pi} f(\theta) d\theta, \quad A_n = \frac{1}{\pi r^n} \int_0^{2\pi} f(\theta) \cos n\theta \, d\theta,$$

$$B_n = \frac{1}{\pi r^n} \int_0^{2\pi} f(\theta) \sin n\theta \, d\theta, \quad n = 1, 2, \ldots.$$

In our case, the boundary condition $u|_{x^2+y^2=1} = 3y$ can be written as $f(\theta) = 3\sin\theta$, so we have: $A_n = 0, \forall n \geq 0$, $B_n = 0, \forall n \geq 2$ and $B_1 = 3$, and $u(\rho, \theta) = 3\rho\sin\theta$. Since $x = \rho\cos\theta$, $y = \rho\sin\theta$, we have $u(x, y) = 3y$.

(b) The boundary condition $u|_{x^2+y^2=4} = x^2$ can be written as: $u(2, \theta) = 4\cos^2\theta = 2 + 2\cos 2\theta = f(\theta)$. By applying the same formulas as above, we obtain: $A_0 = 2$, $A_2 = \frac{1}{2}$ and $A_n = 0, \forall n \neq 0, 2$, $B_n = 0, \forall n \geq 1$, so $u(\rho, \theta) = 2 + \frac{1}{2}\rho^2\cos 2\theta = 2 + \frac{1}{2}\rho^2(\cos^2\theta - \sin^2\theta) \Rightarrow u(x, y) = 2 + \frac{1}{2}(x^2 - y^2)$.

(c) The boundary condition $u|_{x^2+y^2=16} = y^2 + 3xy$; can be written as: $u(4, \theta) = 16\sin^2\theta + 48\sin\theta\cos\theta = 8 - 8\cos 2\theta + 24\sin 2\theta = f(\theta)$. By applying the same formulas as above, we obtain: $A_0 = 8$, $A_2 = -\frac{1}{2}$, $B_2 = \frac{3}{2}$ and $A_n = 0$, $B_n = 0, \forall n \neq 0, 2$, so $u(\rho, \theta) = 8 - \frac{1}{2}\rho^2\cos 2\theta + \frac{3}{2}\rho^2\sin 2\theta = 8 - \frac{1}{2}\rho^2(\cos^2\theta - \sin^2\theta) + 3\rho^2\cos\theta\sin\theta \Rightarrow u(x, y) = 8 - \frac{1}{2}(x^2 - y^2) + 3xy$.

(d) Since $f(\theta) = \sin 3\theta + 4\cos 2\theta$, we have: $A_2 = 1$, $B_3 = \frac{1}{8}$ and $A_n = 0, \forall n \neq 2$, $B_n = 0, \forall n \neq 3$. Therefore, $u(\rho, \theta) = \rho^2\cos 2\theta + \frac{1}{8}\rho^3\sin 3\theta = \rho^2(\cos^2\theta - \sin^2\theta) + \frac{1}{8}\rho^3(3\cos^2\theta\sin\theta - \sin^3\theta) \Rightarrow u(x, y) = x^2 - y^2 + \frac{3}{8}x^2y - \frac{1}{8}y^3$.

10.8 Chapter 8

1. (a) $\|f - g\|_\infty = \sup_{x\in[0,1]} |x - x^2| = \frac{1}{4}$ $\|f' - g'\|_\infty = \sup_{x\in[0,1]} |1 - 2x| = 1 \Rightarrow d_1(f, g) = \frac{1}{4} + 1 = \frac{5}{4}$.

(b) $\|f - g\|_\infty = \sup_{x\in[0,2]} |x\exp(-x) - 1| = 1$ $\|f' - g'\|_\infty = \sup_{x\in[0,2]} |\exp(-x)(x - 1)| = 1 \Rightarrow d_1(f, g) = 1 + 1 = 2$.

2. (a) $\mathcal{F}[x] = \int_0^{\frac{\pi}{2}} (x^2 + 1)dx = \left(\frac{x^3}{3} + x\right)\Big|_0^{\frac{\pi}{2}} = \frac{\pi^3}{24} + \frac{\pi}{2}$.

(b) $\mathcal{F}[\sin x] = \int_0^{\frac{\pi}{2}} (\sin^2 x + \cos^2 x)dx = \int_0^{\frac{\pi}{2}} dx = \frac{\pi}{2}$.

(c) We apply Definition 8.1: $\delta_h \mathcal{F}[y_0] = \varphi_h'(0)$, where $\varphi_h(t) = \mathcal{F}[y_0 + th]$. In the first case, when $y_0 = \cos x$, we have:

$$\varphi_h(t) = \mathcal{F}[\cos x + t\sin 2x] = \int_0^{\frac{\pi}{2}} [(\cos x + t\sin 2x)^2 + (-\sin x + 2t\cos 2x)^2]dx =$$

$$= \int_0^{\frac{\pi}{2}} \left[1 + 2t(\cos x \sin 2x - 2\sin x \cos 2x) + t^2(1 + 3\cos^2 2x)\right]dx =$$

$$= \frac{\pi}{2} + 2t \int_0^{\frac{\pi}{2}} (\cos x \sin 2x - 2\sin x \cos 2x) dx + t^2 \int_0^{\frac{\pi}{2}} (1 + 3\cos^2 2x) dx$$

$$\Rightarrow \varphi_h'(t) = 2 \int_0^{\frac{\pi}{2}} (\cos x \sin 2x - 2\sin x \cos 2x) dx + 2t \int_0^{\frac{\pi}{2}} (1 + 3\cos^2 2x) dx$$

$$\Rightarrow \varphi_h'(0) = 2 \int_0^{\frac{\pi}{2}} (\cos x \sin 2x - 2\sin x \cos 2x) dx =$$

$$= 4 \int_0^{\frac{\pi}{2}} \sin x (1 - \cos^2 x) dx \stackrel{\cos x = t}{=} 4 \int_0^1 (1 - t^2) dt = \frac{8}{3}.$$

In the second case, when $y_0 = \sin 2x$, we have:

$$\varphi_h(t) = \mathcal{F}[(t+1)\sin 2x] = \int_0^{\frac{\pi}{2}} \left[(t+1)^2 \sin^2 2x + 4(t+1)^2 \cos^2 2x\right] dx =$$

$$= (t+1)^2 \int_0^{\frac{\pi}{2}} (1 + 3\cos^2 2x) dx \Rightarrow \varphi_h'(t) = 2(t+1) \int_0^{\frac{\pi}{2}} (1 + 3\cos^2 2x) dx$$

$$\Rightarrow \varphi_h'(0) = 2 \int_0^{\frac{\pi}{2}} (1 + 3\cos^2 2x) dx = \pi + 3 \int_0^{\frac{\pi}{2}} (1 + \cos 4x) = \frac{5\pi}{2}.$$

3.(a) We apply the Euler-Lagrange equation (8.23):

$$\frac{\partial F}{\partial y} - \frac{d}{dx}\left(\frac{\partial F}{\partial y'}\right) = 0.$$

In our case, $F(x, y, y') = (y'^2 + 2y^2 + 4y) \exp x$, so the partial derivatives are: $\frac{\partial F}{\partial y} = (4y + 4)\exp x$, $\frac{\partial F}{\partial y'} = 2y' \exp x$ so the above equation can be written: $(2y+2)\exp x - \frac{d}{dx}(y' \exp x) = 0$, or, equivalently, $(2y+2)\exp x - y'' \exp x - y' \exp x = 0$. We divide the equality by $\exp x$ and obtain:

$$y'' + y' - 2y = 2.$$

The characteristic equation, $\lambda^2 + \lambda - 2 = 0$, has the roots $\lambda_1 = 1$, $\lambda_2 = -2$, so the general solution of the homogeneous equation is $y_{GH} = C_1 \exp x + C_2 \exp(-2x)$. Since the free term, $f(x) = 2$ is a constant and $\lambda = 0$ is not a root of the characteristic equation, we search for a particular solution of the nonhomogeneous equation, $y_P = A$. We get $A = -1$, so the general solution of the above equation is:

$$y = y_{GH} + y_P = C_1 \exp x + C_2 \exp(-2x) - 1.$$

Now, we have to find the constants C_1 and C_2 such that the boundary conditions are satisfied:

$$\left.\begin{array}{l} y(0) = 0 \Rightarrow C_1 + C_2 - 1 = 0 \\ y(1) = e - 1 \Rightarrow C_1 e + C_2 \exp(-2) = e \end{array}\right\} \Rightarrow C_1 = 1, C_2 = 0,$$

so the sought function is $y = \exp x - 1$.

(b) Since $F(x, y, y') = y^2 - 4y \cos x - y'^2$, the partial derivatives are: $\frac{\partial F}{\partial y} = 2y - 4\cos x$, $\frac{\partial F}{\partial y'} = -2y'$, so the Euler-Lagrange equation can be written as: $2y - 4\cos x + 2y'' = 0$, or, equivalently,

$$y'' + y = 2\cos x.$$

The characteristic equation, $\lambda^2 + 1 = 0$, has the complex conjugate roots $\lambda_{1,2} = \pm i$, so the general solution of the homogeneous equation is $y_{GH} = C_1 \cos x + C_2 \sin x$. In this case we have resonance ($\lambda = i$ is a root of the characteristic equation), so we search for a particular solution of the nonhomogeneous equation, $y_P = x(A\cos x + B\sin x)$. By replacing in the above equation, we find $A = 0$, $B = 1$, so the general solution is:

$$y = y_{GH} + y_P = C_1 \cos x + C_2 \sin x + x \sin x.$$

From the boundary conditions we obtain: $y(0) = C_1 = 1$ and $y(\frac{\pi}{2}) = C_2 + \frac{\pi}{2} = \frac{\pi}{2} \Rightarrow C_2 = 0$, so the sought function is $y = \cos x + x \sin x$.

(c) Since $\frac{\partial F}{\partial y} = 12y$, $\frac{\partial F}{\partial y'} = 2x^2 y'$, the Euler-Lagrange equation can be written as: $12y - \frac{d}{dx}(2x^2 y') = 0$, or, equivalently,

$$x^2 y'' + 2xy' - 6y = 0.$$

This is an Euler equation, so we use the change of variable $x = \exp t$ and denote by $\tilde{y}(t) = y(\exp t)$ the new function. The derivatives of y (with respect to x) can be expressed using the derivatives of $\tilde{y}$ (with respect to t) as: $y' = \exp(-t)\tilde{y}'$, $y'' = \exp(-2t)(\tilde{y}'' - \tilde{y}')$ and we obtain a linear equation with constant coefficients in $\tilde{y}(t)$:

$$\tilde{y}'' + \tilde{y}' - 6\tilde{y} = 0.$$

The characteristic equation, $\lambda^2 + \lambda - 6 = 0$, has the roots $\lambda_1 = 2, \lambda_2 = -3$, so the general solution is $\tilde{y} = C_1 \exp 2t + C_2 \exp(-3t)$. Hence, the general solution of the Euler equation is $y = C_1 x^2 + C_2 x^{-3}$. From the boundary conditions we obtain: $y(1) = C_1 + C_2 = 1$ and $y(2) = 4C_1 + \frac{C_2}{8} = 4$ so $C_2 = 0$, $C_1 = 1$ and the sought function is $y = x^2$.

(d) As we can see, the function $F(x, y, y') = xy'^2 - 2y'$ does not depend on y: $\frac{\partial F}{\partial y} = 0$, so the Euler-Lagrange equation becomes: $\frac{d}{dx}(\frac{\partial F}{\partial y'}) = 0$, or, equivalently, $\frac{\partial F}{\partial y'} = C$. But $\frac{\partial F}{\partial y'} = 2xy' - 2$, so we obtain that $xy' = C_1$. From $y' = \frac{C_1}{x}$ we obtain the general solution: $y = C_1 \ln x + C_2$. Now, we use the boundary conditions to find the constants C_1 and C_2: $y(1) = C_2 = 0$, $y(e) = C_1 = 1$, so the sought function is $y = \ln x$.

(e) Since the function $F(x, y, y') = \sqrt{x(1 + y'^2)}$ does not depend on y, from the Euler-Lagrange equation we get $\frac{\partial F}{\partial y'} = C$, so

$$\frac{\sqrt{x}\, y'}{\sqrt{1 + y'^2}} = C.$$

We square this equality and find $y'^2 = \frac{C^2}{x - C^2}$, so, denoting $C_1 = C^2$, we obtain

$$y' = \frac{C_1}{\sqrt{x - C_1}} \Rightarrow y = 2C_1\sqrt{x - C_1} + C_2.$$

From the boundary conditions $y(1) = 0$ and $y(2) = 2$ we find $C_1 = 1$, $C_2 = 0$, so the searched extremal is $y = 2\sqrt{x - 1}$.

(f) In this case, the function $F(x, y, y') = \sqrt{y(1 + y'^2)}$ does not depend on x, so we apply the formula (8.26):

$$F - y'\frac{\partial F}{\partial y'} = C_1 \Rightarrow \sqrt{y(1 + y'^2)} - y'\frac{y'y}{\sqrt{y(1 + y'^2)}} = C_1$$

$$\Rightarrow y = C_1^2(1 + y'^2) \Rightarrow y' = \frac{1}{C_1}\sqrt{y - C_1^2}.$$

This equation can be written $\frac{y'}{2\sqrt{y - C_1^2}} = \frac{1}{2C_1}$ and by integration we get: $\sqrt{y - C_1^2} = \frac{x}{2C_1} + C_2$, so the general solution is

$$y = \left(\frac{x}{2C_1} + C_2\right)^2 + C_1^2.$$

Now, let us use the boundary conditions to find the constants C_1 and C_2:

$$\left.\begin{array}{l} y(0) = C_1^2 + C_2^2 = 1 \\ y(2) = \frac{1}{C_1^2} + \frac{2C_2}{C_1} + C_2^2 + C_1^2 = 2 \end{array}\right\} \Rightarrow C_2 = \frac{C_1}{2} - \frac{1}{2C_1}.$$

By replacing in $C_1^2 + C_2^2 = 1$ we obtain the equation $5C_1^4 - 6C_1^2 + 1 = 0$. Its solutions are ± 1 and $\pm\frac{1}{\sqrt{5}}$. For $C_1 = \pm 1$ we obtain $C_2 = 0$, so $y = \frac{x^2}{4} + 1$. For $C_1 = \pm\frac{1}{\sqrt{5}}$ we find $C_2 = \mp\frac{2}{\sqrt{5}}$ and $y = \frac{5}{4}x^2 - 2x + 1$.

(g) In this case, we apply the Euler–Poisson equation (8.35) for $n = 2$:

$$\frac{\partial F}{\partial y} - \frac{d}{dx}\left(\frac{\partial F}{\partial y'}\right) + \frac{d^2}{dx^2}\left(\frac{\partial F}{\partial y''}\right) = 0.$$

In our case, $F(x, y, y', y'') = 4y^2 - 5y'^2 + y''^2$, so the above equation can be written: $8y - \frac{d}{dx}(-10y') + \frac{d^2}{dx^2}(2y'')$, or, equivalently,

$$y^{(4)} + 5y'' + 4y = 0.$$

The characteristic equation associated to this homogeneous equation (of order 4) is $\lambda^4 + 5\lambda^2 + 4 = 0 \Leftrightarrow (\lambda^2 + 1)(\lambda^2 + 4) = 0$, so its roots are: $\lambda_{1,2} = \pm i$, and $\lambda_{3,4} = \pm 2i$. Hence, the general solution is

$$y = C_1 \cos x + C_2 \sin x + C_3 \cos 2x + C_4 \sin 2x.$$

From the boundary conditions we obtain: $y(0) = C_1 + C_3 = 3$, $y'(0) = C_2 + 2C_4 = 4$, $y(\frac{\pi}{2}) = C_2 - C_3 = 1$ and $y'(\frac{\pi}{2}) = -C_1 - 2C_4 = 0$. The solution of this system of equation is $C_1 = 0$, $C_2 = 4$, $C_3 = 3$, $C_4 = 0$, so the sought function is $y = 4\sin x + 3\cos 2x$.

(h) Here, the Euler-Poisson equation (when $n = 2$) for the function $F(x, y, y', y'') = 2yy' - 2y^2 - y'^2 + y''^2$, can be written:

$$2y' - 4y - \frac{d}{dx}(2y - 2y') + \frac{d^2}{dx^2}(2y'') = 0,$$

or, equivalently,

$$y^{(4)} + y'' - 2y = 0.$$

The characteristic equation associated to this homogeneous equation is $\lambda^4 + \lambda^2 - 2 = 0$. To solve it, we denote $\lambda^2 = t$ and we get: $t^2 + t - 2 = 0$ with the roots 1 and -2. For $\lambda^2 = 1$ we obtain the real roots $\lambda_1 = 1$, $\lambda_2 = -1$, while for $\lambda^2 = -2$ we obtain the complex conjugate roots $\lambda_{3,4} = \pm i\sqrt{2}$. Hence, the general solution is

$$y = C_1 \exp x + C_2 \exp(-x) + C_3 \cos\sqrt{2}x + C_4 \sin\sqrt{2}x,$$

which can be also written in the equivalent form

$$y = K_1 \cosh x + K_2 \sinh x + C_3 \cos \sqrt{2}x + C_4 \sin \sqrt{2}x,$$

From the boundary conditions we obtain:

$$y(0) = K_1 + C_3 = 1$$
$$y'(0) = K_2 + \sqrt{2}C_4 = 0$$
$$y(1) = K_1 \cosh 1 + K_2 \sinh 1 + C_3 \cos \sqrt{2} + C_4 \sin \sqrt{2} = \cosh 1$$
$$y'(1) = K_1 \sinh 1 + K_2 \cosh 1 - C_3\sqrt{2} \sin \sqrt{2} + C_4\sqrt{2} \cos \sqrt{2} = \sinh 1$$

From the first and the second equation we have $K_1 = 1 - C_3$, $K_2 = -\sqrt{2}C_4$ and, after replacing K_1 and K_2 in the last two equations we obtain a homogeneous system (in the unknowns C_3 and C_4) which has the unique solution $C_3 = C_4 = 0$. Hence, $K_1 = 1$, $K_2 = 0$ and the sought function is $y = \cosh x$.

(i) We have $F(x, y, y', y'') = yy' + y'^2 + yy'' + y'y'' + y''^2$, so the Euler-Poisson equation is:

$$y' + y'' - \frac{d}{dx}(y + 2y' + y'') + \frac{d^2}{dx^2}(y + y' + 2y'') = 0$$

$$\Leftrightarrow y' + y'' - y' - 2y'' - y''' + y'' + y''' + y^{(4)} = 0, \text{ or } y^{(4)} = 0.$$

The general solution of this equation is

$$y = C_1 x^3 + C_2 x^2 + C_3 x + C_4$$

From the boundary conditions we have: $y(0) = C_4 = 0$, $y'(0) = C_3 = 0$, $y(1) = C_1 + C_2 + C_3 + C_4 = 1$, $y'(1) = 3C_1 + 2C_2 + C_3 = 2$. Hence, $C_3 = C_4 = C_1 = 0$, $C_2 = 1$ and the solution is $y(x) = x^2$.

(j) Let us write the Euler-Lagrange system (8.42) in our case ($n = 2$, $y = y(x)$, $z = z(x)$ and $F(x, y, z, y', z') = y'^2 + z^2 + z'^2$):

$$\begin{cases} \frac{\partial F}{\partial y} - \frac{d}{dx}\left(\frac{\partial F}{\partial y'}\right) = 0 \\ \frac{\partial F}{\partial z} - \frac{d}{dx}\left(\frac{\partial F}{\partial z'}\right) = 0 \end{cases} \Leftrightarrow \begin{cases} \frac{d}{dx}(2y') = 0 \\ 2z - \frac{d}{dx}(2z') = 0 \end{cases} \Leftrightarrow \begin{cases} y' = C_1 \\ z'' - z = 0 \end{cases}$$

From the first equation we get: $y = C_1 x + C_2$, while the general solution of the second equation is: $z = C_3 \exp x + C_4 \exp(-x)$.

From the boundary conditions we have: $y(1) = C_1 + C_2 = 1$, $y(2) = 2C_1 + C_2 = 2$, so $C_1 = 1$, $C_2 = 0$ and $y = x$.

In the same way, for the function z we get: $z(1) = C_3 e + C_4 e^{-1} = e$ and $z(2) = C_3 e^2 + C_4 e^{-2} = e^2$, so $C_3 = 1$, $C_4 = 0$ and $z = \exp x$.

(k) Since $F(x, y, z, y', z') = 8y\cos x + 2z^2 + 2y'z' + y'^2 - z'^2$, the Euler-Lagrange system is written:

$$\begin{cases} 8\cos x - \frac{d}{dx}(2z' + 2y') = 0 \\ 4z - \frac{d}{dx}(2y' - 2z') = 0 \end{cases} \Leftrightarrow \begin{cases} y'' + z'' = 4\cos x \\ -y'' + z'' + 2z = 0 \end{cases}$$

If we add the equations (and divide everything by 2), we obtain the following equation in z:

$$z'' + z = 2\cos x.$$

The general solution of this equation (see the exercise b)) is

$$z = C_1 \cos x + C_2 \sin x + x \sin x.$$

From the boundary conditions we have: $z(0) = C_1 = 0$, $z(\frac{\pi}{2}) = C_2 + \frac{\pi}{2} = \frac{\pi}{2}$, so $C_2 = 0$ and $z = x \sin x$. Since $y'' = -z'' + 4\cos x$, we get

$$y'' = 2\cos x + x \sin x$$

and we have to integrate twice to obtain the expression of y:

$$y' = 2\sin x + \int x \sin x \, dx = 3\sin x - x\cos x + C_3.$$

$$\Rightarrow y = -4\cos x - x\sin x + C_3 x + C_4.$$

From the boundary conditions we have: $y(0) = -4 + C_4 = 0 \Rightarrow C_4 = 4$, $y(\frac{\pi}{2}) = -\frac{\pi}{2} + C_3 \frac{\pi}{2} + 4 = 4 \Rightarrow C_3 = 1$, so $y = -4\cos x - x\sin x + x + 4$.

4. The work of a force $\mathbf{F} = P(x, y, z)\mathbf{i} + Q(x, y, z)\mathbf{j} + R(x, y, z)\mathbf{k}$ along an oriented curve γ can be calculated as a line integral of the second type: (see [29], 5.2, or [38], 17.2):

$$W = \int_\gamma P(x, y, z)\, dx + Q(x, y, z)\, dy + R(x, y, z)\, dz.$$

We assume that the curve γ is defined by the equations:

$$\gamma : \begin{cases} y = y(x) \\ z = z(x) \end{cases}, \quad x \in [0, 1],$$

where $y(0) = 0$, $y(1) = 1$, $z(0) = 0$, $z(1) = 2/3$ (because γ connects the points $O(0, 0, 0)$ and $A(1, 1, 2/3)$). The work of the force $\mathbf{F} = y^2\mathbf{i} + z\mathbf{j} + x^2\mathbf{k}$ along the curved γ defined above is:

$$W = \int_\gamma y^2 dx + z dy + x^2 dz = \int_0^1 \left[y^2(x) + z(x)y'(x) + x^2 z'(x) \right] dx.$$

We have to find the extremals of the functional

$$\mathcal{F}[y, z] = \int_0^1 \left(y^2 + zy' + x^2 z' \right) dx$$

with the above boundary conditions. The Euler-Lagrange system is:

$$\begin{cases} 2y - \frac{d}{dx}(z) = 0 \\ y' - \frac{d}{dx}(x^2) = 0 \end{cases} \Leftrightarrow \begin{cases} z' = 2y \\ y' = 2x \end{cases}.$$

From the second equation we find $y = x^2 + C_1$ and we replace in the first equation: $z' = 2x^2 + 2C_1 \Rightarrow z = \frac{2}{3}x^3 + 2C_1 x + C_2$. By using the boundary conditions we obtain $C_1 = C_2 = 0$, so the searched curve is:

$$\gamma : \begin{cases} y = x^2 \\ z = \frac{2}{3}x^3 \end{cases}, \quad x \in [0, 1].$$

5. The Euler-Ostrogradsky equation is (see (8.58)):

$$\frac{\partial F}{\partial z} - \frac{\partial}{\partial x}\left(\frac{\partial F}{\partial p}\right) - \frac{\partial}{\partial y}\left(\frac{\partial F}{\partial q}\right) = 0.$$

By using the Monge's notation $p = \frac{\partial z}{\partial x}, q = \frac{\partial z}{\partial y}$, our function is:

$$F(x, y, z, p, q) = q^2 - p^2,$$

so the Euler-Ostrogradsky equation in this case is written:

$$0 - \frac{\partial}{\partial x}(-2p) - \frac{\partial}{\partial y}(2q) = 0,$$

or, equivalently,

$$\frac{\partial^2 z}{\partial x^2} - \frac{\partial^2 z}{\partial y^2} = 0.$$

6. From Theorem 8.3, we have to find the extremals of the auxiliary functional

$$\mathcal{F}[y] + \lambda \mathcal{G}[y] = \int_0^1 \left(y'^2 + x^2 + \lambda y^2\right) dx.$$

The Euler-Lagrange equation for this functional is:

$$2\lambda y - \frac{d}{dx}(2y') = 0 \Rightarrow y'' - \lambda y = 0.$$

To solve this ODE-2, we have to study three cases:

Case I $\lambda = 0 \Rightarrow y = C_1 x + C_2$ and from the boundary conditions $y(0) = y(1) = 0$ we find $C_1 = C_2 = 0$, so $y(x) = 0$, a contradiction because the constraint $\int_0^1 y^2 dx = 2$ is not fulfilled.

Case II $\lambda > 0 \Rightarrow \lambda = \omega^2$, $\omega > 0$, so $y = C_1 \exp x + C_2 \exp(-x)$ and from the boundary conditions $y(0) = y(1) = 0$ we find

$$\begin{cases} C_1 + C_2 = 0 \\ C_1 \exp \omega + C_2 \exp(-\omega) = 0 \end{cases}.$$

The solution of the system is $C_1 = C_2 = 0$, so we obtained again $y(x) = 0$, a contradiction.

Case III $\lambda < 0 \Rightarrow \lambda = -\omega^2$, $\omega > 0$, so $y = C_1 \cos \omega x + C_2 \sin \omega x$. Since $y(0) = 0$ we find $C_1 = 0$ and from $y(1) = 0$ we obtain $C_2 \sin \omega = 0$, so $\omega = k\pi$, $k = 1, 2, \ldots$ (as we saw above, we cannot take $C_2 = C_1 = 0$).

Now, we know that $y = C_2 \sin k\pi x$, let us use the constraint to find C_2:

$$\int_0^1 C_2^2 \sin^2 k\pi x \, dx = 2 \Rightarrow C_2^2 \int_0^1 \frac{1 - \cos 2k\pi x}{2} dx = 2 \Rightarrow C_2^2 = 4.$$

Hence, $C_2 = \pm 2$. The searched extremals are: $y = \pm 2 \sin k\pi x$, $k = 1, 2, \ldots$.

7. From Theorem 8.3, we have to find the extremals of the auxiliary functional

$$\mathcal{F}[y, z] + \lambda \mathcal{G}[y, z] = \int_0^1 \left(y'^2 + z'^2 - 4xz' - 4z + \lambda(y'^2 - xy' - z'^2)\right) dx.$$

Let us write the Euler-Lagrange system for this functional:

$$\begin{cases} 0 - \frac{d}{dx}[2y'(1+\lambda) - \lambda x] = 0 \\ -4 - \frac{d}{dx}(2z' - 4x - 2\lambda z') = 0 \end{cases} \Rightarrow \begin{cases} 2y'(1+\lambda) - \lambda x = C_1 \\ (1-\lambda)z'' = 0, \end{cases}$$

so $\lambda = 1$ or $z'' = 0$.

Case I $\lambda = 1$. Then, from the first equation, we obtain $y' = \frac{x+C_1}{4}$, so $y = \frac{x^2}{8} + \frac{C_1 x}{4} + C_2$. From the boundary conditions $y(0) = 0$ and $y(1) = 1$ we obtain $C_2 = 0$ and $C_1 = \frac{7}{2}$, so $y = \frac{x^2+7x}{8}$.

Now, let us use the constraint to find $z(x)$. Since $y'(x) = \frac{2x+7}{8}$, we get:

$$\int_0^1 \left(\left(\frac{2x+7}{8}\right)^2 - \frac{2x^2+7x}{8} - z'^2 \right) dx = 2,$$

so

$$\frac{1}{64} \int_0^1 \left(-12x^2 - 28x + 49 \right) dx - \int_0^1 z'^2 dx = 2$$

and we obtain that $\int_0^1 z'^2 dx = -\frac{97}{64} < 0$, a contradiction (because $z'^2 \geq 0$), so we have no solutions in this case. Now, let us study the other case.

Case II $z'' = 0$. It follows that z is a linear function: $z = C_2 x + C_3$ and from $z(0) = 0$, $z(1) = 1$ we find $C_3 = 0$, $C_2 = 1$, so $z(x) = x$. From the first equation we obtain that $y' = \frac{\lambda x + C_1}{2(1+\lambda)}$, so

$$y = \frac{\lambda x^2}{4(1+\lambda)} + \frac{C_1 x}{2(1+\lambda)} + C_4.$$

From $y(0) = 0$ we obtain that $C_4 = 0$, from $y(1) = 1$ we get $C_1 = \frac{3\lambda+4}{2}$, so

$$y = \frac{\lambda x^2 + (3\lambda+4)x}{4(1+\lambda)}.$$

In order to find λ, we use the constraint. Since $z' = 1$ and $y' = \frac{2\lambda x + 3\lambda + 4}{4(1+\lambda)}$, the integral can be written:

$$\int_0^1 \left[\left(\frac{2\lambda x + 3\lambda + 4}{4(1+\lambda)}\right)^2 - x\left(\frac{2\lambda x + 3\lambda + 4}{4(1+\lambda)}\right) \right] dx - \int_0^1 dx = 2$$

$$\Leftrightarrow \frac{1}{16(1+\lambda)^2} \int_0^1 \left[-4\lambda(\lambda+2)x^2 - 4(3\lambda+4)x + (3\lambda+4)^2 \right] dx = 3$$

$$\Leftrightarrow \left(-4\lambda(\lambda+2)\frac{x^3}{3} - 4(3\lambda+4)\frac{x^2}{2} + (3\lambda+4)^2 x \right)\Big|_0^1 = 48(1+\lambda)^2$$

and we obtain the equation

$$121\lambda^2 + 242\lambda + 120 = 0 \Leftrightarrow 11^2(\lambda+1)^2 = 1 \Leftrightarrow \lambda + 1 = \pm\frac{1}{11}.$$

Thus, the solutions are $\lambda_1 = -\frac{10}{11}$, and $\lambda_2 = -\frac{12}{11}$ and the searched extremals are: $y = \frac{1}{2}(-5x^2 + 7x)$ and $y = 3x^2 - 2x$.

10.9 Chapter 9

1. We can see that

$$\mathcal{P}(A) = \mathcal{P}(A \cap \overline{B}) + \mathcal{P}(A \cap B)$$
$$\mathcal{P}(B) = \mathcal{P}(\overline{A} \cap B) + \mathcal{P}(A \cap B)$$

   By subtracting these two equalities, we obtain the searched relation.
2. If (A, B) are independent, then $\mathcal{P}(A \cap B) = \mathcal{P}(A)\mathcal{P}(B)$.
   By de Morgan's laws, $\overline{A} \cap \overline{B} = \overline{A \cup B}$, so we have:

$$\mathcal{P}(\overline{A} \cap \overline{B}) = 1 - \mathcal{P}(A \cup B) =$$
$$= 1 - \mathcal{P}(A) - \mathcal{P}(B) + \mathcal{P}(A \cap B) =$$
$$= 1 - \mathcal{P}(A) - \mathcal{P}(B) + \mathcal{P}(A)\mathcal{P}(B) =$$
$$= (1 - \mathcal{P}(A))(1 - \mathcal{P}(B)) =$$
$$= \mathcal{P}(\overline{A})\mathcal{P}(\overline{B}),$$

   hence $(\overline{A}, \overline{B})$ are independent.
   Conversely, if $\overline{A}$ and $\overline{B}$ are independent, then, as we proved above, $A = \overline{\overline{A}}$ and $B = \overline{\overline{B}}$ are independent.
3. Since $P_A(B) = 1/6 = \frac{\mathcal{P}(A \cap B)}{\mathcal{P}(A)}$, we get: $P(A \cap B) = \frac{1}{5} \cdot \frac{1}{6} = \frac{1}{30}$. On the other hand, $P(B) = 1 - P(\overline{B}) = 1 - 5/6 = 1/6$. Thus,

$$\mathcal{P}(A \cup B) = \mathcal{P}(A) + \mathcal{P}(B) - \mathcal{P}(A \cap B) = 1/5 + 1/6 - 1/30 = 1/3.$$

   Since $P(A \cap B) = \frac{1}{30} = P(A)P(B)$, we see that A and B are independent.
4. (a) We apply the "urn model" (sampling with replacement) from Example 9.6. Thus, we use the formula (9.9) with $n = 5$, $p = 0.95$ and $q = 0.05$. Let A_k

denote the event that we get exactly k good products, $k = 0, 1, \ldots, 5$.

$$P(A_3) = \binom{5}{3}(0.95)^3(0.05)^2 = \frac{5!}{3!2!}(0.95)^3(0.05)^2 =$$
$$= 10 \times 0.8574 \times 0.0025 = 0.0214.$$

(b) In this case, the event is $A_1 \cup A_2 \cup A_3 \cup A_4 \cup A_5$, which can also be written $\overline{A_0}$, so we have:

$$P(\overline{A_0}) = 1 - P(A_0) = 1 - (0.05)^5 = 0.9999.$$

(c) $P(A_5) = (0.95)^5 = 0.7538.$

5. (a) We use again the formula (9.9) from Example 9.6. In this case, $n = 10$ and $p = q = 1/2$. Let A_k denote the event that exactly k heads are obtained during the 10 trials. We have to calculate the probability of the event $A_0 \cup A_1 \cup \ldots \cup A_8 = \overline{A_9 \cup A_{10}}$:

$$P(\overline{A_9 \cup A_{10}}) = 1 - \binom{10}{9}\left(\frac{1}{2}\right)^9\left(\frac{1}{2}\right)^1 - \binom{10}{10}\left(\frac{1}{2}\right)^{10}\left(\frac{1}{2}\right)^0 = 0.9892.$$

(b)
$$P(A_7) = \binom{10}{3}\left(\frac{1}{2}\right)^7\left(\frac{1}{2}\right)^3 = \frac{10 \times 9 \times 8}{6}\left(\frac{1}{2}\right)^{10} = \frac{120}{1024} = 0.117.$$

(c) Let H_i denote the event of getting a *head* at the i-th trial, $i = 1, 2, \ldots, 10$. It is easy to see that $\{H_i\}_{i=1,2,\ldots,10}$ are independent, so the required probability is:

$$P(\overline{H_1})P(\overline{H_2})P(\overline{H_3})P(H_4) = \frac{1}{2^4} = 0.0625 \approx 6\%.$$

6. The probability space is isomorphic to the probability space from Example 9.7 (sampling without replacement). Here $a = 7$, $b = 9$ and $n = 3$. Let B_k denote the event that exactly k of the three students are boys.

(a) $P(B_2) = \dfrac{\binom{7}{2} \cdot \binom{9}{1}}{\binom{7+9}{3}} = \dfrac{27}{80} = 0.338.$

(b) $P(B_0) = \dfrac{\binom{7}{0} \cdot \binom{9}{3}}{\binom{7+9}{3}} = \dfrac{3}{20} = 0.15.$

(c) $P(B_1 \cup B_2 \cup B_3) = 1 - P(B_0) = 1 - \dfrac{3}{20} = 0.85.$

7. (a) We have two independent events: A: "the first element is 3" and B: "the second element is 4", each one having the probability $P(A) = P(B) = \frac{1}{10}$.

Thus, we have:

$$P(A \cap B) = P(A)P(B) = \frac{1}{10} \cdot \frac{1}{10} = \frac{1}{100} = 0.01.$$

(b) In the second situation, since $P(B) \neq P_A(B)$, the events A and B are not independent. We have $P(A) = \frac{1}{10}$, but $P_A(B) = \frac{1}{9}$ (after taking out the element 3, there are only 9 possibilities to choose the second element). Therefore, we have:

$$P(A \cap B) = P(A)P_A(B) = \frac{1}{10} \cdot \frac{1}{9} = \frac{1}{90} = 0.011.$$

8. We denote by A_1 the event that a white ball is taken from the first box, and by A_2 the event that a black ball is taken. We have $P(A_1) = \frac{2}{3}$, $P(A_2) = \frac{1}{3}$ and we notice that $\{A_1, A_2\}$ is a partition of Ω: $A_1 \cup A_2 = \Omega$ and $A_1 \cap A_2 = \emptyset$. Let B denote the event of taking a white ball from the second box. If the ball from the first box is white, then the probability of drawing a white ball from the second box is $P_{A_1}(B) = \frac{3}{5}$. Otherwise, if the ball from the first box is black, the probability to get a white ball from the second box is $P_{A_2}(B) = \frac{2}{5}$. We have:

$$P(B) = P(B \cap \Omega) = P((B \cap A_1) \cup (B \cap A_2)) =$$
$$= P((B \cap A_1)) + P((B \cap A_2)) =$$
$$= P_{A_1}(B)P(A_1) + P_{A_2}(B)P(A_2) =$$
$$= \frac{3}{5} \cdot \frac{2}{3} + \frac{2}{5} \cdot \frac{1}{3} = \frac{8}{15} = 0.533.$$

Note that we have just applied the Law of total probability (9.21).

9. (a) Let X denote the event of taking a defective bag, and let A, B, C be the events consisting in taking a sugar bag supplied by the refinery A, B and C respectively. Thus, $A \cup B \cup C = \Omega$ is a partition of the sample space Ω. We apply again the Law of total probability (9.21) and find:

$$P(X) = P_A(X)P(A) + P_B(X)P(B) + P_C(X)P(C) =$$
$$= \frac{2}{100} \cdot \frac{500}{1000} + \frac{1}{100} \cdot \frac{300}{1000} + \frac{1}{100} \cdot \frac{200}{1000} = \frac{15}{1000} = 0.015.$$

(b) We use the Bayes' formula (9.23) to compute the required probability:

$$P_X(B) = \frac{P(X \cap B)}{P(X)} = \frac{P_B(X)P(B)}{P_A(X)P(A) + P_B(X)P(B) + P_C(X)P(C)} = \frac{1}{5}.$$

10. (a) The values of X are: $k = 0, 1, 2, 3$. The probabilities that a child is a boy or a girl are equal, $p = q = \frac{1}{2}$. Consequently, the probability to have exactly

k girls in a family with 3 children is equal to $\binom{3}{k}\left(\frac{1}{2}\right)^k \left(\frac{1}{2}\right)^{3-k}$ i.e. X is a binomial distributed random variable: $X \sim Bin\left(3, \frac{1}{2}\right)$.

$$X \sim \begin{pmatrix} 0 & 1 & 2 & 3 \\ \binom{3}{0}\left(\frac{1}{2}\right)^0 \left(\frac{1}{2}\right)^3 & \binom{3}{1}\left(\frac{1}{2}\right)\left(\frac{1}{2}\right)^2 & \binom{3}{2}\left(\frac{1}{2}\right)^2\left(\frac{1}{2}\right)^1 & \binom{3}{3}\left(\frac{1}{2}\right)^3\left(\frac{1}{2}\right)^0 \end{pmatrix}$$

$$= \begin{pmatrix} 0 & 1 & 2 & 3 \\ \frac{1}{2^3} & \frac{3}{2^3} & \frac{3}{2^3} & \frac{1}{2^3} \end{pmatrix}.$$

(b) Since $F_X(x) = P\{X \leq x\}$, we get:

$$F_X(x) = \begin{cases} 0, & \text{if } x < 0 \\ \frac{1}{2^3} = \frac{1}{8}, & \text{if } x \in [0, 1) \\ \frac{1}{2^3} + \frac{3}{2^3} = \frac{1}{2}, & \text{if } x \in [1, 2) \\ \frac{1}{2^3} + \frac{3}{2^3} + \frac{3}{2^3} = \frac{7}{8}, & \text{if } x \in [2, 3) \\ 1, & \text{if } x \geq 3 \end{cases}$$

(c) The expected value is $E(X) = np = 3 \cdot \frac{1}{2} = 3/2$, and the variance is $Var(X) = npq = 3/4$. The standard deviation is $\sigma_X = \sqrt{3}/2 \approx 0.866$.

11. (a) Since X and Y are independent,

$$P\{X = x_i, Y = y_j\} = P\{X = x_i\}P\{Y = y_j\} = p_i q_j,$$

so we have:

$$XY \sim \begin{pmatrix} x_1 y_1 & x_1 y_2 & x_1 y_3 & x_2 y_1 & x_2 y_2 & x_2 y_3 & x_3 y_1 & x_3 y_2 & x_3 y_3 \\ p_1 q_1 & p_1 q_2 & p_1 q_3 & p_2 q_1 & p_2 q_2 & p_2 q_3 & p_3 q_1 & p_3 q_2 & p_3 q_3 \end{pmatrix}$$

$$= \begin{pmatrix} 1 & 2 & 3 & 2 & 4 & 6 & 3 & 6 & 9 \\ 0.04 & 0.24 & 0.12 & 0.01 & 0.06 & 0.03 & 0.05 & 0.30 & 0.15 \end{pmatrix}$$

$$= \begin{pmatrix} 1 & 2 & 3 & 4 & 6 & 9 \\ 0.04 & 0.25 & 0.17 & 0.06 & 0.33 & 0.15 \end{pmatrix}.$$

We can check our result by summing the probabilities:

$$0.04 + 0.25 + 0.17 + 0.06 + 0.33 + 0.15 = 1.$$

Similarly,

$$X+Y \sim \begin{pmatrix} 2 & 3 & 4 & 3 & 4 & 5 & 4 & 5 & 6 \\ 0.04 & 0.24 & 0.12 & 0.01 & 0.06 & 0.03 & 0.05 & 0.30 & 0.15 \end{pmatrix}.$$

$$= \begin{pmatrix} 2 & 3 & 4 & 5 & 6 \\ 0.04 & 0.25 & 0.23 & 0.33 & 0.15 \end{pmatrix}.$$

Now,

$$3X \sim \begin{pmatrix} 3 & 6 & 9 \\ 0.4 & 0.1 & 0.5 \end{pmatrix},$$

hence

$$Z = 3X + Y \sim \begin{pmatrix} 4 & 5 & 6 & 7 & 8 & 9 & 10 & 11 & 12 \\ 0.04 & 0.24 & 0.12 & 0.01 & 0.06 & 0.03 & 0.05 & 0.30 & 0.15 \end{pmatrix}.$$

(b) The expected value and the variance of X are:

$$E(X) = 1 \cdot 0.4 + 2 \cdot 0.1 + 3 \cdot 0.5 = 2.1 \Rightarrow \mu_X = 2.1$$

$$Var(X) = E(X^2) - E(X)^2 =$$
$$= 1 \cdot 0.4 + 4 \cdot 0.1 + 9 \cdot 0.5 - 2.1^2 = 0.89 \Rightarrow \sigma_X = 0.94$$

The expected value and the variance of Y are:

$$E(Y) = 1 \cdot 0.1 + 2 \cdot 0.6 + 3 \cdot 0.3 = 2.2 \Rightarrow \mu_Y = 2.2$$

$$Var(Y) = E(Y^2) - E(Y)^2 =$$
$$= 1 \cdot 0.1 + 4 \cdot 0.6 + 9 \cdot 0.3 - 2.2^2 = 0.36 \Rightarrow \sigma_Y = 0.6$$

We can compute the mean and the variance of $Z = 3X + Y$ in the same way, since we have calculated the probability mass function of Z, or we can use the properties of the expected value and the variance (Theorems 9.2 and 9.4). Thus, we can write:

$$E(3X + Y) = 3E(X) + E(Y) = 3 \cdot 2.1 + 2.2 = 8.5 \Rightarrow \mu_Z = 8.5$$

and, since X and Y are independent,

$$Var(3X + Y) = 9Var(X) + Var(Y) = 9 \cdot 0.89 + 0.36 = 8.37 \Rightarrow \sigma_Z = 2.89.$$

(c) $P\{\mu_Z - 2\sigma_Z < Z < \mu_Z + 2\sigma_Z\} = P\{2.72 < Z < 12.38\} = 1$, because the values of Z are $4, 5, \ldots, 12$.

Note that the estimation obtained by Chebyshev's inequality (9.133) is far from being a sharp one:

$$P\{\mu_Z - 2\sigma_Z < Z < \mu_Z + 2\sigma_Z\} = P\{|Z - \mu_Z| < 2\sigma_Z\} =$$
$$= 1 - P\{|Z - \mu_Z| \geq 2\sigma_Z\} \geq 1 - \frac{1}{4} = \frac{3}{4}.$$

12. We can see that each of the random variables X and Y takes on the values $1, 2, \ldots, 6$, but always $X \leq Y$. Let $p(i, j) = P\{X = i, Y = j\}$ be the joint probability distribution of X and Y. As one can easily notice, for any $i, j = 1, 2, \ldots, 6$,

$$p(i, j) = \begin{cases} \frac{1}{36}, & \text{if } i = j \\ \frac{1}{18}, & \text{if } i < j \\ 0, & \text{if } i > j, \end{cases}$$

so we can represent the joint distribution function by the matrix:

$$\begin{pmatrix} \frac{1}{36}, & \frac{1}{18}, & \frac{1}{18}, & \frac{1}{18}, & \frac{1}{18}, & \frac{1}{18} \\ 0 & \frac{1}{36}, & \frac{1}{18}, & \frac{1}{18}, & \frac{1}{18}, & \frac{1}{18} \\ 0 & 0 & \frac{1}{36}, & \frac{1}{18}, & \frac{1}{18}, & \frac{1}{18} \\ 0 & 0 & 0 & \frac{1}{36}, & \frac{1}{18}, & \frac{1}{18} \\ 0 & 0 & 0 & 0 & \frac{1}{36}, & \frac{1}{18} \\ 0 & 0 & 0 & 0 & 0 & \frac{1}{36} \end{pmatrix}$$

The individual probability mass functions of X and Y can be calculated by summing the corresponding values on each row or column respectively. For instance,

$$P\{X = 1\} = \sum_{j=1}^{6} P\{X = 1, Y = j\} = \frac{11}{36},$$

$$P\{Y = 2\} = \sum_{i=1}^{6} P\{X = i, Y = 2\} = \frac{3}{36}.$$

Thus, the probability distributions of X and Y are, respectively:

$$X \sim \begin{pmatrix} 1 & 2 & 3 & 4 & 5 & 6 \\ \frac{11}{36} & \frac{9}{36} & \frac{7}{36} & \frac{5}{36} & \frac{3}{36} & \frac{1}{36} \end{pmatrix},$$

$$Y \sim \begin{pmatrix} 1 & 2 & 3 & 4 & 5 & 6 \\ \frac{1}{36} & \frac{3}{36} & \frac{5}{36} & \frac{7}{36} & \frac{9}{36} & \frac{11}{36} \end{pmatrix}.$$

We can see that X and Y are not independent, since

$$\mathcal{P}\{X=1, Y=2\} = \frac{1}{18} \neq \mathcal{P}\{X=1\}\mathcal{P}\{Y=2\} = \frac{11}{36} \cdot \frac{3}{36}.$$

The expected value of X is

$$\mu_X = E(X) = 1 \cdot \frac{11}{36} + 2 \cdot \frac{9}{36} + 3 \cdot \frac{7}{36} + 4 \cdot \frac{5}{36} + 5 \cdot \frac{3}{36} + 6 \cdot \frac{1}{36} = 2.528,$$

and the variance is

$$\sigma_X^2 = Var(X) = E(X^2) - E(X)^2 =$$

$$= 1^2 \cdot \frac{11}{36} + 2^2 \cdot \frac{9}{36} + 9 \cdot \frac{7}{36} + 4^2 \cdot \frac{5}{36} + 5^2 \cdot \frac{3}{36} + 6^2 \cdot \frac{1}{36} - 2.528^2 = 1.971.$$

Similarly, the expected value of Y is

$$\mu_Y = E(Y) = 1 \cdot \frac{1}{36} + 2 \cdot \frac{3}{36} + 3 \cdot \frac{5}{36} + 4 \cdot \frac{7}{36} + 5 \cdot \frac{9}{36} + 6 \cdot \frac{11}{36} = 4.472,$$

and the variance is

$$\sigma_Y^2 = Var(Y) = E(Y^2) - E(Y)^2 =$$

$$= 1^2 \cdot \frac{1}{36} + 2^2 \cdot \frac{3}{36} + 9 \cdot \frac{5}{36} + 4^2 \cdot \frac{7}{36} + 5^2 \cdot \frac{9}{36} + 6^2 \cdot \frac{11}{36} - 4.472^2 = 1.971.$$

The *covariance* of X and Y is given by:

$$Cov(X, Y) = E(XY) - E(X)E(Y) =$$

$$= \sum_{i=1}^{6} \sum_{j=1}^{6} i \cdot j \cdot p(i, j) - E(X)E(Y) =$$

$$= \frac{1}{36}\left(1^2 + 2^2 + \ldots + 6^2\right) + \frac{1}{18}(1 \cdot 2 + 1 \cdot 3 + \cdot + 5 \cdot 6) - E(X)E(Y)$$

$$= 12.25 - 2.528 \cdot 4.472 = 0.945,$$

10.9 Chapter 9

so the *correlation coefficient* of X and Y (see Eq. (9.54)) is

$$Corr(X, Y) = \frac{Cov(X, Y)}{\sqrt{Var(X)Var(Y)}} = \frac{0.945}{1.971} = 0.479.$$

13. (a) The condition that must be fulfilled is $\int_{-\infty}^{\infty} f(x)\, dx = 1$.

$$\int_{-\infty}^{\infty} f(x)\, dx = \int_{0}^{2} Cx^3\, dx = C \frac{x^4}{4}\bigg|_{0}^{2} = 4C = 1 \Rightarrow C = \frac{1}{4}.$$

$$E(X) = \int_{0}^{2} x \cdot \frac{x^3}{4}\, dx = \frac{1}{4} \int_{0}^{2} x^4\, dx = \frac{x^5}{20}\bigg|_{0}^{2} = 1.6.$$

$$E(X^2) = \int_{0}^{2} x^2 \cdot \frac{x^3}{4}\, dx = \frac{1}{4} \int_{0}^{2} x^5\, dx = \frac{x^6}{24}\bigg|_{0}^{2} = 2.667.$$

$$Var(X) = E(X^2) - E(X)^2 = 2.667 - 1.6^2 = 0.107.$$

The cumulative distribution function is

$$F(x) = \int_{-\infty}^{x} f(t)\, dt = \begin{cases} 0, & x \leq 0 \\ \frac{x^4}{16}, & x \in (0, 2) \\ 1, & x \geq 2 \end{cases}$$

(b)

$$\int_{-\infty}^{\infty} f(x)\, dx = \int_{0}^{1} Cx^n\, dx = C \frac{x^{n+1}}{n+1}\bigg|_{0}^{1} = \frac{C}{n+1} = 1 \Rightarrow C = n+1.$$

$$E(X) = (n+1) \int_{0}^{1} x^{n+1}\, dx = (n+1) \frac{x^{n+2}}{n+2}\bigg|_{0}^{1} = \frac{n+1}{n+2}.$$

$$E(X^2) = (n+1) \int_{0}^{1} x^{n+2}\, dx = (n+1) \frac{x^{n+3}}{n+3}\bigg|_{0}^{1} = \frac{n+1}{n+3}.$$

$$Var(X) = E(X^2) - E(X)^2 = \frac{n+1}{n+3} - \left(\frac{n+1}{n+2}\right)^2 = \frac{n+1}{(n+2)^2(n+3)}.$$

The cumulative distribution function is

$$F(x) = \int_{-\infty}^{x} f(t)\, dt = \begin{cases} 0, & x \leq 0 \\ x^{n+1}, & x \in (0, 1) \\ 1, & x \geq 1 \end{cases}$$

(c)

$$\int_{-\infty}^{\infty} f(x)\, dx = \int_{-1}^{1} C(1 - x^2)\, dx = C\left(x - \frac{x^3}{3}\right)\bigg|_{-1}^{1} = \frac{4C}{3} = 1 \Rightarrow C = \frac{3}{4}.$$

$$E(X) = \frac{3}{4} \int_{-1}^{1} (x - x^3)\, dx = 0 \quad \text{(odd function)}.$$

$$E(X^2) = \frac{3}{4} \int_{-1}^{1} (x^2 - x^4)\, dx = \frac{3}{2} \int_{0}^{1} (x^2 - x^4)\, dx = \frac{3}{2}\left(\frac{x^3}{3} - \frac{x^5}{5}\right)\bigg|_{0}^{1} = \frac{1}{5}.$$

$$Var(X) = E(X^2) - E(X)^2 = \frac{1}{5} = 0.2.$$

For any $x \in (-1, 1)$, the distribution function $F(x)$ is:

$$F(x) = \int_{-\infty}^{x} f(t)\, dt = \frac{3}{4} \int_{-1}^{x} (1 - t^2)\, dt = \frac{3}{4}\left(t - \frac{t^3}{3}\right)\bigg|_{-1}^{x} = \frac{-x^3 + 3x + 2}{4},$$

so we have:

$$F(x) = \begin{cases} 0, & x \leq -1 \\ \frac{-x^3 + 3x + 2}{4}, & x \in (-1, 1) \\ 1, & x \geq 1 \end{cases}$$

(d) Using the gamma function, $\Gamma(x) = \int_{0}^{\infty} t^{x-1} \exp(-t)\, dt$, one can write:

$$\int_{-\infty}^{\infty} f(x)\, dx = \int_{0}^{\infty} Cx \exp(-\lambda x)\, dx \overset{\lambda x = t}{=} \frac{C}{\lambda^2} \int_{0}^{\infty} t \exp(-t) =$$

$$= \frac{C}{\lambda^2} \Gamma(2) = \frac{C}{\lambda^2} = 1 \Rightarrow C = \lambda^2.$$

$$E(X) = \lambda^2 \int_{0}^{\infty} x^2 \exp(-\lambda x)\, dx \overset{\lambda x = t}{=} \frac{1}{\lambda} \int_{0}^{\infty} t^2 \exp(-t) =$$

$$= \frac{1}{\lambda} \Gamma(3) = \frac{2!}{\lambda} = \frac{2}{\lambda}.$$

$$E(X^2) = \lambda^2 \int_0^\infty x^3 \exp(-\lambda x)\, dx \stackrel{\lambda x = t}{=} \frac{1}{\lambda^2} \int_0^\infty t^3 \exp(-t) =$$

$$= \frac{1}{\lambda^2} \Gamma(4) = \frac{3!}{\lambda^2} = \frac{6}{\lambda^2},$$

hence the variance is:

$$Var(X) = E(X^2) - E(X)^2 = \frac{6}{\lambda^2} - \left(\frac{2}{\lambda}\right)^2 = \frac{2}{\lambda^2}.$$

For any $x > 0$, the distribution function $F(x)$ is:

$$F(x) = \lambda^2 \int_0^x t \exp(-\lambda t)\, dt = \lambda^2 \left(-\frac{t \exp(-\lambda t)}{\lambda} \bigg|_0^x + \frac{1}{\lambda} \int_0^x \exp(-\lambda t)\, dt \right) =$$

$$= \lambda \left(-x \exp(-\lambda x) - \frac{\exp(-\lambda t)}{\lambda} \bigg|_0^x \right) = 1 - (1 + \lambda x) \exp(-\lambda x),$$

so we have:

$$F(x) = \begin{cases} 0, & x \le 0 \\ 1 - (1 + \lambda x) \exp(-\lambda x), & x > 0. \end{cases}$$

14. (a) $\mathcal{P}\{X > 2\} = 1 - \mathcal{P}\{X \le 2\} = 1 - F(2) = \exp(-4) = 0.018$.
(b) $\mathcal{P}\{1 < X < 3\} = F(3) - F(1) = \exp(-1) - \exp(-9) = 0.368$.
(c) The probability density function is

$$f(x) = F'(x) = \begin{cases} 2x \exp(-x^2), & x \in (0, \infty) \\ 0, & \text{otherwise.} \end{cases}$$

Consequently, the expected value is:

$$E(X) = \int_0^\infty 2x^2 \exp(-x^2)\, dx \stackrel{x^2 = t}{=} \int_0^\infty t^{1/2} \exp(-t)\, dt =$$

$$= \int_0^\infty t^{3/2 - 1} \exp(-t)\, dt = \Gamma\left(\frac{3}{2}\right) = \frac{1}{2} \Gamma\left(\frac{1}{2}\right) = \frac{\sqrt{\pi}}{2}.$$

The second moment, $E(X^2)$ is

$$E(X^2) = \int_0^\infty 2x^3 \exp(-x^2)\, dx \stackrel{x^2 = t}{=} \int_0^\infty t \exp(-t)\, dt = \Gamma(2) = 1,$$

hence
$$Var(X) = E(X^2) - E(X)^2 = 1 - \frac{\pi}{4}.$$

15. (a) Since $f(x)$ is a density function, we have:
$$\int_0^1 (ax^2 + bx)\, dx = a \left.\frac{x^3}{3}\right|_0^1 + b \left.\frac{x^2}{2}\right|_0^1 = \frac{a}{3} + \frac{b}{2} = 1.$$

On the other hand, from the condition $E(X) = 0.6$ we get:
$$E(X) = \int_0^1 (ax^3 + bx^2)\, dx = a \left.\frac{x^4}{4}\right|_0^1 + b \left.\frac{x^3}{3}\right|_0^1 = \frac{a}{4} + \frac{b}{3} = 0.6.$$

Thus, from the linear system
$$\begin{cases} 2a + 3b = 6 \\ 3a + 4b = 7.2 \end{cases}$$

we find $a = -2.4$ and $b = 3.6$.

$$E(X^2) = \int_0^1 (-2.4x^4 + 3.6x^3)\, dx = -2.4 \left.\frac{x^5}{5}\right|_0^1 + 3.6 \left.\frac{x^4}{4}\right|_0^1 = -\frac{2.4}{5} + \frac{3.6}{4} = 0.42,$$

so the variance is
$$Var(X) = E(X^2) - E(X)^2 = 0.42 - 0.36 = 0.06.$$

(b)
$$P\left\{X < \frac{1}{2}\right\} = \int_0^{1/2} (-2.4x^2 + 3.6x)\, dx = -2.4 \cdot \frac{1}{3 \cdot 8} + 3.6 \cdot \frac{1}{2 \cdot 4} = 0.35.$$

(c) For any $x \in (0, 1)$, the cumulative function is:

$$F(x) = \int_0^x (-2.4t^2 + 3.6t)\, dt = -2.4\frac{x^3}{3} + 3.6\frac{x^2}{2} = -0.8x^3 + 1.8x^2,$$

so we have:
$$F(x) = \begin{cases} 0, & x \leq 0 \\ -0.8x^3 + 1.8x^2, & x \in (0, 1) \\ 1 & x \geq 1. \end{cases}$$

16. (a) We have to prove that $\int_{-\infty}^{\infty} f(x)dx = 1$:

$$\int_{-\infty}^{\infty} \frac{k}{\pi(x^2+k^2)} dx = \frac{k}{\pi} \cdot \frac{1}{k} \arctan \frac{x}{k} \Big|_{-\infty}^{\infty} = \frac{1}{\pi}\left[\frac{\pi}{2} + \frac{\pi}{2}\right] = 1,$$

so $f(x)$ is a density function.

(b)
$$F(x) = \int_{-\infty}^{x} \frac{k}{\pi(t^2+k^2)} dt = \frac{1}{\pi} \arctan \frac{t}{k} \Big|_{-\infty}^{x} = \frac{1}{\pi} \arctan \frac{x}{k} + \frac{1}{2}$$

(c) For $k = 1$ we have: $F(x) = \frac{1}{\pi} \arctan x + \frac{1}{2}$.

$$P\{X^2 - 2X - 3 \leq 0\} = P\{-1 \leq X \leq 3\} = F(3) - F(-1) =$$
$$= \frac{1}{\pi}(\arctan 3 - \arctan(-1)) = 0.648.$$

(d) We know that

$$F_X(x) = P\{X \leq x\} = \frac{1}{\pi} \arctan x + \frac{1}{2}, \quad x \in \mathbb{R},$$

and we have to calculate

$$F_Y(x) = P\left\{\frac{1}{X} \leq x\right\}.$$

If $x > 0$, then $\{1/X \leq x\} = \{X \leq 0\} \cup \{X \geq 1/x\}$, hence

$$F_Y(x) = P\{X \geq 0\} + P\left\{X \geq \frac{1}{x}\right\} =$$
$$= F_X(0) + 1 - F_X\left(\frac{1}{x}\right) =$$
$$= \frac{1}{2} + 1 - \frac{1}{\pi} \arctan \frac{1}{x} - \frac{1}{2} =$$
$$= 1 - \frac{1}{\pi}\left(\frac{\pi}{2} - \arctan x\right) =$$
$$= \frac{1}{2} + \frac{1}{\pi} \arctan x = F_X(x).$$

If $x < 0$, then $\{1/X \le x\} = \{1/x \le X < 0\}$, hence

$$F_Y(x) = \mathcal{P}\left\{\frac{1}{x} \le X < 0\right\} = F_X(0) - F_X\left(\frac{1}{x}\right) =$$

$$= \frac{1}{2} - \frac{1}{2} - \frac{1}{\pi}\arctan\frac{1}{x} =$$

$$= -\frac{1}{\pi}\left(-\frac{\pi}{2} - \arctan x\right) =$$

$$= \frac{1}{2} + \frac{1}{\pi}\arctan x = F_X(x).$$

Since $F_Y(0) = \mathcal{P}\{1/X \le 0\} = \mathcal{P}\{X < 0\} = F_X(0) = 1/2$, we obtain that $F_Y(x) = F_X(x)$ for any $x \in \mathbb{R}$, so $Y = 1/X$ is a standard Cauchy distributed random variable.

17. (a) We must have

$$1 = \int_{-\infty}^{\infty} f(x)\,dx = C\int_{10}^{\infty} \frac{1}{x^2}\,dx = -\frac{C}{x}\Big|_{10}^{\infty} = \frac{C}{10},$$

so $C = 10$.

$$\mathcal{P}\{X > 20\} = 1 - \int_{10}^{20} \frac{10}{x^2}\,dx = 1 + \frac{10}{x}\Big|_{10}^{20} = \frac{1}{2}.$$

(b) For $x \le 10$, $F(x) = 0$. For $x > 10$, we have:

$$F(x) = \int_{-\infty}^{x} f(t)\,dt = \int_{10}^{x} \frac{10}{t^2}\,dt = -\frac{10}{t}\Big|_{10}^{x} = 1 - \frac{10}{x},$$

hence

$$F(x) = \begin{cases} 1 - \frac{10}{x}, & x > 10 \\ 0, & \text{otherwise.} \end{cases}$$

(c) Since the integral

$$\int_{-\infty}^{\infty} x f(x)\,dx = \int_{10}^{\infty} \frac{10}{x}\,dx$$

is divergent, the expected value of X does not exists (see also the case of the Cauchy random variable).

(d) The probability that one device is still working after 15 hours is

$$\mathcal{P}\{X \geq 15\} = 1 - F(15) = \frac{10}{15} = \frac{2}{3}.$$

We consider 6 devices, working independently and let Y denote the number of devices that are still functioning after 15 hours. We can see that Y is a binomial random variable, $Y \sim Bin(6, \frac{2}{3})$. Thus

$$Y \sim \begin{pmatrix} 0 & 1 & 2 & \ldots & 6 \\ \left(\frac{1}{3}\right)^6 & \binom{6}{1}\left(\frac{1}{3}\right)^5 \left(\frac{2}{3}\right) & \binom{6}{2}\left(\frac{1}{3}\right)^4 \left(\frac{2}{3}\right)^2 & \ldots & \left(\frac{2}{3}\right)^6 \end{pmatrix}.$$

We have to find the probability that $Y \geq 3$:

$$\mathcal{P}\{Y \geq 3\} = 1 - \mathcal{P}\{Y = 0\} - \mathcal{P}\{Y = 1\} - \mathcal{P}\{Y = 2\}$$

$$= 1 - \frac{1}{3^6} - 6 \cdot \frac{2}{3^6} - 15 \cdot \frac{4}{3^6} = 1 - \frac{73}{3^6} = 0.9.$$

18. If $X \sim \mathcal{N}(\mu, \sigma)$, then $Z = \dfrac{X - \mu}{\sigma} \sim \mathcal{N}(0, 1)$, so we have:

$$\mathcal{P}\{X > 4\} = \mathcal{P}\left\{\frac{X - 10}{6} > \frac{4 - 10}{6}\right\} = \mathcal{P}\{Z > -1\} =$$

$$= \mathcal{P}\{Z < 1\} = \Phi(1) = 0.84.$$

$$\mathcal{P}\{X < 8\} = \mathcal{P}\left\{Z < \frac{8 - 10}{6}\right\} = \mathcal{P}\{Z < -1/3\} = 1 - \Phi(1/3) = 0.37.$$

$$\mathcal{P}\{4 < X < 16\} = \mathcal{P}\left\{\frac{4 - 10}{6} < Z < \frac{16 - 10}{6}\right\} = \mathcal{P}\{-1 < Z < 1\} =$$

$$= \Phi(1) - \Phi(-1) = 2\Phi(1) - 1 = 2 \cdot 0.84 - 1 = 0.68.$$

$$\mathcal{P}\{X < 20\} = \mathcal{P}\left\{Z < \frac{20 - 10}{6}\right\} = \mathcal{P}\{Z < 5/3\} = 0.95.$$

19. We know that $Z = \frac{X-70}{5}$ is a standard normal random variable.

$$0.8 = \mathcal{P}\{|X - 70| < a\} = \mathcal{P}\{-a < X - 70 < a\} =$$

$$= \mathcal{P}\left\{-\frac{a}{5} < Z < \frac{a}{5}\right\} = \Phi\left(\frac{a}{5}\right) - \Phi\left(-\frac{a}{5}\right) =$$

$$= \Phi\left(\frac{a}{5}\right) - 1 + \Phi\left(\frac{a}{5}\right) = 2\Phi\left(\frac{a}{5}\right) - 1$$
$$\Rightarrow \Phi\left(\frac{a}{5}\right) = 0.9 \Rightarrow \frac{a}{5} = 1.28 \Rightarrow a = 6.4.$$

20. Let $Z = \frac{X-5}{\sigma}$. We know that

$$0.8 = \mathcal{P}\{X \le 9\} = \mathcal{P}\left\{Z \le \frac{9-5}{\sigma}\right\} = \Phi\left(\frac{4}{\sigma}\right) \Rightarrow \frac{4}{\sigma} = 0.84$$

so we obtain that $\sigma = 4.76$ and the variance is $\sigma^2 = 22.66$.

21. We denote by X the normal random variable whose values are the lengths of these bars. Thus, $X \sim N(\mu, 36)$ and we know from hypothesis that:

$$0.05 = \mathcal{P}\{X \ge 82\} = 1 - \mathcal{P}\{X < 82\}.$$

Let $Z = \frac{X-\mu}{6}$ denote the reduced normal random variable. We have:

$$0.95 = \mathcal{P}\{X < 82\} = \mathcal{P}\left\{Z = \frac{X-\mu}{6} < \frac{82-\mu}{6}\right\} = \Phi\left(\frac{82-\mu}{6}\right).$$

From Table 9.2 we find: $\frac{82-\mu}{6} = 1.65$, $\mu = 72.1$.

22. Let X_n be the random variable which express the gambler's gain in the n-th bet, $n = 1, 2, \ldots$. Obviously, $X_1, X_2, \ldots$ are independent identically distributed random variables, with the distribution

$$X_n \sim \begin{pmatrix} -1 & -2 & 10 \\ 0.7 & 0.2 & 0.1 \end{pmatrix}.$$

The random variable $Y_n = X_1 + X_2 + \ldots + X_n$ represents the gambler's gain after n bets.

We can see that the gambler's gain after 3 bets is negative if and only if he wins no bet, that is,

$$\mathcal{P}\{Y_3 < 0\} = \mathcal{P}\{X_i \ne 10, i = 1, 2, 3\} = 0.9^3 = 0.729.$$

It is not so easy to compute this probability if we consider 100 bets (instead of 3). But we can estimate it, using the central limit theorem. Since the expected value and the variance of X_n are given by

$$E(X_n) = -1 \cdot 0.7 - 2 \cdot 0.2 + 10 \cdot 0.1 = -0.1,$$
$$Var(X_n) = E(X_n^2) - E(X_n)^2 = 1 \cdot 0.7 + 4 \cdot 0.2 + 100 \cdot 0.1 - 0.01 = 11.49,$$

it follows that, for large values of n, $Y_n \approx \mathcal{N}(-0.1n, 11.49n)$. Thus, in our case, $Y_{100} \approx \mathcal{N}(-10, 1149)$, hence the probability that the gambler's gain will be negative after 100 bets is

$$\mathcal{P}\{Y_{100} \leq -1\} = \mathcal{P}\left\{\frac{Y_{100} + 10}{\sqrt{1149}} \leq \frac{-1 + 10}{\sqrt{1149}}\right\} = \Phi(0.266) = 0.606.$$

Chapter 11
Supplementary Materials

This chapter contains the basic theory in some areas of Linear Algebra, Calculus and Complex Analysis, necessary for a deep understanding of the material presented in the book. The first section contains elementary results on metric, normed, Banach and Hilbert spaces, while the second section provides a brief introduction to Complex Analysis, with a special focus on the calculus of residues.

11.1 Normed, Metric and Hilbert Spaces

11.1.1 Normed Vector Spaces

We assume that the reader has some basic knowledge in Linear Algebra ([18] or [35] for instance) and Calculus ([1, 23, 32], or [38]). In the following we denote by K either the field of real numbers $\mathbb{R}$ or the field of complex numbers $\mathbb{C}$. We begin with normed spaces. The metric spaces will be introduced in Sect. 11.1.3.

Definition 11.1 A *normed space* is a vector space V over K endowed with a mapping $\|\cdot\| : V \to [0, \infty)$, called a *norm* on V, which satisfies the following conditions, for any $x, y, z \in V$ and $\lambda \in K$:

n1. $\|x\| = 0$ if and only if $x = \mathbf{0}$,
n2. $\|x + y\| \leq \|x\| + \|y\|$ (the triangle inequality),
n3. $\|\lambda x\| = |\lambda| \|x\|$.

We usually denote by $(V, \|\cdot\|)$ this normed space.

From n3. one can deduce that $\|-x\| = |-1| \|x\| = \|x\|$.

Definition 11.2 Let V be a vector space over the field K. Suppose that V is equipped with an additional binary operation from $V \times V$ to V, denoted here by "$\cdot$" (if $x, y \in V$, then $x \cdot y \in V$ is the product of x and y). Then V is an *algebra* over K (or a K-*algebra*) if the following identities hold for all vectors $x, y, z \in V$ and all scalars $\alpha, \beta \in K$ (we say that the binary operation is bilinear):

$$(x + y) \cdot z = x \cdot z + y \cdot z$$
$$x \cdot (y + z) = x \cdot y + x \cdot z$$
$$(\alpha x) \cdot (\beta y) = (\alpha \beta)(x \cdot y)$$

If V is a normed vector space and, in addition to the three properties from Definition 11.1, the function $\| \ \|$ also satisfies the condition

n4. $\|x \cdot y\| \leq \|x\| \|y\|$ for all $x, y \in V$,
then V is said to be a *normed K-algebra*.

Different norms can be defined on the same vector space.

Example 11.1 $(K^n, \|\cdot\|_\infty)$, where

$$\|(x_1, x_2, \ldots, x_n)\|_\infty = \max\{|x_i| : i = 1, 2, \ldots, n\},$$

is a normed space.

Example 11.2 $(K^n, \|\cdot\|)$, where

$$\|(x_1, x_2, \ldots, x_n)\| = \sqrt{x_1 \overline{x}_1 + x_2 \overline{x}_2 + \ldots + x_n \overline{x}_n} =$$

$$= \sqrt{|x_1|^2 + |x_2|^2 + \ldots + |x_n|^2}.$$

is a normed space. Here $\overline{x}$ is the complex conjugate of x.

Example 11.3 $M_{m,n}(K)$, the K-vector space of all $m \times n$ matrices with entries in the field K, $A = (a_{i,j})_{i=\overline{1,m}, j=\overline{1,n}}$, with the norm

$$\|A\| = \sqrt{\sum_{i=1}^{n} \sum_{j=1}^{m} |a_{ij}|^2},$$

is a normed space.

Example 11.4 Let A be a nonempty subset of $\mathbb{R}$ and let

$$\mathcal{B}(A) = \{f : A \to K : f \text{ bounded on } A\}$$

11.1 Normed, Metric and Hilbert Spaces

be the K-vector space of all bounded functions defined on A with values in K. Then, one can prove that the mapping:

$$f \to \|f\|_\infty = \sup\{|f(t)| : t \in A\}$$

is a norm on $\mathcal{B}(A)$ (satisfies the conditions n1.–n3). This norm is called the *sup-norm* (also known as the *infinity norm* or the *supremum norm*). It can be also proved that $\mathcal{B}(A)$ is a normed K-algebra (n4. is also fulfilled).

Definition 11.3 We say that a sequence $\{x_n\}$ in a normed space $(V, \|\cdot\|)$ is *convergent* to $x \in V$ if the sequence of real numbers $\{\|x_n - x\|\}_n$ is convergent to 0. We write $x_n \to x$ or $\lim_{n\to\infty} x_n = x$.

Proposition 11.1 *The limit x of a convergent sequence $\{x_n\}$ in a normed space $(V, \|\cdot\|)$ is unique.*

Proof Let us assume that $x_n \to x$ and $x_n \to y$. Thus,

$$\|x - y\| = \|(x - x_n) + (x_n - y)\| \stackrel{n2.}{\leq} \|x - x_n\| + \|x_n - y\| \to 0,$$

when $n \to \infty$, and so the constant sequence $\|x - y\| \to 0$ which implies that $\|x - y\| = 0$, i.e. $x = y$ (see n1.). $\square$

Definition 11.4 Let $(V, \|\cdot\|)$ be a normed space over K. A sequence $\{x_n\}$ is said to be a *Cauchy sequence* in $(V, \|\cdot\|)$ if for any positive real number $\varepsilon > 0$ there exists a natural number $N = N_\varepsilon$, such that $\|x_{n+p} - x_n\| < \varepsilon$ for any $n \geq N$ and for any $p = 1, 2, \ldots$

Remark 11.1 Any convergent sequence $x_n \to x$ is a Cauchy sequence. Indeed, let $\varepsilon > 0$ and let $N = N_\varepsilon \in \mathbb{N}$, such that $\|x_n - x\| < \varepsilon/2$ for any $n \geq N$. Thus, for any $n \geq N$ and for any $p \in \mathbb{N}$, we have:

$$\|x_{n+p} - x_n\| = \|(x_{n+p} - x) + (x - x_n)\| \leq$$
$$\leq \|x_{n+p} - x\| + \|x - x_n\| < \varepsilon/2 + \varepsilon/2 = \varepsilon,$$

The inequality n2. implies: $\|x\| = \|(x - y) + y\| \leq \|x - y\| + \|y\|$, so

$$\|x\| - \|y\| \leq \|x - y\|.$$

We can also write $\|y\| - \|x\| \leq \|y - x\| = \|x - y\|$, hence

$$|\|x\| - \|y\|| \leq \|x - y\|. \tag{11.1}$$

Remark 11.2 If $x_n \to x$, $y_n \to y$ in the normed space $(V, \|\cdot\|)$ and $\alpha_n \to \alpha$ in K, then:

$$\|x_n\| \to \|x\|$$

$$x_n \pm y_n \to x \pm y.$$

$$\alpha_n x_n \to \alpha x.$$

Moreover, if $(V, \|\cdot\|)$ is a K-algebra, then we have also

$$x_n y_n \to xy.$$

Indeed, from (11.1) we can write: $|\|x_n\| - \|x\|| \le \|x_n - x\| \to 0$, so $\|x_n\| \to \|x\|$.

From the triangle inequality n2. we have:

$$\|(x_n \pm y_n) - (x \pm y)\| \le \|(x_n - x) \pm (y_n - y)\| = \|x_n - x\| + \|y_n - y\| \to 0,$$

so $x_n \pm y_n \to x \pm y$. In the same way,

$$\|\alpha_n x_n - \alpha x\| = \|\alpha_n x_n - \alpha x_n + \alpha x_n - \alpha x\| \le \|\alpha_n x_n - \alpha x_n\| + \|\alpha x_n - \alpha x\| =$$

$$= |\alpha_n - \alpha| \|x_n\| + |\alpha| \|x_n - x\| \to 0 \cdot \|x\| + |\alpha| \cdot 0 = 0.$$

To prove the last statement, we write:

$$\|x_n y_n - xy\| = \|(x_n y_n - xy_n) + (xy_n - xy)\| \le \|x_n y_n - xy_n\| + \|xy_n - xy\| \stackrel{n4.}{\le}$$

$$\le \|x_n - x\| \|y_n\| + \|x\| \|y_n - y\| \to 0 \cdot \|y\| + \|x\| \cdot 0 = 0.$$

Definition 11.5 A normed vector space $(V, \|\cdot\|)$ is said to be a *complete normed space* or a *Banach space* if any Cauchy sequence $\{x_n\}$ in V has a limit $x \in V$. A normed K-algebra with this property is said to be a *Banach algebra*.

Definition 11.6 Let $\{x_n\}_{n \ge 1}$ be a sequence in the normed space $(V, \|\cdot\|)$. We say that the series $\sum_{n=1}^{\infty} x_n$ converges if the sequence of partial sums $s_n = x_1 + x_2 + \ldots + x_n$ is convergent to a vector $s \in V$. In this case we say that s is the sum of the series and write $\sum_{n=1}^{\infty} x_n = s$. A series which is not convergent is said to be divergent.

Proposition 11.2 *A normed vector space $(V, \|\cdot\|)$ is complete if and only if any series of vectors $\sum_{n=1}^{\infty} x_n$ with the property that the series $\sum_{n=1}^{\infty} \|x_n\|$ converges (in $\mathbb{R}$), is convergent (in V).*

Proof Suppose that V is complete. Let $\sum_{n=1}^{\infty} x_n$ be a series of vectors such that the sequence of real numbers $\sigma_n = \|x_1\| + \|x_1\| + \ldots + \|x_n\|$ is convergent. We prove that the sequence of vectors $s_n = x_1 + x_2 + \ldots + x_n$ converges to a vector $s \in V$. The sequence $\{\sigma_n\}$ is a Cauchy sequence, so, for any $\varepsilon > 0$, there exists $N = N_\varepsilon$ such that $\sigma_{n+p} - \sigma_n < \varepsilon$, for all $n \geq N$ and $p \geq 1$. It follows that

$$\|s_{n+p} - s_n\| = \|x_{n+1} + x_{n+2} + \ldots + x_{n+p}\| \leq$$
$$\leq \|x_{n+1}\| + \|x_{n+2}\| + \ldots + \|x_{n+p}\| = \sigma_{n+p} - \sigma_n < \varepsilon,$$

for all $n \geq N$ and $p \geq 1$, so $\{s_n\}$ is a Cauchy sequence. Since V is complete, it follows that $\{s_n\}$ converges to a vector $s \in V$, so the series is convergent.

Conversely, suppose that any series $\sum_{n=1}^{\infty} x_n$ such that $\sum_{n=1}^{\infty} \|x_n\|$ converges, is convergent in V. Let $\{a_n\}_{n \geq 1}$ be a Cauchy sequence in V. We prove that $\{a_n\}$ is convergent to a vector $a \in V$. We know that, for any $k = 1, 2, \ldots$, there exists $n_k \in \mathbb{N}$ such that $\|a_m - a_n\| < \frac{1}{2^k}$ for any $n, m \geq n_k$. We can take the natural numbers n_k such that $n_1 < n_2 < \ldots < n_k < \ldots$. Then, we have $\|a_{n_{k+1}} - a_{n_k}\| < \frac{1}{2^k}$, for any $k = 1, 2, \ldots$ and it follows that the series $\sum_{k=1}^{\infty} \|a_{n_{k+1}} - a_{n_k}\|$ converges in $\mathbb{R}$. Therefore, the series $\sum_{k=1}^{\infty} (a_{n_{k+1}} - a_{n_k})$ converges in V. Since the partial sums of this series are

$$s_k = a_{n_2} - a_{n_1} + a_{n_3} - a_{n_2} + \ldots + a_{n_k} - a_{n_{k-1}} = a_{n_k} - a_{n_1},$$

it follows that the subsequence $\{a_{n_k}\}$ is convergent to $a \in V$. But $\{a_n\}$ is a Cauchy sequence, so for any $\varepsilon > 0$ there exists $N = N_\varepsilon$ such that $\|a_m - a_n\| < \varepsilon/2$, for any $m, n \geq N$. Since $\lim_{k \to \infty} a_{n_k} = a$, there exists $n_k \geq N$ such that $\|a_{n_k} - a\| < \varepsilon/2$. Thus, for any $n \geq N$, we have

$$\|a_n - a\| \leq \|a_n - a_{n_k}\| + \|a_{n_k} - a\| < \frac{\varepsilon}{2} + \frac{\varepsilon}{2} = \varepsilon,$$

so the sequence $\{a_n\}$ converges to $a \in V$. $\square$

11.1.2 Sequences and Series of Functions

We consider the normed space $\mathcal{B}(A)$ of the bounded functions defined on $A \subset \mathbb{R}$ with values in K, with the sup-norm defined in Example 11.4. For simplicity, we use the notation $\|\cdot\|$ instead of $\|\cdot\|_\infty$. Thus, for any $f \in \mathcal{B}(A)$,

$$\|f\| = \sup\{|f(t)| : t \in A\}.$$

Theorem 11.1 *The normed vector space* $(\mathcal{B}(A), \|\cdot\|)$ *is a Banach space over* K. *Moreover, as* $\mathcal{B}(A)$ *is normed* K-*algebra, it follows that* $\mathcal{B}(A)$ *is a Banach algebra over* K.

Proof The key of the proof is that K ($= \mathbb{R}$ or $\mathbb{C}$) is a complete field (any Cauchy sequence in K has a limit in K).

Let $\{f_n\}$ be a Cauchy sequence in $(\mathcal{B}(A), \|\cdot\|)$ and $\varepsilon > 0$. We know that there exists a natural number $N = N_\varepsilon$ such that

$$\|f_{n+k} - f_n\| < \varepsilon$$

for any $n \geq N$ and for any $k \in \mathbb{N}$. Consider an element $x \in A$. Since

$$|f_{n+k}(x) - f_n(x)| \leq \|f_{n+k} - f_n\| < \varepsilon \qquad (11.2)$$

for any $n \geq N$ and for any $k \in \mathbb{N}$, the numerical sequence $\{f_n(x)\}$ is a Cauchy sequence in K. Since K is complete, there exists a number (which depends on x), $z = f(x) \in K$ such that $f_n(x)$ is convergent to $f(x)$. Thus, we can define the function $f : A \to K$ such that $\lim_{n \to \infty} f_n(x) = f(x)$, for any $x \in A$. We prove that f is the limit of the sequence $\{f_n\}$, i.e. $\|f_n - f\| \to 0$, when $n \to \infty$. If we make $k \to \infty$ in the inequality (11.2), we find:

$$|f(x) - f_n(x)| \leq \varepsilon \qquad (11.3)$$

for any $n \geq N$ and for any $x \in A$. Thus,

$$\|f_n - f\| \leq \varepsilon \qquad (11.4)$$

for any $n \geq N$, i.e. $\|f_n - f\|_\infty \to 0$, when $n \to \infty$.

It remains to prove that the limit f of the sequence $\{f_n\}$ is bounded. Using (11.3) one can write:

$$|f(x)| \leq |f(x) - f_N(x)| + |f_N(x)| \leq \varepsilon + \|f_N\|,$$

for any $x \in A$, so $f \in \mathcal{B}(A)$. □

In the proof of Theorem 11.1 we used two types of convergence of a sequence of functions: the simple (pointwise) convergence and the uniform convergence. We give the exact definitions below:

Definition 11.7 We say that a sequence of functions $\{f_n\}$, $f_n : A \to K$ is *simply* (pointwise) *convergent* to the function $f : A \to K$ if for any fixed $x \in A$, the sequence of numbers $\{f_n(x)\}$ is convergent to $f(x)$ in K.

We say that the sequence of functions $\{f_n\} \subset \mathcal{B}(A)$ is *uniformly convergent* to the function $f : A \to K$ if the sequence of positive numbers $\|f_n - f\|$ converges to 0, or, equivalently, if the sequence $\{f_n\}$ converges to f in the normed space $\mathcal{B}(A)$.

11.1 Normed, Metric and Hilbert Spaces

It can be proved that the limit f of an uniformly convergent sequence $\{f_n\} \subset \mathcal{B}(A)$ is also a bounded function (see the proof of Theorem 11.1).

Remark 11.3 Obviously, the uniform convergence implies the simple convergence. But the converse it is not true. Indeed, let $f_n : [0, 1] \to \mathbb{R}$, $f_n(t) = t^n$. It is easy to see that $f_n(t) \to f(t) = 0$ for any $t \in [0, 1)$, and $f_n(1) = 1 \to 1$, so $f(1) = 1$. Thus $\|f_n - f\| = 1 \not\to 0$, so $\{f_n\}$ is simply convergent to f but it is not uniformly convergent.

Remark 11.4 Let I be a real interval and $f : I \to \mathbb{C}$ a complex-valued function, $f(t) = u(t) + iv(t)$. The real valued functions $u, v : I \to \mathbb{R}$ are the real part of f, $u = \text{Re } f$, and the imaginary part of f, $v = \text{Im } f$, respectively. The function f is continuous on I if and only if u and v are continuous functions. The function f is differentiable on I if and only if u and v are differentiable on I and in this case we have: $f'(t) = u'(t) + iv'(t)$. Moreover, we say that f is Riemann integrable on I if and only if u and v are Riemann integrable on I and, by definition, $\int_I f(t)dt = \int_I u(t)dt + i \int_I v(t)dt$.

Proposition 11.3 *Let I be a real interval and $f_n : I \to K$, $n = 0, 1, \ldots$ be a sequence of continuous functions which is uniformly convergent to a function $f : I \to K$. Then f is also a continuous function on I.*

Proof Let us fix a $t_0 \in I$ and let $\varepsilon > 0$ be a positive real number. Since $f_n \to f$ uniformly, there exists a number $N = N_\varepsilon \in \mathbb{N}$ such that $\|f_n - f\| < \varepsilon/3$ for any $n \geq N$. This means that $|f_n(t) - f(t)| < \varepsilon/3$ for any $n \geq N$ and for any $t \in I$. In particular, $|f_N(t) - f(t)| < \varepsilon/3$ for any $t \in I$. Since $f_N : I \to K$ is continuous at t_0, there exists $\delta > 0$ such that if $t \in I$ with $|t - t_0| < \delta$, then $|f_N(t) - f_N(t_0)| < \varepsilon/3$. Thus,

$$|f(t) - f(t_0)| \leq |f(t) - f_N(t)| + |f_N(t) - f_N(t_0)| + |f_N(t_0) - f(t_0)| <$$
$$< \varepsilon/3 + \varepsilon/3 + \varepsilon/3 = \varepsilon,$$

for any $t \in I$ with $|t - t_0| < \delta$, so f is continuous at t_0. □

Example 11.5 Consider again the sequence of continuous functions from Remark 11.3: $f_n(t) = t^n$, $t \in [0, 1]$. We saw above that $f_n \to f$ simply, where $f(t) = 0$ for any $t \in [0, 1)$ and $f(1) = 1$. Since f is not a continuous function, by Proposition 11.3 we can deduce that $\{f_n\}$ cannot be uniformly convergent to f.

Proposition 11.4 *Let $f_n : [a, b] \to \mathbb{R}$ be a sequence of continuous functions which is uniformly convergent to $f : [a, b] \to \mathbb{R}$. Then*

(a) f is Riemann integrable on $[a, b]$ and

$$\int_a^b f(t)dt = \lim_{n \to \infty} \int_a^b f_n(t)dt. \tag{11.5}$$

We also say that the limit and the integral are interchangeable:

$$\lim_{n\to\infty}\int_a^b f_n(t)dt = \int_a^b \lim_{n\to\infty} f_n(t)dt$$

(b) *Moreover, if $F_n(t) = \int_a^t f_n(u)du$ and $F(t) = \int_a^t f(u)du$, $t \in [a, b]$, then the sequence of functions $\{F_n\}$ is uniformly convergent to F on $[a, b]$.*

Proof

(a) Take $\varepsilon > 0$ and $N = N_\varepsilon \in \mathbb{N}$ such that $\|f_n - f\| < \varepsilon/(b - a)$ for any $n \geq N$. This means that $|f_n(t) - f(t)| < \varepsilon/(b - a)$ for any $n \geq N$ and for any $t \in [a, b]$. The limit function f is continuous, so it is integrable. For any $n \geq N$, one can write:

$$\int_a^b |f_n(t) - f(t)|\, dt \leq \frac{\varepsilon}{b - a}(b - a) = \varepsilon.$$

Thus,

$$\left|\int_a^b f_n(t)dt - \int_a^b f(t)dt\right| \leq \int_a^b |f_n(t) - f(t)|\, dt \leq \varepsilon$$

for any $n \geq N$, i.e. the sequence of numbers $\int_a^b f_n(t)dt$ is convergent and its limit is $\int_a^b f(t)dt$.

(b) For any $t \in [a, b]$ we can write:

$$|F_n(t) - F(t)| \leq \int_a^t |f_n(u) - f(u)|\, du \leq \|f_n - f\|(b - a).$$

Thus

$$\|F_n - F\| = \sup\{|F_n(t) - F(t)| : t \in [a, b]\} \leq \|f_n - f\|(b - a) \to 0,$$

hence $\{F_n\}$ is uniformly convergent to F.

□

Proposition 11.5 *Let $\{f_n\}$ be a sequence of differentiable functions with continuous derivatives on $[a, b]$ (at $t = a, b$ we speak about side derivatives). We assume that $f_n(t_0) \to f_0$ for a point $t_0 \in [a, b]$ and that $\{f_n'\}$ is uniformly convergent to g. Then $\{f_n\}$ if uniformly convergent to a differentiable function f and $f' = g$ on $[a, b]$.*

Proof For a fixed $n = 0, 1, \ldots$ we write the Newton-Leibniz formula for definite Riemann integrals:

$$f_n(t) - f_n(t_0) = \int_{t_0}^t f_n'(u)\, du. \tag{11.6}$$

11.1 Normed, Metric and Hilbert Spaces

From Proposition 11.4 b) we conclude that the sequence $\{f_n(t) - f_n(t_0)\}$ is uniformly convergent to $h(t)$ on $[a, b]$ with respect to the variable t. Since

$$|f_n(t) - (h(t) + f_0)| \le |f_n(t) - f_n(t_0) - h(t)| + |f_n(t_0) - f_0| \le$$

$$\le \|f_n - f_n(t_0) - h\| + |f_n(t_0) - f_0|.$$

Thus,

$$\|f_n - (h + f_0)\| \le \|f_n - f_n(t_0) - h\| + |f_n(t_0) - f_0| \to 0,$$

so $\{f_n\}$ is uniformly convergent to $f = h + f_0$, where $f_0(x) = f_0$, a constant. Let us use again Proposition 11.4 (a) and make $n \to \infty$ in (11.6). So

$$h(t) + f_0 - f_0 = \int_{t_0}^{t} g(u)\,du.$$

Thus h is a primitive (antiderivative) of g, i.e. $h' = g$ and so $(h(t) + f_0)' = g(t)$, i.e. $f' = g$. $\square$

Definition 11.8 Let $\{f_n\}$, $f_n : A \to K$, be a sequence of functions defined on a nonempty set of real numbers A and $s_n = f_0 + f_1 + \ldots + f_n$ be the "partial sum" of order n. The couple $(\{f_n\}, \{s_n\})$ is called a *series of functions* and is denoted by $\sum_{n=0}^{\infty} f_n$. The set $C \subset A$ of all the points t such that the sequence of numbers $\{s_n(t)\}$ is convergent is called the *set of convergence* of the series. The function $s : C \to K$, $s(t) = \lim_{n \to \infty} s_n(t)$ is called the *sum of the series*. We say that the series is *simply* (pointwise) *convergent* to s and write $s = \sum_{n=0}^{\infty} f_n$ on C. If the sequence $\{s_n\}$ is uniformly convergent to s, then the series is said to be *uniformly convergent*.

Example 11.6 Let us study the geometrical series

$$\sum_{n=0}^{\infty} t^n = 1 + t + t^2 + \ldots + t^n + \ldots$$

It is clear that the set of convergence is $C = (-1, 1)$. Since

$$s_n(t) = 1 + t + \ldots + t^n = \frac{1 - t^{n+1}}{1 - t} \to \frac{1}{1 - t}$$

for any $t \in (-1, 1)$, the sum of the series is the function $s : (-1, 1) \to \mathbb{R}$, $s(t) = \frac{1}{1-t}$. We write:

$$\sum_{k=0}^{\infty} t^n = \frac{1}{1 - t}, \quad t \in (-1, 1).$$

The series is *simply* convergent on $(-1, 1)$; we prove that it is not *uniformly* convergent on this interval. Since

$$|s_n(t) - s(t)| = \frac{|t|^{n+1}}{1-t},$$

we can see that $s_n - s$ is *unbounded* on $(-1, 1)$ (so it is not possible that $\|s_n - s\| \to 0$):

$$\|s_n - s\| = \sup_{t \in (-1,1)} |s_n(t) - s(t)| = \lim_{t \to 1} \frac{|t|^{n+1}}{1-t} = \infty.$$

But the series is uniformly convergent on any interval $[-r, r]$, with $0 < r < 1$: in this case, we have:

$$\|s_n - s\| = \sup_{t \in [-r,r]} |s_n(t) - s(t)| = \sup_{t \in [-r,r]} \frac{|t|^{n+1}}{1-t} = \frac{r^{n+1}}{1-r}.$$

Since $r \in (0, 1)$, we have

$$\lim_{n \to \infty} \|s_n - s\| = \lim_{n \to \infty} \frac{r^{n+1}}{1-r} = 0,$$

hence the series is uniformly convergent on $[-r, r]$.

Definition 11.9 A series of functions $\sum_{n=0}^{\infty} f_n$ is *absolutely convergent* on A if the moduli series $\sum_{n=0}^{\infty} |f_n|$ is convergent on A. If, in addition, the moduli series uniformly converges on A, then the series $\sum_{n=0}^{\infty} f_n$ is said to be *absolutely uniformly convergent*.

Proposition 11.6 *An absolutely uniformly convergent series is also uniformly convergent.*

Proof Let $\sum_{n=0}^{\infty} f_n$ be a series, absolutely uniformly convergent on A. For any $t \in A$, we denote by $s_n(t)$ its partial sums, and by $S_n(t)$ the partial sums of the moduli series, $n = 0, 1, \ldots$:

$$s_n(t) = f_0(t) + f_1(t) + \ldots + f_n(t),$$
$$S_n(t) = |f_0(t)| + |f_1(t)| + \ldots + |f_n(t)|.$$

11.1 Normed, Metric and Hilbert Spaces

Since the sequence $\{S_n\}$ is convergent, it is bounded and it is a Cauchy sequence (see Remark 11.1). Thus, for any $\varepsilon > 0$, there exists $N = N_\varepsilon$ such that

$$\|S_{n+p} - S_n\| < \varepsilon,$$

for any $n \geq N$ and for any $p = 1, 2, \ldots$. But

$$\begin{aligned}
|s_{n+p}(t) - s_n(t)| &= |f_{n+1}(t) + f_{n+2}(t) + \ldots + f_{n+p}(t)| \leq \\
&\leq |f_{n+1}(t)| + |f_{n+2}(t)| + \ldots + |f_{n+p}(t)| = \\
&= S_{n+p}(t) - S_n(t) \leq \|S_{n+p} - S_n\| < \varepsilon,
\end{aligned}$$

for any $n \geq N$, $p = 1, 2, \ldots$ and $t \in A$. It follows that $\|s_{n+p} - s_n\| \leq \varepsilon$ for any $n \geq N$ and $p = 1, 2, \ldots$. Thus, $\{s_n\}$ is a Cauchy sequence in the *complete* space $\mathcal{B}(A)$ (see Theorem 11.1), so it is convergent in the normed space $\mathcal{B}(A)$, which means that the series $\sum_{n=0}^{\infty} f_n$ is uniformly convergent. □

Theorem 11.2 (Weierstrass M-Test) *If $\{M_n\}_{n \geq 0}$ is a sequence of positive numbers such that $|f_n(x)| \leq M_n$ for all $x \in A$ and the series $\sum_{n=0}^{\infty} M_n$ is convergent, then the series $\sum_{n=0}^{\infty} f_n$ is absolutely uniformly convergent on A.*

Proof We follow the same idea as in the proof of Proposition 11.6. Let

$$s_n = |f_0| + |f_1| + \ldots + |f_n|,$$
$$S_n = M_0 + M_1 + \ldots + M_n$$

be the partial sums of order n of the series $\sum_{n=0}^{\infty} |f_n|$ and $\sum_{n=0}^{\infty} M_n$ respectively. Since $(\mathcal{B}(A), \|\cdot\|)$ is a complete space (see Theorem 11.1), it is sufficient to prove that the sequence $\{s_n\}$ is a Cauchy sequence in $\mathcal{B}(A)$. Indeed,

$$\begin{aligned}
|s_{n+p}(t) - s_n(t)| &= |f_{n+1}(t)| + |f_{n+2}(t)| + \ldots + |f_{n+p}(t)| \leq \\
&\leq \|f_{n+1}\| + \|f_{n+2}\| + \ldots + \|f_{n+p}\| \leq \\
&\leq M_{n+1} + M_{n+2} + \ldots + M_{n+p} = S_{n+p} - S_n,
\end{aligned}$$

for any $t \in A$ and we conclude that

$$\|s_{n+p} - s_n\| \leq S_{n+p} - S_n. \tag{11.7}$$

Since $\{S_n\}$ is a Cauchy sequence (of positive numbers), for any $\varepsilon > 0$ there exists $N = N_\varepsilon$ such that $S_{n+p} - S_n < \varepsilon$ for any $n \geq N$ and $p = 1, 2, \ldots$. By (11.7) we obtain that $\|s_{n+p} - s_n\| < \varepsilon$ for any $n \geq N$ and $p = 1, 2, \ldots$. Thus, the sequence of functions $\{s_n\}$ is also a Cauchy sequence, hence it is convergent in $\mathcal{B}(A)$. □

Example 11.7 The series $\sum_{n=1}^{\infty} \dfrac{\cos nx}{n^2}$ is absolutely uniformly convergent because $\left|\dfrac{\cos nx}{n^2}\right| \leq \dfrac{1}{n^2}$ for any $x \in \mathbb{R}$ and the series $\sum_{n=1}^{\infty} \dfrac{1}{n^2}$ is convergent.

Similarly, since $\left|\dfrac{e^{inx}}{n^2}\right| = \dfrac{1}{n^2}$ for any $x \in \mathbb{R}$, we conclude, by Weierstrass M-test that the series $\sum_{n=1}^{\infty} \dfrac{e^{inx}}{n^2} = \sum_{n=1}^{\infty} \dfrac{\cos nx + i \sin nx}{n^2}$ is absolutely uniformly convergent on $\mathbb{R}$.

11.1.3 Metric Spaces. Some Density Theorems

Definition 11.10 A metric space is a nonempty set X with a *metric* on it, i.e. a function that defines the *distance* between any two elements of the set, $d : X \times X \to [0, \infty)$ which satisfies the following conditions for any $x, y, z \in X$:

(d1) $d(x, y) = 0$ if and only if $x = y$,
(d2) $d(x, z) \leq d(x, y) + d(y, z)$ (triangle inequality),
(d3) $d(x, y) = d(y, x)$ (symmetry).

Metric spaces were introduced by M. Fréchet in 1906.

Example 11.8 If $K = \mathbb{R}$ or $\mathbb{C}$ and $V = K^n$, we can define on V the so-called Euclidean metric: for any $x = (x_1, \ldots, x_n)$, $y = (y_1, \ldots, y_n) \in K^n$,

$$d(x, y) = \sqrt{\sum_{k=1}^{n} |x_k - y_k|^2}. \tag{11.8}$$

This is a particular case of the following example.

Example 11.9 Any normed vector space $(V, \|\cdot\|)$ is also a metric space with the metric defined by:

$$d(x, y) = \|x - y\|.$$

11.1 Normed, Metric and Hilbert Spaces

Indeed, the conditions $(d1)$ and $(d3)$ are obviously fulfilled. We prove the condition $(d2)$: for any $x, y, z \in V$ one can write

$$d(x, z) = \|x - z\| = \|(x - y) + (y - z)\| \overset{n2)}{\leq}$$
$$\leq \|x - y\| + \|z - y\| = d(x, y) + d(y, z).$$

Thus, the normed spaces from Examples 11.1, 11.2, 11.3 and 11.4 are also metric spaces.

Let us give now some "geometrical properties" of a metric d. It can be easily proved, by mathematical induction on $n \geq 3$, that for any n points $x_1, x_2, \ldots, x_n \in X$ the following inequality holds:

$$d(x_1, x_n) \leq d(x_1, x_2) + d(x_2, x_3) + \ldots + d(x_{n-1}, x_n). \tag{11.9}$$

This is called the *polygonal inequality* (the length of a side of a polygon is less than the sum of the lengths of the other sides).

Let $x, y, z, w \in X$. We apply (11.9) for $x_1 = x, x_2 = z, x_3 = w, x_4 = y$:

$$d(x, y) - d(z, w) \leq d(x, z) + d(y, w).$$

If we take $x_1 = z, x_2 = x, x_3 = y, x_4 = w$, we find:

$$d(z, w) - d(x, y) \leq d(x, z) + d(y, w).$$

So we obtained the following *quadrilateral inequality*:

$$|d(x, y) - d(z, w)| \leq d(x, z) + d(y, w). \tag{11.10}$$

If $w = y$, we get the following inequality (the difference between the lengths of two sides of a triangle is less than the length of the other side):

$$|d(x, y) - d(y, z)| \leq d(x, z). \tag{11.11}$$

Definition 11.11 Let (X, d) be a metric space and $\{x_n\}_{n \geq 1}$ be a sequence of points in X. We say that $\{x_n\}$ is *convergent* to $x \in X$, and write $x_n \to x$, if the numerical sequence $d(x_n, x) \to 0$. We also write: $\lim_{n \to \infty} x_n = x$.

Proposition 11.7 *Let (X, d) be a metric space and $\{x_n\}_{n \geq 1} \subset X$ be a convergent sequence in X, Then the limit of $\{x_n\}$ is unique.*

Proof Let $\lim_{n \to \infty} x_n = x$ and let $y \in X$ be another limit of $\{x_n\}$. For any $\varepsilon > 0$ we can find a natural number $N = N_\varepsilon$ such that

$$d(x_n, x) < \varepsilon/2, \ d(x_n, y) < \varepsilon/2,$$

for any $n \geq N$. Therefore,
$$d(x, y) \leq d(x, x_n) + d(x_n, y) < \varepsilon.$$

Since ε may be arbitrary small, we see that $d(x, y) = 0$ and from $d1)$ it follows that $x = y$. (see also Proposition 11.1). □

Definition 11.12 We say that a subset A of a metric space (X, d) is an *open set* if, for any $a \in A$, there exists $r > 0$ such that the *open ball*
$$B(a, r) = \{x \in X : d(a, x) < r\}$$
centered at a and of radius r, is contained in A. A subset B of (X, d) is said to be a *closed set* in X if $X \setminus B$ is an open set of X.

Thus A is open if and only if $X \setminus A$ is closed.

Remark 11.5 Any *open ball* $B(a, r) \subset X$ $r > 0$, is an *open* set of X and any *closed ball* $\overline{B}(a, r) \subset X$ is a *closed* set of X, where
$$\overline{B}(a, r) = \{x \in X : d(a, x) \leq r\}.$$

Let y be a point in $B(a, r)$ and let $0 < \rho < r - d(y, a)$. Then $B(y, \rho) \subset B(a, r)$. Indeed, if $x \in B(y, \rho)$, i.e. $d(x, y) < \rho$, then
$$d(x, a) \leq d(x, y) + d(y, a) < \rho + r - \rho = r.$$

Hence $d(x, a) < r$ and so $x \in B(a, r)$.

Let us prove now that $X \setminus \overline{B}(a, r)$ is an open set of X. Let $y \in X \setminus \overline{B}(a, r)$. Then $d(y, a) > r$ and we take $0 \leq \rho \leq d(y, a) - r$. If $x \in B(y, \rho)$, then $d(x, y) < \rho$ and we have:
$$d(a, x) \geq d(a, y) - d(y, x) > r + \rho - \rho = r,$$
so $x \notin \overline{B}(a, r)$ and we proved that $X \setminus \overline{B}(a, r)$ is open, hence $\overline{B}(a, r)$ is closed.

Proposition 11.8 *A subset $B \subset X$ is a closed set of X if and only if any convergent sequence $\{x_n\}_{x \geq 1} \subset B$ has its limit in B.*

Proof

(a) Assume that B is closed and suppose that the sequence $\{x_n\}_{x \geq 1} \subset B$ has the limit $x \notin B$. Since $X \setminus B$ is open and $x \in X \setminus B$, there exists $r > 0$ such that $B(x, r) \subset X \setminus B$. Because $x_n \to x$, there exists $N \in \mathbb{N}$ with $x_n \in B(x, r)$ for any $n \geq N$, so $x_N \notin B$, a contradiction. Hence $x \in B$.

(b) Assume now that any sequence $\{x_n\}_{x \geq 1} \subset B$ has its limit in B. Suppose that B is not closed, i.e. $X \setminus B$ is not open. Thus, there exists $z \in X \setminus B$ such that for any $n = 1, 2, \ldots$, the open ball $B(z, 1/n)$ is not contained in $X \setminus B$, i.e. there

11.1 Normed, Metric and Hilbert Spaces

exists $z_n \in B$ with $d(z_n, z) < 1/n$ for all $n = 1, 2, \ldots$ Thus $z_n \to z \notin B$ and $z_n \in B$ for all $n = 1, 2, \ldots$, a contradiction. Hence B is closed in (X, d).

□

Definition 11.13 Let (X_1, d_1) and (X_2, d_2) be two metric spaces and $a \in X_1$. We say that a function $f : X_1 \to X_2$ is continuous at a if for any $\varepsilon > 0$, there exists a positive number $\delta > 0$ such that, for any $x \in X_1$ with $d_1(x, a) < \delta$ we have $d_2(f(x), f(a)) < \varepsilon$. If f is continuous at any point $a \in X_1$ we say that f is continuous on the whole X_1.

The definition of the continuity of a function can be also stated using sequences as follows:

Proposition 11.9 Let (X_1, d_1) and (X_2, d_2) be two metric spaces and $a \in X_1$. The function $f : X_1 \to X_2$ is continuous at a if and only if for any sequence $x_n \to a$ in (X_1, d_1), we have $f(x_n) \to f(a)$ in (X_2, d_2).

Proof Assume that f is continuous at a and let $\varepsilon > 0$ be a positive number. We know that there exists $\delta_\varepsilon = \delta > 0$ such that $d_1(x, a) < \delta$ implies $d_2(f(x), f(a)) < \varepsilon$. If $\{x_n\}_{n \geq 1}$, is a sequence in X_1 convergent to a, then there exists $N = N_\delta \in \mathbb{N}$, such that $d_1(x_n, a) < \delta$ for any $n \geq N$. So $d_2(f(x_n), f(a)) < \varepsilon$ for any $n \geq N$, i.e. $f(x_n) \to f(a)$.

Conversely, assume that for any sequence $x_n \to a$, we have $f(x_n) \to f(a)$. Suppose that f is not continuous at a. Then there exists $\varepsilon > 0$ such that for any $\delta > 0$, there exists an $x_\delta \in X_1$ with $d_1(x_\delta, a) < \delta$ and $d_2(f(x_\delta), f(a)) > \varepsilon$. We take $\delta = 1/n$, $n = 1, 2, \ldots$ and denote $x_\delta = x_n$. Since $d_1(x_n, a) < 1/n$ we can see that $x_n \to a$ in X_1. But $d_2(f(x_n), f(a)) > \varepsilon$ for any $n = 1, 2, \ldots$, which means that $f(x_n) \not\to f(a)$, a contradiction. Thus f is continuous at a.

□

Example 11.10 Let $f : X \to \mathbb{R}$ be a continuous function defined on a metric space (X, d) with real values, and let r be a real number. Then

$$A_{r_-} = \{x \in X : f(x) < r\},$$

$$A_{r_+} = \{x \in X : f(x) > r\}$$

are open subsets of (X, d), and

$$B_{r_-} = \{x \in X : f(x) \leq r\},$$

$$B_{r_+} = \{x \in X : f(x) \geq r\},$$

$$B_{r_0} = \{x \in X : f(x) = r\},$$

are closed subsets of (X, d). It is sufficient for instance to prove that B_{r_-} is closed. Indeed, let $x_n \to x$, $x_n \in B_{r_-}$ for all $n = 1, 2, \ldots$, i.e. $f(x_n) \leq r$ for any $n = 1, 2, \ldots$ So $f(x_n) \to f(x) \leq r$, i.e. $x \in B_{r_-}$.

Example 11.11 Let (X, d) be a metric space and let $(\mathbb{R}, \rho)$ be the metric space of real numbers with the usual metric: $\rho(x, y) = |x - y|$. Let $a \in X$ be a fixed point in X and $t_a : X \to \mathbb{R}_+$, $t_a(x) = d(a, x)$. Then t_a is continuous on X. Indeed, let $b \in X$ and let $x_n \to b$ be a sequence from X which converges to b. Then

$$|t_a(x_n) - t_a(b)| = |d(x_n, a) - d(b, a)| \stackrel{(11.11)}{\leq} d(x_n, b) \to 0,$$

when $n \to \infty$. So t_a is continuous at b for any $b \in X$, i.e. t_a is continuous on X. In particular, if $(Y, \|\cdot\|)$ is a normed space, then the mapping $x \to \|x\|$ is a continuous mapping. Indeed, we consider the induced metric space (Y, δ), were $\delta(x, y) = \|x - y\|$ and see that $x \to \|x\| = t_0(x)$ is continuous as we have just proved above.

Definition 11.14 Let (X, d) be a metric space and let $\{x_n\}_{n \geq 1} \subset X$ be a sequence in X. We say that $\{x_n\}$ is a Cauchy sequence in (X, d) if for any $\varepsilon > 0$, there exists a natural number $N = N_\varepsilon$, such that $d(x_{n+p}, x_n) < \varepsilon$ for any $n \geq N$ and $p = 1, 2, \ldots$.

Remark 11.6 Like in the case of a normed space, in a metric space (X, d) any convergent sequence $x_n \to x$ is a Cauchy sequence. Indeed, let $\varepsilon > 0$ be a real number and let $N = N_\varepsilon \in \mathbb{N}$, such that $d(x_n, x) < \varepsilon/2$ for any $n \geq N$. Thus,

$$d(x_{n+p}, x_n) \leq d(x_{n+p}, x) + d(x, x_n) < \varepsilon/2 + \varepsilon/2 = \varepsilon$$

for any $n \geq N$ and $p = 1, 2, \ldots$. So, $\{x_n\}$ is a Cauchy sequence in (X, d).

Definition 11.15 Let (X, d) be a metric space. A nonempty subset $A \subset X$ is said to be *bounded* in X if there exists a point $b \in X$ such that the set of real numbers $\{d(x, b) : x \in A\}$ is bounded in $\mathbb{R}$, i.e.

$$d(A, b) = \sup\{d(x, b) : x \in A\} < \infty.$$

Proposition 11.10 *Let A be a bounded subset in the metric space (X, d). Then for any point $y \in X$, $d(A, y) < \infty$.*

Proof If A is bounded, there exist $b \in X$ and $M > 0$ such that $d(x, b) \leq M$ for any $x \in A$. We have:

$$d(x, y) \leq d(x, b) + d(b, y) \leq M + d(b, y)$$

for any $x \in A$. Thus, $d(A, y) = \sup\{d(x, y) : x \in A\} \leq M + d(b, y) < \infty$. □

Proposition 11.11 *Any Cauchy sequence $\{x_n\}_{x \geq 1}$ in a metric space (X, d) is bounded.*

11.1 Normed, Metric and Hilbert Spaces

Proof Let $\varepsilon > 0$, and let $N = N_\varepsilon$ such that $d(x_n, x_N) < \varepsilon$ for any $n \geq N$. We denote

$$M = \max\{\varepsilon, d(x_1, x_N), d(x_2, x_N), \ldots, d(x_{N-1}, x_N)\}.$$

Thus, we have

$$d(x_n, x_N) \leq M$$

for any $n = 1, 2, \ldots, N, N+1, \ldots$, so $\{x_n\}_{x \geq 1}$ is bounded. □

Definition 11.16 A metric space (X, d) is said to be *complete* if any Cauchy sequence $\{x_n\}_{x \geq 1} \subset X$ is convergent to an element $x \in X$.

Example 11.12 We know that $K = \mathbb{R}$ or $\mathbb{C}$ are complete metric spaces with respect to the usual distance $d(x, y) = |x - y|$

By Theorem 11.1 the space $\mathcal{B}(A)$ of all bounded functions on A is a complete metric space with the metric:

$$d(f, g) = \|f - g\| = \sup\{|f(t) - g(t)| : t \in A.\}$$

We prove that $X = K^n$, where $K = \mathbb{R}$ or $\mathbb{C}$, with the distance defined by (11.8), is a complete metric space. Indeed, let us consider Cauchy sequence $\{\mathbf{x}_m\}$, $\mathbf{x}_m = (x_{1,m}, x_{2,m}, \ldots, x_{n,m})$, $m = 0, 1, \ldots$, $i = 1, 2, \ldots, n$. Let $\varepsilon > 0$ be a positive number and let $N = N_\varepsilon$ such that

$$d(\mathbf{x}_{m+k}, \mathbf{x}_m) = \|\mathbf{x}_{m+k} - \mathbf{x}_m\| = \sqrt{\sum_{i=1}^n |x_{i,m+k} - x_{i,m}|^2} < \varepsilon,$$

for any $m \geq N$ and $k = 1, 2, \ldots$. Since, for any $i = 1, \ldots, n$, we have

$$|x_{i,m+k} - x_{i,m}| \leq \|\mathbf{x}_{m+k} - \mathbf{x}_m\| < \varepsilon,$$

it follows that each sequence $\{x_{i,m}\}_{m \geq 0}$, $i = 1, \ldots, n$ is a Cauchy sequence in K. Since K is complete, all the n sequences $\{x_{i,m}\}_{m \geq 0}$ are convergent in K. Let $x_i = \lim_{m \to \infty} x_{i,m}$ and denote $\mathbf{x} = (x_1, \ldots, x_n) \in K^n$. Since, for any $i = 1, \ldots, n$, $|x_{i,m} - x_i| \to 0$ as $m \to \infty$, we have

$$d(\mathbf{x}_m, \mathbf{x}) = \sqrt{\sum_{i=1}^n |x_{i,m} - x_i|^2} \to 0,$$

so the sequence $\{\mathbf{x}_m\}$ is convergent to $\mathbf{x}$ in V.

Definition 11.17 Let (X_1, d_1) and (X_2, d_2) be two metric spaces. A mapping $f : X_1 \to X_2$ such that

$$d_2(f(x), f(y)) = d_1(x, y) \tag{11.12}$$

for any $x, y \in X_1$ is called an *isometry*. Obviously, f is injective. If it is also surjective then the spaces X_1 and X_2 are said to be *isometric*. In this case, $f^{-1} : X_2 \to X_1$ is also an isometry.

Remark 11.7 It can be easily proved that any isometry is continuous: if $x_n \to x$ in X_1 then $f(x_n) \to f(x)$ in X_2. Moreover, if $\{x_n\}$ is a Cauchy sequence in X_1, then $\{f(x_n)\}$ is a Cauchy sequence in X_2.

Example 11.13 Let $X_1 = M_n(K)$, the set of $n \times n$ matrices with entries in the field $K = \mathbb{R}$ or $\mathbb{C}$ and for

$$A = (a_{ij}), B = (b_{ij}) \in X_1, a_{ij}, b_{ij} \in K,$$

we define the following metric:

$$d_1(A, B) = \sqrt{\sum_{k=1}^{n} \sum_{j=1}^{n} |a_{kj} - b_{kj}|^2}.$$

Let $X_2 = K^{n^2}$ be the metric space defined in Example 11.8 with n^2 instead of n. It is easy to see that $f : X_1 \to X_2$,

$$f(A) = (a_{11}, a_{12}, \ldots a_{1n}, a_{21}, a_{22}, \ldots, a_{2n}, \ldots a_{n1}, a_{n2}, \ldots a_{nn})$$

is a bijective isometry from X_1 to X_2.

Definition 11.18 Let (X, d) be a metric space and let Y be a nonempty subset of X. We say that Y is a dense subspace in X if for any $\varepsilon > 0$ and for any $x \in X$, there exists $y \in Y$ such that $d(x, y) < \varepsilon$.

For instance, the field $\mathbb{Q}$ of rational numbers is dense in the field $\mathbb{R}$ of real numbers.

The following result will be useful in proving the most important theorem of this section on the "uniqueness" of the completion of a metric space.

Lemma 11.1 *Let (X_1, d_1), (X_2, d_2), (X_3, d_3), (X_4, d_4) be four metric spaces with the following properties.*

(a) *(X_1, d_1) and (X_2, d_2) are isometric, i.e. there exists a bijective mapping $\varphi : X_1 \to X_2$ such that $d_2(\varphi(x), \varphi(y)) = d_1(x, y)$ for all $x, y \in X_1$.*
(b) *(X_1, d_1) and (X_2, d_2) are embedded in (X_3, d_3) and (X_4, d_4) respectively, i.e. there exist two mappings: $\psi_1 : X_1 \to X_3$ and $\psi_2 : X_2 \to X_4$ such that*

11.1 Normed, Metric and Hilbert Spaces

$d_3(\psi_1(x), \psi_1(y)) = d_1(x, y)$ for all $x, y \in X_1$, and $d_4(\psi_2(x), \psi_2(y)) = d_2(x, y)$ for any $x, y \in X_2$.

(c) The ranges $\psi_1(X_1)$ and $\psi_2(X_2)$ are dense subsets in (X_3, d_3) and (X_4, d_4) respectively.

(d) (X_3, d_3) and (X_4, d_4) are complete metric spaces.

Then, there exists a unique extension $\widetilde{\varphi} : X_3 \to X_4$ of φ such that $\widetilde{\varphi} \circ \psi_1 = \psi_2 \circ \varphi$, i.e. the diagram (11.13) is commutative. In particular $\widetilde{\varphi}$ is a bijective isometry.

$$\begin{array}{ccc} X_1 & \xrightarrow{\psi_1} & X_3 \\ \varphi \downarrow & & \downarrow \widetilde{\varphi} \\ X_2 & \xrightarrow{\psi_2} & X_4 \end{array} \qquad (11.13)$$

Proof First of all, for any $x \in X_1$, we define: $\widetilde{\varphi}(\psi_1(x)) = \psi_2(\varphi(x))$. Let a be in $X_3 \setminus \psi_1(X_1)$ and $n \in \mathbb{N}$. Since $\psi_1(X_1)$ is dense in X_3, there exists $x_n \in X_1$ such that $d_3(\psi_1(x_n), a) < 1/n$. We know that $\widetilde{\varphi}(\psi_1(x_n)) = \psi_2(\varphi(x_n))$ for any $n \in \mathbb{N}$. Since $\psi_1(x_n) \to a$ in (X_3, d_3) it is also a Cauchy sequence in (X_3, d_3) (see Remark 11.6). Since

$$d_4(\widetilde{\varphi}\psi_1(x_{n+k}), \widetilde{\varphi}\psi_1(x_n)) = d_4(\psi_2(\varphi(x_{n+k})), \psi_2(\varphi(x_n))) =$$
$$= d_2(\varphi(x_{n+k}), \varphi(x_n)) = d_1(x_{n+k}, x_n) = d_3(\psi_1(x_{n+k}), \psi_1(x_n)),$$

we see that the sequence $\{\widetilde{\varphi}(\psi_1(x_n))\}_n$ is a Cauchy sequence in (X_4, d_4). Since this last one is a complete metric space, the sequence $\{\widetilde{\varphi}(\psi_1(x_n))\}_n$ is convergent to an element $b \in X_4$. We finally define: $\widetilde{\varphi}(a) = b$. The diagram (11.13) is commutative by the manner we just constructed $\widetilde{\varphi}$. If $\phi : X_3 \to X_4$ is another mapping which extends φ, we see that

$$\phi(\psi_1(x)) = \psi_2(\varphi(x)) = \widetilde{\varphi}(\psi_1(x))$$

for any $x \in X_1$. So ϕ and $\widetilde{\varphi}$ are equal on $\psi_1(X_1)$. Since $\psi_1(X_1)$ is dense in (X_3, d_3), by the same reasoning as above one can prove that

$$\phi(a) = \lim_{n \to \infty} \phi \psi_1(x_n) = \lim_{n \to \infty} \widetilde{\varphi} \psi_1(x_n) = \widetilde{\varphi}(a) = b.$$

Up to now we did not use the fact that $\psi_2(X_2)$ is dense in X_4. We use it now to prove that $\widetilde{\varphi}$ is an onto mapping. Indeed, let $c \in X_4$ and let write it as a limit of a sequence of elements from $\psi_2(\varphi(X_1))$ like we did above for $a \in X_3$. So

$$c = \lim_{n \to \infty} \psi_2(\varphi(z_n)),$$

where $z_n \in X_1$ for all $n = 0, 1, ..$ Since the diagram (11.13) is commutative, we see that

$$c = \lim_{n \to \infty} \widetilde{\varphi}(\psi_1(z_n)).$$

Thus $\{\widetilde{\varphi}(\psi_1(z_n))\}_n$ is a Cauchy sequence in (X_4, d_4), so $\{\psi_1(z_n)\}_n$ is a Cauchy sequence in (X_3, d_3) ($\widetilde{\varphi}$ restricted to $\psi_1(X_1)$ is a bijection onto $\psi_2(X_2)$). If we denote $e = \lim_{n \to \infty} \psi_1(z_n)$, it is not difficult to prove that $c = \widetilde{\varphi}(e)$ (see the definition of $\widetilde{\varphi}(e)$ given at the beginning of the proof). Now, if $c = 0$, since

$$d_4(0, \widetilde{\varphi}(\psi_1(z_n))) = d_2(0, \varphi(z_n)) = d_1(0, z_n) = d_3(0, \psi_1(z_n)),$$

we see that $e = \lim_{n \to \infty} \psi_1(z_n) = 0$ because $\widetilde{\varphi}(\psi_1(z_n)) \to 0$ in X_4.

We prove now that $\widetilde{\varphi}$ is an isometry. Let $a = \lim_{n \to \infty} \psi_1(x_n)$ and $e = \lim_{n \to \infty} \psi_1(z_n)$ be two elements in X_3. By definition and by the continuity of the distance functions (see Example 11.11) we get:

$$d_4(\widetilde{\varphi}(a), \widetilde{\varphi}(e)) = d_4\left(\lim_{n \to \infty} \psi_2\varphi(x_n), \lim_{n \to \infty} \psi_2\varphi(z_n)\right) =$$

$$= \lim_{n \to \infty} d_4(\psi_2\varphi(x_n), \psi_2\varphi(z_n)) = \lim_{n \to \infty} d_2(\varphi(x_n), \varphi(z_n)) =$$

$$= \lim_{n \to \infty} d_1(x_n, z_n) = \lim_{n \to \infty} d_3(\psi_1(x_n), \psi_1(z_n)) =$$

$$= d_3\left(\lim_{n \to \infty} \psi_1(x_n), \lim_{n \to \infty} \psi_1(z_n)\right) = d_3(a, e).$$

Hence $d_4(\widetilde{\varphi}(a), \widetilde{\varphi}(e)) = d_3(a, e)$, i.e. $\widetilde{\varphi}$ is an isometry. □

We recall now some facts from the elementary set theory.

Definition 11.19 Let $\mathcal{C}$ be a nonempty set and let $\mathcal{C} \times \mathcal{C}$ be the set of all ordered pairs (x, y), $x, y \in \mathcal{C}$. A relation $\mathcal{R}$ on $\mathcal{C}$ is a nonempty subset of $\mathcal{C} \times \mathcal{C}$. Instead of writing $(x, y) \in \mathcal{R}$, we write $x \sim y$. We say that $\mathcal{R}$ or "$\sim$" is an equivalence relation on $\mathcal{C}$ if:

(eq1) $x \sim x$ for any $x \in \mathcal{C}$ (reflexivity) i.e. the "diagonal" subset $D = \{(x, x) \in \mathcal{C} \times \mathcal{C} : x \in \mathcal{C}\}$ is contained in $\mathcal{R}$,

(eq2) $x \sim y$ implies $y \sim x$ for $x, y \in \mathcal{C}$ (symmetry), i.e. if $(x, y) \in \mathcal{R}$, then its symmetric (y, x) relative to the diagonal D is also in $\mathcal{R}$,

(eq3) if $x \sim y$ and $y \sim z$, then $x \sim z$ for $x, y, z \in \mathcal{C}$ (transitivity).

Any equivalence relation "$\sim$" on a set $\mathcal{C}$ gives rise to a *partition* of $\mathcal{C}$ into *classes of equivalence*. Namely, x and y belong to the same class $[x]$ if $y \sim x$. So the class

11.1 Normed, Metric and Hilbert Spaces

of equivalence corresponding to x is:

$$[x] = \{y \in C : y \sim x\}. \tag{11.14}$$

It is easy to see that two classes $[x]$ and $[y]$ are identical if and only if $x \sim y$. Thus always the intersection of two distinct classes is the empty set (see $eq\,3$)). It is clear that the union of all distinct classes is the whole C. The set of all equivalence classes of C relative to an equivalence relation "$\sim$" is called the *quotient set* of C with respect to "$\sim$" and we write it as $C/\sim$. The equivalence relation "$\sim$" is in fact the equality in $C/\sim$. It is easy to see that the mapping $x \to [x]$, from C to $C/\sim$ is onto. It is one-to-one if and only if the equivalence relation is the usual equality on C.

Example 11.14 For instance, if $(V, +, \cdot)$ is a vector space over the field $K = \mathbb{R}$ or $\mathbb{C}$ and if W is a vector subspace of it, then the relation $x \sim y$ if and only if $x - y \in W$ is an equivalence relation on V and the quotient $V/\sim$ is usually denoted V/W, the quotient of V relative to W. It is not difficult to prove that the new operations on V/W:

$$[x] + [y] = [x + y], \quad \lambda[x] = [\lambda x], \, x, y \in V, \lambda \in K \tag{11.15}$$

are well defined. Moreover, the addition between classes and the scalar multiplication described in (11.15) endow V/W with a structure of vector space over K.

Theorem 11.3 ((F. Hausdorff) (see [36], 3.81)) *Let (X, d) be an arbitrary metric space. Then there exists a complete metric space $(\overline{X}, \overline{d})$, called the **completion** of (X, d) such that:*

(i) (X, d) *is isometric to a subspace* $(X_1, \overline{d})$ *of* $(\overline{X}, \overline{d})$,
(ii) X_1 *is dense in* $\overline{X}$, *i.e.* $(\overline{X}, \overline{d})$ *is "the least" complete metric space which "contains"* (X, d),
(iii) *Any two complete metric spaces* $(\overline{X}, \overline{d})$ *and* $(\overline{\overline{X}}, \overline{\overline{d}})$ *which verify (i) and (ii) are isometric.*

Proof First of all we try to sketch the construction of $(\overline{X}, \overline{d})$ and to prove (i) and (ii), i.e. how (X, d) can be embedded in $(\overline{X}, \overline{d})$, and why its image there is dense in $\overline{X}$.

For this, let us denote by C the set of all Cauchy sequences $x = \{x_n\}$ in X. We say that two Cauchy sequences $x = \{x_n\}$ and $y = \{y_n\}$ are *equivalent* if $\lim_{n \to \infty} d(x_n, y_n) = 0$ and write this $x \sim y$. It is easy to see that $\sim$ is an equivalence relation. Let us denote $\overline{X} = C/\sim$ the quotient set of C relative to this equivalence relation. We write $[x]$ for the class of Cauchy sequence $x = \{x_n\}$ in $\overline{X}$. We now define a metric $\overline{d}$ on the set $\overline{X}$:

$$\overline{d}([x], [y]) = \lim_{n \to \infty} d(x_n, y_n) \tag{11.16}$$

where $\{x_n\}$ and $\{y_n\}$ are representative Cauchy sequences in the class of x and y respectively. First of all we have to prove that the limit on the right exists. Since $(\mathbb{R}, d_0)$, where $d_0(a, b) = |b - a|$, is a complete metric space, it is sufficient to prove that the sequence $\{r_n = d(x_n, y_n)\}$ of nonnegative real numbers is a Cauchy sequence. Indeed,

$$|r_{n+k} - r_n| = |d(x_{n+k}, y_{n+k}) - d(x_n, y_n)| \leq d(x_{n+k}, x_n) + d(y_{n+k}, y_n)$$

(see formula (11.10)). Let $\varepsilon > 0$ be a positive real number and let $N = N_\varepsilon \in \mathbb{N}$, be a sufficient large natural number, such that

$$d(x_{n+k}, x_n) < \varepsilon/2, \ d(y_{n+k}, y_n) < \varepsilon/2 \tag{11.17}$$

for any $n \geq N$ (we just used the fact that $\{x_n\}$, $\{y_n\}$ are Cauchy sequences). So, from (11.17), we find

$$|r_{n+k} - r_n| < \varepsilon/2 + \varepsilon/2 = \varepsilon$$

and this means that $\{r_n\}$ is a Cauchy sequence in $\mathbb{R}$, i.e. a convergent sequence to a finite nonnegative real number $\overline{d}([x], [y])$. Now we want to prove that this last number does not depend on the representatives $\{x_n\}$ and $\{y_n\}$ of $[x]$ and $[y]$ respectively. Let $\{a_n\} \in [x]$ and $\{b_n\} \in [y]$ be other two representatives of the classes $[x]$ and $[y]$ respectively. We have to prove that $\lim_{n \to \infty} d(x_n, y_n) = \lim_{n \to \infty} d(a_n, b_n)$. Indeed, from the quadrilateral inequality (11.10), we can write:

$$|d(x_n, y_n) - d(a_n, b_n)| \leq d(x_n, a_n) + d(y_n, b_n) \to 0 \tag{11.18}$$

because $\{x_n\} \sim \{a_n\}$ and $\{y_n\} \sim \{b_n\}$. So

$$\lim_{n \to \infty} d(x_n, y_n) = \lim_{n \to \infty} d(a_n, b_n)].$$

We now shall prove that $\overline{d}$ is a distance on $\overline{X}$, i.e. we need to verify the conditions $d1)$, $d2)$ and $d3)$ from Definition 11.10.

$d1)$ $\overline{d}([x], [y]) = 0$ means that $\lim_{n \to \infty} d(x_n, y_n) = 0$, i.e. $[x] = [y]$ as we defined the equality in $\overline{X}$. $d2)$ Let us take $z = \{z_n\}$ another Cauchy sequence in $\mathcal{C}$ and write $[z]$ for its class. Let us consider the triangle inequality in X:

$$d(x_n, z_n) \leq d(x_n, y_n) + d(y_n, z_n)$$

for $n = 0, 1, \ldots$. Take now $n \to \infty$ and find:

$$\overline{d}([x], [z]) \leq \overline{d}([x], [y]) + \overline{d}([y], [z]),$$

11.1 Normed, Metric and Hilbert Spaces

i.e. the triangle inequality in $(\overline{X}, \overline{d})$. The symmetry $d3)$ is obviously satisfied.

For any $x \in X$ we consider the constant sequence $\tilde{x} = \{x_n\}$, where $x_n = x$ for any $n = 0, 1, \ldots$. The set X_1 of all these constant sequences is a subset in $\overline{X}$ because any class $[\tilde{x}]$ contains exactly one sequence of the form $\tilde{y}$, $y \in X$, namely $\tilde{x}$. But the mapping $x \to [\tilde{x}]$ is obvious an isometry from (X, d) to $(X_1, \overline{d})$. So, it remains to prove that X_1 is dense in $\overline{X}$ and that $(\overline{X}, \overline{d})$ is complete. For this, let $[a] = [\{a_n\}]$ be a class of Cauchy sequences in $\overline{X}$ and a positive real number $\varepsilon > 0$. Since $\{a_n\}$ is a Cauchy sequence, there exists $N = N_\varepsilon \in \mathbb{N}$ such that $d(a_{n+k}, a_n) < \varepsilon$ for any $n \geq N$ and for any $k \in \mathbb{N}$. We have:

$$\overline{d}([\tilde{a}_N], [a]) = \lim_{n \to \infty} d(a_N, a_n) \leq \varepsilon.$$

Thus X_1 is dense in $\overline{X}$. To prove that $(\overline{X}, \overline{d})$ is complete, let us take a Cauchy sequence $\{A_n\}_n$ in $\overline{X}$ and a representative $\{a_{n,m}\}_m$ in each class A_n for any $n = 0, 1, \ldots$. Since X_1 is dense in $\overline{X}$ we can find $b_n \in X$ such that $\overline{d}([\tilde{b}_n], A_n) < 1/(n+1)$ for any $n = 0, 1, \ldots$. Let us see that $\{b_n\}$ is a Cauchy sequence in X. Indeed, let $\varepsilon > 0$ be a positive real number and $N = N_\varepsilon$ sufficiently large such that $\overline{d}(A_{n+k}, A_n) < \varepsilon/3$, and $1/n < \varepsilon/3$ for any $n \geq N$ and for any $k \in \mathbb{N}$. Thus,

$$d(b_{n+k}, b_n) = \overline{d}([\tilde{b}_{n+k}], [\tilde{b}_n]) \leq \overline{d}([\tilde{b}_{n+k}], A_{n+k}) + \overline{d}(A_{n+k}, A_n) + \overline{d}(A_n, [\tilde{b}_n])$$
$$< 1/(n+k+1) + \varepsilon/3 + 1/(n+1) \leq \varepsilon/3 + \varepsilon/3 + \varepsilon/3 = \varepsilon.$$

So $d(b_{n+k}, b_n) < \varepsilon$ for any $n \geq N$ and for any $k \in \mathbb{N}$, thus $\{b_n\}$ is a Cauchy sequence in X. Let us denote $B = [\{b_n\}]$ the class of this sequence in $\overline{X}$. Let us prove that $A_n \to B$ in $\overline{X}$. Indeed,

$$\overline{d}(A_n, B) \leq \overline{d}(B, [\tilde{b}_n]) + \overline{d}([\tilde{b}_n], A_n) < \lim_{m \to \infty} d(b_m, b_n) + 1/(n+1) =$$
$$= \lim_{k \to \infty} d(b_{n+k}, b_n) + 1/(n+1).$$

Let $\delta > 0$ be a positive real number and $N_1 = N_{1\delta}$ large enough such that $1/(n+1) < \delta/2$ and $d(b_{n+k}, b_n) < \delta/2$ for any $n \geq N_1$ and for any $k \in \mathbb{N}$ (because $\{b_n\}$ is a Cauchy sequence). Thus

$$\overline{d}(A_n, B) < \delta/2 + \delta/2 = \delta.$$

So $A_n \to B$ in $\overline{X}$ and finally $(\overline{X}, \overline{d})$ is complete.

In order to prove the last statement (iii), we simply apply Lemma 11.1 for $(X_1, d_1) = (X_2, d_2) = (X, d)$, $\varphi =$ the identity mapping, $(X_3, d_3) = (\overline{X}, \overline{d})$ and $(X_4, d_4) = (\overline{\overline{X}}, \overline{\overline{d}})$. □

Remark 11.8 We know that any normed space $(X, \|\cdot\|)$ is also a metric space with $d(x, y) = \|x - y\|$, for all $x, y \in V$. An isometry on normed vector spaces not only preserves the distance, but also it must be a linear mapping. Consequently, if $(X, \|\cdot\|)$ is a normed vector space, then the space $\overline{X} = \mathcal{C}/\sim$ constructed by Theorem 11.3 is a vector space which can be also written as $\overline{X} = \mathcal{C}/W$, where W is the vector subspace of $\mathcal{C}$ generated by all the Cauchy sequence of $\mathcal{C}$ which are convergent to 0 (see (11.15)). It is also a *normed* space, with the norm defined by

$$\|[a]\| = \|[\{a_n\}]\| = \lim_{n \to \infty} \|a_n\|.$$

Let I be a nonempty interval, bounded or not. Let $\mathcal{C}(I)$ be the complex vector space of the functions continuous almost everywhere on I, and consider its subspaces: $R_1(I)$, the space of the functions absolutely integrable on I, and $R_2(I)$, the space of the functions square integrable on I. The mapping

$$f \to \|f\|_1 = \int_I |f(x)|\, dx$$

is "almost" a norm on $R_1(I)$, i.e. it satisfies the conditions $n2)$ and $n3)$ of Definition 11.1 but $n1)$ is not in general satisfied. If we identify with 0 all the functions which are zero almost everywhere on I, i.e. if we substitute $R_1(I)$ with $\widehat{R}_1(I) = R_1(I)/\sim$, where $\sim$ is the equivalence relation of equality almost everywhere, we can see that $\|\cdot\|_1$ is a norm on $\widehat{R}_1(I)$. The same is true for the mapping

$$f \to \|f\|_2 = \sqrt{\int_I |f(x)|^2 dx}.$$

If the interval I is infinite, so the integral is an improper integral, for any function f in $R_1(I)$ (or in $R_2(I)$) and for any $\varepsilon > 0$ there exist a finite interval $[a_\varepsilon, b_\varepsilon] \subset I$ and the functions $g \in R_1(I)$, $h \in R_2(I)$ such that g and h are zero outside the interval $[a_\varepsilon, b_\varepsilon]$, equal to f on $[a_\varepsilon, b_\varepsilon]$ and $\|f - g\|_1 < \varepsilon$ ($\|f - h\|_2 < \varepsilon$). This remark and Lemma 4.9 supply an immediate proof for the following result.

Theorem 11.4 *Let f be a function in $R_1(I)$ (or in $R_2(I)$) and let $\varepsilon > 0$ be a positive number. Then there exists the piecewise linear functions g and h, such that g and h are zero outside a finite interval and $\|f - g\|_1 < \varepsilon$, $\|f - h\|_2 < \varepsilon$.*

We shall see in the following that the above functions g and h can be chosen to be continuous piecewise linear functions. We recall that a function $g : I \to \mathbb{R}$ is *continuous piecewise linear* function if there exists a countable division of I (finite if I is finite)

$$\Delta = \{x_n\}_{n \in \mathbb{Z}} : \ldots \leq x_{-n} \leq x_{-n+1} \leq \ldots \leq x_0 \leq \ldots \leq x_{n-1} \leq x_n \leq \ldots$$

such that g is a linear function on each subinterval $[x_{i-1}, x_i]$, $i \in \mathbb{Z}$, and g is continuous on I. If $g : I \to \mathbb{C}$, we say that it is continuous piecewise linear if its real part Re g and its imaginary part Im g are continuous piecewise linear functions.

11.1 Normed, Metric and Hilbert Spaces

The next theorem easily follows from Lemma 4.8 and Lemma 4.9.

Theorem 11.5 *Let $CPL(I)$ be the complex vector space of all the continuous piecewise linear functions g defined on I, which are zero outside a finite subinterval of I. Then $CPL(I)$ is dense in $R_1(I)$ and in $\widehat{R}_1(I)$.*

11.1.4 The Fields $\mathbb{Q}$, $\mathbb{R}$ and $\mathbb{C}$

We shall use the method presented above on the completion of a metric space by considering quotient spaces in order to construct the ring $\mathbb{Z}$ of integer numbers and the fields $\mathbb{Q}$, $\mathbb{R}$ and $\mathbb{C}$ of rational numbers, real numbers and complex numbers, respectively. Kronecker said that natural numbers were created by God, everything else is the work of men, so we start from the set $\mathbb{N} = \{0, 1, 2, \ldots\}$ of natural numbers.

We consider $\mathbb{N} \times \mathbb{N}$, the set of ordered pairs (n, m), $n, m \in \mathbb{N}$, and denote each pair by $n - m$ instead of (n, m) (this is just a symbolic notation, not a real subtraction). We define on $\mathbb{N} \times \mathbb{N}$ the following equivalence relation $n - m \sim n' - m'$ if and only if $n + m' = n' + m$, (where the addition "+" is the usual operation in $\mathbb{N}$). The equivalence class of $n - m$ is denoted $[n - m]$. The set of these equivalence classes, or the quotient set $\mathbb{N} \times \mathbb{N}/\sim$ is called the set of integer numbers and we denote it by $\mathbb{Z}$. For instance, $-1 = [0 - 1] = [3 - 4] = [71 - 72]$, etc., in $\mathbb{Z}$. We introduce an addition and a multiplication in $\mathbb{Z}$:

$$[n - m] + [n' - m'] = [(n + n') - (m + m')]$$

$$[n - m] \cdot [n' - m'] = [(nn' + mm') - (nm' + n'm)].$$

It is not difficult to prove that these operations are well defined, i.e. their definitions do not depend on the choice of a representative in the class of $[n - m]$ or $[n' - m']$. Since $-[n - m] = [m - n]$, it easy to prove that $(\mathbb{Z}, +)$ is an abelian group and $(\mathbb{Z}, +, \cdot)$ is a commutative ring (see the definitions below). The mapping $n \to [n - 0]$ is an embedding (a one-to-one mapping which preserves addition and multiplication) of $\mathbb{N}$ into $\mathbb{Z}$.

Using the notation: $-n = [0 - n]$ we can identify $(\mathbb{Z}, +, \cdot)$ constructed above with our usual set $\{\ldots, -2, -1, 0, 1, 2, \ldots\}$ with the well known elementary operations of addition, subtraction and multiplication, connected by their common rules of computation.

Before constructing the field $\mathbb{Q}$ of rational numbers, starting from $\mathbb{Z}$, we present some basic notions of algebraic structures (see [24], Ch. 1 and Ch. 2).

Definition 11.20 A *group* is a nonempty set G with an operation "$*$" on it (usually an addition or multiplication) such that the following conditions are satisfied for any

$x, y, z \in G$.

(g1) $(x * y) * z = x * (y * z)$ (associativity),

(g2) there exists $e \in G$ such that $x * e = e * x = x$, for any $x \in G$,

(g3) for any $x \in G$, there exists $x' \in G$ such that $x * x' = x' * x = e$.

The element e is called the *identity* (*neutral*) element and it is unique. The element x' is called the inverse of x and it is unique. We denote the group by $(G, *)$. The group $(G, +)$ is called an additive group and $(G, \cdot)$ is said to be a multiplicative group.

If, in addition, for any $x, y \in G$ we have:

g4) $x * y = y * x$, (commutativity),

then we say that $(G, *)$ is a commutative (abelian) group.

Definition 11.21 A *ring* is a set R with at least two elements, "0" and "1", equipped with two binary operations: "+" (addition) and "$\cdot$" (multiplication) satisfying the following three sets of axioms:

(r1) $(R, +)$ is an abelian group (with the identity element 0);
(r2) $(R, \cdot)$ is a monoid, i.e. the multiplication is associative and has the identity element 1;
(r3) the multiplication is distributive with respect to addition:

$$x \cdot (y + z) = x \cdot y + x \cdot z,$$
$$(x + y) \cdot z = x \cdot z + y \cdot z, \text{ for all } x, y, z \in R.$$

If, in addition, the multiplication is commutative, then $(R, +, \cdot)$ is a *commutative ring*.

Definition 11.22 A *field* is a set F with at least two elements, "0" and "1", equipped with two binary operations: "+" (addition) and "$\cdot$" (multiplication) satisfying the following three sets of axioms:

(f1) $(F, +)$ is an abelian group (with the identity element 0);
(f2) $(F \setminus \{0\}, \cdot)$ is a group (with the identity element 1);
(f3) the multiplication is distributive with respect to addition

If, in addition, the multiplication is commutative, then $(F, +, \cdot)$ is a *commutative field*.

We can construct the field $\mathbb{Q}$ of rational numbers in the same manner as we have just constructed $\mathbb{Z}$ starting from $\mathbb{N} = \{0, 1, 2, \ldots\}$. This time we start from $\mathbb{Z} = \{\ldots, -2, -1, 0, 1, 2, \ldots\}$ and denote $\mathbb{Z}^* = \mathbb{Z} \setminus \{0\}$. On the set of pairs $(a, b) \in \mathbb{Z} \times \mathbb{Z}^*$, written formally a/b, we consider an equivalence relation: $a/b \sim a'/b'$

11.1 Normed, Metric and Hilbert Spaces

if and only if $ab' = a'b$, where we just used the usual multiplication in $\mathbb{Z}$. The quotient set $\mathbb{Z} \times \mathbb{Z}^*/ \sim$ of $\mathbb{Z} \times \mathbb{Z}^*$ relative to this equivalence relation is denoted by $\mathbb{Q}$ and a class of equivalence is denoted by $[a/b]$. Each class $[a/b]$ contains exactly one irreducible fraction n/m, i.e. a pair (n, m) where n and m have no common divisor except 1. Thus, we can write:

$$\mathbb{Q} = \{n/m : n, m \in \mathbb{Z}, m \neq 0, n/m \text{ is irreducible}\}.$$

Since it is difficult to operate only with irreducible fractions, we may consider $\mathbb{Q} = \{a/b : a, b \in \mathbb{Z}, b \neq 0\}$.

The ancient Greeks introduced the set of rational numbers in the following way. Two finite segments of a line (results of two measurements) are said to be *commensurable* if there exists a smaller segment c (a common unit) such that $a = nc$ and $b = mc$, where $n, m \in \{1, 2, \ldots\}$. They associated to the division $a : b = a/b$ the pair (n, m), written as n/m. But the ancient Greeks discovered that not all segments are commensurable, for instance, the diagonal and the side of a square are incommensurable. In Fig. 11.1 a square of side 1m is represented. By Pythagoras' Theorem in the right-angled triangle CDB, the diagonal BD is equal to $\sqrt{2}$.

Numbers like $\sqrt{2}$, i.e., of the form $\sqrt{n}$, where n is not the square of an integer, are not *rational*. Indeed, it can be proved that $\sqrt{2} \notin \mathbb{Q}$. Suppose that $\sqrt{2} = a/b$, where $a, b \in \{1, 2, \ldots\}$ and a, b have no common factor except 1. Taking into account the multiplication rule described bellow, one can see that $2b^2 = a^2$. Since any natural number can be uniquely written as a product of prime numbers (Euclid's Fundamental Theorem of Arithmetic), we see that a is a multiple of 2, so $a = 2c$, where $c \in \mathbb{N}$. Thus $2b^2 = 4c^2$ and so b is also a multiple of 2, i.e. 2 is a common factor of a and b, a contradiction. Hence, the length of the diagonal of a square of side 1m cannot be represented by a rational number. The measurement of this diagonal gives rise to a point on a number line (see Fig. 11.2), but this point does not belong to the set of points Q which represents all the rational numbers.

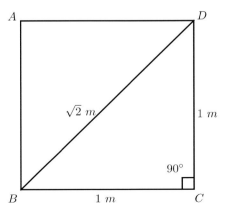

Fig. 11.1 Construction of the irrational number $\sqrt{2}$

Fig. 11.2 The number line $\mathcal{R}$

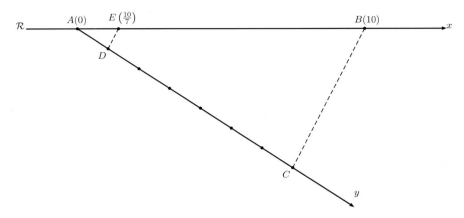

Fig. 11.3 Representation of the rational number 10/7

The rational numbers can be represented using the Thales Theorem. For instance, in Fig. 11.3 the fraction 10/7 is represented.

On the number line $\mathcal{R}$ we take the points $A(0)$ and $B(10)$ (the point A has the coordinate 0: it is the origin; B has the coordinate 10, so the segment $[AB]$ has the length of 10 units) We take an arbitrary half-line $[Ay)$ and an arbitrary point D on it. Let C be the point on the half-line $[Ay)$ such that the length of the segment $[AC]$ is 7 times the length of $[AD]$. Through the point D we construct the parallel DE to the line BC (where $E \in AB$). By Thales Theorem, the coordinate of the point E is equal to $\frac{10}{7}$. In this way we can represent any rational number on the number line $\mathcal{R}$.

We can extend the addition, multiplication, absolute value, norm and the metric functions from $\mathbb{Z}$ to $\mathbb{Q}$ in the following way:

$$\frac{a_1}{b_1} + \frac{a_2}{b_2} = \frac{a_1 b_2 + a_2 b_1}{b_1 b_2},$$

$$\frac{a_1}{b_1} \cdot \frac{a_2}{b_2} = \frac{a_1 a_2}{b_1 b_2},$$

$$\|x\| = |x| = \begin{cases} x, & \text{if } x \geq 0 \\ -x, & \text{if } x < 0 \end{cases}, \text{ for any } x \in \mathbb{Q},$$

$$d(x, y) = |y - x|, \text{ for any } x, y \in \mathbb{Q},$$

11.1 Normed, Metric and Hilbert Spaces

where $x = a/b > 0$ if a and b have the same sign, $x < 0$ if a and b have different signs, and $x = 0$ if $a = 0$.

It can be verified that $(\mathbb{Q}, +, \cdot)$ is indeed a field. This field is called the *field of rational numbers*. There is a natural embedding of the integers in the rational numbers achieved by the one-to-one morphism of rings $\varphi : \mathbb{Z} \to \mathbb{Q}$, $\varphi(a) = \frac{a}{1}$, so we can write $\mathbb{Z} \subset \mathbb{Q}$.

Moreover, we can define on $\mathbb{Q}$ the following relation: for any $x, y \in \mathbb{Q}$, $x \leq y$ if $y - x \geq 0$. It can be verified that this relation satisfies the properties or1) – or6) in the definition bellow, so $(\mathbb{Q}, +, \cdot, \leq)$ is an *ordered field*.

We write $x \geq y$ if $y \leq x$; $x < y$ if $x \leq y$ and $x \neq y$ etc.

Definition 11.23 A *(partial) order* is a binary relation $\leq$ on a set P with the following properties:

(or1) $x \leq x$ for any $x \in P$ (reflexivity),
(or2) $x \leq y$ and $y \leq x$ implies $x = y$, for any $x, y \in P$ (antisymmetry),
(or3) $x \leq y$ and $y \leq z$ implies $x \leq z$, for any $x, y, z \in P$ (transitivity).
In this case, $(P, \leq)$ is said to be a *partially ordered set*.
If, in addition, we have:
(or4) for any $x, y \in \mathbb{Q}$, either $x \leq y$ or $y \leq x$,
then the relation $\leq$ is a *total order* and $(P, \leq)$ is a *totally ordered set*.
If $(P, +, \cdot)$ is a field and the total order relation satisfies the following properties, for all $x, y, z \in P$:
(or5) $x < y$ implies $x + z < y + z$,
(or6) $0 < x$ and $0 < y$ imply $0 < xy$,
then $(P, +, \cdot, \leq)$ is said to be an *ordered field*.

It can be also proved that $(\mathbb{Q}, d)$ is a metric space with the distance defined above, which can also be written as

$$d(x, y) = \begin{cases} x - y, & \text{if } x \geq y, \\ y - x, & \text{if } y > x. \end{cases} \quad (11.19)$$

In this metric space, the addition, multiplication and the mapping $x \to 1/x$, are continuous mappings, i.e. if $d(x_n, x) \to 0$, $d(y_n, y) \to 0$, then $d(x_n + y_n, x + y) \to 0$, $d(x_n y_n, xy) \to 0$, and $d(1/x_n, 1/x) \to 0$, if $x \neq 0$.

Proposition 11.12 *(Archimedes' Axiom for $\mathbb{Q}$) Let $(\mathbb{Q}, +, \cdot, \leq)$ be the ordered field of rational numbers. Then, for any $x \in \mathbb{Q}$, there exists $n \in \mathbb{Z}$ such that $n > x$.*

Proof Let $x = a/b$, with $a, b \in \mathbb{N}^*$. By Euclid's division theorem, there exists the unique natural numbers $q, r \in \mathbb{N}$ such that $a = qb + r$ and $0 \leq r < b$. So $x = a/b = q + r/b < q + 1$. Thus we can take $n = q + 1$. □

Remark 11.9 The metric space $\mathbb{Q}$ is not complete, that is, there exist Cauchy sequences of rational numbers that have no limit in $\mathbb{Q}$. We construct two such sequences below:

Let $\{a_n\}_{n\geq 1}$ and $\{b_n\}_{n\geq 1}$ be the sequences of rational numbers defined in the following way:
$a_1 = 1$, $b_1 = 2$, and, for every $n \geq 1$:

if $\left(\frac{a_n+b_n}{2}\right)^2 > 2$, then $a_{n+1} = a_n$, $b_{n+1} = \frac{a_n+b_n}{2}$;

if $\left(\frac{a_n+b_n}{2}\right)^2 < 2$, then $a_{n+1} = \frac{a_n+b_n}{2}$, $b_{n+1} = b_n$.

(we cannot have $\left(\frac{a_n+b_n}{2}\right)^2 = 2$ because $\sqrt{2} \notin \mathbb{Q}$.)

By mathematical induction it can be easily proved that, for any $n = 1, 2, \ldots$, we have:

$$a_n \leq a_{n+1} < b_{n+1} \leq b_n$$

and

$$d(a_{n+1}, a_n) = a_{n+1} - a_n \leq \frac{1}{2^n}, \quad d(b_{n+1}, b_n) = b_n - b_{n+1} \leq \frac{1}{2^n}.$$

We show that the sequences $\{a_n\}_{n\geq 1}$ and $\{b_n\}_{n\geq 1}$ are Cauchy sequences. For any $\varepsilon > 0$, there exists $N \in \mathbb{N}$ such that $\frac{1}{2^{N-1}} < \varepsilon$. It follows that, for any $n \geq N$ and $p = 1, 2, \ldots$,

$$d(a_{n+p}, a_n) = a_{n+p} - a_n \leq \frac{1}{2^{n+p-1}} + \frac{1}{2^{n+p-2}} + \ldots + \frac{1}{2^n}$$

$$< \frac{1}{2^n}\left(1 + \frac{1}{2} + \frac{1}{2^2} + \ldots\right) = \frac{1}{2^{n-1}} < \varepsilon.$$

By mathematical induction it can be easily proved that, for any $n = 1, 2, \ldots$, we have:

$$2 - \frac{1}{2^{n-2}} < a_n^2 < 2 < b_n^2 < 2 + \frac{1}{2^{n-2}},$$

hence

$$\lim_{n\to\infty} a_n^2 = \lim_{n\to\infty} b_n^2 = 2.$$

Thus, since 2 cannot be the square of a rational number, it follows that $\{a_n\}_{n\geq 1}$ and $\{b_n\}_{n\geq 1}$ have no limit in $\mathbb{Q}$, although they are Cauchy sequences. Hence $\mathbb{Q}$ is not complete.

Using the technique developed in the proof of Theorem 11.3, we construct the completion of $\mathbb{Q}$ relative to the metric d, that is, the field of real numbers $\mathbb{R}$ which can be represented by all the points of the number line $\mathcal{R}$. Thus, we consider the set

11.1 Normed, Metric and Hilbert Spaces

$\mathcal{C}$ of Cauchy sequences with rational terms and the equivalence relation "$\sim$" defined by:

$$x = \{x_n\} \sim y = \{y_n\} \text{ if } \lim_{n\to\infty} (x_n - y_n) = 0.$$

The set $\mathbb{R}$ is the quotient set $\mathcal{C}/\sim$, and a real number r is a class of equivalence $r = [x]$ corresponding to a Cauchy sequence of rational numbers $x = \{x_n\}$. For instance, the sequences $\{a_n\}$ and $\{b_n\}$ from Remark 11.9 belong to the same class of equivalence, and this class corresponds to the real number $r = \sqrt{2}$.

The distance $\overline{d}$ on $\mathbb{R}$ is defined as follows (see (11.16)):

$$\overline{d}([\{x_n\}], [\{y_n\}]) = \lim_{n\to\infty} d(x_n, y_n) = \lim_{n\to\infty} |x_n - y_n|.$$

Obviously, the metric space $(\mathbb{Q}, d)$ is embedded in the *complete* space $(\mathbb{R}, \overline{d})$. Moreover, from Theorem 11.3, we know that $\mathbb{Q}$ is dense in its completion $\mathbb{R}$.

We extend the addition and the multiplication from $\mathbb{Q}$ to $\mathbb{R}$ as follows:

$$[\{x_n\}] + [\{y_n\}] = [\{x_n + y_n\}], \ [\{x_n\}] \cdot [\{y_n\}] = [\{x_n \cdot y_n\}].$$

It can be verified that $(\mathbb{R}, +, \cdot)$ is a field (the *field of real numbers*).

We introduce the following relation of order $\leq$ on $\mathbb{R}$: $[\{x_n\}] \leq [\{y_n\}]$ if either $[\{x_n\}] \sim [\{y_n\}]$ or there exists $N \in \mathbb{N}$ such that $x_n < y_n$ for all $n \geq N$. It can be readily verified that this definition does not depend on the choice of the representative Cauchy sequences $\{x_n\}, \{y_n\}$. The reader can also prove (or1)–(or3) from Definition 11.23, so $\leq$ is an order on $\mathbb{R}$ (which extends to $\mathbb{R}$ the usual relation of order on $\mathbb{Q}$).

We prove that $\leq$ is a *total order* on $\mathbb{R}$ (or4):

Proposition 11.13 *For any $r, s \in \mathbb{R}$, we have either $r \leq s$, or $s \leq r$.*

Proof Let $r = [x] = [\{x_n\}]$ and $s = [y] = [\{y_n\}]$. If $\lim_{n\to\infty}(x_n - y_n) = 0$ then $x \sim y$, so $[x] \leq [y]$ (and also $[y] \leq [x]$; we have $r = s$ in this case).

If $\lim_{n\to\infty}(x_n - y_n) \neq 0$, then there exist $\varepsilon_0 > 0$ and $N_0 \in \mathbb{N}$ such that $|x_n - y_n| > 2\varepsilon_0$ for any $n \geq N_0$ (otherwise, two subsequences $\{x_{n_k}\}$ and $\{y_{m_k}\}$ could be constructed such that $|x_{n_k} - y_{m_k}| < 1/k$, for all $k = 1, 2, \ldots$, which means that $\{x_{n_k}\} \sim \{y_{m_k}\}$, i.e $x \sim y$).

Since $\{x_n\}$ and $\{y_n\}$ are Cauchy sequences, there exists $N \geq N_0$, a natural number such that $|x_n - x_N| < \varepsilon_0$ and $|y_n - y_N| < \varepsilon_0$, for any $n \geq N$.

We know that $|x_N - y_N| > 2\varepsilon_0$. Suppose that $x_N < y_N$ (recall that $\mathbb{Q}$ is a totally ordered set). Then,

$$x_n < x_N + \varepsilon_0 < y_N - \varepsilon_0 < y_n,$$

for all $n \geq N$. According to the definition above, it follows that $[x] \leq [y]$. Similarly, if $x_N > y_N$, we shall obtain that $[y] \leq [x]$, so $(\mathbb{R}, \leq)$ is a totally ordered set. □

One can also prove that $(\mathbb{R}, +, \cdot, \leq)$ satisfies or5) and or6) from Definition 11.23, which means that $\mathbb{R}$ is an *ordered field*. Moreover, the addition, multiplication, the mappings $r \to 1/r$, $r \neq 0$ and $r \to |r|$ are continuous mappings, i.e. if $r_n \to r$ and $s_n \to s$ in $\mathbb{R}$, then $\{r_n + s_n\} \to r + s$, $\{r_n s_n\} \to rs$, $1/r_n \to 1/r$ and $|r_n| \to |r|$. All these statements can be proved by using the density of $\mathbb{Q}$ in $\mathbb{R}$.

As shown above, we can assume that $\mathbb{N} \subset \mathbb{Z} \subset \mathbb{Q} \subset \mathbb{R}$.

Proposition 11.14 (*Archimedes' Axiom for $\mathbb{R}$*) *For any real number $r \in \mathbb{R}$, there exists a positive integer $n \in \mathbb{N}$, such that $|r| < n$.*

Proof Obviously, we can assume that $r > 0$. If $r \in \mathbb{Q}$, the statement follows by Proposition 11.12. Assume that $r \notin \mathbb{Q}$ and let $q \in \mathbb{Q}$ such that $|r - q| < 1$. By Proposition 11.12, there exists $n \in \mathbb{N}$ such that $q + 1 < n$. Hence $r < q + 1 < n$ and the proof is complete. □

Remark 11.10 From Archimedes' Axiom one can deduce that for any real number $x \in \mathbb{R}$ there exists a unique integer $n \in \mathbb{Z}$ (called the *integer part* of x) such that $n \leq x < n + 1$, or, equivalently, $x \in [n, n + 1)$.

Let $\mathcal{R}$ be the number line in Fig. 11.2. To each point $A \in \mathcal{R}$ we associate the length a of the segment OA if A is on the right hand of O, or $-a$ if A is on the left hand of O (for instance B in Fig. 11.2, since the length of $OB = a$). In Fig. 11.2 the point $A \in [4, 5)$. Let us divide this segment $[4, 5)$ into 10 equal parts labeled by $0, 1, 2, \ldots 9$. Let $b_1 \in \{0, 1, \ldots, 9\}$ be the label of the segment $[4 + b_1/10, 4 + (b_1 + 1)/10)$ into which the point A is. We continue to do the same thing with the segment $[4 + b_1/10, 4 + (b_1 + 1)/10)$ instead of $[4, 5)$, and obtain the label $b_2 \in \{0, 1, \ldots, 9\}$. So $A \in [4 + b_1/10 + b_2/10^2, 4 + b_1/10 + (b_2 + 1)/10^2)$, etc. In this way one can obtain a decimal representation of the positive quantity $a = 4.b_1 b_2 \ldots$, where $b_1, b_2, \ldots$ are digits. It is clear that if $b_n = 0$ for any $n > N$, and $b_N \neq 0$,

$$a = 7 + b_1/10 + b_2/10^2 + \ldots + b_N/10^N$$

is a rational number. One can prove that a is in $\mathbb{Q}$ if and only if there exists two natural numbers N and $p \geq 1$ such that $b_{N+i} = b_{N+i+mp}$ for any $m = 1, 2, \ldots$ and for any $i = 1, 2, \ldots p$. This means that a is a simple periodical fraction of period p if $N = 0$, or a mixed periodical fraction if $N > 0$. If $a = K + b_1/10 + b_2/10^2 + \ldots$, where $K \in \mathbb{N}$, we say that the rational number $a_s = K + b_1/10 + b_2/10^2 + \ldots + b_s/10^s$ is the s-*approximate* of a. It is clear that $\lim_{s \to \infty} a_s = a$. Note that all this construction relies on Archimedes' Axiom and Cantor's Intersection Theorem 11.7). Moreover, the number line $\mathcal{R}$ (see Fig. 11.2) is a geometrical model of the ordered field of real numbers. It is "complete", i.e. there is a one-to-one and onto correspondence between all kind of measurements and the points of $\mathcal{R}$. The ordered field of real numbers $\mathbb{R}$ constructed above is another model of the same object.

11.1 Normed, Metric and Hilbert Spaces

Theorem 11.6 (Bolzano–Weierstrass) *If A is a bounded infinite subset of $\mathbb{R}$, then there exists a sequence $\{a_n\}$ with distinct elements of A, which is convergent in $\mathbb{R}$.*

Proof Let $p_0, q_0 \in \mathbb{Q}$ such that $A \subset [p_0, q_0]$ and let $c_0 = (p_0 + q_0)/2$ be the midpoint of the interval $[p_0, q_0]$. Since $A \subset [p_0, q_0]$ is infinite, at least one of the intervals $[p_0, c_0]$ and $[c_0, q_0]$ contains an infinite number of elements of A. We denote by $[p_1, q_1]$ this interval and choose an element $a_1 \in A \cap [p_1, q_1]$. In the same way, (at least) one of the intervals $[p_1, (p_1 + q_1)/2]$ and $[(p_1 + q_1)/2, q_1]$ must contain an infinite number of elements of A and we denote it by $[p_2, q_2]$. We choose $a_2 \in [p_2, q_2] \cap A \setminus \{a_1\}$ and remark that $|a_2 - a_1| \leq q_1 - p_1 = (q_0 - p_0)/2$.

Assume that we constructed $[p_{n-1}, q_{n-1}]$ and $a_{n-1} \in [p_{n-1}, q_{n-1}] \cap A \setminus \{a_1, a_2, \ldots, a_{n-2}\}$. At least one of the intervals $[p_{n-1}, (p_{n-1} + q_{n-1})/2]$ and $[(p_{n-1} + q_{n-1})/2, q_{n-1}]$ must contain an infinite number of elements of A, so we denote it by $[p_n, q_n]$ and choose an element $a_n \in [p_n, q_n] \cap A \setminus \{a_1, a_2, \ldots, a_{n-1}\}$, etc. In this way we construct a sequence $\{a_n\}$ of distinct elements of A with the property that

$$|a_{n+1} - a_n| \leq q_n - p_n = \frac{q_0 - p_0}{2^n}, \quad n = 1, 2, \ldots$$

We prove that $\{a_n\}$ is a Cauchy sequence. Thus, we can write:

$$|a_{n+k} - a_n| \leq |a_{n+k} - a_{n+k-1}| + |a_{n+k-1} - a_{n+k-2}| + \ldots + |a_{n+1} - a_n| \leq$$

$$\leq (q_0 - p_0)\left(\frac{1}{2^{n+k-1}} + \frac{1}{2^{n+k-2}} + \ldots + \frac{1}{2^n}\right) <$$

$$< \frac{q_0 - p_0}{2^n}\left(1 + \frac{1}{2} + \frac{1}{2^2} + \ldots\right) = \frac{q_0 - p_0}{2^{n-1}}, \quad \text{for all } n, k = 1, 2, \ldots.$$

For any $\varepsilon > 0$, there exists $N = N_\varepsilon \in \mathbb{N}$ such that $\frac{q_0 - p_0}{2^{N-1}} < \varepsilon$. It follows that $|a_{n+k} - a_n| < \varepsilon$ for any $n \geq N$, $k \in \mathbb{N}$. Hence $\{a_n\}$ is a Cauchy sequence and it is convergent in $\mathbb{R}$ (since $\mathbb{R}$ is a complete space). □

Theorem 11.7 (Cantor's Intersection Theorem) *Let $\{a_n\}, \{b_n\}$ be two sequences in $\mathbb{R}$ such that:*

(i) $a_0 \leq a_1 \leq \ldots \leq a_n \leq \ldots \leq b_n \leq \ldots \leq b_1 \leq b_0$, and

(ii) $\lim_{n \to \infty}(b_n - a_n) = 0$.

Then the intersection of the nested intervals $[a_n, b_n]$, $n = 1, 2, \ldots$, contains exactly one point:

$$\bigcap_{n \geq 0} [a_n, b_n] = \{c\}.$$

As a consequence, $\lim_{n \to \infty} a_n = \lim_{n \to \infty} b_n = c$.

Proof Assume that both sequences $\{a_n\}$ and $\{b_n\}$ have an infinite number of distinct terms (a monotone sequence with a finite number of distinct terms is constant for $n \geq N$ with N large enough, and the statement is obvious in this case). Since both sequences are bounded, we can apply Bolzano-Weierstrass Theorem (Theorem 11.6) and find two convergent subsequences $a_{k_n} \to a$ and $b_{l_n} \to b$. Since $\{a_n\}$ is an increasing sequence and $\{b_n\}$ is a decreasing sequence, it follows that $a_n \to a$, $a_n \leq a$, for all n and $b_n \to b$, $b_n \geq b$, for all n. Thus, one can write:

$$0 \leq b - a \leq b_n - a_n \to 0.$$

Hence $a = b = c$, $a_n \leq c \leq b_n$ for all $n \geq 1$ and the proof is complete. □

Definition 11.24 Let A be a nonempty subset of $\mathbb{R}$. We say that A is upper bounded in $\mathbb{R}$ if there exists $M \in \mathbb{R}$ such that $a \leq M$ for any $a \in A$. In this case, M is called an upper bound of A. If for any other upper bound M' of A we have $M \leq M'$, then M is said to be the *least upper bound* of A and is denoted by $M = \sup A$.

We say that A is lower bounded in $\mathbb{R}$ if there exists $m \in \mathbb{R}$ such that $a \geq m$ for any $a \in A$. In this case, m is called a lower bound of A. If for any other lower bound m' of A we have $m \geq m'$, then m is said to be the *greatest lower bound* of A and is denoted by $m = \inf A$.

We remark that an upper bound M of A is the least upper bound of A if for any $\varepsilon > 0$, there exists $a \in A$ with $M - \varepsilon < a \leq M$. In other words, $M = \sup A$ if and only if $x \leq M$ for any $x \in A$ and there exists a sequence $\{a_n\} \subset A$ such that $\lim_{n \to \infty} a_n = M$.

Similarly, m is the greatest lower bound of A, $m = \inf A$, if and only if $x \geq m$ for any $x \in A$ and there exists a sequence $\{b_n\} \subset A$ such that $\lim_{n \to \infty} b_n = m$.

The next theorem proves the existence of the least upper bound (and greatest lower bound, respectively), for any bounded subset of $\mathbb{R}$. We say that $\mathbb{R}$ has "the least upper bound property". The set of rational numbers $\mathbb{Q}$ does not have this property.

Theorem 11.8

(i) Any upper bounded nonempty subset A of $\mathbb{R}$ has a (unique) least upper bound in $\mathbb{R}$.

(ii) Any lower bounded nonempty subset A of $\mathbb{R}$ has a (unique) greatest lower bound in $\mathbb{R}$.

Proof

(i) Let b_0 be an upper bound of A and let $a_0 \in A$. We denote by $c_0 = (a_0 + b_0)/2$ the midpoint of the interval $[a_0, b_0]$. If c_0 is an upper bound of A, then we take $a_1 = a_0$ and $b_1 = c_0$. If $[c_0, b_0]$ contains at least one element of A, we put $a_1 = c_0$ and $b_1 = b_0$. We denote by c_1 the midpoint of the interval $[a_1, b_1]$ and continue the process, obtaining the sequence of nested intervals $[a_n, b_n]$, that is, $[a_{n+1}, b_{n+1}] \subset [a_n, b_n]$ for all $n = 0, 1, \ldots$ with the properties that

11.1 Normed, Metric and Hilbert Spaces

$[a_n, b_n] \cap A \neq \emptyset$, $n = 0, 1, \ldots$, and all b_n are upper bounds of A. Since

$$b_n - a_n = \frac{b_0 - a_0}{2^n} \to 0,$$

we can apply the Cantor's Intersection Theorem (Theorem 11.7) and obtain that

$$\bigcap_{n \geq 0} [a_n, b_n] = \{c\},$$

so $c = \lim_{n \to \infty} a_n = \lim_{n \to \infty} b_n$, and $a_n \leq c \leq b_n$ for all $n = 0, 1, \ldots$. Since all $b_n \geq x$ for all $x \in A$, it follows that $c \geq x$, for all $x \in A$, hence c is an upper bound of A. Since $[a_n, b_n] \cap A \neq \emptyset$ for all $n = 0, 1, \ldots$, we can construct the sequence $\{c_n\}$ with elements in A, such that $a_n \leq c_n \leq b_n$, for all $n = 0, 1, \ldots$. It follows that $\lim_{n \to \infty} c_n = \lim_{n \to \infty} a_n = \lim_{n \to \infty} b_n = c$, hence $c = \sup A$.

(ii) Let $-A = \{-a : a \in A\}$ be the reflection of A with respect to 0. Since A is lower bounded, it follows that $-A$ is upper bounded, so there exists its least upper bound, $c = \inf(-A)$. The real number $-c$ is the greatest lower bound of A.

□

Corollary 11.1 *If $\{s_n\}_{n \geq 1}$ is a bounded increasing sequence of real numbers, $s_n \leq s_{n+1}$ and $s_n \leq M$ for any $n \geq 1$, then $\{s_n\}$ is convergent and its limit is $s = \lim_{n \to \infty} s_n = \sup\{s_n : n = 0, 1, \ldots\}$.*

Proof Theorem 11.8 states that $s = \sup\{s_n : n = 0, 1, \ldots\}$ exists. Let $\varepsilon > 0$ be a positive real number. Since s is the least upper bound of $\{s_n : n = 0, 1, \ldots\}$, it follows that there exists $N \in \mathbb{N}$ such that $s - \varepsilon < s_N \leq s$. But the sequence is increasing and bounded by s, so $s - \varepsilon < s_n \leq s$ for any $n \geq N$, i.e. $|s_n - s| < \varepsilon$ for any $n \geq N$. Hence $\lim_{n \to \infty} s_n = s$. □

Theorem 11.9 (Weierstrass' Theorem) *If $f : [a, b] \to \mathbb{R}$ is a continuous function, then f is bounded and attains its bounds, i.e. there exist $\alpha, \beta \in [a, b]$ such that*

$$f(\alpha) \leq f(x) \leq f(\beta), \ \forall x \in [a, b].$$

Proof Suppose that f is not (upper) bounded, i.e. there exists a sequence $\{x_n\}$ in $[a, b]$ such that $f(x_n) > n$, for all $n = 1, 2, \ldots$. It follows that $\lim_{n \to \infty} f(x_n) = \infty$. But the sequence $\{x_n\} \subset [a, b]$ is bounded, and by Bolzano-Weierstrass Theorem (Theorem 11.6) we know that there is a subsequence $\{x_{k_n}\}_n$ of $\{x_n\}$ such that x_{k_n} is convergent to $z \in [a, b]$. Since f is a continuous function, it follows that $\lim_{n \to \infty} f(x_{k_n}) = f(z)$, so we obtain that $f(z) = \infty$, a contradiction. It follows that f

is upper bounded, and, in a similar manner one can prove that f is lower bounded as well.

Now let us prove that f attains its bounds in the interval $[a, b]$. Let $M = \sup_{x \in [a,b]} f(x)$. Suppose that $f(x) < M$ for all $x \in [a, b]$. It follows that the function $g(x) = \frac{1}{M - f(x)}$ is continuous on $[a, b]$, so, as proved above, it is bounded on $[a, b]$. But we know that, for any $n = 1, 2, \ldots$, there exists $x_n \in [a, b]$ such that $M - \frac{1}{n} < f(x_n) < M$. Thus, we obtain that $g(x_n) = \frac{1}{M - f(x_n)} > n$, for all $n = 1, 2, \ldots$, which means that g is not bounded, a contradiction. It follows that there is at least one point $x_0 \in [a, b]$ such that $f(x_0) = M$. In a similar way one can prove that f attains its lower bound $m = \inf_{x \in [a,b]} f(x)$ on $[a, b]$. □

We constructed the ordered complete field $(\mathbb{R}, +, \cdot, \leq)$ which satisfies Archimedes' Axiom and Cantor's Intersection Theorem.

Starting from $\mathbb{R}$ one can construct the field $\mathbb{C}$ of complex numbers. First of all we see that the equation $X^2 + 1 = 0$ has no root in $\mathbb{R}$, because for any $r \in \mathbb{R}$, $r^2 \geq 0$. Let us denote by

$$\mathbb{R}[X] = \{P(X) = a_0 + a_1 X + \ldots + a_n X^n : a_i \in \mathbb{R}, i = \overline{0, n}, a_n \neq 0 \text{ if } n > 0\}$$

the ring of all the polynomials with coefficients in $\mathbb{R}$. The coefficient $a_n \neq 0$ is called the leading coefficient of $P(X)$ and the nonnegative integer $n = \deg P$ is said to be the degree of the polynomial $P(X)$. The degree of the zero polynomial is defined to be $\deg \mathbf{0} = -\infty$. Given two polynomials $P, Q \in \mathbb{R}[X]$, we have:

$$\deg(P + Q) \leq \max(\deg P, \deg Q),$$

$$\deg(PQ) = \deg P + \deg Q.$$

We say that two polynomials P and Q are *equivalent* ($P \sim Q$) if the polynomial $P - Q$ is divisible by $X^2 + 1$. It is easy to see that "$\sim$" is an equivalence relation (see Definition 11.19) on $\mathbb{R}[X]$. We denote by $\mathbb{C} = \mathbb{R}[X]/\sim$, the quotient set of $\mathbb{R}[X]$ relative to this equivalence relation and $\widehat{P} = \{Q \in \mathbb{R}[X] : Q - P \vdots X^2 + 1\}$ the equivalence class of the polynomial $P \in \mathbb{R}[X]$. Thus, for any classes $\widehat{P}, \widehat{Q} \in \mathbb{C}$, we have $\widehat{P} \sim \widehat{Q}$ if and only if $P \sim Q$.

Proposition 11.15 *Each class $\widehat{P} \in \mathbb{C}$ contains a unique polynomial of degree less than or equal to 1.*

Proof Let us use the Euclidean division algorithm to divide $P(X)$ by $X^2 + 1$: $P(X) = (X^2 + 1)Q(X) + R(X)$, where $\deg R(X) \leq 1$. Thus, $P \sim R$, and R is the unique polynomial of degree less than or equal to 1 with this property, so one can identify the set $\mathbb{C}$ with the set of polynomials of degree at most 1, $\mathbb{C} = \{a + bX : a, b \in \mathbb{R}\}$. □

11.1 Normed, Metric and Hilbert Spaces

The addition, the multiplication and the scalar multiplication (with scalars from $\mathbb{R}$) are defined as usual (see the proof of Theorem 11.11):

$$\widehat{P} + \widehat{Q} = \widehat{P+Q}, \tag{11.20}$$

$$\widehat{P} \cdot \widehat{Q} = \widehat{PQ}, \tag{11.21}$$

$$\alpha \cdot \widehat{P} = \widehat{\alpha P}, \tag{11.22}$$

where $P, Q \in \mathbb{R}[X]$ and $\alpha \in \mathbb{R}$. It is not difficult to prove that these operations are well defined, i.e. their definitions do not depend on the choice of representatives in the classes of P and Q. Moreover, with the operations defined in (11.20)–(11.22) the set $\mathbb{C}$ becomes a field and an $\mathbb{R}$-algebra ($a \to a + 0 \cdot X$ is the embedding of $\mathbb{R}$ in $\mathbb{C}$). As $\mathbb{C} = \{a + bX : a, b \in \mathbb{R}\}$ and $\widehat{X^2 + 1} = \widehat{0}$, or, equivalently, $\widehat{X^2} = \widehat{-1}$ in $\mathbb{C}$, we can rewrite the formulas (11.20)–(11.22) as:

$$(a_1 + b_1 X) + (a_2 + b_2 X) = (a_1 + a_2) + (b_1 + b_2)X, \tag{11.23}$$

$$(a_1 + b_1 X) \cdot (a_2 + b_2 X) = (a_1 a_2 - b_1 b_2) + (a_1 b_2 + b_1 a_2)X, \tag{11.24}$$

$$\alpha \cdot (a_1 + b_1 X) = \alpha a_1 + \alpha b_1 X, \tag{11.25}$$

where $\alpha, a_i, b_i, i = 1, 2$ are real numbers.

For instance:

$$(X^3 + 2\widehat{X^2} + 3X + 1) \cdot (X^4 - 4\widehat{X^3} + 2) \stackrel{X^2 = -1}{=}$$
$$= (-1 + 2X) \cdot (3 + 4X) = -11 + 2X.$$

Obviously, $\widehat{X} \ne \widehat{a}$, for any real number a. Euler denoted $\widehat{X}$ by i (an imaginary quantity, i.e. a number which is not real) and so any complex number is uniquely written as $a + bi$, where $a, b \in \mathbb{R}$ and $i = \sqrt{-1}$. Thus $\mathbb{C} = \mathbb{R}[i]$, where $i^2 = -1$, i.e. the polynomials in the variable i such that $i^2 = -1$. It is easy to see that $\mathbb{C}$ is a vector space of dimension 2 over $\mathbb{R}$ and a canonical basis of it is $\{1, i\}$. For any complex number $z \in \mathbb{C}$, $z = a + bi = a + ib$ with $a, b \in \mathbb{R}$, we say that $a = \operatorname{Re} z$ is the real part of z, and $b = \operatorname{Im} z$ is the imaginary part of z. Note that z is a real number if and only if $y = 0$. We also say that z is *pure imaginary* if $x = 0$. Thus, the Cartesian system of coordinates xOy can be used in the complex plane to represent a complex number $z = a + ib$ by the point of coordinates (a, b) (the horizontal x-axis represents the real part, while the vertical y-axis marks the imaginary part). It can be readily proved that $\mathbb{C}$ is a normed space over $\mathbb{R}$ with the norm defined by the modulus (the absolute value) of a complex number $z = a + ib$:

$$\|z\| = |z| = \sqrt{a^2 + b^2}.$$

The great German mathematician, C. F. Gauss proved that any non-constant polynomial with complex coefficients $P \in \mathbb{C}[X]$ has a complex root. This theorem is known as the *Fundamental Theorem of Algebra* (see Theorem 11.29) and states that $\mathbb{C}$ is *algebraically closed*. It can be also stated as follows: every polynomial of degree $n \geq 1$ with complex coefficients has exactly n complex roots (counted with multiplicity). The equivalence of the two statements becomes evident by using successive polynomial division.

Proposition 11.16 *Let $\mathbb{C}$ be the field of complex numbers.*

(a) *A sequence $\{z_n\}$, $z_n = x_n + iy_n$ is convergent to $z = x + iy$ in $\mathbb{C}$ if and only if $x_n \to x$ and $y_n \to y$ in $\mathbb{R}$.*
(b) *A sequence $\{z_n\}$, $z_n = x_n + iy_n$ is a Cauchy sequence in $\mathbb{C}$ if and only if $\{x_n\}$ and $\{y_n\}$ are Cauchy sequences in $\mathbb{R}$.*
(c) *$\mathbb{C}$ is a complete $\mathbb{R}$-algebra, i.e. a Banach $\mathbb{R}$ (or $\mathbb{C}$)-algebra.*

Proof

(a) $|z_n - z| = \sqrt{|x_n - x|^2 + |y_n - y|^2} \to 0 \iff |x_n - x| \to 0$ and $|y_n - y| \to 0$.

(b) Assume that $\{z_n\}$ is a Cauchy sequence in $\mathbb{C}$ and let $\varepsilon > 0$ be a real positive number. Then, there exists $N = N_\varepsilon \in \mathbb{N}$ such that

$$\left|z_{n+p} - z_n\right| = \sqrt{\left|x_{n+p} - x_n\right|^2 + \left|y_{n+p} - y_n\right|^2} < \varepsilon$$

for all $n \geq N$ and $p \in \mathbb{N}$. It follows that

$$\left|x_{n+p} - x_n\right| < \varepsilon, \quad \left|y_{n+p} - y_n\right| < \varepsilon$$

for all $n \geq N$ and $p \in \mathbb{N}$, i.e. $\{x_n\}, \{y_n\}$ are Cauchy sequences in $\mathbb{R}$.

Conversely, suppose that $\{x_n\}$ and $\{y_n\}$ are Cauchy sequences in $\mathbb{R}$ and let $\varepsilon > 0$ be a positive real number. Thus, we can find a common $N = N_\varepsilon$, sufficiently large, such that

$$\left|x_{n+p} - x_n\right| < \varepsilon/\sqrt{2}, \quad \left|y_{n+p} - y_n\right| < \varepsilon/\sqrt{2}$$

for all $n \geq N$ and $p \in \mathbb{N}$. Hence

$$\left|z_{n+p} - z_n\right| = \sqrt{\left|x_{n+p} - x_n\right|^2 + \left|y_{n+p} - y_n\right|^2} < \varepsilon$$

for all $n \geq N$ and $p \in \mathbb{N}$.

(c) Let $\{z_n\}$, $z_n = x_n + iy_n$ be a Cauchy sequence in $\mathbb{C}$. From b) we see that $\{x_n\}$, $\{y_n\}$ are Cauchy sequences in $\mathbb{R}$. Since $\mathbb{R}$ is complete, $x_n \to x$ and $y_n \to y$ in $\mathbb{R}$. From a) we can conclude that $z_n \to z = x + iy$. Hence $\mathbb{C}$ is complete as a metric field or as an $\mathbb{R}$ (or $\mathbb{C}$)-algebra. □

11.1 Normed, Metric and Hilbert Spaces

For any complex number $z = x + iy$ one can define its (complex) *conjugate*, $\bar{z} = x - iy$, which is, from a geometrical point of view, the reflexion of z about the real axis. We remark that

$$z \cdot \bar{z} = x^2 + y^2 = |z|^2.$$

Moreover, the real part and the imaginary part can be written:

$$\operatorname{Re} z = \frac{z + \bar{z}}{2}, \quad \operatorname{Im} z = \frac{z - \bar{z}}{2i}.$$

The following relations can be also easily proved, for any $z, w \in \mathbb{C}$:

$$\overline{z + w} = \bar{z} + \bar{w},$$
$$\overline{z \cdot w} = \bar{z} \cdot \bar{w},$$
$$\overline{z/w} = \bar{z}/\bar{w}, \text{ if } w \neq 0.$$

Complex numbers are often represented in the *trigonometric* (or *polar*) form:

$$z = r \exp(i\alpha) = r(\cos\alpha + i \sin\alpha), \qquad (11.26)$$

where r is the modulus of $z = x + iy$,

$$r = |z| = \sqrt{x^2 + y^2},$$

and $\alpha \in [0, 2\pi)$ is the *argument* of z, that is, the angle of the radius OM with the positive real axis, denoted by $\alpha = \arg z$ (see Fig. 11.4). Except for 0, whose argument is undetermined, any complex number z is uniquely determined by the pair (r, α). We remark that, if $z_1 = r_1 \exp(i\alpha_1)$ and $z_2 = r_2 \exp(i\alpha_2)$ are two complex

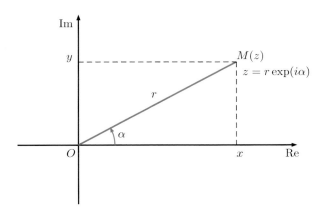

Fig. 11.4 Trigonometric form of complex numbers

numbers ($\neq 0$), then we have:

$$z_1 z_2 = r_1 r_2 \exp[i(\alpha_1 + \alpha_2)], \quad z_1/z_2 = r_1/r_2 \exp[i(\alpha_1 - \alpha_2)].$$

Hence

$$|z_1 z_2| = |z_1||z_2| \text{ and } \arg(z_1 z_2) = \arg z_1 + \arg z_2,$$
$$|z_1/z_2| = |z_1|/|z_2| \text{ and } \arg(z_1/z_2) = \arg z_1 - \arg z_2.$$

11.1.5 Hilbert Spaces

In the Euclidean 3-dimensional space $\mathbb{R}^3$ the *scalar product* (or *dot product*) of two vectors $\vec{a}$ and $\vec{b}$ is defined by

$$\vec{a} \cdot \vec{b} = |\vec{a}| \, |\vec{b}| \cos \theta,$$

where $|\vec{a}|$ is the length of the vector of $\vec{a}$ and $\theta \in [0, \pi]$ is the angle between $\vec{a}$ and $\vec{b}$. Two nonzero vectors are orthogonal if and only if $\vec{a} \cdot \vec{b} = 0$. Let $\{\mathbf{i}, \mathbf{j}, \mathbf{k}\}$ be the canonical base of $\mathbb{R}^3$ and $\vec{a} = a_1 \mathbf{i} + a_2 \mathbf{j} + a_3 \mathbf{k}$, $\vec{b} = b_1 \mathbf{i} + b_2 \mathbf{j} + b_3 \mathbf{k}$. Then, since $\mathbf{i}, \mathbf{j}, \mathbf{k}$ are unit orthogonal vectors, the scalar (dot) product of the vectors $\vec{a}$ and $\vec{b}$ is written:

$$\vec{a} \cdot \vec{b} = a_1 b_1 + a_2 b_2 + a_3 b_3.$$

The *inner product* is a generalization of the dot product.

Definition 11.25 Let V be a vector space over $K = \mathbb{R}$ or $\mathbb{C}$. A mapping $\langle \cdot, \cdot \rangle : V \times V \to K$ which associates to any ordered pair (x, y) of elements of V a real or a complex number $\langle x, y \rangle \in K$ is said to be an **inner product** on V if it satisfies the following properties, for any vectors $x, y, z \in V$ and any scalar $\alpha \in K$:

(p1) $\langle x, y \rangle = \overline{\langle y, x \rangle}$ (conjugate symmetry)
   If $K = \mathbb{R}$, this property becomes symmetry: $\langle x, y \rangle = \langle y, x \rangle$.
(p2) $\langle x + y, z \rangle = \langle x, z \rangle + \langle y, z \rangle$.
(p3) $\langle \alpha x, y \rangle = \alpha \langle x, y \rangle$ (p2) and p3) states the linearity in the first argument).
(p4) $\langle x, x \rangle \geq 0$ for any $x \in V$ (by p1) we know that $\langle x, x \rangle = \overline{\langle x, x \rangle}$, so $\langle x, x \rangle$ is a real number; it must be a nonnegative number)
(p5) $\langle x, x \rangle = 0$ if and only if $x = \mathbf{0}$ (i.e. $\langle x, x \rangle > 0$ for any $x \neq \mathbf{0}$).

A vector space together with an inner product on it is called an *inner product space*.

If only the conditions (p1)–(p4) are satisfied, then $\langle \cdot, \cdot \rangle$ is said to be a *positive semi-definite Hermitian form*.

11.1 Normed, Metric and Hilbert Spaces

The following properties of the inner product can be easily derived from the definition above:

(c1) $\langle \mathbf{0}, x \rangle = \langle x, \mathbf{0} \rangle = 0$ for any $x \in V$ (where $\mathbf{0}$ is the null vector of V, and 0 is the null element of K).
(c2) $\langle x, y + z \rangle = \langle x, y \rangle + \langle x, z \rangle$ for any $x, y, z \in V$;
(c3) $\langle x, \alpha y \rangle = \overline{\alpha} \langle x, y \rangle$ for any $x, y \in V$ and for any $\alpha \in K$, ((c2) and (c3) states the conjugate-linearity in the second argument; if $K = \mathbb{R}$ we have linearity in the second argument).

For any $x \in V$, we denote

$$\|x\| = \sqrt{\langle x, x \rangle}. \tag{11.27}$$

We shall prove that $(V, \|\cdot\|)$ is a normed space over K, i.e. the mapping $\|\cdot\| : V \to [0, \infty)$ satisfies the properties from Definition 11.1. First, we have to prove the following fundamental result.

Theorem 11.10 (Cauchy–Schwarz Inequality) *If $(V, \langle \cdot, \cdot \rangle)$ is an inner product space, then*

$$|\langle x, y \rangle| \le \|x\| \|y\| \tag{11.28}$$

for any $x, y \in V$. The equality takes place if and only if x and y are linearly dependent in V.

Proof We prove the theorem for $K = \mathbb{C}$ (the most general case). Obviously, if $\langle x, y \rangle = 0$, the inequality (11.28) holds true.

Suppose that $\langle x, y \rangle \ne 0$ (hence $x, y \ne 0$). Let α denote the complex number $\alpha = \frac{\langle x, y \rangle}{|\langle x, y \rangle|}$ and let $t \in \mathbb{R}$ be an arbitrary real number. From the definition of the inner product, we have:

$$0 \le \langle tx + \alpha y, tx + \alpha y \rangle =$$
$$= t^2 \|x\|^2 + t\left(\overline{\alpha} \langle x, y \rangle + \alpha \overline{\langle x, y \rangle}\right) + \alpha \overline{\alpha} \|y\|^2 =$$
$$= t^2 \|x\|^2 + 2t|\langle x, y \rangle| + \|y\|^2.$$

If we denote: $A = \|x\|^2$, $B = |\langle x, y \rangle|$ and $C = \|y\|^2$, we can write:

$$At^2 + 2Bt + C \ge 0,$$

for all $t \in \mathbb{R}$. It follows that the discriminant must be either negative, or at most 0: $\Delta = B^2 - AC \le 0$ and so the inequality (11.28) is obtained.

The equality means that $\Delta = B^2 - AC = 0$, so in this case there exists a real number t such that

$$\langle tx + y, tx + y \rangle = At^2 + Bt + C = 0,$$

and by *p5* we obtain that $tx + y = \mathbf{0}$, so x and y are linearly dependent. □

Corollary 11.2 (Triangle Inequality) *If $(V, \langle \cdot, \cdot \rangle)$ is an inner product space, then, for any $x, y \in V$, we have:*

$$\|x + y\| \leq \|x\| + \|y\| \tag{11.29}$$

Proof We can write:

$$\|x+y\|^2 = \langle x+y, x+y \rangle = \langle x, x \rangle + 2\operatorname{Re}\langle x, y \rangle + \langle y, y \rangle \leq$$
$$\leq \|x\|^2 + 2|\langle x, y \rangle| + \|y\|^2 \stackrel{(11.28)}{\leq}$$
$$\leq \|x\|^2 + 2\|x\|\|y\| + \|y\|^2 = (\|x\| + \|y\|)^2,$$

so the inequality (11.29) follows. □

Since

$$\|\lambda x\| = \sqrt{\langle \lambda x, \lambda x \rangle} = \sqrt{\lambda \overline{\lambda} \langle x, x \rangle} = |\lambda| \|x\|,$$

we can see that the conditions $n1 - n3$ from Definition 11.1 are satisfied, so the inner product space $(V, \langle \cdot, \cdot \rangle)$ is also a normed space with the norm defined by (11.27). It is also a metric space with the metric induced by this norm:

$$d(x, y) = \|x - y\| = \sqrt{\langle x - y, x - y \rangle}, \ \forall x, y \in V. \tag{11.30}$$

Proposition 11.17 *Let $(V, \langle \cdot, \cdot \rangle)$ be an inner product space. Then, for any sequences convergent in V, $x_n \to x$, $y_n \to y$, we have:*

$$\langle x_n, y_n \rangle \to \langle x, y \rangle, \ \|x_n\| \to \|x\| \ (in \ K),$$

i.e. the mappings $(x, y) \to \langle x, y \rangle$ and $x \to \|x\|$ are continuous.

Moreover, if $\{x_n\}$, $\{y_n\}$ are two Cauchy sequences in $(V, \|\cdot\|)$, then $\{\langle x_n, y_n \rangle\}_n$ is also a Cauchy sequence in $(K, |\cdot|)$.

Proof Since the sequence $\{x_n\}$, $\{y_n\}$ are convergent, it follows that they are bounded: there is a positive number $M > 0$ such that $\|x_n\| \leq M$, $\|y_n\| \leq M$, for all $n = 1, 2, \ldots$. We can write:

$$|\langle x_n, y_n \rangle - \langle x, y \rangle| = |\langle x_n, y_n \rangle - \langle x, y_n \rangle + \langle x, y_n \rangle - \langle x, y \rangle|$$
$$\leq |\langle x_n - x, y_n \rangle| + |\langle x, y_n - y \rangle| \stackrel{(11.28)}{\leq}$$
$$\leq \|x_n - x\| \|y_n\| + \|y_n - y\| \|x\| \leq$$
$$\leq M \|x_n - x\| + \|x\| \cdot \|y_n - y\| \to 0,$$

11.1 Normed, Metric and Hilbert Spaces

so $\langle x_n, y_n \rangle \to \langle x, y \rangle$ in K. For $x_n = y_n$ we obtain that $\|x_n\| \to \|x\|$.

If $\{x_n\}$, $\{y_n\}$ are two Cauchy sequences in V, then they are bounded (see Proposition 11.11): there is a positive number M such that $\|x_n\| < M$ and $\|y_n\| < M$ for any $n = 1, 2 \ldots$.

Let $\varepsilon > 0$ and $N = N_\varepsilon \in \mathbb{N}$ such that

$$\|x_{n+k} - x_n\| < \frac{\varepsilon}{2M}, \quad \|y_{n+k} - y_n\| < \frac{\varepsilon}{2M}, \quad \text{for all } n \geq N, k \in \mathbb{N}.$$

Thus, by using Cauchy-Schwarz inequality (11.28), we get:

$$|\langle x_{n+k}, y_{n+k} \rangle - \langle x_n, y_n \rangle| = |\langle x_{n+k}, y_{n+k} \rangle - \langle x_n, y_{n+k} \rangle + \langle x_n, y_{n+k} \rangle - \langle x_n, y_n \rangle|$$

$$\leq |\langle x_{n+k} - x_n, y_{n+k} \rangle| + |\langle x_n, y_{n+k} - y_n \rangle| \stackrel{(11.28)}{\leq}$$

$$\leq \|x_{n+k} - x_n\| \|y_{n+k}\| + \|y_{n+k} - y_n\| \|x_n\| <$$

$$< M \cdot \frac{\varepsilon}{2M} + M \cdot \frac{\varepsilon}{2M} = \varepsilon,$$

for any $n \geq N$ and $k = 1, 2, \ldots$. Hence the sequence $\{\langle x_n, y_n \rangle\}_n$ is a Cauchy sequence in K. □

Definition 11.26 An inner product space $(V, \langle \cdot, \cdot \rangle)$ is a *Hilbert space* if it is *complete* with respect to the induced norm $\|x\| = \sqrt{\langle x, x \rangle}$, $x \in V$ (i.e. if every Cauchy sequence is convergent in V).

Example 11.15 Let $K = \mathbb{R}$ or $\mathbb{C}$, $V = K^n$, the n-dimensional arithmetical space over K, and $x = (x_1, x_2, \ldots, x_n)$, $y = (y_1, y_2, \ldots, y_n) \in V$. It is easy to prove that

$$\langle x, y \rangle = \sum_{i=1}^{n} x_i \overline{y}_i \quad (11.31)$$

is an inner product on V and the norm induced by this inner product is

$$\|x\| = \sqrt{\langle x, x \rangle} = \sqrt{\sum_{i=1}^{n} |x_i|^2}. \quad (11.32)$$

Since K is a complete metric space, then V is also a complete metric space (see Example 11.12). Hence $V = K^n$ is a Hilbert space.

Example 11.16 Let $V = l^2$ be the space of all the sequences $\{x_n\}_{n \geq 1}$ with terms in $K = \mathbb{R}$ or $\mathbb{C}$ such that

$$\sum_{n=1}^{\infty} |x_n|^2 < \infty.$$

We define the inner product as:

$$\langle \{x_n\}, \{y_n\} \rangle = \sum_{n=1}^{\infty} x_n \overline{y_n}.$$

It can be proved that the conditions p1)- p5) from Definition 11.25 are satisfied. Moreover, $V = l^2$ is a complete metric space, so it is a Hilbert space. The norm induced by this inner product is

$$\|\{x_n\}\| = \sqrt{\sum_{n=1}^{\infty} |x_n|^2}.$$

Example 11.17 Let $a < b$, $a, b \in \mathbb{R}$ be two real numbers and $V = C[a, b]$ be the K vector space of all continuous functions defined on $[a, b]$ with values in $K = \mathbb{R}$ or $\mathbb{C}$. Then the following formula:

$$\langle f, g \rangle = \int_a^b f(t)\overline{g(t)}dt, \quad f, g \in V, \tag{11.33}$$

defines an inner product on V.

Indeed, the conditions p1)-p4) are obviously satisfied. Let us verify p5). Suppose that $\langle f, f \rangle = \int_a^b |f(t)|^2 dt = 0$ and $f \neq 0$ on $[a, b]$. So there exists $t_0 \in [a, b]$ such that $f(t_0) \neq 0$. Since f is continuous, there exists an interval $[\alpha, \beta] \subset [a, b]$ such that $|f(t)|^2 > 0$ for every $t \in [\alpha, \beta]$ and there exists $c \in [\alpha, \beta]$ such that $\int_\alpha^\beta |f(t)|^2 dt = |f(c)|^2 (\beta - \alpha)$. It follows that

$$0 = \int_a^b |f(t)|^2 dt \geq \int_\alpha^\beta |f(t)|^2 dt = |f(c)|^2 (\beta - \alpha) > 0,$$

a contradiction. Hence $f = 0$ on $[a, b]$ and so V is an inner product space. We remark that it is not a Hilbert space (see [33], 4.5)

Example 11.18 A function $f : [a, b] \to \mathbb{C}$, $f(t) = u(t) + iv(t)$ is said to be *piecewise continuous* if the functions $u, v : [a, b] \to \mathbb{R}$ are continuous on $[a, b]$ possibly except a finite set of points where they still have finite side limits (which may be unequal). We denote by $PC[a, b]$ the $\mathbb{C}$-vector space of all piecewise continuous functions defined on $[a, b]$. The formula (11.33) defines a Hermitian form on $PC[a, b]$ which is not an inner product because p5) is not satisfied. For instance, the function $f(t) = 0$ on $(a, b]$ and $f(a) = 1$ is not equal to zero on $[a, b]$ but $\int_a^b |f(t)|^2 dt = 0$, because f is zero almost everywhere (a.e.) on $[a, b]$.

11.1 Normed, Metric and Hilbert Spaces

We introduce the following equivalence relation on $PC[a, b]$:

$$f \sim g \text{ if } f = g \text{ a.e.}$$

Then, the quotient space $\widehat{PC}[a, b] = PC[a, b]/\sim$ is an inner product space with

$$\langle \widehat{f}, \widehat{g} \rangle = \int_a^b f(t)\overline{g(t)}dt, \tag{11.34}$$

for any $\widehat{f}, \widehat{g} \in \widehat{PC}[a, b]$. The norm induced by this inner product is

$$\|\widehat{f}\| = \sqrt{\int_a^b |f(t)|^2 dt}. \tag{11.35}$$

The corresponding metric in $\widehat{PC}[a, b]$ is given by

$$d(\widehat{f}, \widehat{g}) = \sqrt{\int_a^b |f(t) - g(t)|^2 dt}, \tag{11.36}$$

for any $\widehat{f}, \widehat{g} \in \widehat{PC}[a, b]$.

Definition 11.27 Let $(X_1, \langle \cdot, \cdot \rangle_1)$ and $(X_1, \langle \cdot, \cdot \rangle_2)$ be two inner product spaces over $K = \mathbb{R}$ or $\mathbb{C}$. A linear mapping $F : X_1 \to X_2$ such that

$$\langle x, y \rangle_1 = \langle F(x), F(y) \rangle_2 \tag{11.37}$$

for any $x, y \in X_1$ is said to be an *isometry*.

Any isometry is injective. If F is also surjective, then the spaces X_1 and X_2 are said to be isometric.

Remark 11.11 Let d_1 and d_2 be the metrics induced by the corresponding inner products (see (11.30)) on the spaces X_1 and X_2. It is easy to see that an isometry in the sense of Definition 11.27 is also an isometry in the sense of Definition 11.17. The converse is also true: if $F : X_1 \to X_2$ is a linear mapping such that

$$d_1(x, y) = d_2(F(x), F(y)), \tag{11.38}$$

for any $x, y \in X_1$, then Eq. (11.37) also holds true. To show it, we put $y = \mathbf{0}$ in (11.38) and, since F is a linear mapping, we get:

$$\|x\|_1 = \|F(x)\|_2, \text{ for all } x \in X_1. \tag{11.39}$$

Since $\|x+y\|_1 = \|F(x+y)\|_2 = \|F(x)+F(y)\|_2$, we have:

$$\langle x+y, x+y \rangle_1 = \langle F(x)+F(y), F(x)+F(y) \rangle_2$$

$$\|x\|_1^2 + \langle x, y \rangle_1 + \overline{\langle x, y \rangle_1} + \|y\|_1^2 =$$

$$= \|F(x)\|_2^2 + \langle F(x), F(y) \rangle_2 + \overline{\langle F(x), F(y) \rangle_2} + \|F(x)\|_2^2,$$

and using again (11.39) we find:

$$\langle x, y \rangle_1 + \overline{\langle x, y \rangle_1} = \langle F(x), F(y) \rangle_2 + \overline{\langle F(x), F(y) \rangle_2}, \qquad (11.40)$$

for any $x, y \in X_1$. If $K = \mathbb{R}$, then (11.37) is proved. If $K = \mathbb{C}$, then we obtained only that $\operatorname{Re} \langle x, y \rangle_1 = \operatorname{Re} \langle F(x), F(y) \rangle_2$. If we write (11.40) for ix and y, since $F(ix) = iF(x)$, we obtain:

$$i \langle x, y \rangle_1 - i \overline{\langle x, y \rangle_1} = i \langle F(x), F(y) \rangle_2 - i \overline{\langle F(x), F(y) \rangle_2}, \qquad (11.41)$$

hence $\operatorname{Im} \langle x, y \rangle_1 = \operatorname{Im} \langle F(x), F(y) \rangle_2$ and (11.39) follows.

Theorem 11.3 can be stated for inner product spaces as follows.

Theorem 11.11 *Let* $(X, \langle \cdot, \cdot \rangle)$ *be an inner product space over* $K = \mathbb{R}$ *or* $\mathbb{C}$. *If* X *is not complete, then there exists a Hilbert space* $(\overline{X}, \langle \cdot, \cdot \rangle^-)$, *called the completion of* $(X, \langle \cdot, \cdot \rangle)$, *such that:*

(i) $(X, \langle \cdot, \cdot \rangle)$ *is isometric to an inner product subspace* $X_1 \subset \overline{X}$.
(ii) X_1 *is dense in* $\overline{X}$ *with respect to the corresponding norm,*

$$\|x\|^- = \sqrt{\langle x, x \rangle^-}.$$

(iii) *Any two Hilbert spaces* $(\overline{X}, \langle \cdot, \cdot \rangle^-)$ *and* $(\overline{\overline{X}}, \langle \cdot, \cdot \rangle^=)$ *that satisfy* (i) *and* (ii) *are isometric.*

Proof First of all we construct the quotient space $\overline{X} = \mathcal{C}/\sim$, where $\mathcal{C}$ is the complex vector space of all Cauchy sequences of X, as in the proof of Theorem 11.3 (see also Remark 11.8). We say that two Cauchy sequences $\{x_n\}, \{y_n\}$ are equivalent ($\{x_n\} \sim \{y_n\}$) if $\lim\limits_{n\to\infty} d(x_n, y_n) = 0$, where $d(x, y)$ is the usual metric in an inner product space,

$$d(x, y) = \|x - y\| = \sqrt{\langle x-y, x-y \rangle}, \quad x, y \in X.$$

For any $[x] = [\{x_n\}], [y] = [\{y_n\}] \in \overline{X}$, we define

$$\langle [x], [y] \rangle^- = \lim_{n\to\infty} \langle x_n, y_n \rangle, \qquad (11.42)$$

11.1 Normed, Metric and Hilbert Spaces

and prove that the mapping (11.42) is correctly defined and it is an inner product on $\overline{X}$.

Since $\{x_n\}, \{y_n\}$ are Cauchy sequences, by Proposition 11.17 it follows that $\{\langle x_n, y_n \rangle\}_n$ is a Cauchy sequence in K, which is a complete field, so the sequence $\{\langle x_n, y_n \rangle\}_n$ converges in K.

We show that the definition (11.42) does not depend on the choice of representatives $\{x_n\} \in [x]$ and $\{y_n\} \in [y]$. Let $\{x'_n\} \in [x]$ and $\{y'_n\} \in [y]$ be another two representatives in the class of x and of y respectively. So, $\|x_n - x'_n\| \to 0$ and $\|y_n - y'_n\| \to 0$, when $n \to \infty$. Using the Cauchy-Schwarz inequality (11.28) we can write:

$$|\langle x_n, y_n \rangle - \langle x'_n, y'_n \rangle| = |\langle x_n, y_n \rangle - \langle x'_n, y_n \rangle + \langle x'_n, y_n \rangle - \langle x'_n, y'_n \rangle| \le$$
$$\le |\langle x_n - x'_n, y_n \rangle| + |\langle x'_n, y_n - y'_n \rangle| \le$$
$$\le \|x_n - x'_n\| \|y_n\| + \|y_n - y'_n\| \|x'_n\|.$$

Since $\{y_n\}$ and $\{x'_n\}$ are Cauchy sequences, they are bounded (see Proposition 11.11), so there exists $M > 0$ such that

$$|\langle x_n, y_n \rangle - \langle x'_n, y'_n \rangle| \le M \|x_n - x'_n\| + M \|y_n - y'_n\| \to 0.$$

Hence $\lim_{n \to \infty} \langle x_n, y_n \rangle = \lim_{n \to \infty} \langle x'_n, y'_n \rangle$, so the mapping $\langle \cdot, \cdot \rangle^-$ is well defined.

Now we prove that this mapping is an inner product on $\overline{X}$, i.e. we are going to verify the conditions $p1) - p5)$ from Definition 11.25. Thus, for any $[x], [y], [z] \in \overline{X}$ and $\alpha \in K$, we can write:

$$\langle [y], [x] \rangle^- = \lim_{n \to \infty} \langle y_n, x_n \rangle = \lim_{n \to \infty} \overline{\langle x_n, y_n \rangle} = \overline{\lim_{n \to \infty} \langle x_n, y_n \rangle} = \overline{\langle [x], [y] \rangle^-}.$$

$$\langle [x] + [y], [z] \rangle^- = \lim_{n \to \infty} \langle x_n + y_n, z_n \rangle = \lim_{n \to \infty} \langle x_n, z_n \rangle + \lim_{n \to \infty} \langle y_n, z_n \rangle =$$
$$= \langle [x], [z] \rangle^- + \langle [y], [z] \rangle^-.$$

$$\langle \alpha[x], [y] \rangle^- = \langle [\alpha x], [y] \rangle^- = \lim_{n \to \infty} \langle \alpha x_n, y_n \rangle = \alpha \lim_{n \to \infty} \langle x_n, y_n \rangle = \alpha \langle [x], [y] \rangle^-.$$

$$\langle [x], [x] \rangle^- = \lim_{n \to \infty} \langle x_n, x_n \rangle \in \mathbb{R}_+.$$

$$\langle [x], [x] \rangle^- = \lim_{n \to \infty} \langle x_n, x_n \rangle = 0 \Rightarrow x_n \to \mathbf{0} \text{ in } X \Rightarrow [x] = [\mathbf{0}].$$

So $\langle \cdot, \cdot \rangle^-$ is an inner product on $\overline{X}$. Let $\overline{d} : \overline{X} \times \overline{X} \to \mathbb{R}_+$ be the metric defined by this inner product. Using Proposition 11.17 is easy to show that

$$\overline{d}\,([x], [y]) = \lim_{n\to\infty} d\,(x_n, y_n),$$

for any Cauchy sequences $x = \{x_n\}$, $y = \{y_n\}$ in X.

For every $x \in X$, we denote by $\widetilde{x}$ the constant sequence with all the terms equal to x. Then the class $[\widetilde{x}]$ is formed by the sequences with terms in X that are convergent to x. Let X_1 be the subset of $\overline{X}$ containing all the classes $[\widetilde{x}]$, $x \in X$. Obviously, it is isometric with X. We prove that X_1 is dense in $\overline{X}$. Let $[x] = [\{x_n\}]$ be an element of $\overline{X}$ and $\varepsilon > 0$ be a positive number. We know that $\{x_n\}$ is a Cauchy sequence with all the terms in X, so there exists $N \in \mathbb{N}$ such that $d(x_n, x_N) < \varepsilon$ for all $n > N$. Let $\widetilde{x}_N$ denote the constant sequence with all the terms equal to x_N, hence $[\widetilde{x}_N] \in X_1$ and we have:

$$\overline{d}\,([x], [\widetilde{x}_N]) = \lim_{n\to\infty} d\,(x_n, x_N) \leq \varepsilon,$$

so X_1 is dense in $\overline{X}$.

It can be also proved (as in the proof of Theorem 11.3) that $\overline{X}$ is complete, so it is a Hilbert space. It remains to prove (iii).

Let $(\overline{\overline{X}}, \langle \cdot, \cdot \rangle^=)$ be another Hilbert space that verifies (i) and (ii). So there exists a dense subspace $X_2 \subset \overline{\overline{X}}$ such that X is isometric with X_2, which means that X_1 and X_2 are isometric: there exists a bijective linear mapping $G : X_1 \to X_2$ such that $\overline{d}([\widetilde{x}], [\widetilde{y}]) = \overline{\overline{d}}(G([\widetilde{x}]), G([\widetilde{y}]))$ for any constant sequences $\widetilde{x}, \widetilde{y}$ (where $x, y \in X$ and $\overline{\overline{d}}$ is the metric induced by the inner product on $\overline{\overline{X}}$. We extend G to an isometry $\overline{G}$ defined on the whole space $\overline{X}$.

Let $[x] \in \overline{X}$, where $x = \{x_n\}$ is a Cauchy sequence in X. For every $k = 1, 2, \ldots$, we consider the constant sequences $\widetilde{x}_k = x_k, x_k, \ldots$ and the corresponding classes $[\widetilde{x}_k] \in X_1$. It is easy to see that the sequence $[\widetilde{x}_k]$ converges to $[x]$ in X_1. Since it is a Cauchy sequence, it follows that $G\,([\widetilde{x}_k])$ is also a Cauchy sequence in $\overline{\overline{X}}$, which is a complete space, so there exists the limit of this sequence in $\overline{\overline{X}}$. We define

$$\overline{G}\,([x]) = \lim_{k\to\infty} G\,([\widetilde{x}_k]).$$

The definition of $\overline{G}\,([x])$ does not depend on the choice of the Cauchy sequence $x = \{x_n\}$: If we take $y = \{y_n\}$ another Cauchy sequence in X such that $y \sim x$, then $\lim_{k\to\infty} (x_n - y_n) = 0$, so $\overline{G}\,([x]) = \overline{G}\,([y])$.

Since G is a linear mapping on X, it is easy to see that $\overline{G}$ is a linear mapping on $\overline{X}$. Thus, if $x = \{x_n\}$ and $y = \{y_n\}$ are Cauchy sequences in X, and $\alpha, \beta \in K$, then one can write:

$$\overline{G}(\alpha[x] + \beta[y]) = \overline{G}([\alpha x + \beta y]) = \lim_{k\to\infty} G(\alpha \widetilde{x}_k + \beta \widetilde{y}_k) =$$

$$= \alpha \lim_{k\to\infty} G([\widetilde{x}_k]) + \beta \lim_{k\to\infty} G([\widetilde{y}_k]) = \alpha \overline{G}([x]) + \beta \overline{G}([y]).$$

11.1 Normed, Metric and Hilbert Spaces

Now, let us prove that $\overline{G}$ preserves the scalar product:

$$\overline{\langle \overline{G}([x]), \overline{G}([y]) \rangle} = \left\langle \lim_{k \to \infty} G([\widetilde{x}_k]), \lim_{k \to \infty} G([\widetilde{y}_k]) \right\rangle^{=} = \lim_{k \to \infty} \langle G([\widetilde{x}_k]), G([\widetilde{y}_k]) \rangle^{=}$$
$$= \lim_{k \to \infty} \overline{\langle [\widetilde{x}_k], [\widetilde{y}_k] \rangle} = \overline{\langle [x], [y] \rangle}.$$

Thus $\overline{G}$ is an isometry, so it is one-to-one. We prove that it is also onto. Let z be an element in $\overline{X}$. Since X_2 is dense in $\overline{X}$, there exists a sequence $\{z_k\} \subset X_2$ such that $z_k \to z$ in $\overline{X}$. But $G : X_1 \to X_2$ is bijective, so there is a sequence of constant sequences $[\widetilde{u}_k] \subset X_1$ such that $G([\widetilde{u}_k]) = z_k$. Since $\{z_k\}$ is a Cauchy sequence and G is an isometry, it follows that $[\widetilde{u}_k]$ is a Cauchy sequence, hence it has a limit $[u]$ in the complete space $\overline{X}$ and we have $\overline{G}([u]) = z$. $\square$

Consider now $PC[a, b]$ the vector space of the piecewise continuous functions $f : [a, b] \to \mathbb{C}$ and the quotient space from Example 11.18, $\widehat{PC}[a, b] = PC[a, b]/W$, where W is the subspace of the functions equal to 0 almost everywhere. We have seen that $\widehat{PC}[a, b]$ is an inner product space with

$$\langle \widehat{f}, \widehat{g} \rangle = \int_a^b f(t) \overline{g(t)} dt,$$

for any $\widehat{f}, \widehat{g} \in \widehat{PC}[a, b]$. Let $\|\cdot\|_2$ denote the corresponding norm,

$$\|\widehat{f}\|_2 = \sqrt{\int_a^b |f(t)|^2 \, dt}. \tag{11.43}$$

The corresponding metric in $\widehat{PC}[a, b]$ is given by

$$d(\widehat{f}, \widehat{g}) = \sqrt{\int_a^b |f(t) - g(t)|^2 \, dt},$$

for any $\widehat{f}, \widehat{g} \in \widehat{PC}[a, b]$.

We denote by $H = \overline{\widehat{PC}[a, b]}$ the Hilbert space representing the completion of $\widehat{PC}[a, b]$ (as in Theorem 11.11). The inner product in $H = \overline{\widehat{PC}[a, b]}$ is defined by

$$\langle [f], [g] \rangle = \lim_{n \to \infty} \langle \widehat{f}_n, \widehat{g}_n \rangle, \tag{11.44}$$

for any $[f] = [\{\widehat{f}_n\}], [g] = [\{\widehat{g}_n\}] \in \overline{\widehat{PC}[a, b]}$.

11.1.6 Continuous Functions and Step Functions

Definition 11.28 Let (X_1, d_1) and (X_2, d_2) be two metric spaces. The function $f : X_1 \to X_2$ is *uniformly continuous* on X_1 if for any $\varepsilon > 0$, there exists $\delta_\varepsilon = \delta > 0$ such that for any pair of points $x', x'' \in X_1$ with $d_1(x', x'') < \delta$ one has $d_2(f(x'), f(x'')) < \varepsilon$.

It is clear that any uniformly continuous function on X_1 is also continuous on X_1. Conversely it is not always true. For instance, the continuous function $f : (0, \infty) \to \mathbb{R}$, $f(x) = \ln x$ is not uniformly continuous: we can take the sequences $x'_n = 1/2n$ and $x''_n = 1/n$ with $|x'_n - x''_n| = 1/2n \to 0$, and $|f(x'_n) - f(x''_n)| = \ln 2 \not\to 0$.

Proposition 11.18 *Any continuous function* $f : [a, b] \to K$ *is uniformly continuous on* $[a, b]$.

Proof Suppose that f is not uniformly continuous on $[a, b]$. Thus, there is an $\varepsilon > 0$ such that for any $\delta_n = 1/n$, $n = 1, 2, \ldots$, there exist $x_n, y_n \in [a, b]$ with $|x_n - y_n| < 1/n$ and $|f(x_n) - f(y_n)| > \varepsilon$. The sequence $\{x_n\}$ bounded, so it has a convergent subsequence $\{x_{k_n}\}$, convergent to $x_0 \in [a, b]$ (Bolzano-Weierstrass Theorem 11.6). But

$$\left|y_{k_n} - x_0\right| = \left|(y_{k_n} - x_{k_n}) + (x_{k_n} - x_0)\right| \le \left|y_{k_n} - x_{k_n}\right| + \left|x_{k_n} - x_0\right| <$$

$$< 1/k_n + \left|x_{k_n} - x_0\right| \to 0,$$

so $y_{k_n} \to x_0$. Since f is continuous at x_0, we have $f(x_{k_n}) \to f(x_0)$ and $f(y_{k_n}) \to f(x_0)$, a contradiction, because $|f(x_n) - f(y_n)| > \varepsilon$ for any $n = 1, 2, \ldots$. Hence f is uniformly continuous. $\square$

Definition 11.29 A function $g : [a, b] \to K$, $a \ne b$, is said to be a *step function* on $[a, b]$ if there exists a division $a = x_0 < x_1 < \ldots < x_{n-1} < x_n = b$ of the interval $[a, b]$ and a sequence $c_1, c_2, \ldots c_n$ of numbers in K such that $g(x) = c_i$ for any $x \in [x_{i-1}, x_i)$, $i = 1, 2, \ldots, n$, and $g(b) = c_n$.

The graph of a real step function $g : [a, b] \to \mathbb{R}$ is presented in Fig. 11.5.

Proposition 11.19 *Let* $f : [a, b] \to K$ *be a continuous function and let* $\varepsilon > 0$ *be a positive number. Then there exists a step function* $g : [a, b] \to K$ *such that*

$$\|f - g\|_\infty = \sup_{x \in [a,b]} |f(x) - g(x)| < \varepsilon.$$

Proof Since f is continuous, it is uniformly continuous (see Proposition 11.18). So, there exists $\delta > 0$ such that for any $x', x'' \in [a, b]$ with $|x' - x''| < \delta$, we have $|f(x') - f(x'')| < \varepsilon$. Let $n \in \mathbb{N}$ be large enough such that $(b - a)/n < \delta$ and consider the following division: $x_0 = a$, $x_1 = a + h$, $x_2 = a + 2h, \ldots, x_k = a + kh, \ldots, x_n = a + nh = b$, where $h = (b - a)/n$. Denote $c_i = f(x_{i-1})$ for any

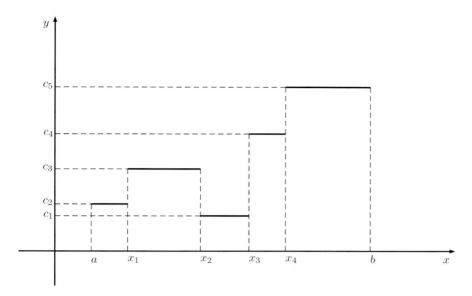

Fig. 11.5 Step function

$i = 1, 2, \ldots, n$. We define the step function $g(x)$ in the following way:

$$g(x) = c_i \text{ for } x \in [x_{i-1}, x_i), \ i = 1, \ldots, n,$$

and $g(b) = c_{n-1}$. So, if $x \in [x_{i-1}, x_i)$, $|x - x_{i-1}| < \delta$, thus $|f(x) - f(x_{i-1})| < \varepsilon$, i.e. $|f(x) - g(x)| < \varepsilon$ and finally we see that $\|f - g\|_\infty < \varepsilon$. □

Proposition 11.19 states that any continuous function $f : [a, b] \to K$ can be well approximated by step functions with respect to the sup-norm or the ∞-norm. This is also true for the norm (11.43) induced by the inner product, because

$$\|f - g\|_2 = \sqrt{\int_a^b |f(t) - g(t)|^2 \, dt} \leq \|f - g\|_\infty \sqrt{b - a}.$$

Corollary 11.3 *Let $f : [a, b] \to K$ be a piecewise continuous function and $\varepsilon > 0$. Then there exists a step function $g : [a, b] \to K$ such that*

$$d(f, g) = \|f - g\|_2 < \varepsilon.$$

Proof If f is continuous on $[a, b]$, then, by Proposition 11.19, there exists a step function $g : [a, b] \to K$ such that $\|f - g\|_\infty < \varepsilon/\sqrt{b - a}$. Thus,

$$d(f, g) = \sqrt{\int_a^b |f(t) - g(t)|^2 \, dt} \leq \|f - g\|_\infty \sqrt{b - a} < \varepsilon. \tag{11.45}$$

Now, let us assume that f is discontinuous at the points $t_1 < t_2 < \ldots < t_m$, where it still has finite side limits. We denote by $t_0 = a$ and $t_{m+1} = b$ and, for every $i = 1, 2, \ldots, m+1$, consider the function $f_i : [t_{i-1}, t_i] \to \mathbb{R}$, $f_i(t) = f(t)$ for $t \in (t_{i-1}, t_i)$, $f_i(t_{i-1}) = \lim_{t \searrow t_{i-1}} f(t)$, $f_i(t_i) = \lim_{t \nearrow t_i} f(t)$. Since f_i is continuous, by (11.45) there exists a step function $g_i : [t_{i-1}, t_i] \to K$ such that

$$d(f_i, g_i) = \sqrt{\int_{t_{i-1}}^{t_i} |f_i(t) - g_i(t)|^2 \, dt} < \frac{\varepsilon}{\sqrt{m+1}}.$$

Consider the step function $g : [a, b] \to K$, $g(x) = g_i(x)$ for any $x \in [t_{i-1}, t_i)$, $i = 1, \ldots, m+1$, $g(b) = g_{m+1}(b)$, and let $\widehat{f}, \widehat{g} \in \widehat{PC}[a, b]$ be the classes of f and g respectively. Then,

$$d(\widehat{f}, \widehat{g}) = \sqrt{\int_a^b |f(t) - g(t)|^2 \, dt} =$$

$$= \sqrt{\sum_{i=1}^{m+1} \int_{t_{i-1}}^{t_i} |f(t) - g(t)|^2 \, dt} < \sqrt{(m+1) \cdot \frac{\varepsilon^2}{m+1}} = \varepsilon.$$

□

Remark 11.12 Let $Step[a, b]$ be the set of all step functions defined on $[a, b]$. It can be readily shown that, for any $f, g \in Step[a, b]$ and $\alpha, \beta \in K$, we have $\alpha f + \beta g \in Step[a, b]$, so $Step[a, b]$ is a vector space (a subspace of the inner product space $\widehat{PC}[a, b]$). Corollary 11.3 states that $Step[a, b]$ is dense in the inner product space $\widehat{PC}[a, b]$ and, implicitly, in its completion, $H = \overline{PC[a, b]}$. The space $C[a, b]$ of all continuous functions is dense in the Hilbert space $L_2[a, b]$, the space of functions $h : [a, b] \to K$, with $|h(t)|^2$ a Lebesgue integrable function (see [33], 3.14). Since any continuous function can be well approximated by step functions (see Proposition 11.19), we see that $Step[a, b]$ is dense in $L_2[a, b]$. By Corollary 11.3 the space $Step[a, b]$ is also dense in the Hilbert space $H = \overline{PC[a, b]}$, so we conclude (by Theorem 11.11) that H and $L_2[a, b]$ are *isometric Hilbert spaces*.

11.1.7 Orthonormal Systems in a Hilbert Space

Definition 11.30 Let $(V, \langle \cdot, \cdot \rangle)$ be an inner product space over $K = \mathbb{R}$ and let $x \neq 0$, $y \neq 0$, be two vectors in V. By the *angle* between x and y we mean that angle $\theta \in [0, \pi]$ whose cosine is the real number

$$\cos \theta = \frac{\langle x, y \rangle}{\|x\| \, \|y\|}, \tag{11.46}$$

11.1 Normed, Metric and Hilbert Spaces

where $\|x\| = \sqrt{(x, x)}$ is the norm induced by the inner product $\langle \cdot, \cdot \rangle$.

We say that x and y are orthogonal (perpendicular) and write $x \perp y$ if $\langle x, y \rangle = 0$ (i.e. if the angle between x and y is $\theta = \pi/2$.)

By Cauchy-Schwarz inequality (11.28), we have $|\cos \theta| = \frac{|(x,y)|}{\|x\|\|y\|} \leq 1$, so the definition formula (11.46) is correct.

Example 11.19 In the 2-dimensional euclidean space $\mathbb{R}^2$ we consider the canonical basis $\{\mathbf{i} = (1, 0), \mathbf{j} = (0, 1)\}$. Let $\vec{a} = a_1\mathbf{i} + a_2\mathbf{j}$ and $\vec{b} = b_1\mathbf{i} + b_2\mathbf{j}$ be two nonzero vectors and $\theta \in [0, \pi]$ be the angle between them. The inner (dot) product is $\vec{a} \cdot \vec{b}$ is the real number

$$\vec{a} \cdot \vec{b} = \|\vec{a}\| \|\vec{b}\| \cos \theta = a_1 b_1 + a_2 b_2,$$

where $\|\vec{a}\| = \sqrt{a_1^2 + a_2^2}$ is the length of the vector $\vec{a}$. Thus

$$\cos \theta = \frac{\vec{a} \cdot \vec{b}}{\|\vec{a}\| \|\vec{b}\|}$$

and we see that (11.46) is a generalization of this geometric formula.

We orthogonally project the vector $\vec{a}$ onto the line spanned by vector $\vec{b}$ (see Fig. 11.6). The magnitude of this orthogonal projection is

$$\|\vec{a}\| \cos \theta = \frac{\vec{a} \cdot \vec{b}}{\|\vec{b}\|}.$$

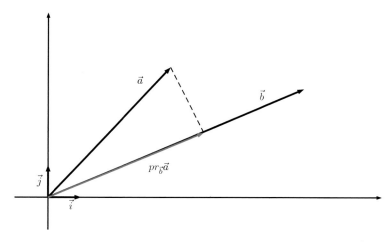

Fig. 11.6 Orthogonal projection of the vector $\vec{a}$ onto the line spanned by the vector $\vec{b}$

By multiplying this number with the unit vector $\frac{\vec{b}}{\|\vec{b}\|}$ we obtain the projection vector,

$$pr_{\vec{b}}\vec{a} = \frac{\vec{a}\cdot\vec{b}}{\|\vec{b}\|} \cdot \frac{\vec{b}}{\|\vec{b}\|} = \frac{\vec{a}\cdot\vec{b}}{\|\vec{b}\|^2}\vec{b}.$$

We can see that the vector $\vec{a} - pr_{\vec{b}}\vec{a}$ is orthogonal to the vector $\vec{b}$.

Definition 11.31 Let $(V, \langle \cdot, \cdot \rangle)$ be an inner product space over $K = \mathbb{R}$ or $\mathbb{C}$. Two vectors $x, y \in V$ are said to be *orthogonal* ($x \perp y$) if $\langle x, y \rangle = 0$.

Definition 11.32 Let $(V, \langle \cdot, \cdot \rangle)$ be an inner product space over $K = \mathbb{R}$ or $\mathbb{C}$ and $x, y \in V$, $y \neq \mathbf{0}$. The orthogonal projection of x on y is the vector

$$pr_y x = \frac{\langle x, y \rangle}{\|y\|^2} y. \tag{11.47}$$

Note that the vectors $x - pr_y x$ and y are orthogonal.

Theorem 11.12 (Pythagoras' Theorem in an Inner Product Space) *Let $(V, \langle \cdot, \cdot \rangle)$ be an inner product space and $\|\cdot\|$ be its corresponding norm. Then, for any orthogonal vectors $x, y \in V$, we have:*

$$\|x + y\|^2 = \|x\|^2 + \|y\|^2. \tag{11.48}$$

The reciprocal holds if and only if $K = \mathbb{R}$.
If $K = \mathbb{C}$, then Eq. (11.48) implies $\operatorname{Re} \langle x, y \rangle = 0$.

Proof

$$\|x+y\|^2 = \langle x+y, x+y \rangle = \|x\|^2 + \|y\|^2 + \langle x+y, x+y \rangle + \overline{\langle x+y, x+y \rangle} =$$
$$= \|x\|^2 + \|y\|^2 + 2\operatorname{Re}\langle x, y \rangle.$$

So $\|x+y\|^2 = \|x\|^2 + \|y\|^2$ if and only if $\operatorname{Re}\langle x, y \rangle = 0$. Since $\langle x, y \rangle = \operatorname{Re}\langle x, y \rangle$ if $K = \mathbb{R}$, the theorem is proved. $\square$

Definition 11.33 A set $\{x_1, x_2, \ldots, x_n, \ldots\}$ (finite or not) of nonzero vectors in an inner product space $(V, \langle \cdot, \cdot \rangle)$ is said to be an *orthogonal* system of vectors if and only if $x_i \perp x_j$ for any $i \neq j$, $i, j = 1, 2, \ldots$ If $\{x_1, x_2, \ldots, x_n, \ldots\}$ is orthogonal and if $\|x_i\| = 1$ for all $i = 1, 2, \ldots$ then the set $\{x_1, x_2, \ldots, x_n, \ldots\}$ is said to be an *orthonormal* system of vectors in V.

Proposition 11.20 *Any orthogonal system of vectors $\{x_1, x_2, \ldots\}$ is linearly independent in V.*

11.1 Normed, Metric and Hilbert Spaces

Proof Take $\alpha_1, \alpha_2, \ldots, \alpha_k \in K$ and $x_{i_1}, x_{i_2}, \ldots, x_{i_k} \in \{x_1, x_2, \ldots\}$ such that

$$\alpha_1 x_{i_1} + \alpha_2 x_{i_2} + \ldots + \alpha_k x_{i_k} = \mathbf{0}$$

Thus, for every $j = 1, 2, \ldots, k$ one can write:

$$\langle \alpha_1 x_{i_1} + \alpha_2 x_{i_2} + \ldots + \alpha_k x_{i_k}, x_{i_j} \rangle = \alpha_j \langle x_{i_j}, x_{i_j} \rangle = 0$$

Since $x_{i_j} \neq \mathbf{0}$, we see that $\alpha_j = 0$ for any $j = 1, 2, \ldots, k$. Hence $\{x_1, x_2, \ldots\}$ is linearly independent in V. □

For any linearly independent set of vectors $S = \{x_1, x_2, \ldots\}$ in V, we denote by $Sp\{x_1, x_2, \ldots\}$ the subspace of V spanned by S, that is, the space of all finite linear combinations with vectors of S:

$$Sp\{x_1, x_2, \ldots\} = \{\alpha_1 x_{i_1} + \ldots + \alpha_k x_{i_k} : \alpha_1, \ldots, \alpha_k \in K, x_{i_1}, \ldots, x_{i_k} \in S\}.$$

Remark 11.13 If $u_1, \ldots, u_n$ are nnonzero orthogonal vectors in V, $W = Sp\{u_1, \ldots, u_n\}$ and $x \notin W$, then the vector $pr_W x$ defined bellow is called the orthogonal projection of x on W:

$$pr_W x = pr_{u_1} x + pr_{u_2} x + \ldots + pr_{u_n} x = \tag{11.49}$$

$$= \frac{\langle x, u_1 \rangle}{\|u_1\|^2} u_1 + \frac{\langle x, u_2 \rangle}{\|u_2\|^2} u_2 + \ldots + \frac{\langle x, u_n \rangle}{\|u_n\|^2} u_n.$$

We can see that

$$\langle x - pr_W x, u_k \rangle = 0, \ k = 1, \ldots, n$$

so $x - pr_W x$ is orthogonal on any vector $y \in W$ (see Fig. 11.7).

Theorem 11.13 (Gram–Schmidt Orthogonalization Process) *Let $(V, \langle \cdot, \cdot \rangle)$ be an inner product space, and let $S = \{v_1, v_2, \ldots\}$ be a linearly independent set of vectors in V. Then there exists an orthonormal system of vectors $E = \{e_1, e_2, \ldots\}$ such that, for any $k = 1, 2, \ldots$,*

$$Sp\{v_1, \ldots, v_k\} = Sp\{e_1, \ldots, e_k\}.$$

Proof First, we construct an orthogonal system of vectors $U = \{u_1, u_2, \ldots\}$ such that $Sp\{v_1, \ldots, v_k\} = Sp\{u_1, \ldots, u_k\}$ for all $k = 1, 2, \ldots$. We take

$$u_1 = v_1,$$
$$u_2 = v_2 - pr_{u_1} v_2$$

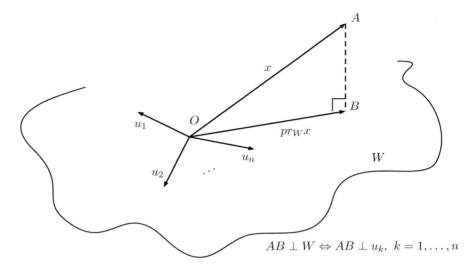

Fig. 11.7 Orthogonal projection of the vector x onto the space W spanned by the vectors $u_1, u_2, \ldots, u_n$

$$u_3 = v_3 - pr_{u_1} v_3 - pr_{u_2} v_3,$$

$$\vdots$$

$$u_n = v_n - pr_{u_1} v_n - pr_{u_2} v_n - \ldots - pr_{u_{n-1}} v_n.$$

It is easy to see that $Sp\{v_1, \ldots, v_k\} = Sp\{u_1, \ldots, u_k\}$, for every $k = 1, 2, \ldots$. Using mathematical induction on k, we can prove that $\langle u_k, u_i \rangle = 0$, for every $i = 1, \ldots, k-1$. Thus, if we assume that $u_1, \ldots, u_{k-1}$ are orthogonal, then the vector $w_k = pr_{u_1} v_k + pr_{u_2} v_k + \ldots + pr_{u_{k-1}} v_k$ is the projection of v_k on the space $Sp\{u_1, \ldots, u_{k-1}\}$, so $u_k = v_k - w_k$ is orthogonal on every u_j, $j = 1, \ldots, k-1$ (see Remark 11.13).

The unit vectors:

$$e_1 = \frac{u_1}{\|u_1\|}, e_2 = \frac{u_2}{\|u_2\|}, \ldots, e_n = \frac{u_n}{\|u_n\|}, \ldots$$

form an orthonormal system in V. □

If the orthogonal system S is finite, $S = \{v_1, v_2, \ldots v_n\}$ and it is a basis of V (i.e. $V = Sp\{v_1, v_2, \ldots v_n\}$), then the orthonormal system constructed by Gram-Schmidt orthogonalization process, $E = \{e_1, e_2, \ldots, e_n\}$ is an *orthonormal basis* of V, so the next corollary follows.

Corollary 11.4 *Any finite dimensional inner product space has an orthonormal basis.*

11.1 Normed, Metric and Hilbert Spaces

Proposition 11.21 *Let $(V, \langle \cdot, \cdot \rangle)$ be an inner product space of dimension $\dim V = n$, and let $E = \{e_1, e_2, \ldots, e_n\}$ be an orthonormal basis of V. Then, for any $x, y \in V$, $x = \alpha_1 e_1 + \ldots + \alpha_n e_n$ and $y = \beta_1 e_1 + \ldots + \beta_n e_n$ we have:*

$$\alpha_k = \langle x, e_k \rangle, \quad k = 1, \ldots, n, \tag{11.50}$$

$$\langle x, y \rangle = \alpha_1 \overline{\beta}_1 + \alpha_2 \overline{\beta}_2 + \ldots + \alpha_n \overline{\beta}_n, \tag{11.51}$$

$$\|x\| = \sqrt{|\alpha_1|^2 + |\alpha_2|^2 + \ldots + |\alpha_n|^2}, \tag{11.52}$$

Moreover, V is isometric as an inner product space with the space K^n with the inner product defined by formula (11.31).

Proof Since $\langle e_i, e_j \rangle = 0$ for $i \neq j$ and $\langle e_i, e_i \rangle = 1$, the formulas (11.50)–(11.52) readily follow.

From (11.51) it follows that the mapping $\Psi : V \to K^n$ defined by $\Psi(x) = (\alpha_1, \alpha_2, \ldots, \alpha_n)$, for $x = \alpha_1 e_1 + \ldots + \alpha_n e_n$, is a bijective isometry of inner product spaces. $\square$

In the following we will focus on infinite dimensional Hilbert spaces.

Definition 11.34 Let $\{e_1, e_2, \ldots\}$ be an orthonormal system of vectors in the Hilbert space H. For any $x \in H$ and $i = 1, 2, \ldots$ we say that the number $\alpha_i = \langle x, e_i \rangle$ is the *Fourier coefficient* of x relative to e_i.

Lemma 11.2 (Best Approximation Lemma) *Let $\{e_1, e_2, \ldots, e_n\}$ be an orthonormal system of vectors in the inner product space $(V, \langle \cdot, \cdot \rangle)$ and let $W = Sp\{e_1, e_2, \ldots, e_n\}$ be the subspace of V spanned by $e_1, e_2, \ldots, e_n$. If $pr_W x$ is the orthogonal projection of x on W,*

$$pr_W x = \langle x, e_1 \rangle e_1 + \langle x, e_2 \rangle e_2 + \ldots + \langle x, e_n \rangle e_n, \tag{11.53}$$

then

$$\|x - pr_W x\| = \inf_{z \in W} \|x - z\| \tag{11.54}$$

for any $x \in V$.

Proof Let us denote by $w = pr_W x$. Since $w \in W$ and $(x - w) \perp W$, we obtain that $(x - w) \perp (w - z)$ for any $z \in W$ (see Fig. 11.8). Thus, from Pythagoras' Theorem 11.12 and using the equality:

$$x - z = (x - w) + (w - z),$$

we obtain:

$$\|x - z\|^2 = \|x - w\|^2 + \|w - z\|^2 \geq \|x - w\|^2,$$

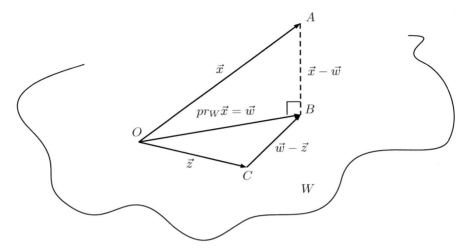

Fig. 11.8 Best approximation lemma

and so $\|x - z\| \geq \|x - w\|$ for any $z \in W$. Thus, $\inf_{z \in W} \|x - z\| \geq \|x - w\|$. Since $w \in W$ the equality (11.54) follows. □

Theorem 11.14 (Bessel's Inequality) *Let $\{e_1, e_2, \ldots, e_n, \ldots\}$ be an orthonormal system of vectors in the Hilbert space $(H, \langle \cdot, \cdot \rangle)$. Then, for every $x \in H$, the series $\sum_{n=1}^{\infty} |\langle x, e_n \rangle|^2$ converges and*

$$\sum_{k=1}^{\infty} |\langle x, e_k \rangle|^2 \leq \|x\|^2. \tag{11.55}$$

Proof For every $n = 1, 2, \ldots$, we denote by $W_n = Sp\{e_1, \ldots, e_n\}$ and

$$w_n = pr_{W_n} x = \langle x, e_1 \rangle e_1 + \ldots + \langle x, e_n \rangle e_n.$$

Since $(x - w_n) \perp W_n$ we see that $x - w_n \perp w_n$ and, from Pythagoras' Theorem 11.12, we find:

$$\|x\|^2 = \|x - w_n\|^2 + \|w_n\|^2. \tag{11.56}$$

Hence

$$\|w_n\|^2 = \sum_{k=1}^{n} |\langle x, e_k \rangle|^2 \leq \|x\|^2. \tag{11.57}$$

11.1 Normed, Metric and Hilbert Spaces

Since the inequality (11.57) holds for every $n = 1, 2, \ldots$, the series $\sum_{n=1}^{\infty} |\langle x, e_n \rangle|^2$ is convergent and the inequality (11.55) follows. □

Definition 11.35 An orthonormal system of vectors in the Hilbert space H is said to be *complete* if no other vector can be adjoined to it such that the resulting set is still orthonormal. A complete orthonormal system is said to be a *Hilbert basis* of H.

This definition can be written in the following equivalent form (see [33], 4.18 for the proof):

Proposition 11.22 *Let $E = \{e_1, \ldots, e_n, \ldots\}$ be an orthonormal system of vectors in the Hilbert space H. E is a complete system if and only if the space of all finite linear combinations with vectors of E:*

$$W = Sp(E) = \{\alpha_1 e_{i_1} + \ldots + \alpha_k e_{i_k} : \alpha_1, \ldots, \alpha_k \in K, x_{i_1}, \ldots, e_{i_k} \in S\}.$$

is dense in H.

Theorem 11.15 *Let $E = \{e_1, \ldots, e_n, \ldots\}$ be an orthonormal system of vectors in the Hilbert space H. Then the following statements are equivalent:*

(i) *E is a Hilbert basis of H.*
(ii) *For any $x \in H$,*

$$x = \sum_{n=1}^{\infty} \langle x, e_n \rangle e_n \tag{11.58}$$

(iii) *For any for any $x \in H$, the Parseval's identity holds true:*

$$\|x\|^2 = \sum_{n=1}^{\infty} |\langle x, e_n \rangle|^2 \tag{11.59}$$

Proof

(i) $\Rightarrow$ (ii) By Proposition 11.22, the system E is complete if and only if $Sp(E)$ is dense in H, that is, for any $x \in H$ and $\varepsilon > 0$, there exists $\alpha_1, \ldots, \alpha_k$ such that

$$\left\| x - \sum_{k=1}^{n} \alpha_k e_k \right\| < \varepsilon.$$

By Lemma 11.2, it follows that

$$\left\| x - \sum_{k=1}^{n} \langle x, e_k \rangle e_k \right\| \leq \left\| x - \sum_{k=1}^{n} \alpha_k e_k \right\| < \varepsilon,$$

so the series $\sum_{n=1}^{\infty} \langle x, e_n \rangle e_n$ converges to x.

$(ii) \Rightarrow (iii)$ Let $s_n = \sum_{k=1}^{n} \langle x, e_k \rangle e_k$, $n = 1, 2, \ldots$ be the partial sums of the series above. Since the series converges to x, we have $\lim_{n \to \infty} s_n = x$, and, by the continuity of the norm, we get $\lim_{n \to \infty} \|s_n\|^2 = \|x\|^2$. But

$$\|s_n\|^2 = \sum_{k=1}^{n} |\langle x, e_k \rangle|^2,$$

hence the Parseval's identity (11.59) follows.

$(iii) \Rightarrow (i)$ Suppose that E is not complete, so there exists $y \in H$, $y \neq \mathbf{0}$, such that $\langle y, e_n \rangle = 0$, for all $n = 1, 2, \ldots$. By the Parseval's identity (11.59), we have

$$\|y\|^2 = \sum_{n=1}^{\infty} |\langle y, e_n \rangle|^2 = 0,$$

hence $y = \mathbf{0}$, a contradiction. It follows that E is a complete orthonormal system (a Hilbert basis). □

Definition 11.36 The series in the formula (11.58) is the *Fourier series* of x relative to the Hilbert basis E. The numbers $\langle x, e_k \rangle$, $k = 1, 2, \ldots$ are the *Fourier coefficients* of x relative to E, and the formula (11.58) is called the Fourier expansion of x relative to E.

We remark that the Fourier expansion of x is unique, i.e. if $x = \sum_{n=1}^{\infty} \alpha_n e_n$, where $\alpha_n \in K$, then, by the continuity of the inner product, $\alpha_n = \langle x, e_n \rangle$ for any $n = 1, 2, \ldots$

In any Hilbert space there exists a complete orthonormal system (which may be countable or not). The proof of this result can be found in [33] (Theorem 4.22).

Theorem 11.16 *Any Hilbert space has a Hilbert basis.*

Consider the piecewise continuous functions $f : \mathbb{R} \to \mathbb{C}$ that are periodic of period 2π. Due to their periodicity, it is sufficient to study these functions on the interval $[-\pi, \pi]$. Thus, we consider the space $PC[-\pi, \pi]$ and the quotient space

11.1 Normed, Metric and Hilbert Spaces

$\widehat{PC}[-\pi, \pi] = PC[-\pi, \pi]/\sim$, where the equivalence relation $\sim$ is defined by: $f \sim g$ if $f = g$ almost everywhere. As shown in Example 11.18, $\widehat{PC}[-\pi, \pi]$ is an inner product space with the inner product defined by

$$\langle \widehat{f}, \widehat{g} \rangle = \langle f, g \rangle = \int_a^b f(t)\overline{g(t)}dt,$$

for any $\widehat{f}, \widehat{g} \in \widehat{PC}[-\pi, \pi]$ classes of equivalence represented by the piecewise continuous functions $f, g \in PC[-\pi, \pi]$. The space $\widehat{PC}[-\pi, \pi]$ is not a Hilbert space (it is not complete), but it can be completed by Theorem 11.11 to a Hilbert space $H = \overline{\widehat{PC}[-\pi, \pi]}$.

Definition 11.37 A trigonometric polynomial (of period 2π) is a finite sum of the form

$$f(t) = a_0 + \sum_{n=1}^{N}(a_n \cos nt + b_n \sin nt), \quad t \in \mathbb{R}, \tag{11.60}$$

where $a_0, a_1, \ldots, a_N$ and $b_1, \ldots, b_N$ are complex numbers.

It is clear that every function of the form (11.60) is periodic of period 2π. We also notice that, using the Euler's formulas

$$\cos x = \frac{e^{ix} + e^{-ix}}{2}, \quad \sin x = \frac{e^{ix} + e^{-ix}}{2i},$$

any trigonometric polynomial (11.60) can be written in the form

$$f(t) = \sum_{n=-N}^{N} c_n e^{int}, \quad t \in \mathbb{R}, \tag{11.61}$$

where $c_{-N}, \ldots, c_N$ are complex numbers.

It can be easily proved that the set of functions

$$U = \left\{ u_n = \frac{1}{\sqrt{2\pi}} e^{int}, \ n \in \mathbb{Z} \right\} \tag{11.62}$$

is an orthonormal system:

$$\langle u_n, u_m \rangle = \frac{1}{2\pi} \int_{-\pi}^{\pi} e^{i(n-m)t} dt = \begin{cases} 0 \text{ if } n \neq m \\ 1 \text{ if } n = m \end{cases}$$

Moreover, this orthonormal system is *complete* in the Hilbert space $L_2[-\pi, \pi]$ containing the (Lebesgue) square integrable functions ([33], 4.24). As shown in

Remark 11.12, the Hilbert space $H = \overline{PC}[-\pi, \pi]$ obtained as the completion of the space $\widehat{PC}[-\pi, \pi]$ can be identified with $L_2[-\pi, \pi]$. Thus, by Theorem 11.15, a piecewise continuous function $f : [-\pi, \pi] \to \mathbb{C}$ is equal to its Fourier series at any point of continuity $t \in (-\pi, \pi)$:

$$f(t) \stackrel{a.e.}{=} \sum_{n \in \mathbb{Z}} \langle f, u_n \rangle u_n = \frac{1}{2\pi} \sum_{n \in \mathbb{Z}} \langle f, e^{int} \rangle e^{int}. \tag{11.63}$$

Let $f : [-\pi, \pi] \to \mathbb{R}$ be a piecewise continuous *real-valued* function. Since $e^{int} = \cos nt + i \sin nt$ (Euler's formula), the Fourier coefficients can be written

$$\langle f, e^{int} \rangle = \int_{-\pi}^{\pi} f(t) e^{-int} dt = \int_{-\pi}^{\pi} f(t) (\cos nt - i \sin nt) dt =$$

$$= \int_{-\pi}^{\pi} f(t) \cos nt \, dt - i \int_{-\pi}^{\pi} f(t) \sin nt \, dt = A_n - i B_n,$$

where $A_n, B_n \in \mathbb{R}$,

$$A_n = \int_{-\pi}^{\pi} f(t) \cos nt \, dt, \quad B_n = \int_{-\pi}^{\pi} f(t) \sin nt \, dt.$$

Thus, the formula (11.63) becomes:

$$f(t) = \frac{1}{2\pi} \sum_{n \in \mathbb{Z}} (A_n - i B_n)(\cos nt + i \sin nt).$$

Since $f(t) \in \mathbb{R}$, we find:

$$f(t) = \frac{1}{2\pi} \sum_{n \in \mathbb{Z}} (A_n \cos nt + B_n \sin nt) =$$

$$= \frac{A_0}{2\pi} + \frac{1}{2\pi} \sum_{n=1}^{\infty} [(A_n + A_{-n}) \cos nt + (B_n - B_{-n}) \sin nt].$$

Since $A_{-n} = A_n$ and $B_{-n} = -B_n$ for any $n \in \mathbb{Z}$, we can write

$$f(t) = \frac{A_0}{2\pi} + \frac{1}{\pi} \sum_{n=1}^{\infty} (A_n \cos nt + B_n \sin nt), \tag{11.64}$$

at any point of continuity $t \in (-\pi, \pi)$.

Finally, we remark that any piecewise continuous function $f(x)$ that is periodic of period $T > 0$ can be transformed into a similar periodic function of period 2π by a simple change of variable: $t = \frac{2\pi}{T}x$. Thus, the function $f(x)$ is periodic of period T if and only if the function $g(t) = f\left(\frac{T}{2\pi}t\right)$ is periodic of period 2π.

11.2 Complex Function Theory

11.2.1 Differentiability of Complex Functions

In this section we shortly review some basic facts on complex functions. Let $f : A \to \mathbb{C}$, be a complex function, where A is a subset of $\mathbb{C}$. Since any $z \in A$, $z = x + iy$, $x, y \in \mathbb{R}$ is completely determined by the two real numbers x and y, we can write

$$w = f(z) = u(x, y) + iv(x, y),$$

where u and v are both real-valued functions of two real variables x and y; $u = \operatorname{Re} f$ is the real part of f, and $v = \operatorname{Im} f$ is the imaginary part of f. So $u, v : D \to \mathbb{R}$, where D is the subset of $\mathbb{R}^2$ defined by

$$D = \{(x, y) \in \mathbb{R}^2 : \text{there exists } z \in A \text{ with } z = x + iy\}$$

and it is usually identified with $A \subset \mathbb{C}$.

The field of complex numbers $\mathbb{C}$ is a Hilbert space with the inner product defined by

$$\langle z_1, z_2 \rangle = z_1 \overline{z_2}.$$

The norm induced by this inner product is

$$\|z\| = |z| = \sqrt{z \cdot \overline{z}} = \sqrt{x^2 + y^2},$$

if $z = x + iy$. We know that $(\mathbb{C}, d)$ is a metric space, with the metric

$$d(z_1, z_2) = |z_1 - z_2|,$$

for any $z_1, z_2 \in \mathbb{C}$. Moreover, $\mathbb{C}$ is a *complete space*.

Let z_0 be a complex number and let $r > 0$ be a positive real number. The subset of $\mathbb{C}$:

$$B(z_0, r) = \{z \in \mathbb{C} : |z - z_0| < r\} \quad (11.65)$$

is called the *open ball centered at z_0 with radius r*. A subset $A \subset \mathbb{C}$ is said to be an *open set* if for any $z_0 \in A$ there exists $r > 0$ such that $B(z_0, r) \subset A$. For instance any open ball is an open set in $\mathbb{C}$.

The subset of $\mathbb{C}$:

$$B[z_0, r] = \{z \in \mathbb{C} : |z - z_0| \leq r\} \tag{11.66}$$

is called the *closed ball centered at z_0 with radius r*.

Let A be a nonempty subset of $\mathbb{C}$. The *boundary* ∂A of A is the set of all points $w \in \mathbb{C}$ such that in any open ball $B(w, r)$, $r > 0$, there exists elements of A as well as elements of $\mathbb{C} \setminus A$. For instance, if $A = B(z_0, r)$, $r > 0$, its boundary ∂A is equal to the set $\{z \in \mathbb{C} : |z - z_0| = r\}$, i.e. the circle centered at z_0 of radius r.

Definition 11.38 Let $I = [a, b]$ be a proper real interval ($a < b$) and let $\gamma : I \to \mathbb{C}$ be a continuous function with complex values. Then, the couple (I, γ) is said to be a *path* (or *curve*) in $\mathbb{C}$. The image of this function, $\gamma(I)$, is said to be the *support* of the path (I, γ). In fact we sometimes identify the path with its support and say that $\gamma(I)$ is a path in $\mathbb{C}$.

A path $\gamma : I \to \mathbb{C}$, $\gamma(t) = x(t) + iy(t)$ such that the functions $x(t)$ and $y(t)$ are continuously differentiable and for any $t \in I$ we have either $x'(t) \neq 0$ or $y'(t) \neq 0$, is said to be a *smooth* path. If this is true on $I \setminus S$, where S is a finite subset of I, then γ is said to be a *piecewise smooth* path.

If $\gamma(a) = \gamma(b)$ then γ is said to be a *closed* path.

If $\gamma(t_1) \neq \gamma(t_2)$ for any $a < t_1 < t_2 < b$, then γ is a *simple* path.

We say that two distinct points $z_1, z_2 \in \mathbb{C}$ are connected by the path (I, γ) if $z_1, z_2 \in \gamma(I)$.

Definition 11.39 A nonempty subset D of $\mathbb{C}$ with at least two distinct elements, such that any two points $z_1, z_2 \in D$ can be connected by a path $\gamma(I) \subset D$ is called a *connected set* of $\mathbb{C}$. If the set $D \subset \mathbb{C}$ is open and connected, then we say that D is a *domain* in $\mathbb{C}$.

Let $A \subset \mathbb{C}$. A point $z_0 \in \mathbb{C}$ is a *limit point* for A if there exists at least one nonconstant sequence $\{z_n\}_{n \geq 1} \subset A$ such that $\lim_{n \to \infty} z_n = z_0$.

Let $f : A \to \mathbb{C}$ be a complex function, $A \subset \mathbb{C}$ and let $z_0 = x_0 + iy_0$ be a limit point of A. We say that f has the limit L at z_0 if for any sequence $\{z_n\}_{n \geq 1} \subset \mathbb{C}$ such that $\lim_{n \to \infty} z_n = z_0$, we have $\lim_{n \to \infty} f(z_n) = L$. In this case, we write $\lim_{z \to z_0} f(z) = L$. If $z_0 \in A$ and $\lim_{z \to z_0} f(z) = f(z_0)$, then f is *continuous* at z_0. Note that $f = u + iv$ is continuous at $z_0 = x_0 + iy_0$ if and only if the real-valued functions $u(x, y)$ and $v(x, y)$ are continuous at (x_0, y_0).

Definition 11.40 Let $D \subset \mathbb{C}$ be a domain, $f : D \to \mathbb{C}$ be a complex function and $z_0 \in D$. We say that f is differentiable at z_0 if the following limit exists in $\mathbb{C}$:

$$\lim_{z \to z_0} \frac{f(z) - f(z_0)}{z - z_0} = f'(z_0). \tag{11.67}$$

The complex number $f'(z_0)$ is the *derivative* of f at z_0.

11.2 Complex Function Theory

If f is differentiable at any point $z_0 \in D$, then we say that f is *holomorphic* on D. A function is said to be *holomorphic* at z_0 if there exists $r > 0$ such that f is holomorphic on $B(z_0, r)$.

A function that is holomorphic on the whole complex plane is called an *entire* function.

It can be easily seen that a complex function which is differentiable at z_0 must be continuous at z_0. It can be also proved that the rules of differentiation of real functions also hold for complex functions:

$$(f + g)' = f' + g', \quad (\alpha f)' = \alpha f',$$

$$(fg)' = f'g + fg', \quad \left(\frac{f}{g}\right)' = \frac{f'g - fg'}{g^2},$$

$$(f \circ g)'(z) = f'(g(z)) \cdot g'(z).$$

The most important formulas of differentiation hold also in the case of complex functions:

$$(\sin z)' = \cos z, \quad (\cos z)' = -\sin z, \quad (\exp(z))' = \exp(z), \text{ etc.}$$

For instance, $(z^2)' = 2z$, because

$$\lim_{z \to z_0} \frac{z^2 - z_0^2}{z - z_0} = \lim_{z \to z_0} (z + z_0) = 2z_0.$$

Theorem 11.17 (Cauchy–Riemann Equations) *If the complex function $f(x + iy) = u(x, y) + iv(x, y)$ is differentiable at the point $z_0 = x_0 + iy_0$, then the real functions $u(x, y)$ and $v(x, y)$ have partial derivatives at (x_0, y_0) and*

$$\begin{cases} \frac{\partial u}{\partial x}(x_0, y_0) = \frac{\partial v}{\partial y}(x_0, y_0), \\ \frac{\partial u}{\partial y}(x_0, y_0) = -\frac{\partial v}{\partial x}(x_0, y_0) \end{cases} \quad (11.68)$$

Moreover,

$$f'(z_0) = \frac{\partial u}{\partial x}(x_0, y_0) + i\frac{\partial v}{\partial x}(x_0, y_0) = \frac{\partial v}{\partial y}(x_0, y_0) - i\frac{\partial u}{\partial y}(x_0, y_0).$$

Proof If f is differentiable at z_0, then the following limit exists:

$$\lim_{z \to z_0} \frac{f(z) - f(z_0)}{z - z_0} = f'(z_0).$$

This means that for any sequence (x_n, y_n) that converges to (x_0, y_0) (in $\mathbb{R}^2$) one has

$$\lim_{(x_n, y_n) \to (x_0, y_0)} \frac{u(x_n, y_n) - u(x_0, y_0) + i[v(x_n, y_n) - v(x_0, y_0)]}{x_n - x_0 + i(y_n - y_0)} = f'(z_0). \tag{11.69}$$

If we take $y_n = y_0$ for any $n = 1, 2, \ldots$ then we get

$$\frac{\partial u}{\partial x}(x_0, y_0) + i \frac{\partial v}{\partial x}(x_0, y_0) = f'(z_0). \tag{11.70}$$

Similarly, if we consider in (11.69) $x_n = x_0$ for any $n = 1, 2, \ldots$ then we find

$$\frac{1}{i}\left[\frac{\partial u}{\partial y}(x_0, y_0) + i \frac{\partial v}{\partial y}(x_0, y_0)\right] = f'(z_0). \tag{11.71}$$

By equating (11.70) and (11.71) we get the Cauchy-Riemann equations (11.68). □

In particular, if $f = u + iv$ is holomorphic at $z_0 = x_0 + iy_0$, then the Cauchy-Riemann equations

$$\frac{\partial u}{\partial x} = \frac{\partial v}{\partial y}, \quad \frac{\partial u}{\partial y} = -\frac{\partial v}{\partial x} \tag{11.72}$$

hold at every point of a small neighborhood of (x_0, y_0). The next theorem states that the converse is also true, provided that the partial derivatives of u and v are continuous (see [36], 10.17)

Theorem 11.18 *If the functions u and v are of class C^1 on some neighborhood D of z_0 and they satisfy the Cauchy-Riemann equations (11.72) in D, then the function $f = u + iv$ is holomorphic at z_0. Moreover, for any $z \in D$ we have:*

$$f'(z) = \frac{\partial u}{\partial x} + i \frac{\partial v}{\partial x} = \frac{\partial v}{\partial y} - i \frac{\partial u}{\partial y}. \tag{11.73}$$

We shall see that if $f(z)$ is a holomorphic function, then its derivative $f'(z)$ is also holomorphic (see Theorem 11.25). Therefore, the functions u and v are twice continuously differentiable, and from the Cauchy-Riemann equations (11.72) we obtain (by differentiation):

$$\begin{cases} \nabla^2 u = \frac{\partial^2 u}{\partial x^2} + \frac{\partial^2 u}{\partial y^2} = 0, \\ \nabla^2 v = \frac{\partial^2 v}{\partial x^2} + \frac{\partial^2 v}{\partial y^2} = 0. \end{cases} \tag{11.74}$$

Thus, the real part and the imaginary part of a holomorphic function satisfy the Laplace's equation. Such functions are called *harmonic functions*. Two harmonic

11.2 Complex Function Theory

functions $u(x, y)$ and $v(x, y)$ are called *conjugate harmonic functions* if they are the real part and the imaginary part of the same holomorphic function $f(z)$, i.e. if they are related by the Cauchy-Riemann equations. Knowing one of the functions u or v, we can determine the other one (up to a constant) such that $f = u + iv$ is a holomorphic function on D.

Suppose that we know the real part $u(x, y)$ and let $z_0 = x_0 + iy_0$ be a point in D. By Cauchy-Riemann equations (11.72) we can write

$$\frac{\partial v}{\partial x}(t, y) = -\frac{\partial u}{\partial y}(t, y),$$

for all $(t, y) \in D$. We integrate with respect to $t \in [x_0, x]$ and find:

$$v(x, y) - v(x_0, y) = -\int_{x_0}^{x} \frac{\partial u}{\partial y}(t, y) \, dt. \tag{11.75}$$

Since, from the same Cauchy-Riemann equations (11.72),

$$\frac{\partial v}{\partial y}(x_0, t) = \frac{\partial u}{\partial x}(x_0, t),$$

we obtain by integration with respect to $t \in [y_0, y]$:

$$v(x_0, y) - v(x_0, y_0) = \int_{y_0}^{y} \frac{\partial u}{\partial x}(x_0, t) \, dt. \tag{11.76}$$

From (11.75) and (11.76) we find that

$$v(x, y) = \int_{y_0}^{y} \frac{\partial u}{\partial x}(x_0, t) \, dt - \int_{x_0}^{x} \frac{\partial u}{\partial y}(t, y) \, dt + C,$$

where $C = v(x_0, y_0)$ is an arbitrary (real) constant.

Example 11.20 Consider $u(x, y) = x^2 - y^2$ and find a holomorphic function $f = u + iv$. In this case, Cauchy-Riemann equations become:

$$\frac{\partial v}{\partial x}(x, y) = 2y$$

and

$$\frac{\partial v}{\partial y}(x, y) = 2x$$

We integrate the first equality with respect to x and obtain:

$$v(x, y) = 2xy + C(y),$$

where $C(y)$ is a function depending only on y (constant with respect to x). Introducing this function $v(x, y)$ in the second relation, we find:

$$2x = 2x + C'(y),$$

so, $C'(y) = 0$, which means that $C(y)$ is a (real) constant, $C(y) = k$. Thus, $v(x, y) = 2xy + k$ and $f(z) = x^2 - y^2 + i(2xy + k) = (x + iy)^2 + ik$.

Any complex number can be written in the trigonometric form

$$z = |z| \exp(i\alpha) = |z|(\cos \alpha + i \sin \alpha), \tag{11.77}$$

where $\alpha = \arg z$ is the argument of z (the angle made by the radius Oz with the real axis Ox).

Let $f(z) = u + iv$ be a holomorphic function on the domain D and $z_0 \in D$. Consider γ a curve in the complex plane passing through the point $M_0(z_0)$ and let $M(z)$ be another point on γ. We denote by $\Gamma = f(\gamma)$ the image of γ through the function f and let $N_0(f(z_0))$ and $N(f(z))$. Let α, α' be the angles made by the x-axis with the tangent line to the curve γ and the secant line M_0M, respectively. Similarly, we denote by β, β' the angles made by the u-axis with the tangent line to the curve Γ and the secant line N_0N, respectively (see Fig. 11.9). Thus, we have:

$$z - z_0 = \left|\overrightarrow{M_0M}\right| \exp(i\alpha') = |z - z_0| \exp(i\alpha')$$

$$f(z) - f(z_0) = \left|\overrightarrow{N_0N}\right| \exp(i\alpha') = |f(z) - f(z_0)| \exp(i\beta')$$

and the formula (11.67) becomes

$$f'(z_0) = \lim_{z \to z_0} \frac{|f(z) - f(z_0)|}{|z - z_0|} \lim_{z \to z_0} \exp\left[i(\alpha' - \beta')\right] = |f'(z_0)| \exp\left[i(\alpha - \beta)\right].$$

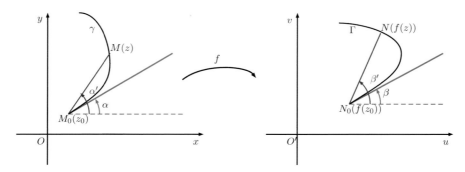

Fig. 11.9 Conformal mapping $f = u + iv$

Therefore, if $f'(z_0) \neq 0$, then

$$\alpha - \beta = \arg f'(z_0),$$

for any curve γ that passes through M_0.

Let γ_1 and γ_2 be two curves intersecting each other at M_0 and let α_1 and α_2 be the angles made by the x-axis with the tangents lines to these curves. We denote by $\Gamma_1 = f(\gamma_1)$ and $\Gamma_2 = f(\gamma_2)$ and by β_1 and β_2, the angles made by the u-axis with the tangents lines to the curves Γ_1 and Γ_2. If $f'(z_0) \neq 0$, then

$$\beta_2 - \beta_1 = \alpha_2 - \alpha_1,$$

which means that the function f *preserves the angles*. A complex function with this property is said to be a *conformal mapping* and we have just proved the following theorem:

Theorem 11.19 *A complex function that is holomorphic on the domain D and $f'(z) \neq 0$ for any $z \in D$, is a conformal mapping.*

11.2.2 Integration of Complex Functions

We know that, if $f : [a, b] \to \mathbb{C}$, $f(t) = u(t) + iv(t)$ is a complex function of a *real* variable such that the functions u, v are integrable on $[a, b]$, then

$$\int_a^b f(t)dt = \int_a^b u(t)dt + i \int_a^b v(t)dt.$$

Here we define the integral of a complex function of a *complex* variable along a path in the complex plane.

Let $\gamma : [a, b] \to \mathbb{C}$ be a piecewise smooth path (or curve),

$$\gamma(t) = x(t) + iy(t), \ t \in [a, b].$$

We usually identify γ with its image the complex plane:

$$\gamma = \{(x(t), y(t)) : t \in [a, b]\}.$$

Before defining the complex integral, we recall some basic facts on line integrals of real functions (see [38], 16.2). If $g : D \to \mathbb{R}^2$ is a continuous (real) function $g(x, y) = (P(x, y), Q(x, y))$, defined on a domain D that contains γ, then the line integral of g along γ is

$$\int_\gamma P(x, y)dx + Q(x, y)dy = \int_a^b \left[P(x(t), y(t))x'(t) + Q(x(t), y(t))y'(t) \right] dt.$$

The length of the curve γ is given by

$$L(\gamma) = \int_\gamma ds = \int_a^b |\gamma'(t)|\, dt = \int_a^b \sqrt{x'(t)^2 + y'(t)^2}\, dt. \tag{11.78}$$

Recall that γ is a *simple, closed* path (curve) if $\gamma(a) = \gamma(b)$ (*closed*), and if $\gamma(t_1) \neq \gamma(t_2)$ for any $t_1, t_2 \in (a, b)$, $t_1 \neq t_2$ (*simple*). Green's Theorem gives a relationship between the line integral along the simple closed curve γ and a double integral over the plane region Ω bounded by γ. Thus, if the functions P and Q are continuously differentiable, then

$$\oint_\gamma P(x,y)\,dx + Q(x,y)\,dy = \iint_\Omega \left(\frac{\partial Q}{\partial x} - \frac{\partial P}{\partial y} \right) dx\,dy, \tag{11.79}$$

where the symbol $\oint$ is used to indicate that the line integral is calculated using the *positive orientation* (counterclockwise) of the closed curve γ.

Definition 11.41 Let $f : D \to \mathbb{C}$ be a continuous function on the domain D, $f(x + iy) = u(x, y) + iv(x, y)$, and let $\gamma : [a, b] \to D$, $\gamma(t) = x(t) + iy(t)$ be a piecewise smooth path in D. The integral of f along γ is

$$\int_\gamma f(z)\,dz = \int_a^b f(\gamma(t))\,\gamma'(t)\,dt, \tag{11.80}$$

or, equivalently,

$$\int_\gamma f(z)\,dz = \int_\gamma (u(x,y) + iv(x,y))\,(dx + i\,dy) = \tag{11.81}$$

$$= \int_\gamma u(x,y)\,dx - v(x,y)\,dy + i \int_\gamma v(x,y)\,dx + u(x,y)\,dy.$$

Properties of the Complex Integral
The following properties easily follow from the definition above and from the properties of the line integral. If $A = \gamma(a)$ and $B = \gamma(b)$ are the endpoints of the curve γ, we also denote the integral $\int_\gamma$ by $\int_{\widehat{AB}}$.

1. $\int_\gamma (\alpha f(z) + \beta g(z))\,dz = \alpha \int_\gamma f(z)\,dz + \beta \int_\gamma g(z)\,dz$, for any $\alpha, \beta \in \mathbb{C}$.

2. $\int_{\widehat{AB}} f(z)\,dz = -\int_{\widehat{BA}} f(z)\,dz$.

3. $\int_{\widehat{AB}} f(z)\,dz = \int_{\widehat{AC}} f(z)\,dz + \int_{\widehat{CB}} f(z)\,dz$, for any $C \in \widehat{AB}$

4. $\left|\int_{\gamma} f(z)\,dz\right| \leq M \cdot L(\gamma)$, where $M = \sup_{\gamma} |f(z)|$.

Let us prove the last property. From the definition formula (11.80) one can write:

$$\left|\int_{\gamma} f(z)\,dz\right| = \left|\int_a^b f(\gamma(t))\,\gamma'(t)\,dt\right| \leq$$

$$\leq \int_a^b |f(\gamma(t))\,\gamma'(t)|\,dt \leq M \int_a^b |\gamma'(t)|\,dt$$

and by the formula (11.78) we obtain that

$$\left|\int_{\gamma} f(z)\,dz\right| \leq M \cdot L(\gamma). \tag{11.82}$$

Example 11.21 Compute the complex integral $I_n = \int_{|z-z_0|=r} \frac{1}{(z-z_0)^n}\,dz$ where $r > 0$ and $n \in \mathbb{Z}$.

The curve γ defined by the equation $|z - z_0| = r$ is the circle with center at z_0 and radius r. It has the parametrization $z = z_0 + r\exp(i\theta)$, where $\theta \in [0, 2\pi]$, or, equivalently,

$$\begin{cases} x = x_0 + r\cos\theta \\ y = y_0 + r\sin\theta \end{cases}, \theta \in [0, 2\pi].$$

We apply the formula (11.80) and obtain:

$$I_n = \int_0^{2\pi} \frac{1}{r^n e^{in\theta}} rie^{i\theta}\,d\theta = r^{1-n}i \cdot \left.\frac{e^{i\theta(1-n)}}{i(1-n)}\right|_0^{2\pi} = 0 \text{ if } n \neq 1,$$

and $I_1 = i\int_0^{2\pi} d\theta = 2\pi i$. Hence

$$\int_{|z-z_0|=r} \frac{1}{(z-z_0)^n}\,dz = \begin{cases} 0, & \text{if } n \neq 1 \\ 2\pi i, & n = 1. \end{cases} \tag{11.83}$$

Theorem 11.20 *If $F : D \to \mathbb{C}$ is a holomorphic function and $\alpha, \beta \in D$, then*

$$\int_\gamma F'(z)\, dz = F(\beta) - F(\alpha), \tag{11.84}$$

for any piecewise smooth path contained in D, $\gamma : [a, b] \to D$, with the endpoints $\gamma(a) = \alpha$, $\gamma(b) = \beta$.

Proof By the formula (11.80) we can write:

$$\int_\gamma F'(z)\, dz = \int_a^b F'(\gamma(t))\, \gamma'(t)\, dt = \int_a^b (F \circ \gamma)'(t)\, dt =$$

$$= F(\gamma(t))\Big|_a^b = F(\beta) - F(\alpha).$$

□

Definition 11.42 Let D be a domain in $\mathbb{C}$ (an open and connected subset of C). Then D is said to be *simply connected* if for any simple closed path γ contained in D, the region of the complex plane bounded by γ is also contained in D (see Fig. 11.10).

A domain which is not simply connected is said to be *multiply-connected*. An multiply-connected domain can be transformed into a simply connected domain by introducing some additional borders (*cuts*). A multiply-connected domain is said to be $n+1$-connected ($n \geq 1$) if it can be transformed into a simply connected domain by n cuts. For instance, the domain in Fig. 11.11 is doubly-connected because it can be transformed into a simply connected domain using one cut.

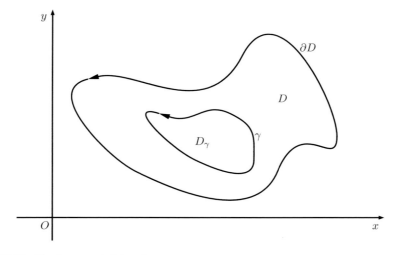

Fig. 11.10 Simply connected domain

Fig. 11.11 Doubly connected domain

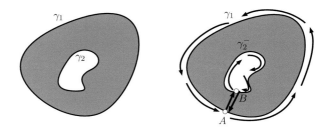

Theorem 11.21 (Cauchy's Theorem for Simply Connected Domains) *Let D be a simply connected domain of $\mathbb{C}$ and let $f : D \to \mathbb{C}$ be a holomorphic function on D. Then, for any piecewise smooth path $\gamma \subset D$, one has*

$$\oint_\gamma f(z)\, dz = 0.$$

Proof Since $f = u + iv$ is holomorphic on D, the functions u and v are of class C^1 on D and satisfy the Cauchy-Riemann equations: $\frac{\partial u}{\partial x} = \frac{\partial v}{\partial y}$, $\frac{\partial u}{\partial y} = -\frac{\partial v}{\partial x}$.

Let D_γ be the domain bounded by the closed path γ (see Fig. 11.10). We apply the Green's Theorem (11.79) for both line integrals in the definition formula (11.81):

$$\int_\gamma f(z)\,dz = \int_\gamma u(x,y)dx - v(x,y)dy + i\int_\gamma v(x,y)dx + u(x,y)dy =$$

$$= \iint_{D_\gamma} \left(-\frac{\partial v}{\partial x} - \frac{\partial u}{\partial y}\right) dx dy + i \iint_{D_\gamma} \left(\frac{\partial u}{\partial x} - \frac{\partial v}{\partial y}\right) dx dy$$

Due to the Cauchy-Riemann equations, both double integrals are equal to 0 and the proof is complete. □

Remark 11.14 The simple connectivity property of D is necessary. Indeed, by (11.83), the following integral is not 0, because the domain $D = \mathbb{C}\setminus\{z_0\}$ is not simple connected:

$$\int_{|z-z_0|=r} \frac{1}{z-z_0}\, dz = 2\pi i. \qquad (11.85)$$

If f is holomorphic on the simply connected domain D, $A(z_1), B(z_2) \in D$ are two points in D and $\gamma_1, \gamma_2 \subset D$ are two different paths from A to B, then, from the Cauchy's Theorem 11.21 and using the properties of the complex integral, one can write:

$$0 = \int_{\gamma_1 \cup \gamma_2^-} f(z)\,dz = \int_{\gamma_1} f(z)\,dz - \int_{\gamma_2} f(z)\,dz$$

where $\widehat{\gamma_2^-} = \widehat{BA}$ is the opposite path of $\gamma_2 = \widehat{AB}$. Hence the next corollary follows.

Corollary 11.5 *If $D \subset \mathbb{C}$ is a simply connected domain and $f : D \to \mathbb{C}$ is holomorphic on D, then, for any points $A(z_1), B(z_2) \in D$, the integral $\int_{\widehat{AB}} f(z)\,dz$ does not depend on the path $\widehat{AB} \subset D$.*

Definition 11.43 Let $F, f : D \to \mathbb{C}$ be two complex functions defined on the domain $D \subset \mathbb{C}$. The function F is said to be a *primitive* of f if F is holomorphic on D and

$$F'(z) = f(z), \quad \text{for all } z \in D.$$

Let D be a simply connected domain and $f : D \to \mathbb{C}$ be a holomorphic function. Let $A(z_0) \in D$ be a fixed point and $M(z)$ be a variable point in D. By the Corollary 11.5 we can see that the integral $\int_{\widehat{AM}} f(\zeta)\,d\zeta$ does not depend on the path $\widehat{AM}$; it depends only on the endpoints z_0, z and, since z_0 is fixed, we can define the function

$$F(z) = \int_{\widehat{AM}} f(\zeta)\,d\zeta = \int_{z_0}^{z} f(\zeta)\,d\zeta. \tag{11.86}$$

We prove that $F(z)$ is holomorphic on D and $F'(z) = f(z)$ for all $z \in D$.

If $f(z) = u(x, y) + iv(x, y)$ and $F(z) = U(x, y) + iV(x, y)$, then, from (11.86) and (11.81), one can write:

$$U(x, y) + iV(x, y) = \int_{\widehat{AM}} u\,dx - v\,dy + i \int_{\widehat{AM}} v\,dx + u\,dy.$$

Since the line integrals above does not depend on the path, we obtain that $dU = u\,dx - v\,dy$ and $dV = v\,dx + u\,dy$, so

$$\frac{\partial U}{\partial x} = u, \quad \frac{\partial U}{\partial y} = -v, \quad \text{and} \quad \frac{\partial V}{\partial x} = v, \quad \frac{\partial V}{\partial y} = u.$$

Thus, since the functions u and v are continuous, we find that U and V are functions of class C^1 that satisfy the Cauchy-Riemann equations

$$\frac{\partial U}{\partial x} = \frac{\partial V}{\partial y}, \quad \frac{\partial U}{\partial y} = -\frac{\partial V}{\partial x}.$$

Therefore, from Theorem 11.18, we obtain that $F(z)$ is a holomorphic function and

$$F'(z) = \frac{\partial U}{\partial x} + i\frac{\partial V}{\partial x} = u + iv = f(z), \quad \text{for all } z \in D.$$

11.2 Complex Function Theory

Remark 11.15 As a matter of fact, we have just proved that any continuous function defined on a simple connected domain, $f : D \to \mathbb{C}$ with the property that $\oint_\gamma f(z)\,dz = 0$ for any closed curve $\gamma \subset D$, has a primitive on D.

The primitive of a complex function is not unique: if $F(z)$ is a primitive of $f(z)$ then any function of the form $\Phi(z) = F(z) + C$ (where C is a complex constant) is also a primitive of $f(z)$. Furthermore, for any primitive $\Phi(z)$ of $f(z)$ we can write:

$$\int_{z_1}^{z_2} f(z)\,dz = \Phi(z_2) - \Phi(z_1).$$

What happens when the domain D is not *simple* connected? Let us consider first the doubly connected domain represented in Fig. 11.11.

Theorem 11.22 (Cauchy's Theorem for a Doubly Connected Domain) *If D is a doubly-connected domain, bounded by the smooth, closed curves γ_1 and γ_2 and f is holomorphic on D, then*

$$\oint_{\gamma_1} f(z)\,dz = \oint_{\gamma_2} f(z)\,dz. \tag{11.87}$$

Proof If we "cut" the domain with the arc $\widehat{AB}$ (see Fig. 11.11), then the doubly-connected domain is transformed into the simply connected domain $D \smallsetminus \widehat{AB}$ with the boundary $\Gamma = \gamma_1 \cup \widehat{AB} \cup \gamma_2^- \cup \widehat{BA}$. By applying the Cauchy Theorem 11.21 one can write:

$$0 = \oint_\Gamma f(z)\,dz = \oint_{\gamma_1} f(z)\,dz + \int_{\widehat{AB}} f(z)\,dz + \oint_{\gamma_2^-} f(z)\,dz + \int_{\widehat{BA}} f(z)\,dz.$$

Since

$$\int_{\widehat{AB}} f(z)\,dz = -\int_{\widehat{BA}} f(z)\,dz \text{ and } \oint_{\gamma_2^-} f(z)\,dz = -\oint_{\gamma_2} f(z)\,dz$$

the equality (11.87) is obtained. □

The general case of an $(n + 1)$-connected domain can be treated in a similar manner (see Fig. 11.12).

Theorem 11.23 (Cauchy's Theorem for an $(n + 1)$-Connected Domain) *Let D be an $(n + 1)$-connected domain, $D = \Omega \smallsetminus \overline{D_1} \smallsetminus \ldots \smallsetminus \overline{D_n}$, where Ω is a simply connected domain with the boundary $\partial \Omega = \gamma_0$ and $D_1, D_2, \ldots, D_n \subset \Omega$ are simply connected, disjoint domains contained in Ω and having the boundaries $\gamma_1, \ldots, \gamma_n$.*

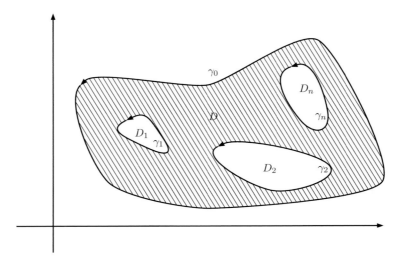

Fig. 11.12 $(n + 1)$-connected domain

Then

$$\oint_{\gamma_0} f(z)\,dz = \sum_{k=1}^{n} \oint_{\gamma_k} f(z)\,dz. \qquad (11.88)$$

The following theorem plays a central role in complex analysis. Known as *Cauchy's integral formula*, it basically states that the values of a holomorphic function inside the domain bounded by a closed curve γ are completely determined by the values of the function on γ. To prove this result we use the following lemma, which is in fact an extension of Cauchy's Theorem 11.21.

Lemma 11.3 *Let $\Omega \subset \mathbb{C}$ be a simply connected domain, and $z_0 \in \Omega$. If $g : \Omega \to \mathbb{C}$ is continuous on Ω and holomorphic on $\Omega \setminus \{z_0\}$, then*

$$\oint_{\gamma} g(z)\,dz = 0.$$

for any closed curve γ surrounding the point z_0 and lying in Ω.

Proof Let $D \subset \mathbb{C}$ be the domain bounded by γ. For any positive number r such that

$$\overline{D_r} = \{z \in \mathbb{C} : |z - z_0| \leq r\} \subset D,$$

we denote by $\gamma_r = \partial D_r$ the circle with center z_0 and radius r. Since the function $g(z)$ is holomorphic on the doubly-connected domain $D \setminus \overline{D_r}$, by Theorem 11.22

one can write:
$$\oint_\gamma g(z)\,dz = \oint_{\gamma_r} g(z)\,dz.$$

Since g is continuous, there exists a positive number $M > 0$ such that
$$|g(z)| \leq M, \quad \forall z \in \overline{D}.$$

It follows that
$$\left|\oint_\gamma g(z)\,dz\right| = \left|\oint_{\gamma_r} g(z)\,dz\right| \leq M \cdot \text{length}(\gamma_r) = M \cdot 2\pi r.$$

Since $r > 0$ can be infinitely small, the integrals above must be 0, hence the proof is complete. □

Theorem 11.24 (Cauchy's Integral Formula) *If $f : \Omega \to \mathbb{C}$ is a holomorphic function on the domain Ω, then, for any $z_0 \in \Omega$, we have*
$$f(z_0) = \frac{1}{2\pi i} \oint_\gamma \frac{f(z)}{z - z_0}\,dz, \tag{11.89}$$

where $\gamma \subset \Omega$ is any simple closed path surrounding the point z_0 such that the domain D bounded by γ is contained in Ω.

Proof Let $r > 0$ be a positive number such that
$$\overline{D_r} = \{z \in \mathbb{C} : |z - z_0| \leq r\} \subset D.$$

We denote by $\gamma_r = \partial D_r$ the circle with center z_0 and radius r. Since the function $\frac{f(z)}{z-z_0}$ is holomorphic on the doubly-connected domain $D \setminus \overline{D_r}$, by Theorem 11.22 one can write:

$$\oint_\gamma \frac{f(z)}{z - z_0}\,dz = \oint_{\gamma_r} \frac{f(z)}{z - z_0}\,dz =$$
$$= \oint_{\gamma_r} \frac{f(z) - f(z_0)}{z - z_0}\,dz + f(z_0) \oint_{\gamma_r} \frac{1}{z - z_0}\,dz \stackrel{(11.85)}{=}$$
$$= \oint_{\gamma_r} \frac{f(z) - f(z_0)}{z - z_0}\,dz + f(z_0) \cdot 2\pi i.$$

Since f is holomorphic, the function
$$\Phi(z) = \begin{cases} \frac{f(z)-f(z_0)}{z-z_0}, & z \in \Omega \setminus \{z_0\} \\ f'(z_0), & z = z_0 \end{cases}$$

is continuous on Ω and holomorphic on $\Omega \smallsetminus \{z_0\}$. By Lemma 11.3 we have

$$\oint_{\gamma_r} \frac{f(z) - f(z_0)}{z - z_0} dz = 0$$

so the proof is complete. □

The next theorem (also referred as the "Cauchy's integral formula") states that a holomorphic function has derivatives of any order, $f^{(n)}$, $n = 1, 2, \ldots$, which can be computed by differentiating n times the function under the integral (11.89). We shall need the following lemma:

Lemma 11.4 *Let γ be a simple closed curve and $D \subset \mathbb{C}$ be the domain bounded by γ. If $\varphi(\zeta)$ is a function continuous on γ, then, for any $n = 1, 2, \ldots$, the function $\widetilde{\varphi}_n : D \to \mathbb{C}$,*

$$\widetilde{\varphi}_n(z) = \oint_\gamma \frac{\varphi(\zeta)}{(\zeta - z)^n} d\zeta$$

is holomorphic on D and

$$\widetilde{\varphi}_n'(z) = n\widetilde{\varphi}_{n+1}(z), \ \forall z \in D. \tag{11.90}$$

Proof We firstly show that $\widetilde{\varphi}_1$ is continuous on D. Let $z_0 \in D$ and $\delta > 0$ such that $B(z_0, \delta) \subset D$. Then, for any $z \in B(z_0, \delta/2)$ and $\zeta \in \gamma$, we have $|\zeta - z| > \delta/2$. Since

$$\widetilde{\varphi}_1(z) - \widetilde{\varphi}_1(z_0) = (z - z_0) \oint_\gamma \frac{\varphi(\zeta)}{(\zeta - z)(\zeta - z_0)} d\zeta, \tag{11.91}$$

and $\varphi(\zeta)$ is continuous, hence bounded on γ, $|\varphi(\zeta)| \leq M$ for all $\zeta \in \gamma$, we obtain:

$$|\widetilde{\varphi}_1(z) - \widetilde{\varphi}_1(z_0)| = |z - z_0| \left| \oint_\gamma \frac{\varphi(\zeta)}{(\zeta - z)(\zeta - z_0)} d\zeta \right|$$

$$\leq |z - z_0| \cdot \frac{2}{\delta^2} \cdot M \cdot L(\gamma),$$

where $L(\gamma)$ is the length of the curve γ. Thus, we obtain that $\widetilde{\varphi}_1$ is continuous on D.

Moreover, from (11.91) we can write:

$$\frac{\widetilde{\varphi}_1(z) - \widetilde{\varphi}_1(z_0)}{z - z_0} = \oint_\gamma \frac{\varphi(\zeta)}{(\zeta - z)(\zeta - z_0)} d\zeta = \widetilde{\psi}_1(z),$$

11.2 Complex Function Theory

where ψ is the function (continuous on γ):

$$\psi(\zeta) = \frac{\varphi(\zeta)}{\zeta - z_0}.$$

Since $\widetilde{\psi}_1(z)$ is continuous, we obtain that

$$\lim_{z \to z_0} \frac{\widetilde{\varphi}_1(z) - \widetilde{\varphi}_1(z_0)}{z - z_0} = \widetilde{\psi}_1(z_0) = \oint_\gamma \frac{\varphi(\zeta)}{(\zeta - z_0)^2} d\zeta = \widetilde{\varphi}_2(z_0),$$

so the function $\widetilde{\varphi}_1(z)$ is holomorphic and satisfies (11.90) (for $n = 1$).

We prove the general result by mathematical induction on n. Suppose that $\widetilde{\varphi}_{n-1}(z)$ is holomorphic and satisfies

$$\widetilde{\varphi}'_{n-1}(z) = (n-1)\widetilde{\varphi}_n(z), \ \forall z \in D.$$

From the identity

$$\widetilde{\varphi}_n(z) - \widetilde{\varphi}_n(z_0) = \left(\oint_\gamma \frac{\varphi(\zeta)\, d\zeta}{(\zeta - z)^{n-1}(\zeta - z_0)} - \oint_\gamma \frac{\varphi(\zeta)\, d\zeta}{(\zeta - z_0)^n} \right) +$$

$$+ (z - z_0) \oint_\gamma \frac{\varphi(\zeta)\, d\zeta}{(\zeta - z)^n (\zeta - z_0)},$$

we find that $\widetilde{\varphi}_n(z)$ is continuous. Indeed, by applying the induction hypothesis to the function $\psi(\zeta) = \frac{\varphi(\zeta)}{\zeta - z_0}$ we find that the first term tends to 0 as $z \to z_0$, and for the second term there is a constant C such that in a certain neighborhood of z_0 we have

$$\left| (z - z_0) \oint_\gamma \frac{\varphi(\zeta)\, d\zeta}{(\zeta - z)^n (\zeta - z_0)} \right| \leq C |z - z_0|.$$

Moreover, if we divide by $z - z_0$ the identity above, we get

$$\lim_{z \to z_0} \frac{\widetilde{\varphi}_n(z) - \widetilde{\varphi}_n(z_0)}{z - z_0} = \lim_{z \to z_0} \frac{\widetilde{\psi}_{n-1}(z) - \widetilde{\psi}_{n-1}(z_0)}{z - z_0} + \lim_{z \to z_0} \widetilde{\psi}_n(z)$$

and by applying the induction hypothesis to the function $\psi(\zeta)$ we obtain:

$$\lim_{z \to z_0} \frac{\widetilde{\varphi}_n(z) - \widetilde{\varphi}_n(z_0)}{z - z_0} = \widetilde{\psi}'_{n-1}(z_0) + \widetilde{\psi}_n(z_0)$$

$$= (n-1)\widetilde{\psi}_n(z_0) + \widetilde{\psi}_n(z_0)$$

$$= n\widetilde{\psi}_n(z_0) = n \oint_\gamma \frac{\varphi(\zeta)\, d\zeta}{(\zeta - z_0)^{n+1}},$$

hence the derivative $\widetilde{\varphi}'_n(z_0)$ exists and $\widetilde{\varphi}'_n(z_0) = n\widetilde{\varphi}_{n+1}(z_0)$, so the proof is complete. □

The following theorem follows directly from Theorem 11.24 and Lemma 11.4. Equation (11.92) is known as the *Cauchy's integral formula* for the derivatives of a holomorphic function.

Theorem 11.25 *If $f : \Omega \to \mathbb{C}$ is a holomorphic function on the domain Ω, then its derivatives $f^{(n)}$, $n = 1, 2, \ldots$, are also holomorphic on Ω and for any $z_0 \in \Omega$, we have*

$$f^{(n)}(z_0) = \frac{n!}{2\pi i} \oint_\gamma \frac{f(z)}{(z-z_0)^{n+1}} dz, \qquad (11.92)$$

where $\gamma \subset \Omega$ is any simple closed path surrounding the point z_0 such that the domain D bounded by γ is contained in Ω.

As noted in Remark 11.15, if $f : \Omega \to \mathbb{C}$ is a continuous function such that $\oint_\gamma f(z) dz = 0$ for any closed curve $\gamma \subset \Omega$, then f has a primitive: there exists a holomorphic function $F : \Omega \to \mathbb{C}$ such that $F' = f$. By Theorem 11.25, the function f is holomorphic, so the following theorem is proved:

Theorem 11.26 (Morera's Theorem) *If $f : \Omega \to \mathbb{C}$ is a continuous function on the domain Ω such that $\oint_\gamma f(z) dz = 0$ for any closed curve $\gamma \subset \Omega$, then f is holomorphic on Ω.*

Theorem 11.27 (Cauchy's Estimates) *Let $f : \Omega \to \mathbb{C}$ be a holomorphic function and $z_0 \in \Omega$, $r > 0$ such that $B[z_0, r] \subset \Omega$. Let γ be the circle $|z - z_0| = r$ and $M > 0$ such that $|f(z)| \leq M$ for all $z \in \gamma$. Then, for every $n = 1, 2, \ldots$, the following inequality holds:*

$$\left|f^{(n)}(z_0)\right| \leq \frac{n! M}{r^n}. \qquad (11.93)$$

Proof By Theorem 11.25 we have:

$$\left|f^{(n)}(z_0)\right| = \frac{n!}{2\pi} \left|\oint_\gamma \frac{f(z)}{(z-z_0)^{n+1}} dz\right|$$

$$\leq \frac{n!}{2\pi} \cdot \frac{M}{r^{n+1}} \cdot 2\pi r = \frac{n! M}{r^n}.$$

□

The next theorem is another classical result following from Theorem 11.25. Recall that a function $f : \mathbb{C} \to \mathbb{C}$ is said to be *entire* if it is holomorphic in the whole complex plane $\mathbb{C}$.

Theorem 11.28 (Liouville's Theorem) *An entire function that is bounded on $\mathbb{C}$ must be a constant.*

Proof Suppose that $|f(z)| \leq M$, for any $z \in \mathbb{C}$. By applying Theorem 11.27 for $n = 1$, we have

$$|f'(z_0)| \leq \frac{M}{r},$$

for any $r > 0$ and $z \in \mathbb{C}$. Hence we obtain that $f'(z) = 0$ for any $z \in \mathbb{C}$ and we conclude that f is constant. □

An immediate and deep application of the Liouville's theorem is the fact that some real functions which are bounded and non-constant on the real line (as $\sin x$, $\cos x$, etc.) are not bounded anymore when they are extended to hololomorphic functions in the whole complex plane. But another important consequence of the Liouville's theorem is the following basic result in Mathematics.

Theorem 11.29 (Fundamental Theorem of Algebra) *Any non-constant polynomial $P(z) = a_0 + a_1 z + \ldots + a_n z^n$ with complex coefficients has at least one root in $\mathbb{C}$. Moreover, a polynomial of degree n has all the n roots in $\mathbb{C}$ and can be written as a product of the form:*

$$P(z) = a_n (z - z_1)(z - z_2) \ldots (z - z_n),$$

where $z_1, z_2, \ldots, z_n \in \mathbb{C}$ are the roots of the polynomial $P(z)$.

Proof Assume the contrary, namely that $P(z)$ has no root in the complex plane. Then $f(z) = \frac{1}{P(z)}$ is holomorphic and bounded, because $\lim_{|z| \to \infty} f(z) = 0$. By Liouville's theorem we find that $f(z)$ is a constant function, i.e. $P(z)$ is constant, which contradicts the hypothesis. Hence $P(z)$ must have at least one root $z_1 \in \mathbb{C}$. Consequently, there exists $Q(z)$ a polynomial of degree $n - 1$ such that $P(z) = (z - z_1) Q(z)$. The proof is completed by induction on n. □

11.2.3 Power Series Representation

A power series is a series of the form

$$\sum_{n=0}^{\infty} c_n (z - a)^n, \qquad (11.94)$$

where $a, c_0, c_1, \ldots$ are complex numbers. For every power series (11.94), there exists $R \in [0, \infty]$ (called the *radius of convergence* of the series) such that:

- the series converges absolutely and uniformly on $B[a, r]$, for any $r < R$,
- the series diverges if $|z - a| > R$.

The radius of convergence R is given by the formula:

$$R = \frac{1}{\limsup_{n \to \infty} \sqrt[n]{|c_n|}}. \tag{11.95}$$

The complex functions that are representable by power series are called *analytic functions*. More exactly, we have the following definition:

Definition 11.44 Let $\Omega \subset \mathbb{C}$ be a domain (an open and connected set). A function $f : \Omega \to \mathbb{C}$ is said to be *analytic* (representable by power series) at $a \in \Omega$ if there exists $r > 0$ such that $B(a, r) \subset \Omega$ and for every $z \in B(a, r)$ one has

$$f(z) = \sum_{n=0}^{\infty} c_n (z - a)^n.$$

A function analytic at every point $a \in \Omega$ is said to be analytic on Ω.

We shall see that a function is analytic at a if and only if it is holomorphic at a, so the two notions are equivalent.

Theorem 11.30 *If $f : \Omega \to \mathbb{C}$ is analytic (representable by power series) on Ω, then it is holomorphic on Ω and its derivative f' is also representable by power series.*

If $a \in \Omega$ and $r > 0$ is a positive number such that $B(a, r) \subset \Omega$ and

$$f(z) = \sum_{n=0}^{\infty} c_n (z - a)^n, \tag{11.96}$$

for all $z \in B(a, r)$, then for these z we also have:

$$f'(z) = \sum_{n=1}^{\infty} n c_n (z - a)^{n-1}. \tag{11.97}$$

Proof From the relation (11.95), the series (11.96) and (11.97) have the same radius of convergence. Since (11.96) is convergent on $B(a, r)$, it follows that (11.97) is also convergent on $B(a, r)$ and we denote its sum by $g(z)$. To simplify calculation, we can take without loss of generality $a = 0$. Let $z_0 \in B(0, r)$ and $\rho > 0$ such that

11.2 Complex Function Theory

$|z_0| < \rho < r$. Then, for any $z \neq z_0$ such that $z \leq \rho$ we can write:

$$\frac{f(z) - f(z_0)}{z - z_0} - g(z_0) = \sum_{n=1}^{\infty} c_n \left(\frac{z^n - z_0^n}{z - z_0} - n z_0^{n-1} \right)$$

$$= \sum_{n=2}^{\infty} c_n \left(z^{n-1} + z^{n-2} z_0 + \ldots + z^2 z_0^{n-3} + z z_0^{n-2} - (n-1) z_0^{n-1} \right)$$

$$= \sum_{n=2}^{\infty} c_n \left[(z^{n-1} - z_0^{n-1}) + z_0 (z^{n-2} - z_0^{n-2}) + \ldots + z_0^{n-2} (z - z_0) \right]$$

$$= (z - z_0) \sum_{n=2}^{\infty} c_n \sum_{k=1}^{n-1} k z^{n-k-1} z_0^{k-1}.$$

Since $|z_0| < \rho$ and $|z| < \rho$, we have:

$$\left| \frac{f(z) - f(z_0)}{z - z_0} - g(z_0) \right| \leq |z - z_0| \sum_{n=2}^{\infty} |c_n| \rho^{n-2} \frac{n(n-1)}{2}$$

and the last series is convergent by the root test (since $\rho < r$). It follows that

$$\lim_{z \to z_0} \left(\frac{f(z) - f(z_0)}{z - z_0} - g(z_0) \right) = 0,$$

so f is differentiable at z_0 and $f'(z_0) = g(z_0)$. □

Since f' satisfies the same hypothesis as f does, we can also apply Theorem 11.30 to f' and so on. It follows that the derivatives $f^{(k)}$, $k = 1, 2, \ldots$ of all orders are representable by power series and, if (11.96) holds for all $z \in B(a, r)$, then we also have

$$f^{(k)}(z) = \sum_{n=k}^{\infty} n(n-1) \ldots (n-k+1) c_n (z-a)^{n-k}, \quad (11.98)$$

for all $z \in B(a, r)$ and $k = 1, 2, \ldots$. Thus, we obtain that the coefficients c_k of the series (11.96) are given by the formula:

$$c_k = \frac{f^{(k)}(a)}{k!} \quad (11.99)$$

for every $k = 0, 1, \ldots$ and the next corollary follows.

Corollary 11.6 *If* $f : \Omega \to \mathbb{C}$ *is analytic (representable by power series) on* Ω *then, for every* $a \in \Omega$ *there is an open disc* $B(a, r) \subset \Omega$ *such that*

$$f(z) = \sum_{n=0}^{\infty} \frac{f^{(n)}(a)}{n!}(z-a)^n, \qquad (11.100)$$

for all $z \in B(a, r)$. *The series* (11.100) *is the* **Taylor series** *of* f.

Now we shall prove the reciprocal of Theorem 11.30 to establish the equivalence of the notions *analytic function* and *holomorphic function*. After proving Theorem 11.31, the words "analytic" and "holomorphic" will be treated like synonyms (as most of the books in complex analysis).

Theorem 11.31 *If* $f : \Omega \to \mathbb{C}$ *is holomorphic on* Ω, *then it is analytic (representable by power series) on* Ω.

Proof Let $a \in \Omega$ and $r > 0$ such that $B[a, r] \subset \Omega$ and let γ denote the circle of radius r centered at a. From the Cauchy's integral formula (11.89), for any $z \in B(a, r)$ we can write:

$$f(z) = \frac{1}{2\pi i} \oint_{\gamma} \frac{f(\zeta)}{\zeta - z} d\zeta = \frac{1}{2\pi i} \oint_{\gamma} \frac{f(\zeta)}{(\zeta - a) - (z - a)} d\zeta$$

$$= \frac{1}{2\pi i} \oint_{\gamma} \frac{f(\zeta)}{\zeta - a} \cdot \frac{1}{1 - \frac{z-a}{\zeta-a}} d\zeta.$$

Since $|\zeta - a| = r$ for any $\zeta \in \gamma$ and $z \in B(a, r)$, we have $\left|\frac{z-a}{\zeta-a}\right| = \rho < 1$, hence

$$\frac{1}{1 - \frac{z-a}{\zeta-a}} = \sum_{n=0}^{\infty} \left(\frac{z-a}{\zeta-a}\right)^n$$

and we have:

$$f(z) = \frac{1}{2\pi i} \oint_{\gamma} \sum_{n=0}^{\infty} \frac{f(\zeta)}{\zeta - a} \left(\frac{z-a}{\zeta-a}\right)^n d\zeta$$

$$= \sum_{n=0}^{\infty} (z-a)^n \frac{1}{2\pi i} \oint_{\gamma} \frac{f(\zeta)}{(\zeta-a)^{n+1}} d\zeta.$$

For the last equality (to interchange the integral and the sum) we used the uniform convergence of the series which results by Weierstrass M-test. Thus, the function f is bounded on γ, $|f(\zeta)| \le M$ for all $\zeta \in \gamma$, so

$$\left|\frac{f(\zeta)}{\zeta - a}\left(\frac{z-a}{\zeta-a}\right)^n\right| \le \frac{M}{r}\rho^n$$

11.2 Complex Function Theory

for all $\zeta \in \gamma$. Since the series $\sum_{n=0}^{\infty} \frac{M}{r}\rho^n$ is convergent, it follows that the series $\sum_{n=0}^{\infty} \frac{f(\zeta)}{\zeta-a}\left(\frac{z-a}{\zeta-a}\right)^n$ uniformly converges. Hence we obtained that

$$f(z) = \sum_{n=0}^{\infty} c_n(z-a)^n,$$

where the coefficients c_n are given by the formula:

$$c_n = \frac{1}{2\pi i} \oint_\gamma \frac{f(\zeta)}{(\zeta-a)^{n+1}} d\zeta \stackrel{(11.92)}{=} \frac{f^{(n)}(a)}{n!}.$$

□

By Theorem 11.31 we can say that a complex function $f(z)$ can be expanded in Taylor series around a point of holomorphy. The functions below are entire functions, so the following formulas hold in the whole complex plane:

$$\exp(z) = 1 + \frac{z}{1!} + \frac{z^2}{2!} + \ldots + \frac{z^n}{n!} + \ldots$$

$$\cos(z) = 1 - \frac{z^2}{2!} + \frac{z^4}{4!} + \ldots + (-1)^n \frac{z^{2n}}{(2n)!} + \ldots$$

$$\sin(z) = z - \frac{z^3}{3!} + \frac{z^5}{5!} + \ldots + (-1)^n \frac{z^{2n+1}}{(2n+1)!} + \ldots$$

$$\cosh(z) = 1 + \frac{z^2}{2!} + \frac{z^4}{4!} + \ldots + \frac{z^{2n}}{(2n)!} + \ldots$$

$$\sinh(z) = z + \frac{z^3}{3!} + \frac{z^5}{5!} + \ldots + \frac{z^{2n+1}}{(2n+1)!} + \ldots$$

An important consequence of Theorem 11.31 is the following fundamental result (see [36], 10.39 a. and b. for a proof).

Theorem 11.32 (Identity Theorem) *If the functions $f, g : \Omega \to \mathbb{C}$ are holomorphic on the domain Ω and $A \subset \Omega$ is a subset of complex numbers with at least one limit point in Ω such that $f(z) = g(z)$ for all $z \in A$, then $f = g$ on the whole domain Ω.*

Let $\Omega \subset \mathbb{C}$ be a simple connected domain, $a \in \Omega$ and $f(z)$ be a complex function which is analytic on $\Omega \setminus a$. Consider the positive numbers $0 < r < R$ such that the circle $|z - a| = R$ is contained in Ω. Since the function f is holomorphic on the doubly connected domain D bounded by the circles $\Gamma : |z - a| = R$ and

$\gamma : |z - a| = r$, from the Cauchy's integral formula (11.89) we can write, for any $z \in D$:

$$f(z) = \frac{1}{2\pi i} \oint_\Gamma \frac{f(\zeta)}{\zeta - z} d\zeta - \frac{1}{2\pi i} \oint_\gamma \frac{f(\zeta)}{\zeta - z} d\zeta$$

$$= \frac{1}{2\pi i} \oint_\Gamma \frac{f(\zeta) d\zeta}{(\zeta - a) - (z - a)} + \frac{1}{2\pi i} \oint_\gamma \frac{f(\zeta) d\zeta}{(z - a) - (\zeta - a)}$$

$$= \frac{1}{2\pi i} \oint_\Gamma \frac{f(\zeta) d\zeta}{(\zeta - a)\left(1 - \frac{z-a}{\zeta-a}\right)} + \frac{1}{2\pi i} \oint_\gamma \frac{f(\zeta) d\zeta}{(z - a)\left(1 - \frac{\zeta-a}{z-a}\right)}.$$

For any $\zeta \in \Gamma$ we have $\left|\frac{z-a}{\zeta-a}\right| = \frac{|z-a|}{R} < 1$ and also, for any $\zeta \in \gamma$ we have $\left|\frac{\zeta-a}{z-a}\right| = \frac{r}{|z-a|} < 1$, so we can write (like in the proof of Theorem 11.31):

$$f(z) = \frac{1}{2\pi i} \oint_\Gamma \frac{f(\zeta)}{\zeta - a} \left[\sum_{n=0}^\infty \left(\frac{z-a}{\zeta-a}\right)^n\right] d\zeta - \frac{1}{2\pi i} \oint_\gamma \frac{f(\zeta)}{z - a} \left[\sum_{n=0}^\infty \left(\frac{\zeta-a}{z-a}\right)^n\right] d\zeta$$

$$= \sum_{n=0}^\infty \left[\frac{1}{2\pi i} \oint_\Gamma \frac{f(\zeta) d\zeta}{(\zeta - a)^{n+1}}\right] (z-a)^n + \sum_{n=1}^\infty \left[\frac{1}{2\pi i} \oint_\gamma f(\zeta)(\zeta - a)^{n-1} d\zeta\right] (z-a)^{-n}$$

$$= \sum_{n=0}^\infty c_n (z - a)^n + \sum_{n=1}^\infty c_{-n}(z - a)^{-n},$$

where

$$c_{-n} = \frac{1}{2\pi i} \oint_\gamma f(\zeta)(\zeta - a)^{n-1} d\zeta = \frac{1}{2\pi i} \oint_C f(\zeta)(\zeta - a)^{n-1} d\zeta, \ n = 1, 2, \ldots,$$

$$c_n = \frac{1}{2\pi i} \oint_\gamma \frac{f(\zeta)}{(\zeta - a)^{n+1}} d\zeta = \frac{1}{2\pi i} \oint_C \frac{f(\zeta)}{(\zeta - a)^{n+1}} d\zeta, \ n = 0, 1, \ldots,$$

where $C \subset D$ is any simple closed curve surrounding the point a.

Hence

$$f(z) = \sum_{n=-\infty}^\infty c_n (z - a)^n, \qquad (11.101)$$

where the coefficients c_n, $n \in \mathbb{Z}$ are given by

$$c_n = \frac{1}{2\pi i} \oint_C \frac{f(\zeta)}{(\zeta - a)^{n+1}} d\zeta. \qquad (11.102)$$

11.2 Complex Function Theory

The series (11.101) is called the *Laurent series* of the function $f(z)$. It is formed by the Taylor part, $\sum_{n=0}^{\infty} c_n(z-a)^n$ and the principal part (with negative powers), $\sum_{n=-1}^{-\infty} c_n(z-a)^n$. If the function is holomorphic on the whole domain Ω, including point a, then all the coefficients of the principal part are 0 and the Laurent series reduces to the Taylor series. Otherwise, if the principal part is not 0, the function is not analytic at a and we say that a is a singular point of $f(z)$ (see the definition below).

Definition 11.45 Consider the domain $\Omega \subset \mathbb{C}$, the function $f : \Omega \to \mathbb{C}$ and $z_0 \in \Omega$.

(i) The point z_0 is a *regular point* of f if the function is analytic at z_0: if there exists $r > 0$ such that f is representable in Taylor series on $B(z_0, r)$:

$$f(z) = \sum_{n=0}^{\infty} c_n(z - z_0)^n, \ z \in B(z_0, r).$$

(ii) The point z_0 is a *singular point* of f if it is not a regular point.
(iii) The point z_0 is a *zero* of the function f if it is a regular point and $f(z_0) = 0$. In this case, there exists a positive integer k such that

$$f(z) = (z - z_0)^k \sum_{n=k}^{\infty} c_n(z - z_0)^{n-k}, \ z \in B(z_0, r),$$

where $c_k \neq 0$. We say that z_0 is a *zero of order k* of the function f and we notice that z_0 is a zero of order k if and only if

$$f^{(n)}(z_0) = 0, \text{ for every } n = 0, 1, \ldots, k-1, \text{ and } f^{(k)}(z_0) \neq 0.$$

(iv) The point z_0 is a *pole* of the function f if the principal part of the Laurent series at z_0 has a finite number of nonzero terms, that is, if there exists $r > 0$ such that

$$f(z) = \frac{c_{-k}}{(z - z_0)^k} + \ldots + \frac{c_{-1}}{z - z_0} + \sum_{n=0}^{\infty} c_n(z - z_0)^n, \ z \in B(z_0, r),$$

where $c_{-k} \neq 0$. In this case, we say that z_0 is a *pole of order k* of the function f.
(v) A singular point z_0 which is not a pole is said to be an *essential singular point* of f. In this case, the principal part of the Laurent series has an infinite number of nonzero terms.

Remark 11.16 From the definition above it is easy to see that z_0 is a zero of order k for the function $f(z)$ if and only if there exists a function $g(z)$ which is analytic at z_0 such that $f(z) = (z - z_0)^k g(z)$ and $g(z_0) \neq 0$.

We can also deduce that z_0 is a pole of order k for the function $f(z)$ if and only if it is a zero of order k for the function $1/f(z)$. Indeed, by the definition we see that z_0 is a pole of order k if and only if the function $g(z) = (z - z_0)^k f(z)$ is analytic at z_0 and $g(z_0) \neq 0$. It follows that the function $1/g$ is analytic at z_0 and that z_0 is a zero of order k for the analytic function $\frac{1}{f(z)} = \frac{(z-z_0)^k}{g(z)}$.

Definition 11.46 A function $f : \Omega \to \mathbb{C}$ is said to be *meromorphic* if the points of the domain Ω are either regular points or poles of f (i.e. $f(z)$ has no essential singular point in Ω).

From Remark 11.16 we deduce that any rational function, $f(z) = \frac{P(z)}{Q(z)}$, where $P(z)$ and $Q(z)$ are polynomials, is a meromorphic function on $\mathbb{C}$.

The function $g(z) = \exp\left(\frac{1}{z-z_0}\right)$ is not a meromorphic function: it has the expansion in Laurent series

$$\exp\left(\frac{1}{z-z_0}\right) = 1 + \frac{1}{1!} \cdot \frac{1}{z-z_0} + \frac{1}{2!} \cdot \frac{1}{(z-z_0)^2} + \ldots + \frac{1}{n!} \cdot \frac{1}{(z-z_0)^n} + \ldots,$$

hence z_0 is an essential singular point for $g(z)$.

11.2.4 Residue Theorem and Applications

Definition 11.47 Let $D = \{z \in \mathbb{C} : 0 < |z - z_0| < r\}$ and let $f(z)$ be a function analytic on D and having a singular point at z_0 (either a pole, or an essential singularity). The *residue* of f at z_0 is the coefficient c_{-1} of $\frac{1}{z-z_0}$ in the Laurent series (11.101). If $\gamma \subset D$ is a simple closed curve surrounding the point z_0, then, by integrating (11.101) we obtain that

$$\oint_\gamma f(z)\,dz = \sum_{n=-\infty}^{\infty} c_n \oint_\gamma (z-z_0)^n\,dz = c_{-1} \cdot 2\pi i,$$

so the residue of f at z_0 is the complex number

$$\operatorname{Res}[f, z_0] = \frac{1}{2\pi i} \oint_\gamma f(z)\,dz. \tag{11.103}$$

Theorem 11.33 (Residue Theorem) *Let $\Omega \subset \mathbb{C}$ be a domain and f be a function analytic on $\Omega \setminus \{z_1, z_2, \ldots, z_n\}$, where $z_1, z_2, \ldots, z_n$ are singular points of $f(z)$.*

11.2 Complex Function Theory

Let Γ be a simple closed curve such that $z_1, z_2, \ldots, z_n$ are contained in the domain D bounded by Γ. Then,

$$\oint_\Gamma f(z)\,dz = 2\pi i \sum_{k=1}^n \text{Res}[f, z_k]. \tag{11.104}$$

Proof For every $k = 1, \ldots, n$ we consider a circle γ_k of center z_k and radius sufficiently small such that $\gamma_1, \ldots, \gamma_n$ are all contained in Ω and do not intersect with each other. Let Δ denote the $n+1$-connected domain bounded by the curves $\Gamma, \gamma_1, \ldots, \gamma_n$. By Cauchy's theorem for multiply connected domains (Theorem 11.23) we obtain that

$$\oint_\Gamma f(z)\,dz = \sum_{k=1}^n \oint_{\gamma_k} f(z)\,dz \stackrel{(11.103)}{=} \sum_{k=1}^n 2\pi i \, \text{Res}[f, z_k]$$

and (11.104) follows. □

Remark 11.17 Suppose that z_0 is a simple pole of the function f. Then, in a neighborhood of z_0 the function has the following expansion in Laurent series:

$$f(z) = \frac{c_{-1}}{z - z_0} + c_0 + c_1(z - z_0) + \ldots + c_n(z - z_0)^n + \ldots$$

It follows that the residue $\text{Res}[f, z_0] = c_{-1}$ is given by the formula:

$$\text{Res}[f, z_0] = \lim_{z \to z_0} [(z - z_0) f(z)]. \tag{11.105}$$

If z_0 is a pole of order k of the function f, then f has the following expansion in Laurent series in a neighborhood of z_0:

$$f(z) = \frac{c_{-k}}{(z - z_0)^k} + \ldots + \frac{c_{-1}}{z - z_0} + c_0 + c_1(z - z_0) + \ldots + c_n(z - z_0)^n + \ldots$$

In order to obtain the residue $\text{Res}[f, z_0] = c_{-1}$ we can multiply this relation by $(z - z_0)^k$, then differentiate it $k-1$ times and find the limit as $z \to z_0$:

$$\text{Res}[f, z_0] = \frac{1}{(k-1)!} \lim_{z \to z_0} \left[(z - z_0)^k f(z)\right]^{(k-1)}. \tag{11.106}$$

Example 11.22 We use the residue theorem to compute the integral:

$$\oint_{|z-i|=2} \frac{dz}{z(z-1)^2}.$$

Notice that the function $f(z) = \frac{1}{z(z-1)^2}$ has two singular points in the open disc bounded by $\gamma : |z - i| = 2$. One is the simple pole $z_1 = 0$ and the other is the double pole $z_2 = 1$.

$$\oint_\gamma \frac{dz}{z(z-1)^2} = 2\pi i \ (\text{Res}[f, 0] + \text{Res}[f, 1])$$

$$= 2\pi i \left(\lim_{z \to 0} z f(z) + \lim_{z \to 1} [(z-1)^2 f(z)]' \right)$$

$$= 2\pi i \left(\lim_{z \to 0} \frac{1}{(z-1)^2} + \lim_{z \to 1} \frac{-1}{z^2} \right) = 0.$$

In the following we apply the residue theorem in the evaluation of some (real) integrals.

Evaluation of Integrals of the Form $I = \int_0^{2\pi} R(\cos t, \sin t) \, dt$

Here $R(u, v) = \frac{P(u,v)}{Q(u,v)}$ is a rational function (P and Q are polynomials with real coefficients).

To compute the integral I, we make the change of variables: $z = e^{it}$ and use the Euler's formulas:

$$I = \frac{1}{i} \int_{|z|=1} R\left(\frac{1}{2}\left(z + \frac{1}{z}\right), \frac{1}{2i}\left(z - \frac{1}{z}\right) \right) \cdot \frac{1}{z} dz.$$

This integral is a complex integral of the rational function:

$$R\left(\frac{1}{2}\left(z + \frac{1}{z}\right), \frac{1}{2i}\left(z - \frac{1}{z}\right) \right) \cdot \frac{1}{z} = \frac{f(z)}{g(z)}.$$

Suppose that the polynomial $g(z)$ has no root of modulus 1 and $z_1, \ldots, z_n$ are the roots of $g(z)$ in the open disc $|z| < 1$. By applying the residue theorem we obtain:

$$I = 2\pi \sum_{k=1}^n \text{Res}\left[\frac{f(z)}{g(z)}, z_k \right].$$

Example 11.23 Compute the integral $a_m = \int_0^{2\pi} \frac{\cos mx}{2 + \sin x} dx, m \in \mathbb{N}$.

11.2 Complex Function Theory

The best idea is to consider also the integrals $b_m = \int_0^{2\pi} \dfrac{\sin mx}{2+\sin x} dx$ and to compute

$$a_m + ib_m = \int_0^{2\pi} \dfrac{\exp(imx)}{2+\sin x} dx.$$

We use the change of variable: $z = e^{ix}$. Thus,

$$a_m + ib_m = \int_{|z|=1} \dfrac{z^m}{2+\frac{z-z^{-1}}{2i}} \cdot \dfrac{dz}{iz} = 2\int_{|z|=1} \dfrac{z^m\,dz}{z^2+4iz-1}.$$

The function $\dfrac{z^m}{z^2+4iz-1}$ has two singular points, $(\sqrt{3}-2)i$ and $-(\sqrt{3}+2)i$, but only the first one has the modulus less than 1, hence

$$a_m + ib_m = 4\pi i\,\mathrm{Res}\left[\dfrac{z^m}{z^2+4iz-1}, (\sqrt{3}-2)i\right] =$$

$$\lim_{z\to(\sqrt{3}-2)i}\left(z-(\sqrt{3}-2)i\right)\dfrac{z^m}{z^2+4iz-1} = \dfrac{2\pi}{\sqrt{3}}(\sqrt{3}-2)^m i^m.$$

Since $a_m = \mathrm{Re}\left(\dfrac{2\pi}{\sqrt{3}}(\sqrt{3}-2)^m i^m\right)$, it follows that $a_{2k+1} = 0$ and $a_{2k} = \dfrac{2\pi(-1)^k}{\sqrt{3}}(\sqrt{3}-2)^{2k}$, for all $k \in \mathbb{N}$.

Integrals of the Form $\int_{-\infty}^{\infty} f(x)\,dx$

These improper integrals of the first type can be easily computed by using residue theorem. First of all we need an auxiliary result.

Lemma 11.5 *Let $f(z)$ be a complex function which is analytic in the upper half-plane $\{z \in \mathbb{C} : \mathrm{Im}\, z > 0\}$, except a finite number of points $z_1, z_2, \ldots, z_n$. We assume that in a neighborhood $|z| > R_0 > 0$ of ∞ the function $f(z)$ is bounded as follows: $|f(z)| < \dfrac{M}{|z|^{1+\delta}}$, where $M > 0$ and $\delta > 0$. Then*

$$\lim_{R\to\infty} \int_{C'_R} f(z)dz = 0, \qquad (11.107)$$

where C'_R is the semicircle $|z| = R$, with $\mathrm{Im}\, z > 0$ (see Fig. 11.13).

Fig. 11.13 The semicircle $|z| = R$, $\text{Im } z > 0$

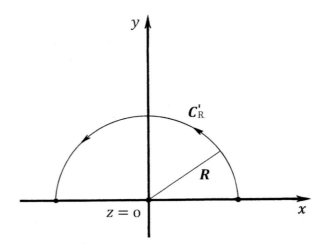

Proof Let us take $R > R_0$ and write:

$$\left| \int_{C'_R} f(z) dz \right| \leq \frac{M}{R^{1+\delta}} \cdot L(C'_R) = \frac{\pi M}{R^\delta},$$

which goes to zero as $R \to \infty$. Hence $\int_{C'_R} f(z) dz \to 0$ when $R \to \infty$ and the lemma is completely proved. $\square$

Theorem 11.34 *Let $f(x)$ be a real function of class C^∞, defined on $(-\infty, \infty)$, and let $f(z)$ be its extension to the upper half-plane $\text{Im } z \geq 0$. We assume that the function $f(z)$ satisfies the conditions of Lemma 11.5. Then the improper integral $\int_{-\infty}^{\infty} f(x) dx$ is convergent and:*

$$\int_{-\infty}^{\infty} f(x) dx = 2\pi i \sum_{k=1}^{n} \text{Res}[f, z_k], \quad (11.108)$$

where $z_1, z_2, \ldots, z_n$ are the singular points of f in the upper half-plane $\text{Im } z > 0$. In particular, if $f(z) = \frac{P(z)}{Q(z)}$ is a rational function which satisfies the supplementary conditions that $\deg Q(z) \geq \deg P(z) + 2$ and the equation $Q(z) = 0$ has no real

11.2 Complex Function Theory

root, then

$$\int_{-\infty}^{\infty} \frac{P(x)}{Q(x)} dx = 2\pi i \sum_{k=1}^{n} \text{Res}[f, z_k] \quad (11.109)$$

Proof Let us take R_0 large enough such that $|z_k| < R_0$ for any $k = 1, 2, \ldots, n$. Now we consider in the upper half-plane the closed contour $[-R, R] \cup C_R'$ (see Fig. 11.13). We apply the residue formula and find:

$$\int_{-R}^{R} f(x)dx + \int_{C_R'} f(z)dz = 2\pi i \sum_{k=1}^{n} \text{Res}[f, z_k].$$

Since the conditions of Lemma 11.5 are satisfied, the second integral tends to 0 as $R \to \infty$, so we obtain the formula (11.108).

The last statement is true because $f(z) = \frac{P(z)}{Q(z)}$ satisfies the conditions of Lemma 11.5. □

We note that a similar result holds for the singular points of f located in the lower half-plane.

Example 11.24 For instance, we know that the integral $I = \int_{-\infty}^{\infty} \frac{dx}{x^4 + 1}$ is convergent but its computation is not so easy. We see that $\left|\frac{1}{z^4+1}\right| \leq \frac{1}{|z|^3}$, for any $z \in \mathbb{C}$ with $|z| \geq 1$ so we can take $R_0 = 1$, $M = 1$ and $\delta = 2 > 0$ in Lemma 11.5. To apply Theorem 11.34, we need to find all the zeros of $z^4 + 1 = 0$ in the half-plane Im $z > 0$. The solutions of the equation $z^4 + 1 = 0$ are $z_{0,1,2,3} = e^{i\frac{\pi + 2k\pi}{4}}$, $k = 0, 1, 2, 3$. But only $z_0 = e^{i\frac{\pi}{4}}$ and $z_1 = e^{i\frac{3\pi}{4}}$ are in the upper half-plane. Thus,

$$I = 2\pi i \left[\text{Res}\left[\frac{1}{z^4 + 1}, e^{i\frac{\pi}{4}}\right] + \text{Res}\left[\frac{1}{z^4 + 1}, e^{i\frac{3\pi}{4}}\right] \right].$$

We can see that it is difficult to apply here the formula (11.105):

$$\text{Res}\left[\frac{1}{z^4 + 1}, e^{i\frac{\pi}{4}}\right] = \lim_{z \to z_0} \left[(z - z_0)\frac{1}{z^4 + 1}\right] = \frac{1}{(z_0 - z_1)(z_0 - z_2)(z_0 - z_3)}.$$

But, if $f(z) = \frac{P(z)}{Q(z)}$ is a rational function and z_0 is a simple pole of it, then, from the formula (11.105) we can write:

$$\text{Res}[f, z_0] = \lim_{z \to z_0} (z - z_0)\frac{P(z)}{Q(z)} = \frac{P(z_0)}{\lim_{z \to z_0} \frac{Q(z) - Q(z_0)}{z - z_0}} = \frac{P(z_0)}{Q'(z_0)}. \quad (11.110)$$

Using this formula we obtain:

$$\operatorname{Res}\left[\frac{1}{z^4+1}, e^{i\frac{\pi}{4}}\right] = \frac{1}{4z^3}\bigg|_{z=e^{i\frac{\pi}{4}}} = \frac{1}{4e^{i\frac{3\pi}{4}}} = \frac{1}{4}\left(-\frac{\sqrt{2}}{2} - i\frac{\sqrt{2}}{2}\right).$$

and

$$\operatorname{Res}\left[\frac{1}{z^4+1}, e^{i\frac{3\pi}{4}}\right] = \frac{1}{4z^3}\bigg|_{z=e^{i\frac{3\pi}{4}}} = \frac{1}{4e^{i\frac{9\pi}{4}}} = \frac{1}{4}\left(\frac{\sqrt{2}}{2} - i\frac{\sqrt{2}}{2}\right).$$

Hence $I = \frac{\pi\sqrt{2}}{2}$.

Integrals of the Form $\displaystyle\int_{-\infty}^{\infty} e^{iax} f(x)\, dx, a \neq 0$

Such integrals appear in the computation of the Fourier transform (see Chap. 5). We begin with an auxiliary result.

Lemma 11.6 (Jordan's Lemma) *Let $f(z)$ be a function which is analytic in the upper half-plane $\operatorname{Im} z \geq 0$, possibly except a finite number of points $z_1, z_2, \ldots, z_n$. For every $R > 0$, we denote by C'_R the semicircle $|z| = R$ in the upper half-plane $\operatorname{Im} z \geq 0$. If $\displaystyle\lim_{R\to\infty} \sup_{z\in C'_R} |f(z)| = 0$, then, for any $a > 0$, we have:*

$$\lim_{R\to\infty} \int_{C'_R} e^{iaz} f(z)\, dz = 0. \tag{11.111}$$

Proof Let $R_0 > 0$ such that the function $f(z)$ is analytic for $|z| \geq R_0$, $\operatorname{Im} z \geq 0$. Then, for every $R \geq R_0$, we can define $\mu(R) = \sup_{z\in C'_R} |f(z)| \in \mathbb{R}$.

Using the change of variable $z = R \cdot e^{i\theta}$, $\theta \in [0, \pi]$, for every $R \geq R_0$ we obtain:

$$\left|\int_{C'_R} e^{iaz} f(z)\, dz\right| \leq \int_{C'_R} \left|e^{iaz}\right| |f(z)|\, |dz| \leq R\mu(R) \int_0^\pi \left|e^{iaRe^{i\theta}}\right| d\theta =$$

$$= R\mu(R) \int_0^\pi \left|e^{iaR\cos\theta - aR\sin\theta}\right| d\theta = R\mu(R) \int_0^\pi e^{-aR\sin\theta}\, d\theta.$$

11.2 Complex Function Theory

Let us evaluate the real integral

$$I = \int_0^{\pi} e^{-aR\sin\theta} d\theta = \int_0^{\frac{\pi}{2}} e^{-aR\sin\theta} d\theta + \int_{\frac{\pi}{2}}^{\pi} e^{-aR\sin\theta} d\theta.$$

Making the change of variable $\eta = \pi - \theta$ in the last integral, we find that

$$\int_{\frac{\pi}{2}}^{\pi} e^{-aR\sin\theta} d\theta = -\int_{\frac{\pi}{2}}^{0} e^{-aR\sin\eta} d\eta = \int_0^{\frac{\pi}{2}} e^{-aR\sin\eta} d\eta,$$

so, $I = 2\int_0^{\frac{\pi}{2}} e^{-aR\sin\theta} d\theta$.

Since $[\tan\theta - \theta]' = \frac{1}{\cos^2\theta} - 1 \geq 0$, the function $\tan\theta - \theta$ is increasing, so $\tan\theta \geq \theta$ if $\theta \in \left(0, \frac{\pi}{2}\right)$. Now $\left[\frac{\sin\theta}{\theta}\right]' = \frac{\theta\cos\theta - \sin\theta}{\theta^2} = \frac{\theta - \tan\theta}{\theta^2\cos\theta} \leq 0$ if $\theta \in \left(0, \frac{\pi}{2}\right)$. Hence the function $\frac{\sin\theta}{\theta}$ is decreasing and its least value on $(0, \frac{\pi}{2}]$ is $\frac{\sin\frac{\pi}{2}}{\frac{\pi}{2}} = \frac{2}{\pi}$, i.e. $\frac{\sin\theta}{\theta} \geq \frac{2}{\pi}$ or $\sin\theta \geq \frac{2}{\pi}\theta$. Therefore,

$$e^{-aR\sin\theta} \leq e^{-\frac{2aR}{\pi}\theta}$$

and so

$$I \leq 2\int_0^{\frac{\pi}{2}} e^{-\frac{2aR}{\pi}\theta} d\theta = -\frac{\pi}{aR} e^{-\frac{2aR}{\pi}\theta} \bigg|_0^{\frac{\pi}{2}} = -\frac{\pi}{aR}\left[e^{-aR} - 1\right].$$

Since

$$\left|\int_{C_R'} e^{iaz} f(z) dz\right| \leq R\mu(R) \int_0^{\pi} e^{-aR\sin\theta} d\theta \leq \mu(R)\frac{\pi}{a}\left[1 - e^{-aR}\right]$$

and $\mu(R) \to 0$, $e^{-aR} \to 0$ when $R \to \infty$, we obtain that

$$\lim_{R\to\infty} \left|\int_{C_R'} e^{iaz} f(z) dz\right| = 0$$

and the lemma is proved. □

Remark 11.18

(a) If $a < 0$ and $f(z)$ satisfies the hypotheses of Lemma 11.6 for the lower half-plane $\operatorname{Im} z \leq 0$, then the formula (11.111) also holds, but in this case C'_R is the semicircle of radius R in the lower half-plane. Indeed, in this case $z = Re^{i\theta}$, $\pi \leq \theta \leq 2\pi$, and

$$\left| \int_{C'_R} e^{iaz} f(z) dz \right| \leq \int_{C'_R} \left| e^{iaz} \right| |f(z)| |dz| \leq R\mu(R) \int_\pi^{2\pi} \left| e^{iaRe^{i\theta}} \right| d\theta =$$

$$= R\mu(R) \int_\pi^{2\pi} e^{-aR\sin\theta} d\theta \stackrel{\theta=\eta+\pi}{=} R\mu(R) \int_0^\pi e^{-\alpha R\sin\eta} d\eta \to 0$$

as $R \to \infty$ (we denoted $a = -\alpha$, $\alpha > 0$).

(b) Similar assertions hold for $a = \pm i\lambda$, $\lambda > 0$, when we integrate on the right ($\operatorname{Re} z \geq 0$, see Fig. 11.14) or on the left ($\operatorname{Re} z \leq 0$) half of the xOy-plane. We leave to the reader to state and prove the corresponding Jordan's lemmas. The Jordan's lemma and its variants are extensively used in operational calculus (Fourier and Laplace transforms).

Theorem 11.35 *Let $f : \mathbb{R} \to \mathbb{R}$ be a function of class C^∞ and let $f(z)$ be its extension to the upper half-plane $\operatorname{Im} z \geq 0$. If $f(z)$ is analytic in the upper half-*

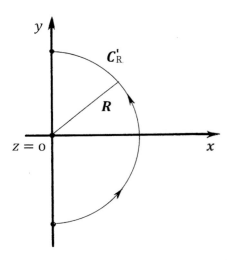

Fig. 11.14 The semicircle $|z| = R$, $\operatorname{Re} z > 0$

plane $\operatorname{Im} z \geq 0$, possibly except a finite number of points $z_1, z_2, \ldots, z_n \notin \mathbb{R}$, and satisfies the conditions from Jordan's lemma, then for any $a > 0$, the integral $\int_{-\infty}^{\infty} e^{iax} f(x) dx$, $a > 0$, is convergent and it can be computed as follows:

$$\int_{-\infty}^{\infty} e^{iax} f(x) dx = 2\pi i \sum_{k=1}^{n} \operatorname{Res}\left[e^{iaz} f(z), z_k\right]. \qquad (11.112)$$

Proof Let R_0 be large enough such that $|z_k| < R_0$, $k = 1, 2, \ldots, n$.

For any $R > R_0$ we consider in the upper half-plane the closed curve $\Gamma = [-R, R] \cup C'_R$, where C'_R is the semicircle $|z| = R$ in the upper half-plane. By applying the residues formula (11.104) we get:

$$\oint_{\Gamma} e^{iaz} f(z) dz = \int_{-R}^{R} e^{iax} f(x) dx + \int_{C'_R} e^{iaz} f(z) dz = 2\pi i \sum_{k=1}^{n} \operatorname{Res}\left[e^{iaz} f(z), z_k\right].$$

We let $R \to \infty$ and apply Jordan's lemma 11.6, so we obtain (11.112). □

Remark 11.19 Let $f(z)$ be a function which is analytic in the lower half-plane $\operatorname{Im} z \leq 0$, possibly except a finite number of points $z_1, z_2, \ldots, z_n \notin \mathbb{R}$. If C'_R denotes the semicircle $|z| = R$ in the lower half-plane $\operatorname{Im} z \leq 0$ and $\lim_{R \to \infty} \sup_{z \in C'_R} |f(z)| = 0$, then, for any $a < 0$, the integral $\int_{-\infty}^{\infty} e^{iax} f(x) dx$ is convergent and and it can be computed by the formula:

$$\int_{-\infty}^{\infty} e^{iax} f(x) dx = -2\pi i \sum_{k=1}^{n} \operatorname{Res}\left[e^{iaz} f(z), z_k\right]. \qquad (11.113)$$

The negative sign that appears in (11.113) is because in the residue formula (11.104) the integral is computed in the direct (trigonometric) sens on the closed contour Γ. Therefore, we have to take $\Gamma = [-R, R]^- \cup C'_R$, where C'_R is the semicircle $z = Re^{i\theta}$, $\theta \in [\pi, 2\pi]$ and $[-R, R]^-$ is the interval considered from R to $-R$.

Example 11.25 In the operational calculus (Fourier transform) we need to compute integrals of the form:

$$I_1 = \int_{-\infty}^{\infty} f(x) \cos nx \, dx, \quad I_2 = \int_{-\infty}^{\infty} f(x) \sin nx \, dx, \quad n = 0, 1, \ldots.$$

For this we consider the integral

$$I = I_1 + iI_2 = \int_{-\infty}^{\infty} e^{inx} f(x)\, dx$$

and apply the formula (11.112).

For instance, let us compute $I_1 = \int_{-\infty}^{\infty} \frac{\cos nx}{x^2 + \lambda^2}\, dx$, where $\lambda > 0$ is a real parameter and $n = 0, 1, 2, \ldots$. We have:

$$I = \int_{-\infty}^{\infty} \frac{e^{inx}}{x^2 + \lambda^2}\, dx = 2\pi i \cdot \operatorname{Res}\left[\frac{e^{inz}}{z^2 + \lambda^2}, i\lambda\right] =$$

$$= 2\pi i \cdot \lim_{z \to i\lambda}\left[(z - i\lambda)\frac{e^{inz}}{z^2 + \lambda^2}\right] = 2\pi i \cdot \frac{e^{-n\lambda}}{2i\lambda} = \frac{\pi}{\lambda}e^{-n\lambda}.$$

Hence $I_1 = \frac{\pi}{\lambda}e^{-n\lambda}$ and $I_2 = \int_{-\infty}^{\infty} \frac{\sin nx}{x^2 + \lambda^2}\, dx = 0$ (this was obvious from the beginning because the function $\frac{\sin nx}{x^2+\lambda^2}$ is odd.

Example 11.26 (Fresnel's Integrals) The following integrals (used in optics) are known as Fresnel's integrals:

$$C(x) = \int_0^x \cos t^2\, dt \text{ and } S(x) = \int_0^x \sin t^2\, dt.$$

We prove that their limits as $x \to \infty$ are both equal to $\sqrt{\frac{\pi}{8}}$, that is,

$$\int_0^{\infty} \cos x^2\, dx = \int_0^{\infty} \sin x^2\, dx = \sqrt{\frac{\pi}{8}}. \tag{11.114}$$

We compute the integral

$$\int_0^{\infty} e^{ix^2}\, dx = \int_0^{\infty} \cos x^2\, dx + i \int_0^{\infty} \sin x^2\, dx.$$

For this, we consider (see Fig. 11.15) for any $r > 0$, the closed curve $\Gamma_r = [OA_r] \cup \gamma_r \cup [B_r O]$, where $A_r(r, 0)$, $B_r\left(\frac{r\sqrt{2}}{2}, \frac{r\sqrt{2}}{2}\right)$ and γ_r is the arc of circle $\widehat{A_r B_r}$ defined by:

$$\gamma_r : z = re^{it},\ t \in \left[0, \frac{\pi}{4}\right].$$

11.2 Complex Function Theory

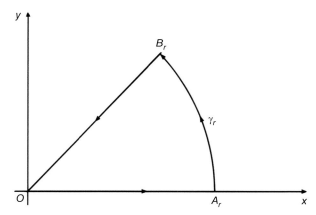

Fig. 11.15 The closed curve $\Gamma_r = [OA_r] \cup \gamma_r \cup [B_r O]$

Since the function $f(z) = e^{iz^2}$ is analytic on the whole complex plane, by Cauchy's Theorem 11.21 it follows that, for any $r > 0$ we can write:

$$0 = \oint_{\Gamma_r} e^{iz^2} dz = \int_{[OA_r]} e^{ix^2} dx + \int_{\gamma_r} e^{iz^2} dz + \int_{[B_r O]} e^{iz^2} dz$$

$$= \int_0^r e^{ix^2} dx + \int_0^{\frac{\pi}{4}} e^{ir^2(\cos t + i \sin t)^2} rie^{it} dt - \int_0^r e^{i\left(\frac{x+ix}{\sqrt{2}}\right)^2} \frac{1+i}{\sqrt{2}} dx.$$

Since $\sin 2t \geq \frac{4t}{\pi}$ for any $t \in \left[0, \frac{\pi}{4}\right]$ (see the proof of Lemma 11.6), we obtain:

$$\left| \int_{\gamma_r} e^{iz^2} dz \right| \leq \int_0^{\frac{\pi}{4}} re^{-r^2 \sin 2t} dt \leq r \int_0^{\frac{\pi}{4}} e^{-\frac{4r^2 t}{\pi}} dt = \frac{\pi}{4r}\left(1 - e^{-r^2}\right) \to 0$$

when $r \to \infty$. Hence

$$\lim_{r \to \infty} \int_{\gamma_r} e^{iz^2} dz = 0$$

and we obtain that

$$\int_0^\infty e^{ix^2} dx = \frac{1+i}{\sqrt{2}} \int_0^\infty e^{-x^2} dx.$$

The last integral (Gaussian integral) equals $\frac{\sqrt{\pi}}{2}$, so

$$\int_0^\infty \cos x^2 dx + i \int_0^\infty \sin x^2 dx = \frac{\sqrt{\pi}}{2\sqrt{2}} + i \frac{\sqrt{\pi}}{2\sqrt{2}}$$

and the equalities (11.114) are proved.

Bibliography

1. T.M. Apostol, *Mathematical Analysis* (Addison-Wesley Publishing Company, 1974)
2. V.I. Arnold, *Ordinary Differential Equations* (Springer, Berlin, 1992)
3. W.E. Boyce, R.C. DiPrima, *Elementary Differential Equations and Boundary Value Problems*, 8th edn. (Wiley, New York, 2005)
4. J.W. Brown, R.V. Churchill, *Fourier Series and Boundary Value Problems*, 5th edn. (McGraw-Hill, New York, 1993)
5. B. van Brunt, *The Calculus of Variations* (Springer, New York, 2004)
6. E.A. Coddington, *An Introduction to Ordinary Differential Equations* (Dover Publications, New York, 1989)
7. E.A. Coddington, N. Levinson, *Theory of Ordinary Differential Equations* (McGraw-Hill, New York, 1955)
8. R. Courant, D. Hilbert, *Methods of Mathematical Physics*, vol. I (John Wiley & Sons, New York, 1966)
9. R. Courant, D. Hilbert, *Methods of Mathematical Physics*, vol. II (John Wiley & Sons, New York, 1966)
10. H. Cramér, *Mathematical Methods in Statistics* (Princeton University Press, Princeton, 1961)
11. A. DasGupta, *Fundamentals of Probability: A First Course* (Springer, Berlin, 2010)
12. G. Doetsch, *Introduction to the Theory and Application of the Laplace Transformation* (Springer, Berlin, 1974)
13. Elsgolts, *Differential Equations and Calculus of Variations* (Mir Publishers, Moscow, 1977)
14. W. Feller, *An Introduction to Probability Theory and Its Applications*, vol. 1 (John Wiley & Sons, 1960)
15. I.M. Gelfand, S.V. Fomin, *Calculus of Variations* (Prentice-Hall, Englewood Cliffs, New Jersey, 1963)
16. B.V. Gnedenko, *The Theory of Probability* (MIR Publishers, Moscow, 1978)
17. E. Hansen *Fourier Transforms: Principles and Applications* (Wiley, New York, 2014)
18. K. Hoffman, R. Kunze, *Linear Algebra*, 2nd edn. (Prentice-Hall, 1971)
19. T. Hsu, *Fourier Series, Fourier Transforms, and Function Spaces: A Second Course in Analysis* (MAA Press, an imprint of the American Mathematical Society, 2020)
20. A. Jeffrey, *Advanced Engineering Mathematics* (Harcourt Academic Press, USA, 2020)
21. E. Kreyszig, H. Kreyszig, *Advanced Engineering Mathematics*, 10th edn. (John Wiley&Sons, 2011)
22. M. Krasnov, A. Kiselev, G. Makarenko, E. Shikin, *Mathematical Analysis for Engineers* (Mir Publishers, Moscow, 1990)
23. S. Lang, *A First Course in Calculus*, 5th edn. (Springer, Berlin, 1986)

24. S. Lang, *Algebra* (Springer, Berlin, 2002)
25. M. Lavrentiev, B. Chabat, *Méthodes de la théorie des fonctions d'une variable complexe* (Springer, Berlin, 2002)
26. P.V. O'Neil, *Advanced Engineering Mathematics* (Thomson, USA, 2007)
27. L.S. Pontryagin, *Ordinary Differential Equations* (Addison-Wesley, Reading, 1962)
28. S.A. Popescu, *Mathematical Analysis I. Differential Calculus* (Conspress, Bucharest, 2009). http://civile-old.utcb.ro/cmat/cursrt/an.htm
29. S.A. Popescu, *Mathematical Analysis II. Integral Calculus* (Conspress, Bucharest, 2011). http://civile-old.utcb.ro/cmat/cursrt/an.htm
30. R. Redheffer, *Differential Equations. Theory and Applications* (Jones & Bartlett Publishers, Boston, 1991)
31. S. Ross, *A First Course in Probability*, 5th edn. (Prentice-Hall, New Jersey, 1998)
32. W. Rudin, *Principles of Mathematical Analysis*, 3rd edn. (Mc Graw-Hill, New York, 1976)
33. W. Rudin, *Real and Complex Analysis*, 3rd edn. (Mc Graw-Hill, New York, 1987)
34. J.L. Schiff, *The Laplace Transform: Theory and Applications* (Springer, Berlin, 1999)
35. G.E. Shilov, *Linear Algebra* (Dover Publications, New York, 1977)
36. G.E. Shilov, *Elementary Real and Complex Analysis* (Dover Publications, New York, 1996)
37. V. Smirnov, *Cours de Mathématiques Supérieures* Tome II, Edition (Mir-Moscou, 1970)
38. J. Stewart, *Calculus Early Transcendentals* (Thomson, Brooks/Cole, 2008)
39. T. Tao, *Analysis I* (Hindustan Book Agency, 2006)
40. T. Tao, *Analysis II* (Hindustan Book Agency, 2006)
41. G.P. Tolstov, *Fourier Series* (Dover, New York, 1976)
42. D.V. Widder, *The Laplace Transform* (Princeton University Press, Princeton, 1946)
43. D.V. Widder, *Advanced calculus* (Prentice-Hall, New Jersey, 1961)
44. D.G. Zill, W.S. Wright *Advanced Engineering Mathematics* (Jones & Bartlett Learning, 2014)

Index

A

Abelian group, 748
Abscissa of convergence, 307
Absolute convergence, 732
Absolutely integrable, 196, 254
Absolutely uniform convergence, 732
Algebraic method for linear systems with constant coefficients, 125
Almost sure convergence, 597
Analytic functions, 804
Archimedes' Axiom, 751
Autonomous system of differential equations, 170

B

Banach algebra, 726
Banach space, 726
Basis of solutions of a linear homogeneous equation, 59
Bayes rule, 509
Beltrami identity, 451
Bernoulli distribution, 536
Bernoulli equation, 20
Bernoulli trial, 495
Bessel's inequality, 211, 780
Binomial distribution, 536
Bolzano-Weierstrass theorem, 755
Brachistochrone, 438, 454

C

Canonical form of elliptic-type PDE, 372
Canonical form of hyperbolic-type PDE, 370
Canonical form of parabolic-type PDE, 371
Cantor's Intersection Theorem, 755
Catenary, 454, 476
Catenoid, 454
Cauchy problem, 4
Cauchy problem for ODE-n, 58
Cauchy problem for ODE-1, 34
Cauchy problem for PDE-1, 181
Cauchy problem for PDE-2, 373
Cauchy problem for systems of differential equations, 103
Cauchy-Riemann equations, 787
Cauchy-Schwarz inequality, 206, 763
Cauchy sequence, 725, 738
Cauchy's integral formula, 799
Cauchy's Theorem for doubly connected domains, 797
Cauchy's Theorem for ($n1$)-connected domains, 798
Cauchy's Theorem for simply connected domains, 795
Characteristic curve of the PDE-2, 368
Characteristic differential equation, 368
Characteristic equation of a linear ODE, 69
Characteristic function of a random variable, 560
Characteristic polynomial of a linear ODE, 69
Chebyshev's inequality, 596
Chi-squared distribution, 588
Circular helix, 481
Clairaut equation, 31
Closed ball, 736, 786
Closed set, 736
Complete metric space, 739
Complete normed space, 726
Completion of a metric space, 743

Completion of an inner product space, 768
Complex form of the Fourier series, 242
Complex Fourier coefficients, 242
Conditional probability, 506
Conformal mapping, 791
Connected set, 786
Conservative field, 171
Continuous almost everywhere, 251
Continuously differentiable, 2
Continuous random variable, 514
Convergence in a metric space, 735
Convergence in a normed space, 725
Convergence in distribution, 597
Convergence in probability, 597
Convolution, 270
Convolution formula for the Fourier transform, 271
Convolution formula for the Laplace transform, 329
Correlation coefficient, 533
Covariance, 532
Critical point of an autonomous system, 173

D

D'Alembert formula, 383
Dense subspace in a metric space, 740
Diagonalizable matrix, 128
Dido's problem, 440, 472, 474
Differential equation, 1
Dirac delta function, 352
Dirichlet integral, 255
Dirichlet kernel, 213
Dirichlet problem for a disk, 425
Dirichlet problem for a rectangle, 423
Discrete Fourier transform, 300
Discrete random variable, 514
Discrete uniform distribution, 535
Distribution function (cumulative distribution function), 514
Domain, 786
Dot product, 762

E

Eigenfunctions, 388
Eigenvalue of a matrix, 126
Eigenvector of a matrix, 126
Elimination method for systems of differential equations, 110
Elliptic-type PDE-2, 367
Entire function, 787
Equivalence class, 742
Equivalence relation on a set, 742

Euclidean space, 762
Euler equation (Euler-Cauchy equation), 93
Euler-Lagrange equation, 450
Euler-Lagrange system, 461
Euler–Ostrogradsky equation, 468
Euler-Poisson equation, 457
Euler-Poisson integral, 257
Euler's formula, 70
Event, 486
Event space, 488
Exact differential equation, 24
Exact differential form, 24
Expansion in Fourier series, 205
Exponential random variable, 572
Extremal, 446

F

Field, 748
Field of complex numbers $\mathbb{C}$, 758
Field of rational numbers $\mathbb{Q}$, 751
Field of real numbers $\mathbb{R}$, 753
Finite vibrating string, 385
First integral of an autonomous system, 170
First variation of a functional, 445
Fourier coefficients, 205
Fourier coefficients in a Hilbert space, 779
Fourier cosine transform, 267
Fourier integral formula, 259
Fourier series, 205
Fourier series of odd/even functions, 220
Fourier sine transform, 267
Fourier transform, 265
Fresnel's integrals, 820
Functional, 435
Functional independence, 173
Function of class C^k, 2
Function of class C^∞, 2
Function of exponential order, 307
Fundamental period, 192
Fundamental set of solutions of a linear homogeneous equation, 59
Fundamental system of solutions, 116
Fundamental Theorem of Algebra, 69, 760, 803
Fundamental Theorem of Calculus, 26

G

Gamma distribution, 584
Gamma function, 253, 291
Gauss error function, 332
Gaussian distribution, 576
Gaussian integral, 257

Index 827

Geodesic curves, 442, 477
Geometric distribution, 543
Gibbs phenomenon, 223
Gram–Schmidt orthogonalization process, 777
Greatest lower bound, 756
Green's theorem, 472, 792
Group, 748

H

Hamilton's principle, 440, 465
Harmonic functions, 422, 788
Harmonic oscillator equation, 48
Heat flow equation, 411
Heaviside step function, 275, 307
Hilbert basis, 235, 781
Hilbert space, 235, 765
Holomorphic function, 787
Homogeneous function, 13
Homogeneous linear equation, 16
Homogeneous linear ODE of order n, 58
Hyperbolic-type PDE-2, 367

I

Image function, 309
Improper integrals, 249
Independent events, 507
Independent random variables, 524
Infinite vibrating string, 381
Inhomogeneous linear equation, 16
Initial conditions, 4
Initial value problem, 4
Inner product, 205, 762
Inner product space, 762
Integrable combinations method, 174
Integrating factor, 27
Inverse Fourier transform, 266
Inverse Laplace transform, 333
Isometric spaces, 740
Isometry, 740
Isometry of Hilbert spaces, 767
Isoperimetric problem, 446

J

Jacobian matrix, 173, 364
Joint distribution function, 555
Joint probability density function, 555
Joint probability distribution, 523
Jordan canonical form, 133
Jordan's lemma, 816

K

Kronecker's symbol $\delta_{i,j}$, 208

L

Lagrange equation, 29
Lagrange's method of variation of constants, 17, 66, 121
Lagrange's multipliers method for functionals, 447
Laguerre's differential equation, 348
Laplace's equation, 360, 422
Laplace transform, 309
Laplacian, 360
Laurent series, 809
Law of conservation of energy, 171, 466
Law of total probability, 509
Least upper bound, 756
Leibniz' formula, 25, 260
Limit point, 786
Linear differential equation of the first order, 16
Linear homogeneous first-order PDE, 180
Linearly correlated, 534
Linear ODE of order n, 57
Linear second-order PDE, 361
Linear system of n differential equations (SLDE-n), 114
Liouville's formula, 119
Liouville's Theorem, 803
Lipschitz condition, 41

M

Markov's inequality, 595
Mellin formula, 333
Meromorphic function, 810
Method of undetermined coefficients, 82, 138
Metric (distance), 734
Metric space, 444, 734

N

Non-homogeneous linear equation, 16
Non-homogeneous linear ODE-n, 64
Norm, 443, 723
Normal distribution, 576
Normal form of a differential equation, 2
Normed K-algebra, 724
Normed space, 443, 723

O

Open ball, 736, 785
Open set, 736, 785
Ordered field, 751
Ordinary differential equation (ODE), 1
Original (function), 306
Orthogonal functions, 207
Orthogonal system, 207, 776
Orthogonal trajectories, 32
Orthogonal vectors, 775
Orthonormal basis, 778
Orthonormal system, 208, 776

P

Parabolic-type PDE-2, 367
Parseval's formula for the Fourier transform, 271
Parseval's identity, 212, 781
Partial differential equation (PDE), 2
Partially ordered set, 751
Partition, 489
Path in $\mathbb{C}$, 786
Periodic function, 192
Picard's method of successive approximations, 35, 104
Piecewise continuous, 195
Piecewise linear function, 231
Piecewise smooth, 197
Plateau's problem, 440, 470
Poisson distribution, 540
Poisson's integral formula, 431
Pole of a complex function, 809
Principle of least action, 440, 465
Probabilistic mathematical model, 496
Probability density function, 545
Probability distribution, 517
Probability function, 492
Probability mass function, 517
Probability space, 492
Pythagoras' Theorem, 776

Q

Quasilinear first-order PDE, 183
Quasilinear second order PDE, 364
Quotient of a vector space relative to a subspace, 743
Quotient set, 743

R

Residue theorem, 810
Riccati equation, 22
Riemann-Lebesgue Lemma, 213

Ring, 748
Ring of integer numbers $\mathbb{Z}$, 747

S

Sample point, 485
Sample space, 485
Scalar product, 762
Separable equation, 7
Series of functions, 731
Simple (pointwise) convergence, 728
Simple random variables, 517
Simply connected domain, 794
Simply supported beam, 404
Sinc function, 274
Singular solutions, 3, 30, 32
Square integrable, 254
Standard deviation, 530
Standard normal distribution, 576
Standard probability space, 496
Step function, 772
Stirling's formula, 503, 594
Student **t**-distribution, 589
Superposition principle, 138
Sup-norm (infinity norm), 725
System of n differential equations (SDE-n), 103

T

Totally ordered set, 751
Triangle inequality, 734
Trigonometric polynomial, 198
Trigonometric series, 198

U

Uniform continuity in metric spaces, 772
Uniform continuous random variable, 568
Uniform convergence, 728
Uniform discrete random variable, 519
Unit step function, 275, 307

V

Variance, 530
Vibrating string equation, 380

W

Wave equation, 380
Weierstrass M-test for improper integrals, 261
Weierstrass M-test for series of functions, 200, 733
Wronskian, 59, 118
Wronski matrix, 118

GPSR Compliance

The European Union's (EU) General Product Safety Regulation (GPSR) is a set of rules that requires consumer products to be safe and our obligations to ensure this.

If you have any concerns about our products, you can contact us on ProductSafety@springernature.com

In case Publisher is established outside the EU, the EU authorized representative is:

Springer Nature Customer Service Center GmbH
Europaplatz 3
69115 Heidelberg, Germany

Batch number: 08011020

Printed by Printforce, the Netherlands